Selected Papers on
Crack Tip Stress Fields

SEM Classic Papers
Volume CP 2

SPIE Milestone Series
Volume MS 138

Selected Papers on
Crack Tip Stress Fields

R. J. Sanford, *Editor*
University of Maryland, College Park
Dept. of Mechanical Engineering

Brian J. Thompson
General Editor, SPIE Milestone Series

THE SOCIETY FOR EXPERIMENTAL MECHANICS
Bethel, Connecticut USA

SPIE OPTICAL ENGINEERING PRESS

SPIE—The International Society for Optical Engineering
Bellingham, Washington USA

Library of Congress Cataloging-in Publication Data

Selected papers on Crack tip stress fields / R.J. Sanford, editor.
      p. cm. — (SEM classic papers series; v. CP 2) (SPIE milestone series ; v. MS 138)
   Includes bibliographical references and index.
   ISBN 0-8194-2621-0 (SPIE) ISBN 0-912053-56-9 (SEM) (alk. paper)
    1. Fracture mechanics—Mathematical models. 2. Strains and stresses—Mathematical models. I. Sanford, R.J. II. Series. III. Series: SPIE milestone series ; v. MS 138.
TA409.S43 1997   97-20468
620.1'126—dc21   CIP

The Society for Experimental Mechanics
ISBN 0-912053-56-9
SPIE—The International Society for Optical Engineering
ISBN 0-8194-2621-0

Copublished by

The Society for Experimental Mechanics (SEM)
7 School Street, Bethel, CT 06801 USA
Telephone 203 790 6373 (Eastern Time)• Fax 203 790 4472
http://www.sem.org/

SPIE—The International Society for Optical Engineering
P.O. Box 10, Bellingham, Washington 98227-0010 USA
Telephone 360 676 3290 (Pacific Time) • Fax 360 647 1445
http://www.spie.org/

This book is a compilation of outstanding papers selected from the world literature on optical and optoelectronic science, engineering, and technology. SPIE Optical Engineering Press and SEM acknowledge with appreciation the authors and publishers who have given their permission to reprint material included in this volume. An earnest effort has been made to contact the copyright holders of all papers reproduced herein.

Copyright © 1997 The Society for Experimental Mechanics (SEM) and The Society of Photo-Optical Instrumentation Engineers (SPIE)

Copying of SEM- and SPIE-copyrighted material in this book for internal or personal use, or the internal or personal use of specific clients, beyond the fair use provisions granted by the U.S. Copyright Law is authorized by SEM and SPIE subject to payment of copying fees. The Transactional Reporting Service base fee for this volume is $10.00 per SEM- or SPIE-copyrighted article and should be paid directly to the Copyright Clearance Center (CCC), 222 Rosewood Drive, Danvers MA 01923. Other copying for republication, resale, advertising, or promotion or any form of systematic or multiple reproduction of any SEM- or SPIE-copyrighted material in this book is prohibited except with permission in writing from the publisher. The CCC fee code for users of the Transactional Reporting Service is 0-8194-2621-0/97/$10.00.

Readers desiring to reproduce materials contained herein copyrighted by other than SEM or SPIE must contact the copyright holder for appropriate permissions.

Printed in the United States of America

Introduction to the Milestone Series

There is no substitute for reading the original papers on any subject even if that subject is mature enough to be critically written up in a textbook or a monograph. Reading a well-written book only serves as a further stimulus to drive the reader to seek the original publications. The problems are, which papers, and in what order?

As a serious student of a field, do you really have to search through all the material for yourself, and read the good with the not-so-good, the important with the not-so-important, and the milestone papers with the merely pedestrian offerings? The answer to all these questions is usually yes, unless the authors of the textbooks or monographs that you study have been very selective in their choices of references and bibliographic listings. Even in that all-too-rare circumstance, the reader is then faced with finding the original publications, many of which may be in obscure or not widely held journals.

From time to time and in many disparate fields, volumes appear that are collections of reprints that represent the milestone papers in the particular field. Some of these volumes have been produced for specific topics in optical science, but as yet no systematic set of volumes has been produced that covers connected areas of optical science and engineering.

The editors of each individual volume in the series have been chosen for their deep knowledge of the world literature in their fields; hence, the selection of reprints chosen for each book has been made with authority and care.

On behalf of SPIE, I thank the individual editors for their diligence, and we all hope that you, the reader, will find these volumes invaluable additions to your own working library.

Brian J. Thompson
University of Rochester

Selected Papers on
Crack Tip Stress Fields

Contents

xi   **Preface**
R.J. Sanford

xiii **Introduction**
R.J. Sanford

Section One
Analytical Methods

3   **Stress analysis of cracks**
Paul C. Paris, George C. Sih (in *Fracture Toughness Testing and Its Applications* 1965)

57 **Westergaard stress functions for several periodic crack problems**
Hiroshi Tada (*Engineering Fracture Mechanics* 1970)

61 **Analysis of an infinite plate containing radial cracks originating at the boundary of an internal circular hole**
O.L. Bowie (*Journal of Mathematics and Physics* 1956)

73 **Rectangular tensile sheet with symmetric edge cracks**
O.L. Bowie (*Journal of Applied Mechanics* 1964)

78 **A modified mapping-collocation technique for accurate calculation of stress intensity factors**
O.L. Bowie, D.M. Neal (*International Journal of Fracture* 1970)

86 **Solution of plane problems of elasticity utilizing partitioning concepts**
O.L. Bowie, C.E. Freese, D.M. Neal (*Journal of Applied Mechanics* 1973)

92 **Elastic field equations for blunt cracks with reference to stress corrosion cracking**
Matthew Creager, Paul C. Paris (*International Journal of Fracture Mechanics* 1967)

98 **The stress intensity factors for cracks in stress gradients**
G.G. Chell (*International Journal of Fracture* 1973)

100 **The stress intensity factors and crack profiles for centre and edge cracks in plates subject to arbitrary stresses**
G.G. Chell (*International Journal of Fracture* 1976)

114 **Crack-tip, stress-intensity factors for plane extension and plate bending problems**
G.C. Sih, P.C. Paris, F. Erdogan (*Journal of Applied Mechanics* 1962)

121 **On cracks in rectilinearly anisotropic bodies**
G.C. Sih, P.C. Paris, G.R. Irwin (*International Journal of Fracture Mechanics* 1965)

136 **Central crack in plane orthotropic rectangular sheet**
O.L. Bowie, C.E. Freese (*International Journal of Fracture Mechanics* 1972)

145 **The stresses around a fault or crack in dissimilar media**
M.L. Williams (*Bulletin of the Seismological Society of America* 1959)

151 **Plane problems of cracks in dissimilar media**
J.R. Rice, G.C. Sih (*Journal of Applied Mechanics* 1965)

157 **Stress-intensity factors for the tension of an eccentrically cracked strip**
M. Isida (*Journal of Applied Mechanics* 1966)

159 **Discussion: Plane strain crack toughness testing of high strength metallic materials (by William F. Brown, Jr., and John E. Srawley)**
C.E. Feddersen (in *Plane Strain Crack Toughness Testing* 1966)

162 **The stress intensity factors of a radial crack in a finite rotating elastic disc**
D.P. Rooke, J. Tweed (*International Journal of Engineering Science* 1972)

168 **The stress intensity factor of an edge crack in a finite rotating elastic disc**
D.P. Rooke, J. Tweed (*International Journal of Engineering Science* 1973)

173 **Stress intensity factors for some through-cracked fastener holes**
A.F. Grandt, Jr. (*International Journal of Fracture* 1975)

185 **Stress intensity factors for hollow circumferentially notched round bars**
D.O. Harris (*Journal of Basic Engineering* 1967)

191 **Three-dimensional stress distribution around an elliptical crack under arbitrary loadings**
M.K. Kassir, G.C. Sih (*Journal of Applied Mechanics* 1966)

202 **Stress intensity factors for penny-shaped cracks. Part 1—Infinite solid**
F.W. Smith, A.S. Kobayashi, A.F. Emery (*Journal of Applied Mechanics* 1967)

208 **Stress intensity factors for semicircular cracks. Part 2—Semi-infinite solid**
F.W. Smith, A.F. Emery, A.S. Kobayashi (*Journal of Applied Mechanics* 1967)

215 **Stress intensity factors for an elliptical crack approaching the surface of a semi-infinite solid**
R.C. Shah, A.S. Kobayashi (*International Journal of Fracture* 1973)

Section Two
Numerical Methods

231 **Stress-intensity factors for a single-edge-notch tension specimen by boundary collocation of a stress function**
Bernard Gross, John E. Srawley, William F. Brown, Jr. (*NASA Technical Note D-2395* 1964)

242 **Stress-intensity factors for single-edge-notch specimens in bending or combined bending and tension by boundary collocation of a stress function**
Bernard Gross, John E. Srawley (*NASA Technical Note D-2603* 1965)

256 **Stress-intensity factors for three-point bend specimens by boundary collocation**
Bernard Gross, John E. Srawley (*NASA Technical Note D-3092* 1965)

269 **Elastic displacements for various edge-cracked plate specimens**
Bernard Gross, Ernest Roberts, Jr., John E. Srawley (*International Journal of Fracture Mechanics* 1968)

280 **Wide range stress intensity factor expressions for ASTM E399 standard fracture toughness specimens**
John E. Srawley (*International Journal of Fracture* 1976)

	282	**Stress intensity factors for deep cracks in bending and compact tension specimens** W.K. Wilson *(Engineering Fracture Mechanics* 1970)
	285	**An improved method of collocation for the stress analysis of cracked plates with various shaped boundaries** J.C. Newman, Jr. *(NASA Technical Note D-6376* 1971)
	316	**On the finite element method in linear fracture mechanics** S.K. Chan, I.S. Tuba, W.K. Wilson *(Engineering Fracture Mechanics* 1970)
	333	**A note on the finite element method in linear fracture mechanics** D.F. Mowbray *(Engineering Fracture Mechanics* 1970)
	337	**Finite elements for determination of crack tip elastic stress intensity factors** Dennis M. Tracey *(Engineering Fracture Mechanics* 1971)
	348	**3-D elastic singularity element for evaluation of K along an arbitrary crack front** D.M. Tracey *(International Journal of Fracture* 1973)
	351	**The computation of stress intensity factors by a special finite element technique** P.F. Walsh *(International Journal of Solids and Structures* 1971)
	361	**On the use of isoparametric finite elements in linear fracture mechanics** Roshdy S. Barsoum *(International Journal for Numerical Methods in Engineering* 1976)
	374	**Crack tip finite elements are unnecessary** R.D. Henshell, K.G. Shaw *(International Journal for Numerical Methods in Engineering* 1975)
	387	**The weight function method for determining stress intensity factors** P.C. Paris, R.M. McMeeking, H. Tada (in *Cracks and Fracture* 1976)
	405	**Numerical evaluation of elastic stress intensity factors by the boundary-integral equation method** Thomas A. Cruse (in *The Surface Crack: Physical Problems and Computational Solutions* 1972)
	423	**Three-dimensional elastic stress analysis of a fracture specimen with an edge crack** T.A. Cruse, W. VanBuren *(International Journal of Fracture Mechanics* 1971)
	438	**Stress-intensity factors for a wide range of semi-elliptical surface cracks in finite-thickness plates** I.S. Raju, J.C. Newman, Jr. *(Engineering Fracture Mechanics* 1979)
	451	**An empirical stress-intensity factor equation for the surface crack** J.C. Newman, Jr., I.S. Raju *(Engineering Fracture Mechanics* 1981)
Section Three **Experimental Methods**	461	**The dynamic stress distribution surrounding a running crack—a photoelastic analysis** A.A. Wells, D. Post *(S.E.S.A. Proceedings* 1958)
	485	**Discussion of "The dynamic stress distribution surrounding a running crack—a photoelastic analysis"** G.R. Irwin *(S.E.S.A. Proceedings* 1958)
	489	**An investigation of propagating cracks by dynamic photoelasticity** W.B. Bradley, A.S. Kobayashi *(Experimental Mechanics* 1970)
	497	**Stress distribution in a tension specimen notched on one edge** J.R. Dixon, J.S. Strannigan, J. McGregor *(Journal of Strain Analysis* 1969)

502 **Photoelastic determination of stress-intensity factors**
R.H. Marloff, M.M. Leven, T.N. Ringler, R.L. Johnson *(Experimental Mechanics* 1971)

513 **An assessment of factors influencing data obtained by the photoelastic stress freezing technique for stress fields near crack tips**
M.A. Schroedl, J.J. McGowan, C.W. Smith *(Engineering Fracture Mechanics* 1972)

524 **A general method for determining mixed-mode stress intensity factors from isochromatic fringe patterns**
Robert J. Sanford, James W. Dally *(Engineering Fracture Mechanics* 1979)

535 **Investigation of the rupture of a Plexiglas plate by means of an optical method involving high-speed filming of the shadows originating around holes drilled in the plate**
Peter Manogg *(International Journal of Fracture Mechanics* 1966)

545 **Analysis of the optical method of caustics for dynamic crack propagation**
Ares J. Rosakis *(Engineering Fracture Mechanics* 1980)

562 **On crack-tip stress state: an experimental evaluation of three-dimensional effects**
Ares J. Rosakis, K. Ravi-Chandar *(International Journal of Solids and Structures* 1986)

576 **An optical method for determining the crack-tip stress intensity factor**
E. Sommer *(Engineering Fracture Mechanics* 1970)

590 **An optical-interference method for experimental stress analysis of cracked structures**
P.B. Crosley, S. Mostovoy, E.J. Ripling *(Engineering Fracture Mechanics* 1971)

604 **The determination of Mode I stress-intensity factors by holographic interferometry**
T.D. Dudderar, H.J. Gorman *(Experimental Mechanics* 1973)

609 **Discussion of "The determination of Mode I stress-intensity factors by holographic interferometry"**
M.E. Fourney *(Experimental Mechanics* 1974)

611 **Fracture analysis by use of acoustic emission**
H.L. Dunegan, D.O. Harris, C.A. Tatro *(Engineering Fracture Mechanics* 1968)

630 **Acousto-elastic measurement of stress and stress intensity factors around crack tips**
A.V. Clark, R.B. Mignogna, R.J. Sanford *(Ultrasonics* 1983)

638 **Strain energy release rate for radial cracks emanating from a pin loaded hole**
D.J. Cartwright, G.A. Ratcliffe *(International Journal of Fracture Mechanics* 1972)

645 **Strain-gage methods for measuring the opening-mode stress-intensity factor, K_I**
J.W. Dally, R.J. Sanford *(Experimental Mechanics* 1987)

653 **Author Index**
655 **Subject Index**

Preface

Fundamental to the theory of linear elastic fracture mechanics is the notion that the functional form of the state of stress at the tip of a crack is an invariant (Irwin, *Journal of Applied Mechanics,* 1957). What is not known *a priori* is the influence of the geometry and applied forces on the magnitude of this stress state. In principle these effects are determined from the solution of boundary value problems in the theory of elasticity. In fracture mechanics theory these geometry/loading effects are embodied in a geometric parameter called the "stress intensity factor." There are a variety of handbooks which tabulate these stress intensity factors for various problems, but the original articles from which the tabulated functions were derived are not as widely available.

This volume consists of original papers and reports which first described solution schemes and/or measurement techniques to determine the stress intensity factor for particular classes of problems. The intent here is not to provide a comprehensive collection of reprints for all of the stress intensity factors tabulated in the handbooks but rather the landmark papers which presented original concepts in mathematics or mechanics from which new solutions can be obtained. The papers are grouped into three categories: analytical, numerical, and experimental.

Not included in this volume are those papers in which the primary focus is the measurement of the crack driving force K_c, as opposed to the geometric factor K. This volume contains 61 papers in over 650 pages and includes several unclassified NASA reports which are frequently cited but generally not readily available outside the United States.

While the selection of papers to be included in this volume unquestionably reflects my own personal choices and biases, I have been privileged to have the guidance and advice of distinguished leaders in the field of fracture mechanics. I wish to express my particular thanks to Drs. George R. Irwin, Albert Kobayashi, and J.C. Newman, Jr., for their review and critique of my list of candidate papers for this volume. I also wish to thank my colleagues and former students now at other institutions for their assistance (and that of their respective libraries) in locating some of the papers included in this collection. Finally, on behalf of researchers, general interest readers, and students, we express our thanks to the authors and publishers of the papers reprinted herein for their contributions to the body of knowledge in the field of fracture mechanics.

R.J. Sanford
Professor Emeritus
University of Maryland
May 1997

Introduction

Section One: Analytical Methods

Irwin's 1957 paper in the *Journal of Applied Mechanics*[1] in which he demonstrated the form invariance of the near-field equations around a crack tip sparked interest within the applied mathematics community to provide the analytical solutions for the stress intensity factor K that Irwin had defined. In a lengthy paper published in 1965 Paris and Sih (p. 3) presented a detailed survey of the then-known solutions for K for both the opening and forward shear modes of deformation. The focus is on the complex variable method for plane stress analysis of crack tip problems, and the paper is considered the primer for anyone interested in this approach. In an appendix Paris and Sih present the fundamentals of Westergaard's unique complex variable formulation[2] of the plane crack problem, and more K solutions, derived using the method, are provided. Regrettably, the underlying Westergaard stress functions which led to the solutions are missing from the paper.

A major contributor of additional Westergaard stress functions is Hiroshi Tada; one example of his many contributions is included in this collection (p. 57). Although the title would indicate that the paper's major focus is on periodic crack problems, the more important contribution is the formulation of Westergaard functions for bodies with internal point loads.

In its early formulation the Westergaard method was limited to the solution of infinite body problems [see the Introduction to a companion volume in this series, *Selected Papers on Foundations of Linear Elastic Fracture Mechanics* (SEM Classic Papers Vol. CP1; SPIE Milestone Series Vol. MS 137), for a more thorough discussion of this limitation and its resolution in later years]. As a consequence, the analytical method of choice for solving problems with finite boundaries during the 1960s and early 1970s was the classical Goursat-Kolossov complex variable formulation of plane elasticity popularized by Muskhelishvili.[3] The next four papers (pp. 61, 73, 78, and 86), by Oscar Bowie and co-authors, are classical examples of this approach. They demonstrate the application of the method to progressively more complex geometries.

Additional topics related to the stress analysis of cracks in isotropic materials are presented in the following four papers. In a still controversial 1967 paper Creager and Paris (p. 92) proposed a modified set of near-field equations for cracks emanating from blunt notches. The next two papers (pp. 98 and 100), by G.G. Chell, examine the problem of arbitrary loading on remote boundaries. In the first paper he combines fundamental principles of energy with finite element results to examine the influence of linear and quadratic stress profiles. In the other he applies Bueckner's weight function principle[4] to determine crack profiles for arbitrary remote stresses.

All of the papers described so far, and for that matter, nearly all of the papers on linear elastic fracture mechanics in general, are directed toward the influence of in-plane forces on the crack tip stress field behavior. Sih, Paris, and Erogan (p. 114) in one grand swoop were able to apply most of these results to cracks in plates undergoing anti-plane bending by comparing the governing differential equations and redefining the stress intensity factor(s) in terms of the customary bending force parameters.

The extension of linear elastic fracture mechanics (LEFM) beyond homogeneous, isotropic materials is of more than academic interest. The emphasis in contemporary materials science on the use of "man-made" materials and bonded interfaces requires an understanding of the nature of the crack tip stress field. To address this issue Sih, Paris, and Irwin (p. 121) extended their earlier work on the complex variable representation of the plane crack problem to anisotropic materials and demonstrated that the characteristic $1/\sqrt{r}$ singularity still persists. They also showed that, in general, the opening and shear modes of deformation are no longer uncoupled. By this we mean that nominal opening mode remote forces will produce local shear stresses at the crack tip and vice versa. As an application of these principles, Bowie and Freese (p. 136) extended their earlier mapping-collocation solution for a central crack in a finite rectangular sheet to include orthotropic material properties.

The first general solution for cracks at the interface of two isotropic materials was due to Williams (p. 145) using the same eigenvalue expansion approach he had used two years earlier to solve the homogeneous problem.[5] An unusual feature of the solution is the presence of a rapidly decaying, oscillating singularity in addition to the classical $1/\sqrt{r}$ singularity. Mathematically, these oscillations lead to material overlap which is, of course, physically impossible. The bimaterial problem was later reexamined by Rice and Sih (p. 151) using the complex variable formulation and the results extended to several important loading cases.

As important as these fundamental papers on the nature of crack tip stress fields were to advancing the discipline of fracture mechanics, the real growth of LEFM was due to its acceptance in practical applications. Toward this end the determination of the stress intensity factor for geometries of engineering interest played a pivotal role. The next series of papers are examples of the application of analytical methods for the solution of some of these practical problems.

Isida (p. 157) answers the question: what is the influence of eccentricity in the crack position in a center-cracked strip? Meanwhile, Feddersen (p. 159) proposed a simple trigonometric function to represent the K-field solution for the finite-width, center-cracked specimen geometry to replace Isida's series solution.

Despite its similarities to the tangent formula of Irwin, his suggestion that there might be a stress function basis for his serendipitous result has not been supported, but not for the lack of trying (see Eftis and Liebowitz, Ref. 6).

Cracks in power generating turbine rotors motivated the solutions by Rooke and Tweed for cracks in rotating disks (pp. 162 and 168). A.F. Grandt, Jr., applied the principles of superposition along with the weight function method to determine the stress intensity factor around fastener holes used extensively in aircraft applications (p. 173). He also examined the case of residual stress around fastener holes using the same analysis. Since his method uses only the residual stress distribution around the unflawed hole, the effects of stress redistribution as the crack advances are not included in the analysis, and the results are suspect. The last of the practical geometries included in this collection is that of the stress intensity factor for a circumferentially notched round bar due to Harris (p. 185).

The final four papers in this section are devoted to analytical methods to address the mathematically challenging problem of the elliptical crack either embedded in an infinite solid (pp. 191 and 202) or interacting with the free surface (pp. 208 and 215). In addition to their historical significance, these papers also serve as benchmark solutions against which contemporary numerical or hybrid solutions can be compared. However, for practical applications the detailed numerical results of Raju and Newman (pp. 438 and 451) are now the accepted standards.

Section Two: Numerical Methods

Despite the advances to the theory of elasticity occasioned by the surge of interest in the fracture mechanics problem, the growth of fracture mechanics into a mature discipline could not be sustained by elegance in mathematics alone. Fortunately, large scale computing power and the FORTRAN programming language became available in the early 1960s. During this same period there was a pressing need within the fracture community for highly accurate elasticity solutions to crack problems for realistic geometries to guide the development of standardized specimens for fracture toughness testing. For its own interests NASA took a lead role in this task. Using their state-of-the-art mainframe computers, NASA engineers combined the just developed general solutions for two-dimensional crack problems with solution schemes for solving large systems of linear equations to generate boundary collocation solutions to a number of practical specimen geometries. These specimens now form the basis for the ASTM fracture toughness testing standards used uniformly throughout the world.

The results of NASA's efforts were presented in report form as NASA Technical Notes, three classical examples of which are included in this volume. In these reports (pp. 231, 242, and 256) the results are presented as tables of values for a range of crack lengths in different geometric shapes. These values were later curve fit to algebraic functions which appeared in the open literature (pp. 269, 280, and 282) and in the ASTM Standard E-399.[7]

A significant advance to the boundary collocation method for crack problems occurred when Newman (p. 285) replaced the Williams stress function with a complex variable stress function that incorporated circular holes. With this new stress function cracks emanating from or approaching holes could be analyzed.

The boundary collocation method had two major drawbacks. First, the approach by its nature requires knowledge of the stress function suitable for the geometry and loading on the model. Second, in the 1960s the method was manpower intensive, with a typical analysis requiring more than a year to complete. (Editor's note: With the present availability of high performance desktop computers and symbolic manipulation software these limitations no longer exist. In fact, I routinely assign a boundary collocation homework problem in my graduate fracture mechanics course.)

To overcome these difficulties it was natural to turn to another computational technique spawned by the availability of the large mainframe computer, i.e., the finite element analysis (FEA) method. The challenges of the FEA approach were to accommodate the extreme strain gradients near the crack tip within a method best suited to smoothly varying stress fields and to extract the stress intensity factor given only nodal information.

One of the earliest examples of the application of the FEA method to fracture mechanics was presented by Chan, Tuba, and Wilson (p. 316) in 1968 (published in 1970). They used a general purpose FEA code with constant strain triangles to model the compact tension specimen geometry. They employed the fundamental definitions of the stress intensity factor and extrapolated to the crack tip to obtain estimates of K. Their paper also describes the importance of mesh refinement in improving the accuracy of the determination. In contrast, Mowbray (p. 333) used the compliance definition of K in combination with linear strain elements. He reported that fine grids were not required but that multiple solutions with increasing crack lengths were needed to generate the compliance values.

Because of the conflict between the lower order gradients permitted with conventional elements and the inherent $1/\sqrt{r}$ strain behavior at a crack tip, attention was soon turned to the development of special elements. Tracey (p. 337) proposed a triangular "singularity" element with the required near-field strain state embodied within it mated to quadrilateral isoparametric elements away from the crack tip. In another paper (p. 348) he extended the concept to three-dimensional elements. Walsh (p. 351) continued Tracey's theme by proposing a band of transitional elements surrounding the singularity elements. These latter elements were constrained to provide compatibility between the inner elements and the conventional isoparametric elements throughout the remainder of the model, thereby removing one of the objections to Tracey's approach.

While the use of singularity elements both improves the accuracy of the stress intensity factor measurement and minimizes the need for extreme mesh refinement, it requires special purpose FEA coding. Barsoum (p. 361) and Henshell and Shaw (p. 374) independently observed that collapsing an 8-noded isoparametric quadrilateral element into a triangular one and moving the mid-side nodes to the quarter point resulted in an element with the required $1/\sqrt{r}$ behavior. With this simple modification any general purpose FEA code that employed 8-noded quadrilateral elements (for 2-D) or 20-noded cubic elements (for 3-D) could be used for fracture mechanics analysis without resorting to fine meshes. It is implicit in these arguments, but never stated, that the size of the singularity (or collapsed node) element cannot exceed the zone of validity of the near-field equations they are meant to model.

Paris, McMeeking, and Tada (p. 387) proposed a novel solution to the fundamental problem of high strain gradients at a crack tip by cutting out (literally) the crack tip entirely! They imposed the theoretical displacements for crack face loading of a Bueckner-type singularity, $r^{-3/2}$, on the small circular boundary created by cutting out the crack tip from the crack problem under consideration and used the resulting FEA solution as input to a Bueckner-type weight function[4] analysis for arbitrary remote boundary loading on the same geometry. Although their example was two dimensional, they suggest that the method would be of more value for three-dimensional crack problems.

The numerical solution of two-dimensional crack problems with the finite element method can now be accomplished with relative ease. However, for three-dimensional problems the complexities of mesh generation and K extraction in three-dimensional space with FEA methods have hampered its widespread implementation. For this class of problems there is a newer numerical method receiving considerable attention within the applied mathematics community. The boundary-integral equation (BIE) method reduces the volume problem of three-dimensional elasticity to a two-dimensional surface (the boundary) problem. In many respects the BIE method is the modern day equivalent of the boundary collocation method with many of the same advantages and constraints. While the literature on this topic is vast, most of it is outside the scope of this volume. However, two early papers on the boundary-integral approach by Cruse are included to illustrate the potential of the method. The first paper (p. 405) is one of the earliest published on the subject within the context of fracture mechanics. It outlines the solution scheme and shows sample results for simple planar models and the three-dimensional surface flaw problem. The second paper, with VanBuren (p. 423), is a detailed analysis of the three-dimensional stress state in the compact tension specimen geometry. This paper was instrumental in understanding the plane stress to plane strain transition behavior of the plastic zone in this specimen geometry.

The final two papers in this section are an application of the three-dimensional finite element method to a problem of great practical importance. The papers are noteworthy not for any innovations but rather for the sheer size of the problem attempted and the volume of data generated. In these papers Raju and Newman collaborate to produce definitive results for the semi-elliptical surface flaw problem in finite thickness plates for a wide range of parametric values. In the first paper (p. 438) the systematic approach to collecting and analyzing the FEA data is described and sample results are shown. In the second paper (p. 451) the large volume of data generated is subjected to a systematic curve-fitting procedure using double-series expansions in terms of dimensionless parameter ratios. The resulting master equations for both remote tension and bending loads are reported to be accurate to within 5% of the individual FEA results over nearly all of the parameter range.

Section Three: Experimental Methods

The advancement of the science of fracture mechanics has had as one of its by-products the involvement of a wide variety of subdisciplines within the applied mechanics community. The experimental mechanics community is no exception. Nearly every major experimental mechanics method has been applied to the fracture problem with particular emphasis on optical methods. The papers presented in this section represent a sampling of these applications.

The fact that Wells and Post (p. 461) would use their novel dynamic polariscope/interferometer to study the running crack problem was no historical accident. At the time the research reported in this paper was conducted both authors were at the Naval Research Laboratory in Washington, D.C., working just a few feet from George Irwin's office. Of the two full-field optical methods described in the paper the photoelastic method was far easier to implement and the Wells and Post dynamic polariscope system was duplicated at several major university laboratories in the years following publication of this paper. The impact of the paper was further multiplied by comments of Dr. Irwin in a prepared Discussion (p. 485) appended to the main paper. Irwin described for the first time the significance of the stress parallel to the crack line, σ_{ox}, on experimental measurements of the stress intensity factor. Equally important, he proposed a simple method for extracting both K and σ_{ox} from the isochromatic fringe loops. Irwin's apogee method was to become the accepted method for the next 20 years for determining these fracture parameters.

One of the laboratories to follow the NRL example was at the University of Washington. Their 16 frame Schardin-type dynamic camera was used for a wide variety of crack propagation studies. The paper included here (p. 489) contains examples of both crack branching and arrest. Also, in this paper Bradley and Kobayashi proposed a modification of Irwin's method for extracting the stress intensity factor.

The simplicity (and popularity) of the photoelastic method lent itself well to static determination of K. Dixon et al. (p. 497) obtained the stress intensity factor in a pin-loaded edge notched specimen. Rather than use a model with an actual crack, they used a notch with a well defined finite root radius and employed the Neuber stress concentration argument proposed by Irwin[8] to compute K. Marloff et al. (p. 502) also used the Neuber analysis as well as classical limit definitions to determine the stress intensity factor in several complex three-dimensional engineering structures using the stress-freezing (a misnomer) properties of certain epoxy compounds. This method takes advantage of the bipolar characteristic of these materials to lock-in the deformed state upon slow cooling from above the critical temperature for the material. The model can then be cut into slices and analyzed plane by plane.

The foremost practitioner of the stress-freezing method has been C.W. Smith at Virginia Polytechnic Institute and State University. Over the years Prof. Smith and his students have performed numerous three-dimensional analyses of cracked bodies. The paper included here is an early one in which Schroedl, McGowan, and Smith (p. 513) discuss some of the subtleties associated with accurate extraction of the stress intensity factor from stress-frozen models.

The final paper in this collection on the photoelastic approach is co-authored by myself and J.W. Dally (p. 524). In the paper we review the previous approaches for determining mixed-mode stress intensity factors from photoelastic models and propose an overdetermined method using the principle of least squares. This method is the experimental equivalent of the boundary collocation method used for numerical solution of fracture problems.

Optical stress analysts had observed that, when viewed in collimated light, the crack tip in transparent models was obscured by a black shadow spot. Peter Manogg (p. 535) at the Ernst Mach Institute investigated the origin of this spot and showed that the stress intensity factor could be computed from the transverse diameter of the spot. This method, now called the method of caustics, has enjoyed wide acceptance in European fracture research. A thorough review by Kalthoff is available in a recent handbook.[9]

Traditionally, a quasi-static analysis has been used to derive the governing equations for optical dynamic fracture experiments even for crack velocities approaching branching speeds. In a definitive paper on the subject Ares Rosakis (p. 545) presented in 1980 an exact elasto-dynamic analysis of the method of caustics and showed that the use of quasi-static equations can result in errors of up to 40%. In a totally different application of the method of caustics Rosakis and Ravi-Chandar (p. 562) exploited the sensitivity of the caustic diameter to plane stress vs plane strain conditions. In a series of very carefully conducted experiments they measured the caustic diameter for increasingly small data acquisition regions around the crack tip. When this region was less than one-half the plate thickness, the experimental patterns deviated from their theoretical predictions. By this method they defined the region of three-dimensional influence at a crack tip that is now the accepted limit for data acquisition.

The crack opening profile of a specimen is uniquely related to the magnitude of the applied stress intensity factor. In glass and transparent epoxy specimens Newton's air gap fringes are produced by light rays reflecting off the partially reflecting opposite faces of the crack. By observing these fringes either from the front plane of the specimen (Sommer, p. 576) or from the edge of the plate (Crosley et al., p. 590) accurate determinations of the stress intensity factor can be made despite the low loads these materials will support.

It is well known that multiple beam interferometry is far more sensitive than photoelasticity for measuring the state of stress near a crack tip, i.e., it produces vastly more fringes. However, the order of magnitude increase in difficulty associated with interferometry is hardly justified in all but extreme cases. On the other hand, the holographic interferometer is relatively easy to construct in those laboratories equipped for such measurements. In experiments in one such laboratory, Bell Laboratories, Dudderar and Gorman (p. 604) used a PMMA model to produce a large number of interference fringes around a crack tip and from them determined the stress intensity factor. In a discussion (p. 609) of that paper M.E. Fourney observed that the region of data acquisition far exceeded the near-field region and compared their results with published results of Theocaris[10] on the size of the plastic zone in experiments with the same material. In their closure Dudderar and Gorman correctly observe that the plastic zone described by Theocaris is, in fact, an optical caustic of the type first described by Manogg.

From the list of papers described so far in this section one might be led to the conclusion that only optical measurements can be correlated to fracture parameters, but this is not the case. In the remainder of this section papers in which non-optical methods of experimental mechanics are used to determine the stress intensity factor are presented. In the first paper (p. 611) acoustic emissions from growing cracks are counted and correlated with the applied stress intensity level. The authors report that there appears to be a power-law relation between the measured emission count and the stress intensity factor. These results have received little corroboration and the role of acoustic emission in fracture mechanics appears best suited to crack growth monitoring. The next paper (p. 630) uses the acoustic birefringence analog of optical photoelasticity to measure the shear stress near a crack tip and, from that, the stress intensity factor.

The determination of K by non-optical means does not require special purpose electronics. Instrumentation available in most industrial and materials testing laboratories will suffice. Cartwright and Radcliffe (p. 638) used traditional extensometers and load cells to determine the compliance of a pin-load hole containing a crack. Although the measurements must be made with some precision, the technique described in this paper can be applied to nearly any manufactured part containing an in-service crack. The final paper in this collection (p. 645) demonstrates that a single electrical resistance strain gauge and commercial instrumentation can be used to measure the stress intensity factor provided the gauge is placed in an unconventional but well defined location relative to the crack tip.

References

1. Irwin, G.R., "Analysis of Stresses and Strains Near the End of a Crack Traversing a Plate," J. Applied Mechanics **24**, 361-364 (1957).
2. Westergaard, H.M., "Bearing Pressures and Cracks," Trans. ASME **61**, A49-A53 (1939).
3. Muskhelishvili, N.I., *Some Basic Problems of the Mathematical Theory of Elasticity*, Noordhoff Ltd., Netherlands (1953).
4. Bueckner, H.F., "A Novel Principle for the Computation of Stress Intensity Factors," Zeitschrift für Angewandte Mathematik und Mechanik **50**(9), 529-546 (1970).
5. Williams, M.L., "On the Stress Distribution at the Base of a Stationary Crack," J. Applied Mechanics **24**, 109-114 (1957).
6. Eftis, J., and Liebowitz, H., "On the Modified Westergaard Equations for Certain Plane Crack Problems," Int'l J. Fracture Mechanics **8**(4), 383-392 (1972).
7. ASTM Standard E 399, "Standard Test Method for Plane-Strain Fracture Toughness Testing of Metallic Materials," American Society for Testing and Materials, Philadelphia, PA, USA.
8. Irwin, G.R., "Fracture Mechanics," in *Proc. Symposium on Naval Structural Mechanics*, J.N. Goodier and N.J. Hoff, editors, Pergamon Press, Oxford, pp. 557-591 (1960).
9. Kalthoff, J.F., "Shadow Optical Method of Caustics," Chap. 9 in *Handbook on Experimental Mechanics,* 2nd Revised Edition, A.S. Kobayashi, editor, VCH Publishers, Inc., New York, pp. 407-476 (1993).
10. Theocaris, P.S., "Local Yielding Around a Crack Tip in Plexiglas," J. Applied Mechanics **37**(2), 409-415 (1970).

Section One
Analytical Methods

STRESS ANALYSIS OF CRACKS

By Paul C. Paris[1] and George C. Sih[1]

Synopsis

A general survey of the results of elastic stress analyses of cracked bodies is the basic objective of this work. The stress-intensity-factor method of representing results is stressed and compared with other similar methods. All three modes of crack-surface displacements are considered, as well as specialized results applicable to plate and shell bending. Results for various media (for example, anisotropic, viscoelastic, or nonhomogeneous) are contrasted with the analysis of homogeneous isotropic media. The accuracy of the representation of the crack-tip stress fields by stress-intensity factor methods is discussed, pointing out some limitations of applicability. Methods of estimating and approximate analysis for stress-intensity factors in complicated practical circumstances are also discussed.

The redistribution of stresses in bodies caused by the introduction of a crack is one of the essential features which should be incorporated into an analysis of strength of structures with flaws. Moreover, the high elevation of stresses near the tip of a crack should receive the utmost attention, since it is at that point that additional growth of the crack takes place. As a consequence, it is the purpose of this paper to present a summary of current knowledge of crack-tip stress fields and of the means of determination of the intensity of those fields.

Small amounts of plasticity and other nonlinear effects may be viewed as taking place well within the crack-tip stress field and hence may be neglected in this presentation of the gross features of those fields. It is the subject of other discussions to assess the effects caused by the fields, for example, the plasticity within them and other requirements of formulation of a complete theory of fracture behavior.

In his now famous paper, Griffith (1)[2] made use of the stress solution provided by Inglis (2) for a flat plate under uniform tension with an elliptical hole which could be degenerated into a crack. However, neither Griffith nor his predecessors had the knowledge of stress fields near cracks which is now available, so as a consequence, he devised an energy-rate analysis of equilibrium of cracks in brittle materials. Sneddon (3) was the first to give stress-field expansions for crack tips for two individual examples; however, it was only later that Irwin (4,5) and Williams (6) recognized the general applicability of these field equations and extended them to the most general case for an isotropic elastic body (5). It is this analysis to which initial attention shall be given.

---

[1] Associate professor of mechanics, Lehigh University, Bethlehem, Pa.

[2] The boldface numbers in parentheses refer to the list of references appended to this paper.

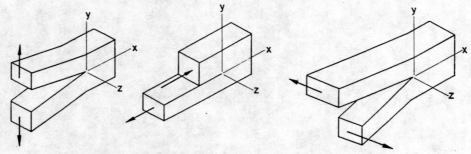

Fig. 1—The Basic Modes of Crack Surface Displacements.

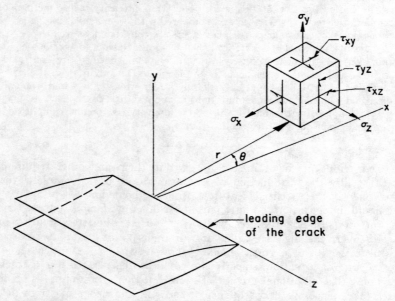

Fig. 2—Coordinates Measured from the Leading Edge of a Crack and the Stress Components in the Crack Tip Stress Field.

Crack-Tip Stress Fields for Isotropic Elastic Bodies

The surfaces of a crack, since they are stress-free boundaries of the body near the crack tip, are the dominating influence on the distributions of stresses in that vicinity. Other remote boundaries and loading forces affect only the intensity of the local stress field.

The stress fields near crack tips can be divided into three basic types, each associated with a local mode of deformation as illustrated in Fig. 1. The opening mode, I, is associated with local displacement in which the crack surfaces move directly apart (symmetric with respect to the x-y and x-z planes). The edge-sliding mode, II, is characterized by displacements in which the crack surfaces slide over one another perpendicular to the leading edge of the crack (symmetric with respect to the x-y plane and skew-symmetric with respect to the

x-z plane). Mode III, tearing, finds the crack surfaces sliding with respect to one another parallel to the leading edge (skew-symmetric with respect to the x-y and x-z planes). The superposition of these three modes is sufficient to describe the most general case of crack-tip deformation and stress fields.

The most direct approach to determination of the stress and displacement fields associated with each mode follows in the manner of Irwin (4,7), based on the method of Westergaard (8). Modes I and II can be analyzed as plane-extensional problems of the theory of elasticity which are subdivided as symmetric and skew-symmetric, respectively, with respect to the crack plane. Mode III can be regarded as the pure shear (or torsion) problem. Referring to Fig. 2 for notation, the resulting stress and displacement fields are given below (a full derivation is found in Appendix I).

Mode I:[3]

$$\begin{aligned}
\sigma_x &= \frac{K_\mathrm{I}}{(2\pi r)^{1/2}} \cos\frac{\theta}{2}\left[1 - \sin\frac{\theta}{2}\sin\frac{3\theta}{2}\right] \\
\sigma_y &= \frac{K_\mathrm{I}}{(2\pi r)^{1/2}} \cos\frac{\theta}{2}\left[1 + \sin\frac{\theta}{2}\sin\frac{3\theta}{2}\right] \\
\tau_{xy} &= \frac{K_\mathrm{I}}{(2\pi r)^{1/2}} \sin\frac{\theta}{2}\cos\frac{\theta}{2}\cos\frac{3\theta}{2} \\
\sigma_z &= \nu(\sigma_x + \sigma_y), \quad \tau_{xz} = \tau_{yz} = 0 \\
u &= \frac{K_\mathrm{I}}{G}[r/(2\pi)]^{1/2}\cos\frac{\theta}{2} \\
&\quad \cdot \left[1 - 2\nu + \sin^2\frac{\theta}{2}\right] \\
v &= \frac{K_\mathrm{I}}{G}[r/(2\pi)]^{1/2}\sin\frac{\theta}{2} \\
&\quad \cdot \left[2 - 2\nu - \cos^2\frac{\theta}{2}\right] \\
w &= 0
\end{aligned} \quad \ldots(1)$$

[3] See Appendix III for explanation of mathematical symbols.

Mode II:

$$\begin{aligned}
\sigma_x &= -\frac{K_\mathrm{II}}{(2\pi r)^{1/2}} \sin\frac{\theta}{2}\left[2 + \cos\frac{\theta}{2}\cos\frac{3\theta}{2}\right] \\
\sigma_y &= \frac{K_\mathrm{II}}{(2\pi r)^{1/2}} \sin\frac{\theta}{2}\cos\frac{\theta}{2}\cos\frac{3\theta}{2} \\
\tau_{xy} &= \frac{K_\mathrm{II}}{(2\pi r)^{1/2}} \cos\frac{\theta}{2}\left[1 - \sin\frac{\theta}{2}\sin\frac{3\theta}{2}\right] \\
\sigma_z &= \nu(\sigma_x + \sigma_y), \quad \tau_{xz} = \tau_{yz} = 0 \\
u &= \frac{K_\mathrm{II}}{G}[r/(2\pi)]^{1/2}\sin\frac{\theta}{2} \\
&\quad \cdot \left[2 - 2\nu + \cos^2\frac{\theta}{2}\right] \\
v &= \frac{K_\mathrm{II}}{G}[r/(2\pi)]^{1/2}\cos\frac{\theta}{2} \\
&\quad \cdot \left[-1 + 2\nu + \sin^2\frac{\theta}{2}\right] \\
w &= 0
\end{aligned} \quad \ldots(2)$$

Mode III:

$$\begin{aligned}
\tau_{xz} &= -\frac{K_\mathrm{III}}{(2\pi r)^{1/2}} \sin\frac{\theta}{2} \\
\tau_{yz} &= \frac{K_\mathrm{III}}{(2\pi r)^{1/2}} \cos\frac{\theta}{2} \\
\sigma_x &= \sigma_y = \sigma_z = \tau_{xy} = 0 \\
w &= \frac{K_\mathrm{III}}{G}[(2r)/\pi]^{1/2}\sin\frac{\theta}{2} \\
u &= v = 0
\end{aligned} \quad \ldots(3)$$

Equations 1 and 2 have been written for the case of plane strain (that is, $w = 0$) but can be changed to plane stress easily by taking $\sigma_z = 0$ and replacing Poisson's ratio, ν, in the displacements with an appropriate value. Equations 1, 2, and 3 have been obtained by neglecting higher-order terms in r. Hence, they can be regarded as a good approximation in the region where r is small compared to other planar (x-y plane) dimensions of a body such as crack length and exact in the limit as r approaches zero.

The parameters, K_I, K_{II}, and K_{III} in the equations are stress-intensity factors[4] for the corresponding three types of stress and displacement fields. It is important to notice that the stress-intensity factors are not dependent on the coordinates, r and θ; hence they control the intensity of the stress fields but not the distribution for each mode. From dimensional considerations of Eqs 1, 2, and 3, it can be observed that

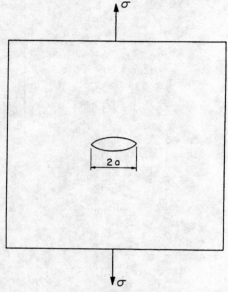

FIG. 3—An Infinite Cracked Sheet with Uniform Normal Stress at Infinity.

the stress-intensity factors must contain the magnitude of loading forces linearly for linear elastic bodies and must also depend upon the configuration of the body including the crack size. Consequently, stress-intensity factors may be physically interpreted as parameters which reflect the redistribution of stress in a body due to the introduction of a crack, and in particular they indicate the type (mode) and magnitude of force transmission through the crack tip region.

ELEMENTARY DIMENSIONAL CONSIDERATIONS FOR DETERMINATION OF STRESS-INTENSITY FACTORS

An infinite plate subjected to uniform tensile stress, σ, into which a transverse crack of length, $2a$, has been introduced, is shown in Fig. 3. As a two-dimensional problem of theory of elasticity, only two characteristic dimensions are present, σ and a. Moreover, this configura-

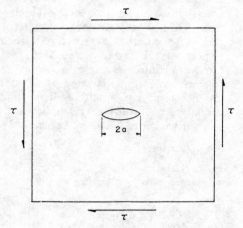

FIG. 4—An Infinite Cracked Sheet with Uniform In-Plane Shear at Infinity.

tion is symmetric with respect to the crack plane; therefore, only the first-mode fields are present. Then, simply from dimensional consideration (9) with Eqs 1, the only possibility is:

$$\left.\begin{array}{l}K_I = C_1\sigma a^{1/2}\\K_{II} = K_{III} = 0\end{array}\right\}\quad\ldots\ldots\ldots(4)$$

Hence, observations of symmetry and dimensional analysis can aid in the determination of stress-intensity factors. Though C_1 is undetermined by such considerations, later results will show it to be $\pi^{1/2}$, (see Eq 17). However, even if C_1 is left undetermined, the fracture-

[4] These stress-intensity factors differ by a factor of $\pi^{1/2}$ with earlier definitions of them.

size effect can be predicted for this configuration, since[5] as $K_I \rightarrow K_{Ic}$, then

$$\sigma a^{1/2} = \text{constant} \quad \ldots \ldots \ldots (5)$$

By similar considerations of the plane-extensional problem of a plate under shear, as shown in Fig. 4, the stress-intensity factors are

$$\left.\begin{array}{l} K_{II} = \tau(\pi a)^{1/2} \\ K_I = K_{III} = 0 \end{array}\right\} \ldots \ldots (6)$$

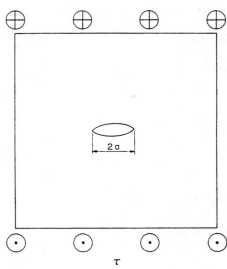

FIG. 5—An Infinite Body with a "Tunnel Crack" Subjected to Out-of-Plane Shear at Infinity.

Moreover, analogous results may be obtained for the problem shown in Fig. 5, that is, an infinity body with shear applied parallel to a tunnel crack of width, $2a$. They are:

$$\left.\begin{array}{l} K_{III} = \tau(\pi a)^{1/2} \\ K_I = K_{II} = 0 \end{array}\right\} \ldots \ldots (7)$$

Though these are relatively interesting examples, more complicated configurations are of practical importance; consequently, more powerful methods of analysis will be cited.

---

[5] $K_I \rightarrow K_{Ic}$, as a fracture criterion, is discussed in many other papers in this symposium.

Stress-intensity factors can be determined from the limiting values of elastic stress-concentration factors (7) as the root radius, p, of the notch approaches zero. Consider a symmetrically loaded notch whereupon the tip will be embedded within a mode I stress field. The maximum stress, σ_o will occur directly ahead of the notch. Again, dimensional considerations of Eqs 1 lead to

$$K_I = C_2 \sigma_o(p)^{1/2} \ldots \ldots \ldots (8)$$

In the limiting case, the notch approaches a crack, as $p \rightarrow 0$, or

$$K_I = \lim_{p \to 0} \frac{\pi^{1/2}}{2} \sigma_o p^{1/2} \ldots \ldots (9)$$

The constant has been evaluated from Eq 4 and the stress-concentration solution for an elliptical hole in the configuration shown in Fig. 3, which is:

$$\sigma_o = \sigma[1 + 2(a/p)^{1/2}] \ldots \ldots (10)$$

A multitude of stress-concentration solutions available in the works of Neuber (10), Peterson (11), Savin (12), Isida (13) and others can be used to determine stress-intensity factors for many configurations. Formulas corresponding to Eq 9 can be as easily derived for modes II and III. They appear in Appendix II.

From the above dimensional considerations, it is evident that the appearance of the $1/r^{1/2}$ type of singularity in the stress-field equations (Eqs 1, 2 and 3) is a controlling feature in fracture-size effects, the relationship of stress concentrations to stress intensity factors, and, as will be noted later, extension of fracture mechanics concepts to other than isotropic, elastic media.

STRESS-INTENSITY FACTORS FROM WESTERGAARD STRESS FUNCTIONS

Several sources (4,5,7,8) give Westergaard stress functions, Z, for crack problems. A discussion of the analysis of plane problems with this type of stress function is given in Appendix I.

For each of the three modes of crack-tip stress fields the Westergaard stress function in the neighborhood of the crack tip takes the form

$$Z = \frac{f(\zeta)}{\zeta^{1/2}}, \qquad \zeta = re^{i\theta} \quad \ldots (11)$$

where $f(\zeta)$ must be well behaved in that vicinity in order to ensure stress-free crack surfaces.[6] Hence in the region close to the crack tip, that is, $|\zeta| \to 0$, it is

As an example, consider a plate with an infinite periodic array of cracks along a line with uniform tension, σ; the half period is b and the half-crack length, a, as shown in Fig. 6. The stress function for this configuration is (4):

$$Z_1 = \frac{\sigma \sin \frac{\pi z}{2b}}{\left[\left(\sin \frac{\pi z}{2b}\right)^2 - \left(\sin \frac{\pi a}{2b}\right)^2\right]^{1/2}} \quad \ldots (16)$$

In order to move the crack tip to the origin, substitute $z = x + iy = a + \zeta$ and trigonometric identities. Eliminating

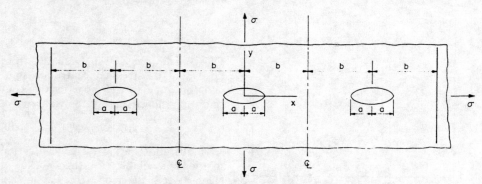

FIG. 6—A Periodic Array of Cracks Along a Line in a Sheet with Uniform Stress at Infinity.

permissible to represent the stress function as (5):

$$Z\big|_{|\zeta|\to 0} = \frac{f(0)}{\zeta^{1/2}} \quad \ldots \ldots (12)$$

for mode I stress fields (see Appendix I). Comparing σ_y along the x-axis as computed from Eq 12 and as given in Eq 1 leads to:

$$K_\mathrm{I} = \lim_{|\zeta|\to 0} (2\pi\zeta)^{1/2} Z_\mathrm{I} \quad \ldots (13)$$

In a similar fashion for the other modes:

$$K_\mathrm{II} = \lim_{|\zeta|\to 0} (2\pi\zeta)^{1/2} Z_\mathrm{II} \quad \ldots (14)$$

$$K_\mathrm{III} = \lim_{|\zeta|\to 0} (2\pi\zeta)^{1/2} Z_\mathrm{III} \quad \ldots (15)$$

---
[6] Simple poles away from the crack tip will appear at locations of concentrated forces.

terms of the order of ζ compared to terms of the order of a, the limiting process in Eq 13 leads to:

$$\left.\begin{array}{l} K_\mathrm{I} = \sigma(\pi a)^{1/2}\left(\dfrac{2b}{\pi a}\tan\dfrac{\pi a}{2b}\right)^{1/2} \\ K_\mathrm{II} = K_\mathrm{III} = 0 \end{array}\right\} \ldots(17)$$

Referring to Fig. 6, the indicated axes of symmetry are lines devoid of shear stress. Subtracting a uniform normal stress, σ, in the horizontal direction leads to no change in K_I and leaves only small self-equilibrating normal stresses, σ_x, along these lines provided a is small compared to b. Consequently, it is regarded as permissible to cut the sheet along these lines and to use Eq 17 as an approximate solution for finite width strips with central cracks provided a is less than $b/2$. Results computed for strips by Isida (13) and Kobayashi (14,15), which are accurate to much larger relative values of a, indicate that this practice is sound (within 7 per cent) (see Table 1).

Similarly, cutting the problem in Fig. 6 along the y-axis and similar lines leads to an approximate solution, Eq 17, for double-edge-notched strips which is acceptably accurate if a is greater than $b/2$ (within 2 per cent) (see Table 3). Bowie (16) has calculated results for edge-notched strips which verify this accuracy.

The configuration shown in Fig. 6 with the applied stress, σ, replaced by in plane-shear stress, τ, leads to:

$$Z_{II} = \frac{\tau \sin \frac{\pi z}{2b}}{\left[\left(\sin \frac{\pi z}{2b}\right)^2 - \left(\sin \frac{\pi a}{2b}\right)^2\right]^{1/2}} \quad (18)$$

Making use of Eq 14 results in:

$$\left. \begin{array}{c} K_{II} = \tau(\pi a)^{1/2} \left(\dfrac{2b}{\pi a} \tan \dfrac{\pi a}{2b}\right)^{1/2} \\ (K_I = K_{III} = 0) \end{array} \right\} \quad (19)$$

In a like fashion, all results such as Eqs 16 and 17 for symmetric problems, mode I, are analogous to the corresponding mode II problem Eqs 18 and 19, obtained by rotation of boundary forces or stresses through 90 deg in plane, or both, when treating extension of infinite plates and certain other cases.

Moreover, for the corresponding mode III problem, with the stress, σ, replaced by out-of-plane shear, τ, for a body of infinite extent in all directions, the stress function is identical to Eq 18 and the stress-intensity factor is:

$$\left. \begin{array}{c} K_{III} = \tau(\pi a)^{1/2} \left(\dfrac{2b}{\pi a} \tan \dfrac{\pi a}{2b}\right)^{1/2} \\ (K_I = K_{II} = 0) \end{array} \right\} \quad (20)$$

It can be noted that the above examples of stress-intensity factors from Westergaard stress functions, Eqs 17, 19 and 20, lead to the results in earlier examples, Eqs 4, 6, and 7, if b becomes very large compared to a.

Westergaard stress functions are available for many problems and with some experience it is easy to add solutions, but there are limitations to the scope of the method. The most serious drawback is that the method is normally restricted to infinite plane (two-dimensional) bodies with cracks along a single straight line. Another more versatile approach to plane problems is available.

Stress-Intensity Factors from General Complex Stress Functions

A complex stress-function approach developed by Muskhelishvili (17) and others has some advantages over the Westergaard method by treating a broader class of plane extensional problems.

An Airy stress function, Φ, must satisfy the boundary conditions of a problem and the biharmonic equation, that is (see Appendix I),

$$\nabla^4 \Phi = 0 \quad (21)$$

The general solution to Eq 21 may be expressed as (17)

$$\Phi = \mathrm{Re}\,[\bar{z}\phi(z) + \chi(z)] \quad (22)$$

From this form for Φ, the sum of the normal stresses becomes

$$\sigma_x + \sigma_y = 4\,\mathrm{Re}\,[\phi'(z)] \quad (23)$$

Defining a complex stress-intensity factor (18) by

$$K = K_I - iK_{II} \quad (24)$$

Eqs 1, 2, and 24 may be combined to give the same stress combination in the vicinity of a crack tip. The result is

$$\sigma_x + \sigma_y = \mathrm{Re}\left[\frac{\sqrt{2}}{(\pi \zeta)^{1/2}} K\right] \quad (25)$$

for a crack tip at z_1 and for corresponding coordinate directions, that is,

$$\zeta = z - z_1 \quad \ldots \ldots (26)$$

Substitution of Eq 26 into Eq 25 and comparison of the result with Eq 23 lead to

$$K = K_I - iK_{II} = 2(2\pi)^{1/2} \lim_{z \to z_1} \cdot (z - z_1)^{1/2} \phi'(z) \ldots (27)$$

The function, $\phi(z)$, has been determined for a large number of crack problems (12,17–20), since with this technique conformal mapping of holes into cracks is permitted.

For a mapping function, $z = w(\eta)$, Eq 27 becomes

$$K = 2\sqrt{2\pi} \lim_{\eta \to \eta_1} (w(\eta) - w(\eta_1))^{1/2} \frac{\phi'(\eta)}{w'(\eta)} \ldots (28)$$

The mapping of a crack of length, $2a$, into a circular hole of unit radius is given by

$$z = w(\eta) = \frac{a}{2}\left(\eta + \frac{1}{\eta}\right) \ldots \ldots (29)$$

For this mapping, Eq 28 simplifies to

$$K = 2(\pi/a)^{1/2} \phi'(1) \ldots \ldots (30)$$

The example of a single concentrated force, F (per unit thickness), on a crack surface with arbitrary inclination, as shown in Fig. 7, is solved by (17,18):

$$\phi(\eta) = \frac{Fa}{4\pi(a^2 - b^2)^{1/2}} \left\{ -\frac{1}{\eta} + \left(\frac{\eta_o}{\eta_o - \eta}\right) \right.$$
$$\cdot \left[\left(\eta + \frac{1}{\eta}\right) - \left(\eta_o + \frac{1}{\eta_o}\right)\right] + \left(\eta_o - \frac{1}{\eta_o}\right)$$
$$\left. \cdot \left[\frac{\kappa}{1 + \kappa} \log \eta - \log(\eta_o - \eta)\right] \right\} \ldots (31)$$

where η_o corresponds to $z = b$, $F = P - iQ$, and κ is an elastic constant, which for plane strain is $\kappa = 3.4\nu$.

Using Eq 30 with Eq 31, the stress-intensity factors are:

$$K_I = \frac{P}{2(\pi a)^{1/2}} \left(\frac{a+b}{a-b}\right)^{1/2}$$
$$+ \frac{Q}{2(\pi a)^{1/2}} \left(\frac{\kappa - 1}{\kappa + 1}\right)$$
$$K_{II} = \frac{-P}{2(\pi a)^{1/2}} \left(\frac{\kappa - 1}{\kappa + 1}\right)$$
$$+ \frac{Q}{2(\pi a)^{1/2}} \left(\frac{a+b}{a-b}\right)^{1/2} \quad \right\} \ldots (32)$$

The concentrated force results, Eqs 32, provide the Green's functions to

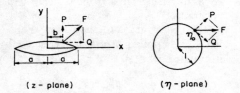

FIG. 7—A Concentrated Force (Per Unit Thickness) on the Surface of a Crack in an Infinite Sheet.

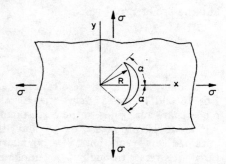

FIG. 8—A Curved Crack in an Infinite Sheet Subjected to Uniform Biaxial Tension.

solve any single straight-crack problem in an infinite plane from a knowledge of the stresses on the prospective crack surface with the crack absent, that is, $\sigma_y(x,0)$ and $\tau_{xy}(x,0)$. The solution is

$$K_I = \frac{1}{(\pi a)^{1/2}} \int_{-a}^{a} \sigma_y(x,0) \left(\frac{a+x}{a-x}\right)^{1/2} dx$$
$$K_{II} = \frac{1}{(\pi a)^{1/2}} \int_{-a}^{a} \tau_{xy}(x,0) \left(\frac{a+x}{a-x}\right)^{1/2} dx \quad \right\} \ldots (33)$$

In order further to illustrate the versatility of the complex stress-function method, the problem of a crack of radius R, subtending an arc of angle, 2α, symmetrically with respect to the x-axis in an infinite sheet subjected to uniform biaxial tension may be treated, see Fig. 8. For this case, Muskhelishvili (17) gives

$$\phi'(z) = \frac{\sigma(R)^{1/2}}{2\left(1 + \sin^2 \frac{\alpha}{2}\right)}$$

$$\cdot \left\{ \frac{\frac{z}{R} - \cos \alpha}{\left[1 - 2\frac{z}{R}\cos \alpha + \frac{z^2}{R^2}\right]^{1/2}} + \sin^2 \frac{\alpha}{2} \right\} \ldots (34)$$

Relocation of a crack tip on the x-axis, as required by Eq 27, may be accomplished by the substitution:

$$\frac{\hat{z}}{R} = ie^{i\alpha}\left(\frac{z}{R} - i - \sin \alpha \cos \alpha\right) \ldots (35)$$

whereupon Eqs 27, 34, and 35 give

$$\left.\begin{array}{l} K_{\mathrm{I}} = \dfrac{\sigma(\pi R)^{1/2}}{\left(1 + \sin^2 \dfrac{\alpha}{2}\right)} \\[6pt] \qquad \cdot \left(\dfrac{\sin \alpha (1 + \cos \alpha)}{2}\right)^{1/2} \\[10pt] K_{\mathrm{II}} = \dfrac{\sigma(\pi R)^{1/2}}{\left(1 + \sin^2 \dfrac{\alpha}{2}\right)} \\[6pt] \qquad \cdot \left(\dfrac{\sin \alpha (1 - \cos \alpha)}{2}\right)^{1/2} \end{array}\right\} \ldots (36)$$

Other notable examples of stress-intensity factors for rather complicated cases of plane extension have been provided (18, 21–23), using this and similar methods. The power of this method for plane extension has been sufficiently illustrated, consequently, additional examples will be removed to Appendix II.

A similar complex-variable approach has been developed to determine stress-intensity factors in prismatic bars (with prismatic cracks) subjected to torsion and flexure (24–26). This type of configuration leads to mode III stress-

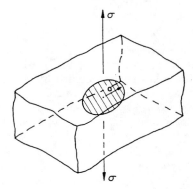

Fig. 9—A "Penny-Shaped" (Circular Disk) Crack in an Infinite Body Subjected to Uniform Tension.

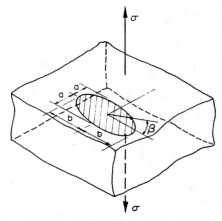

Fig. 10—An Elliptical Crack in an Infinite Body Subjected to Uniform Tension.

intensity factors, some of which will also be tabulated in Appendix II.

STRESS-INTENSITY FACTORS FOR SOME
THREE-DIMENSIONAL CASES

Using a method employing Fourier transforms, Sneddon (3) treated the case of a circular disk crack of radius, a, in an infinite solid subjected to uniform tension, σ, normal to the crack plane, see Fig. 9. His results for crack-tip stress field expansions lead to:

$$K_1 = \frac{2}{(\pi)^{1/2}} \sigma(a)^{1/2} \ldots\ldots\ldots(37)$$

(by symmetry $K_{II} = K_{III} = 0$)

The analysis of stresses near ellipsoidal cavities in infinite bodies subjected to tension has been discussed by Sadowsky (27) and Green (28). However, difficulties arise in the stresses computed from their results near the crack edge when the ellipsoid is degenerated into a crack, see Fig. 10. Subsequently, Irwin (29) calculated the stress-intensity factor at any location on the crack border, described by the angle, β, by comparing Green's results for displacements with Eq 1. The formulas obtained are:

$$K_1 = \frac{\sigma(\pi a)^{1/2}}{\Phi_o} \left(\sin^2 \beta + \frac{a^2}{b^2} \cos^2 \beta \right)^{1/4} \ldots(38)$$

(by symmetry $K_{II} = K_{III} = 0$)

where Φ_o is the elliptic integral[7]

$$\Phi_o = \int_0^{\pi/2} \left[1 - \left(\frac{b^2 - a^2}{b^2} \right) \sin^2 \theta \right]^{1/2} d\theta \ldots(39)$$

Notice that for $b = \infty$, $\beta = \pi/2$, Eqs 38 and 39 reduce to Eq 4 or, for $b = a$, to Eq 37, with corresponding changes from Fig. 10 to Fig. 3 or Fig. 9.

Though the above results for three-dimensional problems are of extreme practical interest, the mathematical difficulty in attempting other such solutions is so great that a discussion of the possible methods would be of little interest. However, in practical application of results it must be kept in mind that all bodies are really three-dimensional and often the cracks which must be analyzed do not suit the idealized results exactly as presented here. Nevertheless, the results which are presented form the basis for sensible judgments from which three-dimensional effects may be assessed.

Moreover, as a prime example of the fact that three-dimensional effects are always present and yet may most often be justifiably neglected, consider sheet of finite thickness with a through-crack. If the sheet were infinitely thick, plane strain would apply; or, if infinitely thin, then plane stress would apply. But with finite thickness, a mixed situation of plane stress near the surfaces of the plate and plane strain in the interior occurs in the crack-tip stress field. Consequently, the stress-intensity factors computed for plane problems represent only their values averaged through the thickness. Therefore, considering that plane-stress versus plane-strain displacement fields differ by a factor of $(1 - \nu^2)$, the actual values of stress-intensity factors for a straight-through-crack can vary by $(1 - \nu^2)^{1/2}$ (or less) from the surface to the interior. The values at the surface are a maximum of 5 per cent less than computed values and, correspondingly, a maximum of 3 per cent more in the interior (for $\nu = 0.3$). Though crack-tip plasticity further complicates the situation, it is partially for this reason that the crack often begins to grow in the interior of a plate rather than at the surface to form a "tongue." Even though this effect is frequently observed, ignoring it leads to a desirable level of accuracy of computed values of stress-intensity factors in developing fracture criteria.[8]

EDGE CRACKS IN SEMI-INFINITE BODIES

The plane-extensional problem of an edge notch, a, into a semi-infinite plane subjected to tension, σ, has been discussed by several authors (30–32,16)

---

[7] Values of elliptic integrals are to be found in many mathematical tables.

[8] So-called pop-in tests actually make direct use of this effect.

(see Fig. 11). With dimensional analysis leading again to Eqs 4, with C_1 left unknown, the task is merely to evaluate that constant. However, formidable methods must be employed to obtain the effect of the free surface of the half-plane. These methods use series-type mapping functions with the complex variable stress-function method (16,30) and dual integral equations resulting from a Green's function approach (31,32) or both. The results may be

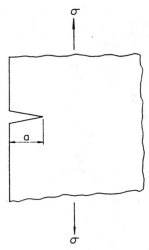

FIG. 11—An Edge Crack in a Semi-Infinite Sheet Subjected to Tension.

computed to any desired degree of accuracy and (within 1 per cent of each other) they are:

$$K_I = 1.12\sigma(\pi a)^{1/2} \brace (K_{II} = K_{III} = 0) \quad \ldots \ldots (40)$$

Comparison of this result with either Eq 4 or Eq 17 leads to the conclusion that the free surface correction factor is 1.12 for edge notches perpendicular to uniform tension.

On the other hand, for the analogous mode III case, Eq 7 and Fig. 5 with the introduction of a free surface perpendicular to the crack plane along the center-line of the crack, no correction is required (26).[9] Therefore, corresponding to Fig. 12, the stress-intensity factor is:

$$K_{III} = \tau(\pi a)^{1/2} \brace (K_I = K_{II} = 0) \quad \ldots \ldots (41)$$

There is no directly analogous mode II case corresponding to Figs. 11 and 12.

With these examples and their results, the methods of determination of "closed form" stress-intensity factors for some basic configurations have been

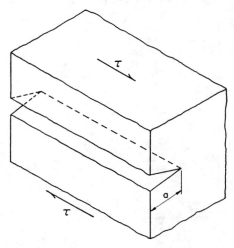

FIG. 12—An Edge Crack in a Semi-Infinite Body Subjected to Shear.

illustrated. Subsequently, some other types of problems which have not lent themselves to closed form solutions bear discussion.

Two-Dimensional Problems of Plate Strips with Transverse Cracks

The class of two-dimensional problems of plate strips with transverse internal, edge, and dual colinear edge cracks subjected to tension and in plane bending is of great practical interest for fracture testing procedures. However, closed-form solutions for such problems are not

---

[9] Unpublished results of G. Sih.

available and many of the approximate solutions in the literature are of doubtful accuracy. Therefore, it is important not only to cite these results but to give estimates of their accuracy.

The limitations on use of the so called "tangent" formula, Eq 17, for centrally cracked strips and double-edge-notched strips subjected to tension were already discussed. The work cited (13–16) which

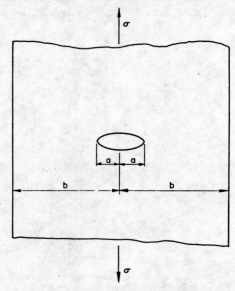

Fig. 13—A Central Crack in a Strip Subjected to Tension.

evaluated those limitations was from direct attacks on the strip problems.

One of the most formidable approaches to this class of problems is found in the work of Isida (13,34–36). Isida has extensively developed mapping functions for strip problems for determination of stress concentrations at the tips of round-ended cracks of end radius, p. His results are presented in the form (13)

$$\sigma_o = \sigma_{max} = \frac{2\sigma a^{1/2}}{p^{1/2}} f(\lambda) \quad \ldots \ldots (42)$$

where λ is the ratio of crack length to strip width. The function $f(\lambda)$ is obtained as a power series as a result of using power-series mapping and stress functions. The form of Eq 42 lends itself to direct substitution into Eq 9 or alternately, to techniques developed by Kobayashi (14). The resulting stress-intensity factors can be computed to any degree of accuracy by Isida's methods, provided the power series employed in the analysis converge, which they do for relatively large variations in λ. Within this minor limitation,

TABLE 1—CORRECTION FACTORS FOR A CENTRALLY CRACKED FINITE-WIDTH STRIP.

$\lambda = a/b$	$[2b/\pi a \tan \pi a/2b]^{1/2}$ a	$f(\lambda)$
0.074	1.00	1.00
0.207	1.02	1.03
0.275	1.03	1.05
0.337	1.05	1.09
0.410	1.08	1.13
0.466	1.11	1.18
0.535	1.15	1.25
0.592	1.20	1.33

a Eq 17.
b Isida (13).
Isida's values agree within 1 per cent of an approximation by Greenspan (100).

Isida's results lead to accuracies of within 1 or 2 per cent.

Isida has computed results in the form of Eq 42 for a variety of problems (13) of special interest in fracture testing such as the case of the centrally notched strip in tension, as shown in Fig. 13. Upon substitution of Eq 42 into Eq 9, it can be noted by comparing the result with Eq 17 that $f(\lambda)$ corresponds to the exact correction factor for the stress-intensity factor of a finite width strip whose approximate form is $[(2b/\pi a) \tan (\pi a/2b)]^{1/2}$. Table 1 compares the two to illustrate the accuracy of Eq 17.

Bueckner[10] (36) has developed integral equation procedures and solved many

[10] H. F. Bueckner in internal reports of the General Electric Co., Schenectady, N. Y.

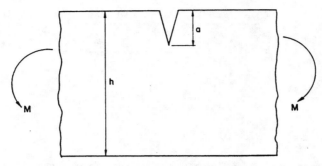

FIG. 14—An Edge Crack in a Strip Subjected to In-Plane Bending.

TABLE 2—STRESS-INTENSITY FACTOR COEFFICIENTS FOR NOTCHED BEAMS.

a/h	0.05	0.1	0.2	0.3	0.4	0.5	0.6 (and larger)
$g(a/h)$	0.36	0.49	0.60	0.66	0.69	0.72	0.73

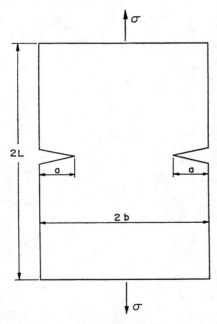

FIG. 15—Double-Symmetric Edge Cracks in Strip of Finite Length Subjected to Tension.

crack problems. He obtained (37) the solution to a strip with a single edge notch subjected to bending, see Fig. 14. The results so reported obviously lack the correction factor for a free surface, for small crack sizes discussed in conjunction with Eq 40, which is a 12 per cent error. However, as noted following Eq 17, the effect of the crack's emanation from a free edge disappears with deepening cracks; consequently, the error should diminish. The results are expressed as follows:

$$K_\mathrm{I} = \frac{6M}{(h-a)^{3/2}} g(a/h) \quad (K_\mathrm{II} = K_\mathrm{III} = 0) \quad \quad (43)$$

where $g(a/h)$ is given in Table 2.

The values in Table 2 suit the limiting case of deep notches as determined from Neuber's results (10). Therefore, it might be presumed that Table 2 reports values with errors of far less than 12 per cent for a/h greater than 0.2. Several recent papers on notch-bending analysis agree with the values in Table 2 and these recent results also claim agreement with "compliance calibrations" for a/h in the normal testing range.

Bowie developed polynomial mapping functions for use with the complex stress function technique to solve plane problems, such as cracks emanating from circular holes (38) and the double-edge-notched strip in tension (16). The latter example, as illustrated in Fig. 15, provides an indication of the validity of

TABLE 3—CORRECTION FACTORS FOR A DOUBLE EDGE-NOTCHED STRIP.[a]

a/b	h(a/b), (L/b = 1.00)	h(a/b), (L/b = 3.00)	$[(2b/\pi a) \tan (\pi a/2b)]^{1/2} h(a/b)$ (L/b → ∞)
0.1	1.13	1.12	1.12
0.2	1.13	1.11	1.12
0.3	1.14	1.09	1.13
0.4	1.16	1.06	1.14
0.5	1.14	1.02	1.15
0.6	1.10	1.01	1.22
0.7	1.02	1.00	1.34
0.8	1.01	1.00	1.57
0.9	1.00	1.00	2.09

[a] The last column agrees within 1 per cent with a similar formula proposed by G. R. Irwin on the basis of estimating the various effects. It is

$$K_I = \sigma (\pi a)^{1/2} \left[\frac{2b}{\pi a} \left(\tan \frac{\pi a}{2b} + 0.1 \sin \frac{\pi a}{b} \right) \right]^{1/2}$$

employing Eq 17 for this configuration. Comparing Bowie's results with Eq 17 is most lucidly accomplished using a correction factor, $h(a/b)$, on Eq 17, or

$$K_I = \sigma(\pi a)^{1/2} \left(\frac{2b}{\pi a} \tan \frac{\pi a}{2b} \right)^{1/2} h(a/b) \quad (44)$$
$$(K_{II} = K_{III} = 0)$$

for which his computed values are given in Table 3.

From Table 3 it can be immediately observed that for low a/b values, the correction factor of 1.12 for a crack from a free surface, as illustrated by Eq 40, is present. As a/b increases, its effect disappears and Eq 17 applies as noted previously. The last column of Table 3 combines the two effects, that is, the free surface and the finite width strip, to give the complete correction factor (within 1 per cent) for all values of a/b. From this study it can be noticed that using Eq 40 for $a/b < 0.5$ and Eq 17 for $a/b > 0.5$ results in errors of less than 3 per cent for the configuration shown in Fig. 15, provided that $L/b > 3$. As a consequence, it has been illustrated that basic solutions such as Eqs 17 and 40 can often be used with proper judgment to provide approximate analyses of more difficult situations such as are shown in Fig. 15.

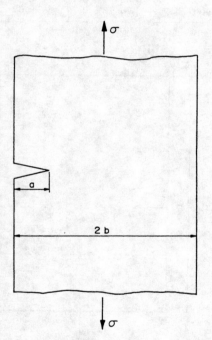

FIG. 16—A Single Edge-Cracked Strip Subjected to Tension.

Collocation procedures for strips of finite length have been developed by Kobayashi (15) and Gross.[11] As an ex-

---
[11] B. Gross, J. E. Srawley, and W. F. Brown, Jr., "Stress Intensity Factors for a Single Edge Notched Tension Specimen by Boundary Collocation of a Stress Function," unpublished report from NASA, Lewis Research Center.

ample of the method, Kobayashi treated the strip configuration in Fig. 13 using the general complex stress functions of Muskhelishvili (17), collocating equally space points on the sides and ends of the strip. He observed agreement within about 5 per cent of Isida's results as given in Table 1.

Gross treated the single-edge-notched strip using Williams' (6) eigenfunction representation of the Airy stress function. The configuration is shown in Fig. 16. He found that collocation at 20 or more boundary points was required to obtain convergence. His results can be stated in the form:

$$K_I = \sigma(a\pi)^{1/2} k(a/b) \quad \quad (45)$$
$$(K_{II} = K_{III} = 0)$$

where $k(a/b)$ is given as a correction factor for this strip problem in Table 4.

By comparison of Gross's results (Fig. 16) with Bowie's double-edge-notched specimen results (Fig. 15) (columns 2 and 3 of Table 4), the apparently large influence of bending due to the lack of symmetry in the single-edge-notch case is observed. Gross's results reportedly agree with experimentally measured values (that is, compliance measurements) within a few percentage points for $0.40 < a/b < 1.00$. Moreover, new results by Bowie (16) shown in column 4 of Table 4 assures the accuracy for this configuration.

Following the procedures of Kobayashi and Gross, it is a straightforward matter to solve additional problems. Moreover, similar numerical procedures based on collocation of boundary conditions in the mean, using other representations of the Airy stress function, or energy methods, or both, are available for development.

TABLE 4—CORRECTION FACTORS FOR A SINGLE EDGE-NOTCHED STRIP.

a/b	$k(a/b)^a$	$[(2b/\pi a) \tan (\pi a/2b)] h(a/b)^b$	$k(a/b)^c$
0.10	1.14	1.12	1.15
0.20	1.19	1.12	1.20
0.30	1.29	1.13	1.29
0.40	1.37	1.14	1.37
0.50	1.50	1.15	1.51
0.60	1.66	1.22	1.68
0.70	1.87	1.34	1.89
0.80	2.12	1.57	2.14
0.90	2.44	2.09	2.46
1.00	2.82	...	2.86

[a] Gross[11]
[b] Table 3 and Eq 44
[c] Bowie (16)

Reinforced Plane Sheets

Many conventional structures are fabricated from plane sheets (plates) with reinforcing stiffeners or doubler plates attached by riveting, welding, and other means. Often the attachments are designed as crack-arrestors in order to provide so called "fail-safe" structures.

In order to analyze some of these configurations, it is appropriate to determine stress-intensity factors for cracks in sheets with stiffeners perpendicular to the cracks. Romualdi (39,40) and Paris (41,42) provided some early attempts to estimate the effect of rivet forces tending to hold a crack closed. Sanders (43) discussed the problem of action of an integral stiffener crossing the center of a crack. Isida (13,44) extended his methods to give results for centrally cracked strips with integrally reinforced edges and to infinite sheets with a periodic array of cracks along a line with interspersed integral stiffeners. Greif (45) has solved the problem of a single crack and an integral stiffener (passing outside the crack) in an infinite sheet, and in a continuation of that work the riveted stiffener has been treated.[12] Moreover, Terry (46) has analyzed some similar riveted and welded stiffener problems, as an extension of work by Erdogan (21). Cracks within one sheet of a riveted doubler-plated area of a structure were treated by Paris (41). Since

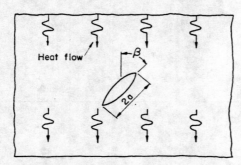

Fig. 17—An Insulated Crack Disturbing Uniform Heat Flow in a Sheet.

this class of problems is difficult to formulate, the methods employed are rather obtuse and specialized. Consequently, they will not be described here other than to remark that the most general approaches available are those of Isida (13), Greif (45) and Terry (46).

Thermal Stresses

It has been shown that the crack-tip stress field equations for isotropic bodies, Eqs 1, 2, and 3, also provide the proper field equations for thermal stress states (47) (with the unlikely exception of the crack tip as a point source of heat). Therefore, the concept of stress-in-

[12] Private communication from J. L. Sanders, Jr.

tensity factors is in general applicable to thermal stress problems.

As an example, consider the case of uniform heat flow in a sheet, with an undisturbed temperature gradient, ∇T, at an angle, β, with respect to a crack of length, $2a$, acting as an insulator, as shown in Fig. 17. Florence and Goodier (48) have provided the complex stress function for this configuration. It is:

$$\phi(\eta) = \frac{iE\alpha a^2 \nabla T}{8} \sin \beta \log \eta \quad \ldots (46)$$

as a consequence of similarity of the re-

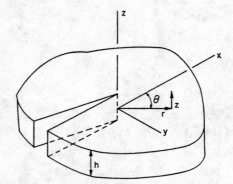

Fig. 18—Coordinates Used in a Cracked Plate Which Will Be Subjected to Transverse Bending.

sulting crack-tip stress field equation with ordinary (isothermal) plane extension, Eqs 29 and 30 may be applied to Eq 46 which results in:

$$\left.\begin{array}{c} K_{II} = \dfrac{E\alpha a^{3/2} \nabla T}{4} \sin \beta \\ (K_I = K_{III} = 0) \end{array}\right\} \ldots (47)$$

where α is the coefficient of thermal expansion and E is Young's modulus. Other examples (47), will be cited in Appendix II.

Stress-Intensity Factors for the Bending of Plates and Shells

The field equations for the stresses near a sharp notch in a plate subjected to bending were first considered by Williams (49,50) who later applied like methods to a more detailed discussion of cracks (51). Using the classical Kirchhoff theory of plate bending, he obtained the following stress-field equations (see Fig. 18):

$$\begin{aligned}
\sigma_r &= \frac{(7+\nu)}{2(3+\nu)} \frac{K_B}{(2\pi r)^{1/2}} \frac{z}{h} \\
&\quad \cdot \left[\frac{(3+5\nu)}{(7+\nu)} \cos\frac{\theta}{2} - \cos\frac{3\theta}{2} \right] \\
&\quad + \frac{(5+3\nu)}{2(3+\nu)} \frac{K_S}{(2\pi r)^{1/2}} \frac{z}{h} \\
&\quad \cdot \left[-\frac{(3+5\nu)}{(5+3\nu)} \sin\frac{\theta}{2} + \sin\frac{3\theta}{2} \right] \\
\sigma_\theta &= \frac{(7+\nu)}{2(3+\nu)} \frac{K_B}{(2\pi r)^{1/2}} \frac{z}{h} \\
&\quad \cdot \left[\frac{(5+3\nu)}{(7+\nu)} \cos\frac{\theta}{2} + \cos\frac{3\theta}{2} \right] \\
&\quad - \frac{(5+3\nu)}{2(3+\nu)} \frac{K_S}{(2\pi r)^{1/2}} \frac{z}{h} \\
&\quad \cdot \left[\sin\frac{\theta}{2} + \sin\frac{3\theta}{2} \right] \\
\tau_{r\theta} &= \frac{(7+\nu)}{2(3+\nu)} \frac{K_B}{(2\pi r)^{1/2}} \frac{z}{h} \\
&\quad \cdot \left[-\frac{(1-\nu)}{(7+\nu)} \sin\frac{\theta}{2} + \sin\frac{3\theta}{2} \right] \\
&\quad + \frac{(5+3\nu)}{2(3+\nu)} \frac{K_S}{(2\pi r)^{1/2}} \frac{z}{h} \\
&\quad \cdot \left[-\frac{(1-\nu)}{(5+3\nu)} \cos\frac{\theta}{2} + \cos\frac{3\theta}{2} \right]
\end{aligned} \quad (48)$$

where the constants in Williams' analysis (51) have been modified in order to define (18,52) the plate bending and plate shearing stress-intensity factors, K_B and K_S, in a manner consistent with (but not quite corresponding to) the first- and second-mode types, K_I and K_{II}, respectively, as defined by Eqs 1 and 2. Though polar instead of rectangular stress components are given for compactness in Eqs 48, the similarity of these results with Eqs 1 and 2 is immediately apparent. This similarity is further clarified upon computing K_B and K_S for some configurations and loadings of interest.

The governing equation for free bending of plates (no transverse loads) by the Kirchhoff theory is:

$$\nabla^4 w = 0 \quad \ldots \ldots \ldots \ldots (49)$$

where w is the transverse displacement. Consequently, an analysis (18) ensues of

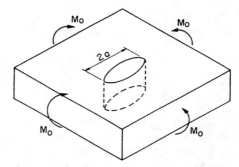

FIG. 19—A Through Crack in an Infinite Plate Subjected to Uniform Biaxial Bending.

an identical nature to Eqs 21–27 which gives:

$$K_B - iK_S = -\frac{(2\pi)^{1/2} Eh(3+\nu)}{(1-\nu^2)}$$
$$\cdot \lim_{z \to z_1} (z - z_1)^{1/2} \phi_B'(z) \ldots (50)$$

where $\phi_B(z)$ is the plate bending stress function discussed extensively by Savin (12).

Furthermore, mapping is again permitted; or, as Eq 30 followed from Eq 27, for the mapping function given by Eq 29, Eq 50 becomes

$$K_B - iK_S = -(\pi/a)^{1/2} \frac{Eh(3+\nu)}{(1-\nu^2)} \phi_B'(1) \ldots (51)$$

For the example of an infinite plate subjected to uniform moment, M_o, all around the boundary, and with a crack of length, $2a$, as in Fig. 19, Savin (12) gives the stress function,

$$\phi_B(\eta) = -\frac{3M_o a(1-\nu)}{Eh^3}$$
$$\cdot \left[\eta + \frac{(1-\nu)}{(3+\nu)}\frac{1}{\eta}\right]\ldots(52)$$

Using Eqs 51 and 52, the result is:

$$\left.\begin{array}{l}K_B = \dfrac{6M_o}{h^2}(\pi a)^{1/2}\\[4pt](K_S = 0)\end{array}\right\}\ldots(53)$$

Since the stress in the surface layer of the plate, σ_o, away from the crack is

$$\sigma_o = \frac{6M_o}{h^2}\ldots\ldots(54)$$

the analogy between Eqs 53 and 4 is evident.

Moreover, Erdogan (52) has shown experimentally that in brittle materials (like Plexiglas) the fracture mechanics concept of K_B reaching a critical value, K_{BC}, is appropriate and analogous to the extensional first-mode case, that is, K_{Ic}. Incidentally, Erdogan (53) also shows that the critical value of stress-intensity factors applies to the extension second mode, that is, K_{IIc}, which again is shown to be analogous to the shear case of bending, that is, K_{Sc}. Consequently, the plate bending and shearing stress-intensity factors as defined in Eqs 48 are of some immediate practical interest.

However, Eqs 1, 2, and 3 were purported to give *all* tip-stress fields for elastic bodies; yet the field for plate bending as predicted by Eqs 48 is not identical to them. This is because the classical Kirchhoff theory of bending is an approximate theory which does not take into account the details of the stress distribution near boundaries nor discontinuities in the plate. The crack-tip and crack-surface boundaries are locations where details are not clear.

Subsequently, Knowles (54) pointed out that using Reissner's (55) more accurate plate theory leads to a correction of Eqs 48 which on the surface of the plate makes them identical to Eqs 1 and 2, except for a constant factor. Moreover, the character and role of K_B and K_S are preserved through this correction. Hence, it is concluded that they are directly proportional to (completely analogous to) their counterparts, K_I and K_{II}, where elastic action is concerned. Williams (56) pointed this out in reference to the experiments by Erdogan (52). This correspondence has also been observed for fatigue crack growth.[13]

Therefore, both theoretical and experimental results for fracture tests have led to:

$$K_B = \frac{(3+\nu)}{(1+\nu)}K_I\ldots\ldots(55)$$

on the surface of the plate. The sensibility of using the Kirchhoff theory to compute K_B values is also clear when it is reasoned that the values of stress-intensity factors reflect the intensity of general transmission of applied loads into the crack-tip region. The general properties of gross-load transmission are unaffected by the boundary layer of about one plate thickness, h, in which the Reissner theory applies. Consequently, Eq 55 is always correct for converting Kirchhoff theory stress-intensity factors, K_B, to the Reissner theory result, K_I, for a given configuration.

Several solutions for K_B and K_S are now available (18) and others can be obtained in a direct fashion using Eqs 50 or their equivalents for other types

---

[13] R. Roberts, Ph.D. dissertation, Lehigh University, 1964.

of stress functions. Some of the available results are tabulated in Appendix II.

The case of general bending and extension of thin shells with cracks has been shown by Sih (57) to give crack-tip stress fields equivalent to combining modes I and II with the bending fields, that is, Eqs 1 and 2, and Eq 48. Modes I and II result from extension of the middle surface of the shell, and the bending fields result from changes in the curvature of the middle surface. Consequently, the stress-intensity factor concept is also of general applicability to shells.

However, computing the values of the stress-intensity factors for particular configurations in shells is very difficult.

Moreover, it may be observed (57) that the extension and bending effects in shells will be coupled, so that the stress-intensity factors resulting from solutions must reflect this coupling. As a consequence of the coupling, formulas for stress-intensity factors will involve many parameters (coupling terms) so that they will, to say the least, be complicated.

Folias (58,59) and Ang (60), noting the similarity of equations for plates on elastic foundations and shallow spherical shells (61), have attempted some problems in these areas. However, no other attempts at the complete solutions to shell problems are known.

On the other hand, some parametric studies of possible shell effects on cracks in cylinders have been attempted in several articles (62–65). The results indicate that the experimental data on failure of cracked shells can in fact be correlated in terms of elastic shell parameters. Hence, it is hopeful that further progress can be made soon toward quantitative prediction of shell effects on an analytical basis.

The problem of crack-arrestor rings on shells is at least another degree more difficult. Nevertheless, since this problem is of prime interest in tear-resistant design, efforts are being made toward empirical methods of design (64,66). The complete analytical solution to such a problem is as yet improbable.

COUPLE-STRESS PROBLEMS WITH CRACKS

Another area analogous to shell problems through having similar governing equations is that of couple stresses (67,68). The formulation of couple-stress problems takes into account the gradients of stresses in terms of couples on infinitesimal elements in order to account for the effects of lattice curvature in crystals, and so forth. Setzer (69) has shown that for extension of cracked plates due to uniform applied stress away from the crack, no modification in the field equations (Eqs 1), nor the stress-intensity factors (for example, Eq 4), is required. However, where the applied stresses away from the crack possess gradients, the values of stress-intensity factors will be modified by factors of the form

$$\left[1 + A_1\left(\frac{l}{a}\right) + A_2\left(\frac{l}{a}\right)^2 + A_3\left(\frac{l}{a}\right)^3 + \cdots \cdot (56)\right]$$

where l is a couple-stress (lattice) parameter or characteristic length of the material. The A_i are of the order of unity or smaller, and l is of the order of lattice dimensions; consequently, these results would be of a greatest interest in analyzing fine cracks in crystals, except for the fact that the methods involved are similar to and may be carried over to the analysis of shells.

ESTIMATION OF STRESS-INTENSITY FACTORS FOR SOME CASES OF PRACTICAL INTEREST

Armed with the principles of linear elastic theory, such as "the principle of superposition," and with an intuitive grasp of a strength-of-materials approach, it is possible to form estimates of stress-intensity factors. This was made partially evident in the case of an embedded elliptical crack in the discussion

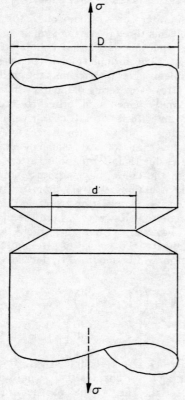

FIG. 20—A Circumferentially Cracked Round Bar Subjected to Tension.

of limiting cases following Eqs 38 and 39. Other situations where limiting cases of different problems are comparable were illustrated in Tables 1, 3, and 4 and examples in the text. Notice especially, as in these tables, that one problem solution often forms an upper or lower bound on the solution of others. These concepts will be employed in examples of estimating to follow.

Consider the configuration of a notched round bar with an outside diameter, D, and notched-section diameter, d, and subjected to extension causing a net-section stress, σ_{net} (see Fig. 20). From dimensional considerations and symmetry, it is noted that the stress-intensity factor may be stated in the form,

$$\left. \begin{array}{l} K_I = \sigma_{net}(\pi D)^{1/2} F(d/D) \\ K_{II} = K_{III} = 0 \end{array} \right\} \quad \ldots (57)$$

where $F(d/D)$ is an unknown dimensionless function of the diameter ratio. The end values (that is, $d/D \to 0$ or 1.0) of the function can be established by examining limiting cases.

As $D \to \infty$, dimensional analysis leads to

$$K_I = C_3 \sigma_{net}(\pi d)^{1/2} \ldots \ldots (58)$$

thus, for small values of d/D,

$$F_u(d/D) = C_3(d/D)^{1/2} \ldots \ldots (59)$$

the value of C_3 is found to be $[1/(2\sqrt{2})]$ using Eq 9 and the stress-concentration solution for the problem given by Neuber (10) and Peterson (11). Since the free surface introduced by the finite diameter of the bar lowers the stress-intensity factor, $F_u(d/D)$ is an upper bound on $F(d/D)$ for all values of d/D.

On the other hand, for $d/D \to 1$, Bowie's solution for the double-edge-notched sheet, Eq 44, simulates the problem upon substituting

$$\left. \begin{array}{l} a = \dfrac{D}{2}\left(1 - \dfrac{d}{D}\right) \\ \dfrac{a}{b} = 1 - \dfrac{d}{D} \\ \sigma = \sigma_{net}\left(\dfrac{d}{D}\right)^2 \end{array} \right\} \quad \ldots (60)$$

The result conforms with Eq. 57, that is,

$$K_I = \sigma_{net}(\pi D)^{1/2}\left[\left(\frac{d}{D}\right)^2 \cdot \left(\frac{1}{\pi}\tan\frac{\pi}{2}\left(1-\frac{d}{D}\right)\right)^{1/2} h\left(1-\frac{d}{D}\right)\right] \quad (61)$$

Consequently,

$$F_L\left(\frac{d}{D}\right) = \left(\frac{d}{D}\right)^2 \cdot \left(\frac{1}{\pi}\tan\frac{\pi}{2}\left(1-\frac{d}{D}\right)\right)^{1/2} h\left(1-\frac{d}{D}\right) \quad (62)$$

where $h(\ )$ is as tabulated in Table 3. This function, $F_L(d/D)$ is a lower bound on $F(d/D)$ for all values of d/D, since the curvature of the bar causes increased crack-tip stress over the flat plate solution as d/D recedes from the value 1.

Finally, from Peterson's (11) stress-concentration values and Eq 9, and other considerations, the maximum value of $F(d/D)$ is estimated to be 0.240. Interpolating between these solutions results in the estimated values in Table 5.

By making use of careful judgment of the limits of applicability of the limiting cases, Eqs 59 and 62 and the analysis of stress concentration (11), the accuracy of $F(d/D)$ in Table 5 can be stated with

TABLE 5—STRESS-INTENSITY FACTOR COEFFICIENTS FOR NOTCHED ROUND BARS.

d/D	$F_L(d/D)$	$F_u(d/D)$	$F(d/D)$
0	0	0	0
0.1	...	0.111	0.111
0.2	0.046	0.158	0.155
0.3	...	0.194	0.185
0.4	0.118	0.223	0.209
0.5	...	0.250	0.227
0.6	0.185	0.274	0.238
0.65	0.203	...	0.240
0.70	0.217	0.296	0.240
0.75	0.226	...	0.237
0.80	0.230	0.317	0.233
0.85	0.224	...	0.225
0.90	0.205	0.336	0.205
0.95	0.162	...	0.162
0.97	0.130	...	0.130
1.00	0	0.353	0

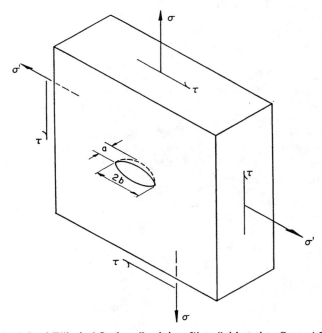

Fig. 21—A Semi-Elliptical Surface Crack in a Plate Subjected to General Extension.

confidence. With a d/D of 0 to 0.4, it is ± 3 per cent; with a d/D of 0.4 to 0.85, it is ± 5 per cent; and with a d/D of 0.85 to 1.0, it is ± 2 per cent. Therefore, a solution with sufficient precision for practical applications has been constructed.

This configuration is often used for fracture testing and a simplified formula is employed, that is (70),

$$K_1 = 0.233\, \sigma_{net}(\pi D)^{1/2} \qquad (63)$$

This formula seems most reasonable since 0.233 agrees with the values of $F(d/D)$ in Table 5 within 5 per cent over the range of d/D from 0.48 to 0.86. Further improvements in the accuracy of the values given would require a full analysis of the problem, such as suggested by Sneddon (71) or Bueckner.[14]

Another configuration, which has been discussed by Irwin (29), is that of a semi-elliptical surface crack in a plate (see Fig. 21). This configuration is both typical of flaws and is used in fracture testing (simulating this type of flaw). If the plate is subjected to general uniform extension by stresses, σ, σ' and τ, the stress, σ', parallel to the crack causes no singularity or no contribution to stress-intensity factors. Consequently σ' will be ignored.

If b/a is large and a/t small compared to 1, the stress-intensity factors at the end of the semi-minor axis, a, can be estimated from Eqs 17 and 20, making use of free-edge corrections as in Eqs 40 and 41. Then, the correction, Φ_o, in Eqs 38 and 39 should be applied as b/a values are reduced toward 1. However, the free-edge correction probably diminishes as b/a approaches 1, and the tangent correction in Eqs 17 and 20 is also an overcorrection in that limit. On the other hand, Eqs 44 and 45 and Table 4 show that single-edge notches induce bending which increases the stress-intensity factor, but less so in this case since the uncracked portion of the plate would inhibit bending. Finally, Table 1 shows the underestimation of the tangent correction as a/t becomes larger. Taking all these factors into account, Eqs 17, 20, 38–42, 44, and 45, and their considerations lead to the approximations:

$$\left. \begin{array}{l} K_1 = \left[1 + 0.12\left(1 - \dfrac{a}{b}\right) \right] \\ \qquad \dfrac{\sigma(\pi a)^{1/2}}{\Phi_o} \left(\dfrac{2t}{\pi a} \tan \dfrac{\pi a}{2t} \right)^{1/2} \\ K_{II} = 0 \\ K_{III} = \dfrac{\tau(\pi a)^{1/2}}{\Phi_o} \left(\dfrac{2t}{\pi a} \tan \dfrac{\pi a}{2t} \right)^{1/2} \end{array} \right\} \quad (64)$$

for the stress-intensity factors at the end of the semi-minor axis, a. For the ranges of b/a from 1 to 10 or more, and of a/t from zero to one half the accuracy is within about ± 5 per cent. Moreover, for b/a up to about 5 and a/t up to three fourths the accuracy is still probably better than ± 10 per cent, considering all the compensating errors. This case has at least provided a classic example of estimating methods using many other solutions for stress-intensity factors to treat an important problem which is all but impossible to solve directly.

A word of warning with complicated cases such as Eqs 64 is in order. If the crack-tip plasticity subtends a major portion (say one half) of the distance between the crack front and the back side of the plate, use of these equations would become indeed doubtful. Moreover, estimation of the amount of plasticity is clearly more complicated here than in other situations, but surely possible. Such estimates are beyond the scope of this discussion and the reader is referred to Ref (72). Moreover, in passing, it is noteworthy that restrictions on crack-tip plastic-zone sizes are always present in making direct

---
[14] Private communication from H. Bueckner.

applications of the elastic analyses (70). For certain situations estimated corrections to the analysis for crack tip plasticity effects have been proposed (41,70,72).

Estimates can be made for stress-intensity factors for quite arbitrary crack-front contours in three-dimensional bodies subjected to uniform tension, σ, perpendicular to the crack plane in the region including the whole crack. Consider the embedded crack whose plan view is shown in Fig. 22. Using the previous results for circular disk cracks, Eq 37, and for tunnel cracks, Eq 4, bounds on the values of stress-intensity factors can be established on the crack

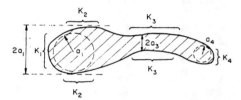

FIG. 22—The Plan View of an Irregular Crack in an Infinite Body.

front for various portions of the contour, where

$$\left. \begin{array}{l} K_\mathrm{I} = K_1 \text{ or } K_2 \text{ or } K_3 \text{ or } K_4 \\ K_\mathrm{II} = 0 \\ K_\mathrm{III} = 0 \end{array} \right\} \quad (65)$$

the value of K_1 will be slightly greater than that for a disk crack of radius, a_1, but far less than a tunnel crack of width, $2a_1$. Therefore, from Eqs 4 and 37,

$$\frac{2}{\pi} \sigma(\pi a_1)^{1/2} < K_1 \ll \sigma(\pi a_1)^{1/2} \quad (66)$$

Since $2/\pi$ is about 0.64, if K_1 is guessed to be

$$K_1 \cong 0.75\, \sigma(\pi a_1)^{1/2} \quad (67)$$

the result is surely within ± 10 per cent along the whole portion of the contour marked, K_1, in Fig. 22. Now, K_2 is closer to the tunnel-crack case or, a guess is

$$K_2 \cong 0.85\, \sigma(\pi a_1)^{1/2} \quad (68)$$

The neck of width, $2a_3$, makes K_3 slightly higher than the comparable tunnel crack, or

$$K_3 \cong 1.05\, \sigma(\pi a_3)^{1/2} \quad (69)$$

Similar to K_1, the guess for K_4 is

$$K_4 \cong 0.75\, \sigma(\pi a_1)^{1/2} \quad (70)$$

These estimates are surely all correct within ± 10 per cent (and probably ± 5 per cent). Moreover, additional refinements are possible, such as noting that K_3 on the upper part of the contour is likely to be about 5 per cent less than on the lower contour in Fig. 22, due to the curvature of the centerline of the neck, $2a_3$.

Corrections can also be added for the proximity to a free surface, such as the tangent correction in Eq 17, or for the emanation of the crack from a free surface, such as Eq 40. The method of estimating has now been sufficiently illustrated to allow direct application to a multitude of examples. In order to develop confidence in estimating procedures, it is suggested that one may, for example, estimate the stress-intensity factor values for an elliptical crack using the above procedure, Eqs 66–70, and compare the results with the exact values, Eq 38.

Stress Fields and Intensity Factors for Homogeneous Anisotropic Media

An interest in stress analysis of cracks for various media, such as anisotropic, viscoelastic, or non-homogeneous materials, stems from two motivations. First, the effects of slight amounts of directionality, creep, and inhomogeneity on the stress distribution and intensity are useful in assessing the limits of applicability of the conceptual model of fracture mechanics based on linear elastic theory. In addition, studies of the stress analysis of these various types of media will provide the basis of extension of fracture mechanics to such materials.

Several authors have treated special cases of crack problems in anisotropic media, such as orthotropy (32,73,74) or particular configurations (75,76). However, the general anisotropic case can be treated in order to determine crack-tip stress fields and to define intensity factors in a manner completely analogous to Eqs 1, 2, and 3. The methods discussed extensively by Lekhnitzki (77) will be employed here.[15]

Hooke's law for a homogeneous (rectilinearly) anisotropic material is:

$$\left.\begin{aligned}\epsilon_x &= a_{11}\sigma_x + a_{12}\sigma_y + a_{13}\sigma_z \\ &\quad + a_{14}\tau_{yz} + a_{15}\tau_{xz} + a_{16}\tau_{xy} \\ \epsilon_y &= a_{21}\sigma_x + \cdots \\ \epsilon_z &= a_{31}\sigma_x + \cdots \\ \gamma_{yz} &= a_{41}\sigma_x + \cdots \\ \gamma_{xz} &= a_{51}\sigma_x + \cdots \\ \gamma_{xy} &= a_{61}\sigma_x + a_{62}\sigma_y + a_{63}\sigma_z \\ &\quad + a_{64}\tau_{yz} + a_{65}\tau_{xz} + a_{66}\tau_{xy}\end{aligned}\right\} \ldots(71)$$

where, from reciprocity

$$a_{ij} = a_{ji}$$

Referring to Fig. 2 for the coordinates and notation with respect to a crack front, the crack-tip stress fields may be resolved from two cases of plane problems which are defined as:

(1) Plane strain, that is, $(\partial u/\partial z) = (\partial v/\partial z) = w = 0$ or $\epsilon_z = \gamma_{yz} = \gamma_{xz} = 0$
(2) Pure shear, that is, $u = v = (\partial w/\partial z) = 0$ or $\epsilon_z = \epsilon_x = \epsilon_y = \tau_{xy} = 0$

---
[15] The mathematical derivation of stress fields leading to Eqs 81–85 are not a requirement of useful interpretation of those results.

The superposition of results from these plane problems will allow treatment of the general case of crack-tip stress fields similar to Eqs 1, 2, and 3.

Plane Strain:

For this case, Hooke's law may be reduced, using the restrictions on strain to eliminate the appearance of z-components of stress, to give:

$$\left.\begin{aligned}\epsilon_x &= A_{11}\sigma_x + A_{12}\sigma_y + A_{16}\tau_{xy} \\ \epsilon_y &= A_{21}\sigma_x + A_{22}\sigma_y + A_{26}\tau_{xy} \\ \gamma_{xy} &= A_{61}\sigma_x + A_{62}\sigma_y + A_{66}\tau_{xy}\end{aligned}\right\} \ldots(72)$$

where again, $A_{ij} = A_{ji}$ and the A_{ij} can be expressed in terms of a_{ij} directly if desired. Using an Airy stress function, U, with stress components defined as the usual second derivatives, equilibrium is automatically satisfied and the compatability equations lead to:

$$D_1 D_2 D_3 D_4 U = 0 \ldots\ldots\ldots(73)$$

where $D_k = (\partial/\partial y) - \mu_k(\partial/\partial x)$ and μ_k are the roots of

$$A_{11}\mu^4 - 2A_{16}\mu^3 + (2A_{12} + A_{66})\mu^2 \\ - 2A_{26}\mu + A_{22} = 0 \ldots(74)$$

These elastic constants, μ_k, are complex or pure imaginary and occur in conjugate pairs (77), that is, $\mu_3 = \bar{\mu}_1$ and $\mu_4 = \bar{\mu}_2$. Defining the complex variables, z_1 and z_2, by

$$\left.\begin{aligned}z_1 &= x + \mu_1 y \\ z_2 &= x + \mu_2 y\end{aligned}\right\} \ldots\ldots\ldots(75)$$

the general solution to Eq 73 can be written, if $\mu_1 = \mu_2$,

$$U = U_1(z_1) + \bar{z}_1 U_2(z_1) + U_3(\bar{z}_1) + z_1 U_3(\bar{z}_1)$$

or, if $\mu_1 \neq \mu_2$,

$$U = U_1(z) + U_2(z_2) + U_3(\bar{z}_1) + U_4(\bar{z}_2) \ldots(76)$$

and with the further restriction that U must be real, they become

$$\left.\begin{aligned}U &= 2\,\text{Re}[U_1(z_1) + \bar{z}_1 U_2(z_1)] \\ U &= 2\,\text{Re}[U_1(z_1) + U_2(z_2)]\end{aligned}\right\} \ldots(77)$$

The similarity of the first case of Eq 77 with Eq 22 is appropriate since for isotropic media, $\mu_1 = \mu_2 = i$. Therefore, the orthotropic case with the crack on a principal plane which leads to $\mu_1 = \mu_2$ has been reduced to the same case as isotropic elasticity with the simple change of variable $z_1 = x + \mu_1 y$. The more general case of anisotropy, the second of Eq 77, or $\mu_1 \neq \mu_2$, will follow in the remaining discussion.

The stress and displacement components are found from the Airy stress function, U, by the usual combination of derivatives which give:

$$\left. \begin{array}{l} \sigma_x = 2\,\text{Re}[\mu_1^2\,U_1''(z_1) + \mu_2^2\,U_2''(z_2)] \\ \sigma_y = 2\,\text{Re}[U_1''(z_1) + U_2''(z_2)] \\ \tau_{xy} = -2\,\text{Re}[\mu_1\,U_1''(z_1) + \mu_2\,U_2''(z_2)] \\ \text{and} \\ u = 2\,\text{Re}[p_1\,U_1'(z_1) + p_2\,U_2'(z_2)] \\ v = 2\,\text{Re}[q_1\,U_1'(z_1) + q_2\,U_2'(z_2)] \\ \text{where:} \\ p_i = A_{11}\mu_i^2 + A_{12} - A_{16}\mu_i \\ q_i = A_{12}\mu_i + (A_{22}/\mu_i) - A_{26}. \end{array} \right\} \quad (78)$$

Therefore, solution to any specific problem is reduced to finding the U_1 and U_2 which satisfy the boundary conditions.

Referring again to Fig. 2 and Eq 75, in the neighborhood of a crack tip, $|z_1|$ and $|z_2|$ are small compared to other planar dimension of problems. Consequently, the stress functions for cracks given by Lekhnitzki (77) may be reduced to the form,

$$\left. \begin{array}{l} U_1''(z_1) = \dfrac{f_1(\mu_1, \mu_2, z_1)}{z_1^{1/2}} \\ U_2''(z_2) = \dfrac{f_2(\mu_1, \mu_2, z_2)}{z_2^{1/2}} \end{array} \right\} \quad \ldots \ldots (79)$$

where f_1 and f_2 are well-behaved in that neighborhood and some restrictions on their form are imposed by the stress-free crack surface boundary conditions. Imposing these conditions, as well as those mentioned earlier and then substituting the variable,

$$z = x + iy = re^{i\theta} \ldots \ldots (80)$$

the crack-tip stress fields are found from Eqs 78–80 and can be stated in the form

$$\left. \begin{array}{l} \sigma_x = \dfrac{K_{Ia}}{(2\pi r)^{1/2}} \text{Re}\left[\dfrac{\mu_1\mu_2}{\mu_1 - \mu_2}\right. \\ \qquad \cdot \left\{ \dfrac{\mu_2}{(\cos\theta + \mu_2 \sin\theta)^{1/2}} \right. \\ \qquad \left. \left. - \dfrac{\mu_1}{(\cos\theta + \mu_1 \sin\theta)^{1/2}} \right\} \right] \\ \qquad + \dfrac{K_{IIa}}{(2\pi r)^{1/2}} \text{Re}\left[\dfrac{1}{\mu_1 - \mu_2}\right. \\ \qquad \cdot \left\{ \dfrac{\mu_2^2}{(\cos\theta + \mu_2 \sin\theta)^{1/2}} \right. \\ \qquad \left. \left. - \dfrac{\mu_1^2}{(\cos\theta + \mu_1 \sin\theta)^{1/2}} \right\} \right] \\ \\ \sigma_y = \dfrac{K_{Ia}}{(2\pi r)^{1/2}} \text{Re}\left[\dfrac{1}{\mu_1 - \mu_2}\right. \\ \qquad \cdot \left\{ \dfrac{\mu_1}{(\cos\theta + \mu_2 \sin\theta)^{1/2}} \right. \\ \qquad \left. \left. - \dfrac{\mu_2}{(\cos\theta + \mu_1 \sin\theta)^{1/2}} \right\} \right] \\ \qquad + \dfrac{K_{IIa}}{(2\pi r)^{1/2}} \text{Re}\left[\dfrac{1}{\mu_1 - \mu_2}\right. \\ \qquad \cdot \left\{ \dfrac{1}{(\cos\theta + \mu_2 \sin\theta)^{1/2}} \right. \\ \qquad \left. \left. - \dfrac{1}{(\cos\theta + \mu_1 \sin\theta)^{1/2}} \right\} \right] \\ \\ \tau_{xy} = \dfrac{K_{Ia}}{(2\pi r)^{1/2}} \text{Re}\left[\dfrac{\mu_1\mu_2}{\mu_1 - \mu_2}\right. \\ \qquad \cdot \left\{ \dfrac{1}{(\cos\theta + \mu_1 \sin\theta)^{1/2}} \right. \\ \qquad \left. \left. - \dfrac{1}{(\cos\theta + \mu_2 \sin\theta)^{1/2}} \right\} \right] \end{array} \right\} \quad (81)$$

$$+ \frac{K_{IIa}}{(2\pi r)^{1/2}} \text{Re} \left[\frac{1}{\mu_1 - \mu_2} \right.$$
$$\left. \cdot \left\{ \frac{\mu_1}{(\cos\theta + \mu_1 \sin\theta)^{1/2}} \right. \right.$$
$$\left. \left. - \frac{\mu_2}{(\cos\theta + \mu_2 \sin\theta)^{1/2}} \right\} \right] \quad ..(81)$$

where higher-order terms in r have been neglected. Reiterating, μ_1 and μ_2 are dimensionless elastic constants. Notice the striking similarity of Eqs 81 with

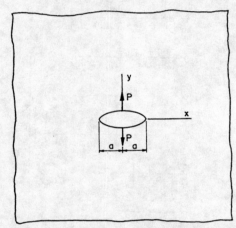

FIG. 23—A Crack in an Infinite Sheet Subjected to Centrally Applied Wedge Forces.

Eqs 1 and 2. The definitions of K_{Ia} and K_{IIa} have been chosen to be identical to K_I and K_{II} for the cases of symmetrical configurations with symmetric or skew-symmetric loadings, respectively.

Consequently, it can be shown that for the general anisotropic problem of the configuration illustrated in Fig. 1:

$$\left. \begin{array}{l} K_{Ia} = \sigma (\pi a)^{1/2} \\ K_{IIa} = 0 \end{array} \right\} \quad \ldots\ldots\ldots (82)$$

and for the problem in Fig. 2,

$$\left. \begin{array}{l} K_{IIa} = \tau (\pi a)^{1/2} \\ K_{Ia} = 0 \end{array} \right\} \quad \ldots\ldots\ldots (83)$$

Moreover, for the symmetrical wedge-force problem, as shown in Fig. 23, the stress functions are:

$$\left. \begin{array}{l} U_1'(\zeta_1) = \dfrac{iP\mu_2}{2\pi(\mu_1 - \mu_2)} \\ \\ \quad \cdot \log\left[\dfrac{\zeta_1 + (\zeta_1^2 - a^2)^{1/2} - ia}{\zeta_1 + (\zeta_1^2 - a^2)^{1/2} + ia}\right] \\ \\ U_2'(\zeta_2) = \dfrac{-iP\mu_1}{2\pi(\mu_1 - \mu_2)} \\ \\ \quad \cdot \log\left[\dfrac{\zeta_2 + (\zeta_2^2 - a^2)^{1/2} - ia}{\zeta_2 + (\zeta_2^2 - a^2)^{1/2} + ia}\right] \\ \\ (\zeta_k = +a + z_k) \end{array} \right\} ..(84)$$

Using Eqs 78–80 with Eqs 84 and comparing results with Eqs 81, it is found that

$$\left. \begin{array}{l} K_{Ia} = \dfrac{P}{(\pi a)^{1/2}} \\ \\ K_{IIa} = 0 \end{array} \right\} \ldots\ldots\ldots(85)$$

Equations 85 can also be obtained from the isotropic case, Eqs 32 or 33, directly. It is therefore easy to add a multitude of examples by simply constructing stress-intensity factors from symmetric and skew-symmetric isotropic counterparts. Attention shall now be turned to the condition of pure shear.

Pure Shear:

For this case the generalized Hooke's law, Eqs 71, may be reduced by the definition of pure shear to:

$$\left. \begin{array}{l} \gamma_{yz} = \dfrac{\partial w}{\partial y} = A_{44}\tau_{yz} + A_{45}\tau_{xz} \\ \\ \gamma_{xz} = \dfrac{\partial w}{\partial x} = A_{54}\tau_{yz} + A_{55}\tau_{xz} \end{array} \right\} \ldots(86)$$

where $A_{54} = A_{45}$. Substituting these expressions into the equilibrium equations, the result is

$$A_{44}\frac{\partial^2 w}{\partial x^2} - 2A_{45}\frac{\partial^2 w}{\partial x \partial y} + A_{55}\frac{\partial^2 w}{\partial y^2} = 0 ..(87)$$

which can be written

$$D_5 D_6 w = 0 \quad \quad (88)$$

where, as previously defined,

$$D_k = \frac{\partial}{\partial y} - \mu_k \frac{\partial}{\partial x}$$

Comparing Eqs 87 and 88, μ_5 and μ_6 are the roots of

$$A_{55}\mu^2 - 2A_{45}\mu + A_{44} = 0 \quad \quad (89)$$

It is observed that these roots are a conjugate pair, that is, $\mu_6 = \bar{\mu}_5$. Defining a complex variable, z_5, by

$$z_5 = x + \mu_5 y \quad \quad (90)$$

the general solution to Eq 88 may be expressed as

$$w = W_1(z_5) + W_2(\bar{z}_5) \quad \quad (91)$$

Since w must be real, for convenience W_2 can be taken as the negative of W_1 or

$$w = 2 \, \text{Im} \, [W_1(z_5)] \quad \quad (92)$$

Referring to Fig. 2 for a description of the coordinates, in order to satisfy the stress-free crack surface conditions, W takes the form,

$$W_1 = A \, (z_5)^{1/2} \quad \quad (93)$$

where A is a real constant in the vicinity of the crack tip. Making use of Eqs 80, 86, 90, 92, and 93, the stress may be written in the form:

$$\left. \begin{array}{l} \tau_{yz} = \dfrac{K_{IIIa}}{(2\pi r)^{1/2}} \, \text{Re} \left[\dfrac{1}{(\cos\theta + \mu_5 \sin\theta)^{1/2}} \right] \\[1em] \tau_{xz} = \dfrac{K_{IIIa}}{(2\pi r)^{1/2}} \, \text{Re} \left[\dfrac{\mu_5}{(\cos\theta + \mu_5 \sin\theta)^{1/2}} \right] \end{array} \right\} \quad (94)$$

where it is necessarily implied that near the crack tip

$$A = \frac{K_{IIIa} (A_{44}A_{55} - A_{45}^2)^{1/2}}{(2\pi)^{1/2}} \quad \quad (95)$$

The anisotropic stress-intensity factor, K_{IIIa}, is defined so that it is also identical to its isotropic counterpart, K_{III}, for all boundary-value problems of pure shear. For example, for the configuration in Fig. 5 the result is

$$K_{IIIa} = K_{III} = \tau \, (\pi a)^{1/2} \quad \quad (96)$$

upon constructing the solution and comparing the result with Eq 7.

Consequently, it has been shown that, for the general homogeneous anisotropic case, the crack-tip stress fields and their intensity factors, the complete analogy with the isotropic case is preserved. By judicious definition of the anisotropic stress-intensity factors, they are identical to those for the isotropic case. The resulting stress field equations (Eqs 81 and 94), when superimposed[16] give the most general state of stress in the neighborhood of a crack tip in an anisotropic body with any configuration or loading.

Perhaps most important of all is the fact that like the isotropic case, the $1/r^{1/2}$ singularity appears in the stress field equations (Eqs 81 and 94). This fact implies that fracture size effects for homogeneous anisotropic media will be identical to the isotropic case.

However, for nonhomogeneous anisotropy, such as polar orthotropy, discussed by Williams (73), singularities other than the $1/r^{1/2}$ type may appear, causing different size effects than the isotropic case.

Cracks in Linear Viscoelastic Media

The deformation of cracks in plane viscous extension has been studied by Berg (78,79). He has shown that in a linear viscous sheet, elliptical holes (including the limiting cases of cracks and circles) always deform into other ellipses for the cases where additional separation is not taking place. The exclusion of separation means that adjacent

---

[16] Components of stress eliminated from the stress-strain laws should be re-introduced. They are derivable directly from the listed components in Eqs 81, 94, and 71.

points on the contour of the hole are remaining adjacent. This assumption may be somewhat restrictive, but it permits the important conclusion of ellipses deforming into ellipses, which in turn allows the use of increments of infinitesimal deformation analysis to provide a stress analysis of this class of problems.

Therefore, for stationary cracks Berg has shown that the treatment by Sih (80) of stress fields near sharp crack tips for arbitrary linear viscoelasticity is in fact pertinent even though "blunting" of the crack tip takes place. Sih has shown that the crack-tip stress fields are as given in Eqs 1, 2, and 3, where the stress-intensity factors are functions of time, that is,

$$\left.\begin{array}{l} K_I = K_I(t) \\ K_{II} = K_{II}(t) \\ K_{III} = K_{III}(t) \end{array}\right\} \dots \dots (97)$$

These stress-intensity factors may be regarded as representing the time history of intensity of a crack-tip stress field of constant spatial distribution.

Treatment of problems of moving (extending) cracks in viscoelastic media is currently unknown. However, they are obviously pertinent to formulating the condition instability of cracks in viscoelastic media where slow growth precedes sudden failure.

On the other hand, the fact that Eqs 1, 2, and 3 have been shown to apply to any crack in a linear viscoelastic body, leads to the conclusion that slight amounts of viscous action may cause time effects but size effect will be identical to the elastic case. Consequently viscous "strain-rate effects" in studies of fracture, see for example Refs (81–83), may be based on the usual elastic stress analysis, that is, Eqs 1, 2, and 3.

Some Special Cases on Nonhomogeneous Media with Cracks

The general problem of nonhomogeneous media with cracks has as yet not been attacked. However, some special cases of practical interest have been treated.

The problem of two semi-infinite half-planes of different material bonded (or welded) together along a line (or plane) containing a crack has received the most attention (84–87). The applications of these analyses include faults in laminations in rock or other materials, cracks formed at steps in the thickness of plates in extension or bending, or both; stresses in glued joints and bond cracks in composite materials.

The stress fields (84,85) for crack tips along such bond lines take the form:

$$\sigma_{ij} = \frac{K}{(2\pi r)^{1/2}} f_{ij}(\epsilon, \theta, \log r) \dots \dots (98)$$

where the terms of the type "$\log r$" are shown (85) to be of little influence on the stress fields. Consequently, the $1/r^{1/2}$ type of singularity is the controlling factor in the stress field. Therefore, again the dimensional character of K is essentially preserved and fracture size effects will be identical to the homogeneous case.

However, Zak (88) observed that for a crack perpendicular to and reaching an interface between two materials, the coefficient, n, of the stress singularity, r^{-n}, will be other than $\frac{1}{2}$. If the new material being entered by the crack has a lower modulus of elasticity, then n will be greater than $\frac{1}{2}$ and vice versa. This seems to indicate a tendency to promote the entering of cracks in hard materials into softer ones due to the increased severity of the type of singularity.

Another implication here is that size effects in transmitting fracture from the harder phases of composite materials to a softer matrix will be different than the case of cracks in homogeneous materials. (More specifically, the stress required for failure should depend inversely upon the size of the hard phase grains to the nth power.)

Inertial Effects on the Stress Field of a Moving Crack

Long before many solutions to elastostatic crack problems were available, Yoffe (89) presented the steady-state solution to a crack of constant length, $2a$, moving through plate subjected to uniform tension, σ. Moreover, she noted that the extending crack tip possessed a stress field of the form,

$$\sigma_{ij} = \frac{\sigma(\pi a)^{1/2}}{(2\pi r)^{1/2}} g_{ij}(\theta, C, E, \nu, \gamma) \ldots (99)$$

where $\sigma(\pi a)^{\frac{1}{2}}$ can be recognized as the stress-intensity factor, C is the crack velocity, and γ is the mass density of the material. Notice that for all values of the crack velocity the $1/r^{1/2}$ singularity is preserved. McClintock (90) obtained similar results for steady-state problems of the mode III variety, pure shear. Both note that g_{ij} is virtually the same as the static case, Eq 1, up to crack speeds, C, of over 0.4 of the shear wave velocity, C_2, where

$$C_2 = \left(\frac{E}{2(1+\nu)\gamma}\right)^{1/2} \ldots (100)$$

However, at some velocity, C, in the neighborhood of $0.5 C_2$, the location of the maximum in the θ-direction stresses changes from $\theta = 0$ deg to an angle of about $\theta = 60$ deg to the crack tip and the distribution of stresses, g_{ij} in general becomes quite different from the static case. The highest triaxiality of stresses near the crack tip shifts from directly ahead of the crack to about $\theta = 60$ deg which is most easily observed from the calculations of Baker (91). Other authors (92–94) have re-emphasized these observations, including the transient states (91).

Experimental photoelastic studies (95) confirm these results and observations of crack branching at velocities near $0.5 C_2$ add further evidence.

In addition, Mott (96) and Roberts (97) studied the acceleration of a crack through dimensional considerations and obtained results tentatively in agreement with those above.

Most important in this discussion of dynamic effects is that the stress fields, Eq 99, are preserved in a form nearly identical to the elastic stress up to very high velocities, that is, $C \rightarrow 0.5 C_2$. Moreover, the $1/r^{1/2}$ singularity appears

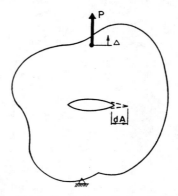

Fig. 24—A Crack in a Body of Arbitrary Shape Subjected to a Load.

for all velocities so that fracture size effects are virtually unchanged.

Energy-Rate Analysis of Crack Extension

Griffith (1) in his original analysis of fracture and later Irwin (98) and Orowan (99) discussed the equilibrium and stability of cracks from an energy-rate viewpoint. Subsequently, Irwin (100,4,5) provided a more detailed study of the energy-rate analysis and its relationship to the crack-tip stress field approach. The details were further generalized and clarified by other authors (20,41,101). The results of these works prove the equivalence of the energy-rate and stress-intensity factor approaches. Application to "compliance calibration" (that is, experimental determination of energy rates) of test configurations is an additional benefit. This discussion will proceed to cover the essential features of energy considerations.

An elastic body subjected to loads and containing an extending crack provides an energy rate (that is, energy per unit of new crack area generated), $\mathcal{G}$, available for the crack-extension process. Referring to Fig. 24, the available energy for an increment of crack extension, dA, is provided from work done by the force, $Pd\Delta$, and the release, $-dV$, in the total strain energy, V, stored in the body (100). Consequently,

$$\mathcal{G} = P\frac{d\Delta}{dA} - \frac{dV}{dA} \quad \ldots \ldots (101)$$

the displacements of a linear elastic body are related to the load by

$$\Delta = \lambda P \ldots \ldots (102)$$

where λ is the compliance (that is, inverse spring constant), which depends upon the configuration, including the crack size, A.

The strain energy in the body is work done in loading, that is,

$$V = \frac{P\Delta}{2} = \frac{\lambda P^2}{2} \ldots \ldots (103)$$

From Eqs 102 and 103 and, using the rule of differentiation,

$$\frac{d}{dA} = \frac{\partial}{\partial A} + \frac{dP}{dA} \cdot \frac{\partial}{\partial P} \ldots \ldots (104)$$

Eq 101 becomes

$$\mathcal{G} = \frac{P^2}{2}\frac{\partial \lambda}{\partial A} \ldots \ldots (105)$$

Terms involving dP cancel in Eq 105. Therefore, the available energy rate, $\mathcal{G}$, for infinitesimal crack extension is independent of the type of load application, for example, fixed grips, constant forces, or intermediate cases. This result applies for an unlimited number of forces on the body (41) and for mixed types of load application (101).

Equation 105 is useful for the experimental determination of energy rates of test configurations. This is accomplished through measurement of the compliance, λ, as a function of crack size, A, in order to compute the derivative in Eq 105. Though this so-called "compliance calibration" method is straightforward in principal, the derivative depends on small changes in λ, which in practice

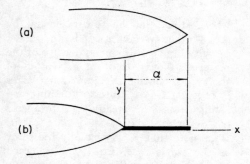

Fig. 25—The Tip of a Crack (a) Which Has Been Pulled Closed (b) Along a Segment Adjacent to the Tip.

require very accurate measurement techniques.

The Equivalence of Energy-Rate and Stress-Intensity Factor Approaches

In the previous section it was noticed that the energy rate, $\mathcal{G}$, is independent of the type of load application. Hence, for convenience in the discussion to follow, the fixed-grip situation may be employed with no loss in generality of results.

If an elastic body is loaded and the grips (load-point displacements) are then fixed, the strain-energy change, dV/dA, is the only contribution to $\mathcal{G}$ (see Eq 101). Under this condition, the work required to close a small segment of the crack, α, as shown in Fig. 25, from the opened position, (a), to the closed position, (b), is identical to the change in the strain energy. The work can be computed as the crack-surface tractions required to close the crack times their closing displacements times one half (since the displacements will be proportional to the tractions). The tractions required are the stresses on the prospective crack surface with the tip at $x = 0$ as in Fig. 25(b). The displacements are the crack-surface displacements of corresponding points in Fig. 25(a).

Therefore, as originally proposed by Irwin (4,5,7), the energy rate, $\mathcal{G}$, can be obtained from these considerations in the form

$$\mathcal{G} = \frac{dV}{dA}\bigg|_{\text{fixed grips}} = \lim_{\alpha \to 0} \frac{2}{\alpha}$$

$$\cdot \int_0^\alpha \left(\frac{\sigma_y v}{2} + \frac{\tau_{yx} u}{2} + \frac{\tau_{yz} w}{2}\right) dx \ldots (106)$$

The stresses, σ_y, τ_{yx} and τ_{yz}, may be obtained from the crack-tip stress field equations, such as Eqs 1, 2, and 3, with $r = x$ and $\theta = 0$. The corresponding displacements are also those of the crack-tip field equations, but with $r = \alpha - x$ and $\theta = \pi$.

For the isotropic case, the result of these substitutions and performance of the integration in Eq 106 leads to (for plane strain):

$$\mathcal{G} = \frac{1-\nu}{2G} K_I^2 + \frac{1-\nu}{2G} K_{II}^2 + \frac{1}{2G} K_{III}^2 \ldots (107)$$

The terms on the right-hand side of Eq 107 indicate that the energy-rate contribution of each mode of crack-tip stress field may be considered separately. Since $E = 2(1 + \nu)G$, the separate contributions are (for plane strain):

$$\left.\begin{array}{l}\mathcal{G}_I = \dfrac{(1-\nu^2)}{E} K_I^2 \\[4pt] \mathcal{G}_{II} = \dfrac{(1-\nu^2)}{E} K_I^2 \\[4pt] \mathcal{G}_{III} = \dfrac{1+\nu}{E} K_{III}^2\end{array}\right\} \ldots (108)$$

where

$$\mathcal{G} = \mathcal{G}_I + \mathcal{G}_{II} + \mathcal{G}_{III} \ldots (109)$$

Equations 108 and 109 also may be adopted to the case of plane stress by appropriately discarding $(1 - \nu^2)$ in the first two of Eqs 108.

TABLE 6—ELASTIC COEFFICIENTS RELATING ENERGY RATES TO STRESS-INTENSITY FACTORS.

$$(\mathcal{G}_i = cK_i^2)$$

(Values of c given below for the case of plane strain)

Mode	Isotropic (5)	Orthotropic (32) $A_{16} = A_{26} = A_{45} = 0$	Anisotropic
I	$\dfrac{(1-\nu^2)}{E}$	$\left(\dfrac{A_{11}A_{22}}{2}\right)^{1/2}\left[\dfrac{A_{22}}{A_{11}} + \dfrac{2A_{12} + A_{66}}{2A_{11}}\right]^{1/2}$	$\dfrac{1}{2} \operatorname{Im}\left[-A_{22}\dfrac{(\mu_1 + \mu_2)}{\mu_1 \mu_2}\right]$
II	$\dfrac{(1-\nu^2)}{E}$	$\dfrac{A_{11}}{2}\left[\dfrac{A_{22}}{A_{11}} + \dfrac{2A_{12} + A_{66}}{2A_{11}}\right]^{1/2}$	$\dfrac{1}{2} \operatorname{Im}[A_{11}(\mu_1 + \mu_2)]$
III	$\dfrac{(1+\nu)}{E}$	$\dfrac{1}{2(A_{44}A_{55})^{1/2}}$	$\dfrac{1}{2}\dfrac{(A_{44}A_{55} - A_{45}^2)^{1/2}}{A_{44}A_{55}}$

As a consequence of Eqs 108 and 109, the direct relationship between energy rates and stress-intensity factors has been illustrated.

Equation 106 can also be used to determine the relationships between energy rates and stress-intensity factors for other elastic media. For example, the relationships for anisotropic media can be obtained by using the appropriate stress fields, Eqs 81 and 94, and corresponding displacements in Eq 106. Table 6 provides the modified elastic coefficients for the equivalent of Eqs 108 for orthotropic (32) and general anisotropic[9] media. Equation 109 applies to orthotropic media, but in its present form not to the general anisotropic case.

However, since cracks normally do not extend in a planar fashion (53) with K_{II} and K_{III} present, or even with K_I present in generally anisotropic media, these relationships are somewhat of academic interest. It is sufficient to have shown the equivalence of the energy-rate and stress-intensity factor approaches, in order that the direct relationship between the Griffith theory and current theories of fracture mechanics be fully understood.

Other Equivalent Methods of Stress Analysis of Cracks and Notches

Several other methods of stress analysis of cracks and notches for incorporation into failure criteria have been proposed. The most notable in the recent literature are those developed by Neuber (10,102), Kuhn (66,103), and Barenblatt (23). Identical to the elastic field approach, each of these methods uses an elastic stress analysis to determine the general character of redistribution of force transmission around cracks. In addition, it is important to note that each of these analyses draws attention to a phenomenon at the crack tip which is regarded as that which precipitates failure.

More specifically, these phenomena are: developing a plastic particle of critical size, developing the ultimate stress at a specific radius from the crack tip, and developing stresses approaching the cohesive bond forces ahead of a crack, respectively. Now, since each of these phenomena occurs imbedded within the elastic crack-tip stress field, their occurrence will always correspond to having that stress field reach a critical value. As a consequence, these and any other methods which draw attention to specific critical phenomena at a crack tip, which proceed to use an essentially elastic stress analysis, will lead to a failure theory equivalent to the current fracture mechanics concept of critical values of stress-intensity factors.

Even though these alternative methods may be regarded as just as true, correct and, useful in a practical sense, the attention that each draws to a specific phenomenon within the crack-tip stress field embodies an assumption which is unnecessarily restrictive in formulating a failure criterion. The strength and generality of fracture mechanics as based on the stress-field approach is in part due to the absence of such an assumption.

On the other hand, this does not mean that the phenomena which do in fact occur within the stress fields near crack tips should be disregarded. Attention to the details of the processes by which materials resist cracking will undoubtedly lead to development of superior materials. Each of the alternative theories of fracture mentioned above (and others) does in fact draw attention to a phenomenon which may be a key feature in the fracture process. Therefore, their high worth in conjunction with and complimentary to the methods of fracture mechanics is clear.

Limitations of the Crack-Tip Stress Field Analyses

In this paper results of linear elastic stress analyses of cracked bodies have been presented for a typical variety of problems which have already been treated. The determination of stress-intensity factors for any particular problem can with time be accomplished. Therefore, the elastic stress analysis is not in itself a real limitation on fracture mechanics.

However, the accomplishment of a stress analysis does represent a delay in the application of fracture mechanics to configurations with cracks which have not yet been treated. Moreover, the accuracy of known solutions for stress-intensity factors represents a temporary limitation on the accuracy of immediate applications. Usually, this limitation is far less severe than others, such as variability of materials, in practical applications. Consequently, the elastic stress analysis itself may be regarded as "exact" and the real limitations of fracture mechanics come only in its application to situations where nonlinearity of material behavior at the crack tip (or elsewhere) disrupts the gross features of the stress distribution.

A certain amount of nonlinear behavior such as plasticity can be tolerated within the crack-tip stress field without grossly affecting the field outside the nonlinear region. Moreover, the disturbances, if embedded within identical fields, will themselves be identical and hence self-compensating in comparisons of fracture strengths. Therefore, it is important to resolve the relative sizes of zones of nonlinearity which can be tolerated within the crack-tip stress fields. This size is of course related to the relative size in which the field equations, such as Eqs 1, apply.

For the configuration shown in Fig. 3, the approximate stress, σ_y, ahead of the crack, obtained by substituting Eq 4 into Eq 1 and setting $\theta = 0$, is:

$$\sigma_{y\,\text{approx}} = \frac{\sigma a^{1/2}}{(2r)^{1/2}} \quad \ldots \ldots (109)$$

The exact stress can be most easily determined by the Westergaard stress function technique, see Appendix I, and is:

$$\sigma_{y\,\text{exact}} = \frac{\sigma(a + r)}{(2ar + r^2)^{1/2}} \quad \ldots \ldots (110)$$

where, in Eqs 109 and 110, r is the distance ahead of the crack tip along the crack line.

Now, taking the ratio of the exact to the approximate stresses gives:

$$\frac{\sigma_{y\,\text{approx}}}{\sigma_{y\,\text{exact}}} = \frac{\left(1 + \dfrac{r}{2a}\right)^{1/2}}{\left(1 + \dfrac{r}{a}\right)} \quad \ldots (111)$$

In a similar fashion, this ratio may also be computed for the configuration shown in Fig. 23 and is:

$$\frac{\sigma_{y\,\text{approx}}}{\sigma_{y\,\text{exact}}} = \left(1 + \frac{r}{a}\right)\left(1 + \frac{r}{2a}\right)^{1/2} \ldots (112)$$

The types of loading in these two configurations, Figs. 3 and 23, are opposite extremes, yet Eqs 111 and 112 show similar deviations of the approximate stresses from the exact, at like values of r/a. Therefore, if the relative tolerable size (compared to crack size, a) of zone of nonlinearity can be established for one configuration it is bound to be applicable to others.

Recent experimental evidence (70) indicates the validity of the elastic stress field approach up to stress levels, σ, of 0.8 of the yield strength, σ_{yp}, for the configuration shown in Fig. 3. For this configuration, the width, ω, of the zone of plasticity is predicted to be (72):

$$\omega = \frac{1}{2}\left(\frac{\sigma}{\sigma_{yp}}\right)^2 a \quad \ldots \ldots (113)$$

Substituting the upper limit of stress, $\sigma = 0.8\ \sigma_{yp}$, mentioned above, the relative size, w/a or r/a, for reasonable accuracy is about 0.3 from Eq 113. For this value of r/a, Eq 111 predicts a deviation of actual stresses from the field equations of about 20 per cent. Thus it appears that the zone of nonlinearity at a crack tip may be fairly sizable, that is, of the order of 0.3 of the crack length (and other planar dimensions such as net-section width), without grossly disturbing the usefulness of the elastic stress field approach. However, a more extensive evaluation of this limitation should be the subject of further research.

In addition to nonlinearity in the region of the crack tip, consideration of other conditions (such as anisotropic and viscous effects having cracks in the bond line between dissimilar materials, thermal stresses, couple stresses, inertial effects of moving cracks) and of all three modes of crack-tip stress fields, has led to positive results. The conclusion is that the current techniques of fracture mechanics may be extended to all of these areas, since similar types of crack-tip stress fields exist for them and the stress-intensity factor methods of assessing failure should apply equally well. At any rate, this conclusion should give full confidence that slight amounts of these effects do not invalidate the useful application of the concepts of fracture mechanics.

As a consequence of the above remarks, it is observed that the only real limitation of elastic stress analysis commences with the advent of sizable zones of nonlinearity that is, plasticity, at the crack tip. The current hope for extension of the applicability of fracture mechanics to such situations lies in developing a full analysis based on the theory of plasticity. This topic is a subject left for other discussions.

Acknowledgment:

The authors gratefully acknowledge the support of this work under the NASA Grant NsG 410 to the Lehigh University Institute of Research. Portions of the analysis were based on previous work supported under grants from the National Science Foundation, G24145, and the Boeing Co.

The many suggestions and data provided by W. F. Brown, O. L. Bowie, and G. R. Irwin were most helpful in preparation of this paper. Appreciation is also expressed for aid in collecting information for Appendix II given by M. Kassir of the Department of Mechanics of Lehigh Univierisity.

APPENDIX I

THE WESTERGAARD METHOD OF STRESS ANALYSIS OF CRACKS

Any elementary text on the theory of elasticity gives a full development of the equations for plane extension. The equilibrium equations are:

$$\left.\begin{array}{l}\dfrac{\partial \sigma_x}{\partial x} + \dfrac{\partial \tau_{xy}}{\partial y} = 0 \\[4pt] \dfrac{\partial \tau_{xy}}{\partial x} + \dfrac{\partial \sigma_y}{\partial y} = 0 \\[4pt] \tau_{xy} = \tau_{yx}\end{array}\right\} \ldots \ldots (114)$$

The strain-displacement relationships and Hooke's law lead to the compatability equation:

$$\nabla^2(\sigma_x + \sigma_y) = \left(\frac{\partial^2}{\partial x^2} + \frac{\partial^2}{\partial y^2}\right)(\sigma_x + \sigma_y) = 0 \ldots (115)$$

The equilibrium equations (114) are automatically satisfied by defining an Airy stress function, Φ, in terms of its relationship to the stresses, that is,

$$\left.\begin{array}{l}\sigma_x = \dfrac{\partial^2 \Phi}{\partial y^2} \\[4pt] \sigma_y = \dfrac{\partial^2 \Phi}{\partial x^2} \\[4pt] \tau_{xy} = \dfrac{-\partial^2 \Phi}{\partial x \partial y}\end{array}\right\} \quad \ldots\ldots (116)$$

Substituting Eq 116 into Eq 115 leads to:

$$\nabla^4 \Phi = \nabla^2(\nabla^2 \Phi) = 0 \ldots\ldots (117)$$

In order to solve a problem, the stress function, Φ, must satisfy Eq 117 and the boundary conditions of that problem.

Choosing the stress function, Φ, to be:

$$\Phi = \psi_1 + x\psi_2 + y\psi_3 \ldots\ldots (118)$$

it will automatically satisfy Eq 117 if the ψ_i are each harmonic, that is,

$$\nabla^2 \psi_i = 0 \ldots\ldots\ldots (119)$$

Define a complex variable, z, by

$$z = x + iy \ldots\ldots\ldots (120)$$

Functions of that complex variable, $\bar{\bar{Z}}(z)$, and its derivatives,

$$\bar{\bar{Z}} = \frac{d\bar{Z}}{dz}, \quad \bar{Z} = \frac{d\bar{Z}}{dz}, \quad Z' = \frac{dZ}{dz} \ldots (121)$$

have harmonic real and imaginary parts, if the function is analytic, for example, if $\bar{Z} = \operatorname{Re} \bar{Z} + i \operatorname{Im} \bar{Z}$, then

$$\nabla^2(\operatorname{Re} \bar{Z}) = \nabla^2(\operatorname{Im} \bar{Z}) = 0 \ldots\ldots (122)$$

This is a result of the Cauchy-Riemann conditions, that is,

$$\left.\begin{array}{l}\dfrac{\partial \operatorname{Re} \bar{Z}}{\partial x} = \dfrac{\partial \operatorname{Im} \bar{Z}}{\partial y} = \operatorname{Re} Z \\[6pt] \dfrac{\partial \operatorname{Im} \bar{Z}}{\partial x} = -\dfrac{\partial \operatorname{Re} \bar{Z}}{\partial y} = \operatorname{Im} Z\end{array}\right\} \ldots (123)$$

Equations 123 may be used to differentiate these functions $\bar{\bar{Z}}$ through Z.

First Mode:

In conformity with Eqs 118–123, Westergaard (8) defined an Airy stress function, Φ, by

$$\Phi_I = \operatorname{Re} \bar{\bar{Z}}_I + y \operatorname{Im} \bar{Z}_I \ldots\ldots (124)$$

which as a consequence automatically satisfies equilibrium and compatability, Eqs 114 and 117.

Using Eqs 116 and 123, the stresses resulting from Φ, as defined in Eq 124, are

$$\left.\begin{array}{l}\sigma_x = \operatorname{Re} Z_I - y \operatorname{Im} Z_I' \\ \sigma_y = \operatorname{Re} Z_I + y \operatorname{Im} Z_I' \\ \tau_{xy} = -y \operatorname{Re} Z_I'\end{array}\right\} \ldots (125)$$

Now any function, Z_I, which is analytic in the region except for a particular branch cut along a portion of the x-axis will have the form

$$Z_I = \frac{g(z)}{[(z+b)(z-a)]^{1/2}} \ldots (126)$$

This will solve crack problems for a crack along the x-axis from $x = -b$ to $x = a$, ($y = 0$), if $g(z)$ is well behaved, since the stresses, σ_y and τ_{xy}, along that interval are zero, provided that

$$\operatorname{Im} g(x) = 0 \text{ (for } -b < x < a) \ldots (127)$$

For example, if the function

$$Z_I = \frac{\sigma z}{(z^2 - a^2)^{1/2}} \ldots\ldots (128)$$

is examined, it solves the problem of a stress-free crack at $-a < x < a$, $y = 0$, and leads to boundary conditions of uniform biaxial stress, σ, at infinity (see Fig. 3).

Now, reverting to the more general case, Eq 126, a substitution of variable

$$\zeta = z - a \ldots\ldots\ldots (129)$$

leads to

$$Z_I = \frac{f(\zeta)}{\zeta^{1/2}} \ldots\ldots (130)$$

where, from Eqs 126 and 127, $f(\zeta)$ is well behaved for small $|\zeta|$ (that is, near the crack tip at $x = a$). Moreover, in that region, as $|\zeta| \to 0$, f may be replaced by a real constant, or Eq 130 may be written

$$Z_I \big|_{|\zeta|\to 0} = \frac{K_I}{(2\pi\zeta)^{1/2}} \ldots\ldots (131)$$

Other stress functions, Z_I, for crack problems, such as Eq 16, will also always lead to this form. Noting that Eq 131 may be substituted into Eqs 125, and using polar coordinates, that is,

$$\zeta = re^{i\theta} \quad \ldots \ldots \ldots (132)$$

the crack-tip stress field is:

$$\left.\begin{aligned}
\sigma_x &= \frac{K_\mathrm{I}}{(2\pi r)^{1/2}} \cos\frac{\theta}{2} \left[1 - \sin\frac{\theta}{2}\sin\frac{3\theta}{2}\right] \\
\sigma_y &= \frac{K_\mathrm{I}}{(2\pi r)^{1/2}} \cos\frac{\theta}{2} \left[1 + \sin\frac{\theta}{2}\sin\frac{3\theta}{2}\right] \\
\tau_{xy} &= \frac{K_\mathrm{I}}{(2\pi r)^{1/2}} \sin\frac{\theta}{2} \cos\frac{\theta}{2} \cos\frac{3\theta}{2}
\end{aligned}\right\} \ldots (133)$$

where, from Eq 131,

$$K_\mathrm{I} = \lim_{|\zeta|\to 0} (2\pi\zeta)^{1/2} Z_\mathrm{I} \ldots \ldots (134)$$

The strain in the y-direction can be written in terms of displacements and stresses by Hooke's law, or

$$\epsilon_y = \frac{\partial v}{\partial y} = \frac{\sigma_y}{E} - \frac{\nu}{E}(\sigma_x + \sigma_z) \ldots (135)$$

For plane strain, Hooke's law ($\epsilon_z = 0$) also leads to

$$\sigma_z = \nu(\sigma_x + \sigma_y) \ldots \ldots \ldots (136)$$

Substituting Eqs 125 and 136 into Eq 135 and integrating lead to

$$v = \frac{1+\nu}{E}[2(1-\nu)\,\mathrm{Im}\,\bar{Z}_\mathrm{I} - y\,\mathrm{Re}\,Z_\mathrm{I}] \ldots (137)$$

Similarly, consideration for ϵ_x gives

$$u = \frac{1+\nu}{E}[(1-2\nu)\,\mathrm{Re}\,\bar{Z}_\mathrm{I} - y\,\mathrm{Im}\,Z_\mathrm{I}] \ldots (138)$$

Substituting Eqs 131 and 132 into Eqs 137 and 138 and noting $E = 2G(1+\nu)$ lead to

$$\left.\begin{aligned}
u &= \frac{K_\mathrm{I}}{G}(r/2\pi)^{1/2}\cos\frac{\theta}{2} \\
&\quad \cdot \left[1 - 2\nu + \sin^2\frac{\theta}{2}\right] \\
v &= \frac{K_\mathrm{I}}{G}(r/2\pi)^{1/2}\sin\frac{\theta}{2} \\
&\quad \cdot \left[2 - 2\nu - \cos^2\frac{\theta}{2}\right]
\end{aligned}\right\} \ldots (139)$$

(for plane strain, $w = 0$)

Equations 133, 134, 136, and 139 are the resulting crack-tip stress and displacement field, that is, Eqs 1 and 13, for the first mode.

Second Mode:

Instead of choosing the Airy stress function as in Eq 124, it is equally permissible to choose the form,

$$\Phi_\mathrm{II} = -y\,\mathrm{Re}\,\bar{Z}_\mathrm{II} \ldots \ldots (140)$$

Repeating all of the operations from Eqs 124–139 and again making use of Eqs 114–123 lead to:

$$\left.\begin{aligned}
\sigma_x &= 2\,\mathrm{Im}\,Z_\mathrm{II} + y\,\mathrm{Re}\,Z_\mathrm{II}' \\
\sigma_y &= -y\,\mathrm{Re}\,Z_\mathrm{II}' \\
\tau_{xy} &= \mathrm{Re}\,Z_\mathrm{II} - y\,\mathrm{Im}\,Z_\mathrm{II}'
\end{aligned}\right\} \ldots (141)$$

and

$$\left.\begin{aligned}
u &= \frac{1+\nu}{E} \\
&\quad \cdot [2(1-\nu)\,\mathrm{Im}\,\bar{Z}_\mathrm{II} + y\,\mathrm{Re}\,Z_\mathrm{II}] \\
v &= \frac{1+\nu}{E} \\
&\quad \cdot [-(1-2\nu)\,\mathrm{Re}\,\bar{Z}_\mathrm{II} - y\,\mathrm{Im}\,Z_\mathrm{II}]
\end{aligned}\right\} \ldots (142)$$

and in the neighborhood of a crack tip, that is, $|\zeta| \to 0$,

$$Z_\mathrm{II}\big|_{|\zeta|\to 0} = \frac{K_\mathrm{II}}{(2\pi\zeta)^{1/2}} \ldots \ldots (143)$$

or

$$K_\mathrm{II} = \lim_{|\zeta|\to 0} (2\pi\zeta)^{1/2} Z_\mathrm{II} \ldots \ldots (144)$$

In addition, near the crack tip, substitution of Eq 143 into Eqs 141 and 142 leads to:

$$\left.\begin{aligned}
\sigma_x &= \frac{-K_\mathrm{II}}{(2\pi r)^{1/2}}\sin\frac{\theta}{2} \\
&\quad \cdot \left[2 + \cos\frac{\theta}{2}\cos\frac{3\theta}{2}\right] \\
\sigma_y &= \frac{K_\mathrm{II}}{(2\pi r)^{1/2}}\cos\frac{\theta}{2}\sin\frac{\theta}{2}\cos\frac{3\theta}{2} \\
\tau_{xy} &= \frac{K_\mathrm{II}}{(2\pi r)^{1/2}}\cos\frac{\theta}{2} \\
&\quad \cdot \left[1 - \sin\frac{\theta}{2}\sin\frac{3\theta}{2}\right]
\end{aligned}\right\} \ldots (145)$$

and, for plane strain,

$$\left.\begin{aligned} u &= \frac{K_{\text{II}}}{G} (r/2\pi)^{1/2} \sin \frac{\theta}{2} \\ &\quad \cdot \left[2 - 2\nu + \cos^2 \frac{\theta}{2} \right] \\ v &= \frac{K_{\text{II}}}{G} (r/2\pi)^{1/2} \cos \frac{\theta}{2} \\ &\quad \cdot \left[1 - 2\nu + \sin^2 \frac{\theta}{2} \right] \end{aligned}\right\} \quad (146)$$

These results are reflected in Eqs 2 and 14, for the second mode.

The first and second modes may be superimposed, since

$$\Phi = \Phi_{\text{I}} + \Phi_{\text{II}} \quad \ldots \ldots (147)$$

is a perfectly permissible Airy stress function, in which case stress and displacement components should simply be added to each other.

Third Mode:

The plane (two-dimensional) problem of pure shear may be specified by:

$$u = 0, \quad v = 0, \quad w = w(x,y) \quad (148)$$

The strain-displacement equations and Hooke's law give (105)

$$\left.\begin{aligned} \gamma_{xz} &= \frac{\partial w}{\partial x} = \frac{\tau_{xz}}{G} \\ \gamma_{yz} &= \frac{\partial w}{\partial y} = \frac{\tau_{yz}}{G} \end{aligned}\right\} \quad \ldots \ldots (149)$$

The stress components, σ_x, σ_y, σ_z, and τ_{xy}, all vanish so the equilibrium equations become

$$\frac{\partial \tau_{xz}}{\partial x} + \frac{\partial \tau_{yz}}{\partial y} = 0 \ldots \ldots (150)$$

which, when combined with Eqs 149, gives

$$\nabla^2 w = 0 \ldots \ldots (151)$$

Choosing

$$w = \frac{1}{G} \text{Im } Z_{\text{III}} \ldots \ldots (152)$$

leads to

$$\left.\begin{aligned} \tau_{xz} &= \text{Im } Z_{\text{III}}' \\ \tau_{yz} &= \text{Re } Z_{\text{III}}' \end{aligned}\right\} \ldots \ldots (153)$$

The stress function, Z_{III}, for a crack along the negative y-axis to the origin, takes the form near the crack tip

$$Z_{\text{III}} \big|_{|\zeta| \to 0} = \frac{K_{\text{III}}}{(2\pi\zeta)^{1/2}} \ldots \ldots (154)$$

Consequently,

$$K_{\text{III}} = \lim_{|\zeta| \to 0} (2\pi\zeta)^{1/2} Z_{\text{III}} \ldots \ldots (155)$$

Moreover, substituting Eq 154 into Eqs 152 and 153 and using Eq 132 lead to

$$\left.\begin{aligned} \tau_{xz} &= -\frac{K_{\text{III}}}{(2\pi r)^{1/2}} \sin \frac{\theta}{2} \\ \tau_{yz} &= \frac{K_{\text{III}}}{(2\pi r)^{1/2}} \cos \frac{\theta}{2} \end{aligned}\right\} \ldots \ldots (156)$$

and

$$w = \frac{K_{\text{III}}}{G} (2r/\pi)^{1/2} \sin \frac{\theta}{2} \ldots \ldots (157)$$

These results are reflected in Eqs 3 and 15 for the third mode.

APPENDIX II

A HANDBOOK OF BASIC SOLUTIONS FOR STRESS-INTENSITY FACTORS AND OTHER FORMULAS

The results to be presented for stress-intensity factors will conform with their definition as implied by Eqs 1–3, 48, 81, and 94. References which contain further results and details will be listed for the readers convenience.

A selection of solutions for stress-intensity factors, in addition to those already listed, will be chosen on the basis of their generality. Since superposition may be used, that is, addition of the stress-intensity factors for each mode, the results which lend themselves to generation of other solutions will be emphasized.

Formulas for Determination of Stress-Intensity Factors from Stress Concentrations:[9]

Mode I:

$$K_I = \lim_{p \to 0} \frac{\pi^{1/2}}{2} \sigma_{max} p^{1/2} \quad \ldots (158)$$

provided $K_{II} = K_{III} = 0$, and where p is the tip radius of the notch and σ_{max} is the

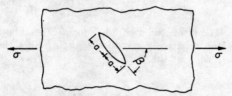

FIG. 26—A Crack in an Infinite Sheet Subjected to Uniform Tension at an Arbitrary Inclination.

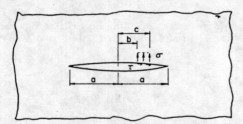

FIG. 27—A Crack in an Infinite Sheet with Uniform Loads on Part of the Crack Surface.

maximum normal stress adjacent to the tip (see Eq 9).

Mode II:

$$K_{II} = \lim_{p \to 0} \pi^{1/2} \sigma_{max} p^{1/2} \quad \ldots (159)$$

provided $K_I = K_{III} = 0$ etc.

Mode III:

$$K_{III} = \lim_{p \to 0} \pi^{1/2} \tau_{max} p^{1/2} \quad \ldots (160)$$

provided $K_I = K_{II} = 0$, and where τ_{max} is the maximum shear stress adjacent to the tip of the notch.

Infinite Sheets Subjected to in-Plane Loads: (See Fig. 26).

$$\left. \begin{array}{l} K_I = \sigma \sin^2 \beta (\pi a)^{1/2} \\ K_{II} = \sigma \sin \beta \cos \beta (\pi a)^{1/2} \\ \phi(\eta) = -\dfrac{\sigma(1 - e^{2i\beta})a}{4\eta} \end{array} \right\} \ldots (161)$$

These cases may be found by the method of Ref (18) or via Eqs 33 or by superimposing results of Eqs 4 and 6. (Note that all other cases of uniform loading at infinity or on the crack surface may be derived from this case by superposition.)

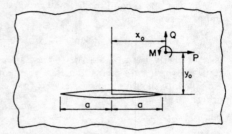

FIG. 28—A Crack in an Infinite Sheet Subjected to an Arbitrary Force and Couple at a Remote Point.

For the right end of the crack in Fig. 27,

$$\left. \begin{array}{l} K_I = \dfrac{\sigma a^{1/2}}{2\pi^{1/2}} \left\{ \sin^{-1} \dfrac{c}{a} - \sin^{-1} \dfrac{b}{a} \right. \\ \left. \quad - \left(1 - \dfrac{c^2}{a^2}\right)^{1/2} + \left(1 - \dfrac{b^2}{a^2}\right)^{1/2} \right\} \\ \quad + \dfrac{\tau(c - b)}{2(\pi a)^{1/2}} \left(\dfrac{\kappa - 1}{\kappa + 1}\right) \\ K_{II} = \dfrac{\sigma(c - b)}{2(\pi a)^{1/2}} \left(\dfrac{\kappa - 1}{\kappa + 1}\right) + \dfrac{\tau a^{1/2}}{2\pi^{1/2}} \\ \quad \cdot \left\{ \sin^{-1} \dfrac{c}{a} - \sin^{-1} \dfrac{b}{a} \right. \\ \left. \quad - \left(1 - \dfrac{c^2}{a^2}\right)^{1/2} + \left(1 - \dfrac{b^2}{a^2}\right)^{1/2} \right\} \end{array} \right\} \ldots (162)$$

where

$\kappa = 3 - 4\nu$ (plane strain)

$\kappa = (3 - \nu)/(1 + \nu)$ (plane stress)

See also Eqs 32.

For the right end of the crack in Fig. 28,

$$K = K_I - iK_{II} = \frac{1}{2(\pi a)^{1/2}(1 + \kappa)}$$
$$\left\{ (P + iQ)\left[\frac{a + z_0}{(z_0^2 - a^2)^{1/2}} - \frac{\kappa(a + \bar{z}_0)}{(\bar{z}_0^2 - a^2)^{1/2}} - 1 + \kappa \right] \right.$$
$$\left. + \frac{a(P - iQ)(\bar{z}_0 - z_0) + ai(1 + \kappa)M}{(\bar{z}_0 - a)(\bar{z}_0^2 - a^2)^{1/2}} \right\} \quad \ldots(163)$$

where

$$z_0 = x_0 + iy_0$$

$$\bar{z}_0 = x_0 - iy_0$$

See Refs (21–23) or Eqs 33.

At the near ends of two equal colinear cracks (see Fig. 29),

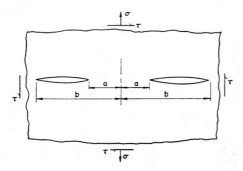

FIG. 29—Two Equal Colinear Cracks in an Infinite Sheet Subjected to Uniform Extension.

$$\left. \begin{array}{l} K_I = \sigma(\pi/a)^{1/2} \dfrac{b^2 \dfrac{E(k)}{K(k)} - a^2}{(b^2 - a^2)^{1/2}} \\[2ex] K_{II} = \tau(\pi/a)^{1/2} \dfrac{b^2 \dfrac{E(k)}{K(k)} - a^2}{(b^2 - a^2)^{1/2}} \end{array} \right\} \ldots(164)$$

At the far ends (see Fig. 29),

$$\left. \begin{array}{l} K_I = \sigma (\pi b)^{1/2} \left(\dfrac{1}{k} - \dfrac{E(k)}{kK(k)}\right) \\[2ex] K_{II} = \tau (\pi b)^{1/2} \left(\dfrac{1}{k} - \dfrac{E(k)}{kK(k)}\right) \end{array} \right\} \ldots(165)$$

where $k = [1 - (a^2/b^2)]^{1/2}$ is the modulus of the complete elliptic integrals $E(k)$ and $K(k)$ of the first and second kind, respectively. See Refs (38,23); and for concentrated forces on the crack surface, see Ref (21).

For an infinite array of cracks at the ends, denoted by e in Fig. 30,

$$\left. \begin{array}{l} K_I = \dfrac{\sigma (4b)^{1/2} \sin \dfrac{\pi c}{2b}}{\left(\cos \dfrac{\pi e}{2b} \left(\sin \dfrac{\pi e}{2b} + \sin \dfrac{\pi c}{2b}\right)\right)^{1/2}} \\[3ex] + \dfrac{P \left(\sin \dfrac{\pi c}{2b}\right)^{1/2}}{\left(b \sin \dfrac{\pi e}{2b} \cos \dfrac{\pi e}{2b} \cdot \left(\sin \dfrac{\pi e}{2b} + \sin \dfrac{\pi c}{2b}\right)\right)^{1/2}} \\[3ex] K_{II} = 0 \end{array} \right\} \ldots(166)$$

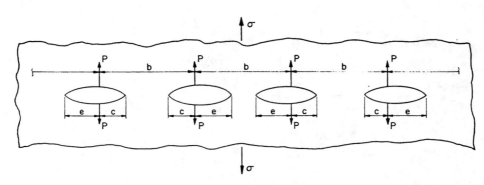

FIG. 30—An Infinite Array of Colinear Cracks in an Infinite Sheet.

For further information see Refs (**4,5**). Note that this result may be used to evaluate eccentrically located cracks in panels.

For the semi-infinite crack in Fig. 31,

$$K_I = \frac{P}{(2\pi c)^{1/2}} \\ K_{II} = \frac{Q}{(2\pi c)^{1/2}} \Biggr\} \quad (167)$$

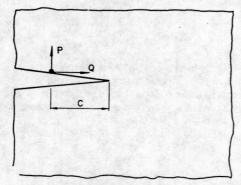

FIG. 31—An Infinite Sheet with a Semi-infinite Crack.

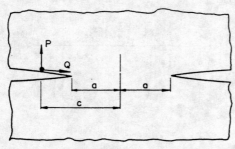

FIG. 32—An Infinite Sheet with Colinear Semi-infinite Cracks with a Concentrated Force.

For further information, see Eqs 32 or Ref (**21**).

For the two semi-infinite cracks in Fig. 32:
At the left crack tip,

$$K_I = \frac{P(c^2 - a^2)^{1/2}}{2(\pi a)^{1/2}(c - a)}$$

$$K_{II} = \frac{Q(c^2 - a^2)^{1/2}}{2(\pi a)^{1/2}(c - a)}$$

At the right crack tip,

$$K_I = \frac{P(c^2 - a^2)^{1/2}}{2(\pi a)^{1/2}(c + a)} \\ K_{II} = \frac{Q(c^2 - a^2)^{1/2}}{2(\pi a)^{1/2}(c + a)} \Biggr\} \quad (168)$$

For further information, see Ref (**21**).

Stress concentrations for deep hyperbolic notches (see Fig. 33 and Ref (**10**)) are as follows.

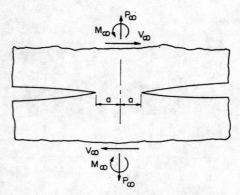

FIG. 33—An Infinite Sheet with Load Transmitted across a Neck Between Hyperbolic Notches (Cracks).

For P_∞ (force per unit thickness) alone:

$$\frac{\sigma_{max}}{\sigma_{net}} = \frac{2\left(\frac{a}{p}\right)^{1/2}\left(1 + \frac{p}{a}\right)}{\left(1 + \frac{p}{a}\right)\tan^{-1}\left(\frac{a}{p}\right)^{1/2} + \left(\frac{p}{a}\right)^{1/2}} \\ \sigma_{net} = \frac{P_\infty}{2a} \Biggr\} \quad (169)$$

For V_∞ alone:

$$\frac{\sigma_{max}}{\tau_{net}} = \frac{\left(\frac{a}{p} + 1\right)^{1/2}}{\left(1 + \frac{p}{a}\right)\tan^{-1}\left(\frac{a}{p}\right)^{1/2} - \left(\frac{p}{a}\right)^{1/2}} \\ \tau_{net} = \frac{V_\infty}{2a} \Biggr\} \quad (170)$$

For M_∞ alone:

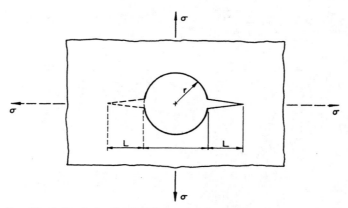

Fig. 34—A Crack (or Cracks) Emanating from a Circular Hole in a Sheet.

TABLE 7—STRESS-INTENSITY FACTOR COEFFICIENTS FOR CRACKS EMANATING FROM A CIRCULAR HOLE.

L/r	F(L/r), One Crack		F(L/r), Two Cracks	
	(uniaxial stress)	(biaxial stress)	(uniaxial stress)	(biaxial stress)
0.00	3.39	2.26	3.39	2.26
0.10	2.73	1.98	2.73	1.98
0.20	2.30	1.82	2.41	1.83
0.30	2.04	1.67	2.15	1.70
0.40	1.86	1.58	1.96	1.61
0.50	1.73	1.49	1.83	1.57
0.60	1.64	1.42	1.71	1.52
0.80	1.47	1.32	1.58	1.43
1.0	1.37	1.22	1.45	1.38
1.5	1.18	1.06	1.29	1.26
2.0	1.06	1.01	1.21	1.20
3.0	0.94	0.93	1.14	1.13
5.0	0.81	0.81	1.07	1.06
10.0	0.75	0.75	1.03	1.03
∞	0.707	0.707	1.00	1.00

$$\left. \begin{array}{l} \dfrac{\sigma_{max}}{\sigma_{net}} = \dfrac{4\left(\dfrac{a}{p}\right)^{1/2}}{3\left[\left(\dfrac{p}{a}\right)^{1/2} + \left(1 - \dfrac{p}{a}\right)\tan^{-1}\left(\dfrac{a}{p}\right)^{1/2}\right]} \\ \sigma_{net} = \dfrac{3M_\infty}{2a^2} \end{array} \right\} \quad (171)$$

Using Eqs 158 and 159, for the right crack crack tip,

$$\left. \begin{array}{l} K_I = \dfrac{P_\infty}{(\pi a)^{1/2}} + \dfrac{2M_\infty}{\pi^{1/2} a^{3/2}} \\ K_{II} = \dfrac{V_\infty}{(\pi a)^{1/2}} \end{array} \right\} \quad (172)$$

For further information, see Refs (**10,42**).
For cracks emanating from a circular hole (see Fig. 34 and Table 7),

$$\left. \begin{array}{l} K_I = \sigma\sqrt{L\pi}\, F\left(\dfrac{L}{r}\right) \\ K_{II} = 0 \end{array} \right\} \quad (173)$$

For further information, see Ref (**38**).
For a circular crack (see Fig. 35): for the crack tip at 0,

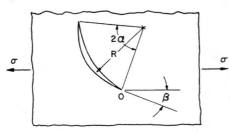

Fig. 35—A Circular-Arc Crack in an Infinite Sheet Subjected to Uniform Tension in an Arbitrary Direction.

$$K_\mathrm{I} = \frac{\sigma(\pi R \sin \alpha)^{1/2}}{2\left(1 + \sin^2 \frac{\alpha}{2}\right)} \left\{\cos \frac{\alpha}{2}\right.$$

$$+ \cos\left(2\beta + \frac{5}{2}\alpha\right)\left[\sin^2 \frac{\alpha}{2}\right]$$

$$- \cos\left(2\beta + \frac{3}{2}\alpha\right)\left[\cos^2 \frac{\alpha}{2}\right.$$

$$\left. - \sin^4 \frac{\alpha}{2}\right] - \sin\left(2\beta + \frac{3}{2}\alpha\right)$$

$$\left.\cdot \left[\sin \alpha \sin^2 \frac{\alpha}{2}\right]\right\}$$

$$K_\mathrm{II} = \frac{\sigma(\pi R \sin \alpha)^{1/2}}{2\left(1 + \sin^2 \frac{\alpha}{2}\right)} \left\{\sin \frac{\alpha}{2}\right.$$

$$+ \sin\left(2\beta + \frac{5}{2}\alpha\right)\left[\sin^2 \frac{\alpha}{2}\right]$$

$$+ \sin\left(2\beta + \frac{3}{2}\alpha\right)\left[\cos^2 \frac{\alpha}{2}\right.$$

$$\left. - \sin \frac{\alpha}{2}\right] - \cos\left(2\beta + \frac{3}{2}\alpha\right)$$

$$\left.\cdot \left[\sin \alpha \sin^2 \frac{\alpha}{2}\right]\right\} \quad \ldots (174)$$

For further information, see Ref (**18**).

Some Cases of Specified Displacements in Infinite Planes:

For an infinite rigid wedge of constant thickness (see Fig. 36),

$$\left.\begin{array}{l} K_\mathrm{I} = \dfrac{Eh}{(2\pi a)^{1/2}} \quad \text{(for plane stress)} \\ K_\mathrm{II} = 0 \end{array}\right\} \ldots (177)$$

For further information, see Ref (**23**), which also has a discussion of other examples of wedging.

When stress is applied first, then the boundaries are clamped, and then a crack is introduced, this sequence (see Fig. 37) results in constant energy release rate, $\mathcal{G}_\mathrm{I}$, or

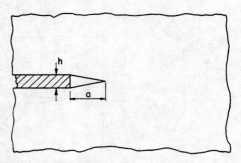

FIG. 36—A Semi-infinite Crack Propped open by a Wedge of Constant Thickness.

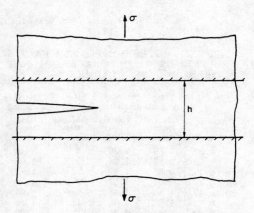

FIG. 37—A Sheet Which is Loaded and Clamped Prior to Introduction of a Crack.

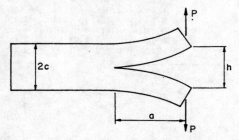

FIG. 38—The Splitting of a Rod of Rectangular Section.

$$\left.\begin{array}{l} K_\mathrm{I} = \dfrac{\sigma h^{1/2}}{2^{1/2}} \\ K_\mathrm{II} = 0 \end{array}\right\} \ldots\ldots\ldots(178)$$

For further information, see Ref (**42**).

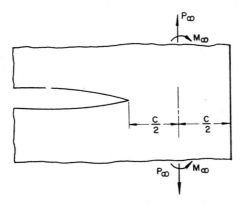

Fig. 39—Semi-finite Notch Approaching the Free Edge of a Half-Plane.

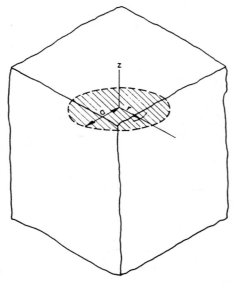

Fig. 40—Axisymmetrical Loading of a Body with a Circular Disk Crack.

A Case of the Splitting of Rods, that is, slender rectangular members ($a \gg 2c$) (see Fig. 38):

Under wedging:

$$\left. \begin{array}{l} K_I = \dfrac{3^{1/2}\, Ehc^{3/2}}{4a^2} \\ K_{II} = 0 \end{array} \right\} \quad \ldots\ldots (179)$$

Under forces:

$$\left. \begin{array}{l} K_I = \dfrac{2(3)^{1/2}\, Pa}{c^{3/2}} \\ K_{II} = 0 \end{array} \right\} \quad \ldots\ldots (180)$$

For further information, see either Ref (**23**) or (**105**).

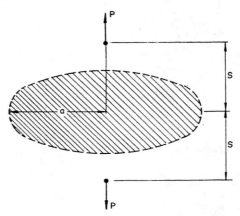

Fig. 41—Concentrated Forces Applied to the Axis of a Circular Disk Crack in an Infinite Body.

A Semi-Infinite Notch Approaching the Free Edge of a Half-Plane (see Fig. 39):

$$\left. \begin{array}{l} K_I = \pi^{1/2}\left(\dfrac{4\pi - 12}{\pi^2 - 8}\right)\dfrac{P_\infty}{c^{1/2}} \\ \quad + \pi^{1/2}\left(\dfrac{4\pi - 8}{\pi^2 - 8}\right)\dfrac{M_\infty}{c^{3/2}} \\ K_{II} = 0 \end{array} \right\} \ldots (181)$$

For further information, see Refs (**10,37**).

Axisymmetrical Loading of a Body with a Circular Disk Crack:

For an axisymmetrical normal pressure distribution, $p(r)$, on both crack surfaces (see Fig. 40),

$$\left. \begin{array}{l} K_I = \dfrac{2}{(\pi a)^{1/2}} \displaystyle\int_0^a \dfrac{r p(r)}{(a^2 - r^2)^{1/2}} dr \\ K_{II} = 0 \end{array} \right\} ..(182)$$

For further information, see Ref. (**23**).

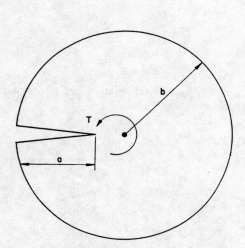

FIG. 42—Torison of a Cylindrical Bar with a Partial Radial Crack.

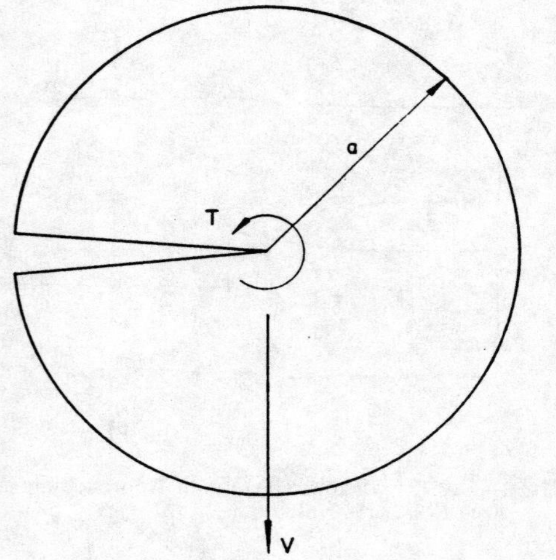

FIG. 43—Torsion and Beam Shear of a Cylindrical Bar with a Radial Crack.

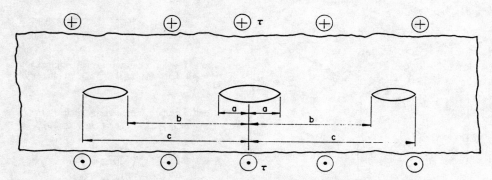

FIG. 44—An Infinite Body with "Tunnel" Cracks Under Longitudinal (Pure) Shear.

Note that with superposition, this enables treatment of all cases of axisymmetrical loading.

For the type of circular disk crack shown in Fig. 41, the K-values are:

$$\left.\begin{aligned} K_{\text{I}} &= \frac{P}{(\pi a)^{3/2}} \frac{\left[1 + \left(\frac{2-\nu}{1-\nu}\right)\frac{s^2}{a^2}\right]}{\left[1 + \frac{s^2}{a^2}\right]^2} \\ K_{\text{II}} &= 0 \end{aligned}\right\} \quad \ldots\ldots (183)$$

Equations 182 may also be used. For further information, see Ref (**23**).

Torsion and Beam Shear of Prismatic Bars with Cracks:

The K-values in prismatic bars under torsion for the type of crack shown in Fig. 42 are:

$$\left.\begin{aligned} K_{\text{I}} &= K_{\text{II}} = 0 \\ K_{\text{III}} &= \frac{(1+\alpha)^{3/2}\,(1-\alpha)^3\,[2(1-\alpha) + \alpha^{1/2}(mJ_0 + J_1)]\pi^{1/2}\,T}{\alpha^2\{2\pi^2 - [2(1-\alpha)^2\,A^2 + \alpha\,(A+B)^2]\}\;a^{5/2}} \end{aligned}\right\} \quad \ldots (184)$$

where:

$\alpha = (b - a)/b$

$$m = \tfrac{1}{2}[\alpha + 1/\alpha]$$
$$J_0 = 4 \arctan (\alpha)^{\frac{1}{2}}$$
$$J_1 = [-(1-\alpha)/4\alpha][4(\alpha)^{\frac{1}{2}} - (1-\alpha)J_0]$$
$$A = 1/\alpha [(1+\alpha)^2 (\arctan (\alpha)^{\frac{1}{2}})/(\alpha)^{\frac{1}{2}} - (1-\alpha)]$$
$$B = [(1-\alpha)/\alpha][2 - \tfrac{3}{4}(1-\alpha)A]$$

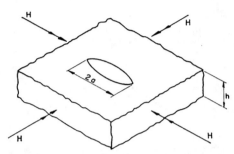

Fig. 45—A Plate Subjected to Uniform Twisting Moments at Infinity.

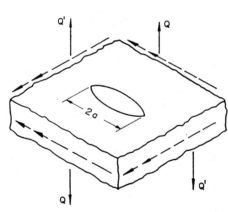

Fig. 46—A Plate with Uniform Shear at Infinity.

For further information, see Ref (**26**). Several other configurations are treated in Refs (**24,25**).

For torsion and beam shear (Fig. 43) the K-values are:

$$\left. \begin{array}{l} K_\mathrm{I} = K_\mathrm{II} = 0 \\ K_\mathrm{III} = 0.969 \dfrac{T}{a^{5/2}} - \left(\dfrac{6.95 + 6.47\nu}{1+\nu}\right)\dfrac{V}{a^{3/2}} \end{array} \right\} \ldots (185)$$

For further information, see Refs (**26,24**).

Cracks under Longitudinal (Pure) Shear:

The K_I and K_II values for the three cracks shown in Fig. 44 are all zero; the K_III values vary with the location of the crack tips as follows:

For the crack tips at *a*:
$$K_\mathrm{III} = \pm \left(\frac{(c^2-a^2)}{(b^2-a^2)}\right)^{1/2} \frac{E(k)}{K(k)} \cdot \tau(\pi a)^{1/2}$$

For the crack tips at *b*:
$$K_\mathrm{III} = \pm \left(\frac{(b^2-a^2)}{(c^2-b^2)}\right)^{1/2} \cdot \left[1 - \left(\frac{c^2-a^2}{b^2-a^2}\right)\frac{E(k)}{K(k)}\right]\tau(\pi b)^{1/2}$$

For the crack tips at *c*:
$$K_\mathrm{III} = \pm \left(\frac{(c^2-a^2)}{(c^2-b^2)}\right)^{1/2} \cdot \left[1 - \frac{(Ek)}{K(k)}\right]\tau(\pi c)^{1/2}$$

$$\bigg\} \ldots (186)$$

where
$$k = \left(\frac{(c^2-a^2)}{(c^2-b^2)}\right)^{1/2}$$

is the modulus of the complete elliptic integrals, $E(k)$ and $K(k)$ of the first and second kind, respectively.

For further information, see Ref (**106**).

The Flexure of Infinite Plates:

A plate subjected to pure twisting moment (per unit length), H, at infinity (see Fig. 45) gives:

$$\phi_B(\eta) = -\frac{iHa}{2D(3+\nu)\eta}$$

or

$$\left. \begin{array}{l} K_B = 0 \\ K_S = \dfrac{6H}{h^2}(\pi a)^{1/2} \end{array} \right\} \ldots (187)$$

For further information, see Ref (**18**).

Uniform shear (per unit length), Q, at

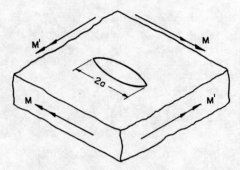

FIG. 47—A Plate with Uniform Bending Moments at Infinity.

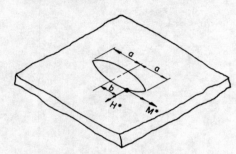

FIG. 48—A Concentrated Couple of Arbitrary Direction on a Crack Surface in an Infinite Plate.

infinity (moments required for equilibrium shown dotted in Fig. 46), gives:

$$\phi_B(\eta) = \frac{Qa^2}{12D}\left[\eta^2 + \left(\frac{1-\nu}{3+\nu}\right)\frac{3}{\eta^2}\right]$$

or

$$\left.\begin{array}{l} K_B = 0 \\ K_S = \dfrac{8\pi^{1/2}Qa^{3/2}}{h^2} \end{array}\right\} \ldots\ldots(188)$$

The results are independent of Q'.

For further information, see Ref (18).

For uniform moments at infinity (see Fig. 47),

$$\left.\begin{array}{l} K_B = \dfrac{6M}{h^2}(\pi a)^{1/2} \\ K_S = 0 \end{array}\right\} \ldots\ldots(189)$$

The results are independent of M'.

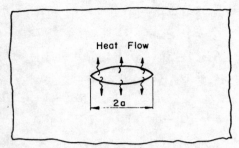

FIG. 49—A Sheet with Uniform Temperature Imposed on the Crack Surface.

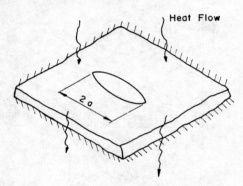

FIG. 50—A Clamped Plate with a Uniform Thermal Gradient Through the Thickness.

For further information, see Ref (18).

For a concentrated couple on the crack surface (see Fig. 48),

$$\left.\begin{array}{l} K_B = \dfrac{3M^*}{h^2(\pi a)^{1/2}}\left(\dfrac{a+b}{a-b}\right)^{1/2} \\ \qquad + \dfrac{3H^*}{2h^2(\pi a)^{1/2}}(1+\nu) \\ K_S = \dfrac{3M^*}{2h^2(\pi a)^{1/2}}(1+\nu) \\ \qquad - \dfrac{3H^*}{h^2(\pi a)^{1/2}}\left(\dfrac{a+b}{a-b}\right)^{1/2} \end{array}\right\}\ldots(190)$$

Thermal Stress Problems:

A plate with uniform temperature supplied on the crack surface (see Fig. 49) gives:

$$\left.\begin{array}{l} K_{\rm I} = -\dfrac{E\alpha\, a^{1/2}\, q}{\pi^{1/2}(1+\kappa)\mu} \\ K_{\rm II} = 0 \end{array}\right\}\ldots\ldots(191)$$

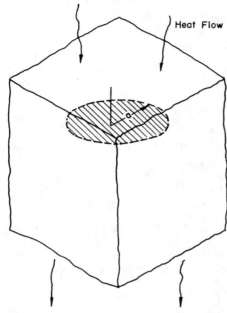

FIG. 51—An Infinite Body with a Uniform Thermal Gradient Normal to an Insulated Circular Disk Crack.

where

μ = thermal conductivity
q = rate of total heat per unit thickness supplied to the plate

For further information, see Ref (**47**). Note that this case has significance for high pressure gas escaping through a crack.

A clamped plate with a thermal gradient through the thickness (see Fig. 50) gives:

$$\left. \begin{array}{l} K_B = \dfrac{\alpha E h \nabla T (\pi a)^{1/2}}{2(1-\nu)} \\ K_S = 0 \end{array} \right\} \quad \ldots \ldots (192)$$

For further information, see Ref (**47**).

An infinite body with a circular disk crack perpendicular to a thermal gradient (see Fig. 51) gives:

$$\left. \begin{array}{l} K_{II} = \dfrac{E \alpha \nabla T a^{3/2}}{3\pi(1-\nu)} \\ K_I = K_{III} = 0 \end{array} \right\} \quad \ldots \ldots (193)$$

For further information, see Ref (**107**).

APPENDIX III

NOTATION

A	= Crack-surface area	G	= Shear modulus of elasticity
a	= Half-crack length of internal crack or crack depth of surface crack	$\mathcal{G}_I, \mathcal{G}_{II}, \mathcal{G}_{III}$	= Energy rates associated with each mode of cracking
		h	= Depth of a beam or plate
A_{ij}	= Elastic compliance coefficients for Hooke's law of plane problems of anisotropic media	i	= $(-1)^{1/2}$
		$K = K_I - iK_{II}$	= Stress-intensity factor with complex representation
a_{ij}	= Elastic compliance coefficients for the general Hooke's law of anisotropic media	K_I, K_{II}, K_{III}	= Stress-intensity factors for each mode (subscript "a" indicates anisotropic type)
b	= Half-width of a strip	K_1, K_2, K_3, K_4	= Stress intensity at various points on a crack contour
C_1, C_2	= Constants		
C	= Crack velocity	K_B, K_S	= Plate and shell bending and shearing stress-intensity factors
D_k	= Complex differential operators ($K = 1, \cdots, 6$)		
		L	= Half-length of a strip
D	= Round bar diameter	l	= A couple-stress elastic constant
d	= Notch diameter	M	= Applied bending moment (per unit thickness)
E	= Modulus of elasticity		
F	= $P - iQ$, force with complex representation (per unit thickness)	P	= Force (per unit thickness)
		p	= Crack-tip radius
		p_i, q_i	= Anisotropic elastic constants

R	= Radius of a curved crack	η	= Complex variable in the mapped plane
r	= Radial coordinate from a crack tip	θ	= Angular coordinate measured from the crack plane
T	= Temperature or torque	κ	= An elastic constant for plane stress or strain (see Eqs 162)
t	= Sheet thickness (or time)		
U, U_1, U_2	= Stress functions for anisotropic media	λ	= Compliance of a linear-elastic body
u, v, w	= Displacement components	μ, μ_k	= Elastic constants for anisotropic media ($k = 1, \cdots, 6$)
V	= Strain energy		
Z, Z_I, Z_{II}, Z_{III}	= Westergaard stress functions	ν	= Poisson's ratio
$\bar{\bar{Z}}, \bar{Z}, Z, Z'$	= Successive derivatives of a Westergaard stress function	σ, τ	= Normal and shear stress (applied at infinity)
z	= Complex variable	σ_o	= Maximum stress at a notch
z_1, z_2, z_3	= Modified complex variables for anisotropic analysis	σ_{net}	= Net-section stress (average)
		$\sigma_{ij} - \sigma_x, \sigma_y, \sigma_z$, $\tau_{xy}, \tau_{yz}, \tau_{xz}$	Rectangular components of stress
α	= An angle (or closing segment of a crack)	$\sigma_r, \sigma_\theta, \tau_{r\theta}$	= Polar components of stress
β	= An angle	Φ	= Airy stress function
γ	= Mass density	ϕ	= A complex stress function for plane stress or strain
$\gamma_{xy}, \gamma_{yz}, \gamma_{zz}$	= Shear strain components		
Δ	= Displacement	Φ_o	= An elliptic integral
∇	= Gradient $[(\partial/\partial x) + (\partial/\partial y)]$	ϕ_B	= A complex stress function for plate bending
∇^2, ∇^4	= Harmonic and biharmonic operators		
ϵ	= The bi-elastic constant of joined halfplanes	χ	= A complex stress function for plane stress or strain
$\epsilon_x, \epsilon_y, \epsilon_z$	= Normal strain components	ψ_i	= Harmonic functions
ζ	= A complex variable (origin at the crack tip)	$F(\ ), g(\ ),$ $h(\ ), k(\ )$	= A function of
ζ_1, ζ_2	= Modified complex variables for anisotropic media	Re, Im	= Real and imaginary parts of complex functions

---

References

(1) A. A. Griffith, "The Phenomena of Rupture and Flow in Solids," *Transactions*, Royal Soc. London, Vol. 221, 1920.

(2) C. E. Inglis, "Stresses in a Plate Due to the Presence of Cracks and Sharp Corners," *Proceedings*, Inst. Naval Architects, Vol. 60, 1913.

(3) I. N. Sneddon, "The Distribution of Stress in the Neighborhood of a Crack in an Elastic Solid," *Proceedings*, Royal Soc. London, Vol. A-187, 1946.

(4) G. R. Irwin, "Analysis of Stresses and Strains Near the End of a Crack Traversing a Plate," *Transactions*, Am. Soc. Mechanical Engrs., *Journal of Applied Mechanics*, 1957.

(5) G. R. Irwin, "Fracture," *Handbuch der Physik*, Vol. VI, Springer, Berlin, 1958.

(6) M. L. Williams, "On the Stress Distribution at the Base of a Stationary Crack," *Transactions*, Am. Soc. Mechanical Engrs., *Journal of Applied Mechanics*, 1957.

(7) G. R. Irwin, "Fracture Mechanics," *Structural Mechanics*, Pergamon Press, New York, N. Y., 1960.

(8) H. M. Westergaard, "Bearing Pressures and Cracks," *Transactions*, Am. Soc. Mechanical Engrs., *Journal of Applied Mechanics*, 1939.

(9) P. C. Paris, "Stress-Intensity-Factors by Dimensional Analysis," Lehigh University, Inst. of Research Report, 1961.

(10) H. Neuber, *Kerbspannungslehre*, Springer, Berlin, 1937 and 1958; English translation available from Edwards Bros., Ann Arbor, Mich.

(11) R. E. Peterson, *Stress Concentration Design Factors*, John Wiley & Sons, Inc., New York, N. Y., 1953.

(12) G. Savin, *Stress Concentration Around*

Holes, Pergamon Press, New York, N. Y., 1961.
(13) M. Isida, "The Effect of Longitudinal Stiffeners in a Cracked Plate Under Tension," *Proceedings*, Fourth U.S. Congress of Applied Mechanics, 1962.
(14) A. S. Kobayashi and R. G. Forman, "On the Axial Rigidity of a Perforated Strip and the Strain Energy Release Rate in a Centrally Notched Strip Subjected to Uniaxial Tension," *Preprint No. 63-WA-29*, Am. Soc. Mechanical Engrs., 1963.
(15) A. S. Kobayashi, R. B. Cherepy, and W. C. Kinsel, "A Numerical Procedure for Estimating the Stress Intensity Factor of a Crack in a Finite Plate," *Preprint No. 63-WA-24*, Am. Soc. Mechanical Engrs., 1963.
(16) O. L. Bowie, "Rectangular Tensile Sheet with Symmetric Edge Cracks," *Transactions*, Am. Soc. Mechanical Engrs., *Journal of Applied Mechanics*, Vol. 31, Ser. E, No. 2, June, 1964, pp. 208–212.
(17) N. I. Muskhelishvili, *Some Basic Problems of Mathematical Theory of Elasticity* (published in Russian in 1933); English translation, P. Noordhoff and Co., 1953.
(18) G. Sih, P. Paris, and F. Erdogan, "Crack Tip Stress Intensity Fractors for Plane Extension and Plate Bending Problems," *Transactions*, Am. Soc. Mechanical Engrs., *Journal of Applied Mechanics*, 1962.
(19) I. S. Sokolnikoff, *Mathematical Theory of Elasticity*, McGraw-Hill Book Co., Inc., New York, N. Y., 1956.
(20) J. L. Sanders, Jr. "On the Griffith-Irwin Fracture Theory," *Transactions*, Am. Soc. Mechanical Engrs., *Journal of Applied Mechanics*, 1960.
(21) F. E. Erdogan, "On the Stress Distribution in Plates with Colinear Cuts Under Arbitrary Loads," *Proceedings*, Fourth U. S. National Congress of Applied Mechanics, 1962.
(22) G. Sih, "Application of Muskhelishvili's Method to Fracture Mechanics," *Transactions*, The Chinese Assn. for Advanced Studies, November, 1962.
(23) G. I. Barenblatt, "Mathematical Theory of Equilibrium Cracks in Brittle Fracture," *Advances in Applied Mechanics*, Vol. VII, Academic Press, New York, 1962.
(24) G. Sih, "On Crack Tip Stress-Intensity-Factors for Cylindrical Bars under Torsion," *Journal of the Aerospaces Sciences*, 1962.
(25) G. Sih, "The Flexural Stress Distribution Near a Sharp Crack," *Journal*, Am. Inst. Aeronautics and Astronautics, 1963.
(26) G. Sih, "Strength of Stress Singularities at Crack Tips for Flexural and Torsional Problems," *Transactions*, Am. Soc. Mechanical Engrs., *Journal of Applied Mechanics*, 1963.
(27) M. A. Sadowsky and E. G. Sternberg, "Stress Concentration Around a Triaxial Ellipsoidal Cavity," *Transactions*, Am. Soc. Mechanical Engrs., *Journal of Applied Mechanics*, 1949.
(28) A. E. Green and I. N. Sneddon, "The Stress Distribution in the Neighborhood of a Flat Elliptical Crack in an Elastic Solid," *Proceedings*, Cambridge Philosophical Soc., Vol. 46, 1950.
(29) G. R. Irwin, "The Crack Extension Force for a Part Through Crack in a Plate," *Transactions*, Am. Soc. Mechanical Engrs., *Journal of Applied Mechanics* 1962.
(30) L. A. Wigglesworth, "Stress Distribution in a Notched Plate," *Mathematika*, Vol. 4, 1957.
(31) G. R. Irwin, "The Crack Extension Force for a Crack at a Free Surface Boundary," *Report No. 5120*, Naval Research Lab., 1958.
(32) G. R. Irwin, "Analytical Aspects of Crack Stress Field Problems," *T&AM Report No. 213*, University of Illinois, March, 1962.
(33) M. Isida, "On the Tension of a Strip with a Central Elliptical Hole," *Transactions*, Japan Soc. Mechanical Engrs., Vol. 21, 1955.
(34) M. Isida, "On the In-Plane Bending of a Strip with a Central Elliptical Hole," *Transactions*, Japan Soc. Mechanical Engrs., Vol. 22, 1956.
(35) M. Isida, "On the Tension of a Semi-Infinite Plate with an Elliptical Hole," *Transactions*, Japan Soc. Mechanical Engrs., Vol. 22, 1956.
(36) H. F. Bueckner, "Some Stress Singularities and Their Computation by Means of Integral Equations," *Boundary Value Problems in Differential Equations*, edited by R. E. Langer, University of Wisconsin Press, Madison, Wis., 1960.
(37) D. H. Winne and B. M. Wundt, "Application of the Griffith-Irwin Theory of Crack Propagation to Bursting Behavior of Disks," *Transactions*, Am. Soc. Mechanical Engrs., Vol. 80, 1958.
(38) O. L. Bowie, "Analysis of an Infinite Plate Containing Radial Cracks Originating from the Boundary of an Internal

Circular Hole," *Journal of Mathematics and Physics*, Vol. 35, 1956.

(39) J. P. Romualdi, J. T. Frasier, and G. R. Irwin, "Crack Extension Force near a Riveted Stringer," *Report No. 4956*, Naval Research Lab., May, 1957.

(40) J. P. Romualdi and P. H. Sanders, "Fracture Arrest by Riveted Stiffeners," *Proceedings*, Fourth Midwest Conference on Solid Mechanics, University of Texas Press, 1959/1960.

(41) P. C. Paris, "The Mechanics of Fracture Propagation and Solutions to Fracture Arrestor Problems," *Document No. D2-2195*, Boeing Co., 1957.

(42) P. C. Paris, *A Short Course in Fracture Mechanics*, University of Washington Press, 1960.

(43) J. L. Sanders, Jr., "Effect of a Stringer on the Stress Concentration Factor Due to a Crack in a Thin Sheet," *NASA Tech. Rep. R-13*, Nat. Aeronautics and Space Administration, 1959.

(44) M. Isida, "Stress Concentration Due to a Central Transverse Crack in a Strip Reinforced on Either Side," *Journal*, Japan Soc. Aero-Space Sciences, Vol. 10, 1962.

(45) R. Greif and J. L. Sanders, Jr., "The Effect of a Stringer on the Stress in a Cracked Sheet," Harvard University, Cambridge, Mass., June, 1963; also ASME *Preprint No. 64WA/APM*, to be published in *Journal of Applied Mechanics*.

(46) T. Terry, "Analysis of a Reinforced Infinite Plate Containing a Single Crack," Ph.D. dissertation, Lehigh University, September, 1963.

(47) G. Sih, "On the Singular Character of Thermal-Stresses Near a Crack Tip," *Transactions*, Am. Soc. Mechanical Engrs., *Journal of Applied Mechanics*, Vol. 29, 1962.

(48) A. L. Florence and J. N. Goodier, "Thermal Stress Due to Disturbance of Uniform Heat Flow by an Ovaloid Hole," *Transactions*, Am. Soc. Mechanical Engrs., *Journal of Applied Mechanics*, Vol. 27, 1960.

(49) M. L. Williams, "Surface Stress Singularities Resulting from Various Boundary Conditions in Angular Corners of Plates under Bending," *Proceedings*, First U. S. National Congress of Applied Mechanics, June, 1951.

(50) M. L. Williams and R. H. Owens, "Stress Singularities in Angular Corners of Plates Having Linear Flexural Rigidities for Various Boundary Conditions," *Proceedings* Second, U. S. National Congress of Applied Mechanics, June, 1954.

(51) M. L. Williams, "The Bending Stress Distribution at the Base of a Stationary Crack," *Transactions*, Am. Soc. Mechanical Engrs., *Journal of Applied Mechanics*, 1961.

(52) F. Erdogan, O. Tuncel, and P. Paris, "An Experimental Investigation of the Crack Tip Stress Intensity Factors in Plates under Cylindrical Bending," *Transactions*, Am. Soc. Mechanical Engrs., *Journal of Basic Engineering*, 1962.

(53) F. Erdogan and G. Sih, "On Crack Extension in Plates under Plane Loading and Transverse Shear," *Transactions*, Am. Soc. Mechanical Engrs., *Journal of Basic Engineering*, 1963.

(54) J. K. Knowles and N. M. Wang, "On the Bending of an Elastic Plate Containing a Crack," *GALCIT SM 60-11*, California Inst. of Technology, July, 1960.

(55) E. Reissner, "The Effect of Transverse Shear Deformation on the Bending of Elastic Plates," *Transactions*, Am. Soc. Mechanical Engrs., *Journal of Applied Mechanics*, Vol. 12, 1945.

(56) M. L. Williams, published discussion of Ref (52). See also Ref (74).

(57) G. Sih and D. Setzer, published discussion of Ref (58), *Transactions*, Am. Soc. Mechanical Engrs., *Journal of Applied Mechanics*, 1964.

(58) E. S. Folias and M. L. Williams, "The Bending Stress in a Cracked Plate on an Elastic Foundation," *Journal of Applied Mechanics*, 1963.

(59) E. S. Folias, Ph.D. dissertation, California Inst. of Technology, 1963.

(60) D. D. Ang, E. S. Folias, and M. L. Williams, "The Effect of Initial Spherical Curvature on the Stresses near a Crack Point," *GALCIT SM 62-4*, California Inst. of Technology, May, 1962.

(61) E. Reissner, "Stress and Displacements of Shallow Spherical Shells," *Journal of Mathematics and Physics*, Vol. 25, 1964.

(62) E. E. Sechler and M. L. Williams, "The Critical Crack Length in Pressurized Cylinders," *GALCIT 96*, California Inst. of Technology, Final Report, September, 1959.

(63) M. L. Williams, "Some Observations Regarding the Stress Field near the Point of a Crack," *Proceedings*, Crack Propagation Conference (Cranfield, England), September, 1961.

(64) J. I. Bluhm and M. M. Mardirosian, "Fracture Arrest Capabilities of An-

nularly Reinforced Cylindrical Pressure Vessels," *Experimental Mechanics*, Vol. 3, 1963.
(65) R. W. Peters and P. Kuhn, "Bursting Strength of Unstiffened Pressure Cylinders with Slits," *NASA TN 3993*, Nat. Aeronautics and Space Administration, April, 1956.
(66) P Kuhn, "The Prediction of Notch and Crack Strength under Static and Fatigue Loading," presented at the SAE-ASME Meeting, New York, N. Y., April, 1964.
(67) Z. and F. Cosserat, *Théorie des Corps Déformables*, A. Hermann et Fils, Paris, 1909.
(68) R. D. Mindlin, "Influence of Couple Stresses on Stress Concentrations," *Experimental Mechanics*, Vol. 3, 163.
(69) D. Setzer, "An Elasto-Static Couple Stress Problem: Extension of an Elastic Body Containing a Finite Crack," Ph.D. dissertation, Lehigh University, December, 1963.
(70) Anon., "Screening Tests for High-Strength Alloys Using Sharply Notched Cylindrical Specimens," Fourth Report of Special ASTM Committee, *Materials Research & Standards*, Vol. 2, 1962.
(71) I. N. Sneddon, "Crack Problems in Mathematical Theory of Elasticity," *Report No. ERD-126/1*, North Carolina State College, May, 1961.
(72) G. R. Irwin and F. A. McClintock, "Plasticity Modifications to Crack Stress Analysis," see p. 84.
(73) R. L. Chapkis and M. L. Williams, "Stress Singularities for a Sharp-Notched Polarly Orthotropic Plate," *Proceedings*, Third U. S. Congress of Applied Mechanics, 1958.
(74) D. D. Ang and M. L. Williams, "Combined Stresses in an Orthotropic Plate Having a Finite Crack," *Journal of Applied Mechanics*, 1961.
(75) T. J. Willmore, "The Distribution of Stress in the Neighborhood of a Crack," *Quarterly Journal of Mechanics and Applied Mathematics*, Vol. II, Part I, 1949.
(76) G. C. Sih and P. C. Paris, "The Stress Distribution and Intensity Factors for a Crack Tip in an Anisotropic Plate Subjected to Extension," Lehigh University, Inst. of Research Report, October, 1961.
(77) S. G. Lekhnitzki, *Anisotropic Plates*, translated by E. Stowell, Am. Iron and Steel Inst., New York, N. Y., 1956. (See alternatively a newer translation available through Holden-Day Book Co., San Francisco, Calif.)
(78) C. A. Berg, "The Influence of Viscous Deformation on Brittle Fracture," D.Sc. dissertation, Massachusetts Inst. of Technology, June, 1962.
(79) C. A. Berg, "The Motion of Cracks in Plane Viscous Deformation," *Proceedings*, Fourth U.S. National Congress of Applied Mechanics, 1962.
(80) G. C. Sih, "Viscoelastic Stress Distribution near the End of a Stationary Crack," Lehigh University, Inst. of Research Report, July, 1962.
(81) J. Krafft and G. R. Irwin, "Crack Velocity Considerations," see p. 114.
(82) J. Krafft and A. M. Sullivan, "Effects of Speed and Temperature upon Crack Toughness and Yield Strength in Mild Steel," *Transactions*, Am. Soc. Metals, Vol. 56, 1963.
(83) G. R. Irwin, "Crack Toughness Testing of Strain Rate Sensitive Materials," *Transactions*, ASME, *Journal of Basic Engineering*, 1964.
(84) M. L. Williams, "The Stresses Around a Fault or Crack in Dissimilar Media," *Bulletin*, Seismological Soc. Am., Vol. 49, 1959.
(85) F. Erdogan, "Stress Distribution in a Non-Homogeneous Elastic Plane with Cracks," *Transactions*, ASME, *Journal of Applied Mechanics*, Vol. 30, 1963.
(86) G. Sih and J. Rice, "The Bending of Plates of Dissimilar Materials with Cracks," *Transactions*, ASME, *Journal of Applied Mechanics*, Vol. 31, 1964. See also: J. Rice and G. Sih, "Plane Problems of Cracks in Dissimilar Media," Lehigh University, Inst. of Research Report, April, 1964.
(87) F. Erdogan and L. Y. Bahar, "On the Stress Distribution in Bonded Dissimilar Materials with a Circular Cavity," to be published in *Journal of Applied Mechanics*, 1964–1965.
(88) A. R. Zak and M. L. Williams, "Crack Point Stress Singularities at a Bimaterial Interface," *GALCIT SM42-1*, California Inst. of Technolgoy, January, 1962.
(89) E. Yoffe, "The Moving Griffith Crack," *Philosophical Mag.*, Ser. 7, Vol. 42, 1951.
(90) F. McClintock and P. Sukhatme, "Traveling Cracks in Elastic Materials under Longitudinal Shear," *Journal of Mechanics and Physics of Solids*, Vol. 8, 1960.
(91) B. R. Baker, "Dynamic Stresses Created by a Moving Crack," *Journal of Applied Mechanics*, Vol. 29, 1962.

(92) J. W. Craggs, "On the Propagation of a Crack in an Elastic-Brittle Material," *Journal of Mechanics and Physics of Solids*, Vol. 8, 1960.

(93) D. D. Ang, "Some Radiation Problems in Elastodynamics," Ph.D. dissertation, California Inst. of Technology, 1958.

(94) B. Cotterell, "On the Nature of Moving Cracks," *Journal of Applied Mechanics*, 1964.

(95) A. A. Wells and D. Post, "The Dynamic Stress Distribution Surrounding a Running Crack—a Photoelastic Analysis," *NRL Report No. 4935*, Naval Research Lab., April, 1957.

(96) N. F. Mott, "Fracture of Metals: Theoretical Considerations," *Engineering*, Vol. 165, 1948.

(97) D. K. Roberts and A. A. Wells, "The Velocity of Brittle Fracture," *Engineering*, Vol. 178, 1954.

(98) G. R. Irwin and J. Kies, "Fracturing and Fracture Dynamics," *Welding Research Journal* (Research Supplement), 1952. See also references to earlier work in this paper.

(99) E. Orowan, "Fundamentals of Brittle Behavior of Metals," *Fatigue and Fracture of Metals*, John Wiley & Sons, Inc., New York, N. Y., 1952. See also earlier work referenced therein.

(100) G. R. Irwin, "A Critical Energy Rate Analysis of Fracture Strength," *Welding Journal* (Research Supplement), 1954.

(101) H. F. Bueckner, "The Propagation of Cracks and Energy of Elastic Deformation," *Transactions*, Am. Soc. Mechanical Engrs., *Journal of Applied Mechanics*, 1958.

(102) J. L. Sanders, "On the Griffith-Irwin Fracture Theory," *Transactions*, Am. Soc. Mechanical Engrs., *Journal of Applied Mechanics*, Ser. E, Vol. 82, 1960.

(103) H. Neuber, discussion in Ref (7), *Structural Mechanics*, Pergamon Press, New York, N. Y., 1960.

(104) P. Kuhn and I. E. Figge, "Unified Notch-Strength Analysis for Wrought Aluminum Alloys," *NASA TN D-1259*, Nat. Aeronautics and Space Administration, 1962.

(105) J. J. Gilman, *Fracture*, edited by Averbach et al, John Wiley & Sons, Inc., New York, N. Y., 1959.

(106) G. C. Sih, "Boundary Problems for Longitudinal Shear Cracks," *Proceedings*, Second Conference on Theoretical and Applied Mechanics, Pergamon Press, New York, N. Y., 1964.

(107) A. L. Florence and J. N. Goodier, "The Linear Thermoelastic Problem of Heat Flow Disturbed by a Penny-shaped Insulated Crack," *International Journal of Engineering Science*, Vol. 1, No. 4, 1963.

DISCUSSION

H. F. Bueckner[1]—It might be well to review briefly the stress-analysis situation regarding the notched round bar in tension, since this geometry appears to be important from the standpoint of fracture testing. Several treatments of this problem can be found in the literature and are compared in Table 8, which gives the coefficient, $F(d/D)$, of Eq 57 in the paper, as derived from various curvature. These values agree well with those reported by Irwin[5] who used the same procedure. The results obtained by this extrapolation method are, of course, only at best as accurate as the stress-concentration factors from which they are derived. The second estimate was made using Neuber's formula for deep notches in combination with a computation for a notch in an elastic half plane.[6]

TABLE 8—COEFFICIENTS FOR COMPUTATION OF THE STRESS-INTENSITY FACTOR, K_I FOR A NOTCHED ROUND BAR.
$[K_I = F(d/D)\sigma_N (\pi D)^{1/2}]$

Notch Depth, d/D	$F(d/D)$ as given by:				
	Lubahn[a]	Irwin[b]	Wundt[c]	Paris[d]	Present Solution[e]
0.5	0.230	0.224	0.239	0.227	0.240
0.6	0.234	0.232	0.252	0.238	0.255
0.707	0.229	0.233	0.258	0.240	0.259
0.8	0.217	0.224	0.250	0.233	0.251
0.9	0.195	0.199	0.210	0.205	0.210

[a] See footnote 2.
[b] See footnote 5.
[c] See footnote 3.
[d] From Table 5 of the paper.
[e] To be published.

references. The values used by Lubahn[2] and by Wundt[3] represent my first and second estimates, respectively. The first estimate was based on extrapolation of Peterson's published stress-concentration factors[4] to a vanishingly small radius of

Recently I completed a more rigorous analysis[7] of the problem using a certain singular integral equation, the kernel of which is found by means of Fourier transforms. The coefficients derived from this solution are shown in the last column

---

[1] General Electric Co., Large Steam Turbine-Generator Dept., Schenectady, N. Y.

[2] J. D. Lubahn, "Experimental Determination of Energy Release Rate," *Proceedings*, Am. Soc. Testing Mats., Vol. 59, 1959, p. 885.

[3] B. M. Wundt, "A Unified Interpretation of Room Temperature Strength of Notch Specimens as Influenced by Size," *ASME Paper No. 59*, MET 9, 1959.

[4] R. E. Peterson, "*Stress Concentration Design Factors*," John Wiley & Sons, Inc., New York, N. Y., 1953.

[5] G. R. Irwin, "Supplement to: Notes for May, 1961 meeting of ASTM Committee for Fracture Testing of High-Strength Metallic Materials."

[6] H. F. Bueckner, "Some Stress Singularities and Their Computation by Means of Integral Equations," *Boundary Problems in Differential Equations*, edited by R. E. Langer, University of Wisconsin Press, Madison, Wisc., 1960, pp. 215–230.

[7] To be published.

of Table 8 and are considered to provide values of K_I having an accuracy within 1 per cent. It is interesting to note that the coefficients obtained by the present analysis agree well with those used by Wundt, but are higher than those given by Irwin or by Paris. Considering, for example, a notched round bar with $d/D = 0.707$, a geometry commonly used in fracture testing, the ASTM Special Committee[8] gives the following expression for the stress-intensity factor:

$$K_I = 0.414\, \sigma_N\, (D)^{1/2}$$

The coefficient in this equation corresponds to the value 0.233 in Eq 57 of the paper. The present analysis yields a value of 0.259, about 10 per cent higher than that given by the above expression.

---

[8] "Screening Tests for High-Strength Alloys Using Sharply Notched Cylindrical Specimens," Fourth Report of a Special ASTM Committee, *Materials Research & Standards*, Vol. 2, March, 1962, pp. 196–203.

WESTERGAARD STRESS FUNCTIONS FOR SEVERAL PERIODIC CRACK PROBLEMS

The crack problems with the geometry and loading configurations which are shown in Figs. 1, 2 and 3 are considered to be of practical interest in the fracture analysis of a plate with riveted or welded stiffeners. The Westergaard type stress functions $Z(\zeta)$, where $\zeta = x + iy$, may be useful to the problems [1].

In this brief note the expressions of $Z(\zeta)$, together with $\bar{Z}(\zeta) = \int Z(\zeta) d\zeta$, which are necessary to calculate the displacements, and K, the crack tip stress intensity factors, are given in closed form. All the expressions are given for the case of plane strain.

The Green's functions used here are given by Irwin [2] as

$$Z_0(\zeta) = \frac{2P}{\pi} f(\zeta, a, b) \tag{1}$$

where

$$f(\zeta, a, b) = \frac{\sqrt{a^2 - b^2}}{(\zeta^2 - b^2)\sqrt{1 - (a/\zeta)^2}} = f_1(\zeta, a, b) \text{ for the case 1.}$$

$$= \left(\frac{\pi}{W} \cos \frac{\pi b}{W}\right) f_1\left(\sin \frac{\pi \zeta}{W}, \sin \frac{\pi a}{W}, \sin \frac{\pi b}{W}\right) \text{ for the cases 2 and 3.}$$

The stress functions of the problems are calculated by

$$Z(\zeta) = \int_0^a \frac{2Q(b) db}{\pi} f(\zeta, a, b) db \tag{2}$$

where $Q(x) = \sigma_y(x, 0)$ is the distribution of σ_y which could be on the crack line in absence of the crack. The functions $\bar{Z}(\zeta)$ and the stress intensity factors are calculated by

$$\bar{Z}(\zeta) = \int Z(\zeta) d\zeta \tag{3}$$

and

$$K = \lim_{\zeta \to 0} \sqrt{2\pi\zeta} Z(a + \zeta). \tag{4}$$

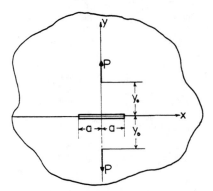

Fig. 1. Single pair of forces acting on a central crack in an infinite plate.

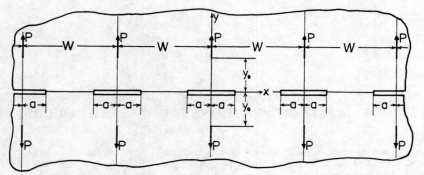

Fig. 2. The force system of Fig. 1 repeated with a constant periodic spacing, W.

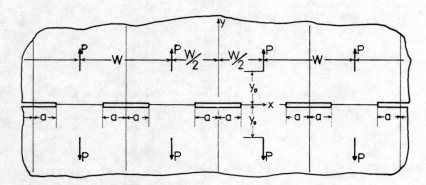

Fig. 3. The repeated cracks of Fig. 2 with the line of action of the forces centered between adjacent cracks.

To avoid the lengthy expressions of the results, it is convenient to use the formal operator, $1 - y_0/[2(1-\nu)] \times \partial/\partial y_0$, which appears in the expression for $Q(x)$ in the case 1 and to write the results in the following forms.

$$Q(x) = \frac{P}{\pi}\left\{1 - \frac{y_0}{2(1-\nu)}\frac{\partial}{\partial y_0}\right\} q(x, y_0) \tag{5}$$

$$Z(\zeta) = \frac{P}{\pi}\left\{1 - \frac{y_0}{2(1-\nu)}\frac{\partial}{\partial y_0}\right\} [F(\zeta, y_0, a) - F(\zeta, y_0, 0)] \tag{6}$$

$$\bar{Z}(\zeta) = \frac{P}{\pi}\left\{1 - \frac{y_0}{2(1-\nu)}\frac{\partial}{\partial y_0}\right\} [\bar{F}(\zeta, y_0, a) - \bar{F}(\zeta, y_0, 0)] \tag{7}$$

$$K = \frac{P}{\pi}\sqrt{\pi a}\left\{1 - \frac{y_0}{2(1-\nu)}\frac{\partial}{\partial y_0}\right\} k(a, y_0). \tag{8}$$

(1) *A central crack of length $2a$ in an infinite plate with opposing concentrated forces P acting at $(0 \pm y_0)$* (*Fig. 1*)

For this case, $Q(x) = Q_1(x)$ is known[3] as

$$Q_1(x) = \frac{P}{\pi}\left\{1 - \frac{y_0}{2(1-\nu)}\frac{\partial}{\partial y_0}\right\} \frac{y_0}{x^2 + y_0^2}.$$

The results are given by

$$q_1(x, y_0) = \frac{y_0}{x^2 + y_0^2} \tag{5-1}$$

$$F_1(\zeta, y_0, a) = \frac{\sqrt{a^2 + y_0^2}}{\zeta^2 + y_0^2} \frac{1}{\sqrt{1 - (a/\zeta)^2}} \tag{6-1}$$

$$\bar{F}_1(\zeta, y_0, a) = \arctan \sqrt{\frac{\zeta^2 - a^2}{a^2 + y_0^2}} \tag{7-1}$$

$$k_1(a, y_0) = \frac{1}{\sqrt{a^2 + y_0^2}}. \tag{8-1}$$

(2) *The case in which the situation 1 is repeated along the x-axis with the period* W *(Fig. 2)*
$Q(x) = Q_2(x)$ is calculated from $Q_1(x)$ as

$$Q_2(x) = \sum_{n=-\infty}^{\infty} Q_1(x - nW)$$

or

$$q_2(x, y_0) = \sum_{n=-\infty}^{\infty} q_1(x - nW, y_0).$$

Using the relations

$$q_1(x, y_0) = \frac{i}{2}\left(\frac{1}{x + iy_0} - \frac{1}{x - iy_0}\right)$$

and

$$\cot z = \frac{1}{z} + \sum_{n=1}^{\infty} \frac{2z}{z^2 - n^2\pi^2}$$

$q_2(x, y_0)$ can be obtained in closed form and the results are as follows.

$$q_2(x, y_0) = \left(\frac{\pi}{W} \cosh \frac{\pi}{W} y_0\right) q_1\left(\sin \frac{\pi x}{W}, \sinh \frac{\pi}{W} y_0\right) \tag{5-2}$$

$$F_2(\zeta, y_0, a) = \left(\frac{\pi}{W} \cosh \frac{\pi}{W} y_0\right) F_1\left(\sin \frac{\pi \zeta}{W}, \sinh \frac{\pi}{W} y_0, \sin \frac{\pi a}{W}\right) \tag{6-2}$$

$$\bar{F}_2(\zeta, y_0, a) = \arctan \left\{ \frac{\cosh \frac{\pi}{W} y_0}{\cos \frac{\pi}{W} \zeta} \sqrt{\frac{\left(\sin \frac{\pi}{W} \zeta\right)^2 - \left(\sin \frac{\pi a}{W}\right)^2}{\left(\sin \frac{\pi a}{W}\right)^2 + \left(\sinh \frac{\pi}{W} y_0\right)^2}} \right\} \tag{7-2}$$

$$k_2(a, y_0) = \sqrt{\frac{W}{\pi a} \tan \frac{\pi a}{W}} \left(\frac{\pi}{W} \cosh \frac{\pi}{W} y_0\right) k_1\left(\sin \frac{\pi a}{W}, \sinh \frac{\pi y_0}{W}\right). \tag{8-2}$$

(3) *The case where the phase of the line of action of forces is different by* W/2 *from the situation 2 (Fig. 3)*
$Q_3(x)$ is calculated directly from $Q_2(x)$ by

$$Q_3(x) = Q_2\left(x - \frac{W}{2}\right)$$

or

$$q_3(x, y_0) = q_2\left(x - \frac{W}{2}, y_0\right).$$

The results are

$$q_3(x, y_0) = \left(\frac{\pi}{W} \cosh \frac{\pi}{W} y_0\right) q_1\left(\cos \frac{\pi x}{W}, \sinh \frac{\pi}{W} y_0\right) \tag{5-3}$$

$$F_3(\zeta, y_0, a) = \left(\frac{\pi}{W} \sinh \frac{\pi}{W} y_0\right) \frac{\sqrt{\left(\cos \frac{\pi a}{W}\right)^2 + \left(\sinh \frac{\pi}{W} y_0\right)^2}}{\left(\cos \frac{\pi a}{W}\right)^2 + \left(\sinh \frac{\pi}{W} y_0\right)^2} \cdot \frac{1}{\sqrt{1 - \left(\frac{\sin \frac{\pi a}{W}}{\sin \frac{\pi \zeta}{W}}\right)^2}} \tag{6-3}$$

$$\bar{F}_3(\zeta, y_0, a) = \arctan\left\{ \frac{\sinh\frac{\pi}{W}y_0}{\cos\frac{\pi}{W}\zeta} \sqrt{\frac{\left(\sin\frac{\pi\zeta}{W}\right)^2 - \left(\sin\frac{\pi a}{W}\right)^2}{\left(\cos\frac{\pi a}{W}\right)^2 + \left(\sinh\frac{\pi}{W}y_0\right)^2}} \right\} \tag{7-3}$$

$$k_3(a, y_0) = \sqrt{\frac{W}{\pi a}\tan\frac{\pi a}{W}} \left(\frac{\pi}{W}\sinh\frac{\pi}{W}y_0\right) k_1\left(\cos\frac{\pi a}{W}, \sinh\frac{\pi}{W}y_0\right). \tag{8-3}$$

Department of Mechanics,   HIROSHI TADA
Lehigh University,
Bethlehem,
Pa. 18015, U.S.A.

REFERENCES

[1] H. M. Westergaard, Bearing pressures and cracks. *Trans. ASME J. appl. Mech.* A49 (1939).
[2] G. R. Irwin, Analysis of stresses and strains near the end of a crack traversing a plate. *Trans. ASME J. appl. Mech.* **24**, 361 (1957).
[3] A. E. H. Love, *A Treatise on the Mathematical Theory of Elasticity*, p. 209, Dover, New York (1944).

ANALYSIS OF AN INFINITE PLATE CONTAINING RADIAL CRACKS ORIGINATING AT THE BOUNDARY OF AN INTERNAL CIRCULAR HOLE

By O. L. Bowie

1. Introduction. Considerable advance has been made in recent years in the application of energy type of theories in determining the influence of cracks in the specimen geometry on the strength of the specimen. In order to apply theories such as that developed by A. A. Griffith [1]*, it is necessary to calculate the elastic strain energy of the system. Although only the boundary stresses and displacements are actually necessary for this calculation, one must nevertheless formally solve the problem as a whole to obtain this information.

We shall consider the solution of the class of plane problems in elasticity corresponding to a distribution of radial cracks, equal and finite in length, originating at the boundary surface of a circular hole in an infinite plate under the two load systems shown in Figure 1. The geometry of the internal boundary, τ, can be conveniently described by considering the plate as the complex Z-plane where $Z = x + iy = re^{i\theta}$. Then, if the center of the hole is chosen as $Z = 0$, we specify that radial cracks of equal length, L, lie along $\theta = 0, 2\pi/k, \cdots, (k-1)2\pi/k$ where $k \geq 1$ is an integer which specifies the number of cracks.

In addition to the plane stress applications of the case of uniform (all-around) tension at infinity, the analysis by a slight modification enables us to study the plane strain problem corresponding to internal pressure acting in hollow cylinders of large wall thickness with longitudinal cracks originating along the inside surface. Indeed, this latter problem is solved by a modification of the manner in which the elastic constants enter in the displacement relations and by the superposition of the simple solution corresponding to uniform hydrostatic pressure.

2. Stress Problem Formulated in Terms of Complex Variables. Due to the irregular geometry of the internal boundary, it would appear that the problem described above can be most conveniently handled by the complex variable method of Mushelisvili [2]. This method depends upon the representation of the well-known Airy's stress function, $U(x, y)$, in terms of two analytic functions of the complex variable, Z, namely, $\varphi(Z)$ and $\psi(Z)$, where

$$U(x,y) = \text{Re}\left[\bar{Z}\varphi(Z) + \int^{z} \psi(Z)\, dZ \right]. \tag{1}$$

With this representation, the stress components in rectangular coordinates can be written as

$$\sigma_y + \sigma_x = 2[\varphi'(Z) + \overline{\varphi'(Z)}] = 4\,\text{Re}[\varphi'(Z)] \tag{2}$$

* Numbers in brackets refer to the references at the end of the paper.

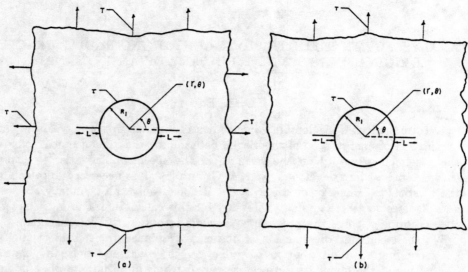

Fig. 1. Geometry and loading for the case of two cracks, $(k = 2)$

and

$$\sigma_y - \sigma_x + 2i\tau_{xy} = 2[\bar{Z}\varphi''(Z) + \psi'(Z)], \qquad (3)$$

where the prime notation denotes differentiation with respect to Z and the bars denote complex conjugates.

It is convenient for the purpose of enforcing boundary conditions to introduce an auxiliary complex plane, the ζ-plane, such that the unit circle, $\zeta = \sigma = e^{i\varphi}$ (where φ is used here to denote angular measure in the ζ-plane), and its exterior are mapped into τ and its exterior, respectively, by the analytic function

$$Z = \omega(\zeta). \qquad (4)$$

The stress functions $\varphi(Z)$ and $\psi(Z)$ will be considered as functions of the parameter ζ. The necessity for introducing considerable new notation can be avoided by designating $\varphi(Z) = \varphi[\omega(\zeta)]$ as $\varphi(\zeta)$, etc., which leads to such relationships as $\varphi'(Z) = \varphi'(\zeta)/\omega'(\zeta)$, etc. The analyticity of $\varphi(\zeta)$ and $\psi(\zeta)$ which follows from their definitions contributes significantly to the advantage gained by using the complex variable method.

The mathematical analysis of the problem requires the determination of the functions $\varphi(\zeta)$ and $\psi(\zeta)$ which are analytic for $|\zeta| > 1$ (with the exception of the point at infinity) and lead to the proper loading conditions at infinity and on the internal boundary τ. The forms of $\varphi(\zeta)$ and $\psi(\zeta)$ can be chosen a priori to yield the proper loading conditions at infinity. The condition that τ be load-free can be written as

$$\varphi(\sigma) + \omega(\sigma)\overline{\varphi'(\sigma)/\omega'(\sigma)} + \overline{\psi(\sigma)} = 0 \qquad (5)$$

3. Description of the Mapping Function. The mapping function can be determined in a fairly easy manner by considering separately the transformation between two upper half planes which carries the real axis of one into the real axis interrupted by vertical slits (cracks) of finite length of the other and the transformation carrying each of the above regions into a circle and its exterior. The mapping function (4) is the product transformation and can be expressed in differential form as

$$dZ/Z = (1 - \zeta^{-k})d\zeta/\zeta(1 + 2\epsilon\zeta^{-k} + \zeta^{-2k})^{\frac{1}{2}}. \qquad (6)$$

In (6), ϵ is a real parameter such that $0 \leq |\epsilon| \leq 1$ and the denominator is considered positive at $\zeta = 1$ in order to define the desired branch of the multi-valued mapping function. By varying ϵ, the crack depth can be adjusted to assigned values. It is evident from symmetry considerations that the unit circle is mapped by (6) into a circular boundary interrupted by k symmetrically distributed radial cracks of equal depth.

It is possible to determine the closed form of the mapping function by quadrature. For $k = 1$, it can be shown that

$$Z = \omega(\zeta) = C[\zeta + \zeta^{-1} + (1 + \epsilon) + (1 + \zeta^{-1})(\zeta^2 + 2\epsilon\zeta + 1)^{\frac{1}{2}}, \qquad (7)$$

where C is a constant of proportionality. The closed form of the mapping function increases in complexity with larger integers, k.

For purposes of the subsequent stress analysis, it is desirable to find a series representation of (6) converging on and exterior to the unit circle. The form of such a series is evidently

$$Z = \omega(\zeta) = C\left[\zeta + \sum_{n=1}^{\infty} A_n \zeta^{1-kn}\right], \qquad (8)$$

where the A_n's are real. The A_n's may be obtained numerically from simple recursive formulae determined by expanding both sides of (6) in series form and equating coefficients of equal powers of ζ. The convergence of (8) on and exterior to the unit circle can be studied by examining the coefficients A_n. It can be shown[1] that $\lim_{n\to\infty} A_n = 0$; thus, using a well-known theorem found in [3], the series (8) converges exterior to the unit circle and at all points on the unit circle except at the roots of

$$\zeta^{2k} + 2\epsilon\zeta^k + 1 = 0. \qquad (9)$$

4. Polynomial Approximation of the Mapping Function. A mathematical difficulty common to most problems in the stress analysis of regions involving irregularities in boundary contours is the determination a priori of the forms of singularities of the stress function associated with the irregularity. Often, the issue is avoided by assuming that the singularities are of such a nature that the stress functions can still be expanded in an infinite series, converging in the region of the problem at all but the singular points of the boundary. Although results which are undoubtedly valid are usually obtained when this argument is used, it is virtually impossible to verify rigorously the convergence assumed.

---

[1] The proof is particularly simple for $k = 1$. From (7), it can be seen that the A_n's behave essentially as the Legendre polynomials. The proof is more difficult for $k \geq 2$.

In this paper, the method of dealing with the singularities of the stress functions will depend on the representation of the mapping function by polynomial approximations. By using polynomial approximations of the internal boundary, an accurate description of the stress distribution at the crack roots can be obtained by introducing cusps in the polynomial mappings to describe the crack roots. On the other hand, the process circumvents a rigorous consideration of the singularities of the stress functions which occur at the roots of (9), i.e., physically at the points of juncture of the cracks and the circle. Convergence of polynomial approximations of the exact problem can be considered a matter of choosing a sufficiently accurate polynomial approximation of the mapping function such that a closer approximation will not effect an improvement in the specified accuracy of the desired information.

The existence of cusps at locations corresponding to the crack roots is ensured in polynomial mapping approximations by demanding that

$$dZ/d\zeta = \omega'(\zeta) = (1 - \zeta^{-k})g(\zeta), \tag{10}$$

where $g(\zeta)$ is a polynomial with coefficients chosen so that the roots of $g(\zeta) = 0$ fall inside the unit circle. Due to the convergence of (8) at all but a finite number of points on the unit circle, suitable polynomial approximations,

$$Z = \omega(\zeta) = C\left[\zeta + \sum_{n=1}^{N} \epsilon_n \zeta^{1-kn}\right], \tag{11}$$

can be obtained by setting $\epsilon_n \cong A_n$; modifications of the ϵ_n's being made to satisfy (10).

5. "All-Around" Tension at Infinity. Consider now the case of uniform tension at infinity illustrated in Figure 1a. It can easily be shown[2] that the loading condition $\sigma_x = \sigma_y = T$ on $|Z| = R$ where R is very large is satisfied by choosing $\varphi(\zeta)$ and $\psi(\zeta)$ such that they approach $CT\zeta/2$ and $CT\gamma_0\zeta^{-1}$, respectively, for large $|\zeta|$. Therefore, let us assume that $\varphi(\zeta)$ is a polynomial of the form

$$\varphi(\zeta) = CT\left[\zeta/2 + \sum_{n=1}^{N} \alpha_n \zeta^{1-kn}\right]. \tag{12}$$

Such indeed[3] would be the form of $\varphi(\zeta)$ were there no boundary irregularities implied by the mapping function (11).

Next, we write the boundary condition (5) as

$$\omega'(\sigma)\psi(\sigma) = -\omega'(\sigma)\overline{\varphi(\sigma)} - \overline{\omega(\sigma)}\varphi'(\sigma). \tag{13}$$

The function $\omega'(\zeta)\psi(\zeta)$ is analytic exterior to the unit circle and, with $\varphi(\zeta)$ assumed as (12), is given as a continuous function on the unit circle by (13). Thus, if the coefficients α_n can be chosen so that the coefficients of all positive powers of ζ in the Laurent expansion of

$$\omega'(\zeta)\varphi(1/\zeta) + \omega(1/\zeta)\varphi'(\zeta) \tag{14}$$

---

[2] A detailed discussion of the stress functions for infinite regions can be found in [4].
[3] E.g., the case for $k = 2, N = 2$ has been discussed by Morkovin [5] in some detail.

vanish, we can determine $\psi(\zeta)$ explicitly. By multiplying both sides of (13) by $1/2\pi i(\sigma - \zeta)$ and integrating around the unit circle, we obtain by a well-known theorem[4]

$$\omega'(\zeta)\psi(\zeta) = -\omega'(\zeta)\varphi(1/\zeta) - \omega(1/\zeta)\varphi'(\zeta) \tag{15}$$

It is interesting to note that the cusp roots are reflected by singularities in $\psi(\zeta)$ in the form of simple poles.

To complete the analysis it is necessary to verify that the coefficients α_n can be determined to meet the condition set forth above. If the coefficients of all positive powers of ζ in the Laurent expansion of (14) are equated to zero, the following system of linear simultaneous equations results:

$$\alpha_p + \sum_{n=1}^{N-p} \alpha_{p+n}\epsilon_n(1 - nk) + \sum_{n=1}^{N-p} \epsilon_{p+n}\alpha_n(1 - nk) + \epsilon_p/2 = 0, \tag{16}$$
$$p = 1, 2, \cdots, N.$$

The conditions can therefore be met by the solution of (16) for the N coefficients, α_n, $n = 1, 2, \cdots, N$.

It is evident that the state of stress is unaffected by the addition of a linear expression of the form $DiZ + a + ib$ to $\varphi(Z)$ or a complex constant $\alpha + i\beta$ to $\psi(Z)$. Thus, for $k = 1$, the quantity α_1 determined by (16) plays no role in the actual stresses. Similarly, identical satisfaction of the boundary condition (5) can be realized by proper adjustments of the coefficients of the above arbitrary additions.

6. Tension in One Direction at Infinity. For the case of tension in one direction at infinity (illustrated in Figure 1b), the analysis can be carried out in a manner similar to that of the preceding section. It can be shown that the loading condition at infinity, $\sigma y = T$, is satisfied by requiring the stress functions $\varphi(\zeta)$ and $\psi(\zeta)$ to approach $CT\zeta/4$ and $CT\zeta/2$, respectively, for large $|\zeta|$. The stress function $\varphi(\zeta)$ is again assumed to have the form of a polynomial

$$\varphi(\zeta) = CT\left[\zeta/4 + \sum_{n=1}^{kN} \alpha_n \zeta^{1-n}\right]. \tag{17}$$

By an argument similar to that of the preceding section, it can again be shown that

$$\omega'(\zeta)\psi(\zeta) = -\omega'(\zeta)\varphi(1/\zeta) - \omega(1/\zeta)\varphi'(\zeta) \tag{18}$$

provided that the coefficients α_n are chosen so that

$$-\omega'(\zeta)\varphi(1/\zeta) - \omega(1/\zeta)\varphi'(\zeta) \approx C^2T\zeta/2, \quad \text{for large} \quad |\zeta|. \tag{19}$$

Due to the relative lack of symmetry in this case, it is difficult to present the linear systems of simultaneous equations for the determination of the α_n's in compact form for arbitrary integers k. Therefore, only the systems for the single

[4] Reference [4], p. 145.

crack ($k = 1$) and the two crack ($k = 2$) cases will be explicitly formulated.

For $k = 1$,

$$\alpha_p + \sum_{n=1}^{N-p} \alpha_{n+p}\epsilon_n(1-n) + \sum_{n=1}^{N-p} \epsilon_{n+p}\alpha_n(1-n) + \epsilon_p/4 = \begin{cases} 0, & p \neq 2 \\ -\tfrac{1}{2}, & p = 2 \end{cases} \quad (20)$$
$$p = 1, 2, \cdots, N$$

For $k = 2$,

$$\alpha_{2p} + \sum_{n=1}^{N-p} \alpha_{2(n+p)}\epsilon_n(1-2n)$$
$$+ \sum_{n=1}^{N-p} \epsilon_{n+p}\alpha_{2n}(1-2n) + \epsilon_p/4 = \begin{cases} 0, p > 1 \\ -1/2, p = 1 \end{cases} \quad (21)$$
$$\alpha_{2p-1} = 0, \qquad p = 1, 2, \cdots, N,$$

7. Stability Condition of the Griffith Theory for Brittle Failure. An application of the preceding results is the determination of the critical applied load for which the radial cracks begin to spread. A purely elastic criterion is found in the Griffith theory of the rupture of brittle materials [1]. The Griffith hypothesis, when applied to the present problem, leads to the following stability condition:

$$\frac{dV}{dL} dL = -2khG dL, \qquad (22)$$

where

$V = V_c - V_0 =$ reduction of the potential energy due to the presence of the radial cracks,
$V_c =$ potential energy of the system with radial cracks,
$V_0 =$ potential energy of the system without cracks,
$k =$ number of radial cracks,
$h =$ thickness of the specimen (axially),
$G =$ surface tension per unit area of the crack surface,
$L =$ length of the radial crack(s),
$d =$ notation for differential.

The critical applied load can be determined from (22) once the reduction of potential energy, V, has been found as a function of L.

8. Calculation of the Reduction in Potential Energy, V, for the Case of "all-around" Tension at Infinity. In order to compare the difference in potential energies in the physical systems with and without cracks, it is convenient and more motivating to refer to the exact geometry rather than polynomial approximations. Accordingly, the calculation of V first will be carried out in terms of the stress functions as referred to the original Z-coordinate system and the exact boundary shape.

For the case of uniform tension at infinity, the potential energy of the system with cracks, V_c, under the assumption of plane stress, reduces to the calculation of

$$V_c = \lim_{R \to \infty} \left\{ -\tfrac{1}{2} h \int_0^{2\pi} (\sigma_r U_r + \tau_{r\theta} U_\theta) R \, d\theta \right\} \tag{23}$$

where σ_r, $\tau_{r\theta}$, U_r and U_θ are stress and displacement components in polar coordinates. The integrand in (23) can be expressed in terms of the original stress functions $\varphi(Z)$ and $\psi(Z)$ defined by (2) and (3) from the relations

$$\sigma_r - i\tau_{r\theta} = \varphi'(+Z) \overline{\varphi'(Z)} - e^{2i\theta}[\overline{Z}\varphi''(Z) + \psi'(Z)] \tag{24}$$

and

$$2\mu(U_r + iU_\theta) = e^{-i\theta}[\eta\varphi(Z) - Z\overline{\varphi'(Z)} - \overline{\psi(Z)}], \tag{25}$$

where, in terms of Young's Modulus, E, and Poisson's ratio, ν,

$$\mu = E/2(1 + \nu) \tag{26}$$

and

$$\eta = \begin{cases} (3 - \nu)/(1 + \nu), & \text{for plane stress} \\ 3 - 4\nu, & \text{for plane strain.} \end{cases} \tag{27}$$

The stress functions $\varphi(Z)$ and $\psi(Z)$ are certainly analytic exterior to τ except for the point at infinity. Therefore, for the case of uniform tension at infinity the stress functions may be expanded in series form,

$$\varphi(Z) = T[Z/2 + a_1 Z^{1-k} + \cdots] \tag{28}$$

and

$$\psi(Z) = T[b_0 Z^{-1} + b_1 Z^{-1-k} + \cdots]. \tag{29}$$

The expansions (28) and (29), valid for large R, can be substituted into (23) to determine V_c. It follows, after some algebra, that

$$V_c = \lim_{R \to \infty} \left\{ -\frac{\pi T^2 h}{E}[(1-\nu)R^2 - 2\nu b_0] \right\}. \tag{30}$$

In order to obtain the potential energy, V_0, for the uncracked plate, the stress analysis for a concentric ring with inner radius R_1 and outer radius R loaded on the outer boundary by the same applied load (as the system with cracks) was carried out. This analysis was straightforward and the details will not be included here. The results lead to the value of the potential energy of the uncracked plate.

$$V_0 = \lim_{R \to \infty} \left\{ -\frac{\pi T^2 h (R^2 + b_0)^2 [(1-\nu)R^2 + 2\nu R_1^2]}{E(R^2 - R_1^2)^2} \right\} \tag{31}$$

In the limit, therefore, the reduction in potential energy for the case of "all-around" tension at infinity for the plane stress assumption is given by

$$V = 2\pi T^2 h(b_0 + R_1^2)/E. \tag{32}$$

We now must interpret this result in terms of the previous analysis based on polynomial approximations of the problem. The stress function $\psi(\zeta)$ given by (15) for polynomial approximations of the geometry can be expanded in a series of the form

$$\psi(\zeta) = CT[\gamma_0' + \gamma_0 \zeta^{-1} + \gamma_1 \zeta^{-1-k} + \cdots]. \tag{33}$$

On the other hand, a series expansion of $\psi(\zeta)$ for the exact geometry which is certainly valid for $|\zeta| > 1$ can be found by substituting the exact mapping (6) into (29). If we assume that the polynomial approximation can be made to converge to the exact solution, we find by the comparison of the coefficients of ζ^{-1} in the expansions of $\psi(\zeta)$ that

$$C^2 \gamma_0 \to b_0 \tag{34}$$

The hitherto unspecified constant C occurring in the mapping function will now be chosen so that the radius of the circular hole in the physical plane is the unit of length. Thus, if $\sigma = \sigma_1$ is that point on the unit circle in the ζ-plane which corresponds to the junction of the crack and the circle in the Z-plane, then C is chosen so that

$$R_1 = |\omega(\sigma_1)| = 1. \tag{35}$$

Thus, if a sufficiently accurate polynomial mapping is taken, we can write (32) within an arbitrarily specified accuracy as

$$V = 2\pi T^2 h\{\gamma_0/[\omega(\sigma_1)/C]^2 + 1\}/E. \tag{36}$$

The reduction in potential energy can be determined as a function of the crack length, L, by noting that L, measured in units of the radius of the circular hole, can be written as

$$L = \omega(1)/\omega(\sigma_1) - 1. \tag{37}$$

Thus, for the case of plane stress and "all-around" tension at infinity,

$$V = 2\pi T^2 h f(L)/E, \tag{38}$$

where,

$$f(L) = \gamma_0[(L+1)/\omega(1)/C]^2 + 1. \tag{39}$$

In (39), it should be noted that both γ_0 and $\omega(1)/C$ are functions of L.

For plane strain, the second form of η in (27) must be used for the calculation of the displacements. Otherwise, the calculation is identical with that for the case of plane stress. The reduction in potential energy for the case of plane strain is given by

$$V = 2\pi T^2 h(1 - \nu^2) f(L)/E \tag{40}$$

where $f(L)$ is given again by (39).

9. The Reduction in Potential Energy for the Case of Simple Tension at Infinity. The calculation of V for the case of simple tension at infinity, $\sigma y = T$, can be carried out in a manner similar to the preceding section. For the case of plane stress, we find

$$V = 2\pi T^2 hg(L)/E, \tag{41}$$

where

$$g(L) = (\gamma_0 + 2\alpha_2 - A)[(L + 1)/\omega(1)/C]^2/2 + \tfrac{3}{4}. \tag{42}$$

In (42), the quantities γ_0, α_2, and A are defined by

$$CT\gamma_0 = \text{coefficient of } \zeta^{-1} \text{ in the Laurent expansion of } \psi(\zeta),$$

$$\alpha_2 = \text{coefficient of } \zeta^{-1} \text{ in the Laurent expansion of } \varphi(\zeta), \tag{43}$$

$$CA = \text{coefficient of } \zeta^{-1} \text{ in the Laurent expansion of } \omega(\zeta).$$

10. The critical load according to the Griffith Hypothesis. The critical value of the applied load, T_c, can now be calculated for the cases considered by substituting (38), (40), or (41) into the stability condition, (22). The results can be summarized as follows:

For plane stress,

$$T_c = [-kEG/\pi f'(L)]^{\frac{1}{2}} \quad \text{for uniform tension at infinity,} \tag{44}$$

and

$$T_c = [-kEG/\pi g'(L)]^{\frac{1}{2}} \quad \text{for simple tension at infinity.} \tag{45}$$

For plane strain,

$$T_c = [-kEG/\pi(1 - \nu^2)f'(L)]^{\frac{1}{2}} \quad \text{for uniform tension at infinity.} \tag{46}$$

According to the Griffith hypothesis, the crack(s) become unstable if $T \geq T_c$.

The plane strain criterion (46) is intended for the study of radial cracks in cylinders under internal pressure, P. Such a load system is obtained simply by superimposing a hydrostatic pressure on the load system corresponding to the uniform tension case for $T = P$. Since it can be shown that the superposition of a hydrostatic stress state does not affect the critical stress as calculated above, the critical internal pressure, P_c, is also given by (46).

11. Numerical Results for the Case of "All-Around" Tension at Infinity." The analysis carried out in the previous sections was numerically evaluated for the cases of a single crack ($k = 1$) and two cracks ($k = 2$). In particular, the stress analysis was carried out for nine values of the crack length-determining parameter, ϵ. From these results, the function $f(L)$ was found in tabular form. These data, in turn, were numerically differentiated with respect to L to find $f'(L)$. The critical stresses according to (44) or (46) were then determined.

For example, when $k = 2$ the coefficient γ_0 in (39) is given by

$$\gamma_0 = -\left[1 + 2\sum_{n=1}^{N} \epsilon_n \alpha_n (1 - 2n)\right]. \tag{47}$$

It was found that γ_0 can be obtained to three significant figure accuracy for the crack lengths chosen if an average of thirty terms of the polynomial approximation of the mapping function is retained. Actually, only a few terms of the polynomial are required for moderately large crack lengths, whereas the convergence is poor for small lengths. Illustrative results for the case of $k = 2$ are listed in Table 1.

TABLE 1

ϵ	L	$\omega(1)/c$	γ_0	$f(L)$
−1.000	0.000	1.000	−1.000	0.000
−0.866	0.303	1.259	−1.075	−0.152
−0.707	0.497	1.383	−1.160	−0.359
−0.500	0.732	1.500	−1.267	−0.689
0.000	1.414	1.707	−1.509	−2.018
+0.500	2.732	1.866	−1.752	−6.008
+0.707	4.027	1.924	−1.852	−11.640
+0.866	6.596	1.966	−1.933	−27.860
+1.000	∞	2.000	−2.000	−∞

For large values of L, the solution approaches the form corresponding to the case of a single crack of length $2(1 + L)$ in an infinite plate. Thus, for large L

$$f(L) \approx -(L + 1)^2/2 + 1. \tag{48}$$

Similar computations were made for the case of a single crack ($k = 1$). The form of $f(L)$ for large L in this case was found to be

$$f(L) \approx -(L + 2)^2/8 + 1. \tag{49}$$

The variation of the critical load, T_c, was then found by differentiating $f(L)$ numerically and substituting into (44) or (46). For the case of plane strain, the variation of the critical load with crack length is shown in Figure 2.

12. Numerical Results for the Case of Simple Tension at Infinity. In a similar manner the cases for $k = 1$ and $k = 2$ were evaluated for the loading $\sigma_y = T$ at infinity. For example, to calculate $g(L)$ in (42) the quantity γ_0 for $k = 2$ is determined from

$$\gamma_0 = \epsilon_1/2 - \left[1/2 + 2\sum_{n=1}^{N} \epsilon_n \alpha_{2n}(1 - 2n)\right] \tag{50}$$

Illustrative data for $k = 2$ is shown in Table 2.

For large crack lengths $g(L)$ approaches the respective forms of $f(L)$ in (48) and (49), which is consistent with the well-known result that a single crack in an infinite sheet is unaffected by a tension in the direction of the crack. The critical stresses, determined by substituting the function $g'(L)$ into (45), are shown as a function of L in Figure 3.

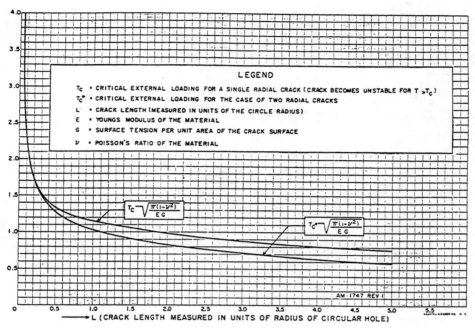

Fig. 2. Variation of the critical load with crack length for the case of "all-around" tension at infinity.

TABLE 2

ϵ	ϵ_1	α_2	γ_0	$g(L)$
−1.000	0.00	−0.50	−0.50	0.00
−0.866	0.07	−0.61	−0.62	−0.28
−0.707	0.15	−0.69	−0.74	−0.57
−0.500	0.25	−0.75	−0.87	−1.00
0.000	0.50	−0.83	−1.13	−2.53
+0.500	0.75	−0.82	−1.33	−6.69
+0.707	0.85	−0.80	−1.41	−12.40
+0.866	0.93	−0.78	−1.46	−28.80
+1.000	1.00	−0.75	−1.50	−∞

13. Observations. For large crack lengths, say $L > 1$, (in units of the radius of the inner hole) the effect of the stress field caused by the circular hole is negligible insofar as the critical stress is concerned. In fact, the solution for a single crack in an infinite plate can be used provided that the single crack is compared with a crack of length $2(1 + L)$ for $k = 2$ and a length of $L + 2$ for $k = 1$.

On the other hand, for very small crack lengths the critical load appears to be governed primarily by the local stress field of the hole. This observation is made on the basis of the following argument: Since the tangential stress in the immediate vicinity of the hole where the crack is to be placed is $3T$ for the

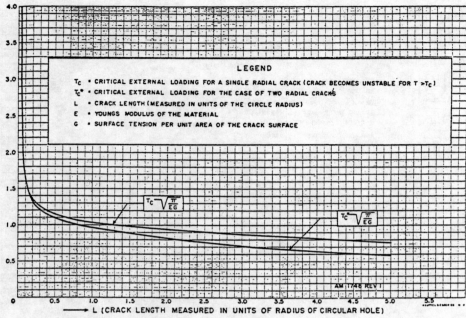

Fig. 3. Variation of the critical load with crack length for the case of simple tension at infinity.

case of simple tension and $2T$ for the case of all-around tension, one would expect that the ratio of the critical load for simple tension to the critical load for all-around tension would be two-thirds if governed by the local stress field. A comparison of the data of Figure 2 and Figure 3 indicates that such indeed is the case as the crack length approaches zero. (Note that the factor $(1 - \nu^2)$ must be neglected to make a consistent plane stress comparison.)

For crack lengths intermediate to those above, the behavior of the critical load does not appear capable of such simple physical arguments.

Finally, the critical internal pressure for the plane strain problem of a cylinder of very large wall thickness can be found simply by replacing T_c by P_c in Figure 2.

BIBLIOGRAPHY

[1] GRIFFITH, A. A., Phil. Trans. Royal Soc., London, A Vol. 221, p. 163 (1921)
[2] MUSHELISVILI, Mathematische Annalen, Vol. 107, pp. 282–312 (1932–33)
[3] TITCHMARSH, E. C., Theory of Functions, Oxford, p. 218 (1939)
[4] SOKOLNIKOFF, I. S., "Mathematical Theory of Elasticity", (Brown University Lecture Notes), pp. 243–319, (1941)
[5] MORKOVIN, V., Quart. Appl. Math., Vol. II, No. 4, pp. 350–352 (1944)
[6] OROWAN, E., Fatigue and Fracture of Metals, Wiley & Sons, Inc., New York, pp. 139–167, (1950)

WATERTOWN ARSENAL, WATERTOWN, MASS.

(Received August 17, 1954)

O. L. BOWIE
U. S. Army Materials Research Agency,
Watertown, Mass.

Rectangular Tensile Sheet With Symmetric Edge Cracks

Complex variable methods are applied to the plane elastic problem of a rectangular tensile sheet with symmetric edge cracks. A mapping function is used which is sufficiently flexible to account for both variations of crack length and length/width ratios of the sheet on the stresses local to the crack tips. The stress-intensity factors are determined numerically for varying crack lengths in sheet with 1×1 and 3×1 length/width ratios and compared with Irwin's earlier approximation derived from Westergaard's collinear crack solution. The results indicate that the approximation is valid for deep cracks but in error by 13 percent for small crack depths.

A MODEL frequently used in fracture mechanics is the rectangular tensile sheet with edge cracks, Fig. 1. The plane elastic problem has not been solved previously although an approximation introduced by Irwin, e.g. [1],[1] is used to estimate the elastic stresses local to the crack tip. This approximation depends on the utilization of Westergaard's [2] solution for a series of equally spaced collinear cracks in an infinite sheet. Load-free conditions on the lateral edges of a strip defined by the perpendicular bisectors of two consecutive cracks are partially satisfied by Westergaard's solution and it is argued physically that the effect of the error is negligible in the vicinity of the crack tip. Tait [3] has considered the infinite strip of finite width with edge cracks; however, the mixed boundary conditions he considers on the lateral edges lead back to the Westergaard solution. Owing to the controversy as to the accuracy of Irwin's approximation for this and related problems, a procedure for an accurate stress analysis of this problem class would seem appropriate.

In the present paper, the analysis is carried out using the complex variable approach with a complex mapping function to describe the geometry. The success of this approach for fairly complicated geometries involving cracks was previously demonstrated by the author [4] by a solution for a plate with radial cracks originating at the boundary of an internal circular hole. The recent trend in solving the problem class of two-dimensional cracks is the formulation of the problem in terms of an integral equation. The latter formulations, although mathematically elegant, usually require a numerical solution of the integral equation. The mapping approach is therefore reintroduced as an effective means for solving this problem class. The flexibility of the approach is illustrated in the present problem by the relative simplicity in which the additional parameter, length/width ratio for finite rectangular sheet, can be included in the analysis.

Initial Formulation

The rectangular sheet under tension weakened by symmetric edge cracks, Fig. 1, will be considered as lying in the complex Z-plane, $Z = x + iy$, with the center of the sheet described by $Z = 0$.

The complex variable methods of Muskhelishvili [5] depend on the representation of the well-known Airy stress function, $U(x, y)$, in terms of two analytic functions of the complex variable Z, namely $\varphi(Z)$ and $\psi(Z)$, where

---
[1] Numbers in brackets designate References at end of paper.
Presented at the Summer Conference of the Applied Mechanics Division, Boulder, Colo., June 9–11, 1964, of THE AMERICAN SOCIETY OF MECHANICAL ENGINEERS.
Discussion of this paper should be addressed to the Editorial Department, ASME, United Engineering Center, 345 East 47th Street, New York, N. Y. 10017, and will be accepted until July 10, 1964. Discussion received after the closing date will be returned. Manuscript received by ASME Applied Mechanics Division, May 27, 1963. Paper No. 64—APM-3.

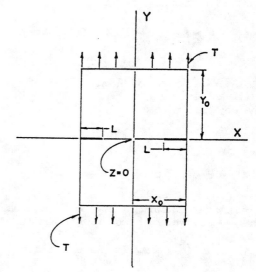

Fig. 1 Z-plane

$$U(x, y) = \text{Re}\left[\bar{Z}\varphi(Z) + \int^Z \psi(Z)dZ\right] \quad (1)$$

To facilitate the consideration of boundary conditions, an auxiliary complex plane, the ζ-plane, is introduced and a functional relationship

$$Z = \omega(\zeta) \quad (2)$$

is found such that the unit circle, $\zeta = \sigma = e^{i\theta}$, and its interior in the ζ-plane map into the boundary and interior, respectively, of the region in Fig. 1. The mapping function, $\omega(\zeta)$, is analytic interior to the unit circle but contains singularities necessary to the description of corner points on the unit circle itself.

The stress functions $\varphi(Z)$ and $\psi(Z)$ can be considered as functions of ζ. New notation can be minimized by designating $\varphi(Z) \equiv \varphi[\omega(\zeta)]$ as $\varphi(\zeta)$, and so on, which leads to such definitions as $\varphi'(Z) = \varphi'(\zeta)/\omega'(\zeta)$, and so forth. (Primes are used to denote differentiation.) Thus, the stresses and displacements in rectangular coordinates can be written as

$$\sigma_y + \sigma_x = 4 \text{ Re } [\varphi'(\zeta)/\omega'(\zeta)] \quad (3)$$

$$\sigma_y - \sigma_x + 2i\tau_{xy} = 2\{\overline{\omega(\zeta)}[\varphi'(\zeta)/\omega'(\zeta)]' + \psi'(\zeta)\}/\omega'(\zeta) \quad (4)$$

$$2\mu(u + iv) = \eta\varphi(\zeta) - \omega(\zeta)\overline{\varphi'(\zeta)/\omega'(\zeta)} - \overline{\psi(\zeta)} \quad (5)$$

where μ and η are constants depending on the material, and bars denote complex conjugates.

The loading conditions in Fig. 1 can be expressed conveniently in terms of the force resultant. Along an arc of the material with element ds, we denote the horizontal and vertical forces by Xds and Yds, respectively. If the arc is taken as the boundary of Fig. 1, then s can be considered as a function of σ. Then

$$\varphi(\sigma) + \omega(\sigma)\overline{\varphi'(\sigma)/\omega'(\sigma)} + \overline{\psi(\sigma)} = i\int^s (X + iY)ds = g(\sigma) \quad (6)$$

The solution requires the determination of the functions $\varphi(\zeta)$ and $\psi(\zeta)$ which are analytic for $|\zeta| < 1$ and satisfy the loading condition (6). The problem will be considered both in terms of power-series expansions of the exact solution and an alternative polynomial approximation of the exact geometry.

The Mapping Function

By application of the Schwartz-Christoffel transformation, it was found that the required mapping function corresponds to an appropriate branch of

$$Z = \omega(\zeta) = -i\int_0^\zeta [(\zeta^2 + 1)/(1 - 2\zeta^2\cos 2\alpha + \zeta^4)^{1/2}(1 - 2\zeta^2\cos 2\beta + \zeta^4)^{1/2}]d\zeta \quad (7)$$

In (7), $0 < \alpha < \beta < \pi/2$, and these two parameters can be varied to obtain desired ratios y_0/x_0 and L/x_0. The choice of branch was made by defining

$$\omega(1) = iy_0 \quad (8)$$

Thus

$$\omega(e^{i\alpha}) = -x_0 + iy_0, \quad \omega(e^{i\beta}) = -x_0, \quad \omega(i) = -x_0 + L,$$

and so on.

By considering the mapping function on the unit circle, it can be shown by well-known substitutions used in elliptic integrals that

$$x_0 = \frac{1}{2\sin\beta}\int_0^{\pi/2}\frac{d\varphi}{(1 - \lambda^2\sin^2\varphi)^{1/2}} = \frac{1}{2\sin\beta}K(\lambda) \quad (9)$$

$$y_0 = \frac{1}{2\sin\beta}\int_1^{1/\lambda}\frac{ds}{(s^2 - 1)^{1/2}(1 - \lambda^2s^2)^{1/2}}$$

$$= \frac{1}{2\sin\beta}K[(1 - \lambda^2)^{1/2}] \quad (10)$$

$$L = x_0 - \frac{1}{2\sin\beta}\int_0^\beta\frac{d\varphi}{(1 - \lambda^2\sin^2\varphi)^{1/2}}$$

$$= x_0 - \frac{1}{2\sin\beta}F(\lambda, \beta) \quad (11)$$

where

$$\lambda = (\sin\alpha)/\sin\beta \quad (12)$$

These integrals are standard elliptic integrals of the first kind and are well-tabulated. A fixed y_0/x_0 ratio determines λ from (9) and (10). A fixed L/x_0 ratio then determines β and hence α. This computation was carried out by interpolation from existing tables.

The mapping function has eight branch points falling on the unit circle, Fig. 2. In addition to these corner-describing singularities, the crack tips are described by the roots of $\omega'(\sigma) = 0$ which occur at $\zeta = \pm i$.

It is convenient to consider the mapping function on an appropriate Riemann surface. The branch cuts can be chosen as the two intervals of the unit circle shown as the wavy arcs in Fig. 2. The top sheet of the surface will be defined as including S_ζ^+ which maps into the physical region occupied by the sheet. Clearly the mapping function can be continued analytically across the uncut

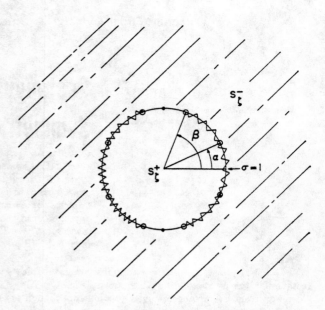

Fig. 2 ζ-plane

intervals of the unit circle into S_ζ^- in the top sheet. By our choice of branch cuts, the uncut intervals on the unit circle correspond to the crack intervals situated along the real axis in the Z-plane. Thus, by Schwartz's reflection principle, the analytic continuation of $\omega(\zeta)$ into S_ζ^- is defined as

$$\omega(\zeta) = \overline{\omega(1/\bar\zeta)}, \quad \zeta\epsilon S_\zeta^- \quad (13)$$

The mapping function can be expressed in series form as

$$Z = \omega(\zeta) = -i\sum_{n=1}^\infty A_n\zeta^{2n-1}, \quad \zeta\epsilon S_\zeta^+ \quad (14)$$

where the mapping coefficients A_n are real. Although the A_n can be related to integrals involving the elliptic functions, it is simpler computationally to calculate them from easily derived recursive formulas [6].

Stresses in Neighborhood of Crack Tips

The character of the stresses in the neighborhood of the crack tips will now be examined in terms of the present formulation. The structure of the stresses in terms of the Muskhelishvili approach has been examined previously by Sih, et al. [7], for the case of single-valued analytic mapping functions. For the exact mapping function of this problem, it is necessary to extend this argument to mapping functions with branch-point type of singularities on the boundary of definition.

An argument particularly useful for analyzing the situation for the exact mapping function will now be carried out utilizing the elegant extension concept of Muskhelishvili. In brief, the function $\varphi(\zeta)$ is extended into S_ζ^- by defining

$$\varphi(\zeta) = -\omega(\zeta)\overline{\varphi'}(1/\zeta)/\overline{\omega'}(1/\zeta) - \overline{\psi}(1/\zeta), \quad \zeta\epsilon S_\zeta^- \quad (15)$$

where the bar notation is defined by

$$\bar f(1/\zeta) = \overline{f(1/\bar\zeta)} \quad (16)$$

In (16), $\omega(\zeta)$ for $\zeta\epsilon S_\zeta^-$ will be considered defined by (14). The form of the foregoing extension for the case of polynomial mapping functions of the circular regions is due to Kartzivadze [8].

The extended definition of $\varphi(\zeta)$ is clearly analytic for all finite points in S_ζ^-. Its structure at infinity is determined by the structures of the defining functions. The function $\psi(\zeta)$ can now be expressed as

$$\psi(\zeta) = -\bar\varphi(1/\zeta) - \bar\omega(1/\zeta)\varphi'(\zeta)/\omega'(\zeta), \quad \zeta\epsilon S_\zeta^+ \quad (17)$$

Thus, the boundary condition (6) can be replaced by

$$\varphi^+(\sigma) - \varphi^-(\sigma) + [\omega^+(\sigma) - \omega^-(\sigma)]\overline{\varphi'^+(\sigma)}/\overline{\omega'^+(\sigma)} = g(\sigma) \quad (18)$$

where the notations $f^+(\sigma)$ and $f^-(\sigma)$ are defined as the values of $f(\zeta)$ as $\zeta \to \sigma$ through S_ζ^+ and S_ζ^-, respectively.

It will now be assumed that $\varphi'^+(\sigma)$ is bounded on intervals of the unit circle corresponding to unloaded portions of the original geometry. (The validity of this assumption can be verified by an *a posteriori* argument.) It then follows that, on the uncut intervals of the unit circle in the present problem, equation (18) reduces to

$$\varphi^+(\sigma) - \varphi^-(\sigma) = g(\sigma) \quad (19)$$

In the present problem $g(\sigma)$ can be considered as vanishing on one of the intervals of the unit circle corresponding to a crack description by a suitable choice of $s = 0$. On this interval, therefore $\varphi^+ = \varphi^-$. Thus $\varphi(\zeta)$ can be considered as analytically continued across this interval by (15). In particular, $\varphi(\zeta)$ is analytic on this interval of the unit circle. The analyticity of $\varphi(\zeta)$ and $\omega(\zeta)$ on this interval in turn establishes the analyticity of $\overline{\omega'(\zeta)\psi(\zeta)}$ at the same points from (17).

The foregoing argument is purposely restrictive for the sake of brevity. The conclusion is valid when $g(\sigma)$ is constant on a crack interval. This can be verified by altering the definition of (15) and utilizing the independence of the stress state to constants added to $\varphi(\zeta)$ and $\psi(\zeta)$. Generalization of the argument to the case of applied loads acting on the cracks is straightforward. For most practical load systems, the structure of $\varphi(\zeta)$ on the crack interval will differ from an analytic function by a Cauchy integral with a density function $g(\sigma)$.

Having established the properties of $\varphi(\zeta)$ and $\psi(\zeta)$ on the intervals of the unit circle containing the crack tip, it is now a straightforward matter to determine the structure of the stresses in the vicinity of the crack tip. Let σ_0 denote a point on the unit circle corresponding to a crack tip in the physical plane. Then the expansions

$$Z = \omega(\sigma_0) + \omega''(\sigma_0)(\zeta - \sigma_0)^2/2! + \ldots \quad (20)$$

$$\varphi(\zeta) = \varphi(\sigma_0) + \varphi'(\sigma_0)(\zeta - \sigma_0) + \ldots \quad (21)$$

are valid in a complete neighborhood of $\zeta = \sigma_0$. If

$$Z - \omega(\sigma_0) = re^{it} \quad (22)$$

then by reversion of series

$$\zeta - \sigma_0 \approx (2r)^{1/2} e^{it/2}/[\omega''(\sigma_0)]^{1/2} \quad (23)$$

From (3), (20), (21), and (23), in the vicinity of the crack tip

$$\sigma_y + \sigma_x \approx 4\,\text{Re}\{\varphi'(\sigma_0)e^{-it/2}/[2r\omega''(\sigma_0)]^{1/2}\} \quad (24)$$

Similarly, from (4), (17), (20), (21), and (23),

$$\sigma_y - \sigma_x + 2i\tau_{xy} \approx \frac{2e^{-it/2}}{[2r\omega''(\sigma_0)]^{1/2}} \left\{ \frac{\overline{\varphi'(\sigma_0)}}{\sigma_0^2} - \frac{\overline{\varphi'(\sigma_0)}}{2\sigma_0^4} - \frac{\varphi'(\sigma_0)}{2}e^{-2it} \right\} \quad (25)$$

In the conventional description, the stress-intensity factor K is defined by

$$\sigma_{\max} = \sigma_y \approx \frac{K}{(2r)^{1/2}}(\cos t/2)[1 + (\sin t/2)(\sin 3t/2)] \quad (26)$$

In the present problem, choosing the crack tip corresponding to $\sigma_0 = i$ leads to

$$K = 2\varphi'(i)/[\omega''(i)]^{1/2} \quad (27)$$

The final result is obviously valid for polynomial approximations of the exact mapping function as well, since no contradictions of the local behavior would be introduced in the argument.

Determination of $\varphi'(i)$ by Power Series

In this section, we consider a power series development of the solution for the determination of $\varphi'(i)$. The following representations are assumed:

$$\varphi(\zeta) = T\omega(\zeta)/4 + iT\sum_{n=1}^{\infty} \alpha_n \zeta^{2n-1}, \quad \zeta \epsilon S_\zeta^+ \quad (28)$$

$$\psi(\zeta) = T\omega(\zeta)/2 + iAT\zeta/(\zeta^2 + 1) + iT\sum_{n=1}^{\infty} \beta_n \zeta^{2n-1}, \quad \zeta \epsilon S_\zeta^+ \quad (29)$$

The first term of each of these representations corresponds to the elementary solution were there no cracks in the geometry. The coefficients A, α_n, and β_n are real from symmetry considerations. The simple poles at $\zeta = \pm i$ in $\psi(\zeta)$ are consistent with the structure indicated in (17); in fact,

$$A = 2i\omega(i)\varphi'(i)/T\omega''(i) \quad (30)$$

It will be sufficient for the present purpose to determine the α_n since

$$\varphi'(i) = -iT\sum_{n=1}^{\infty}(2n-1)\alpha_n(-1)^n \quad (31)$$

It is convenient to express the boundary condition (6) as

$$\overline{\omega'(\sigma)}\varphi(\sigma) + \omega(\sigma)\overline{\varphi'(\sigma)} + \overline{\omega'(\sigma)\psi(\sigma)} = \overline{\omega'(\sigma)}g(\sigma) \quad (32)$$

We write the Fourier expansion

$$g(\sigma) - \frac{T}{2}[\omega(\sigma) + \overline{\omega(\sigma)}] = iT\sum_{n=-\infty}^{\infty} C_{2n-1}\sigma^{2n-1} \quad (33)$$

where

$$C_{2k-1} = +\frac{2}{(2k-1)\pi}\int_\beta^{\pi/2} x(\theta)\cos(2k-1)\theta\,d\theta, \quad k = 1, 2, \ldots. \quad (34)$$

$$C_{-2k-1} = -C_{2k+1}, \quad k = 0, 1, 2, \ldots.$$

Inserting the appropriate series expansions into (32) and equating properly coefficients of equal powers of σ, from the positive powers of σ one finds

$$\sum_{n=1}^{\infty}(2n-1)[\alpha_n A_{n+p-1} + A_n \alpha_{n+p-1}] = +d_p$$

$$p = 1, 2, \ldots. \quad (35)$$

$$d_p = \sum_{n=1}^{\infty}(2n-1)A_n C_{2(p+n)-3}, \quad p = 1, 2, \ldots. \quad (36)$$

The solution of the infinite linear system (35) determines the coefficients α_n.

Truncation of System

The initial numerical procedure carried out followed the conventional study of the numerical convergence of $\varphi'(i)$ for sequences of truncations of the system (35). For the parameter range, large y_0/x_0 in conjunction with small values of L/x_0, this procedure proved fairly effective. Beyond this range, however, the rate of convergence presented serious practical difficulties in the determination of accurate values of $\varphi'(i)$. A plausible explanation of the difficulty can be found by a study of the convergence of the series form of the mapping function. Examination of $\omega_T'(i)$ and $\omega_T''(i)$, where $\omega_T(\zeta)$ denotes the truncated mapping series, indicates that a substantial error in these quantities can occur by indiscriminate truncation. On the other hand, from physical considerations, the solution should be sensitive to the representation of the geometry in the vicinity of the crack root.

An effective alternate plan was found whereby truncations of (14) which preserve key properties of the geometry local to the crack tips, e.g., at $\zeta = \pm i$, were selected. In particular, consider

$$\omega_T(\zeta) = -i \sum_{n=1}^{M+2} \epsilon_n \zeta^{2n-1} \quad (37)$$

where

$$\epsilon_n = A_n, \quad n = 1, 2, \ldots, M; \quad \epsilon_{M+1} = -R; \quad \epsilon_{M+2} = -S$$

$$2(2M+1)R = S_M - 2(M+1)T_M - (-1)^M Q \quad (38)$$

$$2(2M+3)S = S_M - 2MT_M - (-1)^M Q$$

$$T_M = (-1)^M \sum_{n=1}^{M} (2n-1)A_n(-1)^n,$$

$$S_M = (-1)^M \sum_{n=1}^{M} (2n-1)(2n-2)A_n(-1)^n$$

and

$$Q = \omega''(i) = -\tfrac{1}{2} \cos \alpha \cos \beta \quad (39)$$

With this definition of $\omega_T(\zeta)$, satisfaction of $\omega_T'(i) = 0$ and $\omega_T''(i) = Q$ is assured. The choice of M is made from an examination of the partial sums of the exact mapping function and selecting those values of M for which $\omega'(i) \approx 0$ and $\omega''(i) \approx Q$ simultaneously. Thus, corrections of the geometry in the vicinity of the crack tips can be made with little disturbance of the overall configuration.

Modification of the previous setup is minor. The major change involves the assumed form of $\varphi(\zeta)$. In particular, we assume

$$\varphi(\zeta) = T\omega_T(\zeta)/4 + iT \sum_{n=1}^{\infty} C_{2n-2} \zeta^{2n-1} + iT \sum_{n=1}^{M+2} \alpha_n' \zeta^{2n-1} \quad (40)$$

The coefficients C_{2k-1} are calculated from (34) with $\dot{x}(\theta)$ now computed from $Z = \omega_T(\zeta)$. The system for the determination of the α_n can be written as

$$\sum_{n=1}^{M+3-p} (2n-1)[\epsilon_n \alpha_{n'+p-1} + \alpha_n' \epsilon_{n+p-1} + C_{2n-1} \epsilon_{n+p-1}] = 0$$

$$p = 1, 2, \ldots, M+2 \quad (41)$$

Calculation of $\varphi'(i)$ requires the evaluation of

$$\varphi'(i) = -iT \left[\sum_{n=1}^{\infty} (-1)^n (2n-1) C_{2n-1} \right.$$
$$\left. + \sum_{n=1}^{M+2} (-1)^n (2n-1) \alpha_n' \right] = -iT[\Sigma_1 + \Sigma_2] \quad (42)$$

Evaluation of Σ_1 can be made directly from the relation

$$\Sigma_1 = -\frac{1}{\pi} \int_{\beta}^{\pi/2} \dot{x}(\theta) d\theta / \cos \theta \quad (43)$$

Evaluation of Σ_2 requires the solution of the system of equations (41).

Numerical Results

The analysis was carried out numerically for the two cases $y_0/x_0 = 1.00$ and $y_0/x_0 = 3.00$. Both the conventional plan for the exact mapping function and the alternative plan of the preceding section were used in the study of $y_0/x_0 = 3.00$. The latter plan was used exclusively in the study of $y_0/x_0 = 1.00$.

The numerical results for $y_0/x_0 = 1.00$ are contained in Table 1. The values of α and β have been rounded to four decimal places for convenience of presentation. The term Σ_1 which is the dominant contribution to $\varphi'(i)/T$ can be considered as accurately computed to ± 0.001. Evaluation of Σ_2 depended on the solution of the system (41). In general, three values of M were selected for each set of the parameters to study the convergence. For the most part it was possible to select $0 < M_1 < 50$, $50 < M_2 < 100$, $100 < M_3 < 150$. A study of the convergence of Σ_2 with respect to M leads to the assertion that the $i\varphi'(i)/T$ are numerically correct to within 1 percent.

The numerical results for $y_0/x_0 = 3.00$ are contained in Table 2. For the first four sets of parameters, the exact system (35) was considered. Systematic truncation of (35) was carried out successfully for the shorter crack lengths to yield reliably accurate values of $\varphi'(i)$. As the crack length increased this plan became uneconomical and was abandoned in favor of the alternative truncation scheme. It was necessary to find Σ_2 by extrapolation for the largest crack length shown in Table 2 owing to the limitations of the digital computer used in solving the systems. Again, $i\varphi'(i)/T$ in Table 2 can be considered accurate to 1 percent.

Discussion

A matter of considerable interest is the comparison of the calculated results with the stress-intensity factor, denoted here as

Table 1 Evaluation of K/T for $y_0/x_0 = 1.00$

α	β	x_0	L	L/x_0	Σ_1	Σ_2	$i\varphi'(i)/T$	K/T	K_A/T	K/K_A
.7703	1.3963	.9413	.1246	.1324	.346	.055	.401	.401	.356	1.13
.7519	1.3090	.9598	.1896	.1975	.348	.057	.405	.498	.442	1.13
.7268	1.2217	.9865	.2577	.2612	.353	.063	.416	.595	.522	1.14
.6956	1.11345	1.0229	.3308	.3234	.359	.069	.428	.690	.602	1.15
.6591	1.0472	1.0705	.4109	.3838	.372	.074	.446	.793	.685	1.16
.6178	.9599	1.1317	.5007	.4424	.387	.076	.463	.895	.775	1.16
.5236	.7854	1.3110	.7269	.5544	.434	.070	.504	1.12	.995	1.13
.4718	.6981	1.4422	.8769	.6080	.470	.063	.533	1.25	1.14	1.10
.3614	.5236	1.8541	1.3185	.7111	.575	.050	.625	1.59	1.56	1.02
.3035	.4363	2.1936	1.6691	.7609	.683	.041	.724	1.90	1.88	1.01

Table 2 Evaluation of K/T for $y_0/x_0 = 3.00$

α	β	x_0	L	L/x_0	Σ_1	Σ_2	$i\varphi'(1)/T$	K/T	K_A/T	K/K_A
.0354	1.3963	.7978	.0887	.1112	--	--	.283	.334	.299	1.12
.0347	1.3090	.8134	.1356	.1667	--	--	.287	.413	.373	1.11
.0337	1.2217	.8361	.1858	.2222	--	--	.293	.485	.440	1.10
.0311	1.0472	.9072	.3024	.3333	--	--	.311	.622	.577	1.08
.0294	.9599	.9591	.3730	.3889	.297	.027	.324	.694	.654	1.06
.0254	.7854	1.1111	.5556	.5001	.342	.019	.361	.858	.841	1.02
.0152	.4363	1.8590	1.3441	.7230	.575	.015*	.590	1.59	1.59	1.00

*Obtained by extrapolation.

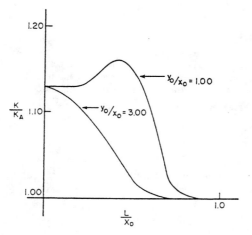

Fig. 3 Ratio of stress-intensity factors

K_A, derived by Irwin [1] from the Westergaard solution. In the present notation

$$K_A/T = \left[\frac{2x_0}{\pi} \tan \frac{\pi L}{2x_0}\right]^{1/2} \qquad (44)$$

The appropriate K_A are listed in Tables 1 and 2 along with the ratios K/K_A. Fig. 3 indicates that the approximation K_A is particularly good for deep cracks while for short cracks K_A differs from the exact solution by approximately 13 percent.

The results of the comparison are consistent for the limiting cases of crack lengths with the following argument. For very short crack lengths the stress intensity should coincide with the corresponding solution for an edge crack in an elastic half-plane. The latter solution was found by Irwin [9] to be $K/T = 1.1\sqrt{L}$. Bueckner [10] subsequently found the more accurate result, $K/T = 1.13\sqrt{L}$. As $L \to 0$ in (44), $K/T \to \sqrt{L}$. Thus, for small crack depths, the 13 percent error in K_A is consistent with Bueckner's result. For very deep cracks it is a straightforward matter to show that K_A approaches the limiting value of Neuber's solution [11] for hyperbolic notches. The foregoing limit arguments are clearly valid for both y_0/x_0 ratios considered. The y_0/x_0 ratio has an effect on the stress-intensity factors for intermediate values of L/x_0 as is indicated in Fig. 3. On the other hand, it is reasonable to assume from the results that the ratio $y_0/x_0 = 3.00$ is effectively infinite as far as effects on K are concerned.

Finally, the mapping approach used in this analysis can be effectively used for other useful examples in this problem class. It is recommended that convergence difficulties be minimized by truncation of the series form of the exact mapping function in the manner described whereby key properties of the geometry local to the crack tip are preserved.

Acknowledgments

The author is indebted to Mr. Donald Neal of the Army Materials Research Agency for programming the analysis for a digital computer. Also, several helpful suggestions by a referee have been included in this paper.

References

1 G. R. Irwin, *Handbuch der Physik*, J. Springer, Berlin, Germany, 1958, pp. 551–590.
2 H. M. Westergaard, "Bearing Pressures and Cracks," TRANS. ASME, vol. 61, 1939, pp. A-49 to A-53.
3 R. J. Tait, "Crack in Strip Finite Width," see I. N. Sneddon, "Crack Problems in the Mathematical Theory of Elasticity," North Carolina State College, Rayleigh, N. C., May 1961.
4 O. L. Bowie, "Analysis of an Infinite Plate Containing Radial Cracks Originating at the Boundary of an Internal Circular Hole," *Journal of Mathematics and Physics*, vol. 25, 1956, pp 60–71.
5 N. I. Muskhelishvili, *Some Basic Problems of the Mathematical Theory of Elasticity*, P. Noordhoff, Limited, Groningen, Holland, 1953.
6 O. L. Bowie, "Rectangular Tensile Bar Weakened by Surface Cracks, Part I, Stress Analysis," Watertown Arsenal Laboratories, WAL TR 811.8/5, March 1963.
7 G. C. Sih, P. C. Paris, and F. Erdogan, "Crack-Tip, Stress Intensity Factors for Plane Extension and Plate Bending Problems," JOURNAL OF APPLIED MECHANICS, vol. 29, TRANS. ASME, vol. 84, Series E, 1962, pp. 306–312.
8 I. N. Kartzivadze, "The Fundamental Problems of the Theory of Elasticity for the Elastic Circle," *Comptes rendus de l'académie des sciences, de l' U.R.S.S. 20*, vol. 12, 1943, pp. 95–104.
9 G. R. Irwin, "The Crack Extension-force for a Crack at a Free Surface Boundary," Naval Research Laboratory, Washington, D. C., NRL Report 5120, 1958.
10 H. F. Bueckner, "Some Stress Singularities and Their Computation by Means of Integral Equations," *Boundary Problems in Differential Equations*, University of Wisconsin Press, Madison, Wis., 1960.
11 H. Neuber, *Kerbspannungslehre: Grundlagen für Genane Spannungsrechnung*, J. Springer, Berlin, Germany, 1937.

A Modified Mapping-Collocation Technique for Accurate Calculation of Stress Intensity Factors

O. L. BOWIE AND D. M. NEAL

Army Materials and Mechanics Research Center, Watertown, Massachusetts

(Received November 20, 1969)

ABSTRACT
A technique combining the advantages of conformal mapping and boundary collocation arguments for calculating stress intensity factors for cracks in plane problems is described. The difficulty of finding the mapping function on a rigidly prescribed parameter region is avoided at the expense of using boundary collocation methods on part of the boundary. Conventional collocation arguments are modified by prescribing stress, force, and moment conditions in a least-square collocation sense. These pseudo-redundant conditions provide a reasonable basis for estimation of the effects of inaccuracy of the boundary conditions. The technique is applied to the problem of a circular disk with an internal crack under a loading of external hydrostatic tension.

Introduction

Methods for calculating stress intensity factors for cracks have important applications in fracture mechanics. For plane problems, the authors have previously developed the Muskhelishvili conformal mapping method [1] into an effective technique for a certain class of crack problems [2], [3]. At about the same time, boundary collocation methods were reintroduced and applied to crack problems by several authors [4], [5].

The boundary collocation methods depend on the selection of a class of stress functions satisfying the loading conditions on the crack and then matching boundary conditions at selected points on the remaining portion of the boundary. The computational simplicity of this approach is attractive; on the other hand, serious difficulties arise in assessment of accuracy. Convergence to the "correct" solution must be assessed on the basis of estimating the effect of the off-point residual errors in the boundary conditions. In fact, "apparent" convergence to incorrect values is quite possible if, for example, the class of stress functions chosen were incomplete.

The mapping technique has its difficulties too. It is frequently very difficult to find accurate polynomial mappings of the physical region onto a suitable parametric region. On the other hand, the ensuing stress analysis is well understood. Assessment of accuracy is related to the approximation of geometry rather than the effect of residual errors in the boundary conditions.

The technique proposed in this paper is a natural compromise of the two methods. The simple form of a mapping function carrying a circle and its exterior in the parameter plane into a crack and its exterior, respectively, will be used. The remaining portion of the boundary in the physical plane will correspond to a directly calculable curve in the auxiliary plane. The continuation arguments of Muskhelishvili are then employed to describe stress functions with, e.g., "traction-free" conditions on the crack. Collocation methods can then be introduced to satisfy the conditions on the remaining portions of the boundary.

This plane eliminates the difficulty of finding accurate polynomial approximations of the exact geometry in terms of a rigidly specified parameter domain. On the other hand, much of the mathematical insight provided by the complex variable formulation is preserved. In fact, it will be shown that an effective modification of the conventional boundary collocation procedure is suggested for the reduction of residual errors intermediate to the points of collocation.

Stress Function for a Circular Disk Containing a Pressurized Internal Crack

The model problem chosen to illustrate the technique is the plane problem corresponding to a circular disk containing an internal crack with loading shown in Fig. 1. It is obvious that the loading in Fig. 1 leads to the same K_I (stress intensity factor) as that for a pressurized internal crack. Such a problem illustrates the practical difficulties of employing the strict mapping technique. Since the region is doubly connected, a natural choice of the parameter domain is a concentric ring. Although such a mapping function is known in terms of elliptic functions, [6], the conversion to accurate polynomial approximations necessary to the stress analysis is not easy.

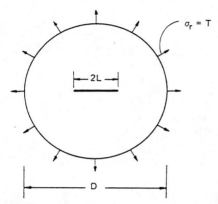

Figure 1. Circular disk with internal crack loaded by uniform external tension, T.

The first step in our procedure is to introduce a limited form of mapping. The physical region in Fig. 1 will be considered as defined in the complex Z-plane. Introducing an auxiliary ζ-plane, we consider the simple mapping

$$Z = \omega(\zeta) = (L/2)(\zeta + \zeta^{-1}). \tag{1}$$

The unit circle, $\zeta = \sigma = e^{i\lambda}$, and its exterior in the ζ-plane map into the crack and its exterior, respectively, in the Z-plane. Clearly the image points in the physical plane can be related to the parameter plane by

$$\zeta = (Z/L) + [(Z/L)^2 - 1]^{\frac{1}{2}}. \tag{2}$$

In particular, the boundary $Z = (D/2)e^{i\theta}$ will correspond to a closed curve τ exterior to the unit circle in the ζ-plane (Fig. 2).

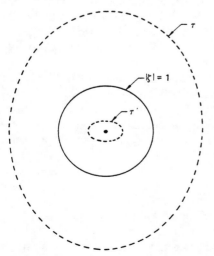

Figure 2. Region defined in the parameter plane.

The stress functions $\phi(Z)$ and $\psi(Z)$ in the Muskhelishvili notation can be considered as analytic functions of ζ. Furthermore, we adopt the notation $\phi'(Z) = \Phi(Z)$, $\psi'(Z) = \Psi(Z)$, $\phi[\omega(\zeta)] = \phi(\zeta)$, $\Phi(\zeta) = \phi'(\zeta)/\omega'(\zeta)$, etc., where primes denote differentiation. Then,

$$\sigma_y + \sigma_x = 4\,\text{Re}\,\{\Phi(\zeta)\} = 4\,\text{Re}\,\{\phi'(\zeta)/\omega'(\zeta)\}$$

$$\sigma_y - \sigma_x + 2i\tau_{xy} = 2\{\overline{\omega(\zeta)}\,\Phi'(\zeta)/\omega'(\zeta) + \Psi(\zeta)\}$$

$$= 2\{\overline{\omega(\zeta)}[\phi'(\zeta)/\omega'(\zeta)]' + \psi'(\zeta)\}/\omega'(\zeta).$$

If we denote by $X_n \mathrm{d}s$ and $Y_n \mathrm{d}s$ the horizontal and vertical forces acting on an element of arc $\mathrm{d}s$ with normal n, the force resultant along the arc can be written as

$$\phi(\zeta) + \omega(\zeta)\overline{\phi'(\zeta)}/\overline{\omega'(\zeta)} + \overline{\psi(\zeta)} = i\int^s (X_n + iY_n)\mathrm{d}s = f_1(s) + if_2(s). \tag{4}$$

Similarly, the moment (with respect to the origin) of the same forces is given by

$$M_0 = \int^s (xY_n - yX_n)\mathrm{d}s = \text{Re}\left\{\int^\zeta \psi(\zeta)\omega'(\zeta)\mathrm{d}\zeta - \omega(\zeta)\psi(\zeta) - |\omega(\zeta)|^2 \Phi(\zeta)\right\}. \tag{5}$$

The condition that the crack be traction-free can be handled effectively by using the extension concept of Muskhelishvili. If S_ζ^+ and S_ζ^- denote the interior and exterior, respectively, of the unit circle in the ζ-plane, the function $\phi(\zeta)$ will be extended into S_ζ^+ by defining

$$\phi(\zeta) = -\omega(\zeta)\overline{\phi'}(1/\zeta)/\overline{\omega'}(1/\zeta) - \overline{\psi}(1/\zeta), \qquad \zeta \in S_\zeta^+ \tag{6}$$

where the bar notation is defined by

$$\overline{f}(1/\zeta) \equiv \overline{f(1/\overline{\zeta})}. \tag{7}$$

The function $\psi(\zeta)$ can now be expressed as

$$\psi(\zeta) = -\overline{\phi}(1/\zeta) - \overline{\omega}(1/\zeta)\phi'(\zeta)/\omega'(\zeta), \qquad \zeta \in S_\zeta^-. \tag{8}$$

Similarly, one finds

$$\omega'(\zeta)\Psi(\zeta) = \zeta^{-2}\overline{\omega}(1/\zeta)\{\Phi(\zeta) + \Phi(1/\zeta)\} - \overline{\omega}(1/\zeta)\Phi'(\zeta), \qquad \zeta \in S_\zeta^- \tag{9}$$

From (4) and (8) evaluated on the unit circle $\zeta = \sigma$, it follows that the resultant force on the crack is identically zero as a function of σ provided the extended definition of $\phi(\zeta)$ is continuous across the unit circle. Thus, let τ' be the closed curve interior to the unit circle obtained by inversion of τ with respect to the unit circle. Then, if $\phi(\zeta)$ is considered as analytic in the doubly connected region enclosed by $\tau + \tau'$, traction-free conditions on the crack are automatically satisfied.

The determination of $\phi(\zeta)$ requires a form of representation and the satisfaction of conditions on τ corresponding to the tractions on $|Z| = D/2$. It will be assumed that $\phi(\zeta)$ can be represented in the form of a Laurent series. This appears to be a reasonable assumption although the boundaries τ' and τ are not circular. There is no *a priori* reason to suspect that the region of convergence of such a series could not extend over the desired parameter range in the present problem. Thus, taking into account obvious stress symmetries,

$$\phi(\zeta) = T\sum_\infty^\infty \alpha_n \zeta^{2n+1} \tag{10}$$

where the α_n's are real and must be determined from the boundary conditions on τ.

It is interesting to compare (10) with the class of stress functions used in [5] for the problem of an internal crack in a rectangular strip under tension. In [5] a generalized Westergaard's stress function, $\phi'_w(Z)$ was chosen of the form

$$\phi'_w(Z) = \sum_{m=0}^{N} C'_m Z^m / [Z^2 - L^2]^{\frac{1}{2}}.$$

This corresponds essentially to

$$\phi_w(\zeta) \approx \sum_{m=0}^{N} C'_m (\zeta + \zeta^{-1})^m \tag{12}$$

in the present formulation. Since (12) represents a restricted version of (10), the question of completeness can certainly be raised. This does not preclude the possibility of accuracy in an asymptotic sense in [5].

Modified Collocation and the Boundary Conditions

For the loading in Fig. 1, we have the stress boundary conditions

$$\sigma_r = T, \quad \tau_{r\theta} = 0 \quad \text{on} \quad |Z| = D/2 \tag{13}$$

and traction-free conditions on the crack. Thus, from (3) and (9)

$$\Phi(\zeta) + \overline{\Phi(\zeta)} - e^{2i\theta} \{\overline{\omega(\zeta)} \Phi'(\zeta) + \zeta^{-2} \overline{\omega}'(1/\zeta)[\overline{\Phi}(1/\zeta) + \Phi(\zeta)] - \overline{\omega}(1/\zeta)\Phi'(\zeta)\}/\omega'(\zeta) = T \quad \zeta \in \tau. \tag{14}$$

The boundary condition in terms of the force resultant (4) becomes

$$\phi(\zeta) - \phi(1/\bar{\zeta}) + [\omega(\zeta) - \omega(1/\bar{\zeta})] \overline{\phi'(\zeta)/\omega'(\zeta)} = T\omega(\zeta), \quad \zeta \in \tau. \tag{15}$$

Only the interval $0 \leq \theta \leq \pi/2$ need be considered explicitly since the symmetry of the stress function guarantees satisfaction of conditions on the remaining interval. Since boundary collocation will be used, the parametric nature of the Muskhelishvili formulation can be utilized by referring to both coordinate systems interchangeably. Thus, e.g., $\omega(\zeta)$ can be programmed in terms of either coordinates (whichever is more convenient), *etc.*

(There is an interesting parallelism here with a method by Bowie [7] for handling edge notches in a semi-infinite region. In [7], the reflection principle was used to ensure traction-free conditions along the real axis by the analyticity of the extended function $\phi(Z)$. Conditions on the notch and its reflected image paralleled the present conditions on τ and τ'. It was shown in [7] that a strict Fourier argument could be made even though explicit boundary conditions were imposed only on the notch interval. It was necessary to recognize that the assumed analytic continuation of $\phi(Z)$ implied conditions on the reflected image of the notch so that a "complete" interval was specified in a Fourier sense. In the present case, a similar argument can be made to justify the apparent inconsistency of determining the coefficients of a Laurent expansion from explicit conditions on only one of the "boundaries".)

To apply the boundary collocation argument, it is necessary to consider truncations of (10), e.g.,

$$\phi(\zeta) = T \sum_{-M}^{N} \alpha_n \zeta^{2n+1} \tag{16}$$

or

$$\omega'(\zeta) \Phi(\zeta) = T \sum_{-M}^{N} (2n+1) \alpha_n \zeta^{2n} \tag{17}$$

Then, if (17) is substituted into (14), there are $M + N + 1$ independent α_n's available to satisfy (14). If the interval $0 \leq \theta \leq \pi/2$ is subdivided into a set of discrete points, the matching of (14) corresponds to two real conditions at each point. If the number of conditions is matched by a consistent choice of M and N, a linear system of equations can then be solved for the α_n's. Of course, the number of conditions can exceed the degrees of freedom if one resorts to a least-square minimization of the total error summed over the discrete points.

When the conventional collocation argument is applied to the stress boundary conditions, a major difficulty becomes immediately obvious. For a fixed M/N ratio in the truncation of (17), apparent convergence to a value of K_I can be found by successively increasing the system size. However, by simply altering the M/N ratio, "convergence" to different values of K_I is found. Furthermore, an examination of the magnitudes of error intermediate to the points of collocation is often inconclusive.

A suitable error measure for boundary collocation is clearly suggested by Saint Venant's principle. The overall effects of boundary errors should be minimized if the resultant force and moment conditions are collocated. If $f_1 + if_2$ is matched at the points of collocation, then the errors in stress boundary conditions correspond to self-equilibrating distributions of loading on the intervals between successive collocation points. A measure of the moment error is found conveniently in the present formulation. Along an arc L

$$M_0 = \int_L (xY_n - yX_n)ds = -\int_L (x\,df_1 + y\,df_2) = -(xf_1 + yf_2)_L + \int_L (f_1\,dx + f_2\,dy). \tag{18}$$

It is clear from (18) that off-point errors in f_1 and f_2 can have an accumulated effect on moment accuracy due to the second term.

The following modification of the conventional boundary collocation plan is therefore recommended: *At each boundary point of collocation, we impose the five conditions corresponding to $M_0, f_1 + if_2$, and the normal and tangential components of the applied stress.* Since these conditions are conveniently expressed by the Muskhelishvili formulation, it is an easy matter to write these conditions in terms of the coefficients α_n of the truncated series (17).

In applying the procedure, it is generally advisable to utilize least-square collocation for economy and also from certain considerations of the system. If strictly continuous arguments were being used, there is obvious redundancy in the five conditions above. Although this redundancy is removed when discrete considerations are made, nevertheless, for large systems a weakness in the determinant for the full system appears probable. This difficulty can be minimized by controlling the degrees of freedom as compared with the number of conditions and minimizing the error in a least-square collocation sense.

Numerical Results for a Circular Disk with an Internal Crack

The stress intensity factor, K_I, in the conventional notation [8] reduces to

$$K_I = 2\pi^{\frac{1}{2}}\phi'(1)/\{\omega''(1)\}^{\frac{1}{2}} = 2\pi^{\frac{1}{2}}\phi'(1)/L^{\frac{1}{2}}. \tag{19}$$

Thus,

$$K_I = 2(\pi/L)^{\frac{1}{2}} \sum_{-M}^{N} (2n+1)\alpha_n. \tag{20}$$

Furthermore, it is well known that

$$\phi(\zeta) \to (TL/8)(\zeta - 3\zeta^{-1}), \quad L/D \approx 0. \tag{21}$$

Therefore,

$$K_I \approx T(\pi L)^{\frac{1}{2}}, \quad L/D \approx 0. \tag{22}$$

Initially, the conventional boundary collocation approach using the stress boundary condition (14) was attempted. When arbitrary M/N ratios were chosen, very poor results were obtained even in the range $0 < L/D < 0.05$ where (22) can be expected to hold. A tedious trail and error process based on a careful variation of M and N guided by (21) yielded some reliable results in this range. However, this uneconomical approach was abandoned in favor of the process described in the preceding section.

The analysis was then set up using (4), (14), and (15). For the radial loading in Fig. 1, $M_0 = 0$. However, since (4) is valid up to a constant of integration, the matter of consistency with the assumed form (16) must be considered. One can simply carry an unknown constant C in (4) and treat it as an unknown in the matching of (4). The usual constant of integration in (15) is zero to be consistent with the assumed form of $\phi(\zeta)$ in (16).

The computations were carried out using double precision. Furthermore, for accuracy in solving the system of equations it was found advisable to scale the unknowns by the substitution

$$\alpha'_n = \alpha_n (D/2)^{|2n|}, \qquad n = 0, \pm 1, \ldots .\tag{23}$$

The interval $0 \leq \theta \leq \pi/2$ was usually divided into 40 stations with equal intervals and approximately 80 degrees of freedom were considered. Least-square collocation was obviously used.

The results obtained are remarkably reliable even for very deep cracks. The values listed in Table 1 for K_I can be considered as accurate to within an error of less than one percent. This estimate is based on examination of the off-point errors and the stability of the answers with variations of M/N ratios, degrees of freedom, number of stations, etc.

Examination of the influence of the individual conditions was also made. In contrast to the poor solutions obtained by using the stress conditions alone, it was found that very stable values of K_I were obtained by using the condition on $f_1 + if_2$ alone. This is consistent with the Saint Venant basis for our argument. The complete system is most reasonably necessary when accuracy is affected by local stress irregularities.

Comparison of Results with an Elliptical Disk with Internal Crack

In Table 1, the present results are compared with the well-known secant correction for an infinite strip [9]. These values are obtained from

$$K_I^{(a)}/T(\pi L)^{\frac{1}{2}} = [\sec(\pi L/D)]^{\frac{1}{2}} .\tag{24}$$

Although (24) is only an approximation, it is frequently accepted on the basis of its agreement with numerical results obtained independently by several investigators.

TABLE 1.

Stress intensity factors K_I, for a circular disk with a pressurized internal crack, $\sigma = -T_1$

D	$D/2L$	$K_I/T(\pi L)^{\frac{1}{2}}$	$K_I^{(a)}/T(\pi L)^{\frac{1}{2}}$	$K_I^{(b)}/T(\pi L)^{\frac{1}{2}}$
100.0	25.00	1.00	1.001	1.00
40.0	10.00	1.02	1.006	1.02
20.0	5.00	1.06	1.025	1.06
10.0	2.50	1.24	1.112	1.28
6.0	1.50	1.74	1.414	2.11
5.4	1.35	1.98	1.589	2.64
5.0	1.25	2.24	1.799	3.43
4.8	1.20	2.43	1.966	4.09
4.6	1.15	2.71	2.217	5.19
4.4	1.10	3.17	2.651	7.61

(a) Secant correction for infinite strip.
(b) Elliptic disk with internal crack.

As $D/2L$ approaches unity, the secant formula increasingly underestimates the K_I in the present problem. Let us consider our present configuration as obtained by cutting away the appropriate material from the infinite strip. Then it is reasonable to hypothesize that the greater K_I values are due to an increase in flexibility with greater bending across the sections on the real axis. If this hypothesis were true, then even greater values of K_I would be obtained by considering an elliptical outer boundary with a major axis of D and a minor axis less than D. This will now be substantiated.

Although a solution for a region bounded by two confocal ellipses has been given by Sheremetjier [10], a direct solution in terms of the present formulation is possible. The mapping function (1) carries the unit circle into a crack of length $2L$ and carries the circle $|\zeta| = \rho_0 > 1$, where

$$2\rho_0 = D/L + [(D/L)^2 - 4]^{\frac{1}{2}} ,\tag{25}$$

into an ellipse with major axis, D, and a minor axis of $L(\rho_0 - \rho_0^{-1})$. Although the eccentricity of the ellipse varies with D/L, the solution for this configuration is still of interest for our purpose.

The stress analysis for external hydrostatic tension again involves a series expansion (10). Substituting (10) into (15) yields the following conditions:

$$\alpha_K(\rho_0^{2K+1} - \rho_0^{-2K-1}) - \alpha_{K-1}(\rho_0^{2K-3} - \rho_0^{-2K-1})$$
$$- \alpha_{-K}(2K-1)(\rho_0^{-2K+1} - \rho_0^{-2K-1}) + \alpha_{-K-1}(2K+1)(\rho_0^{-2K-1} - \rho_0^{-2K-3})$$

$$\begin{aligned}
&= L/2\rho_0, & K &= -1 \\
&= (L/2)(\rho_0 - \rho_0^{-3}), & K &= 0 \\
&= -L/2\rho_0, & K &= 1 \\
&= 0, & K &= \pm 2, \pm 3, \ldots
\end{aligned} \qquad (26)$$

An unusual situation arises in the solution of (26). If N of the conditions are chosen, a linear system of N equations in $N+1$ unknowns results. On the other hand, the structure of (10) is such that no additional relationship is provided by the usual arguments of single-valuedness of stress resultant, *etc.* There remains only the condition of series convergence! For large K, it is clear from (26) that

$$\alpha_K/\alpha_{K-1} \approx \rho_0^{-4} \quad \text{for} \quad K \gg 1. \qquad (27)$$

On the other hand,

$$\alpha_{-K-1}/\alpha_{-K} \approx 1 - \alpha_{K-1}[(1-2K) + 4K\rho_0^{-2}]/\alpha_{-K}, \qquad K \gg 1. \qquad (28)$$

The convergence problem is associated with the behavior of the coefficients α_{-K} by inspection of (28).

The plan of solution by truncation is now fairly evident. To the system formed by the conditions $K = 0, \pm 1, \pm 2, \ldots, \pm M$, we add the condition

$$\alpha_{-M-1} = 0. \qquad (29)$$

The resulting system is a $(2M+2) \times (2M+2)$ linear system. Convergence with respect to M is extremely rapid.

The corresponding stress intensity values are listed in Table 1 as $K_I^{(b)}$. The values exceed the corresponding K_I values as was predicted. It is apparent that for $D/2L > 2.5$, the ellipse closely approximates a circular region and the agreement is good. As $D/2L \to 1$, the ellipse becomes extremely flat and there are no grounds to make any comparisons.

Observations

The modified mapping-collocation technique proposed here appears to be particularly well suited to handling the troublesome problems of internal cracks. Preservation of the Muskhelishvili concepts provides a basis for selecting a class of stress functions with a reasonable guarantee of completeness. In the present problem, for example, the *a priori* expansion in a Laurent series was clearly justified by the excellent matching of the boundary conditions intermediate to the points of collocation.

The modified collocation argument involving the use of pseudo-redundant conditions provides the analyst with a reasonable basis for estimating the effects of inaccuracies in the boundary conditions. For example, it was not surprising that collocation of $f_1 + if_2$ in the present problem yielded excellent values for the stress intensity factors in the range considered. On the other hand it is not difficult to recognize situations when the information desired should involve full use of the five conditions outlined above.

The basic philosophy of the approach can be carried over with suitable modifications to a wide class of problems.

REFERENCES

[1] N. I. Muskhelishvili, *Some Basic Problems of the Mathematical Theory of Elasticity*, P. Noordhoff, Limited, Groningen, Holland, (1953).
[2] O. L. Bowie, Rectangular Tensile Sheet with Symmetric Edge Cracks, *Journal of Applied Mechanics*, 31, Trans. ASME, 86, E (1964) 208–212.
[3] O. L. Bowie and D. M. Neal, "Single Edge Crack in Rectangular Tensile Sheet", *Journal of Applied Mechanics*, 32, Trans. ASME, E, 708–709, (1965).
[4] B. Gross, J. E. Srawley, and W. F. Brown, Jr., *Stress Intensity Factors for a Single-Edge-Notch Tension Specimen by Boundary Collocation of a Stress Function*, NASA TN D-2395, Lewis Research Center, (1964).
[5] A. S. Kobayashi, *Method of Collocation Applied to Edge-Notched Finite Strip Subjected to Uniaxial Tension and Pure Bending*, Boeiing Company, Seattle, Washington, Document No. D2-23551, (1964).
[6] F. Nehari, *Conformal Mapping*, McGraw-Hill Book Company, Inc., New York, (1952).
[7] O. L. Bowie, Analysis of Edge Notches in a Semi-infinite Region, *Journal of Math. and Phys.*, 45, 4 (1966) 356–366.
[8] P. C. Paris and G. C. Sih, "Stress Analysis of Cracks," Fracture Toughness Testing and Its Applications, ASTM STP 381, *Am. Soc. Testing Mats.*, 32, (1965).
[9] W. F. Brown, Jr., and J. E. Srawley, *Plane Strain Crack Toughness Testing of High Strength Metallic Materials*, ASTM STP 410, Philadelphia, Pa., (1966).
[10] M. P. Sheremetjiev, The Elastic Equilibrium of an Elliptic Ring, *Prikl. Mat. i Mech.* XVII, No. 1 (1953) 107–113.

RÉSUMÉ

On décrit une technique de calcul des facteurs d'intensité de contraintes pour des fissures en état plan, qui combine les avantages de la méthode de la représentation conforme et des méthodes de fixation des conditions aux limites.

La difficulté que l'on rencontre à trouver la fonction de représentation qui correspond à une région à paramètres imposés est levée par l'emploi des méthodes de fixation des conditions aux limites sur une partie d'un contour. Le traitement conventionnel de ces méthodes est modifié en imposant les conditions de contraintes, de forces et de moments en un ajustement par moindres carrés.

Les conditions pseudo-redondantes ainsi réunies procurent une base d'appréciation des effets d'une inexactitude dans la définition du contour.

La technique est appliquée au problème du disque circulaire comportant une fissure interne et soumis à l'action d'une tension extérieure uniforme.

ZUSAMMENFASSUNG

Es wird ein Verfahren zur Berechnung von Spannungsintensitätsfaktoren für Riße in einem ebenen Zustand beschrieben, welches sowohl die Vorteile der Methode der konformen Darstellung als auch die der Verfahren zur Bestimmung der Grenzbedingungen miteinander verbindet.

Die Anwendung der Verfahren der Festlegung von Grenzbedingungen für einen Teil der Außenlinie ermöglicht es die Schwierigkeiten zu umgehen, welche sich dann ergeben wenn man versucht die darstellende Funktion für einen Bereich mit streng auferlegten Parameter zu bestimmen. Die konventionelle Behandlung dieser Verfahren wird dadurch abgeändert, daß die Bedingungen für Spannungen, Kräfte und Momente im Sinne der kleinsten Quadratzahlen auferlegt werden.

Diese pseudo-überflüssige Bedingungen ergeben eine Basis zur Beurteilung der Auswirkung einer Ungenauigkeit in der Definition des Umrisses.

Diese Methode wird auf das Problem einer runden Scheibe mit inneren Rissen, welche der Wirkung von äußeren Spannungen unterworfen ist, angewendet.

O. L. BOWIE
C. E. FREESE
D. M. NEAL

Army Materials and
Mechanics Research Center,
Watertown, Mass.

Solution of Plane Problems of Elasticity Utilizing Partitioning Concepts

A partitioning plan combined with the modified mapping-collocation method is presented for the solution of awkward configurations in two-dimensional problems of elasticity. It is shown that continuation arguments taken from analytic function theory can be applied in the discrete to "stitch" several power series expansions of the stress function in appropriate subregions of the geometry. The effectiveness of such a plan is illustrated by several numerical examples.

Introduction

THE evolution of mathematical methodology toward the optimum utilization of computer technology is a continuing matter. An example is provided by the approaches currently used for the solution of plane problems of elasticity. Mathematically, with the adoption of the Airy stress function, the problem reduces to the solution of the biharmonic partial differential equation with boundary properties consistent with the prescribed loading conditions. The classical approaches of course are well known. Specific techniques include power series, integral equations, conformal mapping, transform methods, etc. Although many elegant and important solutions have been found by the classical approaches, a great many practical design configurations have not been amenable to solution by such arguments. In recent years more versatile approaches which utilize the numerical capabilities of the computer have been devised for effective numerical solutions of most plane problems of elasticity.

Recently, the authors have proposed a compromise plan of solution for the plane problem which preserves the essential mathematical properties of the solution predicted by classical analysis yet allows for a nonrigorous use of numerical computation. This plan, the *MMC* (modified mapping-collocation) technique was proposed and developed [1-4][1] for the solution of both isotropic and anisotropic problems of plane elasticity. The *MMC* technique utilizes the complex variable formulation of Muskhelishvili [5] with power series representations of the corresponding analytic stress functions. A limited use of conformal mapping is made for an accurate description of sensitive portions of the geometry. In general, however, the overall geometry of the parameter region is irregular in shape (in contrast to the rigidly prescribed parameter regions familiar to the classical mapping arguments). Simple mapping functions can be employed thus eliminating the major criticism of the mapping approach for irregularly shaped geometries. The critical analytical properties of the solution are preserved by using the classical arguments in the stress-function representation. The boundary conditions are enforced by a modified boundary collocation argument. Included in this modification is a heuristic appeal to Saint Venant's principle to assess the influence of residual errors in the boundary conditions.

In this paper, we shall extend the flexibility of the *MMC* technique to geometries with overall configurations previously awkward for single power series representation of the solution. Here, the notion of geometrical effects is used somewhat loosely for certainly there is an interrelationship with the prescribed boundary-value problem. When the complex variable formulation is introduced, the corresponding analytic functions describing the solution must in general admit mathematical singularities in the complex plane of reference. In many problems it can be assumed that these singularities are sufficiently remote from the physical region so that single series expansions are valid. On the other hand, there is a large class of problems where troublesome singularities can be anticipated, e.g., near sources of high stress concentration due to local irregularities in geometry, etc. In the original *MMC* plan where truncations of power series are used, the existence of such unfavorable singularities is indicated by the

---

[1] Numbers in brackets designate References at end of paper.
Contributed by the Applied Mechanics Division for publication (without presentation) in the JOURNAL OF APPLIED MECHANICS.
Discussion on this paper should be addressed to the Editorial Department, ASME, United Engineering Center, 345 East 47th Street, New York, N. Y. 10017, and will be accepted until October 20, 1973. Discussion received after this date will be returned. Manuscript received by ASME Applied Mechanics Division, March, 1972. Paper No. 73-APM-C.

difficulty numerically in matching boundary conditions even by increasing the number of terms.

To overcome the difficulty just described, we propose to consider the physical region as subdivided into several regions with separate expansions for each region. The solutions for the partitioned regions can then be "stitched" by applying well-known properties of analytic functions in a least-square collocated sense. The procedures will be illustrated by the solution of several crack problems. The application of the arguments to problems of stress concentration will be evident.

Preliminary Discussion of Partitioning Plan

First, let us summarize briefly the notation of the Muskhelishvili [5] formulation. In terms of the two analytic functions $\phi(Z)$ and $\psi(Z)$ of the complex variable $Z = x + iy$, the stresses and displacements in rectangular coordinates can be written as

$$\sigma_y + \sigma_x = 2[\phi'(Z) + \overline{\phi'(Z)}]$$
$$\sigma_y - \sigma_x + 2i\tau_{xy} = 2[\bar{Z}\phi''(Z) + \psi'(Z)] \tag{1}$$

and

$$2\mu(u + iv) = \eta\phi(Z) - Z\overline{\phi'(Z)} - \overline{\psi(Z)} \tag{2}$$

where primes denote differentiation, bars denote complex conjugates, $\mu = E/2(1 + \nu)$ and $\eta = 3 - 4\nu$ (plane strain) and $\eta = (3 - \nu)/(1 + \nu)$ (generalized plane stress). E and ν are Young's modulus and Poisson's ratio. If $X_n ds$ and $Y_n ds$ denote the horizontal and vertical components, respectively, of the force acting on an element of arc, ds, then

$$\phi(Z) + Z\overline{\phi'(Z)} + \overline{\psi(Z)} = i\int^s (X_n + iY_n)ds = f_1(s) + if_2(s) \tag{3}$$

Conformal mapping procedures involve the introduction of a complex parameter plane, the ζ-plane, and the conformal transformation

$$Z = \omega(\zeta) \tag{4}$$

The stresses, displacements, etc., can be written in terms of analytic functions of ζ in the familiar manner

$$\sigma_y + \sigma_x = 4\,\text{Re}\,\{\phi'(\zeta)/\omega'(\zeta)\} \tag{5}$$

$$\sigma_y - \sigma_x + 2i\tau_{xy} = 2\{\overline{\omega(\zeta)}[\phi'(\zeta)/\omega'(\zeta)]' + \psi'(\zeta)\}/\omega'(\zeta)$$

$$2\mu(u + iv) = \eta\phi(\zeta) - \omega(\zeta)\overline{\phi'(\zeta)}/\overline{\omega'(\zeta)} - \overline{\psi(\zeta)} \tag{6}$$

$$\phi(\zeta) + \omega(\zeta)\overline{\phi'(\zeta)}/\overline{\omega'(\zeta)} + \overline{\psi(\zeta)} = f_1(s) + if_2(s) \tag{7}$$

Let us now consider the two-dimensional physical region illustrated in Fig. 1(a) and the problem of determining the two analytic functions $\phi(Z)$ and $\psi(Z)$ defined in this region which satisfy prescribed loading conditions on the boundary. Representation of each of the stress functions by a single series expansion relies on the physical region lying within the circle of convergence of each of the series. It is clear that there are problems for which intrinsic mathematical singularities can exist such that single series expansions are not valid or affect convergence rates detrimentally.

Consider now a partitioning of the region by τ into the two subregions R_1 and R_2. Furthermore, we define

$$\phi(Z) = \phi_1(Z) = \sum_0^\infty a_n(Z - Z_1)^n \quad z \in R_1$$
$$= \phi_2(Z) = \sum_0^\infty b_n(Z - Z_2)^n \quad Z \in R_2 \tag{8}$$

$$\psi(Z) = \psi_1(Z) = \sum_0^\infty c_n(Z - Z_1)^n \quad Z \in R_1$$
$$= \psi_2(Z) = \sum_0^\infty d_n(Z - Z_2)^n \quad Z \in R_2 \tag{9}$$

It is well known that $\phi(Z)$ and $\psi(Z)$ are uniquely determined to within a rigid-body motion by the prescribed loading conditions on the boundary $S_1 + S_2$. Clearly, the boundary conditions on $S_1(S_2)$ alone are not sufficient to determine ϕ_1 and ψ_1 (ϕ_2 and ψ_2). In addition we must require that $\phi_1(Z)$ and $\phi_2(Z)$ are one function $\phi(Z)$. Similarly, $\psi_1(Z)$ and $\psi_2(Z)$ must be elements of the singular function $\psi(Z)$. The conditions under which these elements are analytic continuations of the same analytic function will henceforth be referred to as "stitching" conditions.

If strictly continuous arguments are used in the solution process, a simple requirement for stitching conditions follows directly from a well-known property of analytic functions. If $\phi_1(Z)$ and $\phi_2(Z)$ are analytic in R_1 and R_2, respectively, then a necessary and sufficient condition that $\phi_1(Z)$ and $\phi_2(Z)$ be the elements of a single analytic function $\phi(Z)$ (and similarly for $\psi(Z)$) is

$$\phi_1(Z) = \phi_2(Z), \quad Z \in \tau$$
$$\psi_1(Z) = \psi_2(Z), \quad Z \in \tau \tag{10}$$

Actually (13) can be replaced by continuity across any finite subinterval of τ or, even more generally, continuity across any finite arc common to the regions of convergence of the series in (8) and (9).

In principle, a partitioning plan of solution is evident. The region is partitioned by the analyst in a manner consistent with bypassing anticipated convergence difficulties and local expansions are set up. The set of conditions corresponding to the stitching conditions and the boundary conditions corresponding to each partitioned region will determine the elements of $\phi(Z)$ and $\psi(Z)$. Clearly, the previous argument extends to any finite number of partitioned elements. Furthermore, any of the elements can be individually handled by the mapping techniques in the MMC method.

Although the concept of partitioning the physical region geometrically serves as a convenient guide to the approach, it is not strictly necessary. Alternatively, the procedure can be considered as a set of expansions with overlapping circles of convergence familiar to the classical demonstration of analytic continuation. The latter viewpoint is particularly useful in the choice of expansions for regions for which mathematical singularities in the solution can be anticipated.

Stitching Conditions and Collocation

In practice, we shall be concerned with a pointwise satisfaction of the stitching conditions. Since the previous arguments have relied on continuity on a continuous interval, it is reasonable to reconsider, at least heuristically, the implications of (10) when applied in a discrete sense. It can be seen immediately that the situation is somewhat unsatisfactory. Since the derivatives of the stress functions appear in (1)–(3), continuity of $\phi(Z)$ and $\psi(Z)$ across discrete points of τ does not assure continuous stresses or even continuous displacements at these points.

Let us require as a minimum the continuity of $u + iv$ and $f_1 + if_2$ at the discrete points on τ. The physical implications of continuity of the displacements are obvious. The implications of the continuity of the force conditions are illustrated in Fig. 1(b). Consider the interval AB on τ between two successive collocation stations. Let C_1 and C_2 be curves lying in R_1 and R_2, respectively. If the force conditions across τ are continuous at A and B, then the total force resultant over the closed contour C_1 and C_2 is zero, as it should be for equilibrium.

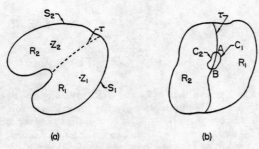

Fig. 1 Example of partitioning

From (2) and (3), continuity of $u + iv$ and $f_1 + if_2$ across τ is equivalent to the corresponding continuity of $\phi(Z)$ and $\bar{Z}\phi'(Z) + \psi(Z)$. It is suggested therefore, as a minimum, for discrete applications, that the stitching condition (10) be replaced by

$$\phi_1(Z_k) = \phi_2(Z_k), \quad Z_k \in \tau$$
$$Z_k\phi_1'(Z_k) + \psi_1(Z_k) = \bar{Z}_k\phi_2'(Z_k) + \psi_2(Z_k), \quad Z_k \in \tau \quad (11)$$

It is obvious that (10) and (11) are equivalent when Z_k represents a continuous description of τ.

Throughout the remainder of this paper we shall adopt the stitching condition (11) unless stated otherwise. This condition does not assure stress continuity across τ. Although this matter can be corrected by additional conditions on the derivatives of the stress functions, the disadvantages of the corresponding computational burden must be weighed. If the stitching locus is chosen in such a manner that it is somewhat remote to the region of interest, it is reasonable to assume that to a certain extent local inaccuracies in stress continuity can be tolerated.

Central Crack in a Rectangular Panel

In this section, we consider the problem of a central crack in a rectangular panel, Fig. 2. This problem has been solved previously by several methods including the MMC method [2]. In a single power series expansion of the solution, convergence difficulties increase for $h/W > 2.5$ particularly for the deeper cracks. In fact in most previous methods accuracy breaks down at about $2L/W = 0.9$ (if not before). Recently, Tada [6] found a solution for the deeper cracks by a careful numerical analysis of a dual integral equation formulation.

The partitioning corresponding to EF and GH with the corresponding regions I, II, III is shown in Fig. 2. The height of region II, namely, $2a$, will be considered as a parameter in the numerical solution to study the effects of partition geometry. It is evident that if the solution symmetry with respect to the x-axis of region II is built into the corresponding stress-function representation, it will be sufficient to consider only regions I and II explicitly.

In Region II: The basic MMC plan can be used to describe the solution in $EFHG$. Thus the mapping function

$$Z = \omega(\zeta) = (L/2)(\zeta + \zeta^{-1}) \quad (12)$$

is introduced which carries the unit circle $|\zeta| = 1$ and its exterior into the crack and its exterior, respectively. The outer boundary of the parameter region corresponding to the points on the rectangle $EFHG$ can be found from

$$\zeta = (Z/L) + [(Z/L)^2 - 1]^{1/2} \quad (13)$$

As in [2], we assume

$$\psi(\zeta) = -\bar{\phi}(1/\zeta) - \bar{\omega}(1/\zeta)\phi'(\zeta)/\omega'(\zeta) \quad (14)$$

where the bar notation is defined by

$$\bar{f}(1/\zeta) = \overline{f(1/\bar{\zeta})} \quad (15)$$

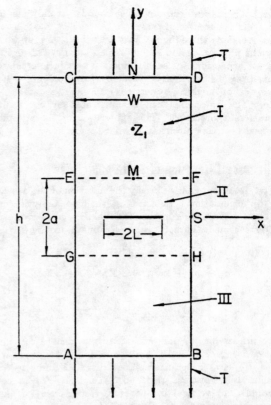

Fig. 2 Central crack in a rectangular plate

With the extended stress function $\phi(\zeta)$ assumed as a Laurent series with real coefficients,

$$\phi(\zeta) = T \sum_{-\infty}^{\infty} \alpha_n \zeta^{2n+1} \quad (16)$$

load-free conditions on the crack and the symmetries of the solution in region II are assured. Thus only the two intervals SF and FM in region II require further consideration. On SF, the resultant force condition (7) becomes

$$\phi(\zeta) - \phi(1/\bar{\zeta}) + [\omega(\zeta) - \omega(1/\bar{\zeta})]\overline{\phi'(\zeta)/\omega'(\zeta)}$$
$$= f_1 + if_2 = Tw/2 \quad (17)$$

On FM, we impose the stitching condition (11) in conjunction with the solution representation in region I. In region I: For the region $ECDF$ we assume a series solution expanded about the center of the region $Z_1 = (h + 2a)i/4$. Thus, taking into account the stress symmetries with respect to the y-axis,

$$\phi(Z) = T\left\{\sum_0^\infty a_{2n+1}(Z - Z_1)^{2n+1} + i\sum_0^\infty a_{2n}(Z - Z_1)^{2n}\right\}$$
$$\psi(Z) = T\left\{\sum_0^\infty b_{2n+1}(Z - Z_1)^{2n+1} + i\sum_0^\infty b_{2n}(Z - Z_1)^{2n}\right\} \quad (18)$$

where the coefficients in (18) can be considered as real.

Since symmetry is included in (18), the boundary conditions need explicit consideration only on FD and DN and the stitching conditions only on FM. The force boundary conditions become

$$\phi(Z) + Z\overline{\phi'(Z)} + \overline{\psi(Z)} = TW/2 \text{ along } FD$$
$$= T(W - X) \text{ along } DN \quad (19)$$

The representation in region II is used for the calculation of the stress-intensity factor K_1 at the crack tip, $Z = L$ (or $\zeta = 1$). If

Table 1 H as a function of $2L/W$

$2L/W$	H	H^*
0.80	1.81	1.85
0.90	2.60	2.61
0.92	2.90	2.92
0.94	3.35	3.37
0.96	4.10	4.13
0.98	5.74	5.84

K_1 is defined in the conventional manner, it is well known that

$$K_1 = 2\sqrt{\pi}\phi'(1)/[\omega''(1)]^{1/2} \quad (20)$$

It is actually more meaningful to consider the dimensionless function H where

$$H = K_1/T\sqrt{\pi L} \quad (21)$$

Numerical studies were carried out for the case $h/W = 3.0$. Stations were chosen along the appropriate boundary and stitching intervals and the corresponding conditions were imposed on truncated forms of the previous series representations. The ratio of the number of conditions to the number of coefficients was maintained at approximately two-to-one and the conditions were satisfied in a least-square sense. In order to study the simplest model of the partitioning approach, the stitching conditions were taken as (11) and only $f_1 + if_2$ was considered in the boundary conditions.

The partitioning defined by $2a/W = 1$ was first considered. For crack length ratios $2L/W < 0.8$ accurate values of the stress-intensity factors were found using 20 terms in the expansion for region II, i.e., equation (16), and 10 terms for each expansion in (18). For $2L/W \geq 0.8$, it was evident from the force and stitching residual errors that convergence in this range becomes difficult with this choice of partitioning. This difficulty was anticipated on the basis of previous experience with the MMC for a central crack with $h/W = 1.0$ (which is of course region II with this choice of partitioning).

When the partitioning was altered by the choice $2a/W = 0.2$, the matchup of force and stitching conditions was dramatically improved for even the very deep crack lengths. With 40 terms of (16) and 17 each for (18), the results for several deep crack lengths are listed in Table 1.

The validity of the results can be examined by a comparison with H^*, where

$$H^* = 0.826(1 - 2L/W)^{-1/2} \quad (22)$$

Koiter [7] has shown that H^* is the exact limit of H as $2L/W \to 1$ for large h/W. The accuracy of solution for the range of crack depths in Table 1 is remarkable particularly in view of the relatively crude stitching and boundary conditions which were used. The apparent trend of inaccuracy for $2L/W \geq 0.98$ can easily be rectified by adding higher ordered stitching and boundary conditions as argued in the basic MMC plan.

Central Crack in a Composite Panel

We now consider the somewhat more complicated problem of a central crack in a rectangular composite panel as illustrated in Fig. 3. The elastic properties, characterized by Poisson's ratio and Young's modulus, will be denoted by (ν_1, E_1) in the central panel, R_0, and by (ν_2, E_2) in the adjacent panels R_1 and R_2.

The loading conditions are taken as follows: The crack and the sides AA' and DD' will be considered as traction-free. The ends will be considered as free from shear and constrained to a uniform vertical displacement, v_0; thus

$$\begin{aligned} v = v_0, & \quad \tau_{xy} = 0 \text{ on } AD \\ v = -v_0, & \quad \tau_{xy} = 0 \text{ on } A'D' \end{aligned} \quad (23)$$

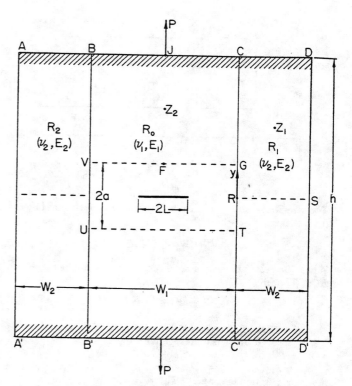

Fig. 3 Central crack in a rectangular composite panel

Across the intersections BB' and CC', we impose continuity of displacements, $u + iv$, and continuity of forces. The latter condition is equivalent to demanding continuity of $f_1 + if_2$ across the intersections. Although the conditions across BB' and CC' coincide with (11) they must not be considered as stitching conditions in the sense of this paper. When the material constants in adjacent panels do not coincide, the stress functions do not continue across the common boundary.

The partitioning plan chosen for this problem is indicated in Fig. 3. Although there are 7 regions indicated, this number can be reduced to 3 by utilizing obvious stress symmetries. The analysis in region $UVGT$ can again be handled by mapping. If symmetry is imposed in the representation, it is sufficient in this region to consider stitching along FG and continuity of $u + iv$ and $f_1 + if_2$ along GR. The solution in the region $VBCG$ can be represented in the manner indicated in (18). Utilizing symmetry in this region, we need only consider stitching across FG, continuity of $u + iv$ and $f_1 + if_2$ across GC and the boundary conditions (23) on JC. It was found desirable, particularly for the deeper cracks, to partition $CDD'C'$ as shown. Expansions for the stress functions were made about Z_1 and symmetry across the real axis was accounted for by imposing the conditions $\tau_{xy} = 0$ and $v = 0$ on RS. The conditions on the remaining sides of $RCDS$ are obvious.

Stitching conditions (11) were adopted and again $2a/h = 0.2$ proved to be an effective choice for the partition geometry particularly for the deeper cracks. There appears to be a natural stress singularity at such points as D due to the junction of the load-free and mixed boundary conditions assumed. This difficulty can be bypassed with a negligible effect on accuracy by considering stations near D but not on D.

In the numerical calculations, the elastic constants were chosen as $\nu_1 = \nu_2 = 0.3$ and three ratios $E_1/E_2 = 3, 1, 1/3$ were considered. It was considered preferable to present the results in terms of an averaged applied stress rather than in terms of the displacement parameter v_0. Thus the averaged applied stress σ_A acting on AD is

Table 2 Values of H for $W_2/W_1 = 0.5$, $h/W_1 = 2.0$

E_1/E_2 \ $2L/W_1$	0.0	0.1	0.2	0.3	0.4	0.5	0.6	0.7	0.8
1/3	1.00	1.00	0.99	0.99	0.98	0.97	0.95	0.93	0.89
1	1.00	1.00	1.00	1.01	1.02	1.03	1.05	1.07	1.10
3	1.00	1.00	1.01	1.03	1.06	1.10	1.16	1.24	1.37

$$\sigma_A = P/(W_1 + 2W_2) = \left\{\int_A^D \sigma_y dx\right\}/(W_1 + 2W_2) \quad (24)$$

The stress-intensity factor can be conveniently normalized by the introduction of σ_M where

$$\sigma_M = P/[W_1 + 2(E_2/E_1)W_2] \quad (25)$$

When the center panel has no crack, σ_M is the applied stress on BC. Thus, if

$$H = K_1/\sigma_M\sqrt{\pi L} \quad (26)$$

then $H \to 1$ as $2L/W_1 \to 0$ independent of the E_2/E_1 ratio.

Values of H for a square configuration with $W_2/W_1 = 0.5$, $h/W_1 = 2.0$ are presented in Table 2.

When $E_1/E_2 = 1$ the problem reduces to a central crack in a homogeneous square panel. The corresponding values of H in Table 2 were checked to three significant figures with a previous solution carried out by the MMC method without partitioning.

The distribution of shear along RC is illustrated in Fig. 4. For a configuration with $W_2/W_1 = 0.5$, $2L/W_1 = 0.6$, and $h/W_1 = 2.0$, τ_{xy}/σ_A is plotted versus y/h where y is defined in Fig. 3.

Although accurate results for the range of parameters in Table 2 and Fig. 4 were obtained by the simpler forms of stitching and boundary conditions, it is obvious that as $2L/W_1 \to 1$, refinement of these conditions would be warranted particularly along RS and FG.

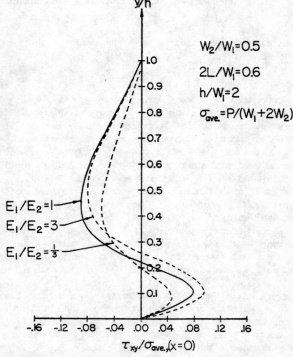

Fig. 4 Shear distribution along RC

Edge Crack in a Rectangular Panel

As a final illustration of the partitioning method, we consider the problem of an edge crack in a rectangular panel with restrained ends, Fig. 5. The crack and the sides BC and AD are considered as traction-free. The end conditions are specified in terms of the displacements, in particular,

$$\begin{array}{lll} v = 0, & u = u_0 \text{ (constant)} & \text{on } CD \\ v = 0, & u = -u_0 & \text{on } AB \end{array} \quad (27)$$

This problem was chosen because of the sensitivity of the solution with respect to W/h for the loading condition (27). The solution of the problem for constant tension on the ends is well known [8]. For this latter case, the rectangle can be considered as an infinite strip for $W/h \geq 3.0$. This is not true for the loading conditions (27). Although the solution for very large W/h-values approaches the tension solution for an infinite strip in the limit, the convergence is very slow as will be shown numerically. We find that the partitioning approach can be used effectively to study the solution of the present problem for large W/h ratios.

For brevity, the details of the formulation will be omitted. The partitioning plan is illustrated in Fig. 5. By utilizing the symmetry of the solution with respect to the y-axis, only regions I, II, III were explicitly considered. The solution in region I, $MNSR$, was formulated in the MMC manner using mapping. In regions II and III, Taylor expansions of $\phi(Z)$ and $\psi(Z)$ were taken about the centers of each region. The boundary and stitching conditions for each region are obvious and again can be handled by the least-square collocation plan of the MMC method.

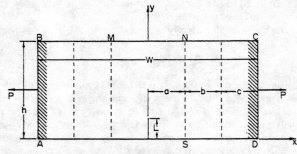

Fig. 5 Edge crack in a rectangular panel

In the numerical calculations, Poisson's ratio was chosen as $\nu = 1/4$. Again, the results are presented in terms of an averaged applied stress rather than the displacement parameter u_0. The averaged applied stress, σ_A, in this case is

$$\sigma_A = (1/h)\int_D^C \sigma_x dy \quad (28)$$

For the range of crack depths considered, it was sufficient to choose $a/h = 1.0$ and $b = c$. For deeper cracks of course a smaller value of a/h would be more effective as we have indicated in the previous examples. Calculations for much larger values of W/h (not included in the present data) indicate that b/h and c/h ratios should not exceed 5.0.

Table 3 H for an edge crack in a rectangular panel

$$H = K_1/\sigma_A\sqrt{\pi L}$$

L/h \ W/h	1	2	3	4	8	12	16	∞[a]
0.05	1.02	1.11	1.13	1.13	1.13	1.13	1.13	1.13
0.10	1.00	1.12	1.15	1.16	1.17	1.18	1.18	1.19
0.20	0.95	1.14	1.21	1.25	1.30	1.32	1.34	1.37
0.30	0.91	1.17	1.30	1.37	1.50	1.55	1.58	1.66
0.40	0.92	1.22	1.40	1.52	1.76	1.86	1.92	2.12
0.50	0.97	1.30	1.53	1.70	2.09	2.28	2.39	2.82
0.60	1.08	1.42	1.69	1.91	2.49	2.83	3.05	...
0.70	1.27	1.61	1.91	2.17	2.97	3.51	3.90	...

[a] Limiting case of uniform tension, e.g., reference [8].

The results for H, where

$$H = K_1/\sigma_A\sqrt{\pi L} \tag{29}$$

are presented in Table 3.

Observations

It has been demonstrated by the several numerical solutions in this paper that the partitioning concepts originally proposed can be effectively incorporated into the MMC plan of solution. Configurations which previously were awkward to analyze by an overall power series expansion can now be handled by this plan. The stitching conditions based on continuation arguments of analytic function theory have been shown to be remarkably effective even when applied in a discrete sense for this class of problems. The latter feature is of course essential to the computational feasibility of the plan.

Although only the simple stitching conditions (11) have been used in the numerical solutions in this paper, it is evident that stronger conditions can be imposed by requiring continuity of first, second, etc., derivatives across the partitioning intervals. The requirement of such refinements of the stitching conditions depends on the accuracy of the information desired and the proximity of the stitching interval to the region of interest.

A direction of future investigation with many interesting possibilities is provided by (10) when τ is chosen as simply a finite arc common to the regions of convergence of the series expansions. The problem of the central crack in a rectangular panel was reexamined with the stitching interval EF, Fig. 2, replaced by a subinterval of length $W/10$. With 10 stations and the requirements of continuity of $\phi(\zeta)$ and $\psi(\zeta)$ and their first derivatives across this interval, the solution for $2L/W = 0.6$, $h/W = 3.0$ was found accurately to three significant figures. This result confirms a previous remark we have made earlier, namely, that the concept of partitioning of the geometry is convenient, yet, not necessary. The possibility of using a set of expansions with overlapping circles of convergence with the choice of common stitching intervals as pieces of boundary arc is clearly a feasible solution plan.

The partitioning concepts just described apply also to a much wider class of problems than the plane problems of elasticity. For example, the problems of torsion and plate bending can be formulated in terms of analytic functions. The MMC plan with partitioning can obviously be modified to handle such problems.

References

1 Bowie, O. L., and Neal, D. M., "A Modified Mapping-Collocation Technique for Accurate Calculation of Stress-Intensity Factors," *International Journal of Fracture Mechanics*, Vol. 6, 1970, pp. 199–206.

2 Bowie, O. L., and Neal, D. M., "A Note on the Central Crack in a Uniformly Stressed Strip," *Engineering Fracture Mechanics*, Vol. 2, 1970, pp. 181–182.

3 Bowie, O. L., and Freese, C. E., "Central Crack in Rectangular Sheet With Rectilinear Anisotropy," Army Symposium on Solid Mechanics, AMMRC, Watertown, Mass., 1970; to appear in *International Journal of Fracture Mechanics*.

4 Bowie, O. L., and Freese, C. E., "Elastic Analysis for a Radial Crack in a Circular Ring," 4th National Symposium on Fracture Mechanics, Carnegie-Mellon, Pittsburgh, Pa., 1970; to appear in *Engineering Fracture Mechanics*.

5 Muskhelishvili, N. I., *Some Basic Problems of Mathematical Theory of Elasticity*, Nordhoff, Gröningen, Holland, 1953.

6 Tada, H., "A Note on the Finite Width Corrections to the Stress-Intensity Factor," *Engineering Fracture Mechanics*, Vol. 3, 1971, pp. 345–347.

7 Koiter, W. T., "Note on the Stress-Intensity Factors for Sheet Strips With Cracks Under Tensile Loads," Delft Technical University, Department of Mechanical Engineering Report No. 314, 1965.

8 Gross, B., Srawley, J. E., and Brown, W. F., "Stress-Intensity Factors for a Single-Edge Notch Tension Specimen by Boundary Collocation of a Stress Function," NASA TND-2395, Lewis Research Center, Cleveland, Ohio, 1964.

9 Bowie, O. L., "Analysis of Edge Notches in a Semi-Infinite Region," *Journal of Mathematics and Physics*, Vol. 45, 1966, pp. 356–366.

ELASTIC FIELD EQUATIONS FOR BLUNT CRACKS WITH REFERENCE TO STRESS CORROSION CRACKING

Matthew Creager[*] and Paul C. Paris[**]

ABSTRACT

The elastic stress field equations for blunt cracks are derived and presented in a form equivalent to the usual sharp crack tip stress fields. These stress field equations are employed in analyzing a dissolution model for the arrest of stress corrosion cracking by crack tip blunting, which is often observed with the arrest of stress corrosion cracks.

Stress corrosion, the growth of cracks due to the combined and inter-related action of stress and environment, is a highly complex phenomena in all its aspects. In general, it involves the diffusion of an environment into a crack which, in some way, attacks the highly stressed material in the vicinity of the crack tip causing the crack to grow. There have been recent experimental observations in which the arrest of a stress corrosion crack was accompanied by the apparent blunting of the crack tip. See Figures 1 and 2. These observations suggest that an investigation of the

Figure 1: 12% Ni - 5% C_r - 3% Mo - Maraging steel in 3-1/2% NaCl solution at K_i equal to 9500 psi $\sqrt{in}$. exposure time in excess of 300 hours. Courtesy of Floyd Brown, U.S. Naval Research Laboratory.

stability of the shape of the crack tip surface may be of great interest.

In order to attempt to relate the conditions of attack to the stress conditions for blunting, a very simplified mechanical view of the process si-

---

[*] Research Assistant, Department of Mechanics, Lehigh University, Bethlehem, Pennsylvania.
[**] Professor of Mechanics, Department of Mechanics, Lehigh University, Bethlehem, Pennsylvania.

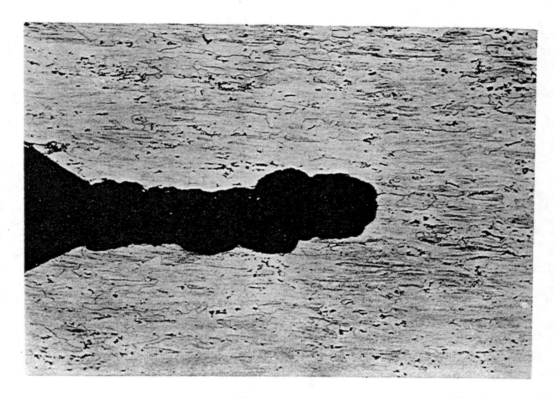

Figure 2: 2024-T351 aluminum alloy in 3-1/2% NaCl solution at K_i equal to 5360 psi $\sqrt{in}$. exposure time in excess of 1000 hours. Courtesy of J. Mulherin, U.S. Army, Frankford Arsenal.

milar to that of Charles and Hillig[1] will be adopted here. The usual continuum model of a crack is a planar void of material, and the corresponding mathematical model is a plane of discontinuity. However, since the chemical attack of a material by the environment at the crack tip is being considered, it is relevant to have as a physical model of the crack, a void that is not the usual plane ending with zero radius of curvature, but a narrow volume with a finite curvature at the tip. This type of blunt crack or notch is conveniently represented mathematically by an elliptical or hyperbolic cylinder, void of material, in which the radius of curvature at the tip is small in comparison to the major dimensions of the void.

It is then of interest to explore the nature of the stress distribution about this blunt crack or notch. As is usual in Fracture Mechanics, an elastic analysis will be attempted which may later be discussed in the light of the effects of nonlinear material behavior near the crack tip. Consequently, the elastic stress distribution in the neighborhood of elliptical holes and hyperbolic notches will be presented. Since regions near the crack tip are of special interest, it is advantageous to expand the expressions for the stresses as a power series in terms of radial distance from some point near the tip and to discard all the second order terms. This is most appropriately done by expanding the expressions for the stresses using the origin chosen in Figure 3, since certain simplifications result. Note that the origin is a distance of $\rho/2$ away from the crack tip, where ρ is the radius of curvature at the crack tip. When ρ/a (a is half the crack length) is small compared to one, the origin is to a very close approximation the focal point of the ellipse or hyperbola that represents the surface of the crack. The results of this expansion for both the elliptic hole and the hy-

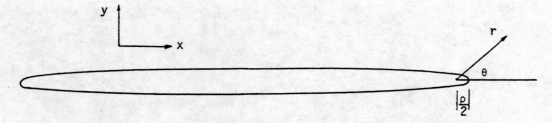

Figure 3: Coordinate system for stress field.

perbolic notch are identical and as follows [2]:

MODE I - OPENING MODE (PLANE-EXTENSION SYMMETRICAL MODE)

$$\sigma_x = \frac{K_I}{(2\pi r)^{1/2}} \cos\frac{\theta}{2}\left[1 - \sin\frac{\theta}{2}\sin\frac{3}{2}\theta\right] - \frac{K_I}{(2\pi r)^{1/2}} \frac{\rho}{2r} \cos\frac{3}{2}\theta$$

$$\sigma_y = \frac{K_I}{(2\pi r)^{1/2}} \cos\frac{\theta}{2}\left[1 + \sin\frac{\theta}{2}\sin\frac{3}{2}\theta\right] + \frac{K_I}{(2\pi r)^{1/2}} \frac{\rho}{2r} \cos\frac{3}{2}\theta \quad (1)$$

$$\tau_{xy} = \frac{K_I}{(2\pi r)^{1/2}} \sin\frac{\theta}{2}\cos\frac{\theta}{2}\cos\frac{3}{2}\theta - \frac{K_I}{(2\pi r)^{1/2}} \frac{\rho}{2r} \sin\frac{3}{2}\theta$$

MODE II - EDGE SLIDING MODE (PLANE-EXTENSION SKEW SYMMETRICAL MODE)

$$\sigma_x = -\frac{K_{II}}{(2\pi r)^{1/2}} \sin\frac{\theta}{2}\left[2 + \cos\frac{\theta}{2}\cos\frac{3}{2}\theta\right] + \frac{K_{II}}{(2\pi r)^{1/2}} \frac{\rho}{2r} \sin\frac{3}{2}\theta$$

$$\sigma_y = \frac{K_{II}}{(2\pi r)^{1/2}} \sin\frac{\theta}{2}\cos\frac{\theta}{2}\cos\frac{3}{2}\theta - \frac{K_{II}}{(2\pi r)^{1/2}} \frac{\rho}{2r} \sin\frac{3}{2}\theta \quad (2)$$

$$\tau_{xy} = \frac{K_{II}}{(2\pi r)^{1/2}} \cos\frac{\theta}{2}\left[1 - \sin\frac{\theta}{2}\sin\frac{3\theta}{2}\right] - \frac{K_{II}}{(2\pi r)^{1/2}} \frac{\rho}{2r} \cos\frac{3}{2}\theta$$

MODE III - TEARING MODE (ANTI-PLANE MODE)

$$\tau_{xz} = -\frac{K_{III}}{(2\pi r)^{1/2}} \sin\frac{\theta}{2}$$

$$\tau_{yz} = \frac{K_{III}}{(2\pi r)^{1/2}} \cos\frac{\theta}{2} \quad (3)$$

These field equations are similar to those for the "mathematically sharp" plane crack. In fact, the relationship describing the third mode stress state is exactly the same as that for the third mode stress state of a plane crack. Moreover, the first and second mode stress state relations differ from their plane crack counterparts by a single additional term dependent upon the radius of curvature at the tip. In these blunt crack stress field equations, the additional term in each may be neglected when ρ/r is negligible compared to one. The remaining terms are the usual sharp crack tip stress field equations which are consequently seen to be applicable to the region $\rho \ll r \ll a$. It is therefore implied that the tips of blunt cracks are imbedded within the usual crack tip stress field and that these usual field equations are disturbed only for mode I and mode II stress states and only in the immediate vicinity of the crack tip. In addition, it is interesting to note that the hydrostatic stress distribution, $\sigma_x + \sigma_y$, is the same for a crack with a finite curvature at the tip and a sharp crack.

For the stress corrosion mechanism under discussion, the crack shape will be considered to be an ellipse and the stress conditions will initially be taken to be entirely elastic. Furthermore, it will be envisioned that the reaction between the solid material and the environment is simply one in which the solid is made to dissolve. Dissolution in this context can be generally viewed as any local degradation in the material's ability to transmit stress. As a consequence, the velocity of dissolution will be considered to be a function of the stress at the material-environment interface.

It is of interest to examine the conditions under which the crack contour could be considered stable. That is, when the shape of the crack at the tip is maintained rather than blunting or sharpening, causing a tendency to arrest or accelerate.

The case where the loading creates a mode I type deformation is of greatest practical interest and therefore will be considered. Referring to Figure 4, $V(\sigma)$ is the dissolution velocity normal to the surface. Note that

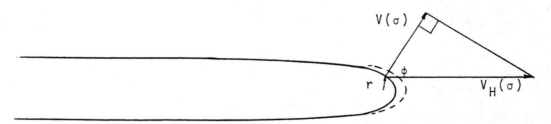

Figure 4: Crack extension due to stress corrosion.

for a constant shape of the whole contour near the tip, it is required that the apparent horizontal velocity, V_H, be constant for all points on the contour. This leads to

$$V_H = \frac{V(\sigma)}{\cos \phi} = \text{const} = C_1 \qquad (4)$$

From simple geometry, it follows that

$$V_H = \frac{V(\sigma)}{\cos \frac{\theta}{2}} = C_1 \qquad (5)$$

The normal stress at the surface is zero so that the stress sum, $\sigma_x + \sigma_y$, equation (1), gives the tangential stress σ_t:

$$\sigma_t = \frac{2K_I}{(2\pi r)^{1/2}} \cos \frac{\theta}{2} \qquad (6)$$

Therefore, for the condition of constant shape of the contour, combining equations (5) and (6) gives

$$V(\sigma_t) = C_1 \frac{(\pi \rho)^{1/4}}{K_1^{1/2}} \sigma_t^{1/2} \tag{7}$$

Consequently, if $V(\sigma)$ is more highly dependent on stress than $\sigma^{1/2}$, the crack will tend to sharpen. Alternately, a consideration of the sign of the derivative of $V(\sigma)/\sigma^{1/2}$ will indicate the tendency for accelerated crack growth or crack arrest. Clearly the stress analysis for this model has not been very exacting physically in that effects of plasticity have been ignored. However, since plasticity effects will serve to reduce the gradient of stress along the crack surface, the above analysis represents a limiting case and indicates the minimum stress dependency of the functional relationship for dissolution velocity with contour stability.

As indicated previously by Figures 1 and 2, there have been a number of reported cases where blunting, as might be expected from the above model, has been observed in specimens where crack growth due to stress corrosion cracking has been arrested. At this time, whether this blunting has occurred before or after arrest has not been properly ascertained, but the relationship of this phenomenon and the above model surely warrants further investigation.

It should be pointed out that in addition to their use in the analysis of this particular stress corrosion mode, the stress field equations given by equations (1), (2) and (3) may be used in a number of other applications. For example, they provide a convenient method for determining stress intensity factors using photoelastic methods where the difficulty of working with a plane crack in a brittle photoelastic material can be easily avoided by using a blunt crack.

CONCLUSIONS

1. The elastic stress field equations for blunt cracks are presented in a form equivalent to the usual sharp crack tip stress fields for each of the three modes.

2. These elastic stress field equations are incorporated in showing that a stress corrosion cracking model, based upon material dissolution (or degradation), suggests that the dissolution velocity should be more strongly dependent upon stress than a one-half power law for accelerated stress corrosion cracking to occur.

3. The previous statement is equally correct if there is a plastic zone in the vicinity of the crack tip and therefore may represent a "minimum criteria" for stress corrosion cracking with material dissolution.

4. The stress field equations presented have other applications as well.

Received on June 1, 1967.

REFERENCES

1. R.J. Charles and W.B. Hillig — The Kinetics of Glass Failure by Stress Corrosion, Symposium on the Mechanical Strength of Glass and Ways of Improving It, Union Scientifique Continentale du Verre, 1961.

2. M. Creager — The Elastic Stress Field Near the Tip of a Blunt Crack, Master's Thesis, Lehigh University, 1966.

RÉSUMÉ - Les équations de champ de tension élastique pour des fissures épointées sont dérivées et présentées de la même façon que celles qui décrivent le champ de tension des pointes de fissures aiguës. Ces équations de champ de tension sont employées dans l'analyse d'un modèle de dissolution décrivant l'arrêt du craquement corrosif sous tension obtenu par l'émoussement des pointes de fissures, qu'on observe souvent dans l'arrêt du craquement corrosif sous tension.

ZUSAMMENFASSUNG - Es wird die Spannungsverteilung vor einem Riss mit einem Kruemmungsradius groesser Null fuer den elastischen Fall abgeleitet und in einer den ueblichen Gleichungen fuer den scharfen Riss aehnlichen Form dargestellt. Diese Spannungsgleichungen werden angewandt, um ein Modell fuer das Anhalten eines Bruchs durch die Abstumpfung der Riss-spitze zu untersuchen. Bei der Spannungskorrosion ist dieses Anhalten eines Bruches haeufig von einer Abstumpfung begleitet.

THE STRESS INTENSITY FACTORS FOR CRACKS IN STRESS GRADIENTS

G. C. Chell
Materials Division, Central Electricity Research Laboratories
Kelvin Avenue, Leatherhead, Surrey, UK
tel: Leatherhead 74488

Cracks occurring in components in service are generally subject to complex loading systems resulting in a stress gradient across the plane containing the crack. To aid the estimation of stress intensity factors for such cracks the compliance functions for centre and edge cracked plates subject to constant, linear, and quadratic stress variations have been determined. A numerical method developed by Hayes [1] from the theoretical work of Bueckner [2] was used. This method has the advantage that it involves only the stresses in the plane of the crack and hence the external loading necessary to induce these stresses do not have to be defined.

The change in strain energy U of a body containing a crack can be written as (Bueckner [2])

$$U = 1/2 \int_{C_1+C_2} T_i \phi_i \, ds$$

Where $-T_i$ represents the tractive components of stress σ_{ij} (due to external loads, body forces, and prescribed deflections) that act on the surfaces defining the crack when it is absent from the body; ϕ_i is the resulting displacement of the surfaces of the crack due *only* to tractions $-T_i$ acting over them and with no external loads or body forces and the prescribed deflections set to zero. The integral is over the upper and lower surfaces C_1 and C_2 of the crack. The strain energy release rate G is given by

$$G = \partial U/\partial a = K^2/E$$

where K is the stress intensity factor and E Young's modulus. Hence

$$K = \sqrt{\left\{ -\frac{E}{2} \frac{\partial}{\partial a} \int_{C_1+C_2} T_i \phi_i \, ds \right\}} \tag{1}$$

Considering centre and edge cracked plates with stresses acting normal to the plane of the crack, equation (1) becomes

$$K^c = \sqrt{\left\{ -E \frac{\partial}{\partial a} \int_{-a}^{a} \sigma(x) \phi(x) \, ds \right\}} \tag{2}$$

and

$$K^e = \sqrt{\left\{ -E \frac{\partial}{\partial a} \int_{0}^{a} \sigma(x) \phi(x) \, ds \right\}} \tag{3}$$

where superscript c denotes centre and e edge, and the crack lies in the regions $-a<x<a$ and $0<x<a$ respectively. The stress intensities defined in (2) and (3) are appropriate for mode I.

Using a finite element stress analysis program K^c and K^e were evaluated for $\sigma(x)$ of the form

$$\sigma_0(x) = \sigma, \text{ a constant} \tag{4a}$$

$$\sigma_1(x) = \sigma(x/w) \tag{4b}$$

$$\sigma_2(x) = \sigma(x/w)^2 \tag{4c}$$

In (4) $2w$ is the width of the plate for a centre crack, the centre of the crack being at origin of coordinates; w is the width of the plate for an edge crack and the origin of coordinates was taken at the crack opening; and σ, a characteristic stress, is defined as the value of the stress at $x = w$.

The stress intensity factors for cracks of lengths $0.2 \leq a/w \leq 0.7$ were calculated. The results are presented in the form of compliance functions $Y_n^{c,e}(a/w)$ where

$$Y_n^{c,e}(a/w) = K_n^{c,e}/\sigma\sqrt{a}$$

Suffix n denotes the compliance function corresponding to the stress $\sigma_n(x)$ defined in 4(a), (b), or (c). The results have been extrapolated below $a/w = 0.2$ so that they agree with the known values of $Y_n^{c,e}(a/w = 0)$. For example, as $a/w \to 0$ then $k \to 1.12\sqrt{(\pi a)}\sigma$ for an edge crack and $Y_o^c \to 1.992$. To aid any stress intensity calculation the computed compliance functions were fitted by a polynomial of order 5 by expressing them in the form

$$Y(a/w) = \sum_{m=0}^{5} A_m (a/w)^m \qquad (5)$$

and determining the coefficients A_m using a least squares fit procedure. The results are displayed in Table 1. Y_o^e, Y_o^e, and $Y_o^e - 2Y_m^e$ when compared with the results of Brown and Srawley [3] for centre and edge cracked plates in tension, and an edge cracked plate subject to a pure bending moment, are within 4% of their values. The compliance function for a centre cracked plate subject to a linear stress is for the high stress end of the crack.

Using the results given above the stress intensity factor for a crack can be obtained from the stress normal to the proposed cracked surface in the uncracked body by expressing the stress in quadratic form as

$$\sigma(x) = A + B(x/w) + C(x/w)^2.$$

Then, due to the superposition of stress, the stress intensity factor is given directly by

$$K^{c,e} = \sqrt{a}\,[AY_o^{c,e}(a/w) + BY_1^{c,e}(a/w) + CY_2^{c,e}(a/w)]$$

since the functions $Y_n^{c,e}$ are known. Although the compliance functions are for cracks in plates it could be argued that, to first order, they can be used for other geometries provided no pronounced geometrical variations occur close to the crack. In determining $\sigma(x)$ the correct body geometry as well as operating loads are used. Thus $\sigma(x)$ already contains information concerning the geometry and it might be expected that a crack in a plate subject to $\sigma(x)$ will have a similar stress intensity factor to the crack in the actual body.

It must be pointed out that the compliance functions calculated above are for cracks in bodies subject to loads and/or body forces. They are not generally suitable for calculations where the stresses are thermally induced or result solely from prescribed deflections but provides an upper bound for these cases.

Acknowledgement: This work was carried out at the Central Electricity Research Laboratories and the paper is published by permission of the Central Electricity Generating Board.

REFERENCES

[1] D. J. Hayes, *International Journal of Fracture Mechanics* 8 (1972) 157-165.

[2] H. F. Bueckner, *Transactions of the American Society of Mechanical Engineers* 80 (1958) 1225

[3] W. F. Brown and J. E. Srawley, *Plane Strain Crack Toughness Testing of High Strength Metallic Materials*, ASTM STP 410, American Society for Testing and Materials, Philadelphia (1966).

26 February 1973

Table 1. Polynomial Coefficients of Compliance Functions

$$Y_n^{c,e}(a/w) = K_n^{c,e}/\sigma\sqrt{a} = \sum_{m=0}^{5} A_m (a/w)^m$$

	Centre Cracked			Edge Cracked		
n	0	1	2	0	1	2
A_0	1.772	0.0	0.0	1.992	0.0	0.0
A_1	0.223	0.837	0.0118	3.081	1.418	0.0312
A_2	-1.597	0.608	0.585	-21.13	-4.120	0.255
A_3	9.802	-4.128	1.228	125.8	25.06	4.179
A_4	-15.08	8.142	-2.217	-235.4	-47.77	-9.811
A_5	10.02	-4.515	1.982	181.2	41.22	12.10

The stress intensity factors and crack profiles for centre and edge cracks in plates subject to arbitrary stresses

G. G. CHELL

Central Electricity Research Laboratories, Leatherhead, Surrey KT22 7SE, U.K.

(Received September 7, 1974; in revised form July 16, 1975)

ABSTRACT

Expressions are given for the stress intensity factors for centre and edge cracks in plates subject to an arbitrary stress. The stress intensity factors for cracks subject to a stress expressed in the form of a polynomial of order 8 are given and their application to cracks in service components are indicated. The results are in good agreement with known solutions. Expressions for the crack profiles of centre and edge cracks subject to arbitrary stress are constructed. They predict profiles in good agreement with the results of a finite element stress analysis.

1. Introduction

Many stress intensity factor solutions are now available in the literature. Tada, Paris and Irwin [1] have documented a large number of crack solutions including both the stress intensity factor and, in some cases, specific displacements due to the crack. In a recent paper [2] the author gave approximate expressions for the stress intensity factors for centre and edge cracks subject to arbitrary stresses. The centre crack expression was approximated by the result for an infinite array of colinear cracks. For large cracks and/or skew-symmetric stresses this approximation introduced appreciable errors. In this paper a more accurate expression for the centre crack is given which agrees well with the known solutions of Isida and Itagaki [3] and Benthem and Koiter [4] for a linear stress. The edge crack solutions were in excellent agreement with the known solution for a linear stress obtained from the pure bend compliance functions given in Brown and Srawley [5]. When more severe stressing cases are considered, good agreement is also obtained with Bueckner's [6] recently given expression for the stress intensity factor for an edge crack subject to an arbitrary stress. The accuracy of these expressions having been demonstrated, in this paper the calculations are extended to enable steeper stress gradients to be handled.

Except for the case of cracks in infinite bodies very little work has been done to obtain solutions for the profiles of cracks although expressions for the relative displacements between the crack surfaces are available [1], [7]. While for elastic solids these solutions are generally of academic interest only, there are cases when they are important. For instance, if a crack propagates into a highly compressive stress analytical solutions may predict a negative displacement between the crack surfaces. This is the case for a centre crack in a plate subject to a pure bending moment. This situation is physically impossible, although, through lack of the appropriate correct solutions, the analytical solutions are accepted. Knowledge of crack profiles can provide a means of correction for this physically unacceptable phenomena. Of more importance is, that the presence of plastic relaxation at the crack tip can cause blunting. In some mathematical models (the Bilby, Cottrell and Swinden [8] model for example) the displacement at the crack tip is an important parameter in predicting the fracture of materials. This displacement is only available analytically if the crack profile is known in detail. With these points in mind expressions for the profile of cracks in bodies subject to an arbitrary stress are constructed from the known solution for cracks in infinite bodies.

2. Analysis

2.1. Stress intensity factors

Approximate expressions for the stress intensity factors for cracks in plates subject to an arbitrary stress have been given by the author [2]. The centre crack expression was assumed to be the same as that for an infinite array of colinear cracks but this can lead to errors [9]. It is, however, possible to construct a solution for a centre crack from the semi-empirical solution given by Chell [10] for a symmetrical, partially loaded crack. This was

$$K^c = \sigma a^{\frac{1}{2}} Y_0^c(a/w)(1-z/w)\frac{2}{\pi}\cos^{-1}\left(\frac{z/a - z/w}{1 - z/w}\right) \tag{1}$$

where K^c is the stress intensity factor and superscript c denotes centre. σ is a uniform stress which loads the crack surface between $|x|=z$ and $|x|=a$, with the centre of the crack as the origin of coordinates (Fig. 1a). The crack is of length $2a$, the plate of width $2w$ and $Y_0^c(a/w)$ is the compliance function corresponding to a uniform stress acting over the whole surface of the crack. Using (1) and performing a Green's function analysis similar to that of [11] the stress intensity factor for a centre crack subject to symmetric arbitrary stress, $\sigma(|x|)$, can be written as

$$K^c = a^{\frac{1}{2}} Y_0^c(a/w) \int_0^a H^c(a, x)\sigma(x)dx \tag{2}$$

where

$$H^c(a, x) = [\partial h^c(a, x)]/\partial x \tag{3}$$

and

$$h^c(a, x) = 1 - (1 - x/w)\frac{2}{\pi}\cos^{-1}\left(\frac{x/a - x/w}{1 - x/w}\right).$$

For an arbitrary stress an approximate solution can be constructed as

$$K^c(\pm a) = \tfrac{1}{2}[a^{\frac{1}{2}} Y_0^c(a/w)] \int_{-a}^a (1 + x/a) H^c(a, |x|)\sigma(\pm x)dx \tag{4}$$

where it is understood that in evaluating $H^c(a, x)$ the derivative of $h^c(a, x)$ is taken and then x replaced by its modulus. $K^c(\pm a)$ corresponds to the stress intensity factor at the crack tips at $x = \pm a$, respectively. In the limit $a/w \to 0$ (4) reduces to the solution for a crack in an infinite body

$$K(\pm a) = (\pi a)^{-\frac{1}{2}} \int_{-a}^a \frac{a+x}{(a^2 - x^2)^{\frac{1}{2}}} \sigma(\pm x)dx. \tag{5}$$

For an edge crack Chell [2] found

$$K^e = a^{\frac{1}{2}} Y_0^e(a/w) \int_0^a H^e(a, x)\sigma(x)dx \tag{6}$$

$$H^e(a, x) = [\partial h^e(a, x)]/\partial x \tag{7}$$

where

$$h^e(a, x) = 1 - (1 - x/w)^2 \frac{2}{\pi}\cos^{-1}\left(\frac{x/a - x/w}{1 - x/w}\right)$$

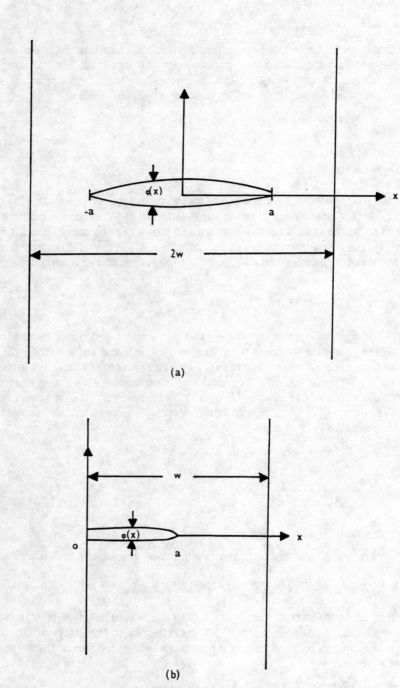

Figure 1. Coordinate system and crack geometry for centre (a) and edge (b) cracked plates.

The coordinate system is shown in Fig. 1b.

The function $G(a, x) = a^{\frac{1}{2}} Y(a/w) H(a, x)$ represents, in a limited sense, the Green's function. It is sometimes referred to as a weight function [6]. For certain geometries solutions for $G(a, x)$ already exist in the literature, for example [11], [6], and [1] has compiled many more. The weight function can be determined using several methods, for example, as the solution to a homogeneous integral equation [6, 12–14] or from finite element computations [7]. In this paper a simple analytical form is used in approximate representation of a one-parameter family of weight functions. The weight or Green's function has the property [6]

$$G(a, a-\Delta)\Delta^{\frac{1}{2}} \to 1 \text{ as } \Delta \to 0$$

where Δ is the distance from the crack tip. Using (3) and (7) the approximate formulae give

$$G^{c,e}(a, a-\Delta)\Delta^{\frac{1}{2}} \to Y^{c,e}(a/w)/[(1-a/w)^n/\pi]^{\frac{1}{2}}, \text{ as } \Delta \to 0$$

where $n=1$ for centre and 3 for edge. This implies that

$$Y^{c,e}(a/w) = [\pi/(1-a/w)^n]^{\frac{1}{2}}. \tag{8}$$

The accuracy of (8) has been discussed previously [10]. For a centre crack it is exact when $a/w=0$, 95% accurate when $a/w=0.1$, 83% accurate for $a/w=0.5$ and 81% accurate when $a/w=0.7$. For an edge crack the equivalent percentage accuracies are 90%, 99%, 100% and 96%.

2.2. Crack profiles

Using the results of [1] and [7] and Eqns. (4) and (6) it can be shown that the relative displacement, $\phi(x)$, between the crack surfaces is given by

$$\phi^c(\pm x) = E^{-1} \int_x^a Y_0^c(x'/w) x'^{\frac{1}{2}} H^c(x', x) \{(1+x/x') K^c(\pm x') + (1-x/x') K^c(\mp x')\} dx'. \tag{9}$$

For an edge crack

$$\phi^e(x) = 2E^{-1} \int_x^a Y_0^e(x'/w) x'^{\frac{1}{2}} H^e(x', x) K^e(x') dx'. \tag{10}$$

Equations (9) and (10) appertain to plane stress, the plane strain equivalent is obtained by replacing E by $E/(1-v^2)$ where v is Poisson's ratio. From (9) and (10) the displacement, $\phi(x=0)$, is

$$\phi^c(0) = E^{-1} \int_0^a Y_0^c(x'/w) x'^{\frac{1}{2}} \left(\frac{1}{w} + \frac{2}{\pi x'}(1-x'/w)\right) (K^c(x') + K^c(-x')) dx' \tag{11}$$

and

$$\phi^e(0) = 2E^{-1} \int_0^a Y_0^e(x'/w) x'^{\frac{1}{2}} \left(\frac{2}{w} + \frac{2}{\pi x'}(1-x'/w)\right) K^e(x') dx' \tag{12}$$

which is the external crack opening displacement.

If $\sigma(x) = \sigma_0$, a uniform stress, then near the crack tip, $x = a - \Delta$, where $\Delta/a \ll 0$, (9) and (10) become, after some algebra,

$$\phi^c(a-\Delta) = \frac{4}{E} \frac{Y_0^c(a/w)}{\pi} [2\Delta(1-a/w)]^{\frac{1}{2}} K^c(a) \tag{13}$$

and

$$\phi^e(a-\Delta) = \frac{4}{E} \frac{Y_0^e(a/w)}{\pi} [2\Delta(1-a/w)^3]^{\frac{1}{2}} K^e(a). \tag{14}$$

Comparing these with the standard results of elasticity theory [15], namely

$$\phi(x \to a) = \frac{8}{E}\left(\frac{\varDelta}{2\pi}\right)^{\frac{1}{2}} K(a) \tag{15}$$

then the approximation in the semi-empirical formula for $G(a, x)$ implied by (8) is again obtained. The accuracy of (8) shows the accuracy to be expected from calculations involving localised loading of the crack surfaces in the vicinity of the crack tip.

3. Calculations

3.1. Stress intensity factors

The stress intensity factors for centre and edge cracks subject to a stress of the form

$$\sigma_n(x) = \sigma_0(x/w)^n \tag{16}$$

where n is an integer and $n < 4$, have been determined by Chell [2]. The centre crack results, however, were calculated by assuming an infinite array of colinear cracks. For large cracks and skew-symmetric stresses this approximation produced appreciable errors. Equation (5) offers an alternative expression for the stress intensity factor to that used by Chell [2]. The calculations of the latter have, therefore, been extended up to $n=8$ using Eqns. (4) and (6). A more convenient form for these equations can be obtained by integrating by parts and writing $h(a, x)$ explicitly thus

$$K^c = \tfrac{1}{2} a^{\frac{1}{2}} Y_0^c(a/w) \left[2\sigma(0) + \int_0^a (1-x/w) \frac{2}{\pi} \cos^{-1}\left(\frac{x/a - x/w}{1 - x/w}\right) \frac{\partial}{\partial x} \right.$$
$$\left. \times [(1+x/a)\sigma(x) + (1-x/a)\sigma(-x)] \, dx \right] \tag{17}$$

and

$$K^e = a^{\frac{1}{2}} Y_0^e(a/w) \left[\sigma(0) + \int_0^a (1-x/w)^2 \frac{2}{\pi} \cos^{-1}\left(\frac{x/a - x/w}{1 - x/w}\right) \frac{\partial \sigma}{\partial x} \, dx \right]. \tag{18}$$

Writing $K_n = \sigma_0 a^{\frac{1}{2}} Y_0(a/w) y_n(a/w)$ as the stress intensity factor corresponding to the stress $\sigma_n(x)$, $n \geqslant 1$, then

$$y_n^c(a/w) = \int_0^a (1-x/w) \frac{2}{\pi} \cos^{-1}\left(\frac{x/a - x/w}{1 - x/w}\right) \frac{(n+1)x^n}{aw^n} \, dx, \quad n \text{ odd} \tag{19a}$$

$$y_n^c(a/w) = (a/w) y_{n-1}^c(a/w) \quad n \text{ even} \tag{19b}$$

and

$$y_n^e = \int_0^a (1-x/w)^2 \frac{2}{\pi} \cos^{-1}\left(\frac{x/a - x/w}{1 - x/w}\right) \frac{nx^{n-1}}{w^n} \, dx \tag{20}$$

The integrals appearing in (19) and (20) have been solved for $n \geqslant 1$. The results for $y_n(a/w)$ have been represented to within 1% over the range $0 \leqslant a/w \leqslant 0.7$ by simple polynomials and are displayed in Table 1.
Consider a stress of the form

$$\sigma_n(x) = \sigma_0 \, \text{sign}(x)(x/w)^n \tag{21}$$

then

$$y_n^c(a/w) = \int_0^a (1-x/w) \frac{2}{\pi} \cos^{-1}\left(\frac{x/a - x/w}{1 - x/w}\right) \frac{(n+1)x^n}{aw^n} \, dx, \quad n \text{ even} \tag{22a}$$

$$y_n^c(a/w) = (a/w) y_{n-1}^c(a/w), \quad n \text{ odd}. \tag{22b}$$

TABLE 1

	Centre crack			Edge crack		
n	A_0	A_1	A_2	A_0	A_1	A_2
1	0.5	−0.132	−0.0267	0.6366	−0.365	0.0581
2	0.5	−0.132	−0.0267	0.5	−0.4185	0.0802
3	0.375	−0.1305	−0.0317	0.4244	−0.4217	0.0936
4	0.375	−0.1305	−0.0317	0.375	−0.4072	0.0934
5	0.3125	−0.1225	−0.03167	0.3395	−0.3912	0.0928
6	0.3125	−0.1225	−0.03167	0.3125	−0.3777	0.0956
7	0.2734	−0.1152	−0.0294	0.2910	−0.3628	0.0939
8	0.2734	−0.1152	−0.0294	0.2734	−0.350	0.0933

$K_n = \sigma_0 a^{\frac{1}{2}} Y_0(a/w) y_n(a/w)$

$y_n(a/w) = (a/w)^n (A_0 + A_1(a/w) + A_2(a/w)^2)$.

Stress intensity factors K_n corresponding to a stress of the form $\sigma_0(x/w)^n$. Y_0 is the compliance function corresponding to the uniform stress σ_0 and the relevant centre or edge crack geometry.

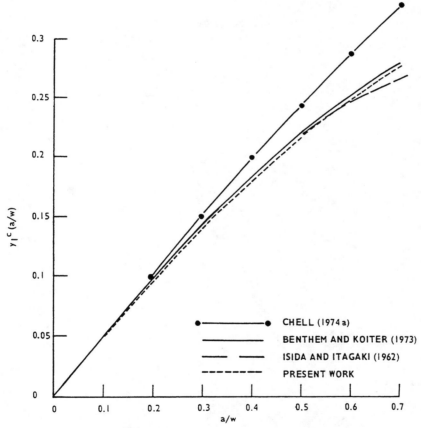

Figure 2. Centre crack compliance functions for linear stress variation.

The integrals appearing in (22) have been performed and the results for $y_n^c(a/w)$ fitted to within 1% over the range $0 \leqslant a/w \leqslant 0.7$ by a simple polynomial. The results are shown in Table 2.

The edge crack expression (6) has already been shown to be in excellent agreement with known solutions [2].

TABLE 2

Centre crack

n	A_0	A_1	A_2
1	0.6366	−0.113	−0.01889
2	0.4244	−0.1328	−0.0317
3	0.4244	−0.1328	−0.0317
4	0.3395	−0.127	−0.0311
5	0.3395	−0.127	−0.0311
6	0.2910	−0.1197	−0.0289
7	0.2910	−0.1197	−0.0289
8	0.2587	−0.1128	−0.0272

$K_n = \sigma_0 a^{\frac{1}{2}} Y_0(a/w) y_n(a/w)$

$y_n(a/w) = (a/w)^n (A_0 + A_1(a/w) + A_2(a/w)^2)$.

Stress intensity factors K_n corresponding to a stress of the form $\sigma_0 \, \text{sign}(x) \, (x/w)^n$. Y_0 is the centre crack compliance function corresponding to the uniform stress σ_0.

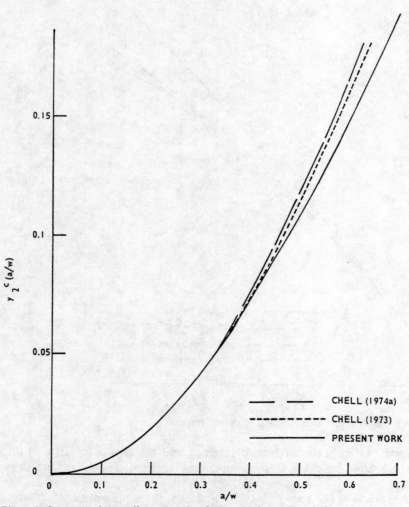

Figure 3. Centre crack compliance function for quadratic stress variation.

The centre crack expression (4) can be compared with the known solutions of Isida and Itagaki [3] and Benthem and Koiter [4] for a centre cracked plate subject to a pure bending moment and Chell [16] for a centre crack subject to a quadratic stress. (The latter was determined using a finite element stress analysis program and the results are probably only accurate to within 5% at most.) The pure bending moment results in a linear stress variation across the crack plane of the form

$$\sigma_1(x) = M/[B(2w)^2](x/w)$$

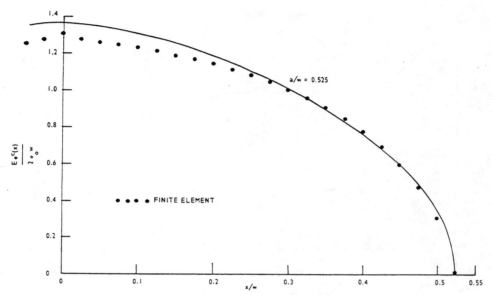

Figure 4. Crack profile for a centre cracked plate subject to a uniform stress.

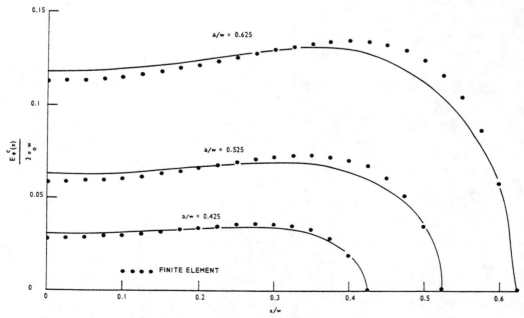

Figure 5. Crack profile for a centre cracked plate subject to a quadratic stress.

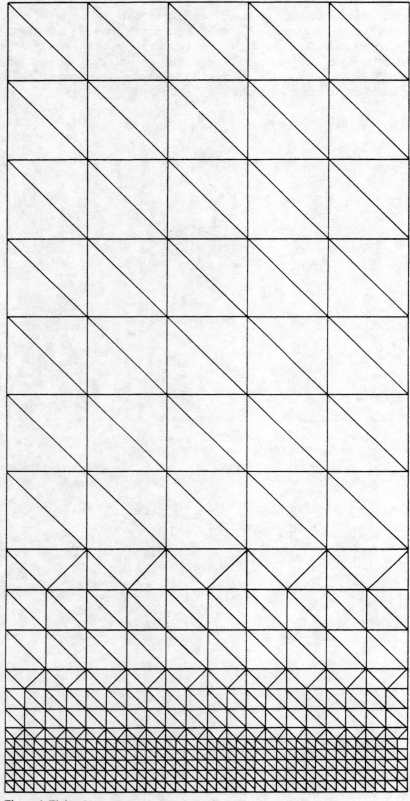

Figure 6. Finite element representation used for all cases considered.

where M is the external bending moment and B the plate thickness. The results of the calculations are shown in Figs. 2 and 3. Clearly very good agreement is obtained between the analytical expression (4) and the known solutions. For the linear stress variation (4) predicts values whose accuracy is superior to the results obtained previously by Chell [2] using the colinear crack approximation. In the limit, $a/w \to 1$, (19a) gives $y_n^c(1) = \frac{1}{3}$, in exact agreement with the ratio of Benthem and Koiters' [4] pure bending and uniform stress compliance functions when $a/w = 1$. In the quadratic stressing case (Fig. 3) the finite element results, colinear crack approximation and (19b) are in good agreement up to $a/w = 0.4$. Above this value the predictions of (19b) are below those of the other two sets of results but are probably the most accurate of the three.

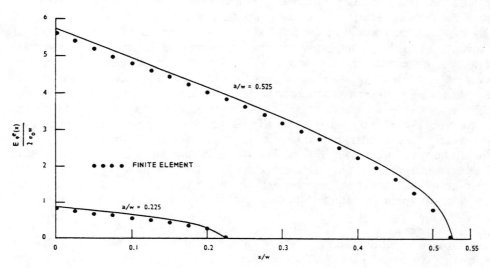

Figure 7. Crack profile for an edge cracked plate subject to a uniform stress.

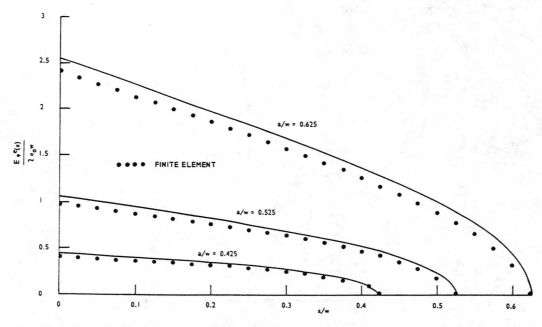

Figure 8. Crack profile for an edge cracked plate subject to a linear stress.

3.2. Crack profiles

Equations (9) and (10) represent expressions for the relative displacement between the surfaces of a centre and edge crack. For a given loading the stress intensity factor can be determined and hence the displacements calculated. Such calculations have been performed for a centre crack subject to a uniform and quadratic stress. The results are displayed in Figs. 4 and 5 together with the displacements obtained from a finite element stress analysis program. The mesh used is shown in Fig. 6 and the displacements were obtained by simulating the required stress by suitably loading the nodes lying on the crack surface. Owing to symmetry the mesh shown could be used to represent a quarter of the plate analysed. Good agreement was obtained between the analytical and computed values of the displacement for the uniform stressing case, although the latter are lower near the centre of the crack. For the quadratic stressing case the two sets of results are qualitatively similar, but quantitatively differ, for instance, near the centre of the crack. Exact agreement between the two sets of values cannot be expected for the relatively coarse mesh used. Since the mesh used was the same as that used in previous stress intensity factor calculations by the author [16] and the accuracy of the results was about 5%, then, since the displacement is approximately proportional to K^2, the finite element values are probably only accurate to about 10%. On the whole the analytical values lie well within this error bar. Similar calculations for an edge crack loaded by a uniform, linear and quadratic stress are shown in Figs. 7–9. Again the computed and analytical values agree quantitatively for all three stressing cases but differ by as much as 15% for the quadratic stressing case. The differences arise from the coarseness of the finite element mesh and the approximate nature of the analytical expressions near to the crack tip ((13) and (14)).

For the centre crack subject to uniform tension and the edge crack subject to tension and bend the relative displacements between the crack surfaces at $x=0$ are within approximately 5% of the displacements calculated from the expressions given in [1].

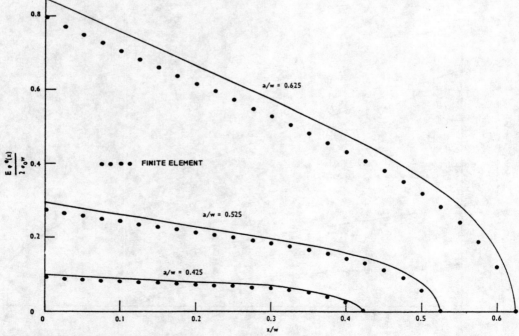

Figure 9. Crack profile for an edge cracked plate subject to a quadratic stress.

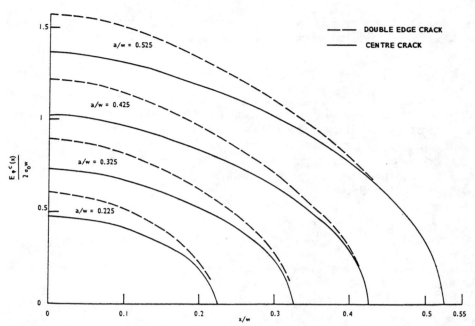

Figure 10. Comparison of centre and double edge crack profiles.

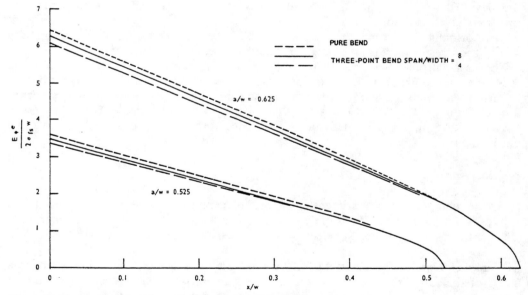

Figure 11. Comparison of crack profiles for the three bend specimens.

In Fig. 10 the effect of specimen geometry is displayed by comparing the calculated results for a centre and double edge cracked plate. In the double-edge crack calculations $H^c(a, x)$ for the centre crack was substituted for $H^e(a, x)$ in (6). Near the crack tip the two profiles are similar and this reflects the similar values of the two compliances, especially for the larger crack lengths. Near $x/w = 0$ the double edge crack displacement is greater than the centre crack displacement, a result arising from the presence of the free surface at $x/w = 0$ in the double edge case. In Fig. 11 the three bend geometries are compared. Again the crack profiles are similar near the crack tip but the pure bend edge crack produces the largest esternal opening displacement. $\sigma_{fs} = 6M/Bw^2$, is the outer fibre stress.

4. Discussion

The stress intensity factors calculated in this paper extend and improve on those previously determined by the author [2]. The results enable the stress intensity factors for cracks which experience very severe stress gradients to be calculated. The stress intensity factor can be determined by substituting the appropriate stress, $\sigma(x)$, into (4) or (6) and performing the integrals or by a simpler calculation using the compliances given in Tables 1 or 2. The latter is possible by firstly expressing the stress $\sigma(x)$ in a polynomial in x/w to order 8 or less, thus

$$\sigma(x) = \sigma_0 \sum_{n=0}^{m<8} B_n^*(x/w)^n \qquad (23)$$

where B_n^* are dimensionless coefficients, and σ_0 is a characteristic stress. Then, by the superposition of stresses, the stress intensity factor is given as

$$K = \sigma_0 a^{\frac{1}{2}} Y_0(a/w) \sum_{n=0}^{m<8} y_n(a/w) B_n^* \qquad (24)$$

where Y_0, the uniform stress compliance function, is obtainable from Brown and Srawley [5] and the $y_n(a/w)$ are easily calculated from the simple polynomial expressions given in Tables 1 and 2. The results given in Table 2 for the symmetric form of the normally skew-symmetric stresses and the skew-symmetric form of the normally symmetric stresses increase the range of stressing cases that can be handled.

The crack profiles, determined using (9) and (10) are mainly of academic interest. It is important however to know the accuracy of the expressions because they provide a means of determining accurate solutions to the Bilby, Cottrell and Swinden [8] model of a crack with yielding ahead of its tip. One of the important parameters in this model is the crack opening displacement and a knowledge of the crack profile enables this to be determined.

5. Conclusions

(i) Expressions for the stress intensity factors for centre and edge cracks subject to an arbitrary stress are given. They supplement the solutions already available in the literature. The centre crack expression supersedes that previously given by the author (2).

(ii) Simple, polynomial expressions for the stress intensity factors for cracks subject to a stress of the form $\sigma(x) = \sigma_0(x/w)^n$ are given where σ_0 is a characteristic stress and n an integer less than, or equal to, 8. These enable the stress intensity factors for cracks in severe stress gradients to be easily determined by expressing the stress in polynomial form.

(iii) Expressions for the relative displacement between the surfaces of centre and edge cracks subject to an arbitrary stress are given. Relatively good agreement is obtained between these and the displacements obtained from a finite element stress analysis program.

(iv) For very localised loading near the crack tip the approximate solutions given will be accurate to within 20% for a centre crack and 10% for an edge crack in the range $0 < a/w < 0.7$.

Acknowledgement

This work was carried out at the Central Electricity Research Laboratories and is published by permission of the Central Electricity Generating Board.

REFERENCES

[1] H. Tada, P. Paris and G. Irwin, *The Stress Analysis of Cracks Handbook*, Del Research Corporation, Hellertown, Pennsylvania (1973).
[2] G. G. Chell, *Eng. Frac. Mech.*, 7 (1975) 137.
[3] M. Isida and Y. Itagaki, Proc. 4th U.S. National Congress of Applied Mechanics, (1962) 955.
[4] J. P. Benthem and W. T. Koiter, *Methods of Analysis and Solutions to Crack Problems*, (Edited by G. C. Sih), Noordhoff, Leyden (1973) 131.
[5] W. F. Brown and J. E. Srawley, Plane Strain Crack Toughness Testing of High Strength Metallic Materials, *ASTM STP* 410, ASTM Philadelphia (1967).
[6] H. F. Bueckner, *Methods of Analysis and Solutions of Crack Problems*, (Edited by G. C. Sih), Noordhoff, Leyden (1973) 239.
[7] J. R. Rice, *Int. J. Solids Structures*, 8 (1972) 751.
[8] B. A. Bilby, A. H. Cottrell and K. H. Swinden, *Proc. Roy. Soc.*, A272 (1963) 304.
[9] R. V. Goldstein, I. N. Ryskov and R. L. Salganik, *Int. J. Fracture Mech.*, 6 (1970) 104.
[10] G. G. Chell, *Mat. Sci. and Eng.*, 17 (1975) 227.
[11] A. F. Emery, G. E. Walker and J. A. Williams, ASME publication 68-WA/MET-19 (1968).
[12] L. A. Wigglesworth, *Mathematika*, 4 (1957) 76 and *ibid.*, 5 (1958) 67.
[13] H. F. Bueckner and I. Giaever, *ZAMM* 46 (1966) 265.
[14] V. V. Panasyuk, *Limiting equilibrium of brittle solids with Fracture*, (p. 184). Translation from Russian, distributed in 1971, by Management Information Services, P.O. Box 5129, Detroit, Michigan 48236.
[15] P. Paris and G. C. Sih, Am. Soc. Testing Mat., *ASTM STP* 381 (1965) 30.
[16] G. G. Chell, *Int. J. Fracture Mechs.*, 9 (1973) 338.

RÉSUMÉ

On présente des expressions pour des facteurs d'intensité de contraintes relatives à des fissures centrales et des fissures latérales dans des tôles sujettes à des contraintes arbitraires. Les facteurs d'intensité des contraintes correspondant à des contraintes exprimées par une loi polynômique de l'ordre de 8 sont établies, et l'on indique leur possibilité d'application à des fissures dans des composants en service.

Les résultats obtenus sont en bon accord avec les solutions connues. On établit des expressions permettant de décrire les profils de fissuration dans le cas de fissures centrales et latérales sujettes à des mises en charge arbitraires. Les prédictions de ces expressions sont en bon accord avec les résultats d'une analyse des contraintes par éléments finis.

Crack-Tip, Stress-Intensity Factors for Plane Extension and Plate Bending Problems

G. C. SIH
Assistant Professor of Mechanics,
Lehigh University,
Bethlehem, Pa.

P. C. PARIS[1]
Instructor of Mechanics,
Lehigh University,
Bethlehem, Pa.

F. ERDOGAN
Associate Professor
of Mechanical Engineering,
Lehigh University,
Bethlehem, Pa.

A complex variable method for evaluating the strength of stress singularities at crack tips in plane problems and plate bending problems is derived. The results of these evaluations give Irwin's stress-intensity factors for plane problems and analogous quantities for bending problems, a form familiar to the practitioner of "fracture mechanics." The methods derived are integrated with the complex variable approach of Muskhelishvili to obtain the stress-intensity factors for various basic examples applicable to the extension and bending of plates with through-the-thickness cracks. The results suggest the possibility of extension of the Griffith-Irwin fracture theory to arbitrary plane extensional and/or bending problems in plates.

THE concept of crack-tip, stress-intensity factors (which are in fact the strength of stress singularities at crack tips) applied to predictions of the static strength of cracked bodies has been developed by Irwin [1, 2, 3][2] for plane extension, symmetric with respect to the crack. Stress-intensity factors have also been shown [4] to control the rate of crack propagation under cyclic loading in such situations. However, extension of these applications to other than symmetric plane problems and/or bending problems is yet to be accomplished.

Both Irwin [1, 2, 3] and Williams [5] have obtained the form of the elastic stress distribution in the vicinity of a crack tip in extensional problems. Williams [6], in a recent work, extends his analysis to thin plates subjected to bending out of the plane. In each case it is shown that the significant stresses in the vicinity of the crack tip are those associated with the singularity of stress of the order $(r)^{-1/2}$, where r is radial distance from the crack tip.

Moreover, the distribution of stress in each, extension or bending, is unique; i.e., its functional form in terms of co-ordinates measured from the crack tip is always the same. The intensity of the local field of stress may be represented in terms of two parameters. Irwin's stress-intensity factors are themselves a convenient form of these parameters which are now in use [4] for extensional problems. Hence, for bending, Williams' results will be modified to define stress-intensity factors in bending in a manner consistent with Irwin's definitions.

Then it will be shown that the complex variable technique of Muskhelishvili [7] and its extension to plate bending problems [8] can be conveniently incorporated into a new method for computing the stress-intensity factors for various configurations.

Stress Field Near Crack-Tip and Stress-Intensity Factors

In plane extension, the stress distribution in the vicinity of the end of a crack has been shown by Irwin to always take the form:

$$\sigma_y = \frac{k_1}{(2r)^{1/2}} \cos \frac{\theta}{2} \left[1 + \sin \frac{\theta}{2} \sin \frac{3\theta}{2} \right]$$
$$+ \frac{k_2}{(2r)^{1/2}} \sin \frac{\theta}{2} \cos \frac{\theta}{2} \cos \frac{3\theta}{2} \quad (Cont.)$$

---

[1] Assistant Professor of Civil Engineering, University of Washington, Seattle, Wash. (on leave 1960–1961); and Consultant to The Boeing Airplane Company.
[2] Numbers in brackets designate References at end of paper.
Presented at the West Coast Conference of the Applied Mechanics Division, Seattle, Wash., August 28–30, 1961, of THE AMERICAN SOCIETY OF MECHANICAL ENGINEERS.
Discussion of this paper should be addressed to the Editorial Department, ASME, United Engineering Center, 345 East 47th Street, New York 17, N. Y., and will be accepted until July 10, 1962. Discussion received after the closing date will be returned. Manuscript received by ASME Applied Mechanics Division, May 16, 1961. Paper No. 61—APMW-29.

Nomenclature

- a = half crack length of a straight crack
- b = dimension locating a concentrated force on a crack surface
- b_m ($m = 1, 2$) = constants determined by loading and configuration
- D = flexural rigidity of a plate
- E = Young's modulus of elasticity
- F = concentrated extensional force on a crack surface (per unit thickness)
- G = shear modulus of elasticity
- h = thickness of a plate
- i = $(-1)^{1/2}$
- k_1, k_2 = crack tip stress-intensity factors for symmetrical and skew-symmetrical stress distributions, respectively, encountered in extensional problems
- k = $k_1 - ik_2$
- K_1, K_2 = crack tip stress-intensity factors for symmetrical and skew-symmetrical stress distributions, respectively, encountered in transverse bending problems
- K = $K_1 - iK_2$
- $M_r, M_\theta, M_{r\theta}$ = bending moments and twisting moment, respectively, all per unit length with reference to (r, θ)-directions
- M_0, Q_0 = bending and shearing force per unit length, respectively, applied at a plate boundary.
- P, Q = x and y-components of a force, F

$$\sigma_x = \frac{k_1}{(2r)^{1/2}} \cos \frac{\theta}{2} \left[1 - \sin \frac{\theta}{2} \sin \frac{3\theta}{2} \right]$$
$$- \frac{k_2}{(2r)^{1/2}} \sin \frac{\theta}{2} \left[2 + \cos \frac{\theta}{2} \cos \frac{3\theta}{2} \right] \quad (1)$$

$$\tau_{xy} = \frac{k_1}{(2r)^{1/2}} \sin \frac{\theta}{2} \cos \frac{\theta}{2} \cos \frac{3\theta}{2}$$
$$+ \frac{k_2}{(2r)^{1/2}} \cos \frac{\theta}{2} \left[1 - \sin \frac{\theta}{2} \sin \frac{3\theta}{2} \right]$$

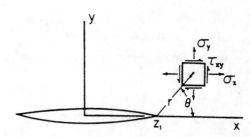

Fig. 1 Notation for rectangular stress components near crack tip

where higher-order terms in r have been neglected. Fig. 1 gives the notation used in this representation. The stress-intensity factors, k_1 and k_2, are the parameters (dependent on the configuration and loading) which control the magnitude of the symmetric and skew-symmetric components of this stress distribution.

In the case of plate bending, Williams' results for the stresses are similar, i.e.,

$$\sigma_r = \frac{12 M_r \delta}{h^3} = \frac{1}{r^{1/2}} \left\{ \left[\cos \frac{3\theta}{2} - \frac{3 + 5\nu}{7 + \nu} \cos \frac{\theta}{2} \right] \frac{3 b_1 G \delta}{2} \right.$$
$$\left. + \left[-\sin \frac{3\theta}{2} + \frac{3 + 5\nu}{5 + 3\nu} \sin \frac{\theta}{2} \right] \frac{3 b_2 G \delta}{2} \right\}$$

$$\sigma_\theta = \frac{12 M_\theta \delta}{h^3} = \frac{1}{r^{1/2}} \left\{ \left[-\cos \frac{3\theta}{2} - \frac{5 + 3\nu}{7 + \nu} \cos \frac{\theta}{2} \right] \frac{3 b_1 G \delta}{2} \right. \quad (2)$$
$$\left. + \left[\sin \frac{3\theta}{2} + \sin \frac{\theta}{2} \right] \frac{3 b_2 G \delta}{2} \right\}$$

$$\tau_{r\theta} = \frac{12 M_{r\theta} \delta}{h^3} = \frac{2}{r^{1/2}} \left\{ \left[-\sin \frac{3\theta}{2} + \frac{1 - \nu}{7 + \nu} \sin \frac{\theta}{2} \right] \frac{3 b_1 G \delta}{2} \right.$$
$$\left. + \left[-\cos \frac{3\theta}{2} + \frac{1 - \nu}{5 + 3\nu} \cos \frac{\theta}{2} \right] \frac{3 b_2 G \delta}{2} \right\}$$

where δ is a co-ordinate perpendicular to the middle plane of the plate, and again higher order terms in r have been neglected, Fig. 2.

For convenience, as previously mentioned, the constants, b_1 and b_2, in Williams' expression, equations (2), are redefined by

$$K_1 = -\frac{3\sqrt{2}(3 + \nu) G h b_1}{7 + \nu}$$

and
$$K_2 = -\frac{3\sqrt{2}(3 + \nu) G h b_2}{5 + 3\nu} \quad (3)$$

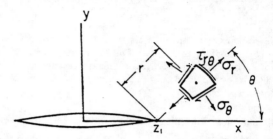

Fig. 2 Notation for polar stress components near crack tip

where K_1 and K_2 will henceforth be called the bending stress-intensity factors. The desirable character of this choice will be obviated as the analysis proceeds.

Now, for the purpose of subsequently developing methods for determining the stress-intensity factors for individual problems, the sum of the normal stresses (an invariant) will be useful. Forming these from equations (1), (2), and (3), they are, for extension:

$$\sigma_x + \sigma_y = k_1 \left[\frac{2}{r} \right]^{1/2} \cos \frac{\theta}{2} - k_2 \left[\frac{2}{r} \right]^{1/2} \sin \frac{\theta}{2} \quad (4)$$

and for bending:

$$\sigma_r + \sigma_\theta = \frac{12(M_r + M_\theta)\delta}{h^3}$$
$$= \frac{2\sqrt{2}(1 + \nu)\delta}{(3 + \nu)h} \left[\frac{K_1}{\sqrt{r}} \cos \frac{\theta}{2} - \frac{K_2}{\sqrt{r}} \sin \frac{\theta}{2} \right]$$

In both cases the stress-intensity factors may be represented as the real and negative imaginary parts of a complex constant in accordance with

$$k = k_1 - i k_2$$
$$K = K_1 - i K_2 \quad (5)$$

Nomenclature

r, θ = polar co-ordinates measured from a crack tip and from line of extension of crack, respectively

w = transverse deflection of a plate in bending, or Airy stress function for extension

x, y = rectangular co-ordinates in middle plane of a plate

z = complex variable $(x + iy)$

α = half-angle subtended by a curved crack

β = angle between direction of an applied shear force, Q_0, or angle between a field of uniform extensional stress and crack plane

δ = co-ordinate perpendicular to middle plane of a plate

ζ = complex variable in a mapped plane

κ = elastic constant which takes values $\kappa = 3 - 4\nu$ for plane strain and $\kappa = (3 - \nu)/(1 + \nu)$ for plane stress

ν = Poisson's ratio

σ = uniform extensional stress at infinity

$\sigma_r, \sigma_\theta, \tau_{r\theta}$ = bending-stress components with reference to (r, θ)-directions

$\sigma_x, \sigma_y, \tau_{xy}$ = components of stress

$\varphi(z), \psi(z), \chi(z), \varphi(\zeta)$ = Muskhelishvili's analytic stress functions of complex variables z and ζ, respectively

φ_e, φ_b = Muskhelishvili's functions for extension and bending, respectively

$\omega(\zeta)$ = a mapping function

Re$\{\ \}$ = real part of

', prime = derivative with respect to argument

Also introducing the complex variable

$$z = x + iy = z_1 + re^{i\theta}$$

where z_1 is the location of the crack tip, equations (4) may be written, for extension:

$$\sigma_x + \sigma_y = \text{Re}\left\{k\left[\frac{2}{z-z_1}\right]^{1/2}\right\}$$

and for bending,

$$M_r + M_\theta = \frac{(1+\nu)h^2}{3\sqrt{2}(3+\nu)}\text{Re}\left\{K\left[\frac{1}{z-z_1}\right]^{1/2}\right\} \quad (6)$$

The similarity of these results is evident; however, two points are to be kept in mind: (a) The stress fields from which they were derived are actually quite different (compare [5] and [6]), and (b), they are valid only in the immediate vicinity of the crack tip, as z approaches z_1.

Application of Complex Variable Stress-Function Technique

A complex variable method of solving boundary-value problems governed by the biharmonic equation,

$$\nabla^4 w = 0 \quad (7)$$

has been developed principally by Muskhelishvili [7] (a résumé is presented in Sokolnikoff's work [9]). The application to many examples of extensional problems is treated in [7] and bending problems are given attention by Savin [8]. The biharmonic equation, equation (7), governs both extensional problems and bending problems (where no transverse loads are present). w is the Airy stress function for extension and the transverse deflection for bending.

In either case the general solution to equation (7) may be written

$$w = \text{Re}\{\bar{z}\varphi(z) + \chi(z)\} \quad (8)$$

where

$$\chi(z) = \int \psi(z)dz$$

Further, $\varphi(z)$ and $\psi(z)$ are analytic functions of the complex variable z. The solution to an individual extension or bending problem is reduced to finding the functions, $\varphi(z)$ and $\psi(z)$, which satisfy the boundary conditions of that problem.

Without going into the details of determining $\varphi(z)$ and $\psi(z)$, a method of determining stress-intensity factors is to be developed. Hence it is expedient to only state the well-known [7, 8] relationships between the stress functions and the quantities in the left-hand members of equations (6). They are, for extension,

$$\sigma_x + \sigma_y = 4\,\text{Re}\{\varphi_e{}'(z)\}$$

and for bending, \hspace{2cm} (9)

$$M_r + M_\theta = -4D(1+\nu)\,\text{Re}\{\varphi_b{}'(z)\}$$

where $D = Eh^3/[12(1-\nu^2)]$ is the flexural rigidity of the plate.

In the vicinity of a crack tip $\varphi'(z)$ takes the form of a complex constant divided by $(z - z_1)^{1/2}$, reference [7] (pp. 496–498). Therefore, equating the results of equations (6) and (9) and recalling that equations (6) are valid only as z approaches z_1, the stress-intensity factors may be evaluated from, for extension,

$$k = 2\sqrt{2}\lim_{z\to z_1}(z-z_1)^{1/2}\varphi_e{}'(z)$$

and for bending, \hspace{2cm} (10)

$$K = -\frac{12\sqrt{2}\,D(3+\nu)}{h^2}\lim_{z\to z_1}(z-z_1)^{1/2}\varphi_b{}'(z)$$

Now, equations (10) illustrate the fact that the stress-intensity factors may be found simply from a knowledge of $\varphi'(z)$ in the vicinity of the crack tip. Other methods can be derived for like determinations; for example, expansions for the stress or moments in the vicinity of the crack tip in a particular problem yield the stress-intensity factors directly. However, it is the authors' opinion that it is usually much easier to perform the indicated operation of equations (10) than to compute and expand the stress for the multitude of problems where the functions $\varphi(z)$ and $\psi(z)$ are already known [7–13]. Moreover, in solving new problems it is somewhat easier to simply find $\varphi(z)$ and ignore $\psi(z)$ if stress-intensity factors are the only desired result. The power of the technique of equations (10) is best illustrated by the simplicity of providing examples for stress-intensity factors heretofore not available in the literature.

Curved Crack in Sheet Under Biaxial Tension

Fig. 3 shows the problem of an infinite sheet under uniform biaxial tension with a circular-sector crack of unit radius about the origin. For this problem Muskhelishvili [7] gives the derivative of the stress function,

$$\varphi_e{}'(z) = \frac{\sigma}{2\left(1+\sin^2\frac{\alpha}{2}\right)}\left[\frac{z-\cos\alpha}{(1-2z\cos\alpha+z^2)^{1/2}}+\sin^2\frac{\alpha}{2}\right] \quad (11)$$

In order to treat the problem by making use of equations (10), the co-ordinates must first be rotated such that the crack tip of interest is parallel to the x-axis. A convenient transformation for this purpose (which also relocates the crack tip on the x-axis) is

$$z = ie^{i\alpha}(\bar{z} - i - \sin\alpha\cos\alpha) \quad (12)$$

Substituting equation (12) into (11) and the result subsequently into the first of equations (10) gives

$$k_1 = \frac{\sigma}{1+\sin^2\frac{\alpha}{2}}\left[\frac{\sin\alpha(1+\cos\alpha)}{2}\right]^{1/2}$$

and \hspace{2cm} (13)

$$k_2 = \frac{\sigma}{1+\sin^2\frac{\alpha}{2}}\left[\frac{\sin\alpha(1-\cos\alpha)}{2}\right]^{1/2}$$

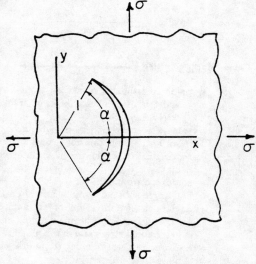

Fig. 3 Infinite sheet containing curved crack of unit radius

Conformal Mapping

In many extensional and bending problems the use of conformal mapping is an expedient. Let the mapping function be defined (with the usual restrictions as to analyticity and single-valuedness) by

$$z = \omega(\zeta) \tag{14}$$

Then

$$\varphi'(z) = \frac{d\varphi}{d\zeta} \cdot \frac{d\zeta}{dz} = \frac{\varphi'(\zeta)}{\omega'(\zeta)} \tag{15}$$

Now, corresponding to the crack tip z_1 in the z-plane there will be a point ζ_1 in the ζ-plane. Thus equations (10) may be written, for extension:

$$k = 2\sqrt{2} \lim_{\zeta \to \zeta_1} [\omega(\zeta) - \omega(\zeta_1)]^{1/2} \frac{\varphi_e'(\zeta)}{\omega'(\zeta)}$$

and for bending: (16)

$$K = -\frac{12\sqrt{2}\, D(3 + \nu)}{h^2} \lim_{\zeta \to \zeta_1} [\omega(\zeta) - \omega(\zeta_1)]^{1/2} \frac{\varphi_b'(\zeta)}{\omega'(\zeta)}$$

Furthermore, for mappings which transform the "sharp" crack tip into a smooth curve an additional simplification results. For example, the mapping,

$$z = \omega(\zeta) = \frac{a}{2}\left(\zeta + \frac{1}{\zeta}\right) \tag{17}$$

transforms a straight crack of length, $2a$, in the x-direction, located centrally at the origin in the z-plane, into a circular hole of unit radius in the ζ-plane. Upon noting that the point corresponding to the crack tip, $z = a$, is at $\zeta = 1$ in the ζ-plane, equations (16) and (17) may be combined to give, for extension,

$$k = \frac{2}{a^{1/2}} \varphi_e'(\zeta)|_{\zeta = 1}$$

and for bending: (18)

$$K = -\frac{12D(3 + \nu)}{h^2 a^{1/2}} \varphi_b'(\zeta)|_{\zeta = 1}$$

since with a circular boundary present $\varphi'(\zeta)$ has no singularity at $\zeta = 1$. Equations (18) will be used to find the stress-intensity factors in the examples to follow.

Concentrated Extensional Force on Crack Surface

The extensional problem of an infinite sheet with a crack along the x-axis of length, $2a$, centered at the origin and having an arbitrary extensional force (per unit sheet thickness), F, applied to its upper surface at $x = b$ is of special interest, since it may be used as a Green's function to form the solution to other problems. For this problem, Fig. 4, the stress function is

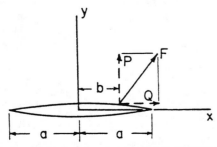

Fig. 4 Crack whose surface is subjected to concentrated force

$$\varphi_e(\zeta) = \frac{Fa}{2\pi(a^2 - b^2)^{1/2}} \left\{ -\frac{1}{\zeta} + \left(\frac{\zeta_0}{\zeta_0 - \zeta}\right) \right.$$
$$\left[\left(\zeta + \frac{1}{\zeta}\right) - \left(\zeta_0 + \frac{1}{\zeta_0}\right) \right]$$
$$\left. + \left(\zeta_0 + \frac{1}{\zeta_0}\right)\left[\frac{\kappa}{1 + \kappa} \log \zeta - \log(\zeta_0 - \zeta)\right]\right\} \tag{19}$$

where the crack has been mapped into a unit circle, by equation (17), on which $\zeta = \zeta_0$ corresponds to $x = b$ on the upper surface of the crack. The concentrated force may be expressed in terms of its y and x-components, P and Q, respectively, whereupon

$$F = P - iQ \tag{20}$$

Using the first of equations (18) to evaluate the stress-intensity factors (at the right end of the crack):

$$k = \frac{P - iQ}{2\pi a^{1/2}} \left[\left(\frac{a + b}{a - b}\right)^{1/2} + i\left(\frac{\kappa - 1}{\kappa + 1}\right)\right]$$

or

$$k_1 = \frac{P}{2\pi a^{1/2}}\left(\frac{a + b}{a - b}\right)^{1/2} + \frac{Q}{2\pi a^{1/2}}\left(\frac{\kappa - 1}{\kappa + 1}\right) \tag{21}$$

and

$$k_2 = -\frac{P}{2\pi a^{1/2}}\left(\frac{\kappa - 1}{\kappa + 1}\right) + \frac{Q}{2\pi a^{1/2}}\left(\frac{a + b}{a - b}\right)^{1/2}$$

The use of equations (21) is self-evident in handling problems of an infinite sheet with a crack on whose surface tractions are applied.

In addition, problems of an infinite sheet with a crack whose surface is free from tractions may be attacked by superposition and resolved to the solution of the problem of a crack with surface tractions. This is accomplished by solving the original problem for the stresses at the crack site, $\sigma_y(x, 0)$ and $\tau_{xy}(x, 0)$, with the crack absent and then superimposing tractions equal and opposite to the crack-site stresses. The result is

$$k_1 = \frac{1}{\pi a^{1/2}} \int_{-a}^{a} \sigma_y(x, 0) \left(\frac{a + x}{a - x}\right)^{1/2} dx$$
$$k_2 = \frac{1}{\pi a^{1/2}} \int_{-a}^{a} \tau_{xy}(x, 0) \left(\frac{a + x}{a - x}\right)^{1/2} dx \tag{22}$$

Equations (22) represent a simple direct method of determining the stress-intensity factors for a crack in an infinite plate with arbitrary extensional loading.

Example of Bending of Infinite Plate Containing Crack

Consider the problem of an infinite plate containing a crack of length $2a$, where a bending moment, M_0, is applied all around the boundary of the plate at infinity, Fig. 5. Savin [8] gives the Muskhelishvili function for this problem using equation (17):

$$\varphi_b(\zeta) = -\frac{M_0 a}{4D(1 + \nu)}\left[\zeta + \frac{1 - \nu}{3 + \nu}\frac{1}{\zeta}\right] \tag{23}$$

Introducing equation (23) into the second of equations (18) results in

$$K_1 = \frac{6M_0}{h^2} a^{1/2} \tag{24}$$

$$K_2 = 0$$

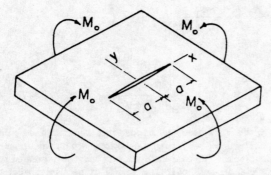

Fig. 5 Infinite plate containing crack and subjected to uniform bending

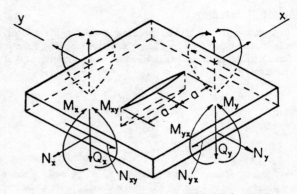

Fig. 6 Element containing crack removed from sheet structure

Upon observing that $6M_0/h^2$ is the stress in the surface layer of the plate at infinity, this result is similar to the "corresponding" extensional problem; i.e., uniform extensional stress, σ, at infinity which gives [1] $k_1 = \sigma a^{1/2}$. The convenient simplicity and familiarity of this result are in part the justification for defining the bending stress intensity factors as in equations (3). Similar simplicity and analogies with extension problems occur in other bending problems, as will be seen in the discussions to follow.

The three preceding examples of determination of stress-intensity factors, equations (13), (21), and (24), are sufficient to illustrate the techniques developed here, equations (10) and (18). It is easy to add more examples; therefore, for the use of the engineer an Appendix is attached to this work giving some additional results in compact form. It is now more pertinent to discuss the significance of these new results with regard to their possible applications.

Extension of Griffith-Irwin Theory for Thin Sheet Structures

Thin plates (and shells with radii of curvature which are very large compared to the crack length) which contain a crack are often the subject of inquiries on their remaining strength. The aircraft industry is one example of interest in such problems. To date, the Griffith-Irwin theory has enjoyed a unique amount of success in treating many of these problems. Yet, so far it has been almost entirely restricted in application to plane extensional problems, symmetric with respect to the crack.

An example of a more general problem is shown in Fig. 6. This element could be a portion of a sheet in a structure, where the introduction of the crack will have little influence on the loading forces if the element is large in planar dimensions compared to the crack length. Presuming the loading is known from a structural analysis, the stress-intensity factors may be found by superposition of those for simpler individual loadings. The stress-intensity factors for each of the individual loadings is given in the Appendix.

Hence, for this example, Fig. 6,

$$k_1 = \frac{N_y}{h} a^{1/2}$$

$$k_2 = \frac{N_{xy}}{h} a^{1/2}$$

$$K_1 = \frac{6M_y}{h^2} a^{1/2} + \frac{8\nu Q_x a^{3/2}}{h^2} \quad (25)$$

$$K_2 = \frac{6M_{xy}}{h^2} a^{1/2} + \frac{8 Q_y a^{3/2}}{h^2}$$

The simple extension of the Griffith-Irwin theory to such a problem is:

The combination of stress-intensity factors present will cause unstable crack extension upon reaching some critical value; i.e., the criterion takes the form

$$f(k_1, k_2, K_1, K_2) = f_{cr} \quad (26)$$

However, some doubtful assumptions remain. For example, interference between crack surfaces has been ignored. Moreover, with bending present, crack extension is likely to favor one side of the sheet more than the other.

Therefore this work is offered as a step toward more general application of Griffith-Irwin concepts, but it should be recognized that practical application must be accompanied by validation or modification of some of the assumptions present.

In addition, it is possible that such a generalization of the Griffith-Irwin concepts may carry over into the field of crack extension under cyclic loading [4]. Current studies are underway in an attempt to apply these ideas.

Conclusions

The complex variable approach presented gives a convenient way to determine crack-tip, stress-intensity factors for plane extension and plate bending problems, provided the Muskhelishvili stress function, $\varphi(z)$, is known in either the actual or a mapped plane. Since Muskhelishvili's methods are well known for their power and direct approach for multiply connected regions, their incorporation is a natural one in analyzing cracks.

Examples have been given of the computation of stress-intensity factors for a few particular problems, which illustrate the power of the technique presented. These examples and other results are tabulated in the Appendix for the use of the engineer in fracture analyses. Thus the most important result may not necessarily be the technique of determining stress-intensity factors, but the primary contribution here is perhaps that the way is cleared for more general application of the Griffith-Irwin fracture concepts as indicated.

Acknowledgment

The authors gratefully acknowledge the encouragement and financial support of the Boeing Airplane Company, Transport Division, in pursuance of this work, which was conducted under their grant to the Lehigh University Institute of Research.

References

1 G. R. Irwin, "Analysis of Stresses and Strains Near the End of a Crack Traversing a Plate," JOURNAL OF APPLIED MECHANICS, vol. 24. TRANS. ASME, vol. 79, 1957, pp. 361-364.

2 G. R. Irwin, "Fracture," *Handbuch der Physik*, vol. 6, Springer, 1958, pp. 551-590.

3 G. R. Irwin, "Fracture Mechanics," *Structural Mechanics*, Pergamon Press, New York, N. Y., 1960, pp. 557-592.

4 W. E. Anderson and P. C. Paris, "Fracture Mechanics Applied to the Evaluation of Aircraft Material Performance," *Metals Engineering Quarterly*, vol. 1, 1961, pp. 33–44.
5 M. L. Williams, "On the Stress Distribution at the Base of a Stationary Crack," JOURNAL OF APPLIED MECHANICS, vol. 24, TRANS. ASME, vol. 79, 1957, pp. 109–114.
6 M. L. Williams, "The Bending Stress Distribution at the Base of a Stationary Crack," JOURNAL OF APPLIED MECHANICS, vol. 28, TRANS. ASME, Series E, vol. 82, March, 1961, pp. 78–82.
7 N. I. Muskhelishvili, "Some Basic Problems of Mathematical Theory of Elasticity," 1933, English translation, P. Noordhoff and Company, New York, N. Y., 1953.
8 G. N. Savin, "Spannungoerhöhung am Rande von Löchern," VEB Verlag Technik, Berlin, Germany, 1956, German translation.
9 I. S. Sokolnikoff, "Mathematical Theory of Elasticity," McGraw-Hill Book Company, Inc., New York, N. Y., 1956, pp. 288–289.
10 J. L. Sanders, Jr., "On the Griffith-Irwin Fracture Theory," JOURNAL OF APPLIED MECHANICS, vol. 27, TRANS. ASME, Series E, vol. 82, 1960, pp. 352.
11 Yi-Yuan Yu, "The Influence of a Small Hole or Rigid Inclusion on the Transverse Flexure of Thin Plates," *Proceedings of Second U. S. National Congress of Applied Mechanics*, 1954, pp. 381–387.
12 M. Goland, "The Influence of the Shape and Rigidity of an Elastic Inclusion on the Transverse Flexure of Thin Plates," JOURNAL OF APPLIED MECHANICS, vol. 10, Trans. ASME, vol. 65, 1943, p. A-69.
13 J. N. Goodier, "The Influence of Circular and Elliptical Holes on the Transverse Flexure of Elastic Plates," *Philosophical Magazine*, series 7, vol. 22, 1936, pp. 68–80.

APPENDIX

The following tabulation of results is presented to supply the reader with the stress-intensity factors for various basic configurations and loadings not available in previous literature. For the sake of brevity, the few examples chosen will be those which can be easily modified and/or superimposed to give solutions to a variety of special case problems which may be of engineering significance. With each result a single reference will be listed where more information on that example may be obtained.

Infinite Plates Subjected to Extensional Loading

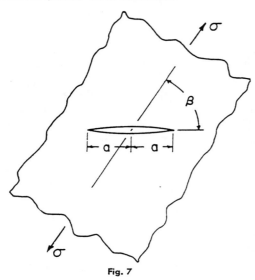

Fig. 7

Reference [7]

$$\varphi_e = -\frac{\sigma(1 - e^{2i\beta})a}{4\zeta}$$

$$k_1 = \sigma(\sin^2 \beta)a^{1/2}$$

$$k_2 = \sigma(\sin \beta \cos \beta)a^{1/2}$$

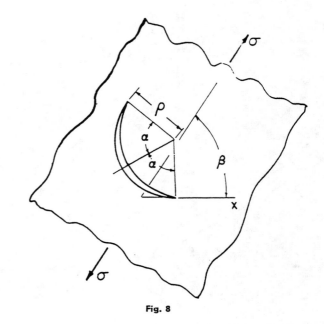

Fig. 8

Reference [7]

$$k_1 = \frac{\sigma(\rho \sin \alpha)^{1/2}}{2\left(1 + \sin^2 \frac{\alpha}{2}\right)} \left\{ \cos \frac{\alpha}{2} + \cos\left(2\beta + \frac{5}{2}\alpha\right)\left[\sin^2 \frac{\alpha}{2}\right] \right.$$
$$- \cos\left(2\beta + \frac{3}{2}\alpha\right)\left[\cos^2 \frac{\alpha}{2} - \sin^4 \frac{\alpha}{2}\right]$$
$$\left. - \sin\left(2\beta + \frac{3}{2}\alpha\right)\left[\sin \alpha \sin^2 \frac{\alpha}{2}\right] \right\}$$

$$k_2 = \frac{\sigma(\rho \sin \alpha)^{1/2}}{2\left(1 + \sin^2 \frac{\alpha}{2}\right)} \left\{ \sin \frac{\alpha}{2} + \sin\left(2\beta + \frac{5}{2}\alpha\right)\left[\sin^2 \frac{\alpha}{2}\right] \right.$$
$$+ \sin\left(2\beta + \frac{3}{2}\alpha\right)\left[\cos^2 \frac{\alpha}{2} - \sin^4 \frac{\alpha}{2}\right]$$
$$\left. - \cos\left(2\beta + \frac{3}{2}\alpha\right)\left[\sin \alpha \sin^2 \frac{\alpha}{2}\right] \right\}$$

Infinite Plates Subjected to Bending (and Shearing)

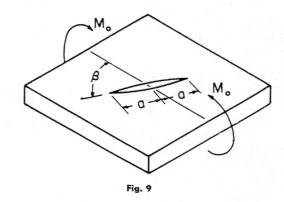

Fig. 9

Reference [8]

$$\varphi_b = -\frac{M_0(\sin^2 \beta)a}{4D(1 + \nu)}\left[\zeta + \frac{1-\nu}{3+\nu}\frac{1}{\zeta}\right] - \frac{iM_0(\sin 2\beta)a}{4D(3+\nu)}\left[\frac{1}{\zeta}\right]$$

$$K_1 = \frac{6M_0}{h^2}(\sin^2\beta)a^{1/2}$$

$$K_2 = \frac{6M_0}{h^2}(\sin\beta\cos\beta)a^{1/2}$$

$$\varphi_b = -\frac{Q_0 a^2}{12D}\left[\zeta^2 e^{-i\beta} + \left(\frac{1-\nu}{3+\nu}\right)\frac{e^{-i\beta} + 2e^{i\beta}}{\zeta^2}\right]$$

$$K_1 = \frac{8\nu Q_0 a^{1/2}}{h^2}\cos\beta$$

$$K_2 = \frac{8Q_0 a^{1/2}}{h^2}\sin\beta$$

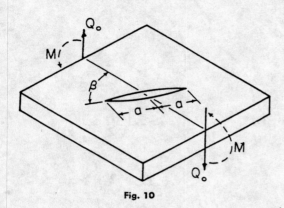

Fig. 10

Reference [11]

ON CRACKS IN RECTILINEARLY ANISOTROPIC BODIES

G. C. Sih*
P. C. Paris**
G. R. Irwin***

ABSTRACT

The general equations for crack-tip stress fields in anisotropic bodies are derived making use of a complex variable approach. The stress-intensity-factors, which permit concise representation of the conditions for crack extension, are defined and are evaluated directly from stress functions. Some individual boundary value problem solutions are given in closed form and discussed with reference to their companion solutions for isotropic bodies.

It is found that an elastic stress singularity of the order $r^{-\frac{1}{2}}$ is always present at the crack tip in a body with rectilinear anisotropy (r being the radial distance from the crack front). This result and some additional consideration of the crack-tip stress fields imply that it is possible to extend current fracture mechanics methods to the representation of fracture conditions for anisotropic bodies with cracklike imperfections.

INTRODUCTION

Recent work [1-3]*** in the mechanics of fracture points out the desirability of a knowledge of the elastic energy release rate, i.e., the crack extension force, and the character of the stress field surrounding a crack tip in analyzing the strength of cracked bodies. The objective of this paper is to provide a discussion of the energy rates, stress fields and the like for various cases of anisotropic elastic bodies which might be of interest.

Wood, laminates, reinforced concrete, and some special types of metals systems with controlled grain orientation are often orthotropic and at least rectilinearly anisotropic from point to point, if regarded as homogeneous media. Plates with stiffners or corregations may also be treated as orthotropic bodies in some cases. The solutions to problems similar to the Yoffe moving crack, a constant length 2-dimensional crack moving with a fixed speed parallel to its length, resemble orthotropic material crack problems in form. Gilman[4] has pointed out the usefulness of energy rate analyses in determining the surface energy of brittle crystals by cleaving them. Most crystals are orthotropic in nature, however, some must be considered as anisotropic media of a more general form[5]. Morever, orthotropic bodies where a crack is not associated with a plane of elastic symmetry may be conveniently treated as a crack problem in a generally anisotropic body.

As has been observed earlier[1], when attention is focussed on the local fields of elastic stress and displacement associated with the singularity near a crack tip, it is always possible to reduce the consideration of the most general situation to a sum of plane and antiplane**** problems. This is equally true in anisotropic elastic media upon centering that attention

---

* Professor of Mechanics, Lehigh University, Bethlehem Pennsylvania.
** Program Director, National Science Foundation, Washington, D.C., (on leave from Lehigh University 1964-1965).
*** Superintendent of Mechanics Division, U.S. Naval Research Laboratories, Washington, D.C.
**** The term antiplane was introduced by L.N.G. Filon[6] to describe such problems as the torsion and flexure of bars by transverse loads.

on a sufficiently small segment of a crack front region. The resulting simplification is most useful. Indeed, the analysis of the resulting plane and antiplane problems of anisotropic elasticity is in a sense less difficult than the isotropic case, since the bi-harmonic and harmonic equations are replaced with homogeneous partial differential equations of identical order but with unequal roots of their characteristic equations.

It is also notable that the aforementioned differential equations are of such a form that stress functions from isotropic elasticity may be employed in the anisotropic case with at most slight modifications. For example, the Westergaard stress function approach[1] may be directly applied to orthotropic problems using exactly the same stress functions as the isotropic case for individual examples. Similar advantages pertain to the more general complex variable approach[7] adopted in this paper.

In treating the crack tip fields by reductions to plane and antiplane problems and subsequently to three characteristic modes[1], a choice for convenience can be made in defining these modes and their corresponding stress-intensity-factors. Defining the basic modes as those which lead to the vanishing of certain stress components ahead of the crack on the crack plane seems most appropriate. This choice results in symmetrically loaded extension being associated with the first mode, skew-symmetrical extension with the second mode and antiplane loadings with the third mode. Moreover, a subsequent judicious selection of the definitions of anisotropic stress intensity factors makes their resulting forms identical to those for most corresponding plane isotropic problems, (i.e., all those with the absence of or self-equilibrating loads on the crack surface). Consequently, these multiple simplifications permit the direct use of the concepts and results from the isotropic case in the general anisotropic case.

An exception is the plane problem of a non-equilibrated concentrated force on a crack surface. The solution to this problem is given to illustrate the influence of elastic constants on stress-intensity-factors, as well as to provide the Green's function for forces on the crack's surface. Moreover, in an analysis of the virtual work of crack extension, in order to obtain the "generalized forces" of crack extension in terms of stress intensity factors, the elastic constants again appear. For the orthotropic and isotropic cases the elastic energy rates associated with virtual crack extension may be simply algebraically added for loadings resulting in the simultaneous presence of more than one mode. However, in the general anisotropic case cross-influences between displacements of various modes will appear in the energy rates. The cross-influences will be illustrated for the two modes associated with plane problems.

The stress singularities associated with cracks in isotropic bodies are always[1] of the order $r^{-\frac{1}{2}}$, where r is the distance measured from the crack front. Chapkis and Williams[8] observed stress singularities of other orders in bodies which are polarly orthotropic about the crack tip. Nevertheless, the $r^{-\frac{1}{2}}$ singularity will be shown to prevail in the general case of rectilinear anisotropy.

Although many authors[7-14] have found expressions for stresses in anisotropic bodies with cracks, none has previously examined the nature of the local crack-tip stress field in a general sense. Ang and Williams[13] did at least examine local stresses for a particular example of uniform extension and bending of an orthotropic plate. Hence, results will be given in some detail for the general character and distribution of crack-tip stress fields. Finally, possible direct application of the concepts of current fracture criteria to anisotropic media will be suggested.

PLANE AND ANTIPLANE PROBLEMS OF ANISOTROPIC ELASTICITY

The basic equations which describe the deformation of anisotropic bodies are the same as those for isotropic bodies except for the adoption of a generalized Hooke's law[*]:

$$\epsilon_i = \sum_{j=1}^{6} a_{ij}\sigma_j, \quad a_{ij} = a_{ji}, \quad (i = 1, 2\ldots 6) \tag{1}$$

where the number of independent elastic constants, a_{ij}, is related to the elastic symmetry properties of the material.

The assumptions of certain states of deformation (and stress) which lead to plane and antiplane problems shall be those which are most convenient in analyzing the stresses and displacements near cracks. In all of the discussions to follow, the z-axis is taken to be parallel to the leading edge of a crack during deformation. The x and y axes are directed parallel and normal to the crack surface, respectively, as in Fig. 1.

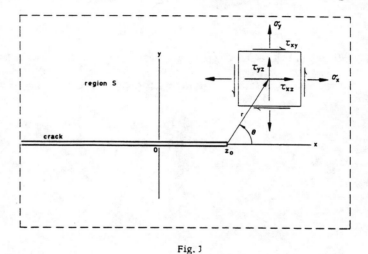

Fig. 1

A brief exposition of the complex variable formulation of plane and antiplane problems of anisotropic elasticity will be given:

(a) *Plane Problems*

The elastic body, considered in problems of plane extension, has at each point a plane of symmetry parallel to the surface of a thin plate in plane stress or normal to the generator of a cylinder of infinite length in plane strain. It follows that the formulation of this class of problems can be reduced to a dependency upon six independent elastic constants[7]

$$a_{ij} \quad (i, j = 1, 2, 6) \tag{2}$$

---

[*] The notation

$$\epsilon_1 = \epsilon_x, \; \epsilon_2 = \epsilon_y, \ldots, \epsilon_6 = \gamma_{xy}$$
$$\sigma_1 = \sigma_x, \; \sigma_2 = \sigma_y, \ldots, \sigma_6 = \tau_{xy}$$

has been adopted in eq. (1).

for plane stress and a corresponding set of

$$b_{ij} = a_{ij} - \frac{a_{i2} a_{j3}}{a_{33}} \quad (i, j = 1, 2, 6) \tag{3}$$

for plane strain. From a mathematical standpoint, the plane stress and plane strain problems are identical except for the values of elastic constants entering into the reduced strain-stress relations, (1). Henceforth it is understood that a_{ij} and b_{ij} for i, j = 1, 2, 6 are the reduced elastic constants used in the plane stress and plane strain problems, respectively.

It has been shown[8] that the problems of two-dimensional anisotropic elasticity can be conveniently formulated in terms of the analytic functions, $\phi_j(z_j)$, of the complex variable, $z_j = x_j + iy_j$ (j = 1, 2), where

$$x_j = x + \alpha_j y, \quad y_j = \beta_j y \quad (j = 1, 2)$$

The parameters α_j and β_j are the real and imaginary parts of μ_j, i.e., $\mu_j = \alpha_j + i\beta_j$, as determined from

$$a_{11} \mu^4 - 2a_{16} \mu^3 + (2a_{12} + a_{66})\mu^2 - 2a_{26}\mu + a_{22} = 0 \tag{4}$$

The roots μ_j of eq. (4) are always complex or purely imaginary and will occur in conjugate pairs as μ_1, $\bar{\mu}_1$ and μ_2, $\bar{\mu}_2$. In the case of distinct roots, the general expressions for the components of stresses and displacements are

$$\sigma_x = 2 \, \text{Re} \left[\mu_1^2 \phi_1'(z_1) + \mu_2^2 \phi_2'(z_2) \right]$$
$$\sigma_y = 2 \, \text{Re} \left[\phi_1'(z_1) + \phi_2'(z_2) \right]$$
$$\tau_{xy} = -2 \, \text{Re} \left[\mu_1 \phi'(z_1) + \mu_2 \phi'(z_2) \right] \tag{5}$$

and

$$u = 2 \, \text{Re} \left[p_1 \phi_1(z_1) + p_2 \phi_2(z_2) \right]$$
$$v = 2 \, \text{Re} \left[q_1 \phi_1(z_1) + q_2 \phi_2(z_2) \right] \tag{6}$$

where:

$$p_j = a_{11} \mu_j^2 + a_{12} - a_{16} \mu_j$$
$$q_j = a_{12} \mu_j + \frac{a_{22}}{\mu_j} - a_{26} \tag{7}$$

Thus, the solution of a plane problem of anisotropic elasticity requires the determination of two functions, $\phi_1(z_1)$ and $\phi_2(z_2)$ which satisfy the boundary conditions on the contour of the crack region under consideration. It may be noted that the structure of $\phi_j(z_j)$ (j = 1, 2) does not change when the constants a_{ij} are replaced by b_{ij}; only the numerical values of the complex roots μ_j and the constants, p_j, q_j, vary.

(b) *Antiplane Problems*

The antiplane problem (i.e., including torsion and/or flexure) of an anisotropic cylinder containing cracks in a region of cross section S can be reduced to the corresponding problem for an isotropic cylinder which has a region of cross section S_3 obtained from S by means of the affine transformation $x_3 = x + \alpha_3 y$ and $y_3 = \beta_3 y$. If there is a two-fold symmetry axis or its equivalent in the z-direction coinciding with the generators of the cylinder, the elastic displacements associated with this problem without body forces are such that

$$u = v = 0, \quad w = w(x, y) \tag{8}$$

The stress-strain law simplifies to

$$\begin{aligned}\tau_{xz} &= c_{45} \gamma_{yz} + c_{55} \gamma_{xz} \\ \tau_{yz} &= c_{44} \gamma_{yz} + c_{45} \gamma_{xz}\end{aligned} \tag{9}$$

the elastic constants c_{ij} (i, j = 4, 5) are defined as

$$\begin{aligned}\Delta_3 c_{ij} &= -a_{ij}, \quad (i \neq j) \\ &= \frac{a_{44} a_{55}}{a_{ij}}, \quad (i = j)\end{aligned} \tag{10}$$

for values of i, j = 4, 5 and

$$\Delta_3 = a_{44} a_{55} - a_{45}^2$$

Making use of (9), the strain-displacement relationships and the equation of equilibrium, the displacement w satisfies the equation

$$c_{55} \frac{\partial^2 w}{\partial x^2} + 2 c_{45} \frac{\partial^2 w}{\partial x \partial y} + c_{44} \frac{\partial^2 w}{\partial y^2} = 0 \tag{11}$$

whose solution can be expressed in terms of a stress function, $\phi_3(z_3)$. The complex variable is $z_3 = x + \mu_3 y$, where $\mu_3 = \alpha_3 + i\beta_3$ and its conjugate $\bar{\mu}_3 = \alpha_3 - i\beta_3$ are the roots of

$$c_{44} \mu^2 + 2 c_{45} \mu + c_{55} = 0 \tag{12}$$

The nontrivial stresses and displacement can be expressed in terms of $\phi_3(z_3)$ by

$$\begin{aligned}\tau_{xz} &= 2 \operatorname{Re}\left[\mu_3 \phi'_3(z_3)\right] \\ \tau_{yz} &= -2 \operatorname{Re}\left[\phi'_3(z_3)\right]\end{aligned} \tag{13}$$

and

$$w = 2 \operatorname{Re}\left[\frac{\mu_3}{c_{55} + \mu_3 c_{45}} \phi_3(z_3)\right] \tag{14}$$

This problem is somewhat simpler than the plane problem since it only requires the knowledge of a single function satisfying the appropriate boundary conditions.

STRESS AND DISPLACEMENT FIELDS NEAR THE CRACK TIP

A knowledge of the stress fields in the neighborhood of the crack tip is essential in analyzing the fracture strength of cracked bodies. In particular, the local intensity of the stress fields plays a dominant role in fracture[1].

The stress field near crack tips in an anisotropic body can be divided into local modes of deformation, but the degree of simplification thus achieved is less than for the isotropic case. This is because the crack surface displacements in general anisotropic media depend upon the directional properties of the material and do not necessarily occur in a planar fashion. However, the general state of stress and displacement near a crack tip can still be considered as the sum of three individual boundary problems; namely, those corresponding to inplane-symmetric loading, inplane-skewsymmetric loading and antiplane shear. It should be mentioned that if the material is orthotropic in nature with one of the preferred directions alined with the crack, it is then also possible to have three associated independent modes of deformation similar to those in the isotropic material[1].

Let the region occupied by the plane of $z = x + iy$ with a straight crack on the x-axis, Fig. 1, be denoted by S. (The complex variable, $z = x + iy$ should not be confused with the coordinate axis, z.) As evidenced from eqs. (5) and (13), the stresses depend on three functions $\phi_j'(z_j)$ of the ordinary complex variables $z_j = x_j + iy_j$, $(j = 1, 2, 3)$. Here, the functions $\phi_j'(z_j)$ must be determined, not in the region S, but in the three regions, S_j $(j = 1, 2, 3)$, which are obtained from S by means of the affine transformations, $x_j = x + \alpha_j y$ and $y_j = \beta_j y$. For problems involving lines of discontinuities, $\phi_j'(z_j)$ are sectionally holomorphic functions in the sense of the definition given by Muskhelishvili[15]. That is, close to a crack tip at z_o, $\phi_j'(z_j)$ may be written as

$$\phi_j'(z_j) = \frac{\psi_j^{(1)}(z_j)}{\sqrt{z_j - z_o}} + \psi_j^{(2)}(z_j), \quad (j = 1, 2, 3) \tag{15}$$

where $\psi_j^{(\ell)}(z_j)$, are holomorphic functions at the crack tip z_o, i.e.,

$$\psi_j^{(\ell)}(z_j) = \sum_{n=0}^{\infty} \lambda_{jn}^{(\ell)}(z_j - z_o)^n, \quad (\ell = 1, 2 \text{ and } j = 1, 2, 3) \tag{16}$$

The character of the crack tip stress and displacement fields may be examined by introducing polar coordinates r and θ such that (see Fig. 1),

$$z_j - z_o = re^{i\theta} \tag{17}$$

In (17), r is the radial distance from the crack tip and θ is the angle between the radius vector and the extension of the crack plane. Substituting (17) into (15), the stress functions near the crack tip may be approximated by

$$\phi_j'(z_j) = \frac{\lambda_{jo}^{(1)}}{\sqrt{r(\cos\theta + \mu_j \sin\theta)}} + O(r^{\frac{1}{2}}) \tag{18}$$

The higher order terms in r may be disregarded in (18) as r becomes small compared to other planar dimensions. The constants $\lambda_{jo}^{(1)}$ are in fact a form of the stress intensity factors. However, they are redefined as k_j $(j = 1, 2, 3)$ in a manner consistent with those for the isotropic case, where

$$\lambda_{10}^{(1)} = -\frac{\mu_2}{2\sqrt{2}(\mu_1 - \mu_2)}(k_1 + \frac{k_2}{\mu_2})$$

$$\lambda_{20}^{(1)} = \frac{\mu_1}{2\sqrt{2}\,(\mu_1 - \mu_2)} \left(k_1 + \frac{k_2}{\mu_1}\right)$$

$$\lambda_{30}^{(1)} = -\frac{k_3}{2\sqrt{2}} \tag{19}$$

The stresses and displacements in a small region surrounding the crack tip may now be obtained without difficulty. They are:

(a) *Plane Extension*

1. Symmetric loading.

$$\sigma_x = \frac{k_1}{\sqrt{2r}} \, \text{Re}\left[\frac{\mu_1 \mu_2}{\mu_1 - \mu_2} \left(\frac{\mu_2}{\sqrt{\cos\theta + \mu_2 \sin\theta}} - \frac{\mu_1}{\sqrt{\cos\theta + \mu_1 \sin\theta}}\right)\right]$$

$$\sigma_y = \frac{k_1}{\sqrt{2r}} \, \text{Re}\left[\frac{1}{\mu_1 - \mu_2} \left(\frac{\mu_1}{\sqrt{\cos\theta + \mu_2 \sin\theta}} - \frac{\mu_2}{\sqrt{\cos\theta + \mu_1 \sin\theta}}\right)\right]$$

$$\tau_{xy} = \frac{k_1}{\sqrt{2r}} \, \text{Re}\left[\frac{\mu_1 \mu_2}{\mu_1 - \mu_2} \left(\frac{1}{\sqrt{\cos\theta + \mu_1 \sin\theta}} - \frac{1}{\sqrt{\cos\theta + \mu_2 \sin\theta}}\right)\right]$$

and

$$u = k_1 \sqrt{2r} \, \text{Re}\left[\frac{1}{\mu_1 - \mu_2} \left(\mu_1 p_2 \sqrt{\cos\theta + \mu_2 \sin\theta} - \mu_2 p_1 \sqrt{\cos\theta + \mu_1 \sin\theta}\right)\right]$$

$$v = k_1 \sqrt{2r} \, \text{Re}\left[\frac{1}{\mu_1 - \mu_2} \left(\mu_1 q_2 \sqrt{\cos\theta + \mu_2 \sin\theta} - \mu_2 q_2 \sqrt{\cos\theta + \mu_1 \sin\theta}\right)\right] \tag{20}$$

2. Skew-symmetric loading.

$$\sigma_x = \frac{k_2}{\sqrt{2r}} \, \text{Re}\left[\frac{1}{\mu_1 - \mu_2} \left(\frac{\mu_2^2}{\sqrt{\cos\theta + \mu_2 \sin\theta}} - \frac{\mu_1^2}{\sqrt{\cos\theta + \mu_1 \sin\theta}}\right)\right]$$

$$\sigma_y = \frac{k_2}{\sqrt{2r}} \, \text{Re}\left[\frac{1}{\mu_1 - \mu_2} \left(\frac{1}{\sqrt{\cos\theta + \mu_2 \sin\theta}} - \frac{1}{\sqrt{\cos\theta + \mu_1 \sin\theta}}\right)\right]$$

$$\tau_{xy} = \frac{k_2}{\sqrt{2r}} \, \text{Re}\left[\frac{1}{\mu_1 - \mu_2} \left(\frac{\mu_1}{\sqrt{\cos\theta + \mu_1 \sin\theta}} - \frac{\mu_2}{\sqrt{\cos\theta + \mu_2 \sin\theta}}\right)\right] \tag{21}$$

and

$$u = k_2 \sqrt{2r} \, \text{Re}\left[\frac{1}{\mu_1 - \mu_2} \left(p_2 \sqrt{\cos\theta + \mu_2 \sin\theta} - p_1 \sqrt{\cos\theta + \mu_1 \sin\theta}\right)\right]$$

$$v = k_2 \sqrt{2r} \, \text{Re}\left[\frac{1}{\mu_1 - \mu_2} \left(q_2 \sqrt{\cos\theta + \mu_2 \sin\theta} - q_1 \sqrt{\cos\theta + \mu_1 \sin\theta}\right)\right]$$

(b) *Antiplane Shear*

$$\tau_{xz} = -\frac{k_3}{\sqrt{2r}} \operatorname{Re}\left[\frac{\mu_3}{\sqrt{\cos\theta + \mu_3 \sin\theta}}\right] \quad (22)$$

$$\tau_{yz} = \frac{k_3}{\sqrt{2r}} \operatorname{Re}\left[\frac{1}{\sqrt{\cos\theta + \mu_3 \sin\theta}}\right]$$

and

$$w = k_3 \sqrt{2r}\, \operatorname{Re}\left[\frac{\sqrt{\cos\theta + \mu_3 \sin\theta}}{c_{45} + \mu_3 c_{44}}\right]$$

The first important observation which may be gained from (20)-(22) is that the stress singularity at the crack tip is of the order of $r^{-\frac{1}{2}}$. Moreover, the angular distribution of local stresses, i.e., their variation with θ, depends upon the material properties through μ_j. The dependency of the crack tip stress distribution on remote boundary conditions is like the isotropic case where the crack geometry and the applied loads affect only the intensity of the stresses, k_j. The conclusions which can be drawn from this knowledge will be elaborated upon in a subsequent section.

STRESS-INTENSITY FACTORS AND ENERGY RATES

A method of determining the stress-intensity factors k_j in generally anisotropic media can now be developed. Since the functions $\phi'_j(z_j)$ each have a removable singularity at the crack tip, z_o, then $\lambda_{jo}^{(1)}$ or k_j in (19) may be interpreted as the strength of the stress singularity.

Inserting (17) and (19) into (18) and recalling that (18) is exact in the limit as z_j approach z_o, the stress intensity factors, k_j, may be evaluated from the formulas:

$$k_1 + \frac{k_2}{\mu_2} = -2\sqrt{2}\left(\frac{\mu_1 - \mu_2}{\mu_2}\right) \lim_{z_1 \to z_o} (z_1 - z_o)^{\frac{1}{2}} \phi'_1(z_1)$$

or (23)

$$k_1 + \frac{k_2}{\mu_1} = 2\sqrt{2}\left(\frac{\mu_1 - \mu_2}{\mu_1}\right) \lim_{z_2 \to z_o} (z_2 - z_o)^{\frac{1}{2}} \phi'_2(z_2)$$

and

$$k_3 = -2\sqrt{2} \lim_{z_3 \to z_o} (z_1 - z_o)^{\frac{1}{2}} \phi'_3(z_3) \quad (24)$$

Therefore, k_j may always be formed directly from $\phi'_j(z_j)$ in the vicinity of the crack tip.

In addition, the stress intensity factors, as found from (23) and (24), may be related to the energy release rates, $\mathfrak{I}_j$. The relationships among the k_j and $\mathfrak{I}_j$ (j = 1, 2, 3) form the basis of the equivalence of the Griffith energy theory and the current stress intensity factor approach in fracture mechanics. As noted earlier, crack extension is not always collinear with the line of the crack itself. Mathematical difficulties involved in angled crack problems seem to prohibit any current solutions of an analytic nature. However, as a matter of resolvable interest, assume that a closure of a crack segment of a distance, δ, takes place. Under these conditions the energy rates, $\mathfrak{I}_j$, may be computed from the integrals*

$$\mathfrak{I}_j = \lim_{\delta \to 0} \frac{1}{\delta} \int_0^\delta \sigma_j(\delta - r, 0) \cdot u_j(r, \pi) dr, \quad (j = 1, 2, 3) \tag{25}$$

Substituting the appropriate stress and displacement expressions given by (20) through (22) into eqs. (25), it is found that

$$\mathfrak{I}_1 = -\frac{\pi k_1^2}{2} a_{22} \operatorname{Im}\left[\frac{\mu_1 + \mu_2}{\mu_1 \mu_2}\right]$$

$$\mathfrak{I}_2 = \frac{\pi k_2^2}{2} a_{11} \operatorname{Im}\left[\mu_1 + \mu_2\right]$$

$$\mathfrak{I}_3 = \frac{\pi k_3^2}{2} \frac{\operatorname{Im}\left[c_{45} + \mu_3 c_{44}\right]}{c_{44} c_{55}} \tag{26}$$

These latter expressions are the energy rates for each mode treated separately, i.e., in the absence of the other two modes. If more than one mode is present at a crack tip, cross-influences of displacements cause the appearance of cross product terms of k_j in the $\mathfrak{I}_j$. For example, if the first two modes are present together, then:

$$\mathfrak{I}_1 = -\frac{\pi k_1}{2} a_{22} \operatorname{Im}\left[\frac{k_1(\mu_1 + \mu_2) + k_2}{\mu_1 \mu_2}\right]$$

$$\mathfrak{I}_2 = \frac{\pi k_2}{2} a_{11} \operatorname{Im}\left[k_2(\mu_1 + \mu_2) + k_1 \mu_1 \mu_2\right]$$

where (26a)

$$\mathfrak{I}_{total} = \mathfrak{I}_1 + \mathfrak{I}_2$$

On the other hand if the material is orthotropic with the crack on one plane of symmetry, the three basic modes[1] are conveniently independent and (26) become

$$\mathfrak{I}_1 = \pi k_1^2 \sqrt{\frac{a_{11} a_{22}}{2}} \left[\sqrt{\frac{a_{22}}{a_{11}}} + \frac{2a_{12} + a_{66}}{2a_{11}}\right]^{\frac{1}{2}}$$

$$\mathfrak{I}_2 = \pi k_2^2 \frac{a_{11}}{\sqrt{2}} \left[\sqrt{\frac{a_{22}}{a_{11}}} + \frac{2a_{12} + a_{66}}{2a_{11}}\right]^{\frac{1}{2}}$$

---

* For compactness, the notation $\sigma_1 = \sigma_y$, $\sigma_2 = \tau_{xy}$, $\sigma_3 = \tau_{yz}$, $u_1 = v$, $u_2 = u$, $u_3 = w$ is introduced.

$$\Im_3 = \pi k_3^2 \frac{1}{2\sqrt{c_{44} c_{55}}} \tag{27}$$

where

$$\Im_{total} = \Im_1 + \Im_2 + \Im_3$$

For an isotropic material, (27) reduce to the usual results[1].

Therefore, once the stress-intensity-factors, k_j, are known, the energy rates, $\Im_j$, can be determined from (26) and (27). It remains to determine the stress-intensity-factors for some example configurations of basic interest with anisotropic media.

EXTENSION OF AN ANISOTROPIC PLATE WITH A CRACK

The problem of stretching of an anisotropic plate containing a traction free crack of length, 2a, will be considered. Let the stress state at infinity be uniform tension of intensity σ^∞ acting at an angle, α, to the x-axis, as in Fig. 2. For this problem, the stress functions, $\phi_j(z_j)$ (j = 1, 2), are of the type,[9]:

$$\phi_1(z_1) = \frac{\sigma^\infty a^2}{4(\mu_1 - \mu_2)} \left[\frac{2\mu_2 \sin^2 \alpha + \sin 2\alpha}{z_1 + \sqrt{z_1^2 - a^2}} \right] + \Gamma_1 z_1$$

$$\phi_2(z_2) = -\frac{\sigma^\infty a^2}{4(\mu_1 - \mu_2)} \left[\frac{2\mu_1 \sin^2 \alpha + \sin 2\alpha}{z_2 + \sqrt{z_2^2 - a^2}} \right] + \Gamma_2 z_2 \tag{28}$$

where the Γ_j are constants.

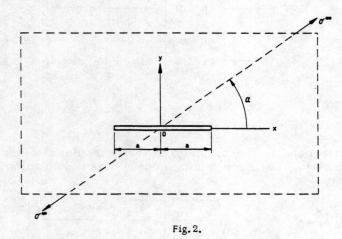

Fig. 2.

Substituting either $\phi_1(z_1)$ or $\phi_2(z_2)$ from (28) into the first or second of eqs. (23) gives the same results for k_1 and k_2:

$$k_1 = \sigma^\infty a^{\frac{1}{2}} \sin^2 \alpha$$
$$k_2 = \sigma^\infty a^{\frac{1}{2}} \sin \alpha \cos \alpha \tag{29}$$

The preceding equations show that the influence of the applied load and the geometric size of the crack on the intensity of the local stresses is identical with the isotropic case[2]. This is in general true for self-balancing loads on the crack surface. The effect of anisotropy on the stress-intensity-factors for problems in which the crack surface loads are not self-equilibrated will be discussed in a later example.

A LONGITUDINAL SHEAR CRACK IN AN ANISOTROPIC CYLINDER

Consider the stress state in a cylindrical body produced by antiplane shear loads uniformly distributed and directed along the longitudinal axis. The cylinder is first assumed infinite in longitudinal extent, then in other dimensions. A uniform shear, τ^∞, inclined at an angle α to the x-axis, is applied at infinity. The anisotropic cylinder contains a crack of length, 2a, as shown in Fig. 3. The stress function, $\phi_3(z_3)$, may be derived in

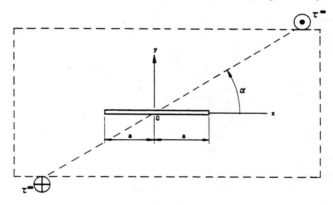

Fig. 3.

the same way as that of the preceding example. It takes the form

$$\phi_3(z_3) = \frac{\tau^\infty a^2}{2}\left[\frac{\sin \alpha}{z_3 + \sqrt{z_3^2 - a^2}}\right] + \Gamma_3 z_3 \tag{30}$$

k_3 is obtained using (24):

$$k_3 = \tau^\infty a^{\frac{1}{2}} \sin \alpha \tag{31}$$

Again, the stress intensity factor, (31), is identical to its isotropic counterpart, (21) in Ref. 3.

CONCENTRATED FORCE ON THE CRACK SURFACE IN AN ANISOTROPIC PLATE

Fig. 4 shows an infinite sheet possessing rectilinear anisotropy with a crack of length, 2a, along the x-axis. A concentrated force, P, per unit thickness is applied normal to the upper surface of the crack at z = b.

Muskhelishvili[15] has obtained the solution of the analogous problem for an isotropic plate containing a finite crack with a uniformly distributed pressure on a portion of its surface. By taking limiting values of the stress functions, the case of a concentrated normal force applied to the crack contour may be found. His solution was based on certain properties of integrals of the Cauchy type, taken along the boundary of the unit circle in

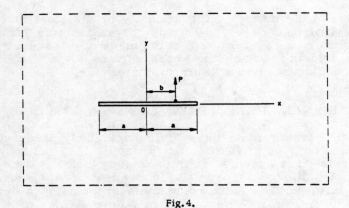

Fig. 4.

a mapped plane. The method can also be adapted for crack problems of anisotropic elasticity. Since such adaptations have already been described in detail by several authors[2,9,15], only the final results will be given here. Furthermore, if stress intensity factors are the only required information, it suffices to find $\phi_1(z_1)$ only, as $\phi_2(z_2)$ yields no additional information on them. For the present problem, Fig. 4, $\Phi_1(\zeta_1)$, which corresponds to $\phi_1(z_1)$ in the mapped plane, is

$$\Phi_1(\rho_1) = - \frac{P a \mu_2}{4\pi(\mu_1 - \mu_2)} (a^2 - b^2)^{\frac{1}{2}} \left\{ -\frac{1}{\zeta_1} + \frac{\zeta_b}{\zeta_b - \zeta_1} \left[(\zeta_1 + \frac{1}{\zeta_1}) - (\zeta_b + \frac{1}{\zeta_b}) \right] + (\zeta_b - \frac{1}{\zeta_b}) \left[\Omega \log \zeta_1 - \log(\zeta_b - \zeta_1) \right] \right\} \quad (32)$$

where the function for mapping the crack problem into the interior of a unit circle is employed, i.e.,

$$z_j = \omega(\zeta_j) = \frac{a}{2} (\zeta_j + \frac{1}{\zeta_j}) \quad (33)$$

In (32) the point ζ_b corresponds to $z = b$, the location of the force, P. The quantity Ω in eq. (32) is a function of the elastic constants given by

$$\Omega = \frac{\mu_1 - \mu_2}{\mu_2} \left(\frac{\Delta_1 + i \Delta_2}{\Delta} \right) \quad (34)$$

in which

$$\Delta = \frac{a_{11} a_{22} \beta_1 \beta_2}{(\alpha_1^2 + \beta_1^2)(\alpha_2^2 + \beta_2^2)} \left[(\alpha_2 - \alpha_1)^4 + 2(\alpha_2 - \alpha_1)^2 (\beta_2^2 + \beta_1^2) + (\beta_2^2 - \beta_1^2)^2 \right] \quad (35)$$

and Δ_j (j = 1, 2) stand for

$$\Delta_1 = \frac{i}{2} \left\{ p_2^{(2)} \left[\alpha_2 q_1^{(1)} - \alpha_1 q_2^{(1)} \right] + q_2^{(2)} \left[\alpha_1 p_2^{(1)} - \alpha_2 p_1^{(1)} \right] + \beta_2 \left[p_1^{(1)} q_2^{(1)} - p_2^{(1)} q_1^{(1)} \right] \right\} \quad (36)$$

$$\Delta_2 = \frac{i}{2}\left\{q_2^{(1)}\left[\beta_1 p_2^{(2)} - \beta_2 p_1^{(2)}\right] + p_2^{(1)}\left[\beta_2 q_1^{(2)} - \beta_1 q_2^{(2)}\right]\right.$$
$$\left. + \alpha_2\left[p_1^{(2)}q_2^{(2)} - p_2^{(2)}q_1^{(2)}\right]\right\} \tag{37}$$

The latter equations are expressed in terms of the real and imaginary parts of p_j and q_j ($j = 1, 2$) as defined in eq. (7), i.e.,

$$p_j = p_j^{(1)} + i\, p_j^{(2)}, \quad q_j = q_j^{(1)} + i q_j^{(2)}$$

Using the first of (23) with modification for mapping, exactly as in Ref. 2, and (32), the stress intensity factors for this problem may be evaluated from

$$k_1 + \frac{k_2}{\mu_2} = \frac{P}{2\pi a^{\frac{1}{2}}}\left[\left(\frac{a+b}{a-b}\right)^{\frac{1}{2}} - i(2\Omega - 1)\right] \tag{38}$$

With the aid of (7) and (34)-(37), the crack tip stress intensity factors, k_j, ($j = 1, 2$) can be obtained from (38). The general results are too cumbersome to be given here, but in the case of orthotropy, i.e., for $\mu_1 = \alpha_0 + i\beta_0$ and $\mu_2 = -\alpha_0 + i\beta_0$,

$$\Omega = -\frac{1}{8}\left(\frac{\mu_1 - \mu_2}{\mu_2}\right)\left\{\frac{i}{\alpha_0 \beta_0}\left[\frac{a_{12}}{a_{11}} + (\alpha_0^2 - \beta_0^2)\right] + 1\right\} \tag{39}$$

Putting (39) into (38), k_j ($j = 1, 2$) are found:

$$k_1 = \frac{P}{2\pi a^{\frac{1}{2}}}\left[\left(\frac{a+b}{a-b}\right)^{\frac{1}{2}} - \frac{1}{2}\left(\frac{\alpha_0}{\beta_0}\right)\right]$$

$$k_2 = -\frac{P}{2\pi a^{\frac{1}{2}}}\left\{\frac{\alpha_0^2}{2\beta_0} + \frac{1}{2\beta_0}\left[\frac{a_{12}}{a_{11}} + (\alpha_0^2 + \beta_0^2)\right]\right\} \tag{40}$$

which agrees with the isotropic solution when $\alpha_0 = 0$, $\beta_0 = 1$ and $a_{12}/a_{11} = -\nu$ for plane stress (see (21) in Ref. 2).

As should have been expected, the magnitude of the stress field depends upon the material properties when the resultant force on the crack surface does not vanish.

APPLICATION OF FRACTURE MECHANICS THEORIES TO ANISOTROPIC BODIES

Essentially, Irwin's extensions[1] of Griffith's approach to fracture are centered upon a proper stress analysis of each example considered in order to examine size effects and in order to compare the strengths of different cracked configurations. It has been noted that an elastic stress analysis is adequate for this purpose[1-3]. Hence, it seems reasonable to presume that similar assumptions might be incorporated into a workable theory of strength of anisotropic bodies with crack-like flaws.

Equations (20)-(22) are the desired general equations for the elastic stress field near a crack tip. In order to compare the cracked strength of bodies of the same material, it is of primary importance that the distribution of stress near the crack tip be the same in the bodies. If bodies having different configurations as well as size are compared, (20)-(22) imply that their stress fields will in fact be distributed in the same manner, and will differ only by their intensity factors for a given crack orientation. These intensity factors will depend upon the magnitude of the loads applied and to some measure on the geometric size of the configuration. Therefore, the spirit of the fracture mechanics concepts implies that the strength of cracked anisotropic bodies may be predicted in exactly the same manner as the case of isotropic bodies. The simple extension of the theory of fracture mechanics to anisotropic bodies may be stated as:

"The combination of stress-intensity-factors, computed from (23) and (24), will cause unstable crack propagation upon reaching some critical value," i.e.,

$$f(k_1, k_2, k_3) = f_{cr} \tag{41}$$

It should be reemphasized that for problems involving self-equilibrating loads, the stress intensity factors for both the isotropic and anisotropic materials are identical as illustrated by (29) and (31). Hence, k_j for numerous problems that are already known for the isotropic case[1-3] may be directly employed to attempt to estimate the fracture strength of cracked anisotropic bodies. If the loads on the crack surface are unbalanced, the stress intensity factors will in general be complicated functions of the material constants and may be found from the stress function, $\phi_j(z_j)$, by the procedure described in this paper.

Received August 31, 1965.

REFERENCES

1. G.R. Irwin — Handbuch der Physik, 6, Springer-Verlag, Berlin, 1958, pp. 551-590.
2. G.C. Sih, P.C. Paris, and F. Erdogan — Trans. ASME, J. Appl. Mech., 29, 2, June 1962, pp. 306-312.
3. G.C. Sih — Trans. ASME, J. Appl. Mech., 32, 1, March 1965, pp. 51-58.
4. J.J. Gilman — Fracture, Technology Press and John Wiley and Sons, 1959, pp. 193-224.
5. C. Zener — Elasticity and Anelasticity of Metals, University of Chicago Press, 1948, pp. 7-23.
6. L.N.G. Filon — Proc. Roy. Soc. (London), Series A, 160, 1937, pp. 137-154.
7. S.G. Lekhnitskii — Theory of Elasticity of an Anisotropic Body, Chapter 3, Holden-Day, Inc., 1963.
8. R.L. Chapkis and M.L. Williams — Proc. of Third U.S. Natl. Congr. Appl. Mech., 1958, pp. 281-286.
9. G.N. Savin — Stress Concentration Around Holes, Chapter 3, Pergamon Press, 1961.
10. L.M. Milne-Thompson — Plane Elastic Systems, Chapter 7, Springer-Verlag, Berlin, 1960.
11. T.J. Willmore — Quar. J. Mech. Appl. Math., 2, 1, 1949, pp. 53-63.
12. G.I. Barenblatt and G.P. Cherepanov — PMM (Translated by ASME), 25, 1, 1961, pp. 61-74.
13. D.D. Ang and M.L. Williams — Trans. ASME, J. Appl. Mech., 28, 3, September 1961, pp. 372-378.
14. A.E. Green and W. Zerna — Theoretical Elasticity, Oxford at the Clarendon Press, 1954, pp. 324-374.
15. N.I. Muskhelishvili — Some Basic Problems of Mathematical Theory of Elasticity, Chapters 15 and 18, P. Noordhoff and Co., 1953.

RÉSUMÉ - On dérive l'équation générale des champs de tension d'une pointe de fissure dans un corps anisotrope par une approximation de variable complexe. On a dévini et évalué les facteurs d'intensité de tension qui permettent la représentation concise des conditions pour l'extension d'une fissure, directement des fonctions de tension. On donne quelques solutions de problèmes à valeur limite en forme implicite et on les discute en correspondance avec les solutions analogues pour corps isotropes.

On a trouvé qu'une singularité de tension élastique de l'ordre de $r^{-\frac{1}{2}}$ (où r est la distance radiale depuis le front de la fissure) est toujours présente à la pointe d'une fissure dans un corps rectilinéairement anisotrope. Ce résultat et quelques considérations additionelles des champs de tension de la pointe d'une fissure impliquent la possibilité d'étendre les applications courantes de théories de mécanisme de fracture à la représentation de conditions de fracture de corps anisotropes avec des imperfections en forme de fissures.

ZUSAMMENFASSUNG - Die allgemeinen Gleichungen für Spannungsfelder an der Spitze eines Risses in anisotropischen Körpern werden näherungsweise unter Benützung einer komplexen Variabele abgeleitet. Die Spannungsintensitätsfaktoren, die eine kurzgefasste Darstellung der Bedingungen für Rissausbreitung gestatten, werden definiert und unmittelbar aus Spannungsfunktionen bestimmt. Einige individuellen Lösungen von Grenzwert-Problemen werden in geschlossener Form gegeben und bezüglich ihrer analogen Lösungen für isotrope Körper diskutiert.

Es ergibt sich, dass eine elastische Spannungssingularität der Ordnung $r^{-\frac{1}{2}}$ an der Spitze eines Risses in einem Körper mit rectilinearischer Anisotropie immer da ist. (r ist die Radiusabstand von der Rissfront). Dieses Ergebnis, und einige weiteren Betrachtungen der Spannungsfelder an der Rissspitze weisen darauf hin, dass es möglich ist, die geläufigen Methoden der Bruchmechanik für die Darstellung von Bruchbedingungen anisotropischer Körper mit rissartigen Fehlern zu erweitern.

Central Crack in Plane Orthotropic Rectangular Sheet

O. L. BOWIE AND C. E. FREESE

Army Materials and Mechanics Research Center Watertown, Massachusetts

(Received July 15, 1971)

ABSTRACT

The plane problem of a central crack in a rectangular sheet of orthotropic material is considered. The solution is found by an extension of the modified mapping-collocation technique, originally formulated for plane isotropic analysis. Application of the technique outlined in this paper for plane orthotropic problems to a wider class of geometries and loading is evident. The numerical results indicate a dependence of the orthotropic stress intensity factors on both geometric and elastic constants over a certain parameter range.

1. Introduction

With the increasing use of materials with directional properties, there has been a renewal of interest in analytical solutions in anisotropic elasticity. It is natural too, that the extension and refinement of fracture mechanics concepts is the subject of much recent attention in this direction. The initial groundwork for such an extension was proposed in 1965 by Sih, Paris, and Irwin [1] for rectilinearly anisotropic* bodies. Additional details are included in a survey paper by Sih and Liebowitz [2].

A comprehensive treatment of anisotropic elasticity has been made by Lekhnitskii [3]. Included in [3] is a systematic formulation of the plane theory of rectilinear anisotropy in terms of analytic functions of complex variables. Several of Muskhelishvili's concepts [4] for problem-solving in the realm of isotropic elasticity are paralleled. Additional refinements as well as several numerical solutions are described by Savin [5].

Conceptionally, the analysis for plane problems in anisotropic elasticity is well-defined. On the other hand, as in isotropic analysis, effective numerical procedures are not as obvious. In fact only a few numerical solutions are available. Furthermore there are often difficulties in the carry-over of isotropic techniques such as polynomial mapping to other than the simpler geometries.

Recently the "modified mapping-collocation" technique was developed for isotropic analysis [6]. This technique combines the advantages of conformal mapping methods with boundary collocation arguments and it has been successfully applied to several crack problems in plane isotropy involving difficult geometries, e.g., [6], [7], [8]. This paper will show that the nature of the technique lends itself to extension to problems of plane anisotropy.

The problem of a central crack in a rectangular sheet under uniaxial tension will be considered. The analagous problem for isotropic materials has already been investigated, e.g., [7], [9], [10]. Mapping will be used only in a limited sense; namely, to generate a class of stress functions consistent with traction-free conditions on the crack surface. As in the isotropic applications, boundary conditions on the rectangle are then enforced by collocation. Conventional boundary collocation procedures are modified by prescribing both stress and "force" conditions at the boundary stations. (Force conditions correspond to the well-known Muskhelishvili formulation [4] of the stress boundary problem in terms of the force resultant of the applied tractions as a function of the arc length of the boundary.) In general it has been found more effective to over-specify the number of boundary stations and match the corresponding conditions in a least square sense.

---

* A rectilinearly anisotropic body possessing three planes of elastic symmetry will be referred to as orthotropic.

2. Basic Equations

The basic equations for the plane problem of rectilinear anisotropy will now be briefly summarized using the notation in [2]. The Airy stress function can be expressed in the form

$$U(x, y) = 2 \operatorname{Re}[U_1(z_1) + U_2(z_2)] \qquad (1)$$

where $U_1(z_1)$ and $U_2(z_2)$ are the arbitrary functions of the complex variables $z_1 = x + s_1 y$ and $z_2 = x + s_2 y$, respectively. The complex constants s_1 and s_2 are related to the roots μ_j of the characteristic equation

$$a_{11}\mu_j^4 - 2a_{16}\mu_j^3 + (2a_{12} + a_{66})\mu_j^2 - 2a_{26}\mu_j + a_{22} = 0 \qquad (2)$$

where the a_{ij} are elastic constants occurring in the generalized Hooke's law. Assuming unequal roots μ_j in (2), then

$$s_1 = \mu_1 = \alpha_1 + i\beta_1, \quad s_2 = \mu_2 = \alpha_2 + i\beta_2,$$
$$\mu_3 = \bar{\mu}_1, \quad \mu_4 = \bar{\mu}_2 \qquad (3)$$

where α_j, β_j are real constants. Without loss of generality, we assume

$$\beta_1 > 0, \quad \beta_2 > 0; \quad \beta_1 \neq \beta_2. \qquad (4)$$

In the case of orthotropy, the additional symmetries imply that $a_{16} = a_{26} = 0$ in (2). In the case of isotropy, $s_1 = s_2 = i$ and the assumption in (4) is no longer valid.

The stresses can now be expressed in terms of the well-known Airy function in the usual manner. To simplify notation, the new functions

$$\phi(z_1) = dU_1/dz_1, \quad \psi(z_2) = dU_2/dz_2 \qquad (5)$$

are introduced. Then

$$\sigma_x = 2 \operatorname{Re}[s_1^2 \phi'(z_1) + s_2^2 \psi'(z_2)]$$
$$\sigma_y = 2 \operatorname{Re}[\phi'(z_1) + \psi'(z_2)]$$
$$\sigma_{xy} = -2 \operatorname{Re}[s_1 \phi'(z_1) + s_2 \psi'(z_2)]$$

where primes denote differentiation. (6)

It will be useful to have an expression for the force resultant as a function of arc length on an arbitrary curve. It can be shown [5] if we denote by $X_n ds$ and $Y_n ds$ the horizontal and vertical forces acting on an element of arc ds with normal n, the force resultant along the arc can be written as

$$(1 + is_1)\phi(z_1) + (1 + is_2)\psi(z_2) + (1 + i\bar{s}_1)\overline{\phi(z_1)} + (1 + i\bar{s}_2)\overline{\psi(z_2)} = i\int^s (X_n + iY_n)ds =$$
$$= f_1(s) + if_2(s). \qquad (7)$$

3. Mapping Formulation

In the modified mapping-collocation method, stress functions ensuring traction-free conditions on the crack surfaces are derived by using Muskhelishvili-type extension arguments of analytic functions. The physical region (Figure 1) will be considered as defined in the complex z-plane. Introducing an auxiliary ζ-plane, we consider the simple mapping

$$z = \omega(\zeta) = (L/2)(\zeta + \zeta^{-1}). \qquad (8)$$

This mapping carries the unit circle $\zeta = \sigma$ and its exterior into a slit along the real axis from $z = -L$ to $z = +L$ and its exterior, respectively. Clearly a point z can be related to the auxiliary plane by

$$\zeta = z/L + [(z/L)^2 - 1]^{\frac{1}{2}}. \qquad (9)$$

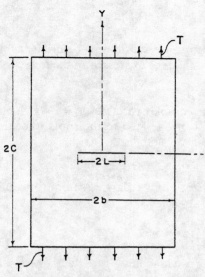

Figure 1. Central crack in rectangular sheet under tension, T.

Consider now the complex variables $z_1 = x + \alpha_1 y + i\beta_1 y$ and $z_2 = x + \alpha_2 y + i\beta_2 y$. Since the crack lies on the real axis, clearly $z = z_1 = z_2$ on the crack. Thus, we set

$$z_1 = \omega(\zeta_1) = (L/2)(\zeta_1 + \zeta_1^{-1})$$
$$z_2 = \omega(\zeta_2) = (L/2)(\zeta_2 + \zeta_2^{-1}) \tag{10}$$

where ζ_1 and ζ_2 are the values of ζ corresponding to z_1 and z_2, respectively. On the unit circle, ζ, ζ_1, and ζ_2 coincide, that is, the three parameter regions corresponding to z, z_1, z_2, on the crack intersect. Otherwise, ζ_1 and ζ_2 are distinct and are found from

$$\zeta_1 = z_1/L + [(z_1/L)^2 - 1]^{\frac{1}{2}}$$
$$\zeta_2 = z_2/L + [(z_2/L)^2 - 1]^{\frac{1}{2}}. \tag{11}$$

For convenience in notation, we denote $\phi(z_1) = \phi[\omega(\zeta_1)] = \phi(\zeta_1)$, $\psi(z_2) = \psi[\omega(\zeta_2)] = \psi(\zeta_2)$, thus, $\phi'(z_1) = \phi'(\zeta_1)/\omega'(\zeta_1)$, etc. The stresses can now be written as

$$\sigma_x = 2\,\mathrm{Re}\,\{s_1^2 \phi'(\zeta_1)/\omega'(\zeta_1) + s_2^2 \psi'(\zeta_2)/\omega'(h_2)\} \quad \text{etc.} \tag{12}$$

The resultant force condition (7) now becomes

$$(1 + is_1)\,\phi(\zeta_1) + (1 + is_2)\,\psi(\zeta_2) + (1 + i\bar{s}_1)\,\overline{\phi(\zeta_1)} + (1 + i\bar{s}_2)\,\overline{\psi(\zeta_2)} = f_1(s) + if_2(s). \tag{13}$$

4. Representation of the Stress Function

To complete the representation of the stress function, we modify the Kartzivadze extension [11] in the isotropic formulation to a comparable result for rectilinear anisotropy. This does not appear to have been done previously.

Let $S_{\zeta 1}^+$ and $S_{\zeta 2}^+$ denote the two parameter regions corresponding to ζ_1 and ζ_2, respectively. Their union, $S_{\zeta 1}^+ + S_{\zeta 2}^+$, will be denoted by S_ζ^+. The region S_ζ^+ is bounded from within by the unit circle in the ζ-plane. The region inside the unit circle will be denoted by S_ζ^-. We introduce $F(\zeta)$, analytic in S_ζ^+ and extend it by definition into S_ζ^-

$$\bar{B}F(\zeta) = \bar{\psi}(1/\zeta) - \bar{C}F(1/\zeta), \qquad \zeta \in S_\zeta^- \tag{14}$$

where $\bar{B}$ and $\bar{C}$ are constants and

$$\bar{F}(1/\zeta) = \overline{F(1/\bar{\zeta})}. \tag{15}$$

Since $1/\bar{\zeta}_2 \in S_\zeta^-$, it follows from (14) that

$$\psi(\zeta_2) = B\bar{F}(1/\zeta_2) + CF(\zeta_2). \tag{16}$$

Furthermore, we assume

$$\phi(\zeta_1) = F(\zeta_1) \qquad \zeta_1 \in S_{\zeta 1}^+. \tag{17}$$

Thus, $\phi(h_1)$ and $\psi(h_2)$ have been defined in terms of $F(\zeta)$ and its extension across the unit circle. Let us now choose

$$\begin{aligned} B &= (\bar{s}_2 - \bar{s}_1)/(s_2 - \bar{s}_2) \\ C &= (\bar{s}_2 - s_1)/(s_2 - \bar{s}_2). \end{aligned} \tag{18}$$

Inserting equations (16)–(18) into (13) and evaluating on the unit circle yields

$$(1 + i\bar{s}_2)(s_2 - s_1)(\bar{s}_2 - s_2)^{-1}[F^+(\sigma) - F^-(\sigma)] + (1 + is_2)(\bar{s}_2 - \bar{s}_1)(s_2 - \bar{s}_2)^{-1}[\overline{F^+(\sigma)} - \overline{F^-(\sigma)}]$$
$$= -i \int (X_n + iY_n)ds = -f_1 - if_2 \tag{19}$$

where $F^+(\sigma)$ and $F^-(\sigma)$ denote $F(\sigma)$ approached through S_ζ^+ and S_ζ^-, respectively.

The necessary and sufficient condition that the crack be traction-free (or the vanishing of the right-hand side of (19) is

$$F^+(\sigma) = F^-(\sigma). \tag{20}$$

Thus, $F(\zeta)$ continues analytically across the unit circle corresponding to unloaded intervals in the physical plane. This is of course the analogous result corresponding to the Kartzivadze extension in the isotropic formulation.

Traction-free conditions on the crack are therefore enforced by considering $F(\zeta)$ as analytic in the annulus corresponding to S_ζ^+ and inversion with respect to the unit circle. A Laurent expansion for $F(\zeta)$ will be assumed even though the annulus is not circular. There is no reason to doubt the validity of such an expansion in the present problem. Therefore,

$$F(\zeta) = T \sum_{-\infty}^{\infty} (a_n + ib_n)\zeta^n. \tag{21}$$

5. Modified Boundary Collocation

Completion of the solution requires the determination of the coefficients in (21) such that the prescribed boundary conditions on the outer boundary are met. This phase of the solution was carried out by the modified boundary collocation plan [6]. First, a finite number of boundary stations were chosen in the z-plane which, in turn, along with the material properties, determine corresponding ζ_1 and ζ_2 locations from (11). Truncated forms of the series (21) were then considered in conjunction with the boundary conditions at these stations.

An important feature of the modified boundary collocation argument is the manner in which boundary conditions are specified. In addition to the specified boundary stresses (or tractions), we specify the resultant force, $(f_1 + if_2)$, and the resultant moment, M_0 (with respect to the origin of the forces f_1 and f_2), at the boundary stations. If strictly continuous arguments were being used, there would be an obvious redundancy in the use of this set of conditions. This redundancy is removed when we consider that these conditions are imposed on a truncated series at a finite number of discrete points. In practice, we choose a sufficient number of boundary stations so that the number of conditions exceeds the degrees of freedom provided by the coefficients of the truncated series by a factor of two. The coefficients are then determined on the basis of satisfying the discrete boundary conditions in a least square sense.

The modified boundary collocation plan was suggested by Saint Venant's principle. If $f_1 + if_2$ is accurately matched at the boundary stations, then the errors in stress boundary conditions correspond to self-equilibrating distributions of loading on the intervals between successive stations. Moment collocation is suggested because of possible build-up of moment error even when $f_1 + if_2$ is collocated. This can be seen by considering the moment, M_0. It can be shown, e.g. [4], by integration by parts along the arc L, that

$$M_0 = \int_L (xY_n - yX_n)ds = -\int_L (\dot{x}df_1 + \dot{y}df_2) = -(xf_1 + yf_2) + \int_L (f_1 dx + f_2 dy). \quad (22)$$

Although f_1 and f_2 are forced to be correct at collocated stations, an accumulated error in M_0 can occur in the last integral in (22). Finally, the conditions corresponding to the local applied tractions are significant when the desired information is substantially influenced by the local boundary conditions, e.g., a crack tip near the boundary.

It was found for the problem type under consideration in this paper, that "force" and stress conditions were sufficient. The "force" conditions in the present formulation become

$$(1+is_1)F(\zeta_1)+(1+is_2)[B\bar{F}(1/\zeta_2)$$
$$+CF(\zeta_2)]+(1+i\bar{s}_1)\overline{F(\zeta_1)}+(1+i\bar{s}_2)[\bar{B}F(1/\bar{\zeta}_2)+\bar{C}\overline{F(\zeta_2)}] = f_1 + if_2. \quad (23)$$

The stresses become

$$\sigma_x = 2\,\text{Re}\,\{s_1^2 F'(\zeta_1)/\omega'(\zeta_1) + s_2^2[-B\bar{F}'(1/\zeta_2)/\zeta_2^2\omega'(\zeta_2) + CF'(\zeta_2)/\omega'(\zeta_2)]\} \quad \text{etc.} \quad (24)$$

6. Stress Intensity Factors

If we assume the in-plane lines of elastic symmetry of the orthotropic plate shown in Figure 1 coincide with the x- and y-axes, then there exist obvious stress symmetries including the vanishing of the shear stresses along the lines $x=0$ and $y=0$. Under these circumstances it is easily shown that

$$F(\zeta) = T[s_2/(s_2-s_1)] \sum_{-\infty}^{\infty} C_n \zeta^{2n+1} \quad (25)$$

where the C_n's are real. Furthermore, when (25) is used, it is sufficient to apply the collocation argument over a quarter of the outer boundary (e.g., in the first quadrant corresponding to the physical plane).

In such a situation, the stress intensity factor K_1 as defined by Sih [2] for "plane symmetric loading" is applicable. In particular,

$$K_1 = 2\sqrt{2}\,[(s_2-s_1)/s_2] \lim_{z_1 \to z_0} (z_1-z_0)^{\frac{1}{2}} \phi'(z_1) \quad (26)$$

where z_0 corresponds to the crack tip. In terms of the present formulation,

$$K_1 = (2/\sqrt{L})[(s_2-s_1)/s_2]F'(1) = (2T/\sqrt{L}) \sum_{-\infty}^{\infty} (2n+1)C_n. \quad (27)$$

It will be convenient to introduce a "finite correction factor", H, where

$$K_1 = TH\sqrt{L}. \quad (28)$$

If the rectangular boundary were infinite in extent, then $H=1$. In general, therefore, the variation of H from unity is a measure of the influence of the outer boundary. Clearly,

$$H = 2L^{-1} \sum_{-\infty}^{\infty} (2n+1)C_n. \quad (29)$$

It is interesting to consider the case when the lines of elastic symmetry do not coincide with the coordinate axes. One can no longer infer all the symmetries leading to (25). Although the exponent of ζ in (25) will still be valid, the more general coefficients $a_n + ib_n$ in (21) must be retained. Boundary conditions on the half panel (first and second quadrants in the physical plane) require consideration due to the loss of symmetry. The shear stresses along the x- and y-axis will no longer vanish except in an integrated sense. The local stress intensity will have two components, K_1 and K_2, which can be calculated from Sih's definitions [2] for "plate symmetric loading" and "plane skew-symmetric loading", respectively.

The remarks of the preceding paragraph have been verified by numerical calculations. The K_2 contribution (shear effect) becomes increasingly small as the crack length is decreased and the solution is effectively that for a crack in an infinite region. This is of course consistent with the solution [3] of the latter problem which indicates no shear ahead of the crack tip independent of the orientation of the material symmetries.

7. Numerical Results

The procedure as outlined above has been programmed for the general case of plane rectilinear anisotropy. For the purpose of numerical illustration, we consider the case of orthotropy with the lines of material symmetry coinciding with the x- and y-axes in Figure 1. Since $a_{16} = a_{26} = 0$ in (2), we are left with the four elastic constants,

$$a_{11} = 1/E_1, \quad a_{22} = 1/E_2, \quad a_{66} = 1/\mu_{12}$$
$$a_{12} = -v_{12}/E_1 = -v_{21}/E_2 \tag{30}$$

in which E_1, E_2 are the Young's moduli, v_{12}, v_{21} the Poisson ratios, and μ_{12} the shear modulus. For concreteness we assume further that the constants s_1 and s_2 are purely imaginary, thus, $s_1 = i\beta_1$ and $s_2 = i\beta_2$ where

$$\beta_1 \beta_2 = (E_1/E_2)^{\frac{1}{2}}$$
$$\beta_1 + \beta_2 = \sqrt{2} \{(E_1/E_2)^{\frac{1}{2}} + E_1/2\mu_{12} - v_{12}\}^{\frac{1}{2}}. \tag{31}$$

The limiting case of isotropy corresponds to $\beta_1 = 1$ and $\beta_2 = 1$. In fact if we choose $\beta_1 = 1$ and $\beta_2 = 1 + \varepsilon$ and consider the previous analysis as $\varepsilon \to 0$, it is possible to recover the equations of the isotropic formulation.

In this paper the primary objective of numerical computation is the illustration of the effectiveness of the procedure. Of additional interest is 1) the degree to which the isotropic stress intensity factors can be approximated by suitable values of the orthotropic constants, and 2) a comparison of the isotropic and orthotropic stress intensity factors in view of remarks made in [1] and [2]. For these purposes, rather than deal with explicit material constants, it was considered sufficient to fix β_1 as unity and vary the parameter β_2.

TABLE 1

Finite correction factors, H, for $c/b = 1$

$\beta_1 = 1$, $\beta_2^2 = E_1/E_2$

L/b \ β_2^2	0.1	0.3	0.5	0.7	0.9	1.1	1.5	2.5	3.5	4.5
0.0	1.00	1.00	1.00	1.00	1.00	1.00	1.00	1.00	1.00	1.00
0.1	1.04	1.02	1.02	1.02	1.01	1.01	1.01	1.01	1.01	1.01
0.2	1.16	1.10	1.08	1.07	1.06	1.05	1.05	1.04	1.03	1.03
0.3	1.34	1.21	1.17	1.15	1.13	1.12	1.10	1.08	1.07	1.07
0.4	1.57	1.37	1.30	1.26	1.23	1.21	1.18	1.15	1.13	1.12
0.5	1.85	1.57	1.46	1.39	1.35	1.32	1.28	1.24	1.22	1.20
0.6	2.20	1.80	1.64	1.56	1.50	1.46	1.41	1.36	1.33	1.31
0.7	2.63	2.06	1.86	1.76	1.69	1.65	1.59	1.53	1.50	1.49
0.8	3.10	2.37	2.17	2.05	1.99	1.95	1.87	1.81	1.80	1.80

The "finite correction factors", H, were calculated for varying values of β_2^2, L/b, and c/b. In Table 1 and Figure 2 the results for a square plate, $c/b = 1$, are presented. It is interesting to compare the results for $\beta_2^2 \approx 1$ with the isotropic solution for a square plate, [7]. The values of H for $\beta_2^2 = 0.9$ agree within one percent over the range of L/b considered. Another interesting result is the behavior of H for large β_2^2. The correction H appears to converge (from above) to

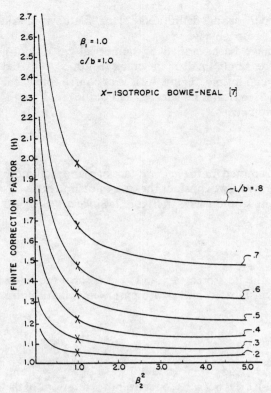

Figure 2. Finite correction factor, H, for square plate of orthotropic material.

TABLE 2

Finite correction factors, H, for $c/b = 1.5$

$\beta_1 = 1$, $\beta_2^2 = E_1/E_2$

L_b \ β_2^2	0.1	0.2	0.3	0.5	0.7	0.9	1.1
0.0	1.00	1.00	1.00	1.00	1.00	1.00	1.00
0.1	1.02	1.01	1.01	1.01	1.01	1.01	1.01
0.2	1.08	1.06	1.05	1.04	1.03	1.03	1.03
0.3	1.18	1.13	1.10	1.08	1.07	1.07	1.07
0.4	1.31	1.22	1.18	1.15	1.13	1.12	1.11
0.5	1.46	1.34	1.28	1.24	1.22	1.21	1.19
0.6	1.64	1.48	1.41	1.36	1.33	1.32	1.31
0.7	1.85	1.66	1.59	1.53	1.51	1.49	1.49
0.8	2.12	1.95	1.88	1.83	1.80	1.80	1.80

the isotropic solution for an infinite strip of finite width, e.g., [7]. The general behavior of the solution is illustrated in Figure 2.

The results for $c/b = 1.5$ and $c/b = 2.0$ are presented in Tables 2 and 3, respectively. Again H is bounded from below by the isotropic solution for an infinite strip of finite width. As c/b increases, the bound is attained for smaller values of β_2^2. This is consistent with the result shown in [7] for isotropy. It was shown there that for $c/b \geq 2.0$ the rectangle could be considered as effectively a semi-infinite strip over most of the range of crack lengths. Convergence with respect to c/b is illustrated in Figure 3 for $\beta_2^2 = 0.4$.

The basic numerical details have been described previously, e.g. [6]. Again the listed values can be considered as accurate to within at least one percent. The method utilizes the intrinsic

TABLE 3

Finite correction factors, H for c/b = 2.0

$\beta_1 = 1$. $\beta_2^2 = E_1/E_2$

L/b \ β_2^2	0.1	0.2	0.3	0.4	0.5
0.0	1.00	1.00	1.00	1.00	1.00
0.1	1.01	1.01	1.01	1.01	1.01
0.2	1.05	1.03	1.03	1.03	1.03
0.3	1.11	1.08	1.07	1.06	1.06
0.4	1.19	1.14	1.12	1.12	1.11
0.5	1.29	1.22	1.20	1.19	1.19
0.6	1.41	1.34	1.31	1.31	1.31
0.7	1.58	1.51	1.49	1.49	1.49
0.8	1.85	1.80	1.80	1.80	1.80

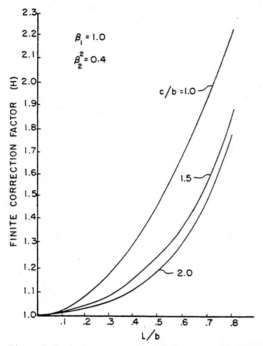

Figure 3. Typical convergence of finite correction factor, *H*, with geometric ratio, *c/b*.

simplicity of the orthotropic solution. The computational time was about a tenth of the time for the corresponding isotropic calculations.

8. Observations

It has been demonstrated that the modified mapping-collocation technique originally developed for isotropic analysis can be suitably extended for the analysis of cracks in plane problems of orthotropy. It is evident that the basic arguments of the technique can be extended to configurations other than cracks. Since the original version of this paper several problems of stress concentration around cutouts in finite geometries of orthotropic material have been analyzed.

One of the most interesting results of the numerical calculation is the clear indication of the dependence of the orthotropic stress intensity factors on the elastic constants. In contrast, for boundary value problems of this type, the isotropic stress intensity factors are independent of the material constants. The equivalence of the anisotropic and isotropic stress intensity factors for boundary value problems involving self-equilibrating load systems on the crack, inferred in [1] and [2], does not appear to be generally true. On the other hand, the results indicate that such an approximation can be made over large practical ranges of geometry. This is consistent with many of the numerical results found in [5]. Savin observed that the disturbance in isotropic stress distributions for problems of stress concentration involving orthotropic materials is quite localized.

Another result of considerable value is the remarkable accuracy to which isotropic solutions can be obtained by setting $\beta_1 \approx 1$ and $\beta_2 \approx 1$. For most practical numerical purposes, there is no need to consider two separate analyses.

Acknowledgment

The authors are indebted to J. I. Bluhm, Chief of the Theoretical and Applied Mechanics Research Laboratory, Army Materials and Mechanics Research Center, for his support and interest in this work. In addition, the authors are indebted to a reviewer for several valuable criticisms of the original manuscript.

REFERENCES

[1] G. C. Sih, P. C. Paris and G. R. Irwin, *Int. Jour. of Fract. Mech.*, 1, 3 (1965) 189–203.
[2] G. C. Sih and H. Liebowitz, *Fracture*, Vol. II, Ed. by H. Liebowitz, Academic Press, New York (1968) 67–190.
[3] S. G. Lekhnitskii, *Theory of Elasticity of an Anisotropic Body*, Holden-Day, Inc., San Francisco (1963).
[4] N. I. Muskhelishvili, *Some Basic Problems of Mathematical Theory of Elasticity*, P. Noordhoff, Groningen (1963).
[5] G. N. Savin, *Stress Concentration Around Holes*, Pergamon Press, New York (1961).
[6] O. L. Bowie and D. M. Neal, A Modified Mapping-Collocation Technique for Accurate Calculation of Stress Intensity Factors, *Int. Journ. of Fract. Mech.*, 6 2 (1970) 199–206.
[7] O. L. Bowie and D. M. Neal, A Note on the Central Crack in a Uniformly Stressed Strip, *Eng. Fract. Mech.*, 2 (1970) 181–182.
[8] O. L. Bowie and C. E. Freese, *Elastic Analysis for a Radial Crack in a Circular Ring*, AMMRC MS70-3, Presented at Fourth National Symposium on Fracture Mechanics, Carnegie-Mellon University, Pittsburgh, Penn. Aug. 24, 1970.
[9] M. Isida, *Crack Tip Stress Intensity Factors for the Tension of an Eccentrically Cracked Strip*, Lehigh University, Depth. of Mech. Report 1965.
[10] W. F. Brown, (Jr) and J. E. Srawley, *Plane Strain Crack Toughness Testing of High Strength Materials*, A.S.T.M. Special Technical Publication No. 410 (1966) 77–79.
[11] I. N. Kartzivadze, *Comptes rendus de l'académie des sciences de l'U.R.S.S.*, 20, V. XII (1943) 95–104.

RÉSUMÉ

On considère le problème plan d'une fissure au centre d'une tôle mince rectangulaire d'un matériau orthotrope.

La solution résulte d'une extension de la technique modifiée de représentation conforme par correspondance point par point, qui fut à l'origine suggérée pour l'analyse de conditions planes et isotropes.

L'application de la technique développée dans le mémoire pour des problèmes plans et orthotropes, à une classe plus large de géométries et de conditions de mise en charge, est évidente.

Les résultats numériques indiquent que les facteurs d'intensité de contrainte en conditions orthotropes dépendent à la fois des constantes géométriques et des constantes élastiques, du moins dans une certaine gamme de leurs valeurs.

ZUSAMMENFASSUNG

Es wird das plane Problem eines Risses im Zentrum eines Feinbleches aus orthotropen Material behandelt.

Die Lösung ergibt sich aus einer Ausweitung des abgewandelten Verfahrens der konformen Darstellung Punkt für Punkt, welche ursprünglich für die Untersuchung der planen und isotropen Bedingungen vorgeschlagen worden war.

Die Anwendung des in diesem Bericht für plane und orthotrope Probleme dargelegten Verfahrens auf andere geometrische Formen und unterschiedliche Belastungsweisen ist selbstverständlich.

Die numerischen Ergebnisse zeigen eine Abhängigkeit der orthotropen Spannungsintensitätsfaktoren von den geometrischen und den elastischen Konstanten über einen gewissen Bereich der Parameterwerte.

THE STRESSES AROUND A FAULT OR CRACK IN DISSIMILAR MEDIA

By M. L. Williams

ABSTRACT

In order to investigate some problems of geophysical interest, the usual consideration of symmetrical or antisymmetrical loading of an isotropic homogeneous plate containing a crack was extended to the case where the alignment of the crack separates two separate isotropic homogeneous regions. It develops that the modulus of the singular behavior of the stress remains proportional to the inverse square root of the distance from the point of the crack, but the stresses possess a sharp oscillatory character of the type $r^{-\frac{1}{2}} \sin(b \log r)$, which seems to be confined quite close to the point, as well as a shear stress along the material joint line as long as the materials are different.

The off-fault areas of high strain energy release reported by St. Amand for the White Wolf fault are qualitatively shown to be expected.

As a logical extension of a previous plane-stress or plane-strain problem[1] which dealt with the stress distribution at the base of a stationary crack in an isotropic homogeneous material, it is proposed to discuss the characteristic behavior of the stress in the vicinity of a crack between the plane-bounding surfaces of two dissimilar materials. The problem was suggested by a possible application to situations in geological investigations dealing with fault lines along the interface between two layers of rock strata, but it may also apply to certain weld joints which, owing to faulty joining techniques, or for that matter applied loading, develop cracks along the original weld line.

The geometry considered is that of a material M_1 occupying the upper half plane and a material M_2 in the lower half plane, joined without residual stress along the positive x-axis, or positive radial direction r measured from the origin along $\psi = 0$. The elastic plane stress solution is therefore desired for unloaded edges along the negative x-axis.

In the previous homogeneous case where $M_1 = M_2$ it was found that the stresses near the base of the crack became (mathematically) infinite according to an inverse square-root law, $\sigma \sim r^{-\frac{1}{2}}$. The interesting question which arises concerns how the character of the stress is changed as a result of the discontinuity of the material properties across the line of crack prolongation.

Outline of Solution

Reviewing the method of solution which is naturally very similar to the homogeneous case, a biharmonic stress function $\chi(r, \psi)$, that is, a solution of

$$\nabla^4 \chi(r, \psi) = 0, \tag{1}$$

is to be found such that the normal stress, σ_ψ, and shear stress, $\tau_{r\psi}$ vanish along $\psi = \pm \pi$, and further that the displacements and stresses are continuous across the material demarcation line $\psi = 0$.

Manuscript received for publication August 8, 1958.

[1] M. L. Williams, "On the Stress Distribution at the Base of a Stationary Crack," *Jour. Applied Mechanics*, March, 1956.

Typical solutions are chosen of the form

$$\chi(r, \psi) = r^{\lambda+1} F(\psi) = r^{\lambda+1} \{a \sin (\lambda + 1)\psi + b \cos (\lambda + 1)\psi$$
$$+ c \sin (\lambda - 1)\psi + d \cos (\lambda - 1)\psi\}, \qquad (2)$$

where the usual relations between stresses, displacements, and stress function are given[2] as

$$\sigma_r = \frac{1}{r^2} \frac{\partial^2 \chi}{\partial \psi^2} + \frac{1}{r} \frac{\partial \chi}{\partial r} = r^{\lambda-1} [F''(\psi) + (\lambda + 1)F(\psi)] \qquad (3)$$

$$\sigma_\theta = \frac{\partial^2 \chi}{\partial r^2} = r^{\lambda-1} \lambda(\lambda + 1)F(\psi) \qquad (4)$$

$$\tau_{r\psi} = -\frac{1}{r} \frac{\partial^2 \chi}{\partial r \partial \theta} + \frac{1}{r^2} \frac{\partial \chi}{\partial \theta} = -\lambda r^{\lambda-1} F'(\psi) \qquad (5)$$

$$u_\psi = \frac{1}{2\mu} r^\lambda \{-F'(\psi) - 4(1 - \sigma) [c \cos (\lambda - 1)\psi - d \sin (\lambda - 1)\psi]\} \qquad (6)$$

$$u_r = \frac{1}{2\mu} r^\lambda \{-(\lambda + 1)F(\psi) + 4(1 - \sigma) [c \sin (\lambda - 1)\psi + d \cos (\lambda - 1)\psi]\}. \qquad (7)$$

Here μ is the shear modulus and σ, in terms of Poisson's ratio, ν, is $\sigma \equiv \nu/(1 + \nu)$. The primes denote differentiation with respect to ψ.

As a matter of notation, let the quantities in the regions M_1 and M_2 have the appropriate subscript $F_1, F_2; \lambda_1, \lambda_2; a_1, a_2; \nu_1, \nu_2$; etc., respectively. With this convention the first four boundary conditions for free edges at $\psi = \pm\pi$ become

$$F_1(\pi) = F_1'(\pi) = F_2(-\pi) = F_2'(-\pi) = 0, \qquad (8)-(11)$$

and upon noting that $\lambda_1 \equiv \lambda_2$ in order that the second four boundary conditions be independent of r, there results

$$F_1(0) = F_2(0) \qquad (12)$$

$$F_1'(0) = F_2'(0) \qquad (13)$$

$$\frac{1}{2\mu_1} [-F_1'(0) - 4c_1(1 - \sigma_1)] = \frac{1}{2\mu_2} [-F_2'(0) - 4c_2(1 - \sigma_2)] \qquad (14)$$

$$\frac{1}{2\mu_1} [-(\lambda + 1)F_1(0) + 4d_1(1 - \sigma_1)] = \frac{1}{2\mu_2} [-(\lambda + 1)F_2(0) + 4d_2(1 - \sigma_2)]. \qquad (15)$$

---

[2] S. Timoshenko and J. N. Goodier, *Theory of Elasticity* (New York: McGraw-Hill, 1951).

Substitution of $F_1(\psi)$ and $F_2(\psi)$ in (8)–(15) leads to the following eight homogeneous linear equations in the eight unknowns $a_1, a_2 \cdots d_1, d_2$.

$$a_1 \sin(\lambda+1)\pi + b_1 \cos(\lambda+1)\pi + c_1 \sin(\lambda-1)\pi + d_1 \cos(\lambda-1)\pi = 0$$

$$-a_2 \sin(\lambda+1)\pi + b_2 \cos(\lambda+1)\pi - c_2 \sin(\lambda-1)\pi + d_2 \cos(\lambda-1)\pi = 0$$

$$a_1(\lambda+1)\cos(\lambda+1)\pi - b_1(\lambda+1)\sin(\lambda+1)\pi + c_1(\lambda-1)\cos(\lambda-1)\pi$$
$$- d_1(\lambda-1)\sin(\lambda-1)\pi = 0$$

$$a_2(\lambda+1)\cos(\lambda+1)\pi + b_2(\lambda+1)\sin(\lambda+1)\pi + c_2(\lambda-1)\cos(\lambda-1)\pi$$
$$+ d_2(\lambda-1)\sin(\lambda-1)\pi = 0$$

$$b_1 + d_1 = b_2 + d_2$$

$$(\lambda+1)a_1 + (\lambda-1)c_1 = (\lambda+1)a_2 + (\lambda-1)c_2$$

$$4(1-\sigma_1)c_1 = 4k(1-\sigma_2)c_2 + (k-1)[(\lambda+1)a_2 + (\lambda-1)c_2]$$

$$4(1-\sigma_1)d_1 = 4k(1-\sigma_2)d_2 - (k-1)(\lambda+1)[b_2 + d_2],$$

where the shear modulus ratio $k \equiv \mu_1/\mu_2$ has been introduced.

A nontrivial solution for the constants exists if the determinant of the eight equations vanish. After some algebraic simplification, the determinant can be written in the form

$$\cot^2 \lambda\pi + \left[\frac{2k(1-\sigma_2) - 2(1-\sigma_1) - (k-1)}{2k(1-\sigma_2) + 2(1-\sigma_1)}\right]^2 = 0. \tag{16}$$

THE HOMOGENEOUS CASE

When the material in both regions is the same, $\sigma_1 = \sigma_2$ and $k = \mu_1/\mu_2 = 1$. Thus (16) reduces to the simple form

$$\cot^2 \lambda\pi = 0, \tag{17}$$

and by inspection the eigen values are

$$\lambda = (2n+1)/2 \qquad n = 0, 1, 2, \cdots \tag{18}$$
$$= 1/2, 3/2, 5/2, \cdots$$

where negative values of n have been excluded so that the physical displacements u_ψ and u_r are finite as the origin is approached, i.e., the stress function and its first derivative exist along the boundary.

Thus an infinite number of λ_n exist with the lowest—the one controlling the stress behavior near the base of the crack, $r \sim 0$—being $\lambda_{\min} = 1/2$. Hence the local stress behavior, from (3) to (5), is of order $\sigma \sim r^{\lambda-1} = r^{-\frac{1}{2}}$.

The Bimaterial Case

It is immediately evident from (16) that no real solutions exist for different materials as this equation is the sum of two positive terms equated to zero, which solution is the homogeneous case. Nothing excludes the admissibility of complex values of the eigen parameter, however, and upon assuming $\lambda \equiv \lambda_r + i\lambda_j$ it is quickly determined that

$$\frac{[\tan^2 \lambda_r\pi + 1]\tanh \lambda_j\pi}{\tan^2 \lambda_r\pi + \tanh^2 \lambda_j\pi} = \pm \frac{2k(1-\sigma_2) - 2(1-\sigma_1) - (k-1)}{2k(1-\sigma_2) + 2(1-\sigma_1)} \quad (19)$$

$$\frac{\tan \lambda_r\pi[1 - \tanh^2 \lambda_j\pi]}{\tan^2 \lambda_r\pi + \tanh^2 \lambda_j\pi} = 0 \ . \quad (20)$$

Two sets of solutions are possible. The first, from (20), if $\tan \lambda_r\pi = 0$ giving

$$\lambda_r = n = 0, 1, 2, 3, \cdots$$

$$\lambda_j = \pm \frac{1}{\pi} \coth^{-1}\left[\frac{2k(1-\sigma_2) - 2(1-\sigma_1) - (k-1)}{2k(1-\sigma_2) + 2(1-\sigma_1)}\right] \neq 0$$

The second solution occurs, from (20), if $\tan \lambda_r\pi = \infty$ giving

$$\lambda_r = 1/2, 3/2, 5/2, \cdots$$

$$\lambda_j = \pm \frac{1}{\pi} \tanh^{-1}\left[\frac{2k(1-\sigma_2) - 2(1-\sigma_1) - (k-1)}{2k(1-\sigma_2) + 2(1-\sigma_1)}\right]$$

which is observed to approach the homogeneous material solution ($\lambda_j \to 0$, as $\sigma_1 \to \sigma_2$ and $k \to 1$).

Discussion of the Solution

The stresses from the second solution are therefore seen to behave, for the minimum eigen value, according to

$$\sigma \sim r^{\lambda-1} \sim r^{(\lambda_r - 1) \pm i\lambda_j}$$

or

$$\sigma \sim r^{-1/2} \begin{pmatrix} \sin \\ \cos \end{pmatrix} (\lambda_j \log r) ,$$

which is of an oscillating character with its maximum modulus determined by $r^{-\frac{1}{2}}$ which is the same bound as determined in the case of homogeneous materials. Indeed, for this special case it was shown that $\lambda_j = 0$ for all roots, and the results are seen to be identical with those of the previous more comprehensive analysis.

Turning to a further consideration of the oscillatory behavior, consider a case when $\nu_1 = \nu_2$. Then

$$|\lambda_j| = \left| \frac{1}{\pi} \tanh^{-1} \frac{1-\nu}{2} \cdot \frac{k-1}{k+1} \right| \lesssim \frac{1}{\pi} \tanh^{-1} \frac{1-\nu}{2} \sim \frac{1}{5}$$

so that the first zero as the origin is approached would be of the order of $|\lambda_j \log r| \sim \pi/2$ or $|\log r| \sim -7$ and hence with the radial dimension expressed in terms of some characteristic distance in the plane of the plate, say R, $r/R \sim e^{-7} \sim 10^{-3}$. A pronounced oscillatory character of the stress, on the basis of this qualitative calculation, would therefore appear to be confined quite close to the base of the crack.

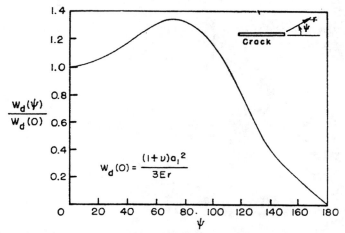

Fig. 1. Distortion energy distribution at constant radius for symmetric fault loading.

The first solution, however, bears further examination owing to the existence of the pure imaginary solution $\lambda = i\lambda_j$ for $n = 0$. The displacements then appear to be finite $u \sim r^\lambda \sim r^{\pm i\lambda_j}$ so that

$$u \sim \binom{\sin}{\cos} (\lambda_j \log r)$$

and stresses of the form

$$\sigma \sim r^{-1} \binom{\sin}{\cos} (\lambda_j \log r) ,$$

which would result in a stronger stress singularity than the $r^{-\frac{1}{2}}$ behavior found in the second solution.

It is believed, nevertheless, that the proper interpretation would require that the displacements tend to zero as the base of the crack is approached. Thus the value of $n = 0$ would be excluded and the two sets of eigen values would be intertwined—the lowest, however, being generated from the second set. The latter set yields the physically expected homogeneous case as the materials M_1 and M_2 become the same.

On this basis, therefore, one would expect very high strain concentrations at the ends of a fault. Indeed, for the elastic homogeneous case it was shown that the strain energy stored in the medium was not uniformly distributed over a peripheral region about the end of the crack, but at a constant radius was one-third higher at ±70 degrees to the left or right of the direction of crack prolongation than directly ahead (fig. 1). Further graphs and additional formulas for symmetrical and antisymmetrical loading are given in detail for the case of similar materials in Williams, *op. cit.* (n. 1).

Dr. Beno Gutenberg has pointed out the interesting feature to the writer that the areas of maximum energy release reported by St. Amand[3] also seems to occur ahead, but to the side, of the fault direction and hence is qualitatively in agreement with the previous calculations.

---

[3] Pierre St. Amand, "Two Proposed Measures of Seismicity," *Bull. Seism. Soc. Am.*, Vol. 46, No. 1, January, 1956.

CALIFORNIA INSTITUTE OF TECHNOLOGY,
PASADENA, CALIFORNIA.

J. R. RICE[1]
Research Associate,
Division of Engineering,
Brown University,
Providence, R. I.

G. C. SIH
Associate Professor of Mechanics,
Lehigh University,
Bethlehem, Pa.
Mem. ASME

Plane Problems of Cracks in Dissimilar Media[2]

The in-plane extension of two dissimilar materials with cracks or fault lines along their common interface is considered. A method is offered for solving such problems by the application of complex variables integrated with the eigenfunction-expansion technique presented in an earlier paper. The solution to any problem is resolved to finding a single complex potential resulting in a marked economy of effort as contrasted with the more laborious conventional methods which have not yielded satisfactory results. Boundary problems are formulated and solutions are given in closed form. The results of these evaluations also give stress-intensity factors (which determine the onset of rapid fracture in the theory of Griffith-Irwin) for plane problems.

A PROBLEM of considerable practical importance is that of two semi-infinite elastic bodies with different elastic properties joined along straight-line segments. The problem represents idealizations of two dissimilar metallic materials welded together with flaws or cracks developed along the original weld line owing to faulty joining techniques. The bonding materials also may be metallic to nonmetallic.

Although a great deal of progress has been made in solving elasticity problems involving lines of discontinuities, mathematical formulation of the problem of cracks between the bonding surfaces of two different materials remains inadequately treated. Recently, several authors have attempted to solve the problem by methods such as the eigenfunction-expansion approach [1, 2],[3] the Hilbert problem [3], and by techniques using integral transforms [4]. However, not one of the foregoing papers has given satisfactory results to the problem. The present investigation therefore is undertaken to give a complete formulation of the "two dissimilar media" crack problem in a manner which is simpler and more thorough than would have been possible by other, hitherto known methods.

In an enlightening paper [1], Williams considered the plane problem of dissimilar materials with a semi-infinite crack. It was discovered for the first time that the stresses possess an oscillatory character of the type $r^{-1/2}$ sin (or cos) of the argument $\epsilon \log r$, where r is the radial distance from the crack tip and ϵ is a function of material constants. This problem was later extended to the case of bending loads by Sih and Rice [2]. While the eigenfunction approach of Williams is an expedient method for determining qualitatively the characteristic behavior of the stress in the vicinity of crack tips, it does not give the solution quantitatively.[4]

Associated with the problem of dissimilar materials having cracks is that of the less complicated one of punches acting on a half-plane. Using the Plemelj formulas and Cauchy integrals, Muskhelishvili [5] has solved the problem of a single punch with straight-line profile pressing on a horizontal base. He showed that the stress changes its sign an infinite number of times underneath the punch. This oscillatory character of the stress is in fact the same as that observed by Williams [1] for the crack problem. Furthermore, upon identifying the width of the punch with the length of the bond line, the Goursat functions for the punch and crack problems take the same form. Similarly, the problem of two collinear punches corresponds to the dissimilar material problem of two semi-infinite planes bonded along two finite line segments. For a detailed account of the similarities between the punch and crack problems, refer to the work of Muskhelishvili [5] in conjunction with that of Erdogan [3]. It should be pointed out that the Hilbert formulation in [3] is based on the condition of free crack surface. A more general application of the problem of linear relationship (or the Hilbert problem) to the problem of straight or circular-arc cracks along the bond line of two different materials will be discussed in a separate paper.

In an effort to obtain a complete solution of the problem, Bahar [4] proposed an alternative method based on integral transforms. He resolved the problem to the solution of simultaneous dual integral equations which in turn were reduced to a system of linear algebraic equations by means of the discontinuous Weber-Sonine-Schafheitlin integrals. In contrast to all the previous results [1, 2, 3, 5], he found that the stresses near the crack

---

[1] Formerly NSF Fellow, Department of Applied Mechanics, Lehigh University, Bethlehem, Pa.

[2] This research was supported by the Office of Naval Research, U. S. Navy, under Contract Nonr-610(06) with the Department of Mechanics, Lehigh University.

[3] Numbers in brackets designate References at end of paper.

Presented at the Applied Mechanics/Fluids Engineering Conference, Washington, D. C., June 7–9, 1965, of THE AMERICAN SOCIETY OF MECHANICAL ENGINEERS.

Discussion of this paper should be addressed to the Editorial Department, ASME, United Engineering Center, 345 East 47th Street, New York, N. Y. 10017, and will be accepted until July 10, 1965. Discussion received after the closing date will be returned. Manuscript received by ASME Applied Mechanics Division, May 5, 1964. Paper No. 65—APM-4.

[4] Note that in [2] the Goursat functions were expressed independently of uncertainties of both the external loads and the crack dimensions.

Nomenclature

a = half crack length
b = dimension
A, B, C = complex constants
E_j = Young's moduli ($j = 1, 2$)
$f(z), g(z), F(z)$ = complex function of z
G_j = shear moduli ($j = 1, 2$)
i = $(-1)^{1/2}$
k_1, k_2 = stress-intensity factors
P, Q = y and x-components of R
R = complex force, $Q + iP$
u_j, v_j = displacements ($j = 1, 2$)
x, y = rectangular coordinates
z = complex variable, $x + iy$
ϵ = bielastic constant
ϵ_x = strain component
η_j = $3 - 4\nu_j$ for plane strain and $(3 - \nu_j)/(1 + \nu_j)$ for plane stress ($j = 1, 2$)
μ = $[G_2(\eta_1 + 1)]/[G_1(\eta_2 + 1)]$
ν_j = Poisson's ratio ($j = 1, 2$)
$(\sigma_x)_j, (\sigma_y)_j, (\tau_{xy})_j$ = stress components ($j = 1, 2$)
$\Phi_j(z), \Psi_j(z)$ = Goursat functions ($j = 1, 2$)
ω_j^∞ = rotation at infinity ($j = 1, 2$)

tip are not oscillatory in character but decay monotonically as r, the radial distance from the crack front, increases. The validity of this result is therefore questioned.

In what follows, it is shown how the complex-variable method combined with eigenfunction expansion in [1, 2] can be applied to formulate the problem of bonded dissimilar elastic planes containing cracks along the bond. Solutions are given in closed form for a number of extensional problems of fundamental interest. In particular, the problem of an isolated complex force, i.e., a force vector having components in the x and y-directions, applied at an arbitrary location on each side of the crack surface is solved. Aside from its application to such problems as wedge loading at an arbitrary angle, the isolated-force solution may be used as Green's functions to obtain the stresses in welded dissimilar plates owing to any arbitrary distribution of tractions on the crack surface.

The results in this paper are also discussed in connection with the Griffith-Irwin theory of fracture. In their theory, the critical length of a crack may be predicted from the crack-tip stress-intensity factors. It is shown that these factors can be evaluated readily from a complex potential function $\Phi(z)$.

Statement of Problem

Let a material with elastic properties E_1 and ν_1 occupy the upper half-plane, $y > 0$, and a material with elastic properties E_2 and ν_2 occupy the lower half-plane, $y < 0$. The two materials are bonded along straight-line segments of the x-axis. In the following, all quantities such as the elastic constants, stresses, and so on, pertaining to the region $y > 0$ and $y < 0$ will be marked with subscripts 1 and 2, respectively.

Muskhelishvili [5] and others have shown that the solution to an individual problem in the plane theory of elasticity can be reduced to finding two complex functions, which satisfy the boundary conditions of that problem. In the case of two different materials, however, the elastic properties are discontinuous across the bond line, and a complete solution to the problem requires the knowledge of four complex functions $\Phi_j(z)$, $\Psi_j(z)$, $j = 1, 2$, of the complex variable $z = x + iy$. The basic equations for two-dimensional isotropic elasticity in the form used by Kolosov-Muskhelishvili are

$$(\sigma_x)_j + (\sigma_y)_j = 4\,\mathrm{Re}[\Phi_j(z)]$$
$$(\sigma_y)_j - (\sigma_x)_j + 2i(\tau_{xy})_j = 2[\bar{z}\Phi_j'(z) + \Psi_j(z)] \quad (1)$$

and

$$2G_j(u_j + iv_j) = \eta_j\int\Phi_j(z)dz - z\bar{\Phi}_j(\bar{z}) - \int\bar{\Psi}_j(\bar{z})d\bar{z} \quad (2)$$

where u_j, v_j are components of displacement, $(\sigma_x)_j$, $(\sigma_y)_j$, $(\tau_{xy})_j$ are components of stress, and G_j is the shear modulus. Also $\eta_j = 3 - 4\nu_j$ for plane strain and $\eta_j = (3 - \nu_j)/(1 + \nu_j)$ for generalized plane stress, ν_j being Poisson's ratio.

Isolated Forces on Surface of a Semi-Infinite Crack

Let the semi-infinite planes of different materials be joined along the positive x-axis, Fig. 1. A line crack is situated along the negative x-axis extending from $x = 0$ to $x = -\infty$ and is opened by a complex force $R = Q + iP$ at $z = -a$ on each side of the crack.

For this problem, the general forms of the Goursat functions are given by equation (41) in [2]. These were derived by expressing the Airy stress function, obtained by the Williams method as a power series in terms of polar coordinates r and θ, in the form

$$\mathrm{Re}\,[\bar{z}\phi_j(z) + \chi_j(z)]$$

The functions $\phi_j'(z)$ and $\chi_j''(z)$ are $\Phi_j(z)$ and $\Psi_j(z)$ in this paper, respectively. Upon defining

$$f(z) = 2\sum_{n=1}^{\infty}[(n - \tfrac{1}{2}) - i\epsilon][(n + \tfrac{1}{2}) - i\epsilon]\bar{B}^{(n)}z^{n-1} \quad (3)$$

it is possible to express the functions $\Phi_j(z)$ and $\Psi_j(z)$ in terms of $f(z)$ alone. The results are

$$\Phi_1(z) = z^{-\frac{1}{2}-i\epsilon}f(z)$$
$$\Psi_1(z) = e^{2\pi\epsilon}z^{-\frac{1}{2}+i\epsilon}\bar{f}(z) - z^{-\frac{1}{2}-i\epsilon}[(\tfrac{1}{2} - i\epsilon)f(z) + zf'(z)] \quad (4)$$

for the region $y > 0$ and

$$\Phi_2(z) = e^{2\pi\epsilon}z^{-\frac{1}{2}-i\epsilon}f(z)$$
$$\Psi_2(z) = z^{-\frac{1}{2}+i\epsilon}\bar{f}(z) - e^{2\pi\epsilon}z^{-\frac{1}{2}-i\epsilon}[(\tfrac{1}{2} - i\epsilon)f(z) + zf'(z)] \quad (5)$$

for the region $y < 0$. In equations (3) through (5), ϵ is defined as the bielastic constant given by (see equation (39) in [2])

$$\epsilon = \frac{1}{2\pi}\log\left[\left(\frac{\eta_1}{G_1} + \frac{1}{G_2}\right)\bigg/\left(\frac{\eta_2}{G_2} + \frac{1}{G_1}\right)\right] \quad (6)$$

The problem is to find the function $f(z)$ such that it is holomorphic in a region close to the crack tip. Outside of this region, $f(z)$ may have poles of the order $1/z$. In effect, this permits loading on the crack surface except for isolated loads near the tip of the crack. Hence, in the proximity of the requisite force R at $z = -a$ in Fig. 1, the Goursat functions, say for $y > 0$, must take the form

$$\Phi_1(z) = -\frac{R}{2\pi}\frac{1}{z + a}$$
$$\Psi_1(z) = \frac{\bar{R}}{2\pi}\frac{1}{z + a} + \frac{R}{2\pi}\frac{a}{(z + a)^2} \quad (7)$$

Equation (7) represents Boussinesq's solution [6] of an isolated force R acting on the boundary of a half-plane, but now expressed in terms of complex potentials. Upon comparing both equations (4) and (7) for the stresses in the neighborhood of the pole at $z = -a$, it is found that

$$f(z) = \frac{P - iQ}{2\pi e^{\pi\epsilon}}\frac{a^{\frac{1}{2}+i\epsilon}}{z + a} \quad (8)$$

Inserting equation (8) into (4) and (5) gives the Goursat functions from which the stresses and displacements can be computed without difficulty.

Goursat Functions for Finite-Crack Problems

The Goursat functions, equations (4) and (5), originally derived for a semi-infinite crack, may be modified to solve the problem of a finite crack, Fig. 2. Since the branch points are now located at $z = \pm a$ (the crack tips), the singular terms $(z - a)^{-\frac{1}{2}-i\epsilon}$ and $(z + a)^{-\frac{1}{2}+i\epsilon}$ must be introduced into the complex potentials $\Phi_j(z)$ and $\Psi_j(z)$. This is accomplished by defining $f(z)$ in equations (4) and (5) as

$$f(z) = (z + a)^{-\frac{1}{2}+i\epsilon}g(z) \quad (9)$$

in which $g(z)$ is well behaved at $z = \pm a$ and it may have poles sufficiently far away from the crack tip when isolated forces are present.

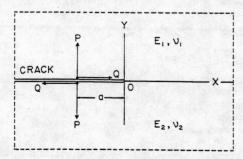

Fig. 1 Isolated forces on a semi-infinite crack

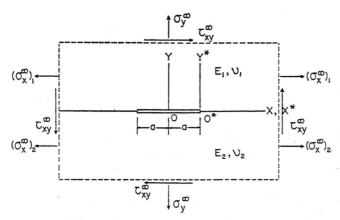

Fig. 2 Infinite plate with a crack subjected to stresses at infinity

In addition, the effect of the translation of coordinate axes on the Goursat functions must be considered. Denote by $\Phi_j(z)$, $\Psi_j(z)$ the functions referred to the axes x, y and $\Phi_j^*(z)$, $\Psi_j^*(z)$ to x^*, y^* (see Fig. 2). Since the stress components, equation (1), are not altered by the translation $z = z^* + a$, it is observed that

$$\Phi_j(z) = \Phi_j^*(z - a)$$
$$\Psi_j(z) = \Psi_j^*(z - a) - a\Phi_j^{*\prime}(z - a) \quad (10)$$

Making use of equations (9) and (10), equations (4) and (5) may be rearranged to read as

$$\Phi_1(z) = g(z)F(z) + A$$
$$\Psi_1(z) = e^{2\pi\epsilon}\bar{g}(z)\bar{F}(z)$$
$$+ \left[\frac{a^2 + 2i\epsilon az}{z^2 - a^2}g(z) - zg'(z)\right]F(z) - (A + \bar{A}) \quad (11)$$

and

$$\Phi_2(z) = e^{2\pi\epsilon}g(z)F(z) + \mu A$$
$$\Psi_2(z) = \bar{g}(z)\bar{F}(z)$$
$$+ e^{2\pi\epsilon}\left[\frac{a^2 + 2i\epsilon az}{z^2 - a^2}g(z) - zg'(z)\right]F(z) - \mu(A + \bar{A}) \quad (12)$$

where

$$\mu = [G_2(\eta_1 + 1)]/[G_1(\eta_2 + 1)]$$

and

$$F(z) = (z^2 - a^2)^{-1/2}\left(\frac{z+a}{z-a}\right)^{i\epsilon} \quad (13)$$

Those terms containing the complex constant A represent the degenerate case of $\Phi_j(z)$ and $\Psi_j(z)$. A detailed derivation is given in Appendix 1.

It can be shown that equations (11) and (12) give zero stresses on the crack surface and continuous stresses across the bond line, $|z| > a$ for $z = \bar{z}$. The continuity of displacements may be verified by first computing for the complex displacements

$$2G_1(u_1 + iv_1) = \eta_1 \int g(z)F(z)dz$$
$$- e^{2\pi\epsilon}\int g(\bar{z})F(\bar{z})d\bar{z} + (\bar{z} - z)\bar{g}(\bar{z})\bar{F}(\bar{z}) \quad (14)$$

in the upper half-plane and

$$2G_2(u_2 + iv_2) = e^{2\pi\epsilon}\eta_2 \int g(z)F(z)dz - \int g(\bar{z})F(\bar{z})d\bar{z}$$
$$- e^{2\pi\epsilon}(\bar{z} - z)\bar{g}(\bar{z})\bar{F}(\bar{z}) \quad (15)$$

in the lower half-plane. From equations (14) and (15), the difference between the displacements for $y > 0$ and $y < 0$ is found to be

$$2[(u_1 + iv_1) - (u_2 + iv_2)] = \left(\frac{\eta_1}{G_1} - e^{2\pi\epsilon}\frac{\eta_2}{G_2}\right)\int g(z)F(z)dz$$
$$+ \left(\frac{1}{G_2} - e^{2\pi\epsilon}\frac{1}{G_1}\right)\int g(\bar{z})F(\bar{z})d\bar{z} \quad (16)$$

on the bond line $z = \bar{z}$. In view of equation (6), this difference is indeed zero when

$$\int g(z)F(z)dz$$

is single-valued on the bond line.

Hence, the Goursat functions given by equations (11) and (12) satisfy all the conditions of the problem of a finite crack between two dissimilar materials. The foregoing analysis may be extended easily to a finite or infinite number of collinear cracks. An example is given in Appendix 2. It is now more pertinent to illustrate the use of this method by finding the constant A and the function $g(z)$ for specific problems.

Extension of Infinite Plate With a Crack

From the point of view of application, the consideration of infinite region is of interest when the crack length is small in comparison with plate dimensions. The geometry of the present problem is shown in Fig. 2, where the plate composed of two different materials is subjected to normal and shear stresses at infinity.

In order for the stresses to be bounded as $z \to \infty$, the function $g(z)$ can at most be linear in z; i.e.,

$$g(z) = Bz + C \quad (17)$$

where B and C are complex constants yet to be found. The physical interpretation of the constant $A = A_1 + iA_2$ in equations (11), (12) and $B = B_1 + iB_2$ in equation (17) is considerably more complicated than in the case of similar material, $\epsilon = 0$. Putting equation (17) into (11) and (12) and letting $z \to \infty$, then by way of equation (1), the stresses at infinity lead to

$$A_1 = \frac{(\sigma_x^\infty)_1 + \sigma_y^\infty}{4} - \frac{\sigma_y^\infty}{1 + e^{2\pi\epsilon}}$$
$$B = B_1 + iB_2 = \frac{\sigma_y^\infty - i\tau_{xy}^\infty}{1 + e^{2\pi\epsilon}} \quad (18)$$

It should be mentioned that the normal stress, σ_x, in the x-direction is discontinuous across the bond line. Thus, it is necessary to distinguish $(\sigma_x^\infty)_1$ in the region $y > 0$ from $(\sigma_x^\infty)_2$ in the region $y < 0$. In fact, it follows directly that they are related to each other by

$$(\sigma_x^\infty)_2 = \mu(\sigma_x^\infty)_1 + \frac{(3+\mu)e^{2\pi\epsilon} - (3\mu+1)}{1 + e^{2\pi\epsilon}}\sigma_y^\infty \quad (19)$$

Alternatively, equation (19) also may be obtained from the conditions of continuity of stresses and displacements across the x-axis along which the component σ_x has a jump (see Appendix 3).

The constant A_2 may be related to the rotation at an infinitely remote part of the x, y-plane as follows:

$$A_2 = \frac{\tau_{xy}^\infty}{1 + e^{2\pi\epsilon}} + \frac{2G_1\omega_1^\infty}{1 + \eta_1} = \frac{1}{\mu}\left(\frac{e^{2\pi\epsilon}}{1 + e^{2\pi\epsilon}}\tau_{xy}^\infty + \frac{2G_2\omega_2^\infty}{1 + \eta_2}\right) \quad (20)$$

in which ω_1^∞ and ω_2^∞ denote the rotations at infinity in the upper and lower half-planes, respectively. After some algebraic manipulations, equation (20) gives

$$\omega_2^\infty - \omega_1^\infty = \left(\frac{G_2 - G_1}{2G_1G_2}\right)\tau_{xy}^\infty \quad (21)$$

In contrast to the homogeneous case ($G_1 = G_2$), where the rotation may be assumed to vanish as it does not affect the stresses, ω_1^∞ and ω_2^∞ in the bimaterial problem cannot be set arbitrarily to zero at the same time unless $\tau_{xy}^\infty = 0$.

Hitherto, no consideration has been given to the condition of single-valuedness of displacements. It is necessary and sufficient for the one-valuedness of $u_j + iv_j$ $(j = 1, 2)$ that the integral

$$\int g(z)F(z)dz$$

in equations (14) and (15) be a single-valued function of z. For $|z| > a$, the function $F(z)$ in equation (13) may be represented by a series of the form

$$F(z) = \frac{1}{z} + \frac{2i\epsilon a}{z^2} + \frac{a^2(1-4\epsilon^2)}{2z^3} + \cdots \quad (22)$$

By virtue of equations (17) and (22)

$$\int g(z)F(z)dz = Bz + (C + 2i\epsilon aB) \log z - \left(\frac{1-4\epsilon^2}{2} a^2 B - 2i\epsilon aC\right) \frac{1}{z} + \cdots$$

For single-valued displacements, i.e., solutions involving no dislocations, the integral should have no logarithmic term. Therefore, the constant C is determined:

$$C = -2i\epsilon aB \quad (23)$$

where B is given by equation (18). The final result in terms of $g(z)$ may be written as

$$g(z) = \frac{\sigma_y^\infty - i\tau_{xy}^\infty}{1 + e^{2\pi\epsilon}} (z - 2i\epsilon a) \quad (24)$$

from which the Goursat functions $\Phi_j(z)$ and $\Psi_j(z)$ may be obtained.

Green's Function

The problem of two semi-infinite planes bonded along the x-axis with a crack of length, $2a$, centered at the origin, Fig. 3, and having two equal and opposite forces $R = Q + iP$ applied at $z = b$ is of fundamental interest, since it may be used as a Green's function to form the solution to other problems.

Replacing the constant a in equation (7) by $-b$ and equating the results in terms of stresses with those obtained from equations (11) and (12) in the vicinity of $z = b$, it gives

$$g(z) = \frac{R}{2\pi i} \frac{e^{-\pi\epsilon}}{z-b} (a^2 - b^2)^{1/2} \left(\frac{a-b}{a+b}\right)^{i\epsilon} \quad (25)$$

where $A = 0$ as the stresses are zero at infinity. The isolated force solution, equation (25), may now be taken as a Green's function to solve problems with any loading desired on the crack surface.

Moreover, by judicious application of the "principle of superposition," the solution of the problem of a crack with surface tractions may be further used to attack all the general problems of an infinite plate with a crack whose surface is free from tractions. First, the stresses $\sigma_y(x, 0)$ and $\tau_{xy}(x, 0)$ on the prospective crack surface with no crack present are computed from the prescribed loading in the original problem. Then, superimposing tractions equal and opposite to those on the prospective crack surface (i.e., to free the crack surface), the result is

$$g(z) = \frac{e^{-\pi\epsilon}}{2\pi} \int_{-a}^{a} [\sigma_y(x, 0) - i\tau_{xy}(x, 0)] \\ \times \left(\frac{a-x}{a+x}\right)^{i\epsilon} \frac{\sqrt{a^2 - x^2}}{z - x} dx \quad (26)$$

Once $g(z)$ is known, the stresses and displacements are completely determined. Hence, equation (26) provides a direct method of solving any problem involving a crack between two bonded dissimilar materials.

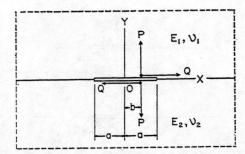

Fig. 3 A finite crack opened by wedge forces

Stress-Intensity Factors in Dissimilar Materials

In a previous paper [2], the stresses in the immediate vicinity of the crack tip were given as a function of r and θ, where r is the distance from the crack front and θ the angle between r and the crack plane. It was found that the singular behavior of the stresses remains proportional to the inverse square root of r (i.e., $r^{-1/2}$) as in the case of homogeneous materials, but the stresses possess a pronounced oscillatory character of the type

$$\sigma \sim r^{-1/2} \binom{\sin}{\cos} (\epsilon \log r) \quad (27)$$

which was first observed by Williams [1]. Equation (27) shows that the stress-intensity factors $k_j(j = 1, 2)$, used in the Griffith-Irwin theory of fracture, can be evaluated in a manner similar to that of the homogeneous case [7]. However, in the bimaterial problem, k_1 and k_2 can no longer be regarded as the crack tip stress-intensity factors for symmetrical and skew-symmetrical stress distributions. This point will be discussed later.

An examination of equations (43) through (44) in [2] indicates that the parameters k_1 and k_2 in general may be considered as the strength of the stress singularities at crack tips. Quantitatively, $k_j(j = 1, 2)$ depend on the external loads and the crack dimensions. For a given problem, the stress-intensity factors may be computed from the complex potential $\Phi_1(z)$. Take the case of a semi-infinite crack with its tip at the origin, Fig. 1. The stress-intensity factors at $z = 0$ are given by[5]

$$k_1(0) - ik_2(0) = 2\sqrt{2} \, e^{\pi\epsilon} \lim_{z \to 0} z^{1/2 + i\epsilon} \Phi_1(z) \quad (28)$$

As a first example, consider the semi-infinite crack problem stated earlier, Fig. 1. The isolated forces P and Q are located at a distance a away from the origin. Using equations (4) and (8), equation (28) becomes

$$k_1(0) = \frac{1}{\pi} \left(\frac{2}{a}\right)^{1/2} [P \cos(\epsilon \log a) + Q \sin(\epsilon \log a)]$$

$$k_2(0) = \frac{1}{\pi} \left(\frac{2}{a}\right)^{1/2} [Q \cos(\epsilon \log a) - P \sin(\epsilon \log a)] \quad (29)$$

Contrary to the conclusion in [3], the stress-intensity factors for a single bond do depend on the bielastic constant ϵ. Thus, the dependency of k_j on the material constants is not a simple matter of identifying it with the number of bond lines. As should have been expected, when $\epsilon = 0$, equation (29) reduces to the solution for the homogeneous material.

Similarly, the evaluation of k_j for a finite crack of length $2a$, Fig. 2, may be carried out by redefining equation (28) in the form

$$k_1(a) - ik_2(a) = 2\sqrt{2} \, e^{\pi\epsilon} \lim_{z \to a} (z - a)^{1/2 + i\epsilon} \Phi_1(z) \quad (30)$$

As a second example, consider a straight crack of length $2a$ along the x-axis in an infinite plate with normal and shear stresses at large distances from the crack, Fig. 2. From equations (11),

---

[5] Equation (48) in [2].

(24), and (30), the stress-intensity factors at $z = a$ are obtained. They are

$$k_1 = \frac{\left\{\begin{array}{l}\sigma[\cos(\epsilon \log 2a) + 2\epsilon \sin(\epsilon \log 2a)] \\ + \tau[\sin(\epsilon \log 2a) - 2\epsilon \cos(\epsilon \log 2a)]\end{array}\right\}}{\cosh \pi\epsilon} a^{1/2}$$

$$k_2 = \frac{\left\{\begin{array}{l}\tau[\cos(\epsilon \log 2a) + 2\epsilon \sin(\epsilon \log 2a)] \\ - \sigma[\sin(\epsilon \log 2a) - 2\epsilon \cos(\epsilon \log 2a)]\end{array}\right\}}{\cosh \pi\epsilon} a^{1/2}$$

(31)

An interesting feature of equation (31) is that both the symmetric and skew-symmetric loadings, σ_y^∞ and τ_{xy}^∞, are intermixed in the expressions for k_1 and k_2. As a result, the $k_j(j = 1, 2)$ do not have the simple physical interpretation as in the homogeneous case where the symmetric and skew-symmetric loads are separately contained in $k_1 = \sigma_y^\infty a^{1/2}$ and $k_2 = \tau_{xy}^\infty a^{1/2}$ for $\epsilon = 0$. When $\epsilon \neq 0$, even if the external loads were symmetric, say $\tau_{xy}^\infty = 0$ in equation (31), more than one stress-intensity factor is involved. Hence, in the application of the Griffith-Irwin theory of fracture, it is necessary to assume that a function of k_1, k_2 will cause the crack to grow upon reaching some critical value. The criterion may be written as

$$f(k_1, k_2) = f_{cr} \tag{32}$$

The specific form of equation (32) must be determined experimentally. Such studies will be left for future investigations.

Conclusions

A simple method for determining the Goursat functions for two dissimilar (or similar) materials bonded along straight-line segments is developed. The unbonded portion of the interface may be regarded as cracklike imperfections. The derivation combines an eigenfunction-expansion method with the complex-function theory of Muskhelishvili. The problem of isolated forces on the crack surface is solved with the aid of Boussinesq's solution.

In general, the results in this paper can be used in any one of the current fracture-mechanics theories. In particular, it is shown that the concept of stress-intensity factor in the Griffith-Irwin theory of fracture may be extended to cracks in dissimilar materials.

The Goursat functions for out-of-plane bending of cracks along the interface of two joined materials may be obtained in the same way. These results will be reported at a later date.

References

1 M. L. Williams, "The Stresses Around a Fault or Crack in Dissimilar Media," *Bulletin of the Seismological Society of America*, vol. 49, 1959, pp. 199–204.

2 G. C. Sih and J. R. Rice, "The Bending of Plates of Dissimilar Materials With Cracks," JOURNAL OF APPLIED MECHANICS, vol. 31, TRANS. ASME, vol. 86, Series E, 1964, pp. 477–482.

3 F. Erdogan, "Stress Distribution in a Nonhomogeneous Elastic Plane With Cracks," JOURNAL OF APPLIED MECHANICS, vol. 30, TRANS. ASME, vol. 85, Series E, 1963, pp. 232–236.

4 L. Y. Bahar, "On an Elastostatic Problem in Nonhomogeneous Media Leading to Coupled Dual Integral Equations," PhD dissertation, Lehigh University, Bethlehem, Pa., 1963.

5 N. I. Muskhelishvili, *Some Basic Problems of the Mathematical Theory of Elasticity*, 1933, English translation, P. Noordoff and Company, New York, N. Y., 1953.

6 J. Boussinesq, "Équilibre d'élasticité," *Comptes Rendus de l'Académie des Sciences*, Paris, France, vol. 114, 1892, pp. 1510–1516.

7 G. C. Sih, P. C. Paris, and F. Erdogan, "Crack-Tip, Stress-Intensity Factors for Plane Extension and Plate Bending Problems," JOURNAL OF APPLIED MECHANICS, vol. 29, TRANS. ASME, vol. 84, Series E, 1962, pp. 306–312.

8 W. T. Koiter, "An Infinite Row of Collinear Cracks in an Infinite Elastic Sheet," *Ingenieur-Archiv*, vol. 28, 1959, pp. 168–172.

APPENDIX 1

Degenerate Case of Goursat Functions

When the Goursat functions $\Phi_j(z)$ and $\Psi_j(z)$ degenerate to the constants A_j and B_j, respectively, equations (1) and (2) may be written as

$$(\sigma_x)_j + (\sigma_y)_j = 2(A_j + \bar{A}_j)$$

$$(\sigma_y)_j - (\sigma_x)_j + 2i(\tau_{xy})_j = 2B_j \tag{33}$$

$$2G_j(u_j + iv_j) = (\eta_j A_j - \bar{A}_j - \bar{B}_j)z$$

where $j = 1, 2$. Now, consider a uniaxial state of stress parallel to the crack surface which is not affected by the presence of the crack. This is given by

$$(\sigma_y)_1 + i(\tau_{xy})_1 = (\sigma_y)_2 + i(\tau_{xy})_2 = 0$$

Hence, equation (33) yields

$$-B_j = A_j + \bar{A}_j, \qquad j = 1, 2 \tag{34}$$

From equation (34) and the continuity of displacements along the bond line, i.e.,

$$u_1 + iv_1 = u_2 + iv_2, \qquad \text{at} \quad y = 0$$

it is found that

$$G_2(\eta_1 + 1)A_1 = G_1(\eta_2 + 1)A_2 \tag{35}$$

To simplify the notation, let $A_1 = A$. Thus, the Goursat functions for the two materials become

$$\Phi_1 = A, \qquad \Psi_1 = -(A + \bar{A}), \quad \text{for} \quad y > 0$$

and

$$\Phi_2 = \frac{G_2}{G_1}\left(\frac{\eta_1 + 1}{\eta_2 + 1}\right)A,$$

$$\Psi_2 = -\frac{G_2}{G_1}\left(\frac{\eta_1 + 1}{\eta_2 + 1}\right)(A + \bar{A}), \quad \text{for} \quad y < 0$$

(36)

APPENDIX 2

An Infinite Row of Collinear Cracks

The problem of an infinite series of equal cracks of length $2a$ along the bond line of two dissimilar materials and spaced at constant intervals $b(>2a)$ may be solved by the method described earlier. Referring to Fig. 4 for notation and the external loads at infinity, analogously to the expressions of (11), (13), and (17),

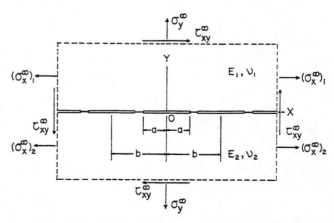

Fig. 4 An infinite series of collinear cracks between two dissimilar materials

$\Phi_1(z)$ is represented in the periodic form, giving stress-free crack surface and uniform stress at infinity,

$$\Phi_1(z) = B \prod_{n=-\infty}^{+\infty} (z - nb - d)F(z - nb) + A$$

$$= B \left(\frac{z-d}{b}\right)\left(\frac{z-a}{b}\right)^{-\frac{1}{2}-i\epsilon}\left(\frac{z+a}{b}\right)^{-\frac{1}{2}+i\epsilon}$$

$$\times \prod_{n=1}^{\infty} \left\{\left[1 - \frac{1}{n^2}\left(\frac{z-d}{b}\right)^2\right]\left[1 - \frac{1}{n^2}\left(\frac{z-a}{b}\right)^2\right]^{-\frac{1}{2}-i\epsilon}\right.$$

$$\left. \times \left[1 - \frac{1}{n^2}\left(\frac{z+a}{b}\right)^2\right]^{-\frac{1}{2}+i\epsilon}\right\} + A$$

One may show that for single-valued displacements, $d = 2i\epsilon a$; and that boundary conditions at infinity are satisfied by expressing $A = A_1 + iA_2$ and $B = B_1 + iB_2$ in terms of the applied stresses and rotation through equations identical to (18) and (20).

Noting that

$$\sin \pi t = \pi t \prod_{n=1}^{\infty} (1 - t^2/n^2),$$

$$\Phi_1(z) = B \sin \frac{\pi(z - 2i\epsilon a)}{b} \left[\sin \frac{\pi(z-a)}{b}\right]^{-\frac{1}{2}-i\epsilon}$$

$$\times \left[\sin \frac{\pi(z+a)}{b}\right]^{-\frac{1}{2}+i\epsilon} + A \quad (37)$$

and the remaining Goursat functions are given by

$$\Psi_1(z) = e^{2\pi\epsilon}[\bar{\Phi}_1(z) - \bar{A}] - [\Phi_1(z) - A + z\Phi_1'(z)] - (A + \bar{A})$$

$$\Phi_2(z) = e^{2\pi\epsilon}[\bar{\Phi}_1(z) - A] + \mu A$$

$$\Psi_2(z) = [\bar{\Phi}_1(z) - \bar{A}] - e^{2\pi\epsilon}[\Phi_1(z) - A + z\Phi_1'(z)] - \mu(A + \bar{A})$$

In the special case when $\epsilon = 0$, the foregoing solution reduces to that obtained by Koiter [8] for two similar materials.

APPENDIX 3

Stress Jump Across Interface

In general, equation (19) may be derived by considering the equilibrium of an element occupying both the region $y > 0$ and $y < 0$, $y = 0$ being the bond line. The stress component σ_x is taken to be discontinuous across the line $y = 0$ and the strain component ϵ_x to be continuous along such a line, i.e.,

$$(\epsilon_x)_1 = (\epsilon_x)_2$$

It follows from the strain-stress relations that

$$(\sigma_x)_2 = \frac{E_2}{E_1}(\sigma_x)_1 + \left[\nu_2 - \frac{E_2}{E_1}\nu_1\right]\sigma_y \quad (38)$$

for plane stress and

$$(\sigma_x)_2 = \frac{E_2}{E_1}\left(\frac{1 - \nu_1^2}{1 - \nu_2^2}\right)(\sigma_x)_1 + \left[\frac{\nu_2}{1 - \nu_2} - \frac{E_2}{E_1}\frac{\nu_1(1 + \nu_1)}{1 - \nu_2^2}\right]\sigma_y \quad (39)$$

for plane strain. Equations (38) and (39) may be made into a single generalization upon defining $\eta_j (j = 1, 2)$ such that $\eta_j = 3 - 4\nu_j$ for plane strain and $\eta_j = (3 - \nu_j)/(1 + \nu_j)$ for plane stress. Moreover, using equation (6), the final result may be put into the form of equation (19).

Stress-Intensity Factors for the Tension of an Eccentrically Cracked Strip

M. ISIDA[1]

Consider the tension of an infinite strip with a transverse crack located eccentrically from the center line. We denote the uniform tensile stress at infinity by T, plate width by $2b$, crack length by $2l$, and distances of the crack center from both straight edges and the plate center line by b_1, b_2, and e, respectively. We take the axes x, y as shown in Fig. 1 and define the following dimensionless coordinates and parameters:

$$\begin{aligned} \epsilon &= \frac{e}{b}, & \lambda &= \frac{l}{b_1} \\ \xi &= \frac{x}{b}, & \eta &= \frac{y}{b} \\ z &= x + iy, & \zeta &= \xi + i\eta = \frac{z}{b} \end{aligned} \quad (1)$$

The plate is assumed to be in a state of generalized plane stress, and the Airy's stress function may be expressed in terms of complex potentials $\varphi(\zeta)$ and $\psi(\zeta)$ as follows:

$$\chi = \frac{T}{2} b^2 \xi^2 + Tb^2 \, \mathrm{Re}\{\bar{\xi}\varphi(\zeta) + \psi(\zeta)\} \quad (2)$$

where the first term in equation (2) corresponds to the uniform tensile stress at infinity and $\varphi(\zeta)$, $\psi(\zeta)$ should be determined from the boundary conditions.

In the present problem, not only stresses, strains, and rotation but also the resultant force and displacements must be single-valued within the plate region. It follows that $\varphi(\zeta)$ and $\psi'(\zeta)$ must be analytic functions and can be expanded in Laurent's series convergent in a region bounded by certain two concentric circles. They are written in the following forms:

$$\begin{aligned} \varphi(\zeta) &= \sum_{n=0}^{\infty} (F_n \zeta^{-(n+1)} + M_n \zeta^{(n+1)}) \\ \psi'(\zeta) &= -D_0 \zeta^{-1} - \sum_{n=0}^{\infty} n D_n \zeta^{-(n+1)} \\ &\quad + \sum_{n=0}^{\infty} (n+2) L_n \zeta^{n+1} \end{aligned} \quad (3)$$

where unknown coefficients F_n, M_n, D_n, and L_n must be determined from the boundary conditions.

First unknown coefficients in (3) must satisfy some relations in order that the straight edges may be free from stress. They were derived by C. B. Ling [1][2] in his paper on a circular hole

---

[1] Visiting Professor of Mechanics, Lehigh University, Bethlehem, Pa.; on leave from Second Airframe Division, National Aerospace Laboratory, Tokyo, Japan.

[2] Numbers in brackets designate References at end of Note.

Manuscript received by ASME Applied Mechanics Division, June 17, 1965; final draft, February 8, 1966.

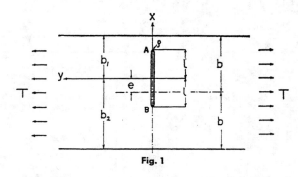

Fig. 1

problem and rewritten in our notations as follows:

$$\begin{aligned} L_n &= \sum_{p=0}^{\infty} (\alpha_p{}^n D_p + \beta_p{}^n F_p) \\ M_n &= \sum_{p=0}^{\infty} (\gamma_p{}^n D_p + \delta_p{}^n F_p) \end{aligned} \quad (4)$$

where the coefficients $\alpha_p{}^n$, $\beta_p{}^n$, $\gamma_p{}^n$, and $\delta_p{}^n$ are given in forms of infinite integrals.

On the other hand, the stress-free conditions for the crack edge were discussed in the author's previous paper and given in the following forms [2]:

$$\begin{aligned} D_{2n} &= \frac{1}{2} a_0{}^{2n} \lambda^{2n+2} + \sum_{p=0}^{\infty} (L_{2p} + M_{2p}) a_{2p}{}^{2n} \lambda^{2n+2p+2} \\ F_{2n} &= -\frac{1}{2} c_0{}^{2n} \lambda^{2n+2} - \sum_{p=0}^{\infty} (L_{2p} + M_{2p}) c_{2p}{}^{2n} \lambda^{2n+2p+2} \\ D_{2n+1} &= \sum_{p=0}^{\infty} (L_{2p+1} + M_{2p+1}) a_{2p+1}{}^{2n+1} \lambda^{2n+2p+4} \\ F_{2n+1} &= -\sum_{p=0}^{\infty} (L_{2p+1} + M_{2p+1}) c_{2p+1}{}^{2n+1} \lambda^{2n+2p+4} \end{aligned} \quad (5)$$

where $a_p{}^n$, $c_p{}^n$ are expressed in linear forms of the expansion coefficients of $(1 - X)^{1/2}$.

The unknown coefficients F_n, M_n, D_n, and L_n are calculated by solving (4) and (5) with the perturbation method, and they are determined in forms of power series of λ.

Now the stress-intensity factors for crack tip points A and B are given by the formula

$$K_{A,B} = 2\sqrt{2}\, T\sqrt{b} \lim_{\zeta \to \pm\lambda} [\sqrt{\zeta \mp \lambda}\, \varphi'(\zeta)] \quad (6)$$

Using equation (3), together with the obtained series expressions for M_n and F_n, the K are given in power series of λ, whose coefficients are also infinite series due to the assumed series expansion for $\varphi(\zeta)$. Those coefficients, however, can be summed up in closed form, and their values are evaluated exactly by using the expressions for $c_{2p}{}^{2n}$ and $c_{2p+1}{}^{2n+1}$ in equation (5) [3].

Analysis is performed for 15 values of ϵ, including the extreme cases when $\epsilon = 0$ and $\epsilon \to 1$, which are the centrally cracked strip and the cracked semi-infinite plate, respectively; and the stress intensity factors are given by the following series:

Table 1 Values of coefficients in correction factors

ϵ	C_2	C_3	C_4	C_5	C_6	C_7	C_8	C_9	C_{10}
0	0.5948	0	0.4812	0	0.3963	0	0.3367	0	0.2972
0.02	0.5726	0.0339	0.4462	0.0315	0.3548	0.0433	0.2917	0.0464	0.2498
0.04	0.5535	0.0639	0.4173	0.0574	0.3234	0.0759	0.2608	0.0788	0.2208
0.06	0.5371	0.0903	0.3936	0.0785	0.2998	0.1003	0.2400	0.1014	0.2035
0.08	0.5231	0.1134	0.3743	0.0958	0.2823	0.1185	0.2263	0.1172	0.1939
0.1	0.5112	0.1335	0.3585	0.1099	0.2694	0.1319	0.2175	0.1281	0.1890
0.2	0.4761	0.1975	0.3155	0.1485	0.2428	0.1576	0.2073	0.1467	0.1904
0.3	0.4635	0.2179	0.3016	0.1571	0.2374	0.1538	0.2083	0.1428	0.1936
0.4	0.4555	0.2126	0.2922	0.1507	0.2292	0.1405	0.2012	0.1310	0.1860
0.5	0.4404	0.1939	0.2754	0.1355	0.2113	0.1236	0.1832	0.1154	0.1677
0.6	0.4123	0.1707	0.2473	0.1192	0.1841	0.1061	0.1574	0.0989	0.1429
0.7	0.3704	0.1495	0.2108	0.1029	0.1529	0.0905	0.1298	0.0841	0.1175
0.8	0.3197	0.1341	0.1735	0.0899	0.1246	0.0783	0.1063	0.0727	0.0969
0.9	0.2729	0.1264	0.1449	0.0814	0.1051	0.0706	0.0910	0.0656	0.0837
1	0.25	0.125	0.1328	0.0781	0.0967	0.0671	0.0836	0.0618	0.0766

ϵ	C_{11}	C_{12}	C_{13}	C_{14}	C_{15}	C_{16}	C_{17}	C_{18}	C_{19}
0	0	0.2713	0	0.2535	0	0.2404	0	0.2300	0
0.02	0.0533	0.2219	0.0576	0.2021	0.0627	0.1873	0.0669	0.1756	0.0711
0.04	0.0878	0.1948	0.0920	0.1774	0.0974	0.1650	0.1011	0.1558	0.1048
0.06	0.1099	0.1810	0.1127	0.1668	0.1167	0.1575	0.1189	0.1512	0.1212
0.08	0.1241	0.1749	0.1251	0.1638	0.1275	0.1570	0.1284	0.1528	0.1294
0.1	0.1331	0.1731	0.1325	0.1644	0.1336	0.1594	0.1334	0.1567	0.1337
0.2	0.1447	0.1817	0.1413	0.1772	0.1396	0.1748	0.1383	0.1735	0.1376
0.3	0.1387	0.1854	0.1355	0.1806	0.1336	0.1776	0.1324	0.1757	0.1318
0.4	0.1266	0.1771	0.1236	0.1715	0.1218	0.1679	0.1205	0.1654	0.1197
0.5	0.1111	0.1585	0.1082	0.1526	0.1063	0.1487	0.1049	0.1459	0.1040
0.6	0.0949	0.1343	0.0921	0.1289	0.0902	0.1252	0.0888	0.1225	0.0877
0.7	0.0804	0.1104	0.0779	0.1060	0.0761	0.1030	0.0747	0.1007	0.0736
0.8	0.0694	0.0915	0.0672	0.0881	0.0655	0.0858	0.0643	0.0840	0.0632
0.9	0.0626	0.0796	0.0606	0.0770	0.0591	0.0752	0.0579	0.0737	0.0570
1	0.0585	0.0724	0.0562	0.0697	0.0544	0.0678	0.0529	0.0662	0.0517

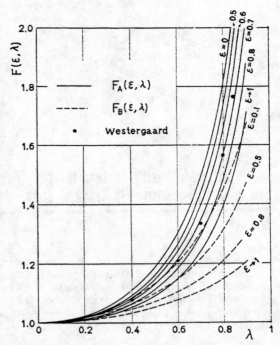

Fig. 2 Variation of correction factors $F_A(\epsilon, \lambda)$ and $F_B(\epsilon, \lambda)$ with ϵ and λ

$$K_A = T\sqrt{l}\, F_A(\epsilon, \lambda), \quad K_B = T\sqrt{l}\, F_B(\epsilon, \lambda)$$
$$F_A(\epsilon, \lambda) = 1 + \sum_{n=2}^{19} C_n \lambda^n, \quad F_B(\epsilon, \lambda) = 1 + \sum_{n=2}^{19} (-1)^n C_n \lambda^n \quad (7)$$

Numerical values of $C_n(\epsilon)$ are shown in Table 1. Convergency of the series in equations (7) is not discussed mathematically, but still those series are regarded to have better convergency than geometrical series of which the common ratios are λ, since absolute values of C_n have the general trend of decreasing with increasing values of n as shown in Table 1. If it is accepted, the errors due to truncating the series would be smaller than $|C_{18}\lambda^{19}/(1-\lambda)|$ and, on this basis, equations (7) are expected to give correct four-digit values of F_A and F_B if $\lambda \leq 0.7$, and correct three and two-digit values for $\lambda = 0.8$ and 0.9, respectively. In Fig. 2, $F_A(\epsilon, \lambda)$ and $F_B(\epsilon, \lambda)$ are plotted against λ for typical values of ϵ by solid and dotted curves, respectively. F_A and F_B start from the same curve when $\epsilon = 0$, the case of a centrally cracked strip. As eccentricity ϵ increases with fixed values of λ, F_A and F_B decrease as shown in the figure, the former gradually and the latter quite rapidly, and converge to the limiting curves for $\epsilon \to 1$, corresponding to the case of a cracked semi-infinite plate. Solid circles in the figure show the values by the tangent formula [4]

$$F_A(0, \lambda) = \left[\left(\frac{2}{\pi\lambda}\right)\left(\tan\frac{\pi\lambda}{2}\right)\right]^{1/2} \quad (8)$$

which is actually for a wide plate with an infinite number of colinear cracks, but often employed as an approximation for a centrally cracked strip, the case of $\epsilon = 0$ in the present problem. Those values are plotted fairly well below the author's exact curve, the discrepancy for $\lambda = 0.5$ being about 10 percent, and by coincidence very close to the curve of $F(0.7, \lambda)$. In Fig 3, F_A are also plotted against ϵ for various values of l/b, and this would be convenient to know the effect of the location of a crack of definite length on the stress-intensity factor.

Acknowledgment

The author wishes to express his sincere thanks to Mr. Y. Itagaki of The National Aerospace Laboratory, Tokyo, Japan, for programming the problem for a Burroughs Datatron 205 computer.

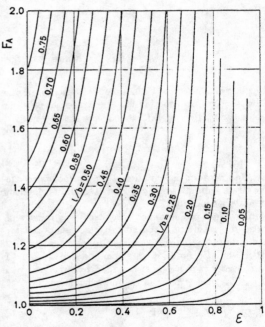

Fig. 3 Effect of location of crack of definite length on F_A

References

1 C. B. Ling, "Stresses in a Perforated Strip," JOURNAL OF APPLIED MECHANICS, vol. 24, TRANS. ASME, vol. 79, 1957, pp. 365–375.

2 M. Isida, "On the Stress Function in the Plane Problems of an Elastic Body Containing a Free Elliptical Hole," *Transactions of the Japan Society of Mechanical Engineers*, vol. 21, no. 107, 1955, pp. 502–506.

3 M. Isida, "On the Tension of a Strip With a Central Elliptical Hole," *Transactions of the Japan Society of Mechanical Engineers*, vol. 21, no. 107, 1955, pp. 507–518.

4 H. M. Westergaard, "Bearing Pressures and Cracks," JOURNAL OF APPLIED MECHANICS, vol. 6, TRANS. ASME, vol. 61, 1939, pp. A-49–A-53.

Reprinted with permission from *Plane Strain Crack Toughness Testing*, ASTM STP 410, pp. 77-79 (1966). ©1966 American Society for Testing and Materials.

Discussion: Plane strain crack toughness testing of high strength metallic materials (by William F. Brown, Jr., and John E. Srawley)

C. E. Feddersen[19]—From the perspective of design applications, the calibration factors (or finite-width corrections) are especially important elements of fracture mechanics theory. These expressions are the scaling transformations by which test data are extrapolated into practical design criteria. Since the precedents set by this committee have a far-reaching effect on the engineering applications, the following criticisms are offered in a constructive sense.

The comments may be resolved into two points about the calibration factors for the center-notch specimens. The first point concerns the leading coefficient of the calibration expression, Y. The coefficient 1.77 appears to be the value of $\sqrt{\pi}$. For clarity of interpretation and for

TABLE 5—*Finite width corrections.*
$Y/\sqrt{\pi} = f(\lambda)$ where $\lambda = 2a/W$

Author	Expression
Brown and Srawley	
3rd degree	$Y/\sqrt{\pi} = 1 + 0.128\lambda - 0.288\lambda^2 + 1.525\lambda^3$
2nd degree	$Y/\sqrt{\pi} = 1 - 0.1\lambda + \lambda^2$
Isida	$Y/\sqrt{\pi} = 1 + 0.5948\lambda^2 + 0.4812\lambda^4 + \cdots + 0.2535\lambda^{14}$
Forman and Kobayashi	$Y/\sqrt{\pi} = [\Sigma F(\lambda)g(\lambda)]^{1/2}$
Dixon	$Y/\sqrt{\pi} = (1 - \lambda^2)^{-1/2}$
Greenspan	$Y/\sqrt{\pi} = (1 - 0.5\lambda^2 - 0.5\lambda^4)^{-1}$
Modified Greenspan (Brossman and Kies)	$Y/\sqrt{\pi} = (1 + 0.5\lambda^4)^{1/2}(1 - 0.5\lambda^2 - 0.5\lambda^4)^{-1}$
Irwin	$Y/\sqrt{\pi} = [(2/\pi\lambda) \tan (\pi\lambda/2)]^{1/2}$

consistency with fracture mechanics theory, it is considered better form to retain the leading coefficient as the symbol $\sqrt{\pi}$.

The second and more important point is concerned with the actual form of the calibration expressions. Consider Table 5, which lists a few of the popular expressions for the center-notch calibration in algebraic form. (Note that the term 1.77 has been transposed in the first expression for purposes of comparison). The Brown and Srawley expressions are considered compact expressions which closely match the Isida equation, now considered the most precise expression for center-notch specimens. The Forman and Kobayashi expression is another formulation which appears to substantiate that of Isida. The formulations of Dixon, Greenspan, and Brossman and Kies are others which have been utilized. The final expression, that of Irwin, is in present usage but is to be superseded by the first listed expressions of Brown and Srawley.

Now a question arises: "How different are these various formula-

---

[19] Battelle Memorial Inst., Columbus, Ohio.

tions?" To answer this, consider Table 6. Here, values of the calibrations are tabulated for discrete values of the aspect ratio, $\lambda = 2a/W$. Note that at the currently recommended aspect ratio of 0.33 for plane strain toughness testing the discrepancies are quite small. However, as the recommended aspect ratio for testing increases to 0.50, the discrepancies increase, and it is very desirable to approximate Isida's work as closely as possible.

TABLE 6—*Comparison of various finite width corrections.*

Aspect Ratio	Brown and Srawley		Isida	Forman and Kobayashi	Dixon	Greenspan	Modified Greenspan (Brossman and Kies)	Irwin
	3rd Degree	2nd Degree						
0	1.000	1.00	1.000	...	1.000	1.000	1.000	1.000
0.1	1.012	1.00	1.006	...	1.005	1.005	1.005	1.001
0.2	1.026	1.02	1.025	...	1.021	1.029	1.029	1.017
0.3	1.053	1.06	1.058	...	1.037	1.051	1.052	1.040
0.4	1.103	1.12	1.109	...	1.091	1.101	1.107	1.076
0.5	1.183	1.20	1.187	...	1.155	1.185	1.205	1.130
0.6	1.303	1.30	1.303	...	1.250	1.324	1.370	1.208
0.7	1.473	1.42	1.487	1.464	1.401	1.574	1.670	1.335
0.8	1.699	1.56	1.799	...	1.667	2.135	2.370	1.565
0.9	1.993	1.72	2.391	...	2.292	3.745	4.320	2.115
1.0	2.365	1.90	3.631	...	∞	∞	∞	∞

TABLE 7—*Comparison of Isida and secant corrections.*

Aspect Ratio	Isida	$(\sec \pi\lambda/2)^{1/2}$	Difference
0	1.000	1.000	0
0.1	1.006	1.006	0
0.2	1.025	1.025	0
0.3	1.058	1.059	0.001
0.4	1.109	1.112	0.003
0.5	1.187	1.189	0.002
0.6	1.303	1.304	0.001
0.7	1.487	1.484	0.003
0.8	1.799	1.796	0.003
0.9	2.391	2.525	0.134
1.0	3.631	∞	∞

There exists a natural trigonometric function which approaches the Isida work more closely over a wider range than do the proposed Brown and Srawley results. This is shown in Table 7. Note that the precision appears to be within 0.3 per cent through an aspect ratio of 0.8. Here it is recommended that the secant expression be used to match Isida's work. In addition to greater accuracy over a wider range, the secant expression is certainly more compact than a second or third degree polynomial.

Now, a more subtle question appears: "From where does the secant correction arise?" With the simplification of notation $\theta = \pi\lambda/2$, we write the secant expression as

$$\left[\sec\theta\right]^{1/2} = \left[\frac{1}{\sin\theta} \cdot \tan\theta\right]^{1/2} \quad \dots \dots \dots \dots \dots \quad (6)$$

and compare it with the current Irwin expression,

$$\left[\frac{1}{\theta} \cdot \tan\theta\right]^{1/2} \quad \dots \dots \dots \dots \dots \dots \dots \quad (7)$$

There is a strong analytical similarity between these expressions through the usual trigonometric approximation $\theta \approx \sin\theta$, at small values of θ. While the Irwin analysis yielding Eq 7 is not questioned, it is suggested that there may exist a closely related stress function which would yield Eq 6.

However, whether it is an exact or approximate equation for matching Isida's work, the concise and accurate nature of the secant equation has considerable merit. The committee is also urged to survey the other calibration equations for simpler, more direct representations. Direct and concise format is certainly advantageous to the committee, as well as invaluable to those who will be applying the committee's developments.

Messrs. Brown and Srawley—For the practical purpose with which we are concerned in our paper we consider it desirable to express K calibrations in a simple standard form wherever possible. These calibrations are interpolation functions which are fitted to a limited number of primary results. The polynomial form permits determination of the coefficients of the interpolation function by a standard least-squares-best-fit computational procedure. It is also a convenient form for computation and manipulation.

We are indebted to Mr. Feddersen for his interesting observation that Isida's results for the center-cracked specimen correspond very closely to $(\sec \pi a/W)^{1/2}$; this is a convenient and compact expression. It seems most unlikely, however, that equally simple forms could be found for other configurations, and we are not aware of any methodical procedure that could be used to search for such forms.

From our point of view the expression of the polynomial coefficients in terms of the factor $\pi^{1/2}$ (1.77 ...) is an unnecessary embellishment. It amounts to the same thing as using the alternate stress intensity factor which is usually, but not always, written as a script K in the literature. The ASTM Special Committee on Fracture Testing (now E-24) decided early in its life to standardize on the K used in our paper in order to avoid ambiguity.

THE STRESS INTENSITY FACTORS OF A RADIAL CRACK IN A FINITE ROTATING ELASTIC DISC

D. P. ROOKE
The Royal Aircraft Establishment, Farnborough, Hants., England

and

J. TWEED
The University of Glasgow, Scotland

Abstract — The elastic problem of a radial crack in a finite rotating elastic disc is reduced to the solution of a Fredholm equation. From this solution the stress intensity factor at both tips and the crack formation energy are derived.

1. INTRODUCTION

IT HAS been shown by Tweed et al.[1] that the stress intensity factors and the strain energy for a radial crack, subject to an arbitrary internal pressure, in a finite disc can be expressed in terms of a function which is the solution of a Fredholm equation. In that paper[1] the plane strain case of a constant internal pressure was considered. We will here consider the pressure which gives results appropriate to the case of a crack in a uniformly rotating disc.

2. BASIC EQUATIONS

Consider a disc of unit radius containing a crack defined, in plane polar word coordinates, by $\theta = 0$, $0 < a \leq r \leq b < 1$ which is subject to an internal pressure $f(r)$. The stress intensity factors, K_a and K_b, and the crack energy, W, under plane strain conditions are given by the following:

$$\left. \begin{array}{ll} \text{for the tip at } r = a & K_a = \sqrt{\dfrac{2}{b-a}} P(a); \\[2ex] \text{for the tip at } r = b & K_b = -\sqrt{\dfrac{2}{b-a}} P(b) \end{array} \right\} \quad (1)$$

and

$$W = -\frac{2(1-\eta^2)}{E} \int_a^b \frac{P(t)\,dt}{\sqrt{(b-t)(t-a)}} \int_a^t f(r)\,dr, \quad (2)$$

where η is Poisson's ratio and E is Young's modulus. $P(t)$ is the solution of the integral equation.

$$P(t) - \int_a^b \frac{M(t,\rho) P(\rho)\,d\rho}{\sqrt{(b-\rho)(\rho-a)}} = S(t); a \leq t \leq b \quad (3)$$

with

$$S(t) = \frac{1}{\pi} \int_a^b \frac{\sqrt{(b-y)(y-a)}}{y-t} f(y)\,dy \quad (4)$$

and

$$M(t,\rho) = \frac{1}{\pi}\{\rho^2[J_3-J_1]+\rho^{-2}[J_3-J_2]+[J_0+J_2-2J_3]+(\rho-\rho^{-1})J\}, \quad (5)$$

where

$$J = \tfrac{1}{2}(a+b-2t) \quad (6)$$

$$J_0 = \frac{\sqrt{ab}}{t}-1 \quad (7)$$

$$J_1 = \frac{\sqrt{ab}}{t}-\frac{\sqrt{(1-a\rho)(1-b\rho)}}{(1-\rho t)}, \quad (8)$$

$$J_2 = \frac{\sqrt{ab}}{t}-\frac{\sqrt{(1-a\rho)(1-b\rho)}}{(1-\rho t)^2}+\frac{\rho(a+b-2ab\rho)}{2(1-\rho t)\sqrt{(1-a\rho)(1-b\rho)}} \quad (9)$$

and

$$J_3 = \frac{\sqrt{ab}}{t}-\frac{\sqrt{(1-a\rho)(1-b\rho)}}{(1-\rho t)^3}+\frac{\rho(2-\rho t)(a+b-2ab\rho)}{2(1-\rho t)^2\sqrt{(1-a\rho)(1-b\rho)}}$$

$$+\frac{\rho^2(a-b)^2}{8(1-\rho t)[(1-a\rho)(1-b\rho)]^{3/2}}. \quad (10)$$

The function $P(t)$ is closely related to the crack shape function $v(r,0)$ since [1], for $a \leq r \leq b$,

$$v(r,0) = -\frac{2(1-\eta^2)}{E}\int_r^b \frac{P(t)\,dt}{\sqrt{(b-t)(t-a)}}. \quad (11)$$

3. ROTATING DISC

For the problem of a crack in a disc, of density ρ_0, rotating at an angular velocity ω the pressure takes the form

$$f(r) = p_0[f_1(r)-\alpha f_2(r)] \quad (12)$$

where

$$p_0 = \frac{\rho_0\omega^2(3-2\eta)R^2}{8(1-\eta)}, \quad \alpha = \frac{1+2\eta}{3-2\eta}$$

$$f_1(r) = 1 \quad \text{and} \quad f_2(r) = r^2/R^2$$

where R is the radius of the disc (numerically equal to unity).

If we let $P_i(t)$, $i = 1, 2$ be defined by

$$P_i(t) - \int_a^b \frac{M(t,\rho)P_i(\rho)\,d\rho}{\sqrt{(b-\rho)(\rho-a)}} = S_i(t), a < t < b \quad (13)$$

where

$$S_i(t) = \frac{1}{\pi} \int_a^b \frac{\sqrt{(b-y)(y-a)}}{y-t} f_i(y) \, dy \tag{14}$$

then it follows from (3) and (4) that

$$P(t) = p_0[P_1(t) - \alpha P_2(t)]. \tag{15}$$

From (1) we have

$$K_a = p_0 \sqrt{\frac{2}{(b-a)}} [P_1(a) - \alpha P_2(a)] \tag{16}$$

and

$$K_b = -p_0 \sqrt{\frac{2}{(b-a)}} [P_1(b) - \alpha P_2(b)]. \tag{17}$$

If K_0 is the stress intensity factor of a Griffith crack of length $(b-a)$ which is opened by a constant pressure p_0, then clearly we may write

$$\frac{K_a}{K_0} = \frac{K_a^{(1)}}{K_0} - \alpha \frac{K_a^{(2)}}{K_0} \quad \text{and} \quad \frac{K_b}{K_0} = \frac{K_b^{(1)}}{K_0} - \alpha \frac{K_b^{(2)}}{K_0} \tag{18}$$

where

$$\frac{K_a^{(i)}}{K_0} = \frac{2}{b-a} P_i(a) \quad \text{and} \quad \frac{K_b^{(i)}}{K_0} = \frac{2}{b-a} P_i(b), i = 1, 2. \tag{19}$$

From (2) and (12) the energy can be written as

$$W = -\frac{2(1-\eta^2)}{E} \int_a^b \frac{P(t)}{\sqrt{(b-t)(t-a)}} p_0 \left[(t-a) - \frac{\alpha(t^3 - a^3)}{3R^2} \right] dt. \tag{20}$$

Since, from (11), crack closure at $r = a$ implies

$$\int_a^b \frac{P(t)}{\sqrt{(b-t)(t-a)}} \, dt = 0 \tag{21}$$

we have,

$$W = -\frac{2(1-\eta^2) p_0^2}{E} \int_a^b \frac{(t - \alpha t^3/3R^2)}{\sqrt{(b-t)(t-a)}} [P_1(t) - \alpha P_2(t)] \, dt. \tag{22}$$

Let W_0 be the energy of a Griffith crack of length $(b-a)$ which is opened by a constant pressure p_0, then

$$W_0 = \frac{\pi(1-\eta^2) p_0^2 (b-a)^2}{4E} \tag{23}$$

and

$$\frac{W}{W_0} = \frac{1}{W_0}\left[W_1 - \alpha W_2 - \frac{\alpha}{3}W_3 + \frac{\alpha^2}{3}W_4\right] \tag{24}$$

where

$$\left.\begin{aligned}\frac{W_1}{W_0} &= -\frac{8}{\pi(b-a)^2}\int_a^b \frac{tP_1(t)\,dt}{\sqrt{(b-t)(t-a)}} \\ \frac{W_2}{W_0} &= -\frac{8}{\pi(b-a)^2}\int_a^b \frac{tP_2(t)\,dt}{\sqrt{(b-t)(t-a)}} \\ \frac{W_3}{W_0} &= -\frac{8}{\pi(b-a)^2 R^2}\int_a^b \frac{t^3 P_1(t)\,dt}{\sqrt{(b-t)(t-a)}} \\ \frac{W_4}{W_0} &= -\frac{8}{\pi(b-a)^2 R^2}\int_a^b \frac{t^3 P_2(t)\,dt}{\sqrt{(b-t)(t-a)}}\end{aligned}\right\} \tag{25}$$

and

In order to evaluate K_a, K_b, and W we require $S_1(t)$ and $S_2(t)$ from (14). $S_1(t)$ which is the same function as $S(t)$, for a constant pressure, in [1] is given by

$$S_1(t) = \frac{a+b-2t}{2}. \tag{26}$$

Elementary integration of (14) for $i = 2$ gives

$$S_2(t) = \frac{1}{R^2}\left[\frac{t^2}{2}(a+b-2t) + \frac{(b-a)^2}{16}(a+b+2t)\right]. \tag{27}$$

4. NUMERICAL PROCEDURE

The integral equations (13) are solved using the Gauss–Chebyshev quadrature formula, i.e. (13) becomes

$$P_i(t_j) - \frac{\pi}{n}\sum_{k=1}^{n} M(t_j, t_k) P_i(t_k) = S_i(t_j) \tag{28}$$

for $i = 1, 2$ and $j, k = 1, 2, \ldots n$ where n is the order of the Gauss–Chebyshev approximation. The arguments t_j are given by

$$t_j = \frac{a+b}{2} + \frac{b-a}{2}\cos\frac{\pi(2j-1)}{2n}. \tag{29}$$

Solution of the set of simultaneous linear equations (28) yields $P_i(t_j)$ for $i = 1$ and 2. From (13) it follows that

$$P_i(a) = S_i(a) + \frac{\pi}{n} \sum_{k=1}^{n} M(a,t_k) P_i(t_k)$$

and

$$P_i(b) = S_i(b) + \frac{\pi}{n} \sum_{k=1}^{n} M(b,t_k) P_i(t_k). \quad (30)$$

Thus K_a/K_0 and K_b/K_0 can now be calculated using (18) and (19). To the same approximation the energy terms are obtained as:

$$\frac{W_1}{W_0} = -\frac{8}{(b-a)^2 n} \sum_{j=1}^{n} t_j P_1(t_j)$$

$$\frac{W_2}{W_0} = -\frac{8}{(b-a)^2 n} \sum_{j=1}^{n} t_j P_2(t_j)$$

$$\frac{W_3}{W_0} = -\frac{8}{(b-a)^2 n R^2} \sum_{j=1}^{n} t_j^3 P_1(t_j) \quad (31)$$

and

$$\frac{W_4}{W_0} = -\frac{8}{(b-a)^2 n R^2} \sum_{j=1}^{n} t_j^3 P_2(t_j).$$

To calculate W_3 and W_4 R is put equal to unity in the above equations.

5. RESULTS

Since the stress intensity factors and the crack energy are physical quantities, the way in which they vary with a and b must be independent of the mathematical technique used to find them. It follows that we may dispense with the restriction $0 < a < b < 1$ which was made at the beginning of [1] only to facilitate the use of the Mellin transform. By using the procedure outlined above, the authors have constructed a comprehensive

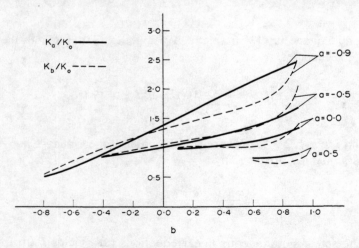

Fig. 1. The variation of K_a/K_0 and K_b/K_0 with a and b for the case in which Poisson's ratio is 0·3.

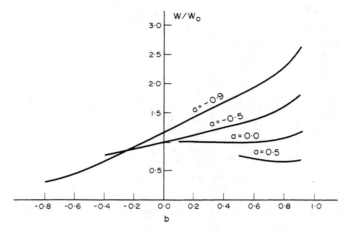

Fig. 2. The variation of W/W_0 with a and b for the case in which Poisson's ratio is 0·3.

set of tables which give the values of $K_a^{(i)}/K_0$, $K_b^{(i)}/K_0$, $i = 1, 2$ and $W^{(j)}/W_0, j = 1, 2, 3, 4$ for values of a and b satisfying the condition $-1 < a < b < 1$. These tables, which are accurate to four significant figures may be obtained from either author. Figures 1 and 2 show the variation of K_a/K_0, K_b/K_0 and W/W_0 with a and b for the special case in which Poisson's ratio is 0·3.

REFERENCES
[1] J. TWEED, S. C. DAS and D. P. ROOKE, *Int. J. Engng Sci.* **10**, 323 (1972).
[2] S. TIMOSHENKO and J. N. GOODIER, *Theory of Elasticity*. McGraw-Hill (1965).
[3] M. ABRAMOWITZ and I. A. STEGUN, *Handbook of Mathematical Functions*. Dover (1965).
[4] I. S. GRADSHTEYN and I. M. RYZHIK, *Tables of Integrals, Series and Products*. Academic Press (1965).

(Received 1 October 1971)

Résumé — Le problème classique d'une fissure radiale dans un disque élastique fini en rotation est ramené à la résolution d'une équation de Fredholm. A partir de cette solution, le facteur d'intensité de contrainte à chaque extrémité et l'énergie de formation de la fissure sont déduits.

Zusammenfassung — Das elastische Problem eines Radialrisses in einer endlichen sich drehenden Scheibe wird auf die Lösung einer Fredholm'schen Gleichung reduziert. Für diese Lösung werden der Spannungsintensitätsfaktor an beiden Spitzen und die Rissbildungsenergie abgeleitet.

Sommario — Il problema elastico di un'incrinatura radiale in un disco elastico finito in rotazione è ridotto alla soluzione di un'equazione di Fredholm. Da questa soluzione il fattore d'intensità delle sollecitazioni ad entrambe le punte e l'energia della formazione dell' incrinatura sono rispettivamente calcolate.

Абстракт — Проблема об упругости для радиальной трещины в конечном, вращающемся, упругом диске сводится к решению уравнения Фредгольма. На основе этого решения выводятся коэффициенты интенсивности напряжения у обеих концов и энергию образования трещины.

THE STRESS INTENSITY FACTOR OF AN EDGE CRACK IN A FINITE ROTATING ELASTIC DISC

D. P. ROOKE[†] and J. TWEED[‡]

(Communicated by I. N. SNEDDON)

Abstract — In this paper the problem of determining the stress intensity factor and crack formation energy of an edge crack in a finite rotating elastic disc is reduced to the solution of a Fredholm equation. Numerical results are given.

1. INTRODUCTION

IN A RECENT paper[1] the authors considered the problem of determining the stress intensity factor and the crack energy of an edge crack in a finite elastic disc when the surfaces of the crack are subjected to a distribution of pressure which is symmetric about the plane of the crack but is otherwise arbitrary. In that paper[1] the plane strain case of a constant internal pressure was considered in detail. We will here consider the pressure which gives results appropriate to the case of a crack in a uniformly rotating disc.

2. BASIC EQUATIONS

Consider a disc of unit radius containing a crack which is defined, in plane polar coordinates (r, ϑ), by the relations $\vartheta = 0$, $0 < b \leq r \leq 1$ and which is subject to an internal pressure $f(r)$. Under plane strain conditions the stress intensity factor K and the crack energy W are given[1] by

$$K = [2/(1-b)]^{1/2} P(b) \tag{2.1}$$

and

$$W = \frac{2(1-\eta^2)}{E} \int_b^1 \frac{P(t)}{[(1-t)(t-b)]^{1/2}} \int_t^1 f(r)\,\mathrm{d}r\,\mathrm{d}t, \tag{2.2}$$

where E is the Young's modulus and η the Poisson's ratio of the material. $P(t)$ is the solution of the integral equation

$$P(t) - \int_b^1 \frac{P(\rho) M(t, \rho)}{[(1-\rho)(\rho-b)]^{1/2}}\,\mathrm{d}\rho = S(t), \quad b < t < 1 \tag{2.3}$$

with

$$S(t) = \frac{1-t}{\pi} \int_b^1 \left[\frac{y-b}{1-y}\right]^{1/2} \frac{f(y)}{y-t}\,\mathrm{d}y \tag{2.4}$$

[†]The Royal Aircraft Establishment, Farnborough, Hants.
[‡]The University of Glasgow, Glasgow, Scotland.

and

$$M(t,\rho) = \frac{1-t}{\pi}\left\{\frac{(\rho^2-1)^2(1-b)}{2(1-\rho t)^2(1-\rho)^{3/2}(1-b\rho)^{1/2}} + \frac{(\rho^2-1)^2(1-b)[4-(3b+1)\rho]}{8(1-\rho t)(1-\rho)^{5/2}(1-b\rho)^{3/2}}\right.$$

$$+ \frac{\rho(\rho^2-1)(1-b\rho)^{1/2}}{(1-\rho t)^2(1-\rho)^{1/2}} + \frac{\rho^2(\rho^2-1)(1-b)}{2(1-\rho t)(1-\rho)^{3/2}(1-b\rho)^{1/2}}$$

$$+ \frac{\rho^2-1}{\rho}\left[1 + \frac{(\rho^2-1)(1-b\rho)^{1/2}}{(1-\rho t)^3(1-\rho)^{1/2}}\right]$$

$$\left. - \frac{\rho(1-b\rho)^{1/2}}{(1-\rho t)(1-\rho)^{1/2}}\right\}. \tag{2.5}$$

3. ROTATING DISC

For the problem of a crack in a disc, of density ρ_0, rotating at an angular velocity ω the pressure takes [2] the form

$$f(r) = p_0[f_1(r) - \alpha f_2(r)] \tag{3.1}$$

where, for a disc of unit radius,

$$p_0 = \frac{\rho_0 \omega^2 (3-2\eta)}{8(1-\eta)}, \quad \alpha = \frac{1+2\eta}{3-2\eta},$$

$$f_1(r) = 1 \quad \text{and} \quad f_2(r) = r^2.$$

If we let $P_k(t)$, $k = 1, 2$, be defined by

$$P_k(t) - \int_b^1 \frac{P_k(\rho)M(t,\rho)}{[(1-\rho)(\rho-b)]^{1/2}} \, d\rho = S_k(t), \quad b < t < 1 \tag{3.2}$$

where

$$S_1(t) = \frac{1-t}{\pi}\int_b^1 \left[\frac{y-b}{1-y}\right]^{1/2}\frac{f_1(y)}{y-t}\,dy = 1-t \tag{3.3}$$

and

$$S_2(t) = \frac{1-t}{\pi}\int_b^1 \left[\frac{y-b}{1-y}\right]^{1/2}\frac{f_2(y)}{y-t}\,dy$$

$$= \frac{(1-t)}{8}\{(1-b)(3+b+4t) + 8t^2\}, \tag{3.4}$$

we see that

$$P(t) = p_0[P_1(t) - \alpha P_2(t)] \tag{3.5}$$

and hence that the stress intensity factor K is given by

$$K = p_0[2/(1-b)]^{1/2}[P_1(b) - \alpha P_2(b)]. \tag{3.6}$$

Similarly, we see that

$$W = \frac{2p_0^2(1-\eta^2)}{E} \int_b^1 \frac{[P_1(t) - \alpha P_2(t)]}{[(1-t)(t-b)]^{1/2}} \left[(1-t) - \frac{\alpha}{3}(1-t^3)\right] dt. \tag{3.7}$$

Let K_0 and W_0 be the stress intensity factor and crack energy respectively, of a Griffith crack of length $2(1-b)$ in an infinite elastic sheet. Then, in the case of a constant pressure p_0, it is found that

$$K_0 = p_0(1-b)^{1/2}$$

and

$$W_0 = \frac{\pi p_0^2(1-b)^2(1-\eta^2)}{E}$$

(see e.g. reference [3]). With this notation equations (3.6) and (3.7) may be written in the form

$$K/K_0 = K_1/K_0 - \alpha K_2/K_0 \tag{3.8}$$

and

$$W/W_0 = W_1/W_0 - \alpha W_2/W_0 - \alpha W_3/W_0 + \alpha^2 W_4/W_0, \tag{3.9}$$

where

$$K_k/K_0 = \frac{2^{1/2}}{1-b} P_k(b), \qquad k = 1, 2 \tag{3.10}$$

$$W_1/W_0 = \frac{2}{\pi(1-b)^2} \int_b^1 \frac{(1-t)^{1/2}}{(t-b)^{1/2}} P_1(t)\, dt$$

$$W_2/W_0 = \frac{2}{\pi(1-b)^2} \int_b^1 \frac{(1-t)^{1/2}}{(t-b)^{1/2}} P_2(t)\, dt$$

$$W_3/W_0 = \frac{2}{\pi(1-b)^2} \int_b^1 \frac{(1-t^3)P_1(t)}{[(t-b)(1-t)]^{1/2}}$$

$$\tag{3.11}$$

and

$$W_4/W_0 = \frac{2}{3\pi(1-b)^2} \int_b^1 \frac{(1-t^3)P_2(t)}{[(t-b)(1-t)]^{1/2}} dt.$$

From Bettz's reciprocal theorem (see [2], p. 339) it follows that W_2 and W_3 are equal.

4. NUMERICAL PROCEDURE

The integral equations (3.2) may be solved by using the Gauss–Chebyshev quadrature formula to turn them into the linear algebraic systems

$$P_k(t_i) - \frac{\pi}{n} \sum_{j=1}^n P_k(t_j) M(t_i, t_j) = S_k(t_i), \tag{4.1}$$

where $k = 1, 2$; $i = 1, 2, \ldots, n$ and the arguments t_i are given by

$$t_i = \frac{1+b}{2} + \frac{1-b}{2} \cos\frac{(2i-1)\pi}{2n}, \tag{4.2}$$

$i = 1, 2, \ldots, n$. Once the equations (4.1) have been solved the quantities $P_k(b)$ $k = 1, 2$ may be obtained from

$$P_k(b) = S_k(b) + \frac{\pi}{n} \sum_{j=1}^{n} P_k(t_j) M(b, t_j), \qquad k = 1, 2. \tag{4.3}$$

K/K_0 can now be calculated by substituting these results into (3.8) and (3.10). Similarly, by making use of the formulae

$$\left.\begin{aligned} W_1/W_0 &= \frac{2}{n(1-b)^2} \sum_{j=1}^{n} (1-t_j) P_1(t_j), \\ W_2/W_0 &= \frac{2}{n(1-b)^2} \sum_{j=1}^{n} (1-t_j) P_2(t_j), \\ W_4/W_0 &= \frac{2}{3n(1-b)^2} \sum_{j=1}^{n} (1-t_j^3) P_2(t_j), \end{aligned}\right\} \tag{4.4}$$

and equation (3.9) we can calculate W/W_0 to the same degree of approximation.

The quantities K/K_0 and W/W_0 have been computed for several values of the crack length $a = 1 - b$ for the cases in which Poisson's ratio is 0.2, 0.3 and 0.4. The results are shown in Fig. 1.

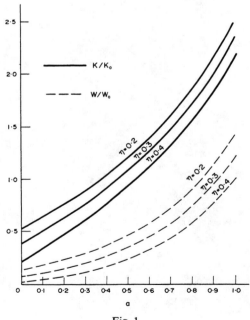

Fig. 1.

REFERENCES
[1] J. TWEED and D. P. ROOKE, *Int. J. Engng Sci.* **11**, 65 (1972).
[2] S. TIMOSHENKO and J. N. GOODIER, *Theory of Elasticity*. McGraw-Hill (1951).
[3] I. N. SNEDDON and H. A. ELLIOTT, *Q. appl. Math.* **4**, 229 (1946).

(Received 8 *May* 1972)

Résumé — Dans cet article, le problème de détermination du facteur d'intensité d'effort et de l'énergie de formation de fissure d'une fissure de lisière dans un disque élastique rotatif fini est réduite à la solution d'une équation de Fredholm. On donne les résultats numériques.

Zusammenfassung — In dieser Arbeit wird das Problem der Bestimmung des Spannungsintensitätsfaktors und der Rissbildungsenergie eines Kantenrisses in einer endlichen rotierenden elastischen Scheibe auf die Lösung einer Fredholm-Gleichung reduziert. Es werden numerische Resultate gegeben.

Sommario — In questo articolo, l'A. riduce a soluzione di un'equazione di Fredholm il problema di determinare il fattore d'intensità delle sollecitazioni e d'energia di formazione di un'incrinatura di una fessura sul bordo di un disco elastico rotante finito. Si presentano risultati numerici.

Абстракт — В этой работе проблема определения коэффициента интенсивности напряжения и энергии формирования трещины для краевой трещины, находящейся в конечном, вращающемся, упругом диске, сводится к решению уравнения Фредгольма. Даны численные результаты.

Stress intensity factors for some through-cracked fastener holes

A. F. GRANDT Jr.

Metals and Ceramics Division, Air Force Materials Laboratory, Wright-Patterson Air Force Base, Dayton, Ohio 45433, U.S.A.

(Received December 12, 1973; in revised form March 22, 1974)

ABSTRACT

A stress intensity factor solution is developed for a large plate containing radial hole cracks loaded with arbitrary crack face pressure. When the pressure is defined as the unflawed hoop stress surrounding a mechanical fastener, stress intensity factor calibrations are readily computed by the linear superposition principle. Results obtained in this manner agree well with previous solutions determined for open holes loaded in remote tension. The potential usefulness of the present analysis is further demonstrated with application to specific fastener configurations, including interference fit fasteners, pin-loaded plates, and cold-worked holes.

1. Introduction

Cracks which emanate from fastener holes represent one of the most common sources for brittle fracture in aircraft structures [1]. In order to determine the severity of such flaws, the design engineer may utilize the stress intensity factor to relate crack length, remote loading, and structural geometry. With the stress intensity factor one may calculate the critical flaw size for a given structure and loading as well as estimate the service life of components containing subcritical flaws.

Recent attempts to reduce the severity of flaws located at fastener holes by introducing controlled residual stress fields have been shown to provide some improvement in service life [2]. Cold-working holes with an oversized mandrel, for example, develops residual compression next to the hole which may retard, or in some cases, stop cracks from growing under subsequent cyclic loading.

To date, analytical consideration of holes which have been subjected to some type of fatigue life extension process has been limited to determining the residual stress field surrounding the hole [3–6]. Corresponding stress intensity factor solutions are presently unavailable for "fatigue improvement" fasteners. In previous work [7] it was shown, however, that the mode I stress intensity factor (K_1) could be estimated for cracked fastener holes from a knowledge of the unflawed stress distribution in the region of the hole. The procedure employed a linear superposition technique which has seen frequent use in the literature [8–14].

In order to apply the linear superposition method, one needs stress intensity factor solutions for flaws subjected to arbitrary crack face pressure. Such solutions are available for central flaws in wide plates [8] and for edge cracked strips [9], [12]. More recently, the central crack solution was modified to estimate K_1 for two diametrically opposed radial cracks emanating from a circular hole [13], [14].

The linear superposition method was applied directly to the fatigue rated fastener hole problem in Ref. [7] by using the single edge crack solution. Although good results were obtained for small flaws ($a/r < 0.15$), it was felt at that time that improved accuracy could be achieved for longer crack lengths by working directly with a solution for cracked holes loaded with arbitrary crack face pressure. With this motivation for the present work, a solution is developed in the next section for single and double radial cracks emanating from a circular hole in a wide plate. The single flaw geometry is shown in Fig. 1, with the double crack configuration chosen as its symmetric counterpart. Once the arbitrary crack face pressure solution is obtained, its utility is demonstrated by application to typical fastener hole configurations through the linear superposition method.

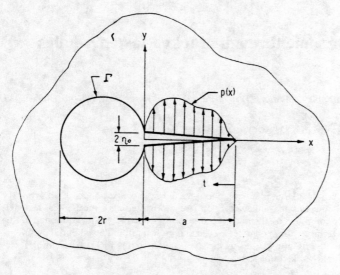

Figure 1. Open hole containing a radial crack subjected to pressure $p(x)$.

2. Development of the fastener hole solution

The analytical approach to determine the stress intensity factor solution for radial cracks subjected to an arbitrary pressure distribution is discussed in the following section. The method is based on work by Rice [15], who has shown that once the displacement field and stress intensity factor are known for one geometry and loading (case 1), K_1 may be obtained for any other symmetric loading applied to the same crack geometry (case 2).

As shown in Fig. 1, the problem to be solved here consists of radial cracks of length a emanating from a circular hole of radius r. The flaws are loaded with an arbitrary pressure $p(x)$ which is perpendicular to the crack faces and symmetric with respect to the x-axis. In addition, the pressure distribution is restricted to mode I loading only (i.e., the crack faces are not allowed to close in compression).

Following Rice [15], the stress intensity factor for a given crack geometry under arbitrary loading (case 2) is given by

$$K_1 = \int_\Gamma \mathbf{s} \cdot \mathbf{h} \, d\Gamma + \int_A \mathbf{f} \cdot \mathbf{h} \, dA . \tag{1}$$

Here $\mathbf{s}$ is the stress vector acting on boundary Γ chosen around the crack tip, and $\mathbf{f}$ is the body force acting in region A defined by Γ. The vector $\mathbf{h}$ is a universal weight function for the given crack geometry determined by the known loading condition (case 1). The weight function is given by

$$\mathbf{h} = \mathbf{h}(x, y, a) = \frac{H}{2K} \frac{\partial \mathbf{u}}{\partial a} \tag{2}$$

where H is the constant

$$H = \begin{cases} E & \text{for plane stress} \\ E/(1-v^2) & \text{for plane strain} \end{cases} \tag{3}$$

Here E is the elastic modulus and v is Poisson's ratio. In Eqn. (2), K and $\mathbf{u}$ are the known stress intensity factor and displacement field for the specified load configuration (case 1).

In order to find the desired stress intensity factor solution to the cracked hole in Fig. 1 (case 2), a remote uniform tensile stress σ was selected as the known (case 1) loading. Crack tip stresses have been obtained for the latter geometry by Bowie [16] and stress intensity factor results given in Ref. [17].

If one chooses the Bowie problem as the case 1 loading, and defines the boundary Γ to consist of the hole perimeter and crack faces as shown in Fig. 1, specific terms in Eqn. (1) become

$$s = \begin{bmatrix} s_x = 0 \\ s_y = p(x) \\ s_x = s_y = 0 \end{bmatrix} \begin{array}{l} \text{along the crack faces} \\ \\ \text{along the hole perimeter} \end{array} \quad (4)$$

$$f = 0 \quad (\text{no body forces in Area } A) \quad (5)$$

$$h_y = \frac{H}{2K_B} \cdot \frac{\partial u_y}{\partial a} = \frac{H}{2K_B} \cdot \frac{\partial \eta}{\partial a} \quad (6)$$

where η is the y-component of the crack surface displacements and K_B is the Bowie solution for the stress intensity factor. For crack lengths up to $a/r = 10$, the Bowie solution may be represented within three percent by the function

$$K_B = \sigma (\pi a)^{\frac{1}{2}} \left[\frac{F_1}{F_2 + a/r} + F_3 \right] \quad (7)$$

where F_1, F_2, and F_3 are the constants given in Table 1 [7]. In this instance Eqn. (1) reduces to

$$K_1 = \frac{H}{K_B} \int_0^a p(x) \frac{\partial \eta}{\partial a} dx . \quad (8)$$

Now, once $\partial \eta / \partial a$ is defined, the desired stress intensity factor solution for arbitrary crack surface pressure is given by Eqn. (8).

TABLE 1

Least squares approximation to Bowie solution for radially cracked holes

$$K_1 = \sigma_0 (\pi a)^{\frac{1}{2}} \left[\frac{F_1}{F_2 + a/r} + F_3 \right]$$

	Single crack	Double crack
F_1	0.8733	0.6865
F_2	0.3245	0.2772
F_3	0.6762	0.9439

Since crack opening displacements were not given in the Bowie analysis of the cracked hole [16], it was necessary to determine η in the present work. Near the crack tip the displacement field is readily determined from the known stress intensity factor expression by the following equation [17].

$$\eta = 4 \frac{K_B}{H} (t/2\pi)^{\frac{1}{2}} . \quad (9)$$

Here t is the distance from the crack tip ($t = a - x$). For large values of t (approximately $t > a/10$), Eqn. (9) loses validity and η must be found by an alternate approach. In this instance, the finite element method was used to determine crack surface displacements for the radially flawed holes. The single crack model consisted of a 6 inch wide by 14 inch long aluminum plate ($E = 10^7$ psi, $\nu = 0.31$) containing a 0.3125 inch diameter hole [18]. The finite element grid used for the double crack problem was 30 inches wide by 20 inches long with a 2.0 inch diameter hole [19]. The unflawed hoop stress determined by both finite element models was well within one percent of the accepted values [20].

In order to compute $\partial \eta / \partial a$, this displacement information must be represented by an appropriate expression for η. Orange [21] has shown that the shape of edge cracks under remote tension or bending may be approximated within three percent by conic sections of the form

$$\left(\frac{\eta}{\eta_0}\right)^2 = \frac{2}{2+m}(t/a) + \frac{m}{2+m}(t/a)^2. \tag{10}$$

Here η_0 is the displacement at the crack mouth ($t=a$) and m is the conic section coefficient found from

$$m = \pi\left[\frac{H\eta_0}{2\sigma a Y}\right]^2 - 2. \tag{11}$$

In this instance Y is the factor defined from Eqn. (7) as

$$Y = \frac{K_B}{\sigma a^{\frac{1}{2}}} = \pi^{\frac{1}{2}}\left[\frac{F_1}{F_2 + a/r} + F_3\right]. \tag{12}$$

Finite element results for the crack mouth displacement η_0 were closely represented by the least squares expression

$$\eta_0 = r\sum_{i=0}^{6} D_i(a/r)^i \tag{13}$$

where D_i are the constants given in Table 2. When the finite element representation lost accuracy for small crack lengths ($a/r < 0.1$), the crack mouth displacements were determined from edge cracked plate results [21] modified to reflect the hole stress concentration.

TABLE 2

Least squares fit of finite element data for crack mouth displacement

$\eta_0/r = \sum_{i=0}^{6} Di(a/r)^i$

Coefficient	single crack	Double crack
D_0	-1.567×10^{-6}	1.548×10^{-5}
D_1	6.269×10^{-4}	5.888×10^{-4}
D_2	-6.500×10^{-4}	-4.497×10^{-4}
D_3	4.466×10^{-4}	3.101×10^{-4}
D_4	-1.725×10^{-4}	-1.162×10^{-4}
D_5	3.485×10^{-5}	2.228×10^{-5}
D_6	-2.900×10^{-6}	-1.694×10^{-6}

Crack surface profiles determined by Eqns. (10)–(13) for the double crack problem are compared in Fig. 2 with results from Eqn. (9) and the finite element data. Similar results were obtained for the single flaw case. Examining Fig. 2, note that Eqn. (10) closely matches Eqn. (9) near the crack tip and forces the displacement curve through the finite element value for η_0 at the edge of the hole ($t=a$). The fact that Eqn. (10) predicts larger values for η near the midpoint of the crack length than given by the finite element data should not be considered significant, since the finite element answers lose accuracy as one approaches the crack tip. Thus, calculating $\partial \eta/\partial a$ from Eqn. (10) and inserting the result into Eqn. (8) completes the desired solution for the cracked holes loaded with arbitrary crack face pressure. The general utility of this solution is demonstrated with several specific examples in the remainder of this paper.

3. Application to typical fastener problems

Employing the linear superposition method, the crack face pressure solution developed in the previous section may be applied to a wide variety of fastener problems. Stress intensity factor calibrations are computed directly from Eqn. (8) by defining $p(x)$ as the hoop stress surrounding

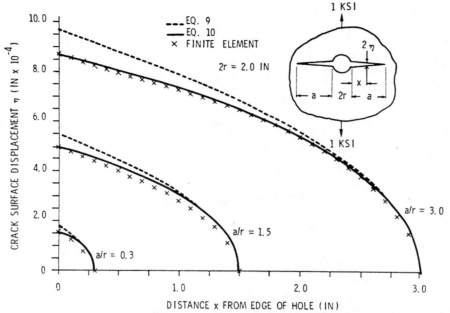

Figure 2. Crack surface profiles for Bowie problem (double crack).

the *unflawed* fastener hole. This approach is used below to calculate K_1 from several typical stress distributions taken from the literature. (Since the author has made no attempt to assess the accuracy of the initial unflawed stress solutions, these examples have been chosen to demonstrate the application of Eqn. (8) through use with the linear superposition method rather than to provide quantitative K_1 solutions.)

3.1. Bowie problem

The Bowie problem provides a convenient test for the accuracy of the present solution. For an *unflawed* hole loaded with remote tension σ, the hoop stress σ_H along a radial line located at an angle θ to the loading axis is given by [20]

$$\sigma_H = \frac{\sigma}{2}\left[1 + \left(\frac{r}{r+x}\right)^2\right] - \frac{\sigma}{2}\left[1 + 3\left(\frac{r}{r+x}\right)^4\right]\cos 2\theta. \tag{14}$$

Letting $\theta = \pi/2$ and taking the resulting expression as $p(x)$, Eqns. (8) and (14) give the stress intensity factor calibrations shown in Figs. 3 and 4 for single and double flaws. Note that the present solution agrees within seven percent of the Bowie analysis throughout the crack length range investigated. An earlier application of the edge crack solution to the Bowie problem [7] gave results which agreed within 10 percent for small cracks ($a/r < 0.15$) but which exceeded the Bowie analysis by as much as 23 percent for the larger flaws shown in Fig. 3. Thus, use of the present analysis rather than the edge crack solution provides considerable improvement in accuracy when the linear superposition method is applied to the Bowie problem.

The remaining difference between the present solution and the Bowie analysis for the open hole problem is most likely due to the approximations employed in the numerical calculations. Possible sources of deviation include the least squares representation of the Bowie coefficients (Eqn. (7)) and the approximations necessary to determine the crack surface displacements (Eqns. (10) and (13)).

3.2. Biaxial loading

The effect of biaxial loading on the open hole problem may be determined by the present analysis. Consider flawed holes loaded with a combination of remote stress σ and $C\sigma$ as shown in Figs. 5 and 6. The unflawed hoop stress perpendicular to the subsequent direction of crack

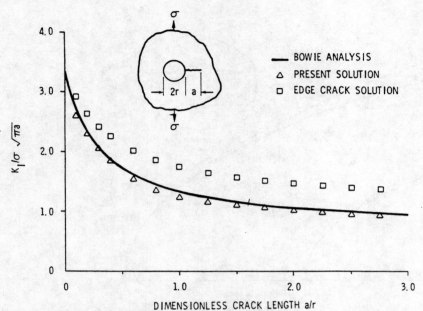

Figure 3. Comparison of stress intensity factor solutions for the Bowie problem (single flaw).

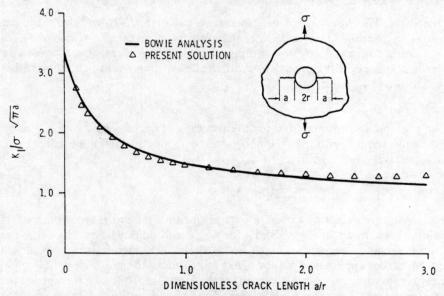

Figure 4. Comparison of stress intensity factor solutions for the Bowie problem (double flaws).

growth may be found by appropriate superposition of specific solutions to Eqn. (14). Defining the crack face pressure $p(x)$ in this manner for $-1.0 \leq C \leq 1.0$ and using the general solution of Eqn. (8) gives the stress intensity factor calibrations shown in Figs. 5 and 6.

Examining these curves, one notices a significant biaxial stress effect in the region near the hole, but that this result rapidly decreases for longer crack lengths ($a/r > 1.0$). The equal biaxial stress case ($C = +1.0$) has been solved by Bowie [16]. A comparison of these results indicates that the present linear superposition method agrees well with Bowie's stress intensity factor analysis for flawed holes loaded in equal biaxial tension.

3.3. Hole loaded in compression

Consider a plate loaded in compression with radial cracks growing in the direction of loading

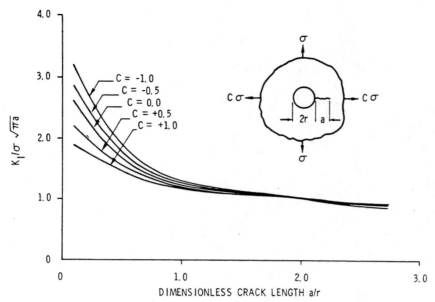

Figure 5. Effect of biaxial stress on flawed hole (single crack).

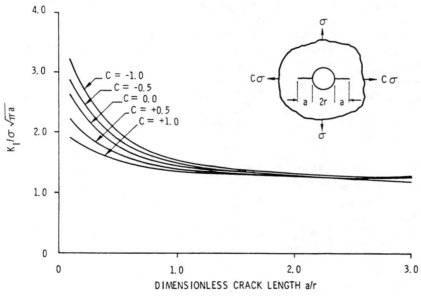

Figure 6. Effect of biaxial stress on flawed hole (double crack).

as shown in Fig. 7. In this case, one notices from Eqn. (14) that the unflawed hoop stress in the direction of crack growth is tensile in the region adjacent to the hole. Taking this unflawed stress distribution as $p(x)$ in Eqn. (8) as before, one obtains the stress intensity factor results for single and double flaws shown in Fig. 7. Note that there is a positive stress intensity factor for small cracks located at holes loaded in remote compression, but that this value rapidly decreases to zero as the flaw extends.

3.4. Interference fit fasteners

Slightly oversize fasteners are commonly employed in structural applications to improve fatigue performance. The finite element method was used in Ref. [4] to study the influence of interference level and material properties on the stress distribution surrounding unflawed

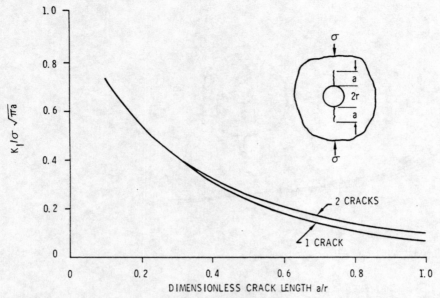

Figure 7. Stress intensity factor calibration for cracked hole in remote compression.

holes. Stress analyses are presented in Ref. [4] for interference fit loading (no remote load) and for combinations of interference and remote load.

The hoop stresses around an unflawed hole in an aluminum plate containing a steel interference fit fastener (0.00125 in. radial interference) were selected as a typical structural application for Eqn. (8). Defining $p(x)$ as a least squares polynomial representation of the finite element hoop stresses along a line perpendicular to the direction of remote loading gives the stress intensity factor calibrations shown in Fig. 8. Note from these curves that a sizeable residual stress intensity factor may exist around interference fit fasteners in the absence of remote loading. This result may have significance for structures which are stored for extended periods of time in an agressive environment.

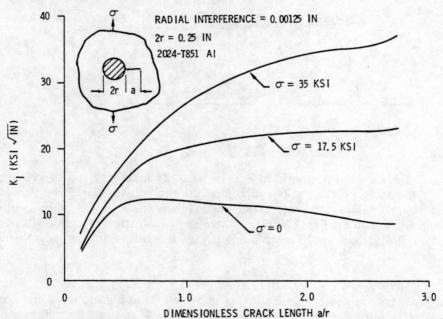

Figure 8. Stress intensity factor calibration for flawed interference fit fastener installation (single crack).

3.5. Pin-loaded holes

The finite element method was used in Ref. [22] to determine the stress field around load transfer fasteners. The problem considered here consisted of a 0.25 in. diameter hole located on the vertical axis of symmetry of a 1.875 in. wide (2 W) by 8 in. long aluminum sheet (thickness = 0.125 in.). The hole was placed 0.625 in. from one end of the sheet and subjected to a uniform 8000 lb. load at the far end. This load was reacted through a rigid pin located in the hole with two values of pin clearance (based on diameters) of 2.0 percent and 0.4 percent. In this instance the maximum tangential stress was found to occur at an 81° angle with the loading axis as shown in Fig. 9.

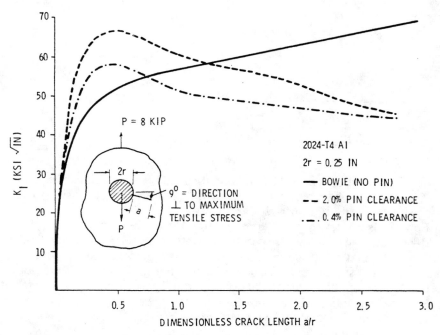

Figure 9. Stress intensity factor solutions for pin-loaded holes (single crack).

Expressing the hoop stress given in Ref. [22] for this 81° radial line with a least squares polynomial expansion, and using the solution of Eqn. (8), one obtains the stress intensity factor predictions of Fig. 9. Although in this instance the crack was assumed to grow along the 81° radial direction, one would expect cracks occurring in service to curve over to a path perpendicular to the loading axis as propagation extends away from the influence of the pin. Since the largest a/W values considered here are less than 0.4, no attempt was made to apply a width correction to the stress intensity factor calculations. The influence of the free boundaries are reflected, however, in the initial finite element stresses used to calculate $K_1(p(x)$ in Eqn. (8)). Note in Fig. 9 that K_1 increases with pin clearance and that initially the stress intensity factor exceeds the Bowie analysis [16] for the open hole (no pin), but decreases as the crack extends. The strain energy release rate found by compliance techniques for a pair of radial cracks emanating from pin-loaded holes [23] confirms the initial magnification in stress intensity factor for pin-loaded holes in wide plates. The fact that K_1 eventually falls below the open hole (Bowie) analysis for longer cracks is also expected from the well known result that crack line loaded flaws show a decreasing stress intensity factor [17].

3.6. Mandrel hole enlargement

The residual stress field caused by pre-expanding fastener holes with an oversized mandrel has been studied by means of an elastic-plastic finite element model in Ref. [5]. The problem considered consisted of a 0.26 in. diameter hole in a 2.5 in. wide aluminum plate (thickness =

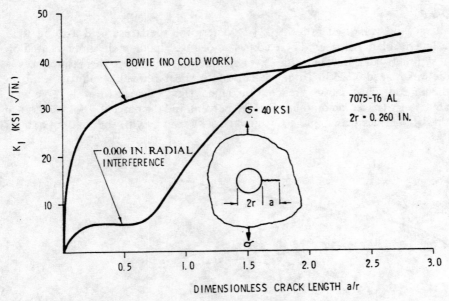

Figure 10. Stress intensity factor calibration for a cold worked hole.

0.25 in.). The hole was radially expanded by 0.006 in. and then unloaded to define the residual stresses and strains. The plate was then remotely loaded with a uniform 40 ksi tensile stress to determine the subsequent redistribution in stress and strain.

Defining $p(x)$ in Eqn. (8) as a least squares representation for the hoop stress distribution surrounding the cold-worked hole gives the stress intensity factor calibration shown in Fig. 10. Note for small crack lengths that the 0.006 in. mandrel hole enlargement causes a dramatic reduction in K_1 when the hole is subsequently loaded in remote tension. This result provides a qualitative explanation for the increased fatigue life observed in Ref. [2] for radially cracked cold-worked holes. Since the stress intensity factor is significantly lowered by the cold-working process, ΔK is reduced, resulting in slower fatigue crack growth rates.

4. Summary and conclusions

A stress intensity factor solution was developed for an infinite plate containing radial hole cracks which are loaded with arbitrary crack face pressure. When the pressure distribution is chosen as the hoop stress surrounding an unflawed fastener hole, stress intensity factor calibrations may be determined for flawed holes by the linear superposition method. Stress intensity factors computed in this manner agree well with previous results obtained for open holes loaded in remote tension. Although the present solution does not reflect finite plate boundaries directly (other than through their effect on the unflawed stress input), reasonable K_1 estimates for moderate width panels should be possbile with standard width corrections. One should be cautious, however, in applying the current results to crack configurations which are strongly influenced by the proximity of free boundaries.

In order to apply the linear superposition method to specific problems, one must first determine the circumferential stresses around the unflawed hole. Although stress analyses for complex problems, such as "fatigue improvement" fasteners in structural applications, will, in general, require sophisticated analysis techniques, the more difficult task of resolving crack stresses has been avoided. In addition, since an entire K_1 calibration curve is obtained from a single unflawed stress solution, computation time and expense is greatly reduced.

The potential utility of the present solution for application to common structural problems was demonstrated with selected examples for interference fit fasteners, pin-loaded plates, and cold-worked holes. Further research should be directed toward securing unflawed stress solutions for specific fracture critical fastener configurations. Once these results are available, the present analysis may be used to study the influence of cracking at these critical locations.

Acknowledgement

Although the author is indebted to many colleagues for advice and assistance, he especially wishes to thank Dr. J. P. Gallagher who suggested the linear superposition method, and to acknowledge the computer programming support of S. LaMacchia and the technical aid of J. Wagner.

REFERENCES

[1] R. J. Gran, F. D. Orazio, P. C. Paris, G. R. Irwin and R. Hertzerg, *Investigation and Analysis Development of Early Life Aircraft Structural Failures*, Technical Report AFFDL-TR-70-149, Wright-Patterson Air Force Base, Ohio, March (1971).
[2] C. F. Tiffany, R. P. Stewart and T. K. Moore, *Fatigue and Stress-Corrosion Test of Selected Fasteners/Hole Processes*, Technical Report ASD-TR-72-111, Wright-Patterson Air Force Base, Ohio, January (1973).
[3] J. A. Regalbuto and O. E. Wheeler, Stress Distribution from Interference Fits and Uniaxial Tension, *Experimental Mechanics*, 10, 7 (1970) 274–280.
[4] M. Allen and J. A. Ellis, *Stress and Strain Distribution in the Vicinity of Interference Fit Fasteners*, Technical Report AFFDL-TR-72-153, Wright-Patterson Air Force Base, Ohio, January (1973).
[5] W. F. Adler and D. M. Dupree, *Stress Analysis of Coldworked Fastener Holes*, Technical Report AFML-TR-74-44, Wright Patterson Air Force Base, Ohio, March 1974.
[6] J. H. Crews, Jr., *An Elastic Analysis of Stresses in a Uniaxially Loaded Sheet Containing an Interference-Fit Bolt*, NASA Technical Note NASA TN D-6955, Langley Research Center, Hampton, Virginia, October 1972.
[7] A. F. Grandt, Jr. and J. P. Gallagher, Procedures for Infinite Life Design of Mechanical Fasteners, *Proceedings of the Air Force Systems Command 1973 Science and Engineering Symposium*, AFSC TR-73-003, Vol. I, Kirtland Air Force Base, New Mexico, October (1973).
[8] P. C. Paris, M. P. Gomez and W. E. Anderson, A Rational Analytic Theory of Fatigue, *The Trend in Engineering*, 13, 1, January (1961). University of Washington.
[9] A. F. Emery, Stress-Intensity Factors for Thermal Stresses in Thick Hollow Cylinders, *Journal of Basic Engineering, Transactions of the ASME, Series D*, March (1966) 45–52.
[10] A. F. Emery and G. E. Walker, Jr., *Stress Intensity Factors for Edge Cracks in Rectangular Plates with Arbitrary Loadings*, ASME Paper No. 68-WA/Met-18.
[11] A. F. Emery, G. E. Walker, Jr. and J. A. Williams, A Green's Function for the Stress Intensity Factors of Edge Cracks and Its Application to Thermal Stresses, *Journal of Basic Engineering, Transactions of the ASME*, Vol. 91, Series D, No. 4, December (1969) 618–624.
[12] H. F. Bueckner, Weight Functions for the Notched Bar, *Zeitschrift für Angewandte Mathematik und Mechanik*, Vol. 51, pp 97–109, (1971).
[13] J. H. Crews, Jr. and N. H. White, *Fatigue Crack Growth From a Circular Hole With and Without High Prior Loading*, Technical Report NASA TN D-6899, Langley Research Center, Hampton, Virginia, September (1972).
[14] R. A. Schmidt, *An Approximate Technique for Obtaining Stress Intensity Factors for Some Difficult Planar Problems*, paper presented at Fracture and Flaws Symposium, Albuquerque, New Mexico, March 2 (1973).
[15] J. R. Rice, Some Remarks on Elastic Crack-Tip Stress Fields, *International Journal of Solids and Structures*, 8, 6 June (1972) 751–758.
[16] O. L. Bowie, Analysis of an Infinite Plate Containing Radial Cracks Originating from the Boundary of an Internal Circular Hole, *Journal of Mathematics and Physics*, 35 (1956) 60–71.
[17] P. C. Paris and G. C. Sih, Stress Analysis of Cracks, *Fracture Toughness Testing and Its Applications, ASTM STP 381, American Society for Testing and Materials*, (1965) 30–81.
[18] Private correspondence with G. F. Zielsdorff, Air Force Flight Dynamics Laboratory, Wright Patterson Air Force Base, Ohio.
[19] Private correspondence with H. Keck, United States Air Force Academy, Colorado Springs, Colorado.
[20] S. Timoshenko and J. N. Goodier, *Theory of Elasticity*, McGraw-Hill Book Company (1951).
[21] T. W. Orange, Crack Shapes and Stress Intensity Factors for Edge-Cracked Specimens, *Stress Analysis and Growth of Cracks, Proceedings of the 1971 National Symposium on Fracture Mechanics, Part I, ASTM STP 513, American Society for Testing and Materials*, (1972) 71–78.
[22] H. G. Harris, I. U. Ojalvo, and R. E. Hooson, *Stress and Deflection Analysis of Mechanically Fastened Joints*, Technical Report AFFDL-TR-70-49, Wright Patterson Air Force Base, Ohio (1970).
[23] D. J. Cartwright and G. A. Ratcliffe, Strain Energy Release Rate for Radial Cracks Emanating from a Pin Loaded Hole, *International Journal of Fracture Mechanics*, 8, 2, June (1972) 175–181.

RÉSUMÉ

Une solution pour déterminer le facteur d'intensité des contraintes est développée dans le cas d'une grande plaque comportant des fissures émanant radialement d'un trou et sollicitées par une pression arbitraire agissant sur leurs faces. Lorsque la pression est définie comme valant la tension de membrane aux alentours d'un rivet et en l'absence de défauts, le facteur d'intensité des contraintes peut être directement calculé par le principe de superposition linéaire des effets. Les résultats obtenus suivant cette approche sont en bon accord avec les solutions précédemment développées pour des trous ouverts sollicités par des efforts de traction agissant à une certaine distance. L'utilité potentielle de l'analyse proposée est en outre démontrée dans le cas de son application à des configurations particulières de rivetage, telles que les rivets à coincement, les tôles sollicitées par des broches, et les trous écrouis par mandrinage.

ZUSAMMENFASSUNG

Man stellt eine Lösung für die Spannungsintensitätsfaktoren auf für den Fall einer großen Platte mit Rissen ausgehend von radialen Löchern mit beliebigen Druck am Rißausgang. Wenn der Druck als die Zylinder, die ein mechanisches Verbindungselement umgibt wenn kein Fehler vorhanden ist, definiert wird, dann kann man die Spannungsintensitätsfaktoren sofort nach dem Verfahren der linearen Überlagerung rechnen. Die so erhaltene Ergebnisse stimmen gut mit den früheren Lösungen für offene Löcher die durch Zugkräfte belastet sind, die in einiger Entfernung angebracht werden. Die große Nützlichkeit dieser Analyse wird weiterhin durch Anwendungen auf verschiedene Verbindungselemente, einschließlich Drucknieten, durch Spindeln belastete Platten und kaltgeformte Löcher, bewiesen.

D. O. HARRIS

Mechanical Engineer,
Lawrence Radiation Laboratory,
University of California,
Livermore, Calif.
Assoc. Mem. ASME

Stress Intensity Factors for Hollow Circumferentially Notched Round Bars

Stress intensity factors are presented for a hollow circumferentially notched round bar subjected to tension, torsion, bending, and transverse shear. These previously undetermined factors are calculated from the limiting values of stress concentration factors for a small notch root radius. The stress intensity factors obtained are compared with results obtained by other investigators, when possible, and a favorable correlation is observed. The accuracy is further indicated by the good agreement of the fracture toughness of 7075-T6 aluminum, determined in a series of tension and bending tests conducted on hollow notched round bars, with the corresponding value obtained by other investigators.

Introduction

THE relation between loading, the body geometry, and the crack tip stress intensity factors is necessary in the application of sharp crack fracture mechanics (as originated by Griffith [1],[1] and later extended by Irwin [2, 3]), to the failure analysis of a structure containing flaws. This paper presents the stress intensity factors for some previously unsolved geometries, as determined from the limiting values of stress concentration factors for a small notch root radius. This method, which is covered by Irwin [3] and Paris and Sih [4], is quite simple when compared with the other mathematical techniques that have been used. Paris and Sih, in their review paper [4], provide a thorough coverage of the techniques available; they also include a convenient collection of previously solved problems, and an extensive list of references in the field.

A serious limitation of the stress concentration factor approach is that an elastic analysis of the notched geometry is necessary. However, a great deal of work has been done on this type of problem, notably by Neuber in *Theory of Notch Stresses* [5]. The present paper gives the stress intensity factors for a hollow circumferentially notched round bar subjected to various loadings, a result readily obtainable from Neuber's work on stress concentration factors.

Determination of Stress Intensity Factors From Neuber's Work

The stress field near an elastic crack tip can be completely characterized by three stress intensity factors [3, 4] corresponding to the three modes of deformation—opening, sliding, and tearing. The stress intensity factors can be found from stress concentration factors through the relations [3, 4]:

$$K_I = \frac{\sqrt{\pi}}{2} \lim_{p \to 0} p^{1/2} \sigma_{\max} = \frac{\sqrt{\pi}}{2} \lim_{p \to 0} p^{1/2} \sigma_{\text{net}} \alpha_i \quad (1)$$

provided $K_{II} = K_{III} = 0$;

$$K_{II} = \sqrt{\pi} \lim_{p \to 0} p^{1/2} \sigma_{\max} = \sqrt{\pi} \lim_{p \to 0} p^{1/2} \sigma_{\text{net}} \alpha_i \quad (2)$$

provided $K_I = K_{III} = 0$;

$$K_{III} = \sqrt{\pi} \lim_{p \to 0} p^{1/2} \tau_{\max} = \sqrt{\pi} \lim_{p \to 0} p^{1/2} \tau_{\text{net}} \alpha_i \quad (3)$$

provided $K_I = K_{II} = 0$; where α_i is the appropriate stress concentration factor.

In some instances, the stress intensity factors can be found more conveniently by comparing the elastic stress field close to the notch root, as given by Neuber, to the field given by the stress intensity equations for the stresses in the vicinity of the crack tip. These latter equations are [3, 4]:

$$\sigma_x = \frac{K_I}{\sqrt{2\pi r}} \cos \frac{\theta}{2} \left[1 - \sin \frac{\theta}{2} \sin \frac{3\theta}{2} \right]$$
$$\quad - \frac{K_{II}}{\sqrt{2\pi r}} \sin \frac{\theta}{2} \left[2 + \cos \frac{\theta}{2} \cos \frac{3\theta}{2} \right]$$

$$\sigma_y = \frac{K_I}{\sqrt{2\pi r}} \cos \frac{\theta}{2} \left[1 + \sin \frac{\theta}{2} \sin \frac{3\theta}{2} \right]$$
$$\quad + \frac{K_{II}}{\sqrt{2\pi r}} \sin \frac{\theta}{2} \cos \frac{\theta}{2} \cos \frac{3\theta}{2}$$

$$\tau_{xy} = \frac{K_I}{\sqrt{2\pi r}} \sin \frac{\theta}{2} \cos \frac{\theta}{2} \cos \frac{3\theta}{2} \quad (4)$$

---

[1] Numbers in brackets designate References at end of paper.
Contributed by the Metals Engineering Division for presentation at the Winter Annual Meeting, New York, N. Y., November 27–December 1, 1966, of THE AMERICAN SOCIETY OF MECHANICAL ENGINEERS. Manuscript received at ASME Headquarters, July 27, 1966. Paper No. 66—WA/Met-19.

Nomenclature

a = radial distance from crack tip to surface of axial hole
c = radius of axial hole
d = notch root diameter of solid round bar
D = outside diameter of round bar
K_I = stress intensity factor, opening mode
K_{II} = stress intensity factor, sliding mode
K_{III} = stress intensity factor, tearing mode

M = bending moment
p = notch root radius
P = tensile force
r = radial coordinate from crack tip
t = crack depth
T = torsional moment
V = transverse shear force
W = wall thickness of hollow round bar
α_k = stress concentration factor for arbitrary notch depth
α_{tk} = stress concentration factor for a deep notch

α_{fk} = stress concentration factor for a shallow notch
θ = angular coordinate from crack tip
ν = Poisson's ratio
σ_{ys} = 0.2 percent offset yield strength
$\sigma_{\max}$ = maximum normal stress adjacent to crack tip
$\tau_{\max}$ = maximum shear stress adjacent to crack tip
ϕ = angular coordinate around bar circumference

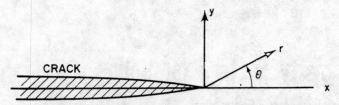

Fig. 1 Coordinate system for stresses in the vicinity of a crack tip

$$+ \frac{K_{II}}{\sqrt{2\pi r}} \cos \frac{\theta}{2} \left[1 - \sin \frac{\theta}{2} \sin \frac{3\theta}{2} \right] \quad (4)$$
(Cont.)

$$\tau_{xz} = -\frac{K_{III}}{\sqrt{2\pi r}} \sin \frac{\theta}{2} \quad \tau_{yz} = \frac{K_{III}}{\sqrt{2\pi r}} \cos \frac{\theta}{2}$$

$\sigma_z = \nu(\sigma_x + \sigma_y)$, plane strain

or

$\sigma_z = 0$, plane stress.

The coordinate system is shown in Fig. 1.

The stress concentration factors were obtained directly from Neuber's work. The procedure followed by Neuber is:

1 The stress field for the deeply notched solid round bar is completely determined by the use of the theory of elasticity. The stress concentration factors, which are always defined as the ratio of the maximum stress to the nominal net stress, are then found from the stress field equations.

2 The stress concentration factor for a shallow notch is determined, usually by approximations.

3 The stress concentration factor for a deep notch with a large axial hole is determined, also usually by approximations.

4 The stress concentration factors for arbitrary dimensions are then estimated by the use of equations presented by Neuber for interpolation between the foregoing three limiting cases.

Thus only the concentration factors for a deep notch are exact, other cases being approximations.

Once the stress concentration factors are found from Neuber's work, the stress intensity factors can be determined from equations (1), (2), and (3).

Calculations

The calculations for determining the stress intensity factors from Neuber's stress concentration factors are quite straightforward. A sample calculation is presented in the Appendix. It was necessary to modify the stress concentration factor for a shallow notch in tension or in bending. For these two cases Neuber gives

$$\alpha_{fk} = 1 + 2(t/p)^{1/2} \quad (5)$$

This gives, from equation (1),

$$K_I = \sigma_{net}(\pi t)^{1/2} \quad (6)$$

However, it is known [3, 4, 6, 7] that

$$K_I = 1.12 \sigma_{net}(\pi t)^{1/2} \quad (7)$$

for an edge crack in an infinite plate. Therefore, to agree with this more accurate expression, the equation for α_{fk} must be modified to

$$\alpha_{fk} = 1 + 2.24(t/p)^{1/2} \quad (8)$$

The stress intensity factors for torsion and tension are quite simple, because they do not vary around the notch root circumference. In the case of bending, K_I for the deep notch, as found from the stress field equations, varies as $\cos \phi$. The angular variations for the shallow notch and for the deep notch with large

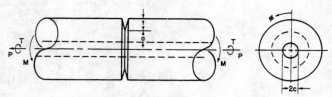

$$K_I = \frac{P}{\pi[(c+a)^2 - c^2]} \sqrt{\pi t} \left\{ 0.80 + \frac{1}{a+c}\left(4 + 1.08 \frac{c}{a}\right) \right\}^{-\frac{1}{2}}$$

$$+ \frac{4M(c+a) \cos \phi}{\pi[(c+a)^4 - c^4]} \sqrt{\pi t} \left\{ 0.80 + \frac{1}{a+c}\left(7.12 + 1.08 \frac{c}{a}\right) \right\}^{-\frac{1}{2}}$$

$$K_{III} = \frac{2T(c+a)}{\pi[(c+a)^4 - c^4]} \sqrt{\pi t} \left\{ 1 + \frac{1}{a+c}\left(\frac{64}{9} + 2.468 \frac{c}{a}\right) \right\}^{-\frac{1}{2}}$$

P – TENSILE LOAD
M – BENDING MOMENT
T – TORSIONAL MOMENT

Fig. 2 Stress intensity factors for notched round bar with axial hole: General case. The angular variation of K_I is assumed to be the same as for deep notch, as discussed in the text.

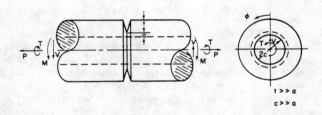

$$K_I = \frac{0.96 P}{(2c+a)\sqrt{\pi a}} + \frac{1.28 M \cos \phi}{c(c+a)\sqrt{\pi a}}$$

$$K_{III} = \frac{4T}{3c(c+a)\sqrt{\pi 3a}} + \frac{2V \sin \phi}{c\sqrt{\pi 3a}}$$

Fig. 3 Stress intensity factors for deeply notched round bar with large axial hole. The angular variation of K_I is assumed to be the same as for deep notch, as discussed in the text.

axial hole are not given by Neuber, but it seems reasonable that these stress intensity factors would vary with ϕ in the same manner as for the deep notch.

The analysis for transverse shear loading is considerably more involved, because, as determined from the elastic stress fields given by Neuber, the type of stress field present at the crack tip is different for the deep notch than for the shallow one. That is, the crack tip stress field for the deep notch is mode II, which is the sliding mode and gives K_{II} as the only nonzero stress intensity factor, whereas a mode III crack tip stress field is present for the two cases of a shallow notch and a deep notch with a large axial hole. Mode III is the tearing mode, and gives K_{III} as the only nonzero stress intensity factor. Thus K_{II} is present for the limiting case of a deep notch, and K_{III} for the other two limiting cases of a shallow notch and a deep notch with a large axial hole. The equations proposed by Neuber for interpolation between the limiting cases can therefore not be used to determine the stress intensity factors for intermediate notch depths. Consequently, the stress intensity factors for transverse shear loading can be obtained by this method only for the three limiting cases. Also, since some question arises as to which mode is present, it is

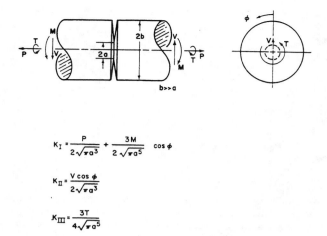

Fig. 4 Stress intensity factors for deeply notched round bars

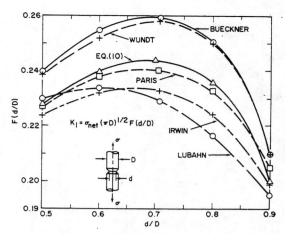

Fig. 5 Various results for K_I for solid notched round bar in tension

preferable to determine the stress intensity factors by comparing the stress field near the notch root, as given by Neuber, to the field given by the stress intensity equations, rather than using the stress concentration factors. Another advantage of comparing the stress fields to find the stress intensity factors is that the angular variation around the notch circumference is simultaneously determined. Therefore, the stress intensity factors for transverse shear loading were determined by comparison of the elastic stress fields. These calculations, while not overly complex, are quite lengthy, and therefore are not included.

Results

The results obtained for the general case are presented in Fig. 2, and special cases are shown in Figs. 3 and 4. These special cases may be of particular interest in certain cases, as discussed in the following section.

Discussion

The results of this investigation are of interest in many respects. Their use makes possible the failure analysis of flawed shafts subjected to various loadings. This is an extension of previous applications of sharp crack fracture mechanics, since the equations for the stress intensity factors for this geometry were not previously known (except for the solid circumferentially notched round bar in tension).

The hollow notched round bar may be quite useful as a fracture toughness specimen. The circumferentially notched round without an axial hole has been widely used as a mode I fracture toughness specimen [8, 9, 10], but, as pointed out by Srawley and Brown [8], it suffers from a high fracture load, compared to other conventional specimens. The axial hole would reduce the necessary load, but plane-strain conditions at the notch tip might be disrupted by the presence of the axial hole.

The hollow notched round in torsion may be a convenient specimen for mode III fracture testing to determine K_{IIIc}, a parameter which would be useful primarily in the failure analysis of flawed shafts in torsion. The presence of a large axial hole would eliminate the problem of crack arrest owing to the propagation of the crack into a decreasing stress field (as may occur in the solid notched round). The geometry which has been analyzed may also be convenient in combined-mode failure testing.

A thorough assessment of the accuracy of the results obtained in this investigation is presently impossible. However, their validity can be estimated by comparing them with results of two special cases reported by other investigators:

1 The solid notched round bar in tension, which corresponds to the case shown in Fig. 2, with $c = 0$ (and $V = M = T = 0$).

2 The single-edge notched plate in tension, which also corresponds to the case shown in Fig. 2, but with $c \to \infty$ (and $V = M = T = 0$).[2]

The circumferentially notched solid round in tension has been the object of several investigations which are conveniently summarized by Bueckner [11], in his discussion of Paris and Sih's paper. This discussion includes experimental results obtained by Lubahn [12], results obtained by Irwin [13] and Wundt [14] from extrapolation of stress concentration factors, the results of an analysis made by Bueckner, and an approximate analysis included in Paris and Sih's paper.

The equation for the stress intensity factor for the solid notched round bar in tension can be expressed as

$$K_I = \sigma_{net}(\pi D)^{1/2} F(d/D). \qquad (9)$$

$F(d/D)$, as obtained in this investigation, can be determined from the equation included in Fig. 2 (with $c = 0$, $a = d/2$, $a + t = D/2$) (see Appendix):

$$F(d/D) = \left[\frac{1}{2} \frac{d}{D} \frac{1 - d/D}{4 - 3.20 d/D}\right]^{1/2} \qquad (10)$$

Table 1 $F(d/D)$ obtained by various investigators for K_I for solid notched round in tension

d/D	Lubahn (Ref. 12)	Irwin (Ref. 13)	Wundt (Ref. 14)	Paris (Ref. 4)	Bueckner (Ref. 11)	Eq. (10)
0.5	0.230	0.224	0.239	0.227	0.240	0.228
0.6	0.234	0.232	0.252	0.238	0.255	0.240
0.707	0.229	0.233	0.258	0.240	0.259	0.244
0.8	0.217	0.224	0.250	0.233	0.251	0.236
0.9	0.195	0.199	0.210	0.205	0.210	0.200

$K_I = \sigma_{net}(\pi D)^{1/2} F(d/D)$

A comparison of this result with those of other investigators is presented in Table 1 and in Fig. 5, for d/D between 0.5 and 0.9. It is seen that equation (10) agrees favorably with the other analyses.

Bueckner claims an accuracy of 1 percent for his results, but the relatively poor agreement with others indicates the possibility of a larger error. The experimental results obtained by Lubahn are subject to relatively large error, especially for large d/D. The results obtained by extrapolation of stress concentration factors

---

[2] The author is indebted to Prof. A. S. Kobayashi, of the University of Washington, Seattle, Wash., for suggesting this comparison.

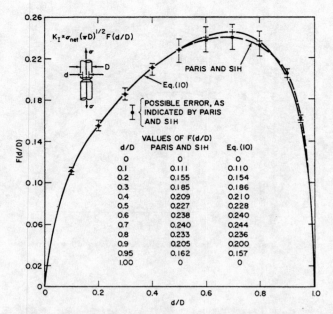

Fig. 6 Comparison of results for stress intensity factor for solid notched round bar in tension with Paris and Sih's values, full range of d/D

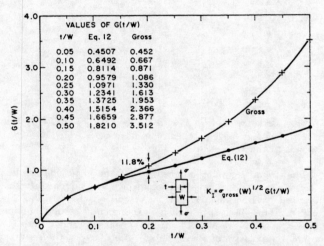

Fig. 7 Comparison of Gross' results for single-edge notched plate with equation (12) for notched round bar with large axial hole

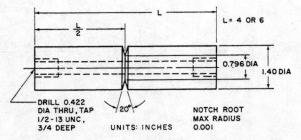

Fig. 8 Sketch of specimen used in experimental work

closely, it will suffice to use only the theoretical results obtained by Gross, Srawley, and Brown [17] for this comparison.

The equation for a single-edge notched plate in tension is conveniently expressed as

$$K_{\mathrm{I}} = \sigma_{\mathrm{gross}} W^{1/2} G(t/W) \qquad (11)$$

where σ_{gross} is the gross nominal stress, t is the crack length, and W is the plate width (equal to $a + t$). $G(t/W)$, as obtained in this investigation, can be found from the equation included in Fig. 2 by letting c approach infinity, putting $t + a = W$, and converting to the gross, rather than the net, stress. The result is

$$G(t/W) = \left[\frac{25\pi t/W}{20 - 13(t/W) - 7(t/W)^2}\right]^{1/2} \qquad (12)$$

A comparison of this result with the theoretical solution obtained in reference [17] is included in Fig. 7. The agreement is favorable for values of t/W up to 0.2, but poor for larger values. A possible explanation of the deviation of the two sets of values is that the single-edge notched plate analysis was based on a uniform stress distribution across the width of the plate at a distance far from the crack, so that, for a fairly deep crack, the ends of the plate may rotate somewhat in the plane of the plate. If this rotation is constrained, as would probably be the case for the hollow notched round with a large axial hole, the stress intensity factor for the crack will be reduced. This could account for the considerably higher values of $G(t/W)$ which were obtained for the single-edge notched plate. The disagreement of 11.8 percent for $t/W = 0.2$ is for the extreme case of c very large, so even if the inaccuracy of the result obtained in this investigation was this large in the extreme case, it is probable that the error would diminish as c became finite. This further indicates the accuracy of the stress intensity factors obtained from Neuber's work.

Neuber [20] has recently proposed a new formula for interpolating between the two limiting cases of the deep and the shallow notch. The new equation is believed to give more accurate results than those obtained through use of the older formula, which was used in this investigation. However, the new formula cannot be used if an axial hole is present. The stress intensity factor for a solid notched round bar in tension, as obtained by use of the two interpolation equations, disagree by 5.3 percent, at the most. The results using the newer, and probably more accurate, formula agree best with Wundt's values.

Based on the limited number of comparisons possible, it may be concluded that the accuracy of the results of this investigation is favorable, even when the older, less accurate interpolation formula is used.

Experimental Work

A limited amount of experimental work has been conducted with the specimen shown in Fig. 8. The results further indicate the accuracy of the stress intensity factor equations that have been obtained. Two series of tests were conducted on specimens made from 7075-T6 aluminum round bar.

(Wundt and Irwin) vary widely, indicating that this method is relatively inaccurate. It is of interest that the result expressed by equation (10), which was also determined from stress concentration factors but with no extrapolation, falls midway between Wundt's and Irwin's values.

The result obtained in this investigation agrees best with those obtained by Paris and Sih. Fig. 6 shows a comparison with their results for the full range of d/D. The estimated error quoted by Paris and Sih is indicated, and the agreement between their results and equation (10) is quite good, being within the quoted range of error for most values of d/D. Thus it may be concluded that the stress intensity factor for a solid notched round in tension which was obtained from Neuber's work is probably nearly as accurate as any other analysis presently available.

As the diameter of the inside hole ($2c$) of the hollow notched round bar in tension becomes very large relative to the other dimensions (a and t), the configuration approaches that of a single-edge notched plate in tension. The single-edge notched plate has been thoroughly investigated both experimentally [15, 16] and theoretically [17, 18, 19]. Since all of these analyses agree very

In the first series of tests the specimens were loaded to failure in uniaxial tension. The fracture toughness, K_{Ic}, was then calculated using the applicable equation included in Fig. 2. No plasticity corrections were made in the calculations, as they would be quite small for the low-toughness material under consideration. (For test no. 1, plasticity corrections would reduce K_{Ic} by less than 2 percent.) The results of the tension tests are presented in Table 2; the average of the fracture toughness values was 35.7 ksi-in.$^{1/2}$.

The second series of tests consisted of loading the 6-in-long specimen to failure in four-point bending. The value of K_{Ic} was calculated from the applicable equation in Fig. 2; the results of the bending tests appear in Table 3. It must be emphasized that the validity of applying sharp crack fracture mechanics to the bending test results is questionable, since the high value of σ_{nc}/σ_{ys} of 1.41 indicates that plasticity effects may be excessive. However, several factors indicate that plasticity effects were minor: The two pieces of the broken specimen could be placed back together and no permanent deformation could be observed; the fracture surfaces were completely flat; the calculated plastic zone size was much smaller than any of the specimen dimensions; and, most importantly, the calculated value of K_{Ic} agreed closely with the value obtained in the tension tests.

The consistency of the results of the two series of tests indicated good accuracy of the loading and measurement systems, as well as consistency in the preparation of the machined notches.

The accuracy of the equations used to calculate K_{Ic} can be assessed by comparison of the results obtained in this investigation with plane-strain fracture toughness values determined by other investigators, whose work is summarized in Table 4. The fact that the average tension value of 35.7 ksi-in.$^{1/2}$ and the bending value of 36.9 ksi-in.$^{1/2}$ fall right in the center of the range of results of other investigators indicates that the accuracy of the equations for the stress intensity factor for a hollow notched round in tension and bending is fairly good. It further indicates that plane-strain conditions existed at the crack tip.

Another point of interest is that the fracture surfaces of the specimens tested were all flat, thus further indicating that plane-strain conditions at the crack tip were not disrupted by the presence of the axial hole (as it was thought might occur).

The good agreement between K_{Ic}-values also indicates that the large inaccuracies suggested by Fig. 7 for a deep notch in tension do not actually occur.

Conclusions

The results of this investigation show that the stress intensity factors for relatively complex geometries can be readily obtained from limiting values of stress concentration factors for a small notch root radius. A comparison of the results with those of other investigators, when possible, indicates favorable accuracy of the stress intensity factor equations which were obtained. The accuracy of the equations for the hollow notched round bar in tension and in bending was further indicated by comparison of the fracture toughness obtained from a fracture test of these specimens with results obtained by other investigators. In conclusion, it appears that the stress intensity factors can be accurately determined in a straightforward manner from the corresponding elastic solution for the notched body under consideration.

Acknowledgment

This work was performed under the auspices of the U. S. Atomic Energy Commission.

References

1 A. A. Griffith, "The Phenomena of Rupture and Flow in Solids," *Philosophical Transactions of the Royal Society of London,* series A, vol. 221, 1920, pp. 163–198.

Table 2 Experimental results of tension tests

Test no.	Specimen length (in.)	Failure load (kips)	Critical net stress σ_{nc} (ksi)	σ_{nc}/σ_{ys}	K_{Ic} (ksi-in$^{1/2}$)
1	4	28.0	79.5	1.10	35.4
2	4	27.7	78.7	1.09	34.9
3	6	28.6	81.2	1.13	36.2
4	6	28.7	81.5	1.13	36.2

σ_{ys} = 72 ksi   Average   35.7

Table 3 Experimental results of bending tests

Test no.	Failure bending moment (kip-in)	σ_{nc} (ksi)	σ_{nc}/σ_{ys}	K_{Ic} (ksi-in$^{1/2}$)
5	4.55	102	1.41	36.9
6	4.55	102	1.41	36.9

Table 4 Summary of results from other investigators

K_{Ic}	Specimen type	Notch preparation	Reference
40	Notched round bar	Unspecified	ASTM Committee [21]
30.9	Double edge notched plate	Machined	Kaufman and Hunsicker [22]
29.8	Center notched plate	Machined	Kaufman and Hunsicker [22]
24.1	Center notched plate	Fatigued	Kaufman and Hunsicker [22]
31.0	Single edge notched plate	Machined	Kaufman and Hunsicker [22]
25.1	Single edge notched plate	Fatigued	Kaufman and Hunsicker [22]
40.2	Single edge notched plate	Machined	Jones and Brown [23]
28.2	Single edge notched plate	Fatigued	Jones and Brown [23]
33.5	Center notched plate	Fatigued	Jones and Brown [23]
35.9	Notched round bar	Unspecified	Irwin [3]
38.8	Notched bend test	Unspecified	Irwin [3]

2 G. R. Irwin, "Fracture," *Handbuch der Physik,* vol. VI, Springer-Verlag, Berlin, Germany, 1960.

3 G. R. Irwin, "Fracture Mechanics," *Structural Mechanics,* edited by Goodier and Hoff, Pergamon Press, New York, N. Y., 1960, pp. 567–591.

4 P. C. Paris and G. C. Sih, "Stress Analysis of Cracks," *Fracture Toughness Testing and Its Applications,* ASTM Special Technical Publication no. 381, Philadelphia, Pa., 1965, pp. 30–81.

5 H. Neuber, *Kerbspannungslehre,* Springer-Verlag, Berlin, Germany; first edition 1937, second edition 1958; English translations available: *Theory of Notch Stresses,* first edition, J. W. Edwards, Ann Arbor, Mich., 1946; second edition, United States Atomic Energy Commission Translation AEC-tr-4547, available from Office of Technical Services, Dept. of Commerce, Washington, D. C.

6 L. A. Wigglesworth, "Stress Distribution in a Notched Plate," *Mathematika,* vol. 4, 1957, pp. 76–96.

7 O. L. Bowie, "Rectangular Tensile Sheet With Symmetric Edge Cracks," *Journal of Applied Mechanics,* vol. 31, TRANS. ASME, vol. 86, Series E, 1964, pp. 208–212.

8 J. E. Srawley and W. F. Brown, "Fracture Toughness Testing Methods," *Fracture Toughness Testing and Its Applications,* ASTM Special Technical Publication no. 381, Philadelphia, Pa., 1965, pp. 133–196.

9 "Screening Tests of High-Strength Alloys Using Sharply Notched Cylindrical Specimens," *Materials Research and Standards,* Fourth Report of ASTM Special Committee on Fracture Testing of High-Strength Materials, vol. 2, no. 3, 1962, pp. 196–203.

10 J. E. Campbell, "Current Methods of Fracture-Toughness Testing of High-Strength Alloys With Emphasis on Plane Strain," DMIC Report no. 207, Defense Metals Information Center, Battelle Memorial Institute, Columbus, Ohio, 1964.

11 H. F. Bueckner, Discussion of Reference 4, *Fracture Toughness Testing and Its Applications,* ASTM Special Technical Publication no. 381, Philadelphia, Pa., 1965, pp. 82–83.

12 J. D. Lubahn, "Experimental Determination of Energy Release Rate for Notch Bending and Notch Tension," *Proceedings of ASTM*, vol. 59, 1959, pp. 885–913.

13 G. R. Irwin, "Supplement to: Notes for May, 1961 Meeting of ASTM Committee for Fracture Testing of High-Strength Metallic Materials," 1961.

14 B. M. Wundt, "A Unified Interpretation of Room Temperature Strength of Notch Specimens as Influenced by Size," ASME Paper No. 59—Met-9.

15 J. E. Srawley, M. H. Jones, and B. Gross, "Experimental Determination of the Dependence of Crack Extension Force on Crack Length for Single-Edge-Notch Tension Specimen," National Aeronautics and Space Administration TN D-2396, August, 1964.

16 A. M. Sullivan, "New Specimen Design for Plane Strain Fracture Toughness Testing," *Materials Research and Standards*, vol. 4, no. 1, 1964, pp. 20–24.

17 B. Gross, J. E. Srawley, and W. F. Brown, "Stress Intensity Factors for a Single-Edge-Notch Tension Specimen by Boundary Collocation of a Stress Function," National Aeronautics and Space Administration TN D-2395, August, 1964.

18 A. S. Kobayashi, "Method of Collocation Applied to Edge-Notched Finite Strip Subjected to Uniaxial Tension and Pure Bending," Boeing Co., Seattle, Wash., Document D2-23551, 1964.

19 O. L. Bowie and D. M. Neal, "Single Edge Crack in Rectangular Sheet," *Journal of Applied Mechanics*, vol. 32, Trans. ASME, vol. 87, Series E, 1965, pp. 708–709.

20 H. Neuber, "Notch Stress Theory," Air Force Materials Laboratory Tech. Rep. AFML-TR-65-225, 1965.

21 "Progress in Measuring Fracture Toughness and Using Fracture Mechanics," *Materials Research and Standards*, Fifth Report of ASTM Special Committee on Fracture Testing of High-Strength Materials, vol. 4, no. 3, 1965, pp. 107–119.

22 J. G. Kaufman and H. Y. Hunsicker, "Fracture Toughness Testing at Alcoa Research Laboratories," *Fracture Toughness Testing and Its Applications*, ASTM Special Technical Publication no. 381, Philadelphia, Pa., 1965, pp. 290–308.

23 M. H. Jones and W. F. Brown, "Acoustic Detection of Crack Initiation in Sharply Notched Specimens," *Materials Research and Standards*, vol. 4, no. 3, 1964, pp. 120–127.

APPENDIX

Sample of Calculations for Determination of Stress Intensity Factors From Stress Concentration Factors

The problem of a solid notched bar subjected to tension as shown in Fig. 9 will be worked out as an example.

Neuber's equations for stress concentration factors[3]

$$\alpha_{tk} = \frac{\frac{a}{p}\sqrt{\frac{a}{p}+1} + \left(\frac{1}{2}+\nu\right)\frac{a}{p} + (1+\nu)\left(1+\sqrt{\frac{a}{p}+1}\right)}{\frac{a}{p} + 2\nu\sqrt{\frac{a}{p}+1} + 2}$$

(13)

$$\alpha_{fk} = 1 + 2.24(t/p)^{1/2} \quad (14)$$

These equations are simplified when we consider a small notch root radius, p. Retaining only predominant terms leads to:

$$\alpha_{tk0} = (a/p)^{1/2} \quad (15)$$

$$\alpha_{fk0} = 2.24(t/p)^{1/2} \quad (16)$$

where the "0" subscript refers to the special case of p small.

---
[3] Modified as discussed in text.

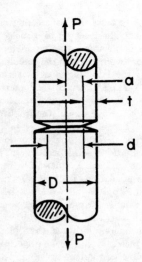

Fig. 9 Solid notched bar subjected to tension

The equation given by Neuber for interpolation between these two limiting cases is:

$$\alpha_k = 1 + \frac{(\alpha_{fk}-1)(\alpha_{tk}-1)}{[(\alpha_{fk}-1)^2 + (\alpha_{tk}-1)^2]^{1/2}} \quad (17)$$

Since the stress concentration factors for sharp cracks are much greater than one, the previous equation can be simplified by dropping negligible terms.

$$\alpha_{k0} = \frac{\alpha_{fk0}\alpha_{tk0}}{(\alpha_{fk0}^2 + \alpha_{tk0}^2)^{1/2}} \quad (18)$$

The general stress concentration factor, α_{k0}, can now be determined by substituting equations (15) and (16) into (18).

$$\alpha_{k0} = \frac{2.24(t/p)^{1/2}(a/p)^{1/2}}{(a/p + 5.02t/p)^{1/2}} = p^{-1/2}\left(\frac{5.02at}{a+5.02at}\right)^{1/2} \quad (19)$$

K_I can now be found by use of equation (1).

$$K_I = \frac{\pi^{1/2}}{2}\lim_{p\to 0} p^{1/2}\sigma_{\max} = \frac{\pi^{1/2}}{2}\lim_{p\to 0}\alpha_{k0}\sigma_{\text{net}}$$

$$= \frac{\pi^{1/2}}{2}\lim_{p\to 0} p^{1/2}p^{-1/2}\left(\frac{5.02at}{a+5.02at}\right)^{1/2} = \sigma_{\text{net}}\left(\frac{\pi at}{4t+0.80a}\right)^{1/2}$$

$$= \sigma_{\text{net}}(\pi t)^{1/2}(0.80 + 4t/a)^{-1/2} \quad (20)$$

To put this in the form used for equation (10), substitute $a = d/2$, $a + t = D/2$.

$$K_I = \sigma_{\text{net}}\left[\frac{\pi\frac{d}{2}\frac{1}{2}(D-d)}{4\left(\frac{1}{2}\right)(D-d) + 0.80\frac{d}{2}}\right]^{1/2}$$

$$= \sigma_{\text{net}}(\pi D)^{1/2}\left[\frac{1}{2}\frac{d}{D}\frac{1-d/D}{4-3.20d/D}\right]^{1/2} \quad (21)$$

M. K. KASSIR
Instructor of Mechanics,
Lehigh University,
Bethlehem, Pa.

G. C. SIH[2]
Visiting Associate Professor,
Firestone Flight Sciences Laboratory,
California Institute of Technology,
Pasadena, Calif. Mem. ASME

Three-Dimensional Stress Distribution Around an Elliptical Crack Under Arbitrary Loadings[1]

As a companion problem to that of a flat elliptical crack subject to a uniform tension perpendicular to the crack plane, this paper deals with the case of arbitrary shear loads. Upon superposition, solutions to problems of an infinite solid containing an elliptical crack subjected to loads of a general nature may be obtained. It is shown that the three-dimensional stresses near the crack border can be expressed explicitly in terms of a convenient set of coordinates r and θ defined in a plane normal to the edge of the crack. In such a plane, the local stresses in a solid are found to have the same angular distribution and inverse square-root stress singularity as those in a two-dimensional body under the action of in-plane stretching and out-of-plane shear. This result will, in general, hold for any plane of discontinuity bounded by a smooth curve. Such information provides a clear interpretation of current fracture-mechanics theories to three dimensions. In particular, stress-intensity factors $k_j (j = 1, 2, 3)$, used in the Griffith-Irwin theory of fracture, are evaluated from the stress equations for determining the fracture strength of elastic solids with cracks or flaws.

Introduction

BECAUSE of mathematical tractability of two-dimensional theory of elasticity, the majority of the literature on the analysis of cracks has been concerned with thin sheet structures. For bodies having planes of discontinuities bounded by closed curves, the only three-dimensional configuration that has received much attention is the so-called "penny-shaped" crack. The distribution of stress in the neighborhood of a flat elliptical crack in an infinite solid is yet to be found.

In the present paper, the general character of the three-dimensional stresses near the border of a plane of discontinuity under arbitrary loadings is examined. Stress solutions are also obtained for the particular case of uniform shear applied to the surface of an "elliptically shaped" crack.

By taking the circular crack as a degenerate case of an oblate spheroid, Sack [1][3] has discussed the three-dimensional aspects of the Griffith theory of brittle fracture. Instead of using curvilinear coordinates, Sneddon [2] has obtained the stresses near such a crack in well-determined directions by application of Hankel transforms. More recently, Keer [3] has calculated the modulus of cohesion in Barenblatt's theory for the penny-shaped crack. The same problem was also studied by Payne [4], Green and Zerna [5], Collins [6], and others. In references [1–6], the round crack is opened by loads applied symmetrically with respect to the crack faces. The case of a penny-shaped crack in an unbounded medium with crack surfaces subjected to uniform shear was treated by Westmann [7]. Florence and Goodier [8] have presented a solution to the problem of a uniform steady heat flow disturbed by an insulated circular crack. Thus a variety of boundary problems of the penny-shaped crack is available for fracture-mechanics applications.

In reality, as pointed out by Irwin [9], the crack is more likely to be in the shape of a plane ellipse. While Sadowsky and Sternberg [10] have considered the stress concentration around a triaxial ellipsoidal cavity in a body of infinite extent, the solution to the problem of a flat elliptical crack given by Green and Sneddon [11] can be used directly to obtain information pertinent in the instability behavior of cracks. The applied loads in [10, 11] are placed symmetrically with respect to the principal planes of the elliptical cavity or the plane of the flat crack. In this paper, the antisymmetric problem of a flat elliptical crack is formulated and solved in closed form by means of harmonic functions. The stress-intensity factors k_2 and k_3 corresponding to the edge-sliding and tearing modes of fracture are determined from the stresses near the crack border. Compared to the relatively simple expression of k_1 [9] for the opening mode, k_2 and k_3 are complicated functions of Poisson's ratio of the material and tabulated elliptic functions of the first and second kinds. In [9], Irwin has derived the k_1-factor based on the solution in [11] from the displacement field with the knowledge that normal tension produces an elliptical crack opening. Actually, the $k_j (j = 1, 2, 3)$ factors can be just as easily obtained from the stress equations. To illustrate this point, the normal component of stress σ_{zz} on the plane $z = 0$ and $\eta = 0$ (outside the ellipse) is quoted from [11]:

$$\sigma_{zz} = \frac{p}{E(k)} \left\{ \frac{ab^2}{\sqrt{Q(\xi)}} - \left[E(u) - \frac{snu\, cnu}{dnu} \right] \right\} \quad (1)$$

where p is the magnitude of the uniform tension normal to the crack plane at $z = 0$; and a, b are, respectively, the major and minor semiaxes of the ellipse. In equation (1),

$$Q(\xi) = \xi(a^2 + \xi)(b^2 + \xi), \qquad E(u) = \int_0^u dn^2 t\, dt$$

and $snu, cnu, \ldots$, are the Jacobian elliptic functions related to ξ by

$$\xi = \frac{a^2\, cn^2 u}{sn^2 u}$$

In the limit as $\xi \to 0$, the crack border, $E(u)$, becomes the com-

---

[1] The research reported in this paper was supported by the U. S. Navy through the Office of Naval Research under Contract Nonr-610(06).
[2] On leave from Lehigh University, Bethlehem, Pa., 1965–1966.
[3] Numbers in brackets designate References at end of paper.
Contributed by the Applied Mechanics Division for publication (without presentation) in the JOURNAL OF APPLIED MECHANICS.
Discussion of this paper should be addressed to the Editorial Department, ASME, United Engineering Center, 345 East 47th Street, New York, N. Y. 10017, and will be accepted until October 15, 1966. Discussion received after the closing date will be returned. Manuscript received by ASME Applied Mechanics Division, May 12, 1965; final draft, November 29, 1965. Paper No. 66—APM-N.

plete elliptic integral of the second kind, $E(k)$, with argument $k = [1 - (b/a)^2]^{1/2}$, and $(snu\, cnu)/dnu$ vanishes. Therefore, the quantity in the bracket in equation (1) remains finite as ξ approaches zero. Moreover, knowing the limiting form of ξ given by equation (55a) with $\theta = 0$, the expansion of σ_{zz} near the crack border is

$$\sigma_{zz} = \frac{p}{E(k)}\left(\frac{b}{a}\right)^{1/2}(a^2\sin^2\phi + b^2\cos^2\phi)^{1/4}\frac{1}{(2r)^{1/2}} + 0(r^{1/2})$$

(2)

where r is the radial distance measured from the crack border, and ϕ is the angle in the parametric equations of an ellipse. It is now obvious that the coefficient of $(2r)^{-1/2}$ gives the stress-intensity factor

$$k_1 = \frac{p}{E(k)}\left(\frac{b}{a}\right)^{1/2}(a^2\sin^2\phi + b^2\cos^2\phi)^{1/4}, \qquad a > b \quad (3)$$

which agrees with that obtained by Irwin [9]. For flat elliptical cracks under shear loads, k_2 and k_3 can be evaluated in the same way. Once k_j ($j = 1, 2, 3$) are known, the energy release rates $\mathcal{G}_j$ ($j = 1, 2, 3$) follow immediately from the expressions originally developed by Irwin [12] for two-dimensional problems of cracks.

Several essential features of the three-dimensional crack problem are discussed in detail. Briefly, the crack-border stress fields are expressed independently of uncertainties of both the external loads and crack geometries. The stress components when expressed in a convenient set of coordinates are shown to have counterparts in two-dimensional analysis of stresses. Of particular interest is the problem of a flat elliptical crack subjected to surface shear, where two different types of stress-intensity factors k_2 and k_3 occur simultaneously. In addition, their magnitudes vary around the border of the crack. The results of the present study provide a better understanding of the Griffith-Irwin theory of fracture in three dimensions.

Equations of Elasticity in Three Dimensions

The mathematical formulation of three-dimensional elastostatic problems involves an appropriate selection of harmonic functions based upon the solution of the Navier displacement equations of equilibrium. In the absence of body forces, Navier's equations become

$$(\lambda + \mu)\nabla\nabla\cdot\mathbf{u} + \mu\nabla^2\mathbf{u} = 0 \quad (4)$$

where $\mathbf{u}$ is the displacement vector, ∇ is the gradient operator, and λ, μ are the Lamé coefficients for an isotropic solid.

Let x, y, z be the rectangular Cartesian coordinates. For problems having a surface of discontinuity, say in the plane $z = 0$, it is convenient to introduce a vector displacement potential $\boldsymbol{\phi}$ and a scalar displacement potential ψ such that

$$\mathbf{u} = \boldsymbol{\phi} + z\nabla\psi \quad (5)$$

in which $\boldsymbol{\phi}$ and ψ are real harmonic functions, i.e.,

$$\nabla^2\boldsymbol{\phi} = 0, \qquad \nabla^2\psi = 0 \quad (6)$$

In addition, it is not difficult to show that equations (4) are satisfied if

$$\frac{\partial\psi}{\partial z} = -\left(\frac{\lambda + \mu}{\lambda + 3\mu}\right)\nabla\cdot\boldsymbol{\phi} \quad (7)$$

The stress tensor $\boldsymbol{\sigma}$ may be obtained from the displacement vector $\mathbf{u}$ by

$$\boldsymbol{\sigma} = \lambda(\nabla\cdot\mathbf{u})\mathbf{I} + \mu(\nabla\mathbf{u} + \mathbf{u}\nabla) \quad (8)$$

Here, $\mathbf{I}$ is the isotropic tensor. Note that

$$\nabla\cdot\mathbf{u} = \left(\frac{2\mu}{\lambda + 3\mu}\right)\nabla\cdot\boldsymbol{\phi} \quad (9)$$

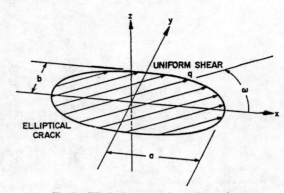

Fig. 1 Elliptical crack under uniform shear

The constants λ, μ are related to Poisson's ratio ν and Young's modulus E as follows:

$$\lambda = \frac{E\nu}{(1 + \nu)(1 - 2\nu)}, \qquad \mu = \frac{E}{2(1 + \nu)} \quad (10)$$

Statement of the Physical Problem

Consider a flat elliptical crack in the xy-plane with major and minor semiaxes a and b, respectively, as shown in Fig. 1. The boundary of the crack is described by the ellipse

$$\frac{x^2}{a^2} + \frac{y^2}{b^2} = 1 \quad (11)$$

in the plane $z = 0$. Equal and opposite shear stresses of magnitude q are applied to the crack surfaces and directed at an angle ω to the x-axis which coincides with the major axis of the ellipse. The stresses at a sufficiently large distance away from the crack are assumed to vanish. In view of the shear load, the entire plane $z = 0$ must be free from normal stress, i.e.,

$$\sigma_{zz} = 0, \quad \text{for} \quad z = 0 \quad (12)$$

On the surfaces of the crack,

$$\left.\begin{array}{l}\tau_{xz} = q\cos\omega \\ \tau_{yz} = q\sin\omega\end{array}\right\} z = 0, \quad \left(\frac{x^2}{a^2} + \frac{y^2}{b^2} \leq 1\right) \quad (13)$$

Exterior to the ellipse, the displacements u, v are zero:

$$u = v = 0, \quad \text{for} \quad z = 0 \quad \text{and} \quad \left(\frac{x^2}{a^2} + \frac{y^2}{b^2} \geq 1\right) \quad (14)$$

If the Cartesian components of $\boldsymbol{\phi}$ are denoted by ϕ_x, ϕ_y, ϕ_z, then from equations (5), (7), and (8), σ_{zz} can be written as

$$\sigma_{zz} = \lambda\left(\frac{\partial\phi_x}{\partial x} + \frac{\partial\phi_y}{\partial y} + \frac{\partial\phi_z}{\partial z}\right) + (\lambda + 2\mu)\frac{\partial}{\partial z}(\psi + \phi_z) \quad (15)$$

From equations (12) and (15), it is seen that the formulation may be further simplified by taking

$$\phi_x = -2(1 - \nu)\frac{\partial f}{\partial z}, \qquad \phi_y = -2(1 - \nu)\frac{\partial g}{\partial z} \quad (16)$$

The functions $f(x, y, z), g(x, y, z)$ satisfy Laplace's equations in three dimensions:

$$\nabla^2 f(x, y, z) = 0, \qquad \nabla^2 g(x, y, z) = 0 \quad (17)$$

The quantity $-2(1 - \nu)$ in equations (16) is introduced merely for the sake of convenience in subsequent work. Making use of equations (10), (15), and (16), the condition, equation (12), is satisfied only if

$$2\nu F = \psi + \phi_z \quad (18)$$

where

$$F = \frac{\partial f}{\partial x} + \frac{\partial g}{\partial y}, \qquad \nabla^2 F(x, y, z) = 0$$

It is now possible to express the functions ψ, ϕ_z in terms of f and g. Combining equations (7), (10), (16), and (18) gives

$$\psi = \frac{\partial f}{\partial x} + \frac{\partial g}{\partial y}, \qquad \phi_z = -(1 - 2\nu)\left(\frac{\partial f}{\partial x} + \frac{\partial g}{\partial y}\right) \quad (19)$$

Hence the problem is reduced to the determination of two real functions f and g which may be evaluated from the remaining boundary conditions.

In the usual notation of u, v, w being the rectangular components of the displacement vector, it is found that

$$u = -2(1 - \nu)\frac{\partial f}{\partial z} + z\frac{\partial F}{\partial x} \quad (20a)$$

$$v = -2(1 - \nu)\frac{\partial g}{\partial z} + z\frac{\partial F}{\partial y} \quad (20b)$$

$$w = -(1 - 2\nu)F + z\frac{\partial F}{\partial z} \quad (20c)$$

Similarly, the components of the stress tensor σ are

$$\frac{\sigma_{xx}}{2\mu} = -2(1 - \nu)\frac{\partial^2 f}{\partial x \partial z} - 2\nu\frac{\partial F}{\partial z} + z\frac{\partial^2 F}{\partial x^2} \quad (21a)$$

$$\frac{\sigma_{yy}}{2\mu} = -2(1 - \nu)\frac{\partial^2 g}{\partial y \partial z} - 2\nu\frac{\partial F}{\partial z} + z\frac{\partial^2 F}{\partial y^2} \quad (21b)$$

$$\frac{\sigma_{zz}}{2\mu} = z\frac{\partial^2 F}{\partial z^2} \quad (21c)$$

$$\frac{\tau_{xy}}{2\mu} = -(1 - \nu)\frac{\partial}{\partial z}\left(\frac{\partial f}{\partial y} + \frac{\partial g}{\partial x}\right) + z\frac{\partial^2 F}{\partial x \partial y} \quad (21d)$$

$$\frac{\tau_{xz}}{2\mu} = -(1 - \nu)\frac{\partial^2 f}{\partial z^2} + \nu\frac{\partial F}{\partial x} + z\frac{\partial^2 F}{\partial x \partial z} \quad (21e)$$

$$\frac{\tau_{yz}}{2\mu} = -(1 - \nu)\frac{\partial^2 g}{\partial z^2} + \nu\frac{\partial F}{\partial y} + z\frac{\partial^2 F}{\partial y \partial z} \quad (21f)$$

Note that there is little resemblance between the formulation of the present problem of uniform shear and that of uniform tension solved by Green and Sneddon [11]. However, the method of solution may follow along the same line of reasoning. This will be done in the next section.

Shear of an Elliptical Crack

The evaluation of the unknown functions f, g is more expedient by application of ellipsoidal coordinates (ξ, η, ζ). From Whittaker and Watson [13], the coordinates (x, y, z) are determinate in terms of (ξ, η, ζ) and vice versa by the relations

$$a^2(a^2 - b^2)x^2 = (a^2 + \xi)(a^2 + \eta)(a^2 + \zeta) \quad (22a)$$

$$b^2(b^2 - a^2)y^2 = (b^2 + \xi)(b^2 + \eta)(b^2 + \zeta) \quad (22b)$$

$$a^2 b^2 z^2 = \xi \eta \zeta \quad (22c)$$

where

$$-a^2 \leq \zeta \leq -b^2 \leq \eta \leq 0 \leq \xi < \infty$$

When $z = 0$, $\xi = 0$ represents the point (x, y) inside the ellipse and $\eta = 0$ the outside.

An appropriate solution to the present problem is [11]:

$$\begin{bmatrix} f(x,y,z) \\ g(x,y,z) \end{bmatrix} = \frac{1}{2}\begin{bmatrix} A \\ B \end{bmatrix}\left\{\int_\xi^\infty \times \left[\frac{x^2}{a^2 + s} + \frac{y^2}{b^2 + s} + \frac{z^2}{s} - 1\right]\frac{ds}{\sqrt{Q(s)}}\right\} \quad (23)$$

where

$$Q(s) = s(a^2 + s)(b^2 + s)$$

and the constants A, B may be evaluated from the boundary conditions, equations (13), as

$$\frac{q \cos \omega}{2\mu} = \nu\left[\frac{\partial^2 f}{\partial x^2} + \frac{\partial^2 g}{\partial x \partial y}\right] - (1 - \nu)\frac{\partial^2 f}{\partial z^2}, \quad \xi = 0 \quad (24a)$$

$$\frac{q \sin \omega}{2\mu} = \nu\left[\frac{\partial^2 f}{\partial x \partial y} + \frac{\partial^2 g}{\partial y^2}\right] - (1 - \nu)\frac{\partial^2 g}{\partial z^2}, \quad \xi = 0 \quad (24b)$$

Substituting equations (23) into equations (24a), (24b) and performing the necessary integrations as worked out in Appendix 1, the result gives

$$4\mu A = \frac{ab^2 k^2 q \cos \omega}{(k^2 - \nu)E(k) + \nu k'^2 K(k)}, \quad k^2 = 1 - \frac{b^2}{a^2} \quad (25a)$$

$$4\mu B = \frac{ab^2 k^2 q \sin \omega}{(k^2 + \nu k'^2)E(k) - \nu k'^2 K(k)} \quad (25b)$$

where

$$k^2 + k'^2 = 1$$

The quantities $K(k)$, $E(k)$ are complete elliptical integrals of the first and second kinds, respectively. While equations (25a) and (25b) give the solution of the problem, there remains the verification of equations (14). Since

$$\begin{bmatrix} \frac{\partial f}{\partial z} \\ \frac{\partial g}{\partial z} \end{bmatrix} = z\begin{bmatrix} A \\ B \end{bmatrix}\int_\xi^\infty \frac{ds}{s\sqrt{Q(s)}} = 0 \quad (26)$$

for $z = 0$ and $\eta = 0$ (outside the ellipse), equations (20a) and (20b) show that the conditions, equation (14), are indeed satisfied. Moreover, for $(x^2/a^2 + y^2/b^2) > 1$, w, τ_{xz}, and τ_{yz} must be continuous across the plane $z = 0$, which may all be satisfied upon observing from equations (23) that the functions f, g and their derivatives with respect to x and/or y are even functions of z.

The solution of the problem is thus complete. Hence the displacements and stresses at any point of the elastic solid with an elliptical crack under shear may be obtained from equations (20) and (21). Of immediate interest are the shear components τ_{xz} and τ_{yz} outside the ellipse ($\eta = 0$) on the plane $z = 0$. Putting equations (23) into equations (21e) and (21f) yields

$$-\frac{\tau_{xz}}{2\mu} = A\left\{2\left[\frac{a^2 + \xi}{\xi(b^2 + \xi)}\right]^{1/2}\left[\frac{1}{a^2 + \xi} + \frac{\nu(b^2 + \zeta)}{(a^2 - b^2)(\xi - \zeta)}\right]\right.$$
$$+ \int_\xi^\infty \left[\nu - 1 - \left(\frac{b^2 + s}{a^2 + s}\right)\right]\frac{ds}{(b^2 + s)\sqrt{Q(s)}}\right\}$$
$$+ \left[-\frac{(a^2 + \zeta)(b^2 + \zeta)}{\xi}\right]^{1/2}\frac{2\nu B}{(a^2 - b^2)(\xi - \zeta)} \quad (27a)$$

$$-\frac{\tau_{yz}}{2\mu} = A\left\{2\left[\frac{b^2 + \xi}{\xi(a^2 + \xi)}\right]^{1/2}\left[\frac{1}{b^2 + \xi} - \frac{\nu(a^2 + \zeta)}{(a^2 - b^2)(\xi - \zeta)}\right]\right.$$
$$+ \int_\xi^\infty \left[\nu - 1 - \left(\frac{a^2 + s}{b^2 + s}\right)\right]\frac{ds}{(a^2 + s)\sqrt{Q(s)}}\right\}$$
$$+ \left[-\frac{(a^2 + \zeta)(b^2 + \zeta)}{\xi}\right]^{1/2}\frac{2\nu B}{(a^2 - b^2)(\xi - \zeta)} \quad (27b)$$

where $-(b^2 + \zeta)$ is positive definite. In the case of a penny-shaped crack, these expressions simplify considerably. When $b \to a$, $\xi \to r^2 - a^2$, $\zeta \to -a^2$, $K \to E \to \pi/2$ and

$$A = \frac{a^3 q \cos \omega}{\pi\mu(2 - \nu)}, \quad B = \frac{a^3 q \sin \omega}{\pi\mu(2 - \nu)} \quad (28)$$

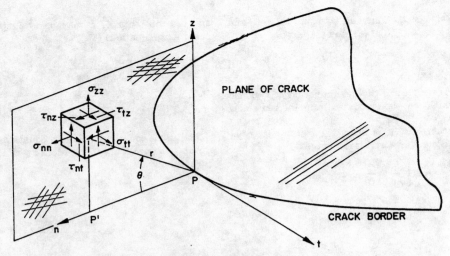

Fig. 2 Rectangular stress components in plane normal to crack border

Equations (27a) and (27b) reduce to

$$\tau_{xz} = \frac{2q \cos \omega}{\pi} \left\{ \sin^{-1}\left(\frac{a}{r}\right) \right.$$
$$\left. - \left[1 - \left(\frac{\nu}{2-\nu}\right)\left(\frac{a}{r}\right)^2\right]\left[\left(\frac{r}{a}\right)^2 - 1\right]^{-1/2} \right\}$$
$$- \frac{4\nu q \sin \omega}{(2-\nu)}\left(\frac{a}{r}\right)^2\left[\left(\frac{r}{a}\right)^2 - 1\right]^{-1/2} \quad (29a)$$

$$\tau_{yz} = \frac{2q \sin \omega}{\pi} \left\{ \sin^{-1}\left(\frac{a}{r}\right) \right.$$
$$\left. - \left[1 - \left(\frac{\nu}{2-\nu}\right)\left(\frac{a}{r}\right)^2\right]\left[\left(\frac{r}{a}\right)^2 - 1\right]^{-1/2} \right\}$$
$$+ \frac{4\nu q \cos \omega}{(2-\nu)}\left(\frac{a}{r}\right)^2\left[\left(\frac{r}{a}\right)^2 - 1\right]^{-1/2} \quad (29b)$$

It can be easily shown that $[\tau_{xz}]_{\omega=0}$ and $[\tau_{yz}]_{\omega=\pi/2}$ check with $-[\tau_{\theta z}]_{\theta=\pi/2}$, equation (23e), and $[\tau_{zr}]_{\theta=0}$, equation (23f), in [7], respectively. The other stress components also agree.

It is now more pertinent to examine the normal and tangential components of τ_{xz} and τ_{yz} near the border of the elliptical crack. Consideration of the equilibrium of a triangular element on the crack border in the plane $z = 0$ leads to the relations

$$\tau_{nz} = \tau_{xz} \cos \beta + \tau_{yz} \sin \beta \quad (30a)$$

$$\tau_{tz} = -\tau_{xz} \sin \beta + \tau_{yz} \cos \beta \quad (30b)$$

where β is the angle between the outward unit normal vector of the crack border and the x-axis. More precisely, β is related to the parametric equations of an ellipse as

$$x = a \sin \phi = (a^2 \sin^2 \phi + b^2 \cos^2 \phi)^{1/2} \sin \beta \quad (31a)$$

$$y = b \cos \phi = (a^2 \sin^2 \phi + b^2 \cos^2 \phi)^{1/2} \cos \beta \quad (31b)$$

In the vicinity of the crack edge, equations (52a) and (55a) in Appendix 2 give

$$\left. \begin{array}{l} \xi = 2ab(a^2 \sin^2 \phi + b^2 \cos^2 \phi)^{-1/2} r \\ \zeta = -(a^2 \sin^2 \phi + b^2 \cos^2 \phi) \end{array} \right\} z = 0, \quad \eta = 0$$

and r is the radial distance normal to the crack border in the plane $z = 0$. From equations (27), (30), and (31), the shear stresses in the neighborhood of the elliptical boundary are

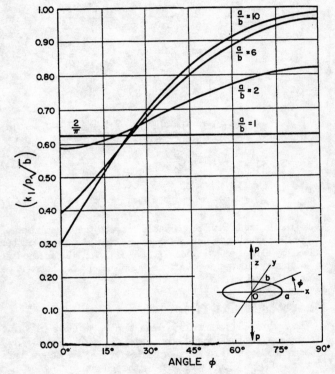

Fig. 3 Stress-intensity factor for symmetric mode

$$\tau_{nz} = -\frac{4\mu}{(ab)^{1/2}}(a^2 \sin^2 \phi + b^2 \cos^2 \phi)^{-1/4}$$
$$\times (aB \sin \phi + bA \cos \phi)\frac{1}{(2r)^{1/2}} + 0(r^{1/2}) \quad (32a)$$

$$\tau_{tz} = \frac{4\mu(1-\nu)}{(ab)^{1/2}}(a^2 \sin^2 \phi + b^2 \cos^2 \phi)^{-1/4}$$
$$\times (aA \sin \phi - bB \cos \phi)\frac{1}{(2r)^{1/2}} + 0(r^{1/2}) \quad (32b)$$

Equations (32) imply that both the edge-sliding and tearing modes of kinematic movements of the crack surfaces exist in the three-dimensional problem of skew-symmetric loading. According to the Griffith-Irwin theory of fracture, these modes of fracture can be best described by their corresponding stress-intensity factors, which may be extracted from equations (32a) and (32b):

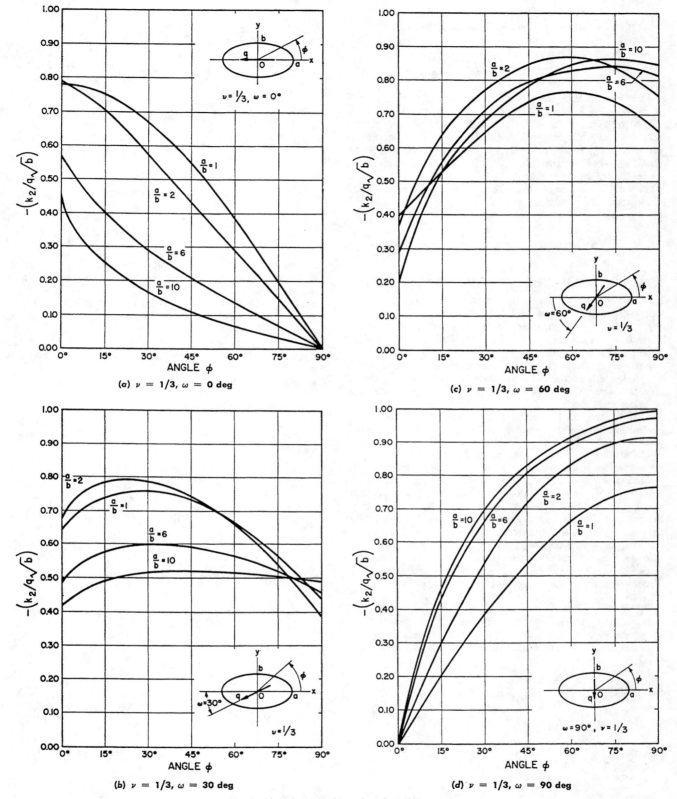

Fig. 4 Stress-intensity factor for edge-sliding mode

$$k_2 = -\frac{4\mu}{(ab)^{3/2}}(a^2\sin^2\phi + b^2\cos^2\phi)^{-1/4}(aB\sin\phi + bA\cos\phi) \quad (33a)$$

$$k_3 = \frac{4\mu(1-\nu)}{(ab)^{3/2}}(a^2\sin^2\phi + b^2\cos^2\phi)^{-1/4}(aA\sin\phi - bB\cos\phi) \quad (33b)$$

where A, B are given by equations (25). In comparison with k_1, equation (3), for the opening mode, the expressions of k_2 and k_3 are considerably more complicated. If the line of action of shear is directed along the major axis of the ellipse, i.e., $\omega = 0$ and $B = 0$, then at $\phi = 0$, $k_3 = 0$ as it should:

$$k_2 = -\frac{bk^2q}{\sqrt{a}[(k^2 - \nu)E(k) + \nu k'^2 K(k)]}, \qquad k_3 = 0 \quad (34)$$

Consequently, at $\phi = \pi/2$, k_2 should vanish, giving a state of pure tearing:

$$k_2 = 0, \qquad k_3 = \frac{(1-\nu)b^{1/2}k^2 q}{(k^2 - \nu)E(k) + \nu k'^2 K(k)} \qquad (35)$$

Similar confirmations may be obtained for values of $\omega = \pi/2$ ($A = 0$) and $\phi = 0, \pi/2$. In general, the variations of k_j ($j = 2, 3$) with the angle ϕ for different values of a/b and ω are plotted in Figs. 4 and 5. For the purpose of comparing the relative magnitude of the k_j-factors along the crack border, Irwin's result [9] of k_1 is also shown in graph form, Fig. 3. A detailed discussion of the results will be given later on.

It should be pointed out that the stress-intensity factors k_2 and k_3 also occur in the limiting case of a penny-shaped crack. Setting $a = b$ for $\phi = 0$, equations (33) reduce to

$$k_2 = -\frac{4}{\pi} \frac{q \cos \omega}{(2-\nu)} \sqrt{a}, \qquad k_3 = -\frac{4(1-\nu)q \sin \omega}{\pi(2-\nu)} \sqrt{a} \qquad (36)$$

Three-Dimensional Stress Distribution Near Crack Border

From the viewpoint of fracture mechanics, a great deal of information may be gained from a study of the stress distribution near the border of irregularities or flaws in an elastic solid. These flaws are three-dimensional in nature and may be idealized as planes of discontinuities bounded by smooth curves, or simply referred to as plane cracks. It is the purpose here to acquire a knowledge of the stress fields associated with such cracks so that the three-dimensional aspects of the current fracture theories may be better understood.

At first sight, the evaluation of the three-dimensional stresses in elliptical coordinates seems to be insurmountable. This task was bypassed by Irwin [9]. Instead, he studied the general shape of the crack opening for determining the stress-intensity factor k_1 for symmetric loading. If the loads are skew symmetric with respect to the crack surfaces, the shape to which the crack would grow is no longer evident. For this reason, k_2 and k_3 can be more readily obtained from the stress equations. The success of finding the crack-border stresses depends largely upon the relationship between the orthogonal system (ξ, η, ζ) and the coordinate system r, θ in a plane normal to the periphery of the crack. In order to preserve continuity, the mathematical details are outlined in the Appendixes. The symmetrical and skew-symmetrical stress distributions about the crack plane are given separately.

Symmetrical Distribution

Substituting the appropriate expressions in Appendix 2 into the stress equations (57) in Appendix 3, the symmetrical stress distribution in the immediate vicinity of the crack border is found:

$$\sigma_{xx} = \frac{abk_1}{(a^2 \sin^2 \phi + b^2 \cos^2 \phi)(2r)^{1/2}} \cos \frac{\theta}{2}$$
$$\times \left\{ \left(\frac{b}{a}\right) \cos^2 \phi \left[1 - \sin \frac{\theta}{2} \sin \frac{3\theta}{2}\right] \right.$$
$$\left. + 2\nu \left(\frac{a}{b}\right) \sin^2 \phi \right\} + 0(r)^{1/2} \quad (37a)$$

$$\sigma_{yy} = \frac{abk_1}{(a^2 \sin^2 \phi + b^2 \cos^2 \phi)(2r)^{1/2}} \cos \frac{\theta}{2}$$
$$\times \left\{ \left(\frac{a}{b}\right) \sin^2 \phi \left[1 - \sin \frac{\theta}{2} \sin \frac{3\theta}{2}\right] \right.$$
$$\left. + 2\nu \left(\frac{b}{a}\right) \cos^2 \phi \right\} + 0(r^{1/2}) \quad (37b)$$

$$\sigma_{zz} = \frac{k_1}{(2r)^{1/2}} \cos \frac{\theta}{2} \left(1 + \sin \frac{\theta}{2} \sin \frac{3\theta}{2}\right) + 0(r^{1/2}) \quad (37c)$$

$$\tau_{xy} = \frac{abk_1 \sin \phi \cos \phi}{(a^2 \sin^2 \phi + b^2 \cos^2 \phi)(2r)^{1/2}} \cos \frac{\theta}{2}$$
$$\times \left[(1-2\nu) - \sin \frac{\theta}{2} \sin \frac{3\theta}{2}\right] + 0(r^{1/2}) \quad (37d)$$

$$\tau_{xz} = \frac{bk_1 \cos \phi}{(a^2 \sin^2 \phi + b^2 \cos^2 \phi)^{1/2}(2r)^{1/2}} \sin \frac{\theta}{2}$$
$$\times \cos \frac{\theta}{2} \cos \frac{3\theta}{2} + 0(r^{1/2}) \quad (37e)$$

$$\tau_{yz} = \frac{ak_1 \sin \phi}{(a^2 \sin^2 \phi + b^2 \cos^2 \phi)^{1/2}(2r)^{1/2}} \sin \frac{\theta}{2}$$
$$\times \cos \frac{\theta}{2} \cos \frac{3\theta}{2} + 0(r^{1/2}) \quad (37f)$$

Referring to Fig. 2, r is the radial distance from the crack front, and θ is the angle between r and the crack plane. Both r and θ lie in a plane normal to the border. In equations (37), the higher-order terms in r have been neglected and the stresses in a solid are shown to possess the same inverse square root singularity as those in two dimensions. However, their intensities governed by k_1 are different. For the flat elliptical crack subjected to tension perpendicular to the crack plane, k_1 in equations (37) is given by equation (3). The critical value of k_1 will presumably determine the onset of rapid fracture [12].

Skew-Symmetrical Distribution

In the same way, the skew-symmetrical stress distribution in the local crack-border regions may be obtained from equations (21), (46), (52a), and (55):

$$\sigma_{xx} = \frac{8\mu(\nu aB \sin \phi + bA \cos \phi)}{(ab)^{3/2}(a^2 \sin^2 \phi + b^2 \cos^2 \phi)^{1/4}(2r)^{1/2}} \sin \frac{\theta}{2}$$
$$- \frac{b^2 k_2 \cos^2 \phi}{(a^2 \sin^2 \phi + b^2 \cos^2 \phi)(2r)^{1/2}} \sin \frac{\theta}{2} \cos \frac{\theta}{2} \cos \frac{3\theta}{2} + 0(r^{1/2})$$
$$(38a)$$

$$\sigma_{yy} = \frac{8\mu(aB \sin \phi + \nu bA \cos \phi)}{(ab)^{3/2}(a^2 \sin^2 \phi + b^2 \cos^2 \phi)^{1/4}(2r)^{1/2}} \sin \frac{\theta}{2}$$
$$- \frac{a^2 k_2 \sin^2 \phi}{(a^2 \sin^2 \phi + b^2 \cos^2 \phi)(2r)^{1/2}} \sin \frac{\theta}{2} \cos \frac{\theta}{2} \cos \frac{3\theta}{2} + 0(r^{1/2})$$
$$(38b)$$

$$\sigma_{zz} = \frac{k_2}{(2r)^{1/2}} \sin \frac{\theta}{2} \cos \frac{\theta}{2} \cos \frac{3\theta}{2} + 0(r)^{1/2} \quad (38c)$$

$$\tau_{xy} = \frac{4\mu(1-\nu)(aA \sin \phi + bB \cos \phi)}{(ab)^{3/2}(a^2 \sin^2 \phi + b^2 \cos^2 \phi)^{1/4}(2r)^{1/2}} \sin \frac{\theta}{2}$$
$$- \frac{abk_2 \sin \phi \cos \phi}{(a^2 \sin^2 \phi + b^2 \cos^2 \phi)(2r)^{1/2}} \sin \frac{\theta}{2} \cos \frac{\theta}{2} \cos \frac{3\theta}{2} + 0(r^{1/2})$$
$$(38d)$$

$$\tau_{xz} = -\frac{4\mu(a^2 \sin^2 \phi + b^2 \cos^2 \phi)^{1/4}}{(ab)^{1/2}(2r)^{1/2}} \left\{ A \cos \frac{\theta}{2} \left[\left(\frac{1-\nu}{ab}\right) \right. \right.$$
$$\left. + \frac{b \cos^2 \phi}{a(a^2 \sin^2 \phi + b^2 \cos^2 \phi)} \cdot \left(\nu - \sin \frac{\theta}{2} \sin \frac{3\theta}{2}\right)\right]$$
$$\left. + \frac{B \sin \phi \cos \phi}{(a^2 \sin^2 \phi + b^2 \cos^2 \phi)} \cos \frac{\theta}{2} \left(\nu - \sin \frac{\theta}{2} \sin \frac{3\theta}{2}\right)\right\} + 0(r^{1/2})$$
$$(38e)$$

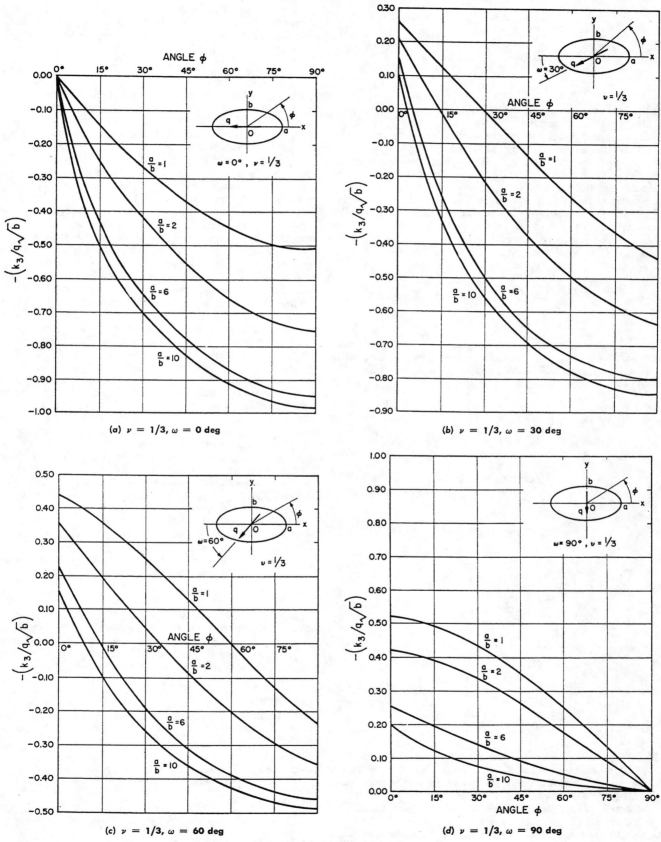

Fig. 5 Stress-intensity factor for tearing mode

$$\tau_{yz} = -\frac{4\mu(a^2 \sin^2\phi + b^2 \cos^2\phi)^{1/4}}{(ab)^{1/2}(2r)^{1/2}} \left\{ B \cos\frac{\theta}{2} \left[\left(\frac{1-\nu}{ab}\right) \right. \right.$$
$$\left. + \frac{a \sin^2\phi}{b(a^2 \sin^2\phi + b^2 \cos^2\phi)} \cdot \left(\nu - \sin\frac{\theta}{2} \sin\frac{3\theta}{2}\right) \right]$$
$$\left. + \frac{A \sin\phi \cos\phi}{(a^2 \sin^2\phi + b^2 \cos^2\phi)} \cos\frac{\theta}{2} \left(\nu - \sin\frac{\theta}{2} \sin\frac{3\theta}{2}\right) \right\} + 0(r^{1/2}) \quad (38f)$$

where the constants A, B, and k_2 are given by equations (25a), (25b), and (33a), respectively.

The angular distribution of the stresses shown by equations (37) and (38) appears to be rather involved and does not offer a simple physical interpretation. Intuitively, however, it seems that the more general three-dimensional crack-border stresses should have counterparts to the corresponding problems of plane extension of plates [14] and torsion [15] (or longitudinal shear [16]) of cylindrical bodies. A close examination of equations (37) and (38) reveals that such counterparts indeed exist. Upon transformation of coordinate axes from (x, y, z) to (n, t, z), Fig. 6, the stress components σ_{xx}, σ_{yy}, τ_{xy} may be combined to yield

$$\sigma_{nn} = \tfrac{1}{2}(\sigma_{xx} + \sigma_{yy}) + \tfrac{1}{2}(\sigma_{xx} - \sigma_{yy}) \cos 2\beta + \tau_{xy} \sin 2\beta \quad (39a)$$

$$\sigma_{tt} = \tfrac{1}{2}(\sigma_{xx} + \sigma_{yy}) + \tfrac{1}{2}(\sigma_{yy} - \sigma_{xx}) \cos 2\beta - \tau_{xy} \sin 2\beta \quad (39b)$$

$$\tau_{nt} = \tfrac{1}{2}(\sigma_{yy} - \sigma_{xx}) \sin 2\beta + \tau_{xy} \cos 2\beta \quad (39c)$$

and the component σ_{zz} remains invariant. Inserting equations (37), (38) into equations (30), (39) and superimposing the results, the most general stress field containing all three stress-intensity factors k_j ($j = 1, 2, 3$) becomes

$$\sigma_{nn} = \frac{k_1}{(2r)^{1/2}} \cos\frac{\theta}{2} \left(1 - \sin\frac{\theta}{2} \sin\frac{3\theta}{2}\right)$$
$$- \frac{k_2}{(2r)^{1/2}} \sin\frac{\theta}{2} \left(2 + \cos\frac{\theta}{2} \cos\frac{3\theta}{2}\right) + 0(r^{1/2}) \quad (40a)$$

$$\sigma_{zz} = \frac{k_1}{(2r)^{1/2}} \cos\frac{\theta}{2} \left(1 + \sin\frac{\theta}{2} \sin\frac{3\theta}{2}\right)$$
$$+ \frac{k_2}{(2r)^{1/2}} \sin\frac{\theta}{2} \cos\frac{\theta}{2} \cos\frac{3\theta}{2} + 0(r^{1/2}) \quad (40b)$$

$$\sigma_{tt} = \frac{k_1}{(2r)^{1/2}} 2\nu \cos\frac{\theta}{2} - \frac{k_2}{(2r)^{1/2}} 2\nu \sin\frac{\theta}{2} + 0(r^{1/2}) \quad (40c)$$

$$\tau_{nt} = -\frac{k_3}{(2r)^{1/2}} \sin\frac{\theta}{2} + 0(r^{1/2}) \quad (40d)$$

$$\tau_{nz} = \frac{k_1}{(2r)^{1/2}} \sin\frac{\theta}{2} \cos\frac{\theta}{2} \cos\frac{3\theta}{2}$$
$$+ \frac{k_2}{(2r)^{1/2}} \cos\frac{\theta}{2} \left(1 - \sin\frac{\theta}{2} \sin\frac{3\theta}{2}\right) + 0(r^{1/2}) \quad (40e)$$

$$\tau_{tz} = \frac{k_3}{(2r)^{1/2}} \cos\frac{\theta}{2} + 0(r)^{1/2} \quad (40f)$$

in which σ_{tt} is equal to Poisson's ratio ν times the sum of σ_{nn} and σ_{zz}, a well-known relationship in two-dimensional problems of plane strain. It is clear that σ_{tt} has no counterpart to two-dimensional analysis of generalized plane stress.

Equations (40) show that the most general stresses in the close neighborhood of the crack outer boundary consist of the linear sum of three separate stress fields, each of which contains a different stress-intensity factor. In fact, as observed by Irwin [12], equations (40d), (40f) may be related to the tearing mode of fracture and the remaining equations pertain to the opening and edge-sliding modes of fracture. These modes are characterized by the basic movements of the crack surfaces as discussed in [12]. It should be emphasized that, although equations (40) are formally the same as those in two-dimensional problems, the intensity factors k_j ($j = 1, 2, 3$) will in general depend upon the curvature of the crack edge for three-dimensional problems. The same applies to the displacements.

Discussion of Results

The significant result is that the magnitude of loading forces and the crack configuration have been shown by equations (40) to be contained in the factors k_j, which are generally known as the crack-border stress-intensity factors. Hence the functional relationship of the stresses in equations (40) to r and θ remains unchanged as the point P of the triply orthogonal system (n, t, z), Fig. 2, traces the periphery of a crack of any arbitrary shape. The system (n, t, z), however, must form a trihedral in such a way that the axes n, t, z are always directed along the binormal, tangent, and principal normal of the curve, respectively. Generally speaking, equations (40) represent the three-dimensional local stress field in a plane normal to the border of any plane of discontinuities with the possible exception of those places where the crack border is sharp in the form of a corner. Stress singularities at such a corner still remain to be investigated.

As mentioned earlier, the instability behavior of cracks is governed by the intensity of the crack-border stress field. To fix ideas, the variations of k_j ($j = 1, 2, 3$) along the border of an elliptical crack are studied. For loads applied symmetrically with respect to the crack plane, Fig. 3 shows that k_1 is greatest at $\phi = \pi/2$ for all ratios of a/b. According to the Griffith-Irwin theory of fracture, crack extension would tend to initiate at $y = b$ in the plane $z = 0$, Fig. 1. When the crack is subjected to uniform shear of magnitude q, which makes an angle ω to the x-axis, k_2 and k_3 are complicated functions of ν, ϕ, ω, and a/b as shown by equations (33). For Poisson's ratio of $\nu = 1/3$, the maximum values of k_2 shift continuously from $\phi = 0$ deg to $\phi = 90$ deg as the shear angle ω is increased from 0 to 90 deg. [See Figs. 4(a) through 4(d).] Similar interpretation may be drawn for the variations of k_3 with the angle ϕ, Figs. 5(a) to 5(d). It should be noticed that k_3 changes sign for values of 0 deg $< \omega <$ 90 deg. This implies that k_2 is dominant at those places where the k_3-curves intersect the ϕ-axis. In general, both k_2 and k_3 are operative along the crack border. Thus crack initiation will depend upon reaching some critical values of the combination of k_2 and k_3.

In the presence of all three factors k_j ($j = 1, 2, 3$), a criterion of fracture in three dimensions may be postulated as follows:

$$f(k_1, k_2, k_3) = f_{cr} \quad (42)$$

Equation (42) states that the combination of all three stress-intensity factors present will cause unstable crack propagation when the function f takes the critical value f_{cr}. Experimental verification of such a statement for internal cracks in three dimensions would be difficult, if not impossible.

References

1 R. A. Sack, "Extension of Griffith's Theory of Rupture to Three Dimensions," *Proceedings of the Physical Society*, London, England, vol. 58, 1946, pp. 729–736.

2 I. N. Sneddon, "The Distribution of Stress in the Neighborhood of a Crack in an Elastic Solid," *Proceedings of the Royal Society*, London, England, Series A, vol. 187, 1946, pp. 229–260.

3 L. M. Keer, "Stress Distribution at the Edge of an Equilibrium Crack," *Journal of Mechanics and Physics of Solids*, vol. 12, 1964, pp. 149–163.

4 L. E. Payne, "On Axially Symmetric Punch, Crack, and Torsion Problems," *Journal of the Society of Industrial Applied Mathematics*, vol. 1, 1953, pp. 53–71.

5 A. E. Green and W. Zerna, *Theoretical Elasticity*, Oxford University Press, Oxford, England, 1954, pp. 177–178.

6 W. D. Collins, "Some Axially Symmetric Stress Distributions in Elastic Solids Containing Penny-Shaped Cracks," *Proceedings of*

the Royal Society, London, England, Series A, vol. 266, 1962, pp. 359–386.

7 R. A. Westmann, "Asymmetric Mixed Boundary-Value Problems of the Elastic Half-Space," JOURNAL OF APPLIED MECHANICS, vol. 32, TRANS. ASME, vol. 87, Series E, 1965, pp. 411–417.

8 A. L. Florence and J. N. Goodier, "The Linear Thermoelastic Problem of Uniform Heat Flow Disturbed by a Penny-Shaped Insulated Crack," *International Journal of Engineering Sciences*, vol. 1, 1963, pp. 533–540.

9 G. R. Irwin, "Crack-Extension Force for a Part-Through Crack in a Plate," JOURNAL OF APPLIED MECHANICS, vol. 29, TRANS. ASME, vol. 84, Series E, 1962, pp. 651–654.

10 M. A. Sadowsky and E. Sternberg, "Stress Concentration Around an Elliptical Cavity in an Infinite Body Under Arbitrary Plane Stress Perpendicular to the Axis of Revolution of Cavity," JOURNAL OF APPLIED MECHANICS, vol. 14, TRANS. ASME, vol. 69, 1947, pp. A-191-A-201.

11 A. E. Green and I. N. Sneddon, "The Distribution of Stress in the Neighborhood of a Flat Elliptical Crack in an Elastic Solid," *Proceedings of the Cambridge Philosophical Society*, vol. 46, 1950, pp. 159–164.

12 G. R. Irwin, "Fracture Mechanics," *Structural Mechanics*, Pergamon Press, London, England, 1960, pp. 560–574.

13 E. T. Whittaker and G. N. Watson, *Modern Analysis*, Cambridge University Press, Cambridge, England, 1962, pp. 548–552.

14 G. C. Sih, P. C. Paris, and F. Erdogan, "Crack-Tip Stress-Intensity Factors for Plane Extension and Plate Bending Problems," JOURNAL OF APPLIED MECHANICS, vol. 29, TRANS. ASME, vol. 84, Series E, 1962, pp. 306–312.

15 G. C. Sih, "Strength of Stress Singularities at Crack Tips for Flexural and Torsional Problems," JOURNAL OF APPLIED MECHANICS, vol. 30, TRANS. ASME, vol. 85, Series E, 1963, pp. 419–425.

16 G. C. Sih, "Stress Distribution Near Internal Crack Tips for Longitudinal Shear Problems," JOURNAL OF APPLIED MECHANICS, vol. 32, TRANS. ASME, vol. 87, Series E, 1965, pp. 51–58.

APPENDIX 1

Evaluation of the Fundamental Integral and Its Partial Derivatives

To find the displacements and stresses at any point of an elastic solid containing a flat elliptical crack, the fundamental integral

$$I = \tfrac{1}{2} \int_\xi^\infty \left[\frac{x^2}{a^2+s} + \frac{y^2}{b^2+s} + \frac{z^2}{s} - 1 \right] \frac{ds}{\sqrt{Q(s)}} \quad (43)$$

must be evaluated. On taking partial derivatives of I with respect to x, y, z, the following relations are needed [13]:

$$\frac{\partial \xi}{\partial x} = \frac{x}{2(a^2+\xi)h_1^2}, \quad \frac{\partial \xi}{\partial y} = \frac{y}{2(b^2+\xi)h_1^2}, \quad \frac{\partial \xi}{\partial z} = \frac{z}{2\xi h_1^2} \quad (44a)$$

$$\frac{\partial \eta}{\partial x} = \frac{x}{2(a^2+\eta)h_2^2}, \quad \frac{\partial \eta}{\partial y} = \frac{y}{2(b^2+\eta)h_2^2}, \quad \frac{\partial \eta}{\partial z} = \frac{z}{2\eta h_2^2} \quad (44b)$$

$$\frac{\partial \zeta}{\partial x} = \frac{x}{2(a^2+\zeta)h_3^2}, \quad \frac{\partial \zeta}{\partial y} = \frac{y}{2(b^2+\zeta)h_3^2}, \quad \frac{\partial \zeta}{\partial z} = \frac{z}{2\zeta h_3^2} \quad (44c)$$

in which

$$h_1^2 = \frac{(\xi-\eta)(\xi-\zeta)}{4\xi(a^2+\xi)(b^2+\xi)}, \quad h_2^2 = \frac{(\eta-\zeta)(\eta-\xi)}{4\eta(a^2+\eta)(b^2+\eta)},$$

$$h_3^2 = \frac{(\zeta-\xi)(\zeta-\eta)}{4\zeta(a^2+\zeta)(b^2+\zeta)} \quad (45)$$

Using equations (43) and (44), the pertinent quantities leading to the displacements and stresses are

$$\frac{\partial I}{\partial x} = x \int_\xi^\infty \frac{ds}{(a^2+s)\sqrt{Q(s)}} = \frac{2x}{a^3 k^2}[u - E(u)] \quad (46a)$$

$$\frac{\partial I}{\partial y} = y \int_\xi^\infty \frac{ds}{(b^2+s)\sqrt{Q(s)}}$$
$$= \frac{2y}{a^3 k^2}\left[\left(\frac{a}{b}\right)^2 E(u) - u - \left(\frac{a^2-b^2}{b^2}\right)\frac{snu\, cnu}{dnu}\right] \quad (46b)$$

$$\frac{\partial I}{\partial z} = z \int_\xi^\infty \frac{ds}{s\sqrt{Q(s)}} = \frac{2z}{ab^2}\left[\frac{snu\, dnu}{cnu} - E(u)\right] \quad (46c)$$

$$\frac{\partial^2 I}{\partial x \partial y} = -\frac{2xy\xi^{1/2}}{(\xi-\eta)(\xi-\zeta)(a^2+\xi)^{1/2}(b^2+\xi)^{1/2}} \quad (46d)$$

$$\frac{\partial^2 I}{\partial y \partial z} = -\frac{2y[\eta\zeta(a^2+\xi)]^{1/2}}{ab(\xi-\eta)(\xi-\zeta)(b^2+\xi)^{1/2}} \quad (46e)$$

$$\frac{\partial^2 I}{\partial z \partial x} = -\frac{2x[\eta\zeta(b^2+\xi)]^{1/2}}{ab(\xi-\eta)(\xi-\zeta)(a^2+\xi)^{1/2}} \quad (46f)$$

$$\frac{\partial^2 I}{\partial x^2} = \int_\xi^\infty \frac{ds}{(a^2+s)\sqrt{Q(s)}} - \frac{2x^2[\xi(b^2+\xi)]^{1/2}}{(\xi-\eta)(\xi-\zeta)(a^2+\xi)^{3/2}} \quad (46g)$$

$$\frac{\partial^2 I}{\partial y^2} = \int_\xi^\infty \frac{ds}{(b^2+s)\sqrt{Q(s)}} - \frac{2y^2\sqrt{Q(\xi)}}{(\xi-\eta)(\xi-\zeta)(b^2+\xi)^2} \quad (46h)$$

$$\frac{\partial^2 I}{\partial z^2} = \frac{2\xi^{1/2}[\xi(a^2b^2-\eta\zeta)-a^2b^2(\eta+\zeta)-(a^2+b^2)\eta\zeta]}{a^2b^2(\xi-\eta)(\xi-\zeta)(a^2+\xi)^{1/2}(b+\xi)^{1/2}}$$
$$- \frac{2}{ab^2}\left[E(u) - \frac{snu\, cnu}{dnu}\right] \quad (46i)$$

$$\frac{\partial^3 I}{\partial x \partial y^2} = \frac{2x\xi^{1/2}}{(\xi-\eta)(\xi-\zeta)(b^2+\xi)^{1/2}} \Bigg\langle \frac{1}{(a^2+\xi)^{1/2}}$$
$$\times \left\{\left(\frac{2y^2}{\eta-\zeta}\right)\left[\frac{\eta(a^2+\eta)}{(\xi-\eta)^2} - \frac{\zeta(a^2+\zeta)}{(\xi-\eta)^2}\right] - 1\right\}$$
$$+ \frac{y^2\xi(a^2+\xi)^{1/2}}{(\xi-\eta)(\xi-\zeta)}\left[2\left(\frac{1}{\xi-\eta}+\frac{1}{\xi-\zeta}\right)\right.$$
$$\left.+\frac{1}{a^2+\xi}+\frac{1}{b^2+\xi}-\frac{1}{\xi}\right]\Bigg\rangle \quad (46j)$$

$$\frac{\partial^3 I}{\partial y \partial x^2} = \frac{2y\xi^{1/2}}{(\xi-\eta)(\xi-\zeta)(a^2+\xi)^{1/2}} \Bigg\langle \frac{1}{(b^2+\xi)^{1/2}}$$
$$\times \left\{\left(\frac{2x^2}{\eta-\zeta}\right)\left[\frac{\eta(a^2+\eta)}{(\xi-\eta)^2} - \frac{\zeta(a^2+\zeta)}{(\xi-\zeta)^2}\right] - 1\right\}$$
$$+ \frac{x^2\xi(b^2+\xi)^{1/2}}{(\xi-\eta)(\xi-\zeta)}\left[2\left(\frac{1}{\xi-\eta}+\frac{1}{\xi-\zeta}\right)\right.$$
$$\left.+\frac{1}{a^2+\xi}+\frac{1}{b^2+\xi}-\frac{1}{\xi}\right]\Bigg\rangle \quad (46k)$$

$$\frac{\partial^3 I}{\partial y \partial z^2} = \frac{2y(a^2+\xi)^{1/2}}{\xi^{1/2}(\xi-\eta)(\xi-\zeta)(b^2+\xi)^{1/2}}$$
$$\times \Bigg\{\left(\frac{2z^2}{\eta-\zeta}\right)\left[\frac{(a^2+\eta)(b^2+\eta)}{(\xi-\eta)^2} - \frac{(a^2+\zeta)(b^2+\zeta)}{(\xi-\zeta)^2}\right] - 1$$
$$+ \frac{z^2(a^2+\xi)^{1/2}(b^2+\xi)^{1/2}}{\xi^{1/2}(\xi-\eta)(\xi-\zeta)}\left[\frac{2(2\xi-\eta-\zeta)\sqrt{Q(\xi)}}{(\xi-\eta)(\xi-\zeta)}\right.$$
$$\left.+\frac{\xi^2+2a^2\xi+a^2b^2}{\sqrt{Q(\xi)}}\right]\Bigg\} \quad (46l)$$

199

$$\frac{\partial^3 I}{\partial z \partial y^2} = \frac{2z\xi^{1/2}(a^2+\xi)^{1/2}}{(\xi-\eta)(\xi-\zeta)(b^2+\xi)^{1/2}}\left\{\frac{2y^2}{(\eta-\zeta)(b^2+\xi)}\right.$$
$$\times\left[\frac{(a^2+\eta)(b^2+\eta)}{(\xi-\eta)^2} - \frac{(a^2+\zeta)(b^2+\zeta)}{(\xi-\zeta)^2}\right] - \frac{1}{\xi}$$
$$+ \frac{y^2(a^2+\xi)}{(\xi-\eta)(\xi-\zeta)}\left[2\left(\frac{1}{\xi-\eta}+\frac{1}{\xi-\zeta}\right)\right.$$
$$\left.\left. - \frac{1}{a^2+\xi}+\frac{3}{b^2+\xi}-\frac{1}{\xi}\right]\right\} \quad (46m)$$

$$\frac{\partial^3 I}{\partial x \partial z^2} = \frac{2x(b^2+\xi)^{1/2}}{\xi^{1/2}(\xi-\eta)(\xi-\zeta)(a^2+\xi)^{1/2}}$$
$$\times\left\{\left(\frac{2z^2}{\eta-\zeta}\right)\left[\frac{(a^2+\eta)(b^2+\eta)}{(\xi-\eta)^2} - \frac{(a^2+\zeta)(b^2+\zeta)}{(\xi-\zeta)^2}\right] - 1\right.$$
$$+ \frac{z^2(a^2+\xi)^{1/2}(b^2+\xi)^{1/2}}{\xi^{1/2}(\xi-\eta)(\xi-\zeta)}\left[\frac{2(2\xi-\eta-\zeta)\sqrt{Q(\xi)}}{(\xi-\eta)(\xi-\zeta)}\right.$$
$$\left.\left. + \frac{\xi^2+2b^2\xi+a^2b^2}{\sqrt{Q(\xi)}}\right]\right\} \quad (46n)$$

$$\frac{\partial^3 I}{\partial z \partial x^2} = \frac{2z\xi^{1/2}(b^2+\xi)^{1/2}}{(\xi-\eta)(\xi-\zeta)(a^2+\xi)^{1/2}}\left\{\frac{2x^2}{(\eta-\zeta)(a^2+\xi)}\right.$$
$$\times\left[\frac{(a^2+\eta)(b^2+\eta)}{(\xi-\eta)^2} - \frac{(a^2+\zeta)(b^2+\zeta)}{(\xi-\zeta)^2}\right]$$
$$- \frac{1}{\xi} + \frac{x^2(b^2+\xi)}{(\xi-\eta)(\xi-\zeta)}\left[2\left(\frac{1}{\xi-\eta}+\frac{1}{\xi-\zeta}\right)\right.$$
$$\left.\left. + \frac{3}{a^2+\xi}-\frac{1}{b^2+\xi}-\frac{1}{\xi}\right]\right\} \quad (46p)$$

$$\frac{\partial^3 I}{\partial x \partial y \partial z} = \frac{2xyz\xi^{1/2}}{(\xi-\eta)(\xi-\zeta)(a^2+\xi)^{1/2}(b^2+\xi)^{1/2}}$$
$$\times\left\{\left(\frac{2}{\eta-\zeta}\right)\left[\frac{(a^2+\eta)(b^2+\eta)}{(\xi-\eta)^2} - \frac{(a^2+\zeta)(b^2+\zeta)}{(\xi-\zeta)^2}\right]\right.$$
$$+ \frac{(a^2+\xi)(b^2+\xi)}{(\xi-\eta)(\xi-\zeta)}\left[2\left(\frac{1}{\xi-\eta}+\frac{1}{\xi-\zeta}\right)\right.$$
$$\left.\left. + \frac{1}{a^2+\xi}+\frac{1}{b^2+\xi}-\frac{1}{\xi}\right]\right\} \quad (46q)$$

$$\frac{\partial^3 I}{\partial x^3} = \frac{2x\xi^{1/2}(b^2+\xi)^{1/2}}{(\xi-\eta)(\xi-\zeta)(a^2+\xi)^{3/2}}$$
$$\times\left\{\left(\frac{2x^2}{\eta-\zeta}\right)\left[\frac{\eta(b^2+\eta)}{(\xi-\eta)^2} - \frac{\zeta(b^2+\zeta)}{(\xi-\zeta)^2}\right] - 3\right.$$
$$+ \frac{x^2\xi(b^2+\xi)}{(\xi-\eta)(\xi-\zeta)}\left[2\left(\frac{1}{\xi-\eta}+\frac{1}{\xi-\zeta}\right)\right.$$
$$\left.\left. + \frac{3}{a^2+\xi}-\frac{1}{b^2+\xi}-\frac{1}{\xi}\right]\right\} \quad (46r)$$

$$\frac{\partial^3 I}{\partial y^3} = \frac{2y\xi^{1/2}(a^2+\xi)^{1/2}}{(\xi-\eta)(\xi-\zeta)(b^2+\xi)^{3/2}}$$
$$\times\left\{\left(\frac{2y^2}{\eta-\zeta}\right)\left[\frac{\eta(a^2+\eta)}{(\xi-\eta)^2} - \frac{\zeta(a^2+\zeta)}{(\xi-\zeta)^2}\right] - 3\right.$$
$$+ \frac{y^2\xi(a^2+\xi)}{(\xi-\eta)(\xi-\zeta)}\left[2\left(\frac{1}{\xi-\eta}+\frac{1}{\xi-\zeta}\right)\right.$$
$$\left.\left. - \frac{1}{a^2+\xi}+\frac{3}{b^2+\xi}-\frac{1}{\xi}\right]\right\} \quad (46s)$$

$$\frac{\partial^3 I}{\partial z^3} = \frac{2z(a^2+\xi)(b^2+\xi)}{a^2b^2(\xi-\eta)(\xi-\zeta)\sqrt{Q(\xi)}}$$
$$\times\left\{\left(\frac{2\xi}{\eta-\zeta}\right)\left[\frac{\zeta(a^2+\eta)(b^2+\eta)}{(\xi-\eta)^2} - \frac{\eta(a^2+\zeta)(b^2+\zeta)}{(\xi-\zeta)^2}\right]\right.$$
$$\left. + a^2b^2\left(\frac{1}{a^2+\xi}+\frac{1}{b^2+\xi}\right)\right\} + \frac{2z\sqrt{Q(\xi)}}{a^2b^2(\xi-\eta)^2(\xi-\zeta)^2}$$
$$\times\left\{[a^2b^2(\eta+\zeta-\xi) + (a^2+b^2)\eta\zeta + \xi\eta\zeta]\right.$$
$$\times\left[2\left(\frac{1}{\xi-\eta}+\frac{1}{\xi-\zeta}\right) + \frac{1}{a^2+\xi}+\frac{1}{b^2+\xi}-\frac{1}{\xi}\right]$$
$$\left. + 2(a^2b^2 - \eta\zeta)\right\} \quad (46t)$$

APPENDIX 2

Limiting Forms of (ξ, η, ζ) to (r, θ)

The derivation of the stress state around the border of an elliptical crack depends upon a knowledge of the limiting forms of (ξ, η, ζ) to (r, θ). To this end, construct an ellipse with semi-axes

$$a' = a + r\cos\theta, \qquad b' = b + r\cos\theta \quad (47)$$

where r is very small in comparison with a and b, Fig. 6. The length PP' (or $r\cos\theta$) represents the projection of the radial distance r in the direction of the binormal, Fig. 2. Accordingly, it is observed from Fig. 6 that the coordinates of any point $P'(x, y)$ on $x^2/a'^2 + y^2/b'^2 = 1$ can be expressed by

$$x = a\cos\phi + r\cos\theta\cos\beta \quad (48a)$$
$$y = b\sin\phi + r\cos\theta\sin\beta \quad (48b)$$

The angle β is related to ϕ by equations (31) in such a way that the line PP' always remains normal to the boundary of the ellipse $x^2/a^2 + y^2/b^2 = 1$. Hence the point (r, θ) in the nz-plane, Fig. 2, is equivalent to (x, y, z), where the expressions for x, y are given by equations (48) and

$$z = r\sin\theta \quad (49)$$

Now, upon adding, equations (22) give

$$x^2 + y^2 + z^2 = \xi + \eta + \zeta + a^2 + b^2 \quad (50)$$

and it follows that

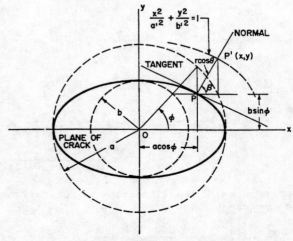

Fig. 6 Polar coordinates near crack edge

$$\xi + \eta + \zeta = -(a^2 \sin^2 \phi + b^2 \cos^2 \phi)$$
$$+ \frac{2abr \cos \theta}{(a^2 \sin^2 \phi + b^2 \cos^2 \phi)^{1/2}} + r^2 \quad (51)$$

Since $\xi = \eta = 0$ corresponds to $r = 0$, i.e., on the border of the ellipse, equation (51) implies

$$\zeta = -(a^2 \sin^2 \phi + b^2 \cos^2 \phi) \quad (52a)$$

and

$$\xi + \eta = \frac{2abr \cos \theta}{(a^2 \sin^2 \phi + b^2 \cos^2 \phi)^{1/2}} + r^2 \quad (52b)$$

Furthermore, combining equations (22c), (49), and (52a), a second equation relating ξ, η is found:

$$\xi \eta = -\frac{(abr \sin \theta)^2}{a^2 \sin^2 \phi + b^2 \cos^2 \phi} \quad (53)$$

Substituting equation (53) into (52b), the quadratic

$$\alpha^2 - \left[\frac{2abr \cos \theta}{(a^2 \sin^2 \phi + b^2 \cos^2 \phi)^{1/2}} + r^2\right] \alpha$$
$$- \frac{(abr \sin \theta)^2}{a^2 \sin^2 \phi + b^2 \cos^2 \phi} = 0 \quad (54)$$

is obtained. The roots are

$$\xi = \frac{2abr}{(a^2 \sin^2 \phi + b^2 \cos^2 \phi)^{1/2}} \left(\cos \frac{\theta}{2}\right)^2 \quad (55a)$$

$$\eta = -\frac{2abr}{(a^2 \sin^2 \phi + b^2 \cos^2 \phi)^{1/2}} \left(\sin \frac{\theta}{2}\right)^2 \quad (55b)$$

where higher-order terms in r have been neglected. Therefore, equations (55) apply only in the immediate vicinity of the crack border.

APPENDIX 3
Stresses and Displacements for Symmetrical Distribution

In order to have a consistent set of notations, the stress and displacement expressions in reference [11] will be slightly modified. Defining

$$h = \frac{C}{2} \int_\xi^\infty \left[\frac{x^2}{a^2 + s} + \frac{y^2}{b^2 + s} + \frac{z^2}{s} - 1\right] \frac{ds}{\sqrt{Q(s)}} \quad (56)$$

the stresses in terms of the function h are

$$\frac{\sigma_{xx}}{2\mu} = \frac{\partial^2 h}{\partial x^2} + 2\nu \frac{\partial^2 h}{\partial y^2} + z \frac{\partial^3 h}{\partial z \partial x^2} \quad (57a)$$

$$\frac{\sigma_{yy}}{2\mu} = \frac{\partial^2 h}{\partial y^2} + 2\nu \frac{\partial^2 h}{\partial x^2} + z \frac{\partial^3 h}{\partial z \partial y^2} \quad (57b)$$

$$\frac{\sigma_{zz}}{2\mu} = -\frac{\partial^2 h}{\partial z^2} + z \frac{\partial^3 h}{\partial z^3} \quad (57c)$$

$$\frac{\tau_{xy}}{2\mu} = (1 - 2\nu) \frac{\partial^2 h}{\partial x \partial y} + z \frac{\partial^3 h}{\partial x \partial y \partial z} \quad (57d)$$

$$\frac{\tau_{yz}}{2\mu} = z \frac{\partial^3 h}{\partial y \partial z^2} \quad (57e)$$

$$\frac{\tau_{zx}}{2\mu} = z \frac{\partial^3 h}{\partial x \partial z^2} \quad (57f)$$

and the displacements are

$$u = (1 - 2\nu) \frac{\partial h}{\partial x} + z \frac{\partial^2 h}{\partial x \partial z}, \quad v = (1 - 2\nu) \frac{\partial h}{\partial y}$$
$$+ z \frac{\partial^2 h}{\partial y \partial z}, \quad w = -2(1 - \nu) \frac{\partial h}{\partial z} + z \frac{\partial^2 h}{\partial z^2} \quad (58)$$

NOTE ADDED IN PROOF: C. M. Segedin ("A Penny-Shaped Crack Under Shear," *Proceedings of the Cambridge Philosophical Society*, vol. 47, 1950, pp. 396–400) has solved the problem of an infinite solid containing a penny-shaped crack under uniform shear prior to the work in [7].

Stress Intensity Factors for Penny-Shaped Cracks
Part 1—Infinite Solid

F. W. SMITH
Assistant Professor,
Department of Mechanical Engineering,
Colorado State University,
Fort Collins, Colo.

A. S. KOBAYASHI
Professor.

A. F. EMERY
Associate Professor.

Department of Mechanical Engineering,
University of Washington,
Seattle, Wash

An expression is developed for the stress intensity factor of a penny-shaped crack in an infinite elastic solid subjected to nonaxisymmetric normal loading. The stress intensity factor can then be determined for penny-shaped cracks in infinite or finite solids subjected to symmetric loading about the plane containing the crack. The singular state associated with the embedded crack with finite, nonaxisymmetric normal loading is that of plane strain. Results are also presented for two problems: A penny-shaped crack subjected to two symmetrically located concentrated forces and a penny-shaped crack in a large beam subjected to pure bending.

Introduction

THE PROBLEM of a penny-shaped crack embedded in an infinite solid and subjected to axisymmetric loading conditions has been considered by several investigators. Sneddon [1][1] solved a problem involving uniform pressure prescribed on the crack surface. Later Olesiak and Sneddon [2] solved the problem with either a prescribed axisymmetric heat flux or temperature on the crack surface. Florence and Goodier [3] solved the problem of an insulated crack perpendicular to uniform heat flux. In addition, the widely used elliptical crack in an infinite solid subjected to uniform applied pressure at the crack surface was solved by Green and Sneddon [4], from which Irwin derived the stress intensity factors for an embedded flaw and a surface flaw of elliptical shape [5].

The objective of this paper is to develop a procedure for solving penny-shaped crack problems in an infinite medium and subjected to nonaxisymmetric loading conditions. The method of approach used follows that of Sneddon [6, 7], which is also well documented by Muki [8] and Westmann [9].

General Equations

Consider a penny-shaped crack embedded in an infinite body. The crack lies inside the circle $r = a$ in the plane $z = 0$ as shown in Fig. 1. The stresses in the neighborhood of the crack are induced by identical distributions of normal stress or displacement applied

[1] Numbers in brackets designate References at end of paper.
Contributed by the Applied Mechanics Division for presentation at the Winter Annual Meeting, Pittsburgh, Pa., November 12-16, 1967, of THE AMERICAN SOCIETY OF MECHANICAL ENGINEERS.
Discussion of this paper should be addressed to the Editorial Department, ASME, United Engineering Center, 345 East 47th Street, New York, N. Y. 10017, and will be accepted until one month after final publication of the paper itself in the JOURNAL OF APPLIED MECHANICS. Manuscript received by ASME Applied Mechanics Division, April 28, 1966; final draft, March 20, 1967. Paper No. 67—WA/APM-1.

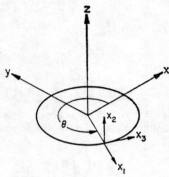

Fig. 1 Local coordinate system near crack tip

to the two surfaces of the crack which are free of shear stresses. On the $z = 0$ plane, therefore, the shear components σ_{rz}, $\sigma_{\theta z}$ are zero, and the displacement component u_z is zero outside the crack. The applied normal stress is assumed to be finite at every point on the crack surface, and the distributions of normal stress are thus expressible in Fourier series.

Following the formulation given by Green and Zerna [10], in the absence of body forces, the complete solution for a restricted class of problems in which the shear stresses are zero on the plane $z = 0$ can be represented by a single harmonic function, $\phi(r, \theta, z)$. The displacement components in terms of this stress function are

$$u_r = \frac{\partial \phi}{\partial r} + (\beta^2 - 1) z \frac{\partial^2 \phi}{\partial r \partial z}$$

$$u_\theta = \frac{1}{r} \frac{\partial \phi}{\partial \theta} + (\beta^2 - 1) \frac{z}{r} \frac{\partial^2 \phi}{\partial \theta \partial z} \qquad (1a)$$

$$u_z = -\beta^2 \frac{\partial \phi}{\partial z} + (\beta^2 - 1) z \frac{\partial^2 \phi}{\partial z^2}$$

where

Nomenclature

a = radius of circular crack
I = moment of inertia of a beam in bending
K_I = plane strain stress intensity factor
$p(r, \theta)$ = prescribed pressure nondimensionalized with respect to μ
r, θ, z = circular cylindrical coordinates
u_i = displacement components nondimensionalized with respect to a
$w(r, \theta)$ = prescribed nondimensional displacement in z-direction
x, y, z = rectangular Cartesian coordinates nondimensionalized with respect to a
x_1, x_2, x_3 = local nondimensional rectangular coordinates
$\beta^2 = 2(1 - \nu)/(1 - 2\nu)$
λ, μ = Lamé constants
ν = Poisson's ratio
ϕ = stress function nondimensionalized with respect to μ
σ_{ij} = stress components nondimensionalized with respect to μ

$$\beta^2 = \frac{\lambda + 2\mu}{\mu} = \frac{2(1-\nu)}{1-2\nu} \quad (1b)$$

The coordinates r, z and the displacement components u_r, u_θ, u_z in these equations are nondimensionalized with respect to the crack radius, a. The stress components are

$$\sigma_{rr} = -2(\beta^2 - 2)\frac{\partial^2 \phi}{\partial z^2} + 2\frac{\partial^2 \phi}{\partial r^2} + 2(\beta^2 - 1)z\frac{\partial^3 \phi}{\partial r^2 \partial z} \quad (2a)$$

$$\sigma_{\theta\theta} = -2(\beta^2 - 2)\frac{\partial^2 \phi}{\partial z^2} + \frac{2}{r}\frac{\partial \phi}{\partial r} + \frac{2}{r^2}\frac{\partial^2 \phi}{\partial \theta^2}$$
$$+ 2(\beta^2 - 1)\frac{z}{r^2}\frac{\partial^3 \phi}{\partial \theta^2 \partial z} + 2(\beta^2 - 1)\frac{z}{r}\frac{\partial^2 \phi}{\partial r \partial z} \quad (2b)$$

$$\sigma_{zz} = -2(\beta^2 - 1)\frac{\partial^2 \phi}{\partial z^2} + 2(\beta^2 - 1)z\frac{\partial^3 \phi}{\partial z^3} \quad (2c)$$

$$\sigma_{r\theta} = \frac{2}{r}\frac{\partial^2 \phi}{\partial r \partial \theta} + 2(\beta^2 - 1)\frac{z}{r}\frac{\partial^3 \phi}{\partial r \partial \theta \partial z}$$
$$- \frac{2}{r^2}\frac{\partial \phi}{\partial \theta} - 2(\beta^2 - 1)\frac{z}{r^2}\frac{\partial^2 \phi}{\partial \theta \partial z} \quad (2d)$$

$$\sigma_{rz} = 2(\beta^2 - 1)z\frac{\partial^3 \phi}{\partial r \partial z^2} \quad (2e)$$

$$\sigma_{\theta z} = 2(\beta^2 - 1)\frac{z}{r}\frac{\partial^3 \phi}{\partial \theta \partial z^2} \quad (2f)$$

These stresses are nondimensionalized with respect to the shear modulus, μ. On the plane $z = 0$, the boundary equations involve

$$\sigma_{rz} = \sigma_{\theta z} = 0 \quad (3a)$$

$$u_z = -\beta^2 \frac{\partial \phi}{\partial z}\bigg|_{z=0} = 0 \quad \text{for} \quad r > 1 \quad (3b)$$

$$\left.\begin{array}{l}\sigma_{zz} = -2(\beta^2 - 1)\dfrac{\partial^2 \phi}{\partial z^2}\bigg|_{z=0} = -p(r, \theta) \quad \text{for} \quad r < 1 \\ \text{when normal stresses are prescribed on the crack surface,} \\ u_z = -\beta^2 \dfrac{\partial \phi}{\partial z}\bigg|_{z=0} = w(r, \theta) \quad \text{for} \quad r < 1 \\ \text{when normal displacements are prescribed on the crack surface.}\end{array}\right\} \quad (3c)$$

The unknown harmonic stress function ϕ is expressed as

$$\phi = \sum_{n=0}^{\infty} \begin{array}{c}\cos n\theta \\ \sin n\theta\end{array} \int_0^\infty \frac{1}{\xi} f_n(\xi) e^{-z\xi} J_n(\xi r) d\xi \quad (4)$$

where $f_n(\xi)$ is an unknown function to be determined from the foregoing boundary conditions. The function, ϕ, in this form is harmonic, and the stresses derived from it approach zero as r and z tend toward infinity. Since the stress function, ϕ, with $\sin n\theta$ or with $\cos n\theta$ has similar properties, only $\cos n\theta$ will be carried through in the following derivations.

If the loading on the crack surface is expressed as a Fourier series, the boundary equation (3c) becomes

$$\sigma_{zz}\bigg|_{z=0} = -p(r, \theta) = -2(\beta^2 - 1)\sum_{n=0}^{\infty} K_n(r)\cos n\theta \quad (5a)$$

where $K_n(r)$ are the Fourier coefficients of the function $p(r, \theta)/2(\beta^2 - 1)$ given by

$$K_0(r) = \frac{1}{2\pi(\beta^2 - 1)}\int_0^\pi p(r, \theta)d\theta$$
$$K_n(r) = \frac{2}{2\pi(\beta^2 - 1)}\int_0^\pi p(r, \theta)\cos n\theta d\theta \quad (5b)$$

Substituting equation (4) into equations (3a), (3b), and (3c), the boundary equations thus obtained constitute a set of dual integral equations in terms of $f_n(\xi)$ for which the solution is given by Sneddon [11] as

$$f_n(\xi) = \sqrt{\frac{2}{\pi}}\xi^{1/2}\int_0^1 \eta^{3/2}J_{n+1/2}(\xi\eta)d\eta \int_0^1 K_n(\zeta\eta)\frac{\zeta^{n+1}}{(1-\zeta^2)^{1/2}}d\zeta \quad (6)$$

Equation (6) with equations (1), (2), and (4) then constitute a solution to the problem.

State of Stress Near Crack Tip

Consider a local coordinate system x_1, x_2, and x_3 near the tip of the crack as shown in Fig. 1. Define the dilatation in the form of

$$\theta = \bar{\theta} + u_{3,3} \quad (7a)$$

where

$$\bar{\theta} = u_{1,1} + u_{2,2} \quad (7b)$$

and the comma denotes covariant differentiation. The stress components near the crack tip may then be expressed in terms of displacements and dilatation as

$$\sigma_{\alpha\beta} = \frac{\lambda}{\mu}\delta_{\alpha\beta}\bar{\theta} + (u_{\alpha,\beta} + u_{\beta,\alpha}) + \frac{\lambda}{\mu}\delta_{\alpha\beta}u_{3,3} \quad (8a)$$

$$\sigma_{\alpha 3} = (u_{\alpha,3} + u_{3,\alpha}) \quad (8b)$$

$$\sigma_{33} = \frac{\lambda\bar{\theta}}{\mu} + \frac{(\lambda + 2\mu)}{\mu}u_{3,3} \quad (8c)$$

where α and β take on values of 1 or 2. The first two terms in equation (8a) correspond to the stress components in a state of plane strain.

It is known that the state of stresses in the vicinity of the crack tip of a penny-shaped crack in an infinite medium subjected to uniaxial tension is in a state of plane strain. It is not known, however, if the state of plane strain is maintained when the crack surface is subjected to a nonaxisymmetric normal stress or displacement loading condition. The generalization of this plane strain condition to such loading has practical significance in that the strain fracture toughness, K_{Ic}, determined by other crack geometries can be used to predict the load at the onset of rapid fracture for all problems involving a penny-shaped crack.

If it could be shown that, in the plane $z = 0$, $u_{3,3}$ and the $\sigma_{\alpha 3}$ are continuous and finite at the crack tip, then the singularity will exist only in the plane strain portion of the expressions. Noting that, in cylindrical polar coordinates, $u_{3,3}$ corresponds to $(1/r) \times (\partial u_\theta/\partial \theta) + (u_r/r)$, it is then necessary to study the order of magnitude of these two terms. In the plane $z = 0$,

$$\frac{1}{r}\frac{\partial u_\theta}{\partial \theta} = -\frac{1}{r^2}\sum_{n=0}^{\infty} n^2 \cos n\theta \int_0^1$$
$$\times \left[\eta^{1/2}g_n(\eta)\int_0^\infty \xi^{-1/2}J_{n+1/2}(\xi\eta)J_n(\xi r)d\xi\right]d\eta \quad (9a)$$

where $g_n(\eta)$ is defined as

$$g_n(\eta) = \frac{2\eta}{\pi}\int_0^1 K_n(\eta\xi)\frac{\xi^{n+1}d\xi}{\sqrt{1-\xi^2}} \quad (9b)$$

The infinite integral is of the Weber-Schafheitlin type (sometimes referred to as Sonine-Schafheitlin integrals) and may be expressed as [12]

$$\int_0^\infty \xi^{-1/2} J_{n+1/2}(\xi\eta) J_n(\xi r) d\xi$$

$$= \begin{cases} \dfrac{\eta^{n+1/2} \Gamma\left(n + \dfrac{1}{2}\right)}{\sqrt{2}\, r^{n+1} \Gamma\left(\dfrac{1}{2}\right) \Gamma\left(n + \dfrac{2}{3}\right)} \\ \quad \times {}_2F_1\left(n + \dfrac{1}{2}, \dfrac{1}{2}; n + \dfrac{3}{2}; \dfrac{\eta^2}{r^2}\right) & r > \eta \\ \\ \dfrac{\Gamma\left(n + \dfrac{1}{2}\right) r^n}{\sqrt{2}\, \Gamma(n+1) \eta^{n+1/2}} & r < \eta \end{cases} \quad (10)$$

For $r < \eta$, equation (10) has no singularity near the crack tip. For $r > \eta$, equation (10) has also no singularity near the crack tip since the hypergeometric function, ${}_2F_1(a, b, c; z)$, is absolutely convergent everywhere including the crack tip. It is also noted that, for $r = 1$, equation (10) can be evaluated by using ${}_2F_1(a, b, c; 1) = \dfrac{\Gamma(c)\Gamma(c - a - b)}{\Gamma(c - a)\Gamma(c - b)}$, and it then represents a finite value. Thus the infinite integral of equation (9a) is finite between the limits of the first integral. The function $g_n(\eta)$ is continuous so that $(1/r)(\partial u_\theta/\partial \theta)$ is finite for all r. For the second term in $u_{3.3}$, the displacement component u_r can be represented as

$$\frac{1}{r} u_r = \sum_{n=0}^\infty \left[\frac{n}{r^2} \int_0^1 \eta^{1/2} g_n(\eta) d\eta \sqrt{\frac{\pi}{2}} \int_0^\infty \xi^{-1/2} J_{n+1/2}(\xi r) \right.$$
$$\times J_n(\xi r) d\xi - \frac{1}{r} \int_0^1 \eta^{1/2} g_n(\eta) d\eta \sqrt{\frac{\pi}{2}} \int_0^\infty \xi^{1/2} J_{n+1/2}(\xi\eta)$$
$$\left. \times J_{n+1}(\xi r) d\xi \right] \cos n\theta \quad (11)$$

The first infinite integral is given by equation (10) and causes no singularities at the crack tip. The second infinite integral is

$$\sqrt{\frac{\pi}{2}} \int_0^\infty \xi^{1/2} J_{n+1/2}(\xi\eta) J_{n+1}(\xi r) d\xi = \begin{cases} \dfrac{\eta^{n+1/2}}{r^{n+1}(r^2 - \eta^2)^{1/2}} & r > \eta \\ 0 & r < \eta \end{cases}$$
$$(12)$$

and thus the second term of equation (11) for $r > 1$ becomes

$$\text{finite value} + \frac{1}{r^{n+2}} g_n(1) \int_{1-\epsilon}^1 \frac{\eta^{n+1} d\eta}{(r^2 - \eta^2)^{1/2}}$$

which is always finite at the crack tip. Thus, for a penny-shaped crack, $u_{3,3}$ is finite at the crack tip.

The stress component σ_{23} which corresponds to $\sigma_{z\theta}$ can be expressed in terms of the infinite integrals discussed earlier and is finite at the crack tip. To complete the reasoning, it can be readily shown that all terms contained in the plane strain portion of equations (8a), (8b), and (8c) are singular at the crack tip.

It has been shown analytically that the plane strain portions of equations (8a), (8b), and (8c) become dominant very near the crack tip; thus the state of stress becomes that of plane strain at the tip of a penny-shaped crack for any nonaxisymmetric continuous distribution of loading on the crack surface. Although this is a generally accepted concept in fracture mechanics, to the writers' knowledge it has never been proved analytically. Irwin [1] deduced this result for elliptical cracks in pure tension based on an argument that the stresses and displacement vary slowly along the crack border.[2] The present argument only requires that certain displacement derivatives be finite at the crack border.

---

[2] Note added in proof. Kassir and Sih have proved the plane-strain concept recently for an elliptical crack under combined loading of uniform tension and shear [13].

Stress Intensity Factor

The stress intensity factor, K_I,[3] for the opening mode of fracture is defined in terms of the stress σ_{zz} on the plane $z = 0$ as

$$K_I = \lim_{\delta \to 0} \sqrt{2\pi\delta}\, \mu \sigma_{zz} \quad (13)$$

where δ is a small distance measured from the crack tip. By using the condition that the state of plane strain prevails in the vicinity of the crack tip, the crack opening shape in terms of the stress intensity factor is

$$u_z\bigg|_{z=0} = \frac{\beta^2}{2(\beta^2 - 1)} (1 - r^2)^{1/2} \frac{K_I}{\mu \sqrt{\pi a}} \quad (14)$$

By calculating separately the opening shape from the exact expression of equations (1a), one obtains

$$u_z\bigg|_{z=0} = \beta^2 (1 - r^2)^{1/2} \sum_{n=0}^\infty g_n(1) \cos n\theta \quad (15)$$

which proves that the crack opening shape near the crack tip will always be elliptical for any possible finite and continuous force boundary condition on the crack surface.

Comparing equations (14) and (15), the stress intensity factor for symmetric loading with respect to θ is

$$K_I = 2\mu \sqrt{a\pi}\, (\beta^2 - 1) \sum_{n=0}^\infty g_n(1) \cos n\theta \quad (16)$$

For skew symmetric loading with respect to θ, similar results are obtained by replacing $\cos n\theta$ with $\sin n\theta$.

Solution for Prescribed Normal Displacements on Crack Surface

This solution represents cooling of an elastic medium with a frictionless, penny-shaped, rigid inclusion of small variable thicknesses. Reversing the problem, the state of stresses can be represented by a penny-shaped crack subjected to prescribe normal displacements on its shear traction-free crack surface.

From equation (3c), the integral equation to be satisfied is

$$\beta^2 \sum_{n=0}^\infty \cos n\theta \int_0^\infty \xi h_n(\xi) J_n(\xi r) d\xi = \begin{cases} w(r, \theta) & r \leq 1 \\ 0 & r \geq 1 \end{cases} \quad (17)$$

where $h_n(\xi)$ is the unknown function for the stress function φ, which now takes the form of

$$\phi = \sum_{n=0}^\infty \cos n\theta \int_0^\infty h_n(\xi) e^{-z\xi} J_n(\xi r) d\xi \quad (18)$$

If the function $w(r, \theta)$ is expressible in the form

$$w(r, \theta) = \beta^2 \sum_{n=0}^\infty W_n(r) \cos n\theta \quad (19)$$

then the unknown function $h_n(\xi)$ can be represented in terms of a Hankel transform[4] of $H(1 - r) W_n(r)$ given by

$$h_n(\xi) = \int_0^1 r W_n(r) J_n(\xi r) dr \quad (20)$$

Equations (18), (19), and (20), with equations (1) and (2), constitute a solution to the problem.

In order to investigate the order of stress singularity in displacement prescribed problems, it is necessary to choose a particular form for $W_n(r)$. A convenient form is

---

[3] Plane strain stress intensity factor.
[4] $H(1 - r)$ is the Heaviside function.

$$W_n(r) = r^n \sum_{p=1,3\ldots}^{\infty} \alpha_{pn}(1 - r^2)^{\gamma p} \qquad (21)$$

It may be shown that, if an expression is derived for the opening shape of a crack subjected to finite surface tractions, the Fourier coefficients of that expression will take on the form of equation (21), with $\gamma = 1/2$. Choosing $W_n(r)$ of the form in equation (21) is then quite reasonable since the special case $\gamma = 1/2$ corresponds to finite loads on the crack surface and an opening shape near the crack tip which is elliptical.

It is possible in displacement prescribed problems that the opening shape near the crack tip might not be elliptical; i.e., γ may not be $1/2$. In the following, γ is allowed to range from zero to one. Substituting equation (21) into equation (20), one obtains

$$h_n(\xi) = \sum_{p=1,3\ldots}^{\infty} \alpha_{pn} \frac{2^{\gamma p}\Gamma(\gamma p + 1)}{\xi^{p+1}} J_{n+\gamma p+1}(\xi) \qquad (22)$$

The stress component σ_{zz} on the plane $z = 0$ now becomes

$$\sigma_{zz} = -2(\beta^2 - 1) \sum_{n=0}^{\infty} \cos n\theta \sum_{p=1,3\ldots}^{\infty} \alpha_{pn} 2^{\gamma p} \Gamma(\gamma p + 1)$$
$$\times \int_0^{\infty} \xi^{1-\gamma p} J_{n+\gamma p+1}(\xi) J_n(\xi r) d\xi \qquad (23)$$

where the infinite integral is of the Weber-Schafheitlin type and, for $r > 1$, is expressible in terms of [12]

$$_2F_1\left(n + \frac{3}{2}, \frac{3}{2}, n + \gamma p + 2; \frac{1}{r^2}\right) \qquad (24)$$

which is absolutely convergent for $\gamma p > 1$. For $0 < \gamma p \leq 1$, equation (24) becomes conditionally convergent and may be transformed to

$$\frac{r^{2(1-\gamma p)}}{(r^2 - 1)^{1-\gamma p}} \,_2F_1\left(\gamma p + \frac{1}{2}, n + \gamma p + \frac{1}{2}, n + \gamma p + 2, \frac{1}{r^2}\right) \qquad (25)$$

where the hypergeometric series is absolutely convergent. This expression agrees with the known fact that singularities in the stresses near the crack tip of order other than $1/(r^2 - 1)^{1/2}$ are possible. In particular, equation (25) allows singularities of $1/(r^2 - 1)$ to any power between zero and one.

Numerical Procedure

So far, only methods of calculating stress intensity factors for embedded circular cracks and related topics have been discussed. In the following, the methods employed to evaluate the stress components will be discussed.

Equations necessary to calculate the stress components may be derived by substituting equations (4), (5), and (6) into equations (2). The task of transforming these equations into numerical results is a rather formidable one. The analysis consists of evaluating integrals similar to those involved in equations (4), (5), and (6), but, before a numerical integration may be applied, the integrands must be calculated.

Both equations (4) and (6) involve Bessel functions which satisfy the following recurrence formulas:

$$J_{n+1}(\xi) = (2n/\xi)J_n - J_{n-1} \qquad (26)$$

When recurring upward, however, the round-off error builds up very rapidly and completely overshadows the correct value when the ratio n/ξ is much greater than 1. This instability was overcome by using a method due to Abramowitz [12], which is based on the fact that, for a given argument, the function $J_n(\xi)$ approaches zero as n becomes large. Thus, by assuming that

$$J_N = 0, \qquad J_{N-1} = 1 \qquad (27)$$

where N is sufficiently large, the two values of equation (27) can then be substituted into equation (26), which is then recurred downward to J_0. Since J_n is a decreasing function of n for $n/\xi > 1$, there is no danger of significant round-off error in this procedure. The results are then normalized by using the following relation:

$$J_0 + 2J_2 + 2J_4 + \ldots\ldots = 1 \qquad (28)$$

The number of accurate decimal places in the result is the same as the accurate number of decimal places in the assumption of $J_N = 0$. It was found that, for the Bessel functions of interest, $N = 2(6 + \xi)$ gave results accurate to at least eight decimal places.

For values of the ratio n/ξ much less than 1, equation (20) may be recurred upward with negligible error. For these cases, J_0 and J_1 were calculated using a formula by Abramowitz [12], which gave results accurate in the seventh decimal place.

With the integrands determined, the infinite integrals similar to that of equation (4) were evaluated for each value of n and then the results were summed.

Since the functions $f_n(\xi)$, $J_n(\xi r)$ and $e^{-z\xi}$ all tend toward zero for large values of the integrating variables ξ, it is only necessary to integrate to some large finite value to obtain a fairly accurate value for the integral. For values of $z < 1$, the effect of $e^{-z\xi}$ is not very strong so Weddle's rule is used in evaluating the integrand at 162 points in the range $0 \leq \xi \leq 102$. For $z > 1$, the function e^{-z} becomes quite strong so the Laguerre integration formula

$$\int_0^{\infty} e^{-x}f(x)dx = \sum_{i=1}^{M} \omega_i f(x_i)$$

is used where $M = 15$, and ω_i and x_i are weighting factors and abscissas, respectively, which minimize the error in the approximation. The weighting factors and abscissas are tabulated in [12]. By comparing with the closed-form solution for the case of a constant load on the crack surface, given by Sneddon [1], it was found that the foregoing integrating scheme worked very well except for values of z close to zero. This difficulty was traced to a term of the type

$$I(2m, 1) = \int_0^{\infty} e^{-z\xi} \cos \xi J_{2m}(\xi r) d\xi \qquad (29)$$

the integrand of which converges very slowly for small z as the integrating variable becomes large. This type of integral has been discussed by Sneddon [1], George [14], and Kobayashi [15] and may be evaluated by recurrence formulas.

Once the integration has been performed, the results must be summed on n. To avoid an excessive use of computer time, it is necessary to limit the number of terms in the series. The greatest number of terms that could be practically summed was six, which gives rather inaccurate results because of the slow convergence of the series. A numerical method to improve the accuracy of the values calculated from six terms was developed. This method is patterned after a method due to Krylov [16] for improving the convergence of a series in which analytical expressions for each term are known.

As a check of this numerical integration procedure, the special case of a constant load on the crack surface was evaluated using Sneddon's [1] closed form for comparison. Reference [17] gives the results for the numerically integrated solution. The agreement at all points was found to be within $3/4$ percent of the maximum value of the function and, for values of z greater than 0.4, was exact to four decimal places. Discrepancies between stresses for z greater than 0.8 and the corresponding original values presented by Sneddon are believed to be due to the inadequacies of computing facilities at the time reference [1] was published.

In the following, the equations derived earlier will be employed to calculate stress intensity factors for two cases in which the load applied to the crack surface is nonaxisymmetric.

Two Concentrated Forces at Equal Radial Distance on Crack Surface

The solution to this problem can be used as a Green's function to solve problems of practical importance which will be described in the following section and in Part 2 (ASME Paper No. 67-WA/APM-2). Consider two concentrated forces symmetrically located with respect to the $\theta = 0$ plane and represented as

$$p(r, \theta) = \frac{\delta(r - r_1)\delta(\theta \pm \theta_1)}{r} \frac{F_0}{a^2\mu} \tag{30}$$

where F_0 is the magnitude of each of the forces located at the points $r_1, \pm\theta_1$; and δ is the Dirac delta function. From equations (5),

$$K_0(r) = \frac{F_0}{2\mu\pi a^2(\beta^2 - 1)} \frac{\delta(r - r_1)}{r}$$

$$K_n(r) = \frac{F_0}{\mu\pi a^2(\beta^2 - 1)} \frac{\delta(r - r_1)}{r} \cos n\theta_1 \tag{31}$$

the function $g_n(1)$ is

$$g_0(1) = \frac{2F_0}{2\mu\pi^2 a^2(\beta^2 - 1)} \frac{1}{(1 - r_1^2)^{1/2}}$$

$$g_n(1) = \frac{4F_0}{2\mu\pi a^2(\beta^2 - 1)} \frac{r_1^n}{(1 - r_1^2)^{1/2}} \cos n\theta_1 \tag{32}$$

Equation (16) gives the stress intensity factor as

$$K_I = \frac{2F_0\sqrt{a}}{\pi^{3/2}a^2}\left[1 + 2\sum_{n=1}^{\infty} r_1^n \cos n\theta_1 \cos n\theta\right]\frac{1}{(1 - r_1^2)^{1/2}} \tag{33}$$

It should be noted that equation (33) reduces to the result derived by Barenblatt [18] for cases with variable axisymmetric loading. It is interesting to note that, if the forces are located at the origin, then the stress intensity factor becomes

$$K_I = \frac{2F_0\sqrt{a}}{\pi^{3/2}a^2}$$

where $2F_0$ is the total force on the crack. If we define an average stress over the crack as

$$P_0 = \frac{2F_0}{\pi a^2}$$

the stress intensity factor is

$$K_I = \frac{1}{\sqrt{\pi}} P_0 \sqrt{a}$$

which is exactly one half of that given by Sneddon [1] for a constant pressure P_0 on the crack surface.

Penny-Shaped Crack in a Large Beam Subjected to Pure Bending

Consider now a large beam subjected to pure bending with a circular crack embedded in it. The crack is located so that a line through its center is parallel to the centerline of the beam as in Fig. 2. The crack is small compared to the size of the beam cross section and must be located such that its center is at least three crack radii away from any surface. This latter restriction assures that the error in the free surface boundary conditions will be less than 1 percent of the maximum load on the crack surface.

The stress intensity factor may be found for this case by the following superposition. The stresses σ_{zz} in a crack free beam are calculated in the region where the crack will be. Then a stress free crack is introduced by superposing the solution for a crack with $-\sigma_{zz}$ prescribed on its surface.

If the crack center is located at a distance b above the neutral

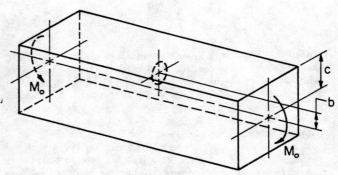

Fig. 2 Penny-shaped crack in a large beam subjected to pure bending

axis as in Fig. 2, the stress in the region where the crack will be is

$$\sigma_{zz} = \frac{M_0 b}{I\mu} - \frac{M_0 a}{I\mu} r \cos\theta \tag{34}$$

where M_0 is the applied moment on the beam. Applying the negative of this to the crack surface, we have

$$p(r, \theta) = \frac{M_0 b}{I\mu} - \frac{M_0 a}{I\mu} r \cos\theta$$

and, from previous sections,

$$K_0(r) = \frac{1}{2(\beta^2 - 1)} \frac{M_0 b}{I\mu}$$

$$K_1(r) = -\frac{1}{2(\beta^2 - 1)} \frac{M_0 a}{I\mu} r$$

The functions $g_n(1)$ are

$$g_0(1) = \frac{2}{\pi} \cdot \frac{1}{2\mu(\beta^2 - 1)} \cdot \frac{M_0 b}{I}$$

$$g_1(1) = -\frac{2}{\pi} \frac{1}{2\mu(\beta^2 - 1)} \frac{2M_0 a}{3I}$$

so the stress intensity factor is

$$K_I = \frac{2}{\sqrt{\pi}} \frac{M_0 b \sqrt{a}}{I}\left[1 - \frac{2}{3}\frac{a}{b}\cos\theta\right] \tag{35}$$

The first term in equation (35) represents the stress intensity factor resulting from the average stress over the crack surface. The second term represents the error that would result if the first term were used alone as an approximation. If the crack is a long way from the neutral axis—that is, a/b is small—the error due to assuming a constant load on the crack is negligible. On the other hand, if the crack is very close to the neutral axis, the error can be very sizable. In the case where $a/b = 1$, the error is 67 percent at the top edge of the crack.

Another interesting case occurs when $a/b = 3/2$. In this case, the lower third of the line $\theta = 0$ extends into the compression portion of the beam. At the crack tip when $\theta = 0$, the stress intensity factor is also zero. This physically means that the compressive stresses at the point $r = 1$, $\theta = 0$ are just sufficient to keep the crack from opening near that point. If the crack were moved further into the compression field, the stress intensity factor and the crack opening displacement would become negative at some points, which physically cannot occur. For this case, the equations do not describe what actually occurs

Conclusions

A procedure was developed for analyzing a problem involving embedded cracks in an infinite elastic solid and subjected to non-axisymmetric loading. In the course of this development, it was found that:

1 The singular state associated with an embedded crack subjected to arbitrary, finite, nonaxisymmetric normal loading is that of plane strain.

2 The opening shape near the border of an embedded circular crack subjected to finite nonaxisymmetric normal loads is always elliptical.

3 A general expression was derived for the stress intensity factor of an embedded circular crack being opened by arbitrary finite nonaxisymmetric loading.

Acknowledgment

The writers wish to thank the Research Computer Laboratory of the University of Washington for providing free computer time for this project. F. W. Smith held a NASA Predoctoral Traineeship during the period this investigation was conducted.

References

1 Sneddon, I. N., "The Distribution of Stress in the Neighborhood of a Crack in an Elastic Solid," *Proceedings of the Royal Society*, London, Series A, Vol. 187, 1946.

2 Olesiak, A., and Sneddon, I. N., "The Distribution of Thermal Stress in an Infinite Elastic Solid Containing a Penny-Shaped Crack," *Archive for Rational Mechanics and Analysis*, 1960, pp. 238–254.

3 Florence, A. L., and Goodier, J. N., "The Linear Thermoelastic Problem of Uniform Heat Flow Disturbed by a Penny-Shaped Insulated Crack," *International Journal of Engineering Sciences*, Vol. 1, 1963, pp. 533–540.

4 Green, A. E., and Sneddon, I. N., "The Distribution of Stress in the Neighborhood of a Flat Elliptical Crack in an Elastic Solid," *Proceedings of the Cambridge Philosophical Society*, Vol. 46, 1950, pp. 159–163.

5 Irwin, G. R., "Crack Extension Force for a Part Through Crack in a Plate," JOURNAL OF APPLIED MECHANICS, Vol. 29, No. 4, TRANS. ASME, Vol. 84, Series E, Dec. 1962, pp. 651–654.

6 Sneddon, I. N., "Crack Problems in the Mathematical Theory of Elasticity," North Carolina State University, Applied Mathematics Research Group, May 15, 1961.

7 Sneddon, I. N., "The Use of Transform Methods in Elasticity," North Carolina State University, Applied Mathematics Research Group, Nov. 6, 1964.

8 Muki, R., "Asymmetric Problems of the Theory of Elasticity for a Semi-Infinite Solid and a Thick Plate," *Progress in Solid Mechanics*, Vol. 1, North-Holland Publishing Co., Amsterdam, 1960, pp. 401–439.

9 Westmann, R. A., "Asymmetric Mixed Boundary-Value Problem of the Elastic Half-Space," JOURNAL OF APPLIED MECHANICS, Vol. 32, No. 2, TRANS. ASME, Vol. 87, Series E, June 1965, pp. 411–417.

10 Green, A. E., and Zerna, W., *Theoretical Elasticity*, Oxford at the Clarendon Press, 1960.

11 Sneddon, I. A., *Fourier Transform*, McGraw-Hill, New York, 1951.

12 Abramowitz, M., and Stegun, I. A., *Handbook of Mathematical Functions*, National Bureau of Standards Applied Mathematics Series, No. 55, 1964.

13 Kassir, M. K., and Sih, G. C., "Three-Dimensional Stress Distribution Around an Elliptical Crack Under Arbitrary Loadings," JOURNAL OF APPLIED MECHANICS, Vol. 33, No. 3, TRANS. ASME, Vol. 88, Series E, Sept. 1966, pp. 601–611.

14 George, D. L., "Numerical Values of Some Integrals Involving Bessel Functions," *Proceedings of the Glasgow Mathematics Association*, Vol. 4, 1962, pp. 87–113.

15 Kobayashi, A. S., "Tables of Some Integrals Involving Bessel Functions," Boeing Company Document No. D2-23879-1, 1965.

16 Kantorovich, L. F., and Krylov, V. I., *Approximate Methods of Higher Analysis*, Wiley, New York, 1964.

17 Smith, F. W., "Stresses Near a Semi-Circular Edge Crack," PhD thesis, University of Washington, Mar. 1966.

18 Barenblatt, C. I., "Mathematical Theory of Equilibrium Cracks," *Advances in Applied Mathematics*, Academic Press, New York, 1962.

F. W. SMITH
Assistant Professor,
Department of Mechanical Engineering,
Colorado State University,
Fort Collins, Colo.

A. F. EMERY
Associate Professor.

A. S. KOBAYASHI
Professor,
Department of Mechanical Engineering,
University of Washington,
Seattle, Wash.

Stress Intensity Factors for Semicircular Cracks

Part 2—Semi-Infinite Solid

The stress intensity factor for a semicircular edge crack is derived. Numerical values for axial, bending, and thermal loads in half spaces and plates are presented. The results show that a magnification of the stress intensity factor of about 20 percent occurs at the free surface.

Introduction

THE THREE-DIMENSIONAL problem of part-through cracks embedded at the surface of semi-infinite solids and subjected to asymmetrical loading conditions is a problem of great practical interest. In 1962, Irwin [1][1] estimated the stress intensity factor and crack-extension force for a part-through crack using the solution of Green and Sneddon [2] for a flat, elliptically shaped crack in an infinite body. He used the two-dimensional edge crack solution to estimate the effect of introducing a free surface perpendicular to the crack surface. At present, there are no other known three-dimensional studies for the part-through crack.

However, two-dimensional edge cracks have been studied extensively. Wigglesworth [3] treated the long notched plate, and Bueckner [4] examined the case of bending. Bowie [5] and Koiter [6] have considered edge cracks in two-dimensional beams. Lachenbruch [7] treated edge cracks in plane strain by using the Schwarz alternating technique.

Application of the alternating technique to three-dimensional problems involving part-through cracks is a natural extension of the two-dimensional edge-crack solution. Whereas the solution for a plane crack of finite length was used in the two-dimensional edge-crack solution, a completely enclosed penny-shaped crack solution described in Part 1 (ASME Paper No. 67—WA APM-1) is utilized in this part-through crack problem.

Formulation

Fig. 1 depicts the semicircular edge crack and the coordinate system. A flat crack described by $0 \leq r \leq 1$, $-\pi/2 \leq \theta \leq \pi/2$ is placed in the plane $z = 0$. It is required that the loading be such that the crack surfaces will spread apart at every point, thus making the crack surface stress free. This requirement does not restrict the loading on the crack surface to be compressive at every point. The loadings are further restricted to be symmetric with respect to the plane $z = 0$, thus implying that the shear stress components σ_{rz} and $\sigma_{\theta z}$ are zero on the plane $z = 0$. The remaining boundary condition to be satisfied on the plane $z = 0$, in

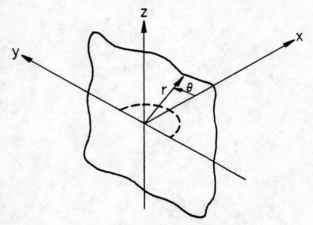

Fig. 1 Semicircular edge crack

[1] Numbers in brackets designate References at end of paper.
Contributed by the Applied Mechanics Division for presentation at the Winter Annual Meeting, Pittsburgh, Pa., November 12–16, 1967, of THE AMERICAN SOCIETY OF MECHANICAL ENGINEERS.
Discussion of this paper should be addressed to the Editorial Department, ASME, United Engineering Center, 345 East 47th Street, New York, N. Y. 10017, and will be accepted until January 15, 1968. Discussion received after the closing date will be returned. Manuscript received by ASME Applied Mechanics Division, April 28, 1966; final draft, March 20, 1967. Paper No. 67—WA/APM-2.

Nomenclature

- a = radius of edge crack
- c = specific heat
- e_{kk} = dilatation
- E = elastic modulus
- h = plate half thickness
- I = moment of inertia
- K_I = stress intensity factor for bending
- k = thermal conductivity
- m = $(3\lambda + 2\mu)\alpha$
- M = bending moment
- p = nondimensional pressure upon crack surface
- r = radius
- t = time
- T = temperature
- u = displacement nondimensionalized with respect to a
- x, y, z = nondimensional rectangular Cartesian coordinates
- α = linear coefficient of thermal expansion
- $\beta^2 = 2(1 - \nu)/(1 - 2\nu)$
- $\bar{\beta} = (3\lambda + 2\mu)/(\lambda + 2\mu)$
- ξ, ζ, η = nondimensional rectangular Cartesian coordinates, Fig. 2
- σ = stress nondimensionalized with respect to μ
- μ, λ = Lamé constants
- χ = surface temperature
- ρ = density
- κ = thermal diffusivity
- ψ = angle measured from crack tip in a plane perpendicular to crack surface
- ψ_0, ψ_1, ψ_2 = functions related to stress intensity factor for semicircular cracks

addition to the prescribed normal stresses on the crack surface, is that of zero normal displacement u_z outside the crack. Also, the plane $\theta = \pm \pi/2$ must be free of all stresses.

The problem may be decomposed into two parts which yield the full solution upon superposition. The first part is the determination of the stresses due to a completely embedded crack, and the second part is the satisfaction of stresses on the bounding plane $\theta = \pm \pi/2$.

The steps necessary to solve the edge crack problem are:

Step 1. Determine the elastic solution with no crack present and calculate the stress σ_{zz} on the plane $z = 0$.

Step 2. Determine the solution for the crack in an infinite body. Calculate the stress $\sigma_{\theta\theta}$ on the plane $\theta = \pm \pi/2$.

Step 3. Determine the solution for a crack-free half space with $-\sigma_{\theta\theta}$ prescribed on the bounding plane. Calculate the stresses σ_{zz} on the plane $z = 0$.

Step 4. Apply the alternating method to Steps 2 and 3 to obtain the solution for the edge crack with the surface load $-\sigma_{zz}$.

Step 5. Superimpose the results of Steps 1 and 4.

The formulation of the embedded crack problem and the numerical methods used in determining the stress components σ_{zz} are given in Part 1 and reference [8]. This paper details only the formulation and necessary methods for Steps 2–5.

To satisfy the requirements of Step 2, it is necessary to calculate the normal stress components on the plane $\theta = \pm \pi/2$. For the examples considered herein, the boundary conditions and crack geometry will be symmetric with respect to the planes $y = 0$ and $z = 0$ (see Fig. 1). Thus the stress component $\sigma_{\theta\theta}$ on the plane $\theta = \pm \pi/2$ is symmetric with respect to both the y and the z-axes. This stress component is evaluated to be

$$\sigma_{\theta\theta} = \sum_{n=0}^{\infty} \cos \frac{n\pi}{2} \int_0^\infty \left\{ \left[-2\xi r^2 + \frac{n(n-1)}{\xi} \right. \right.$$
$$\left. \left. \times (4z\xi - 2) \right] \frac{1}{r^2} J_n(\xi r) - [2 - 4z\xi] \frac{1}{r} J_{n+1}(\xi r) \right\} e^{-z\xi} f_n(\xi) d\xi$$
(1)

for a Poisson's ratio of 0.25, which corresponds to $\beta^2 = 3$.

Solution for Half Space

Step 3 in the solution requires that, on the surface $x = 0$ shown in Fig. 1,

$$\begin{aligned} \sigma_{xy} = \sigma_{xz} &= 0 & x &= 0 \\ \sigma_{xx} &= -\sigma_{\theta\theta}|_{\theta=\pm\pi/2} & x &= 0 \end{aligned}$$
(2)

The solution to this problem may be formulated very simply in terms of Fourier transforms. However, the numerical evaluation of the resulting expressions for the stresses is complicated because of the four repeated integrals which occur. For this reason, an approximate solution is developed using Love's solution [9] for stresses in a half space subjected to a constant pressure on the boundary over a rectangular area. The boundary conditions for this problem are

$$\begin{aligned} \sigma_{\zeta\xi} = \sigma_{\zeta\eta} &= 0 & \zeta &= 0 \\ \sigma_{\zeta\zeta} &= p & \zeta &= 0, |\xi| < a, |\eta| < b \\ \sigma_{\zeta\zeta} &= 0 & \zeta &= 0, |\xi| > a \text{ or } |\eta| > b \end{aligned}$$
(3)

The equations of elasticity are satisfied for the restricted case of zero shear on the plane $z = 0$ if

$$\nabla^2 \chi = \nabla^2 V = 0, \qquad V = \partial \chi / \partial \zeta \tag{4}$$

and

$$\begin{aligned} u_\xi &= -\frac{1}{4\pi\mu} \left(\frac{1}{\beta^2 - 1} \frac{\partial \chi}{\partial \xi} + \zeta \frac{\partial V}{\partial \xi} \right) \\ u_\eta &= -\frac{1}{4\pi\mu} \left(\frac{1}{\beta^2 - 1} \frac{\partial \chi}{\partial \eta} + \zeta \frac{\partial V}{\partial \eta} \right) \\ u_\zeta &= \frac{1}{4\pi\mu} \left(\frac{\beta^2}{\beta^2 - 1} V - \zeta \frac{\partial V}{\partial \zeta} \right) \end{aligned}$$
(5)

The stress component which satisfies the requirements of Step 3 is

$$\sigma_{\eta\eta} = \frac{1}{2\pi} \left(\frac{\beta^2 - 2}{\beta^2 - 1} \frac{\partial V}{\partial \zeta} - \frac{1}{\beta^2 - 1} \frac{\partial^2 \chi}{\partial \eta^2} - \zeta \frac{\partial^2 V}{\partial \eta^2} \right) \tag{6}$$

The solution given by Love [9] which satisfies the boundary conditions of equations (3) is

$$\frac{\partial^2 \chi}{\partial \eta^2} = -p \left[\tan^{-1} \frac{a - \xi}{b - \eta} + \tan^{-1} \frac{a + \xi}{b - \eta} - \tan^{-1} \frac{\zeta(a - \xi)}{(b - \eta)a_1} \right.$$
$$- \tan^{-1} \zeta \frac{(a + \xi)}{(b - \eta)b_2} + \tan^{-1} \frac{a - \xi}{b + \eta} + \tan^{-1} \frac{a + \xi}{b + \eta}$$
$$\left. - \tan^{-1} \zeta \frac{(a - \xi)}{(b + \eta)d_4} - \tan^{-1} \frac{\zeta(a + \xi)}{(b + \eta)c_3} \right] \tag{7}$$

where a and b are the dimensions of the rectangle shown in Fig. 2;

$$\frac{\partial V}{\partial \zeta} = p \left[2\pi - \tan^{-1} \frac{\zeta a_1}{(a - \xi)(b - \eta)} \right.$$
$$- \tan^{-1} \frac{\zeta d_4}{(a - \xi)(b + \eta)} - \tan^{-1} \frac{\zeta b_2}{(a + \xi)(b - \eta)}$$
$$\left. - \tan^{-1} \frac{\zeta c_3}{(a + \xi)(b + \eta)} \right] \tag{8}$$

where the arc tangents have values between $-\pi/2$ and $\pi/2$:

$$\frac{\partial^2 V}{\partial \eta^2} = p \left[\frac{b - \eta}{(b - \eta)^2 + \zeta^2} \left(\frac{a - \xi}{a_1} + \frac{a + \xi}{b_2} \right) \right.$$
$$\left. + \frac{b + \eta}{(b + \eta)^2 + \zeta^2} \left(\frac{a - \xi}{d_4} + \frac{a + \xi}{c_3} \right) \right] \tag{9}$$

The quantities a_1, b_2, c_3, and d_4 are given by

$$\begin{aligned} a_1 &= [(\xi - a)^2 + (\eta - b)^2 + \zeta^2]^{1/2} \\ b_2 &= [(\xi + a)^2 + (\eta - b)^2 + \zeta^2]^{1/2} \\ c_3 &= [(\xi + a)^2 + (\eta + b)^2 + \zeta^2]^{1/2} \\ d_4 &= [(\xi - a)^2 + (\eta + b)^2 + \zeta^2]^{1/2} \end{aligned}$$
(10)

where the positive root is always taken.

The solution given by equations (6–10) is applicable only for a constant load applied on a rectangular area. However, the solution may be used to approximately satisfy the boundary condi-

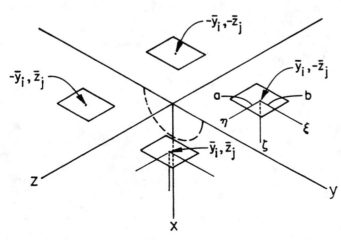

Fig. 2 Symmetrical distribution of rectangular areas used in conjunction with Love's solution

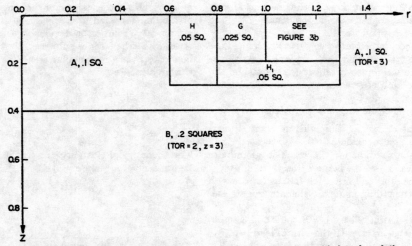

Fig. 3(a) Grid spacing on bounding plane used in conjunction with Love's solution

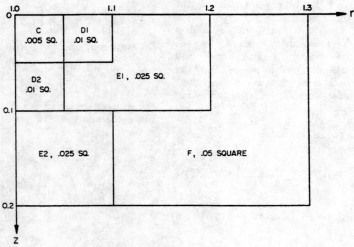

Fig. 3(b)

tions of equations (2) by use of the superposition theorem. Let us first divide the plane $x = 0$ ($\theta = \pm \pi/2$) into many small rectangular sections, denoting the location of each rectangle by the coordinates $\bar{y}_i$, $\bar{z}_j$ of its center. From the boundary condition represented by equation (2), an average stress may be calculated for each rectangle. Due to symmetry of $\sigma_{\theta\theta}$ on the plane $x = 0$, there will be four rectangles, one in each quadrant, which have the same average pressure.

By characterizing equation (6) as

$$\sigma_{\eta\eta} = pF(\xi, \eta, \zeta) \quad (11)$$

the stress σ_{xx} at any point x, y, z due to the pressure $p(\bar{y}_i, \bar{z}_j)$ on the four rectangles can be represented as

$$\sigma_{xx} = p(\bar{y}_i, \bar{z}_j)[F(y - \bar{y}_i, z - \bar{z}_j, x) + F(y + \bar{y}_i, z - \bar{z}_j, x)$$
$$+ F(y - \bar{y}_i, z + \bar{z}_j, x) + F(y + \bar{y}_i, z + \bar{z}_j, x)] \quad (12)$$

Inspection of equations (7–10) reveals the following symmetry properties of F:

$$F(\xi, \eta, \zeta) = F(-\xi, \eta, \zeta) = F(\xi, -\eta, \zeta) = F(-\xi, -\eta, \zeta) \quad (13)$$

The total stress at any point x, y on the crack surface is then

$$\sigma_{xx} = \sum_{i=1}^{M} \sum_{j=1}^{M} 2p(\bar{y}_i, \bar{z}_j)[F(y - \bar{y}_i, \bar{z}_j, x)$$
$$+ F(y + \bar{y}_i, \bar{z}_j, x)] \quad (14)$$

summing over all the rectangles in the first quadrant.

The rectangles for which the average values of normal stress $p(\bar{y}_i, \bar{z}_j)$ are calculated are laid out according to Fig. 3. The outer boundaries of the pressed area are taken to be the lines $r = 2$ and $z = 3$, since the embedded crack solution shows that the stresses beyond these lines are very small. Values of $\sigma_{\theta\theta}$ are calculated at each node point, and the average value of the stress for each rectangle is calculated as the arithmetic average of the values at the corners. The average value for the rectangle near the crack tip is found by expressing the normal stress near the crack tip as

$$\sigma = \frac{C}{\sqrt{r}} \cos \frac{1}{2} \psi$$

which is a result from Sneddon's solution for the crack in simple tension. This expression is then analytically averaged over a square of dimension 0.005. The constant C is found by evaluating $\sigma_{\theta\theta}$ at $r = 1.005, z = 0$ and setting it equal to σ.

To evaluate the stress component σ_{xx}, it is only necessary to calculate the appropriate functions

$$2\{F(y + \bar{y}_i, \bar{z}_j, x) + F(y - \bar{y}_i, \bar{z}_j, x)\}$$

multiply by the appropriate average pressures $p(\bar{y}_i, \bar{z}_j)$, and sum over all the rectangles in Fig. 3.

Thermal Stress Formulation

Since three of the examples involve the state of stress at a crack tip caused by thermal gradients, the necessary thermal stress formulation is included here. The necessary field equations are

$$\sigma_{ij,j} = 0 \qquad (15)$$

$$k\nabla^2 T = \rho c \dot{T} \qquad (16)$$

where k is the thermal conductivity, ρ the density, and c the specific heat in the absence of body forces.

In equation (15), an inertia term $\rho \ddot{u}_i$ and, in equation (16), a coupling term $mT_0 \dot{e}_{kk}$ have been assumed to be negligible. Boley and Weiner [10] show that, if the inertia term is negligible, then the coupling term is also negligible. By a solution including inertia, Boley and Weiner demonstrate, for the case of a half space subjected to the convection boundary condition at the surface, that the inertia term is very small. For a film conductance of 10^5 Btu/hr-ft^2-deg F, which may be taken to be the upper bound of practical values, the error is of the order of 0.001 percent for ordinary materials. Some problems treated in this work will have the surface temperature prescribed which corresponds to the case of infinite film conductance in which inertia cannot be neglected. In these cases, the temperature will be interpreted as the limit of the temperature as the film conductance becomes large, but finite, so that coupling and inertia may be neglected.

Equations (15) and (16) are subject to the assumption that the change in temperature from the reference state be very small. Experimental results show that the temperature change can be fairly large, but the linear theory will still give good results. A paper by Emery, et al. [11], shows good agreement between experiment and theory for temperature differences of 140 F.

From Goodier's solution to the half space [12] with one-dimensional temperature variation, the stresses are found to be

$$\sigma_{xx} = \sigma_{xy} = \sigma_{xz} = \sigma_{yz} = 0 \qquad (17)$$

$$\sigma_{yy} = \sigma_{zz} = -2\bar{\beta} T(x, t) \qquad (18)$$

where $\bar{\beta} = \alpha(3\lambda + 2\mu)/(\lambda + 2\mu)$, and α is the linear coefficient of thermal expansion.

The temperature solution for one dimensional heat flow in a semi-infinite body with temperature prescribed at the surface is

$$T = \frac{2}{\sqrt{\pi}} \int_{\frac{x}{2\sqrt{\kappa t}}}^{\infty} \chi\left(t - \frac{x^2}{4\kappa u^2}\right) e^{-u^2} du \qquad (19)$$

where $\chi(t)$ is the prescribed surface temperature. Substituting equation (19) into equation (18), the desired stress component is found to be

$$\sigma_{zz} = -\frac{4\bar{\beta}}{\sqrt{\pi}} \int_{\frac{x}{2\sqrt{\kappa t}}}^{\infty} \chi\left(t - \frac{x^2}{4\kappa u^2}\right) e^{-u^2} du \qquad (20)$$

If the surface temperature is

$$\chi(\tau) = -T_w \text{ (const)} \qquad \text{for } \tau \geq 0$$

the stress component σ_{zz} is given by

$$\sigma_{zz} = 2\bar{\beta} T_w \text{ erfc } \frac{x}{2\sqrt{\kappa t}} \qquad (21)$$

This is the case of a half-body being cooled due to a step change in surface temperature suddenly applied at $t = 0$.

Consider now a plate, free at the edges and subjected to a temperature which varies only through the thickness. Boley and Weiner [10] give the solution for this case as

$$\sigma_{xx} = \sigma_{xy} = \sigma_{xz} = \sigma_{yz} = 0 \qquad (22)$$

$$\sigma_{yy} = \sigma_{zz} = \frac{E/\mu}{1-\nu}\left[-T + \frac{1}{2h}\int_{-h}^{h} T dx \right. \qquad (22)$$
$$\left. + \frac{3x}{2h^3}\int_{-h}^{h} Tx dx\right] \qquad (Cont.)$$

where h is one half the plate thickness, and the distance x is measured from the center of the plate. This solution was obtained by superimposing on the solution of equation (18) a normal force and a couple so that the moment and average normal stress are zero at the edges.

If the values of time are suitably restricted, the temperature in a plate with the surface temperature of T_w is

$$T = T_w \text{ erfc } \frac{x/a}{2\sqrt{\kappa t}/a} \qquad (23)$$

where a is the crack radius. This solution will be valid as long as the temperature is not appreciably felt at the back side of the plate. If we make the restrictions $\sqrt{\kappa t}/a \leq h/2a$, the temperature change at the back surface will never be larger than 0.5 percent of T_w. Substituting equation (23) into equation (22), the stress is found to be

$$\sigma_{zz} = 2\bar{\beta} T_w \left\{ -\text{erfc } \frac{x}{2\eta} \right.$$
$$+ \frac{4h - 3x}{h^2}\left[h \text{ erfc } \frac{h}{\eta} + \frac{\eta}{\sqrt{\pi}}(1 - e^{-h^2/\eta^2})\right]$$
$$\left. + \frac{3(x-h)}{2h^3}\left[2h^2 \text{ erfc } \frac{h}{\eta} - \frac{2\eta h}{\sqrt{\pi}} e^{-h^2/\eta^2} + \eta^2 \text{ erf } \frac{h}{\eta}\right]\right\} \qquad (24)$$

where $\eta = \sqrt{\kappa t}/a$, and h and x are nondimensional with the characteristic length of the crack radius a.

Examples

1 Crack in Surface of a Half Space, Simple Tension. The problem of a semi-circular edge crack in simple tension is treated by superimposing two solutions. The first is that of a half space in simple tension with no crack. The stresses are

$$\sigma_{xx} = \sigma_{yy} = \sigma_{xy} = \sigma_{xz} = \sigma_{yz} = 0$$
$$\sigma_{zz} = p_0 \qquad (25)$$

The second solution is for the edge crack with a prescribed surface loading of $\sigma_{zz} = -p_0$, which, when superimposed on the first solution, will give the solution for the stress-free crack subjected to pure tension at infinity. The second solution is carried out by the foregoing iteration procedure.

Table 1 of reference [8] gives the results of each iteration for an initial load of unity applied to the crack surface. The value of the stress intensity factor for the complete solution is then the sum of the results for each iteration. Five iterations were carried out, and the values of K_I were plotted against the iteration number. A power law was determined as an upper limit for the value of K_I after each iteration. Using this power law and summing to infinity, an upper bound on the error due to stopping after five iterations was found to be 0.4 percent.

It should be noted that some error occurs due to the approximate nature of the half-space solution, which is manifested by small excursions in the stresses. However, the stress intensity factor is determined by integrals of the stress over the whole crack surface and since, at fairly small distances away from the surface $\theta = \pm\pi/2$, the effects of the grid spacings become negligible, small local errors will have little effect.

Fig. 4 presents the stress intensity factor for the semicircular edge crack in simple tension. We note that the magnification due to the presence of the free surface ranges from about 3 percent at $\theta = 0$ to about 20 percent at the free surface. Thus Irwin's [1] estimate of 10 percent is a good average value.

We note also that, since the stress intensity factor is larger near the surface, the crack would tend to grow from a length to depth ratio of 2.0 into an elliptical shape if subjected to repeated loading.

An estimate of an equilibrium shape ratio can be made with the help of an assumption. The stress intensity factor for an elliptical flat crack in an infinite solid is [1]

$$K_I = K_c \left(\frac{a}{c}\right)^{1/2} \frac{\left[\sin^2 \theta + \left(\frac{c}{a}\right)^2 \cos^2 \theta\right]^{1/4}}{\Phi}$$

where (26)

$$\Phi = \frac{2}{\pi} \int_0^{\pi/2} \left[1 - \left(1 - \frac{a^2}{c^2}\right) \sin^2 \theta\right]^{1/2} d\theta$$

K_c is the stress intensity factor for a penny-shaped crack, and θ is measured from the minor axis. Let us assume that the stress intensity factor for a semielliptical edge crack may be found by replacing K_c with the stress intensity factor found for the semicircular edge crack. The cases for which the stress intensity factors are most constant as θ varies have length to depth ratios in the range 2.6–3.2. This indicates that an equilibrium shape should lie somewhere in this range. Some low-cycle fatigue data exist for pressurized cylinders, which appears to support this estimate. The fatigue cracks in this case were found to have length to depth ratios at failure in the range 3–3.5 [13]. Reference [14] presents a case in which an initial surface flaw was machined with a shape ratio of about 5.5. At an intermediate stage, the shape ratio was about 3.6 and, at fracture, the shape ratio was 2.6.

2 Crack in Top Surface of a Beam in Pure Bending. In this and the other problems involving finite rather than infinite bodies, the crack is taken to be small compared to the dimensions of the body. This is to insure that the free-surface boundary conditions will be met at the surfaces other than the one in which the crack is located. The semicircular edge crack in the top of a beam is shown in Fig. 5. The stress in a crack free beam is

$$\sigma_{zz} = \frac{M_0 c}{I \mu} - \frac{M_0 a}{I \mu} r \cos \theta = \frac{M_0 c}{I \mu} - Qr \cos \theta \quad (27)$$

To introduce a stress-free crack, the foregoing loading must be removed from the crack surface. The first term is a constant, which is the case solved in Example 1. The results of the iterations for the second term are presented in Table 2 of reference [8]. Compared to the simple tension case, the iteration procedure converged very rapidly. The second value of K_I at $\theta = 0$ was about 1 percent of the first value at the same point, while the third value was less than 0.05 percent of the total. The stress in the last iteration was still on the order of 1 percent of the maximum applied.

The rapid convergence here can be attributed to the fact that the stresses applied are high deep in the crack and low near the plane $\theta = \pi/2$. This results in relatively low stresses on the plane $\theta = \pi/2$ to be removed, which, in turn, causes relatively low residual stress on the crack surface.

Fig. 6 presents values of the stress intensity factor calculated from the formula

$$K_I = \frac{2\sqrt{a}}{\sqrt{\pi}} \frac{M_0 c}{I} \left[\psi_0(\theta) - \left(\frac{a}{c}\right) \psi_1(\theta)\right] \quad (28)$$

where $(2\sqrt{a}/\sqrt{\pi})\psi_0$ and $(2\sqrt{a}/\sqrt{\pi})\psi_1$ are the stress intensity factors for an edge crack with applied loads of unity and $r \cos \theta$, respectively. We note from equation (28) that, for very small cracks, the error due to assuming a constant load on the crack surface is very small.

3 Crack in Surface of a Half Space, Prescribed Temperature. If the temperature prescribed at the cracked surface is $-T_w$, the stress which must be freed from the crack surface is

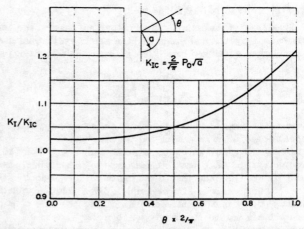

Fig. 4 Stress intensity factor for a semicircular edge crack in simple tension

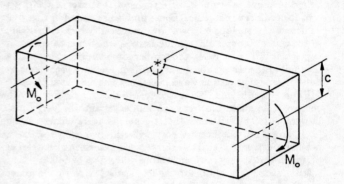

Fig. 5 Edge crack in a beam under pure bending

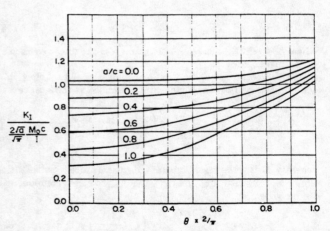

Fig. 6 Stress intensity factor for a semicircular edge crack in top surface of a beam in pure bending

$$\sigma_{zz} = 2\bar{\beta} T_w \, \text{erfc} \, \frac{x}{2\sqrt{\kappa t}} \quad (29)$$

The function, erfc, has the value 1 at $x = 0$ and decreases rapidly as x increases. Because of the improved convergence attained with a function that is small at $x = 0$ and increases with x, the solution for this case was found by applying the function

$$\text{erf} \, \frac{x}{2\sqrt{\kappa t}} = 1 - \text{erfc} \, \frac{x}{2\sqrt{\kappa t}} \quad (30)$$

to the surface of the crack. The function, erf, has this desired property. We note that, when $t = 0$, equation (30) corresponds to the constant-load case which has already been solved. Table 3 of reference [8] presents the results of the iterations for various

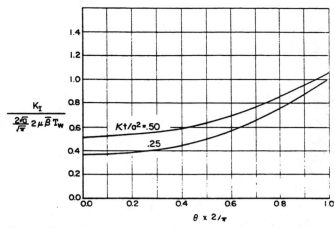

Fig. 7 Stress intensity factor for a semicircular edge crack in low-temperature surface of a half-body

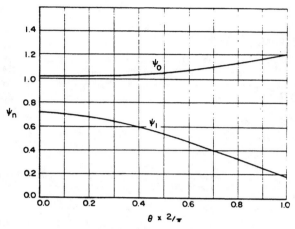

Fig. 8 Functions ψ_0 and ψ_1

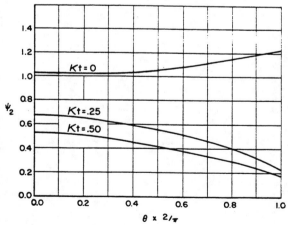

Fig. 9 Function $\psi_2(\theta, \kappa t)$

values of time. Only one iteration was necessary for the larger values of time. Fig. 7 presents values of the stress intensity factor calculated from the formula

$$K_I = 2\mu\bar{\beta}T_w \frac{2\sqrt{a}}{\sqrt{\pi}} [\psi_0(\theta) - \psi_2(\theta, \kappa t)] \quad (31)$$

where $2\sqrt{a}\,\psi_2/\sqrt{\pi}$ is the stress intensity factor for the loading function of equation (30).

Figs. 8 and 9 present curves for the functions ψ_0, ψ_1, and ψ_2. The stress intensity factor for any loading on the crack surface of the form

$$p(r, \theta) = p_0 + p_1 r \cos\theta + p_2 \operatorname{erf}(r\cos\theta/2\sqrt{\kappa t}) \quad (32)$$

may be calculated from these curves using the expression

$$K_I = \frac{2\sqrt{a}\,\mu}{\sqrt{\pi}}(p_0\psi_0 + p_1\psi_1 + p_2\psi_2) \quad (33)$$

The function ψ_2, for values of $\kappa t/a^2$ other than those shown in Fig. 9, may be evaluated approximately by cross-plotting.

4 Crack in Front Surface of a Free Plate, Low Temperature on Front Surface. The stress intensity factor for this case is found as before by superposition to be

$$K_I = \frac{2\sqrt{a}}{\sqrt{\pi}} 2\mu\bar{\beta}T_w \Bigg\{ \psi_0 - \psi_2$$
$$+ \psi_0 \left[-\operatorname{erfc}\frac{h}{\eta} - \frac{\eta}{h\sqrt{\pi}}(4 - e^{-h^2/\eta^2}) \right.$$
$$\left. + \frac{3\eta^2}{2h^2}\operatorname{erf}\frac{h}{\eta} \right] + \psi_1 \left[\frac{3\eta}{h^2\sqrt{\pi}} - \frac{3\eta^2}{2h^3}\operatorname{erf}\frac{h}{\eta}\right] \Bigg\} \quad (34)$$

Fig. 10 presents the results for this case. Approximate results may be obtained for values of the time parameter other than 0.25 and 0.50 by cross-plotting Fig. 9 and evaluating the foregoing expression. When this is done, it is found that K_I at $\theta = \pi/2$ is maximum at about $\kappa t/a^2 = 0.25$ for h/a in the range 2 to 4, while the maximum occurs at about $\kappa t/a^2 = 0.50$ for $h/a = 10$.

5 Crack in Back Surface of a Free Plate, High Temperature on Front Surface. For this case, the stress to be freed from the crack surface is given by equation (24) with the exception of the first term, which is negligible compared to the other terms in the region of the crack. The stress intensity factor is found by superposition to be

$$K_I = \frac{2\sqrt{a}}{\sqrt{\pi}} 2\mu\bar{\beta}T_w \Bigg\{ \psi_0 \left(\operatorname{erfc}\frac{h}{\eta} - \frac{\eta}{h\sqrt{\pi}}(2 + e^{-h^2/\eta^2}) \right.$$
$$\left. + \frac{3\eta^2}{2h^2}\operatorname{erf}\frac{h}{\eta}\right) + \psi_1\left(\frac{3\eta}{h^2\sqrt{\pi}} - \frac{3\eta^2}{2h^3}\operatorname{erf}\frac{h}{\eta}\right)\Bigg\} \quad (35)$$

Numerical evaluation of this function indicates that, for the range of $\sqrt{\kappa t}/a$ for which the solution is valid, K_I is negative. Physically, this means that cracking will not occur due to heating the plate on the opposite side from the crack in the time ranges restricted such that $\sqrt{\kappa t}/a \leq h/2a$. We note that this is contrary to the results found by Emery [11] for a crack at the outer radius of a hollow cylinder heated on the inner surface. In this case, the closed nature of the region produces tensile stresses at the outer surface.

Comparison With Two-Dimensional Results

This section compares the results for circular cracks with the approximately analogous results from the two-dimensional analysis of cracks. The stress intensity factor for a semicircular crack will be comparable with the two-dimensional case only over the limited portion of the crack border near $\theta = 0$. All the comparisons made in the following sections are valid only over this portion.

Cracks in Simple Tension. The comparison between the stress intensity factors for two-dimensional edge crack and three-dimensional semicircular crack under a constant stress P_0 is

Two-dimensional edge crack

(1.1) $\quad P_0\sqrt{a\pi}$

At $\theta = 0$ deg of a semicircular crack

(1.03) $\quad \dfrac{2}{\pi} P_0\sqrt{a\pi}$

The quantities in parentheses are the magnification factors at the root of the crack due to introducing a free surface through the

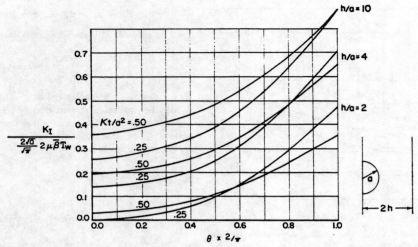

Fig. 10 Stress intensity factor for a semicircular edge crack in low-temperature surface of a free plate

crack. We note that the magnification in the two-dimensional case is higher than in the three-dimensional case.

Cracks Subjected to Bending. The comparison between the two-dimensional and three-dimensional edge cracks due to bending stresses $Qx = Qr \cos \theta$ is

Two-dimensional edge crack

$$(1.36) \quad \frac{1}{2} Q \sqrt{a\pi}$$

At $\theta = 0$ deg of semicircular crack

$$(1.08) \quad \frac{4}{3\pi} Q \sqrt{a\pi}$$

Again, we note that the magnification is greater in the two-dimensional case. The magnifications for the two dimensional cases were derived in reference [7].

Summary and Conclusions

Several special cases of a semicircular edge crack located at a free surface are discussed. Values of the stress intensity factor are presented for cases including both thermal and mechanical loading of a crack either in a semi-infinite solid or in a free plate. Several general conclusions were drawn from this work:

1 The effect of introducing a free surface through a circular crack is to magnify the stress intensity factor. In the case of simple tension loading, the stress intensity factor is magnified by about 20 percent at the free surface.

2 Cracking due to thermal loading will not occur for an edge crack in a free plate heated on the opposite side from the crack for time in the range $0 \leq \kappa t \leq h^2/4$, where κ is thermal diffusivity, and $2h$ is the plate thickness.

3 The replacement of a constant average load over the surface of a circular crack in a case where the load varies rapidly can cause a significant underestimate of the maximum value of the stress intensity factor.

Acknowledgment

The authors wish to thank the Research Computer Laboratory of the University of Washington for providing free computer time for this project. F. W. Smith held a NASA Predoctoral Traineeship during the period that this investigation was conducted.

References

1 Irwin, G. R., "Crack-Extension Force for a Part-Through Crack in a Plate," JOURNAL OF APPLIED MECHANICS, Vol. 29, No. 4, TRANS. ASME, Vol. 84, Series E, Dec. 1962, pp. 651–654.

2 Green, A. E., and Sneddon, I. N., "The Distribution of Stress in the Neighborhood of a Flat Elliptical Crack in a Elastic Solid," *Proceedings of the Cambridge Philosophical Society*, Vol. 46, 1950, pp. 159–163.

3 Wigglesworth, L. A., "Stress Distribution in a Notched Plate," *Mathematika*, Vol. 4, 1957, pp. 76–96.

4 Bueckner, H. F., "Some Stress Singularities and Their Computations by Means of Integral Equations," *Boundary Problems in Differential Equations*, Langer, R. E., ed., Madison, Wis., 1960.

5 Bowie, O. L., "Rectangular Tensile Sheet With Symmetric Edge Cracks," JOURNAL OF APPLIED MECHANICS, Vol. 31, No. 2. TRANS. ASME, Vol. 86, Series E, June 1964, p. 208–212.

6 Koiter, W. T., discussion on "Rectangular Tensile Sheet With Symmetric Edge Cracks," reference [5], JOURNAL OF APPLIED MECHANICS, Vol. 32, No. 1, TRANS. ASME, Vol. 87, Series E, Mar. 1965, p. 237.

7 Lachenbruch, A. H., "Depth and Spacing of Tension Cracks," *Journal of Geophysical Research*, Vol. 66, No. 12, 1961, pp. 4273–4292.

8 Smith, F. W., "Stresses Near a Semi-Circular Edge Crack," PhD thesis, University of Washington, Mar. 1966.

9 Love, A. E. H., "On Stress Produced in a Semi-Infinite Solid by Pressure on Part of the Boundary," *Philosophical Transactions of the Royal Society*, London, Series A, Vol. 228, 1929, pp. 378–395.

10 Boley, B. A., and Weiner, J. H., *Theory of Thermal Stresses* Wiley, New York, 1962.

11 Emery, A. F., Barrett, C. F., and Kobayashi, A. S., "Temperatures and Thermal Stresses in a Partially Filled Annulus," *Journal of the Society for Experimental Stress Analysis*, Vol. 6, 1966, p. 606.

12 Goodier, J. N., "On the Integration of the Thermo-Elastic Equations," *Philosophical Magazine*, Vol. 23, 1937, pp. 1017–1032.

13 Private communication with S. Yukawa, Large Steam Turbine-Generator Department, General Electric Company.

14 Tiffany, C. F., and Masters, J. N., "Applied Fracture Mechanics," presented at the ASTM Symposium on Fracture Toughness Testing and Its Applications, June 1964.

Stress Intensity Factors for an Elliptical Crack Approaching the Surface of a Semi-Infinite Solid

R. C. SHAH [1]) and A. S. KOBAYASHI [2])

(Received Sept. 11, 1970; March 6, 1972)

ABSTRACT

Stress intensity factors for an embedded elliptical crack approaching the free surface of the semi-infinite solid that is subjected to uniform tension perpendicular to the plane of crack are presented in a nondimensional form for various crack aspect ratios and crack distances from the free surface. Stress intensity factors are determined numerically using an alternating technique with two solutions. The first solution involves an elliptical crack in a solid and subjected to normal loading expressible in a polynomial of x and y. The second solution involves stresses in the half space due to prescribed normal and shear stresses on the surface. Effect of the Poisson's ratio on these stress intensity factors is also investigated. Stress intensity factors for a semi-elliptical surface crack in a finite thickness plate are then estimated in a nondimensional form for various crack aspect ratios and crack depth to plate thickness ratios.

Nomenclature

ϕ	suitable harmonic stress function $= \sum_{i=0}^{3} \sum_{j=0}^{3} \Phi_{ij}$; $i+j \leq 3$
λ, μ, ν	ellipsoidal coordinates
a	semi-major axis of ellipse
b	semi-minor axis of ellipse
h	distance from the free surface to the center of ellipse or plate thickness
$\omega(s)$	$1 - \dfrac{x^2}{a^2+s} - \dfrac{y^2}{b^2+s} - \dfrac{z^2}{s}$
$Q(s)$	$s(a^2+s)(b^2+s)$
η	Poisson's ratio
G	shear modulus
u_1	incomplete elliptic integral of the first kind
$K(k)$	complete elliptic integral of the first kind
$E(u_1)$	incomplete elliptic integral of the second kind
$E(k)$	complete elliptic integral of the second kind
$snu_1, cnu_1, dnu_1, dcu_1, cdu_1, ndu_1, ncu_1, sdu_1$	= Jacobian elliptic functions
k, k'	modulus and complimentary modulus of Jacobian elliptic functions, respectively with $k^2 = (1-b^2/a^2)$
θ	angle in the parametric equations of ellipse, $x = a\cos\theta$; $y = b\sin\theta$
β	$90°-\theta$ i.e., $x = a\sin\beta$, $y = b\cos\beta$
K_I	opening mode stress intensity factor
K_{Ie}	stress intensity factor for an elliptical crack embedded in an elastic solid

[1]) Specialist Engineer, Aerospace Group, The Boeing Company, Seattle, Washington.
[2]) Professor, Department of Mechanical Engineering, University of Washington, Seattle, Washington, and also Aerospace Group, The Boeing Company, Seattle.

1. Introduction

The most commonly encountered crack configurations in service failures are surface cracks and embedded cracks which can be approximated by semi-elliptical and elliptical cracks, respectively [1][1]. Due to the complexity of solving a three dimensional elasticity problem, however, only limited analytical work on the elliptical crack problem has been published to date. Utilizing the solution of Green and Sneddon [2] for an elliptical crack embedded in an infinite solid under uniaxial tension, σ, at infinity, Irwin [3] derived the stress intensity factor as

$$K_{Ie} = \frac{\sigma(\pi b)^{\frac{1}{2}}}{E(k)} \left[\frac{b^2}{a^2}\sin^2\beta + \cos^2\beta\right]^{\frac{1}{4}} \tag{1}$$

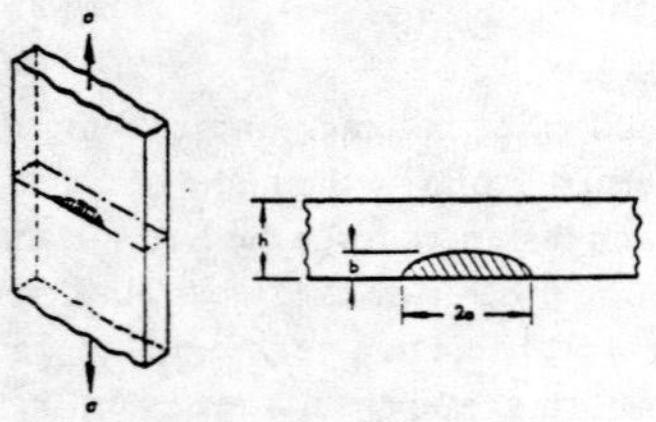

Figure 1. Surface flaw in a finite-thickness plate.

Irwin then estimated that the stress intensity factor for a semi-elliptical surface crack in an elastic plate (shown in Figure 1) as

$$K_I = 1.1\, K_{Ie} \tag{2}$$

where he introduced the constant 1.1 to account for the effects on the stress intensity factor of the two stress-free surfaces. As originally estimated by Irwin [3], experimental data on surface cracks correlate reasonably well when the crack depths are less than half the plate thickness [4, 5]. When the crack becomes deep with respect to the thickness of the plate, experimental results have shown that Irwin's correction of 1.1 for effects of free surfaces tends to underestimate actual stress intensity factor by as much as 90 percent [6]. To obtain better correlations with the experimental results, some efforts in estimating or determining numerically the stress intensity factors for a deep surface flaw have been made by various investigators. Following Irwin [3], the common approach underlying these attempts is to estimate a correction factor to the known solution of the stress intensity factor for an embedded elliptical flaw in a solid. In addition, the deep surface flaw correction factor is considered to be a product of front surface correction and back surface correction and each of these two correction factors is determined separately [7]. Although doubts exist about the independency of these two correction factors, analytical complexity of the original problem forces the use of this assumption at the present time. Work done on the circular crack partially embedded in a plate suggests that the error in assuming independency of these factors is very small (less than 5 percent) [8]. Using this simplification, Kobayashi, and Moss [7] estimated the correction factor for the deep surface flaw. The front surface correction was based on available solutions [9, 10] and the back surface correction was estimated using the approximate solution of two coplanar elliptical flaws in an infinite solid subjected to uniform tension [11]. Smith and Alavi solved numerically the problems of a circular crack embedded in a halfspace [12] and a part circular surface flaw in a halfspace [13]. Thresher and Smith [14] extended the work of Smith and Alavi to solve the problem of a part-circular crack in a plate and obtained the back surface correction for the stress intensity

---

[1]) Numbers in brackets designate references at end of paper.

factor. Their results are based on part-circular flaws and are thus approximate in nature when applied to semi-elliptical surface flaws.

The principal objective of this paper is to present the stress intensity factors for an embedded elliptical crack approaching the free surface of the semi-infinite solid that is subjected to uniform tension perpendicular to the plane of crack at infinity. The theoretical background used in this analysis is partly reported by the authors in a previous publication [15] and the computational procedure is similar to that used by Smith, et al. [9, 12]. The results of the stress intensity factors of this analysis are used, along with the known approximate results of the front surface correction for a semi-elliptical surface flaw in a halfspace [7], to estimate the stress intensity factors for a semi-elliptical surface flaw in a finite thickness plate.

2. Statement of problem

The problem considered is the determination of the elastic state of stress and the stress intensity factor around the periphery of an elliptical crack in a semi-infinite solid shown in Figure 2. The flat elliptical crack is described by

$$\frac{x^2}{a^2} + \frac{y^2}{b^2} \leq 1 \text{ and } z = 0.$$

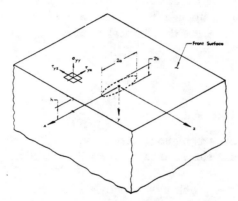

Figure. 2. Elliptical crack in a semi-infinite solid.

The free bounding plane is located at a distance h from the center of the elliptical crack and is perpendicular to the plane of the crack. The semi-infinite solid is subjected to a uniaxial tension, σ, perpendicular to the plane of the crack at infinity. The actual problem solved here is that of a uniformly pressurized crack in a semi-infinite solid. Simple superposition shows that the stress intensity factor for this problem is identical to that of the original problem. The boundary conditions thus are

$$\sigma_{zz} = -\sigma \qquad \left(\frac{x^2}{a^2} + \frac{y^2}{b^2} < 1, z = 0\right) \tag{3a}$$

$$w = 0 \qquad \left(\frac{x^2}{a^2} + \frac{y^2}{b^2} > 1, z = 0\right) \tag{3b}$$

$$\tau_{xz} = \tau_{yz} = 0 \qquad (z = 0) \tag{3c}$$

and

$$\sigma_{yy} = \tau_{yx} = \tau_{yz} = 0 \quad (y = -h) \tag{3d}$$

Also all the stresses must vanish at infinity. The problem can be separated into two parts which can be superposed to yield the solution. The first part is to determine the stresses at $y = -h$ due to an embedded elliptical crack in an infinite solid satisfying the boundary conditions 3a to 3c and the second part is to satisfy the boundary condition 3d, i.e., the bounding plane $y = -h$ be stress free. The two part solution is accomplished by an alternating iterative technique described below.

1. Using the solution of an elliptical crack subjected to constant pressure and embedded in an infinite solid [2, 15], the surface tractions σ_{yy}, τ_{yx}, and τ_{yz} are computed on $y = -h$ plane.

2. In order to free the residual surface tractions acting on $y = -h$ plane, opposing surface tractions are applied on the plane of an uncracked halfspace. Solution based on Love's stress function [16, 17] are used to remove these surface tractions from $y = -h$ plane and to calculate the resulting residual normal stress σ_{zz} on the crack surface.

3. The residual normal stresses σ_{zz} are removed by applying opposing stresses on the crack surface. These opposing stresses are approximated by least square fitting of a polynomial of x and y [15] and are used to compute the surface tractions on $y = -h$ plane and the stress intensity factor around the periphery of ellipse [15].

4. Steps 2 and 3 are repeated until the resulting residual stresses on the crack surface become negligible with comparison to the applied stress. The results of all the iterations are superposed to obtain the stress intensity factor for the elliptical crack in the halfspace.

The two basic theoretical solutions necessary to execute the above procedure are discussed in the following sections.

3. Elliptical crack in an infinite solid

Steps 1 and 3 of the iterative technique require a solution for the stress distribution in a solid containing an elliptical crack subjected to an internal pressure expressible in a polynomial of x and y.

Stress distributions for this problem were derived from the potential function proposed by Segedin [18] and were described in detail by the authors [15]. For the sake of completeness, part of the derivation will be repeated here briefly and formulas for the stresses not previously published will be added.

Consider an elastic solid containing an elliptical crack located in the plane $z = 0$ and with the boundary described as

$$\frac{x^2}{a^2} + \frac{y^2}{b^2} = 1 \tag{4}$$

The crack is opened by applying an internal pressure $p(x, y)$ symmetrically to both surfaces of the crack. The boundary conditions for this problem are as follows

$$\sigma_{zz} = -p(x, y) \quad \left(\frac{x^2}{a^2} + \frac{y^2}{b^2} < 1, \; z = 0\right) \tag{5a}$$

$$w = 0 \quad \left(\frac{x^2}{a^2} + \frac{y^2}{b^2} > 1, \; z = 0\right) \tag{5b}$$

$$\tau_{xz} = \tau_{yz} = 0 \quad (z = 0) \tag{5c}$$

Also, the stresses must vanish at infinity.

For the vanishing shear stresses on $z = 0$ plane and in the absence of body forces, Navier's equations of equilibrium are satisfied by the harmonic function ϕ. Only the relations between ϕ and displacement and stresses pertinent to this descussion are given here. Other relations can be found in Reference 15.

$$w = z \frac{\partial^2 \phi}{\partial z^2} - 2(1-\eta)\frac{\partial \phi}{\partial z} \tag{6a}$$

$$\sigma_{yy} = 2G\left[z\frac{\partial^3 \phi}{\partial y^2 \partial z} + \frac{\partial^2 \phi}{\partial y^2} + 2\eta \frac{\partial^2 \phi}{\partial x^2}\right] \tag{6b}$$

$$\sigma_{zz} = 2G\left[z\frac{\partial^3 \phi}{\partial z^3} - \frac{\partial^2 \phi}{\partial z^2}\right] \tag{6c}$$

$$\tau_{xy} = 2G\left[z\frac{\partial^3 \phi}{\partial x \partial y \partial z} + (1-2\eta)\frac{\partial^2 \phi}{\partial x \partial y}\right] \tag{6d}$$

$$\tau_{xz} = 2Gz\frac{\partial^3 \phi}{\partial x \partial z^2} \tag{6e}$$

$$\tau_{yz} = 2Gz\frac{\partial^3 \phi}{\partial y \partial z^2} \tag{6f}$$

Since it is convenient to solve this boundary value problem with natural coordinates, an ellipsoidal coordinate system λ, μ, and ν is introduced where λ, μ, and ν are the roots of the cubic equation

$$\omega(s) = 1 - \frac{x^2}{a^2+s} - \frac{y^2}{b^2+s} - \frac{z^2}{s} = 0 \tag{7a}$$

with the limits of variation

$$\infty > \lambda \geq 0 \geq \mu \geq -b^2 \geq \nu \geq -a^2 \tag{7b}$$

The cartesian coordinates (x, y, z) are then related to (λ, μ, ν) by

$$a^2(a^2-b^2)x^2 = (a^2+\lambda)(a^2+\mu)(a^2+\nu) \tag{8a}$$

$$b^2(b^2-a^2)y^2 = (b^2+\lambda)(b^2+\mu)(b^2+\nu) \tag{8b}$$

$$a^2 b^2 z^2 = \lambda\mu\nu \tag{8c}$$

In the plane $z = 0$, interior region of the elliptical crack is represented by $\lambda = 0$ and the exterior is given by $\mu = 0$.

From equations 5 and 6 the boundary conditions that still need to be satisfied by the harmonic function ϕ are

$$\frac{\partial^2 \phi}{\partial z^2} = \frac{p(x,y)}{2G} \quad (\lambda = 0) \tag{9a}$$

$$\frac{\partial \phi}{\partial z} = 0 \quad (\mu = 0) \tag{9b}$$

Since the problem of the elliptical crack in a halfspace has symmetry with respect to the y-axis, the pressure distribution, $p(x, y)$, on the elliptical crack in an infinite solid selected contains only the even order terms of x and the distribution $p(x, y)$ is given by

$$p(x, y) = A_{00} + A_{01}y + A_{20}x^2 + A_{02}y^2 + A_{21}x^2y + A_{03}y^3 \tag{10}$$

As described in Reference 15, the voluminous work involved in deriving the necessary equations limited the above polynomial to cubic terms in x and y. The resulting truncation error, as described later, is small due to the use of a least square method to fit the residual surface tractions.

The harmonic function which satisfies both of the boundary conditions of equation 9 is given by [15]

$$\phi(x, y, z) = \Phi_{00} + \Phi_{01} + \Phi_{20} + \Phi_{02} + \Phi_{21} + \Phi_{03} \tag{11a}$$

where

$$\Phi_{00} = C_{00} \int_\lambda^\infty \frac{\omega(s)ds}{\sqrt{Q(s)}} \tag{11b}$$

$$\Phi_{01} = C_{01} y \int_\lambda^\infty \frac{\omega(s)ds}{(b^2+s)\sqrt{Q(s)}} \tag{11c}$$

$$\Phi_{20} = C_{20} \left[\int_\lambda^\infty \frac{\omega^2(s)ds}{(a^2+s)\sqrt{Q(s)}} - 4x^2 \int_\lambda^\infty \frac{\omega(s)ds}{(a^2+s)^2\sqrt{Q(s)}} \right] \tag{11d}$$

$$\Phi_{02} = C_{02} \left[\int_\lambda^\infty \frac{\omega^2(s)ds}{(b^2+s)\sqrt{Q(s)}} - 4y^2 \int_\lambda^\infty \frac{\omega(s)ds}{(b^2+s)^2\sqrt{Q(s)}} \right] \tag{11e}$$

$$\Phi_{21} = C_{21} y \left[\int_\lambda^\infty \frac{\omega^2(s)ds}{(a^2+s)(b^2+s)\sqrt{Q(s)}} - 4x^2 \int_\lambda^\infty \frac{\omega(s)ds}{(a^2+s)^2(b^2+s)\sqrt{Q(s)}} \right] \tag{11f}$$

$$\Phi_{03} = C_{03} y \left[3\int_\lambda^\infty \frac{\omega^2(s)ds}{(b^2+s)^2\sqrt{Q(s)}} - 4y^2 \int_\lambda^\infty \frac{\omega(s)ds}{(b^2+s)^3\sqrt{Q(s)}} \right] \tag{11g}$$

and

$$Q(s) = [s(a^2+s)(b^2+s)] \tag{11h}$$

The unknown constants C_{ij} are a linear function of A_{ij} of the pressure distribution as represented by the matrix equation 17 of Reference 15.

In Reference 15 the stress intensity factor was derived by using the procedure described by Irwin [19] and Kassir and Sih [20]. For the polynomial loading of Equations 10, the stress intensity factor is given by the following equation [15]

$$K_I = \frac{8G}{ab}\left(\frac{\pi}{ab}\right)^{\frac{1}{2}} (a^2 \sin^2\theta + b^2 \cos^2\theta)^{\frac{1}{4}} \left[C_{00} + \frac{C_{01}\sin\theta}{b} \right.$$
$$\left. - \frac{4C_{20}\cos^2\theta}{a^2} - \frac{4C_{02}\sin^2\theta}{b^2} - \frac{4C_{21}\cos^2\theta\sin\theta}{a^2 b} - \frac{4C_{03}\sin^3\theta}{b^3} \right] \tag{12}$$

The surface tractions σ_{yy}, τ_{yx} and τ_{yz} acting on $y = -h$ plane can now be calculated from equations 11 and 6. Partial derivatives needed for these expressions and other details are derived in Reference 21.

4. Surface tractions on the plane of a halfspace

Step 2 to the iterative technique requires a solution for the stresses in the halfspace when its bounding plane is subjected to arbitrary varying normal and shear stresses. The plane $y = -h$ is first divided into many small rectangles (as shown in Figure 3 for one quadrant of the plane) where the rectangle sizes are selected such that the variation of a given surface traction within

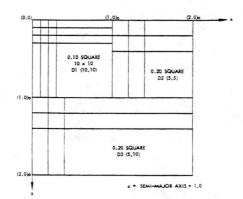

Figure 3. Grid spacing on bounding plane, $y = -h$.

each rectangle is less than 5 percent of its maximum value. The solution for a rectangle on the bounding plane of a halfspace and subjected to constant applied surface tractions is then used to compute the normal stress on the crack surface. The normal stresses are summed for all rectangles to complete step 2 of the iterative process. Shear stresses τ_{xz} and τ_{yz} on the $z = 0$ plane vanish due to symmetry of the problem when all effects from all rectangles are summed.

The solution for a constant normal stress, σ_{yy}, on a rectangular area at the $y = -h$ plane of the halfspace was given by Love [16] and used by Smith et al. [9]. The solutions for constant shear stresses τ_{yx} and τ_{yz} on a rectangular area at the $y = -h$ plane of the halfspace were derived independently by the authors and Smith and Alavi [12] from the workable details of Love [17]. The solutions derived by the authors differ somewhat in the form from those in Reference 12. Details of the derivations are omitted here and only the needed terms for σ_{zz} are given.

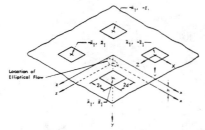

Figure 4. Symmetrical distribution of rectangular areas on the free surface.

Figure 4 shows the coordinate system used for each rectangle in the halfspace solution. Normal stress σ_{zz} due to an applied shear stress, q_X on $Y = 0$ plane described by the following boundary conditions

$$\tau_{YX} = q_X \qquad (|X| < c, Y = 0, |Z| < d) \tag{13a}$$

$$\sigma_{XX} = \sigma_{YY} = \sigma_{ZZ} = \tau_{XZ} = \tau_{YZ} = 0 \quad (Y = 0) \tag{13b}$$

is given by

$$\sigma_{zz} = \frac{1}{\pi}\left[\eta\frac{\partial^2 F}{\partial X\,\partial Y} + \eta\frac{\partial^3 F_1}{\partial X\,\partial Z^2} - \frac{Y}{2}\frac{\partial^3 F}{\partial X\,\partial Z^2}\right] \tag{13c}$$

where

$$\frac{\partial^2 F}{\partial X\,\partial Y} = q_X \log\left[\frac{(d-Z+a_1)(d+Z-c_3)}{(d+Z-d_4)(d-Z+b_2)}\right] \tag{13d}$$

$$\frac{\partial^3 F_1}{\partial X\,\partial Z^2} = -q_X\left[\frac{d-Z}{Y+a_1} + \frac{d-Z}{Y+b_2} - \frac{d+Z}{Y+c_3} + \frac{d+Z}{Y+d_4}\right] \tag{13e}$$

$$\frac{\partial^3 F}{\partial X\,\partial Z^2} = q_X\left[\frac{d-Z}{a_1(Y+a_1)} - \frac{d-Z}{b_2(Y+b_2)} - \frac{d+Z}{c_3(Y+c_3)} + \frac{d+Z}{d_4(Y+d_4)}\right] \tag{13f}$$

a_1, b_2, c_3 and d_4 are given by the following equations.

$$a_1 = [(X-c)^2 + Y^2 + (Z-d)^2]^{\frac{1}{2}} \tag{14a}$$

$$b_2 = [(X+c)^2 + Y^2 + (Z-d)^2]^{\frac{1}{2}} \tag{14b}$$

$$c_3 = [(X+c)^2 + Y^2 + (Z+d)^2]^{\frac{1}{2}} \tag{14c}$$

$$d_4 = [(X-c)^2 + Y^2 + (Z+d)^2]^{\frac{1}{2}} \tag{14d}$$

Normal stress σ_{zz} due to an applied shear stress q_Z on $Y=0$ plane described by the following boundary conditions

$$\tau_{YZ} = q_Z \qquad (|X| < c,\ Y = 0,\ |Z| < d) \tag{15a}$$

$$\sigma_{XX} = \sigma_{YY} = \sigma_{ZZ} = \tau_{YX} = \tau_{XZ} = 0 \quad (Y=0) \tag{15b}$$

is given by

$$\sigma_{zz} = \frac{1}{\pi}\left[\frac{\partial^2 G}{\partial Y\,\partial Z} - \eta\frac{\partial^3 G_1}{\partial X^2\,\partial Z} + \frac{Y}{2}\left(\frac{\partial^3 G}{\partial X^2\,\partial Z} + \frac{\partial^3 G}{\partial Y^2\,\partial Z}\right)\right] \tag{15c}$$

where

$$\frac{\partial^2 G}{\partial Y\,\partial Z} = q_Z \log\left[\frac{(c-X+a_1)(c+X-c_3)}{(c+X-b_2)(c-X+d_4)}\right] \tag{15d}$$

$$\frac{\partial^3 G_1}{\partial X^2\,\partial Z} = -q_Z\left[\frac{c-X}{Y+a_1} + \frac{c+X}{Y+b_2} - \frac{c+X}{Y+c_3} - \frac{c-X}{Y+d_4}\right] \tag{15e}$$

$$\frac{\partial^3 G}{\partial X^2\,\partial Z} = q_Z\left[\frac{c-X}{a_1(Y+a_1)} + \frac{c+X}{b_2(Y+b_2)} - \frac{c+X}{c_3(Y+c_3)} - \frac{c-X}{d_4(Y+d_4)}\right] \tag{15f}$$

$$\frac{\partial^3 G}{\partial Y^2\,\partial Z} = q_Z Y\left[\frac{1}{a_1(c-X+a_1)} + \frac{1}{b_2(c+X-b_2)} - \frac{1}{c_3(c+X-c_3)} - \frac{1}{d_4(c-X+d_4)}\right] \tag{15g}$$

5. Numerical procedure

In order to compute the stresses for the halfspace problem, the plane $y = -h$ is divided into 175 small squares as shown in Figure 3. From symmetry considerations of the problem, squares are taken only in one quadrant of the plane $y = -h$. The outer boundaries of these square grids are taken as lines $x = 2a$ and $z = 2a$ since the embedded crack in the infinite solid solution shows that the stresses σ_{yy}, τ_{yx} and τ_{yz} are very small beyond these lines. As stated before, the sizes of the squares are selected so that the variation of any given stress within the square is less than 5 percent. The stresses, σ_{yy}, τ_{yx} and τ_{yz} are calculated at the center of each square by programming the stress equations. Symmetry properties of the elliptical crack and halfspace solutions are used in evaluating σ_{zz} due to the application of the opposite of the above surface tractions [12]. The stress, σ_{zz}, is then calculated at 96 points on the crack, shown in Figure 5. The opposite of the stress σ_{zz} is least square fitted by a polynomial surface of x and y described by equation 10 and the stress intensity factor is calculated.

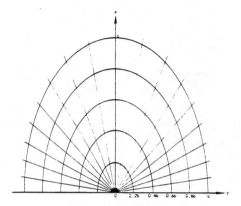

Figure 5. Location of points on crack surface.

The maximum deviation between the calculated stresses and the fitted stresses is less than 5 percent for all values of b/a and b/h ratios except for the following cases. For $b/a = 0.10$ and $b/h = 0.75$, 0.82 and 0.90, and $b/a = 0.20$ and $b/h = 0.90$, the maximum deviation between the calculated stress and the fitted stress is between 8 and 15 percent. Iterations are carried out until the maximum residual stresses on the crack surface are smaller than 0.10 percent of the applied stress. Typically the stress intensity factor for the last iteration is less than 0.20 percent of the stress intensity factor for the applied stress.

The expressions for stresses and stress intensity factors of the elliptical crack involving lengthy expressions of elliptical integrals and Jacobian elliptic functions along with the expressions for stresses of the halfspace problem were programmed for IBM system /360 model 67 computer. Each iteration takes approximately two minutes of CPU time.

To insure that the computer program had no errors, a variety of special cases was run. Some of these special case runs are described in Reference 15. The surface tractions at $y = -h$ plane were computed for a nearly penny-shaped crack ($b/a = 0.99$) subjected to constant and linearly varying pressures. These stresses compared favorably with those computed for these cases by Smith. Also the surface tractions at $y = -h$ plane were computed for various b/a ratios for each component of the pressure expression 10. As expected, stresses diminished to negligible value at a distance few crack lengths away.

6. Results and discussion

Using the aforementioned computer program, computations for the stress intensity factors are carried out for a number of elliptical crack geometries of $b/a = 0.10$ to 0.99 with surface proximities of $b/h = 0.4$ to 0.9. Figure 6a shows the normalized stress intensity factor along the periphery of a nearly circular crack of $b/a = 0.99$ for various b/h ratios and for Poisson's ratio of 0.30. The computed stress intensity factor is normalized with its corresponding stress intensity factor, K_{Ie}, for the elliptical crack in an infinite solid. The crack geometry of $b/a = 0.99$ is sufficiently close to a circular crack so that these results can be compared with those in Reference 12, obtained by a different formulation for Poisson's ratio of 0.25. In theory, all the equations used in this investigation can be reduced to those of a circular crack by letting $b/a = 1.0$. However, the derivations involved for this limiting case become so complex that a numerical approximation by an elliptical crack geometry of $b/a = 0.99$ is used. Results of the stress intensity factors of this investigation for $\eta = 0.25$ checked within two percent of those of Reference 12.

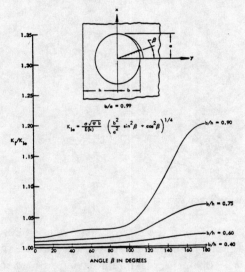

Figure 6a. Stress intensity factor for an elliptical crack approaching free surface ($b/a = 0.99$).

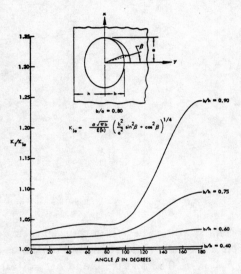

Figure 6b. Stress intensity factor for an elliptical crack approaching free surface ($b/a = 0.80$).

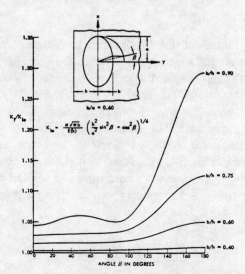

Figure 6c. Stress intensity factor for an elliptical crack approaching free surface ($b/a = 0.60$).

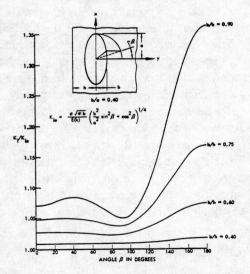

Figure 6d. Stress intensity factor for an elliptical crack approaching free surface ($b/a = 0.40$).

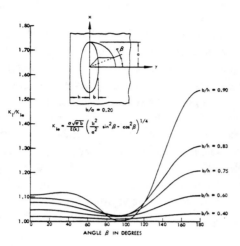
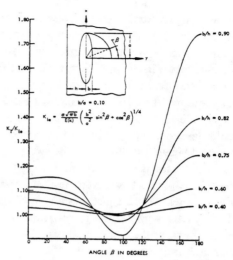

Figure 6e. Stress intensity factor for an elliptical crack approaching free surface ($b/a = 0.20$).

Figure 6f. Stress intensity factor for an elliptical crack approaching free surface ($b/a = 0.10$).

Figures 6b through 6f show the normalized stress intensity factors along the periphery of the elliptical cracks having the aspect ratios, b/a, of 0.80, 0.60, 0.40, 0.20 and 0.10, respectively, for various values of b/h ratios. All computations are carried out for Poisson's ratio of 0.30.

In addition to geometric extreme of $b/a = 1.0$ discussed previously, the stress intensity factors at both ends of an elliptical crack of $b/a = 0.10$ are compared with those for a two dimensional tunnel crack ($b/a = 0.0$) in a halfplane [22]. The stress intensity factors for $b/a = 0.10$ are lower than those for the tunnel crack and the maximum difference between them is less than four percent.

Figure 7a shows the stress intensity factors for $b/a = 0.99$ and 0.60 and $b/h = 0.90$ for Poisson's ratio of 0.4, 0.3 and 0.2. Figure 7b shows the stress intensity factors for $b/a = 0.20$ and $b/h = 0.83$ for Poisson's ratios of 0.4, 0.3 and 0.2. These figures show that the stress intensity

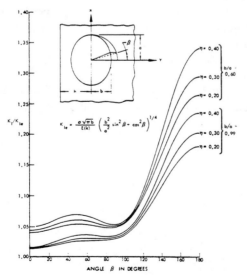
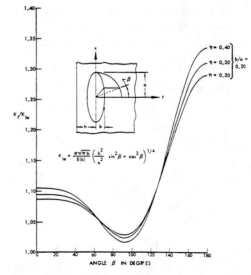

Figure 7a. Effect of Poisson's ratio on stress intensity factor ($b/h = 0.90$).

Figure 7b. Effect of Poisson's ratio on stress intensity factor ($b/h = 0.83$).

factor is higher for the higher Poisson's ratio. Also, as expected, the effect of the Poisson's ratio on the stress intensity factor is higher for the higher aspect ratios.

The dip in the normalized stress intensity factor in the vicinity of $\beta = 90°$ is due to the stress intensity factor itself which is minimum at the ends of the major diameter of the ellipse resulting in minimum free surface effects in this region. The polynomial function of x and y selected for least square fitting for the residual stresses on the crack surface did not fit well for cases of $b/a = 0.20$ and $b/h = 0.90$ and $b/a = 0.10$ and $b/h = 0.90, 0.82$ and 0.75. These fitting errors could have affected the value of the stress intensity factor at $\beta = 90°$ for these cases. From the practical viewpoint, however, the stress intensity factor at $\beta = 90°$ is not significant. The stress intensity factor is maximum at $\beta = 180°$ (the point closest to the free surface) and this is the area of the greatest interest to designers for determining the fracture strength of structures containing cracks. Figure 8 shows a plot of free surface magnification for the maximum stress intensity factor with respect to proximity of the free surface, b/h, for various aspect ratio, b/a. As expected for the elastic analysis the maximum stress intensity factors for all crack geometries approach infinity as $b/h \to 1$.

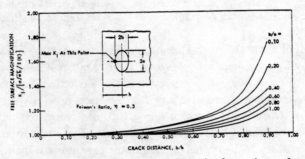

Figure 8. Magnification for maximum stress intensity factor due to free surface.

A more practical application of the results is in the estimation of the stress intensity factors for semi-elliptical surface flaws that are deep with respect to the plate-thickness. Following the procedure described previously [7], the free surface magnifications obtained in this paper are used as the back surface magnification for surface flaws. The stress intensity magnification due to both free surfaces for a surface flaw can be obtained by multiplying this back surface magnification by an estimated front surface magnification, M_F[7].

$$M_F = 1.0 + 0.12(1 - b/2a)^2$$

Figure 9 shows the stress intensity magnification for surface flaws with respect to b/h for various ratios of b/a. These results were then used to calculate critical stress intensity factors from ex-

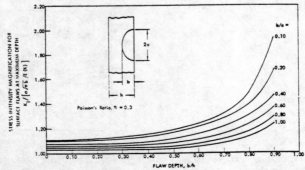

Figure 9. Estimated elastic stress intensity magnification factors for semi-elliptic surface flaws at maximum depth.

perimental data on surface flaws in 0.625 inch thick 2219-T87 aluminum base metal plate tested at three different temperatures [6]. The calculated critical stress intensity factors at each temperature remained nearly constant for wide variations of b/h and b/a ratios, thus providing an indirect experimental verification for the magnification factors in Figure 9.

7. Conclusions

The stress intensity factor for an embedded elliptical crack in a halfspace is determined as a function of position around the crack periphery, crack aspect ratio and crack distance from the free surface. The results show that the effect of the free surface on the stress intensity factor can be as much as 75 percent higher depending on crack aspect ratio and crack distance from the free surface.

Effect of the Poisson's ratio of the material on the stress intensity factor is quite small. Stress intensity factor is shown to be slightly higher for a material with a higher Poisson's ratio than that with a lower Poisson's ratio. Also, the effect of Poisson's ratio on the stress intensity factor is higher for rounder elliptical cracks ($b/a \to 1.0$) than that for flatter elliptical cracks ($b/a \to 0.0$).

Stress intensity magnification factor at the minor axis of a semi-elliptical surface crack in a plate is estimated as a function of crack depth to crack length and crack depth to plate thickness ratio. The results compare well with the experimental data of 2219-T87 aluminum.

Acknowledgements

This investigation is a part of a long term internal research program in fracture mechanics at the Aerospace Group of The Boeing Company. The authors wish to thank Mr. C. F. Tiffany for his encouragement and counsel during this lengthy and sometimes discouraging research problem.

REFERENCES

[1] C. F. Tiffany and J. N. Masters, *Applied Fracture Mechanics*, Fracture Toughness and Its Application, ASTM Special Technical Publication No. 381 (1965) 249–278
[2] A. E. Green and I. N. Sneddon, *The Distribution of Stress in the Neighborhood of a Flat Elliptical Crack in an Elastic Solid*, Proceedings of the Cambridge Philosophical Society, 46 (1950) 159–163
[3] G. R. Irwin, Crack Extension Force for a Part-Through Crack in a Plate, *Journal of Applied Mechanics*, 29, Trans. ASME, 84 (1962) 651–654
[4] C. F. Tiffany, P. M. Lorenz, and L. R. Hall, *Investigation of Plane Strain Flaw Growth in Thick-Walled Tanks*, NASA CR-54837 (February 1966).
[5] L. R. Hall, *Plane Strain Cyclic Flaw Growth in 2014–T62 Aluminum and 6A1-4V(ELI) Titanium*, NASA CR-72396 (November 1968)
[6] J. N. Masters, W. P. Haese and R. W. Finger, *Investigation of Deep Flaws in Thin-Walled Tanks*, NASA CR-72606 (December 1969)
[7] A. S. Kobayashi and W. L. Moss, *Stress Intensity Magnification Factors for Surface-Flawed Tension Plate and Notched Round Tension Bar*, Proceedings of the Second International Conference on Fracture, Brighton, England (1968)
[8] Private communication with R. W. Thresher, Department of Mechanical Engineering. Oregon State University, Corvallis, Oregon.
[9] F. W. Smith A. F. Emery and A. S. Kobayashi, Stress Intensity Factors for Semi-circular Cracks, Part 2 – Semi-Infinite Solid, *Journal of Applied Mechanics*, 34, Trans. ASME, 89 (1967) 953–959.
[10] L. A. Wigglesworth, Stress Distribution in a Notched Plate, *Mathematika*, 4 (1957) 76–96

[11] A. S. Kobayashi, M. Ziv and L. R. Hall, Approximate Stress Intensity Factor for an Embedded Elliptical Crack Near Two Parallel Free Surfaces, *International Journal of Fracture Mechanics*, 1, No. 2 (1965) 81–95
[12] F. W. Smith and M. J. Alavi, Stress Intensity Factors for a Penny Shaped Crack in a Half Space, *Journal of Engineering Fracture Mechanics*, 3, no. 2 (1971) 241–255
[13] F. W. Smith and M. J. Alavi, *Stress Intensity Factors for a Part Circular Surface Flaw*, Proceedings of the First International Conference on Pressure Vessel Technology, Delft, Holland (1969)
[14] R. W. Thresher, and F. W. Smith, Stress Intensity Factors for a Surface Crack in a finite Solid, *Journal of Applied Mechanics*, 39th Trans. of ASME, 95 (1972) 195–200
[15] R. C. Shah, and A. S. Kobayashi, Stress Intensity Factor for an Elliptical Crack Under Arbitrary Normal Loading, *Journal of Engineering Fracture Mechanics*, 3, no. 1 (1971) 71–96
[16] A. E. H. Love, On Stress Produced in a Semi-Infinite Solid by Pressure on Part of the Boundary, Philosophical Transactions of the Royal Society, Series A, 228 (1929) 378–395
[17] A. E. H. Love, *A Treatise on the Mathematical Theory of Elasticity*, Dover Publications, New York (1944) 241–245
[18] C. M. Segedin, A Note on Geometric Discontinuities in Elastostatics, *International Journal of Engineering Science*, 6 (1968) 309–312
[19] G. R. Irwin, *Analytical Aspects of Crack Stress Field Problems*, T. and A. M. Report No. 213, University of Illinois, March 1962
[20] M. K. Kassir and G. C. Sih, Geometric Discontinuities in Elastostatics, *Journal of Mathematics and Mechanics*, 16, No. 9 (1967) 927–948
[21] R. C. Shah and A. S. Kobayashi, *Stress Intensity Factors for an Elliptical Crack Approaching the Surface of Semi-Infinite Solid*, Boeing Company Document No. D-180-14494-1, (1971)
[22] M. Isida and Y. Itagaki, *The Effect of Longitudinal Stiffeners in a Cracked Plate under Tension*, Proceedings of the 4th U.S. National Congress of Applied Mechanics (June 1962) 955–969

Section Two
Numerical Methods

NASA Technical Note D-2395 (August 1964).

STRESS-INTENSITY FACTORS FOR A SINGLE-EDGE-NOTCH TENSION

SPECIMEN BY BOUNDARY COLLOCATION OF A STRESS FUNCTION

by Bernard Gross, John E. Srawley,
and William F. Brown, Jr.

Lewis Research Center

SUMMARY

A boundary value collocation procedure applied to the Williams stress function was employed to determine the elastic stress distribution in the immediate vicinity of the tip of an edge crack in a finite-width specimen subjected to uniform tensile loading. This type of single-edge-notch specimen is particularly suitable for determination of plane strain fracture toughness values. The analytical results are expressed in such a way that the stress intensity factor may be determined from known conditions of specimen geometry and loading.

As the crack length decreased, the results obtained by the collocation procedure approached those derived from a closed solution for an edge crack in a semi-infinite plate. Over a range of ratios of crack length to specimen width between 0.15 and 0.40 the collocation solution yielded results in very good agreement with those derived from experimental compliance measurements.

INTRODUCTION

A method for calculating the stress distribution in a test specimen containing a single-edge crack (sharp notch) and subjected to a uniform tensile load is described herein. The results are particularly useful in determining stress intensity factors K for given conditions of load and geometry and therefore permit the use of the single-edge-notch specimen in fracture toughness testing.

The ASTM Special Committee on Fracture Testing of High Strength Metallic Materials issued a series of reports describing recent developments in fracture toughness testing (refs. 1 to 4). It has been shown that the magnitude of the elastic stress field in the immediate vicinity of a crack but beyond the crack tip plastic zone may be characterized by a single parameter K, the stress intensity factor (refs. 1 and 5). For any given material a characteristic value K_c of the stress intensity factor is assumed to exist that corresponds to the onset of rapid fracture. Like other mechanical properties, K_c is dependent on the strain rate, the temperature, and the testing direction. In the case of sheet and plate materials it is also dependent on the thickness. The value of K_c may be determined from tests on specimens containing sharp notches or cracks, provided that suitable expressions are available that give the stress intensity factor in terms of the specimen geometry and applied loads at fracture instability.

Approximate solutions for K exist in closed form for a number of specimen designs symmetrically notched with respect to the tensile load axis (refs. 5 to 9). The single-edge-notch tension specimen appears to be more efficient, however, than symmetrically notched specimens with respect both to the material and to the loading capacity required (ref. 10). For this reason, the single-edge-notch specimen may be of considerable importance in the determination of K_c for plane strain crack propagation where relatively large cross sections are an inherent requirement of the test. Recent, very careful experimental compliance measurements on the single-edge-notch specimen (ref. 10) provide values of strain energy release rates as a function of crack length from which values of K may be derived. An analytical solution is desirable, however, as an independent check on the experimental procedure; it also has the advantage that the influence of certain geometrical parameters, such as the ratio of height to width V/W, may be rapidly determined without resort to tedious experimental measurements. Furthermore, the method of obtaining an analytical solution is applicable to all combinations of bending and tension applied to a single-edge-notch specimen.

An analytical solution to the stress distribution in the single-edge-notch tension specimen is obtained herein by a boundary value collocation procedure applied to the Williams stress function (ref. 11), which is known to satisfy the boundary conditions along an edge crack. The results are in a form that permits expression of stress intensity factors in terms of the measured quantities of load and specimen dimensions. In addition, the influence of the end effect on the stress intensity factor is determined. The end effect derives from the finite distance between the crack and the uniformly loaded boundary, expressed as V/W. A comparison is made between the present analytical results and a closed solution obtained by Wigglesworth (ref. 12) for an edge crack in a semi-infinite plate. Finally, the collocation solution in terms of the stress intensity factor is compared with experimental results obtained by other investigators for this specimen with strain energy release rate (compliance measurement) experiments.

SYMBOLS

a crack length in single-edge-notch specimen, in.

d_n coefficients of Williams stress function

E Young's modulus, psi

$\mathcal{G}$ strain energy release rate with crack extension; or crack extension force, in.-lb/sq in.

K   stress intensity factor of elastic stress field in vicinity of border of crack, psi $\sqrt{\text{in.}}$

P   load per unit thickness, lb/in.

r, θ angular position coordinates referred to crack tip

V   distance (height) between crack plane and location of uniform stress, in.

W   specimen width, in.

x, y coordinate axes with origin at crack tip, parallel and perpendicular, respectively, to crack plane

σ_0 uniform tensile stress applied to specimen, psi

$\sigma_x, \sigma_y, \tau_{xy}$ stress in x- and y-directions, psi

χ stress function

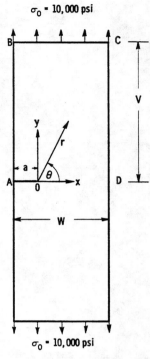

Figure 1. - Specimen geometry and loading assumed for collocation solution.

METHOD

The method of analysis consists in finding a stress function χ satisfying the biharmonic equation $\nabla^4 \chi = 0$ and the boundary conditions at a finite number of stations along the boundaries of the single-edge-notch specimen shown in figure 1. For the present purposes use is made of the Williams stress function (ref. 11) with the correction of a typographical error in that reference:

$$\chi(r,\theta) = \sum_{n=1,2,\ldots}^{\infty} \left\{ (-1)^{n-1} d_{2n-1} r^{n+(1/2)} \left[-\cos\left(n - \frac{3}{2}\right)\theta + \frac{2n-3}{2n+1} \cos\left(n + \frac{1}{2}\right)\theta \right] \right. $$
$$\left. + (-1)^n d_{2n} r^{n+1} \left[-\cos(n-1)\theta + \cos(n+1)\theta \right] \right\} \qquad (1)$$

Because of symmetry (fig. 1) only even terms of the stress function are considered. The stresses in terms of χ obtained by partial differentiation are as follows:

$$\sigma_y = \frac{\partial^2 \chi}{\partial x^2} = \frac{\partial^2 \chi}{\partial r^2} \cos^2\theta - 2 \frac{\partial^2 \chi}{\partial \theta \, \partial r} \frac{\sin\theta \cos\theta}{r} + \frac{\partial \chi}{\partial r} \frac{\sin^2\theta}{r}$$

$$+ 2 \frac{\partial \chi}{\partial \theta} \frac{\sin\theta \cos\theta}{r^2} + \frac{\partial^2 \chi}{\partial \theta^2} \frac{\sin^2\theta}{r^2}$$

$$\sigma_x = \frac{\partial^2 \chi}{\partial y^2} = \frac{\partial^2 \chi}{\partial r^2} \sin^2\theta + 2 \frac{\partial^2 \chi}{\partial \theta \, \partial r} \frac{\sin\theta \cos\theta}{r} + \frac{\partial \chi}{\partial r} \frac{\cos^2\theta}{r}$$

$$- 2 \frac{\partial \chi}{\partial \theta} \frac{\sin\theta \cos\theta}{r^2} + \frac{\partial^2 \chi}{\partial \theta^2} \frac{\cos^2\theta}{r^2}$$

$$-\tau_{xy} = \frac{\partial^2 \chi}{\partial x \, \partial y} = \sin\theta \cos\theta \frac{\partial^2 \chi}{\partial r^2} + \frac{\cos 2\theta}{r} \frac{\partial^2 \chi}{\partial r \, \partial \theta} - \frac{\sin\theta \cos\theta}{r^2} \frac{\partial^2 \chi}{\partial \theta^2}$$

$$- \frac{\sin\theta \cos\theta}{r} \frac{\partial \chi}{\partial r} - \frac{\cos 2\theta}{r^2} \frac{\partial \chi}{\partial \theta}$$

(2)

The Williams stress function is an Airy stress function, which, besides satisfying the biharmonic equation, also satisfies the boundary conditions along the crack surface, namely, that the normal and shearing stresses be zero. Thus, when $\theta = \pm\pi$, equations (1) and (2) give $\sigma_y = 0$, $\tau_{xy} = 0$. The remaining boundary requirements on the stress function for the specimen having the geometry and tractions shown in figure 1 are as follows:

Along boundary A → B:

$$\chi = 0, \quad \frac{\partial \chi}{\partial x} = 0$$

Along boundary B → C:

$$\chi = \sigma_o \left(\frac{x^2}{2} + ax + \frac{a^2}{2} \right), \quad \frac{\partial \chi}{\partial y} = 0$$

(3)

Along boundary C → D:

$$\chi = \frac{\sigma_o W^2}{2}, \quad \frac{\partial \chi}{\partial x} = \sigma_o W$$

Because of symmetry with respect to the crack plane (fig. 1) only half the specimen need be considered.

For the purpose of determining the stress intensity factor as defined in reference 13, which characterizes the stress distribution in the immediate neighborhood of the crack tip ($r \to 0$), only the first coefficient d_1 of the

Williams stress function is necessary, since this term is dominant. As shown later, d_1 is proportional to the stress intensity factor K. Values of d_1 as well as of the other coefficients are obtained by satisfying the boundary conditions (eq. (3)) at a finite number of stations equally spaced along a given boundary for a specimen with the geometry shown in figure 1 that is subjected to a uniform stress of 10,000 psi acting at a distance V from the crack plane. Computations were made for several ratios of crack length to specimen width a/W between 0.04 and 0.5 and for values of V/W ranging from 0.5 to 1.5.

The collocation procedure requires a matrix solution of twice as many equations as the number of boundary stations selected for each combination of the independent variables. This problem was programed for a digital computer with the use of double precision arithmetic (16 significant figures).

In this solution, the number of boundary stations is increased until the first matrix coefficient d_1 converges to a sufficiently stable value. Figure 2, for example, shows the first matrix coefficient as a function of the number of boundary stations for configurations with several V/W ratios at a value of a/W of 1/3. The variation in the first matrix coefficient is not more than ±1 percent when the number of boundary stations is increased from 11 to 23.

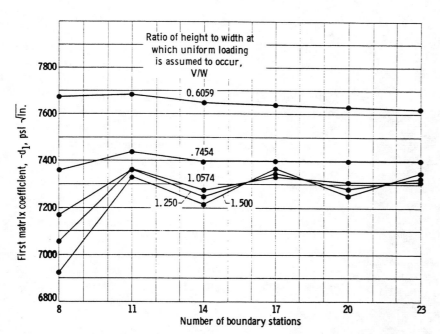

Figure 2. - First matrix coefficient as function of number of boundary stations. Tensile stress applied to specimen, 10,000 psi; specimen width, 1 inch; ratio of crack length to specimen width, 1/3.

The stress-function values at 50 stations along the boundary were computed for several geometries with all d_n coefficients. These values were in good agreement with the prescribed values. The stress-function derivative normal to the boundary, however, showed perturbations near the corners of the specimen. The effect of this variation in terms of stress distribution throughout the specimen could be determined only by additional computation. This additional effort did not appear justified since the first matrix coefficient was, for practical purposes, insensitive to these perturbations.

RESULTS

Stress Intensity Factors

The stress intensity factor K may be derived in terms of the first coefficient of the Williams stress function. The expression for the stress in the y-direction in the immediate vicinity of the crack tip is obtained from the dominant term as follows:

$$\sigma_y = \frac{-d_1}{\sqrt{r}} \cos \frac{\theta}{2} \left(1 + \sin \frac{\theta}{2} \sin \frac{3\theta}{2}\right) \tag{4}$$

The expression for σ_y given in reference 1, based on the Westergaard crack stress analysis, is as follows:

$$\sigma_y = \frac{K}{\sqrt{2\pi r}} \cos \frac{\theta}{2} \left(1 + \sin \frac{\theta}{2} \sin \frac{3\theta}{2}\right) \tag{5}$$

Thus

$$K = -d_1 \sqrt{2\pi} \tag{6}$$

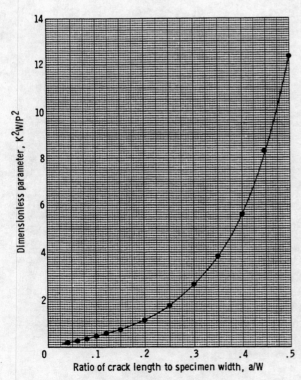

Figure 3. - Collocation results of a plot of dimensionless parameter against ratio of crack length to specimen width in single-edge-notch specimen.

As shown in figure 2, small oscillations sometimes occur in the first matrix coefficient. For this reason stress intensity factors were computed with values of d_1 averaged from 11 to 23 boundary points.

For purposes of fracture toughness testing the results of the collocation procedure are conveniently expressed in the form of a dimensionless parameter involving K and the measured quantities as a function of a/W (ref. 10). Thus,

$$\frac{K^2}{\sigma_o^2 W} = \frac{K^2 W}{P^2} \tag{7}$$

where P is the load per unit thickness. As discussed later, this form is useful for a comparison of the analytical results with experimental compliance calibration data. A curve derived from equation (3) relating K^2W/P^2 to a/W is given in figure 3. This curve applies to any V/W value greater than about 0.8 (see fig. 4).

Influence of End Effects

As an aid to optimizing the specimen design, calculations were made to show how close the assumed position of the uniformly loaded boundary could be to the crack plane without affecting the stress intensity factor. For a given value of a/W the position, of course, is a function of the specimen width, and expressing the first matrix coefficient as a function of the ratio of height to width (see fig. 1, p. 3) is convenient for various values of a/W. According to figure 4, the first matrix coefficient at a/W ratios between 0.15 and 0.5 is essentially constant for ratios of height to width V/W greater than about 0.8.

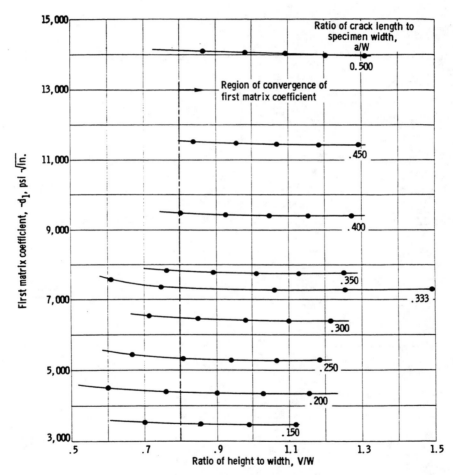

Figure 4. - First matrix coefficient as function of ratio of height to width for various ratios of crack length to specimen width. Tensile stress applied to specimen, 10,000 psi; specimen width, 1 inch.

In a practical test specimen the means by which load is applied introduces nonuniformly stressed regions at the ends of the specimen. The actual specimen length must therefore exceed the minimum length determined from figure 4. In the case of a pin-loaded specimen, for example, the optimum ratio of total length to width is about 4 (ref. 10).

DISCUSSION OF RESULTS

Comparison with Wigglesworth Solution

A check on the validity of the present solution can be obtained by comparison with that reported by Wigglesworth (ref. 12) for an edge crack in a semi-infinite plate. These two solutions should converge as the crack length tends to zero. In the immediate vicinity of the crack the first term of the Wigglesworth solution predominates, and σ_y may be expressed in terms of the coordinate system shown in figure 1 (p. 3) as follows:

$$\sigma_y = 0.793\, \sigma_o \sqrt{\frac{a}{r}} \cos \frac{\theta}{2} \left(1 + \sin \frac{\theta}{2} \sin \frac{3\theta}{2}\right) \tag{8}$$

This equation may be compared with the corresponding expression obtained by the present method:

$$\sigma_y = \frac{-d_1}{\sqrt{r}} \cos \frac{\theta}{2} \left(1 + \sin \frac{\theta}{2} \sin \frac{3\theta}{2}\right) \tag{4}$$

where the first matrix coefficient d_1 depends on σ_o and the specimen dimensions. The respective values of σ_y in any direction near the crack tip may be compared by considering a single-edge-notch specimen of unit width. For the same uniform stress in both cases, the ratio of σ_y obtained from the Wigglesworth solution (eq. (8)) to that computed from equation (4) should ap-

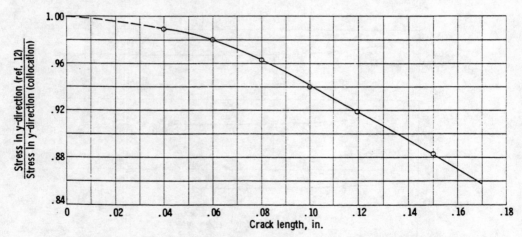

Figure 5. - Comparison of stress ratios in immediate vicinity of crack tip obtained by collocation solution and solution of reference 12.

proach 1 as the crack length decreases. The results presented in figure 5 are in accordance with this behavior.

Comparison with Experimental Results

Two experimental compliance calibrations of single-edge-notch specimens loaded in tension are available for comparison, the earlier by Sullivan (ref. 14) and a more recent one by Srawley, Jones, and Gross (ref. 10). The design of the specimen of reference 14 was loaded through pins separated by a distance less than twice the width, which introduced large end effects that are not accounted for in the analytical solution. The specimen used in reference 10 was of sufficient length that end effects were negligible, since the compliance measurements were made over a gage length of 8 inches on a specimen 3 inches wide that was loaded through pins 10 inches apart. The sufficiency of this gage length was established by preliminary experiments. For this reason better agreement with the analytical results is to be expected from the experiments of reference 10 than from those of reference 14. Furthermore, as discussed in reference 10, those data are expected to be more precise than the data of reference 14 because of differences in specimen size and measurement techniques.

The experimental compliance procedure gives results in terms of the strain energy release rate $\mathscr{G}$. The correct procedure for converting these values of $\mathscr{G}$ to stress intensity factors is not yet completely settled (see ref. 10). For the purposes of comparing the analytical with the experimental results the most reasonable procedure appears to be a conversion on the plane stress basis. Thus,

$$K^2 = E\mathscr{G}$$

or in terms of experimentally measured quantities,

$$\frac{K^2 W}{P^2} = \frac{E \mathscr{G} W}{P^2} \tag{9}$$

where $\mathscr{G}$ is determined by experimental compliance procedures and P is the load per unit specimen thickness.

Ratio of crack length to specimen width, a/W	Dimensionless parameter, $K^2 W/P^2$		
	Experimental results		Collocation results
	Ref. 10	Ref. 14	
0.05	0.314	0.35	0.204
.10	.556	.65	.445
.15	.816	1.00	.758
.20	1.180	1.40	1.180
.25	1.735	1.97	1.768
.30	2.571	2.80	2.603
.35	3.775	4.20	3.813
.40	5.436	6.18	5.596
.45	7.641	8.90	8.276
.50	10.477	12.50	12.399

The comparison between experimental and analytical results is shown in the table. As might be expected from the foregoing discussion of the compliance calibration experiments, the results given in reference 14 are consistently higher than either those obtained by the present collocation solution or those reported in reference 10. In contrast, very good agreement between the analytical solution and the data of reference 10 is noted for values of a/W between 0.15 and 0.40. The differences between these two sets of results at the lower values of a/W are probably associated with uncertainties in the lower range of the experimental data. For the a/W values above 0.4, differences due to bending of the experimental compliance specimen, which are not taken into account by the analytical solution, become important. This bending decreases the eccentricity of loading with respect to the uncracked section, and the compliance for a given slot length is therefore slightly less than if no bending took place.

SUMMARY OF RESULTS

The results of an analytical investigation of the stress intensity factors for a single-edge-notch tension specimen obtained by a boundary value collocation procedure applied to the Williams stress function are as follows:

1. The values of the stress intensity factor were independent of the distance between the uniformly loaded cross section and the notch plane provided that this distance was greater than 80 percent of the width.

2. At small ratios of crack length to specimen width the present results were in good agreement with a closed solution obtained for an edge crack in a semi-infinite plate.

3. When the analytical results were expressed in appropriate dimensionless form, very good agreement was obtained with comparable results obtained from a highly accurate experimental strain energy release rate determination.

Lewis Research Center
   National Aeronautics and Space Administration
      Cleveland, Ohio, May 5, 1964

REFERENCES

1. ASTM Special Committee: Fracture Testing of High-Strength Sheet Materials, pt. II. ASTM Bull. 244, Feb. 1960, pp. 18-28.

2. ASTM Special Committee: The Slow Growth and Rapid Propagation of Cracks. Materials Res. and Standards, vol. 1, no. 5, May 1961, pp. 389-393.

3. ASTM Special Committee: Fracture Testing of High-Strength Sheet Materials. Materials Res. and Standards, vol. 1, no. 11, Nov. 1961, pp. 877-885.

4. ASTM Special Committee: Progress in Measuring Fracture Toughness and Using Fracture Mechanics. Materials Res. and Standards, vol. 4, no. 3, Mar. 1964, pp. 107-119.

5. Irwin, G. R.: Analysis of Stresses and Strains Near the End of a Crack Traversing a Plate. Jour. Appl. Mech., vol. 24, no. 3, Sept. 1957, pp. 361-364.

6. Sneddon, I. N.: The Distribution of Stress in the Neighbourhood of a Crack in an Elastic Solid. Proc. Roy. Soc. (London), Ser. A, vol. 187, no. 1099, Oct. 22, 1946, pp. 229-260.

7. Mendelson, Alexander, and Spero, Samuel W.: Elastic Stress Distribution in a Finite-Width Orthotropic Plate Containing a Crack. NASA TN D-2260, 1964.

8. Bowie, Oscar L.: Rectangular Tensile Sheet with Symmetric Edge Cracks. TR-63-22, Army Materials Res. Agency, Oct. 1963.

9. Irwin, G. R.: Crack-Extension Force for a Part-Through Crack in a Plate. Jour. Appl. Mech. (Trans. ASME), ser. E, vol. 29, no. 4, Dec. 1962, pp. 651-654.

10. Srawley, John E., Jones, Melvin H., and Gross, Bernard: Experimental Determination of the Dependence of Crack Extension Force on Crack Length for a Single-Edge-Notch Tension Specimen. NASA TN D-2396, 1964.

11. Williams, M. L.: On the Stress Distribution at the Base of a Stationary Crack. Jour. Appl. Mech., vol. 24, no. 1, Mar. 1957, pp. 109-114.

12. Wigglesworth, L. A.: Stress Distribution in a Notched Plate. Mathematika, vol. 4, 1957, pp. 76-96.

13. Irwin, G. R.: Fracture. Vol. VI - Elasticity and Plasticity. Encyclopedia of Phys., S. Flügge, ed., Springer-Verlag (Berlin), 1958, pp. 551-590.

14. Sullivan, A. M.: New Specimen Design for Plane-Strain Fracture Toughness Tests. Materials Res. and Standards, vol. 4, no. 1, Jan. 1964, pp. 20-24.

NASA Technical Note D-2603 (January 1965).

STRESS-INTENSITY FACTORS FOR SINGLE-EDGE-NOTCH SPECIMENS

IN BENDING OR COMBINED BENDING AND TENSION BY BOUNDARY

COLLOCATION OF A STRESS FUNCTION

by Bernard Gross and John E. Srawley

Lewis Research Center

SUMMARY

A boundary-value-collocation procedure was used in conjunction with the Williams stress function to determine values of the stress-intensity factor K for single edge cracks of various depths in specimens subjected to pure bending. The results are of use in connection with K_{Ic} crack toughness tests, which utilize rectangular-section crack-notch beam specimens loaded in four-point bending, and are in good agreement with published results derived from experimental compliance measurements. The results are expressed in convenient, compact form in terms of the dimensionless quantity $Y^2 = K^2B^2W^3/M^2$, which is a function of relative crack depth a/W only, where B and W are the specimen width and thickness and M is the applied bending moment.

On the assumption that the condition for a valid K_{Ic} test is that the maximum nominal stress at the crack tip should not exceed the yield strength of the material, the K_{Ic} measurement capacity of bend specimens was estimated as a function of a/W. The measurement capacity is proportional to the yield strength and to the square root of the specimen depth, and it is greatest for a/W in the range 0.2 to 0.3.

Values of K for single-edge-notch specimens subjected to combined bending and tension were obtained by superposition of the present results and those of earlier work for specimens loaded in uniform tension. These values are of interest in connection with the use of single-edge-notch specimens that are off-center pin-loaded in tension. It is shown that the K_{Ic} measurement capacity of such specimens is not very sensitive to the eccentricity of loading.

INTRODUCTION

It was shown previously by Gross, Srawley, and Brown (ref. 1) that the value of the stress-intensity factor K for a single edge crack in a flat plate specimen of finite width could be computed accurately by a boundary-

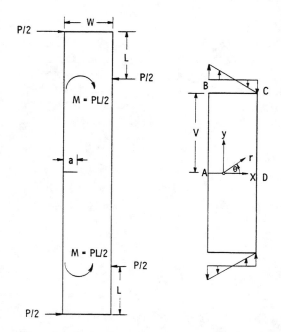

Figure 1. - Single-edge-notch specimen subjected to pure bending.

value-collocation procedure applied to an appropriate stress function. For uniform tensile loading, the computed values of K for various values of the relative crack length a/W were in good agreement with the corresponding values derived from experimental measurements of specimen compliance by Srawley, Jones, and Gross (ref. 2), thus providing confidence in the reliability of the mathematical analysis. The significance of K and its role in the measurement of plane strain crack toughness K_{Ic} are discussed in references 1 and 2, and in greater detail by Srawley and Brown (ref. 3). (All symbols are defined in the appendix).

In the present report, the application of the boundary-value-collocation procedure to the case of single-edge-notch specimens subjected to pure bending is described, and the results are presented. These results are compared with previous results obtained by Bueckner who used a different analytical method (ref. 4), and also with results derived from careful experimental compliance measurements of four-point loaded notched beams by Lubahn (ref. 5). While the results of references 4 and 5 are substantially in agreement, there is sufficient discrepancy between them to warrant a third, independent treatment of the problem in view of the practical importance of the accuracy of K_{Ic} measurements that are conducted with four-point bend specimens.

The use of single-edge-notch specimens loaded in tension through off-center pins has been discussed by Sullivan (ref. 6), and the results of experimental compliance measurements for two positions of the loading pin holes are given in this reference. In the present report, the general case of off-center tension loading is treated by appropriate superposition of the present results for pure bending and the results of reference 1 for uniform tension. This method gives a good approximation to the value of K for any position of the loading pin holes.

In deciding what design of single-edge-notch specimen is to be used for a particular application, an important consideration is the extent to which the K_{Ic} measurement capacity C_{IK} is affected by the design. The K_{Ic} measurement capacity is the largest value of K_{Ic} that could be measured with acceptable accuracy by using a specimen of given width and adequate thickness and a material of given yield strength. For a given specimen design, C_{IK} is proportional to the yield strength and to the square root of the specimen width. In the present report, estimates are obtained of C_{IK} for single-edge-notch specimens loaded in pure bending and in combined bending and tension.

METHOD

The method of analysis consists in finding a stress function χ that satisfies the biharmonic equation $\nabla^4 \chi = 0$ and also the boundary conditions at a finite number of stations along the boundary of a single-edge-notched specimen, such as shown in figure 1. The biharmonic equation and the boundary conditions along the crack are satisfied by the Williams stress function (ref. 7). Because of symmetry (fig. 1) the coefficient of the sine terms in the general stress function must be zero, hence

$$\chi(r,\theta) = \sum_{n=1,2...}^{\infty} \left\{ (-1)^{n-1} d_{2n-1} r^{n+(1/2)} \left[-\cos\left(n - \frac{3}{2}\right)\theta + \frac{2n-3}{2n+1} \cos\left(n + \frac{1}{2}\right)\theta \right] \right.$$
$$\left. + (-1)^n d_{2n} r^{n+1} \left[-\cos(n-1)\theta + \cos(n+1)\theta \right] \right\} \quad (1)$$

The stresses in terms of χ obtained by partial differentiation are as follows:

$$\left.\begin{aligned}
\sigma_y &= \frac{\partial^2 \chi}{\partial x^2} = \frac{\partial^2 \chi}{\partial r^2} \cos^2\theta - 2 \frac{\partial^2 \chi}{\partial \theta \, \partial r} \frac{\sin\theta \cos\theta}{r} + \frac{\partial \chi}{\partial r} \frac{\sin^2\theta}{r} \\
&\qquad + 2 \frac{\partial \chi}{\partial \theta} \frac{\sin\theta \cos\theta}{r^2} + \frac{\partial^2 \chi}{\partial \theta^2} \frac{\sin^2\theta}{r^2} \\
\sigma_x &= \frac{\partial^2 \chi}{\partial y^2} = \frac{\partial^2 \chi}{\partial r^2} \sin^2\theta + 2 \frac{\partial^2 \chi}{\partial \theta \, \partial r} \frac{\sin\theta \cos\theta}{r} + \frac{\partial \chi}{\partial r} \frac{\cos^2\theta}{r} \\
&\qquad - 2 \frac{\partial \chi}{\partial \theta} \frac{\sin\theta \cos\theta}{r^2} + \frac{\partial^2 \chi}{\partial \theta^2} \frac{\cos^2\theta}{r^2} \\
-\tau_{xy} &= \frac{\partial^2 \chi}{\partial x \, \partial y} = \sin\theta \cos\theta \frac{\partial^2 \chi}{\partial r^2} + \frac{\cos 2\theta}{r} \frac{\partial^2 \chi}{\partial r \, \partial \theta} - \frac{\sin\theta \cos\theta}{r^2} \frac{\partial^2 \chi}{\partial \theta^2} \\
&\qquad - \frac{\sin\theta \cos\theta}{r} \frac{\partial \chi}{\partial r} - \frac{\cos 2\theta}{r^2} \frac{\partial \chi}{\partial \theta}
\end{aligned}\right\} \quad (2)$$

The remaining boundary conditions to be satisfied for the case of pure bending (fig. 1) are as follows:

Along

A-B $\quad \chi = 0; \quad \frac{\partial \chi}{\partial x} = 0$

B-C $\quad \chi = -\frac{12M}{BW^3}\left(\frac{a^3}{6} + \frac{a^2 x}{2} + \frac{ax^2}{2} + \frac{x^3}{6}\right) + \frac{6M}{BW^2}\left(\frac{x^2}{2} + ax + \frac{a^2}{2}\right); \quad \frac{\partial \chi}{\partial y} = 0$

C-D $\quad \chi = \frac{M}{B}; \quad \frac{\partial \chi}{\partial x} = 0$

The collocation procedure consists of solving $2m$ simultaneous algebraic equations corresponding to the values of χ and either $\partial \chi/\partial x$ or $\partial \chi/\partial y$ at m selected boundary stations, thus obtaining values for the first $2m$ coefficients in the Williams stress function, the remaining terms being neglected. Only the value of the first coefficient d_1 is needed for the present purpose because this is directly proportional to the stress-intensity factor K. According to Irwin (ref. 8), the stress component σ_y in the immediate vicinity of the crack tip (as r approaches zero) is given by

$$\sigma_y = \frac{K}{\sqrt{2\pi r}} \cos\frac{\theta}{2}\left(1 + \sin\frac{\theta}{2}\sin\frac{3\theta}{2}\right)$$

while in terms of the Williams stress function, as r approaches zero

$$\sigma_y = \frac{-d_1}{\sqrt{r}} \cos\frac{\theta}{2}\left(1 + \sin\frac{\theta}{2}\sin\frac{3\theta}{2}\right)$$

and similarly for the other stress components, hence,

$$K = -\sqrt{2\pi}\, d_1$$

For given values of M, B, W, a, and V (fig. 1), the value of d_1 computed by the collocation procedure will vary somewhat with m but will approach a limit as m is increased. This is illustrated by the example shown in figure 2 for specimens having a/W equal to 0.3 and various values of V/W. For a given V/W, as the number of boundary stations is increased in steps of three, the computed d_1 oscillates about and converges toward a limit that is close to the average of the five computed values. Accordingly, the value of d_1 that was used to obtain K in each case was the average of five

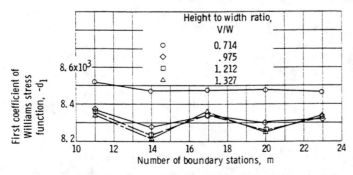

Figure 2. - Value of coefficient d_1 against number of boundary stations. Specimen width, 1 inch; actual crack length, 0.30 inch, applied bending moment, 200 inch-pounds; specimen thickness, 1/16 inch.

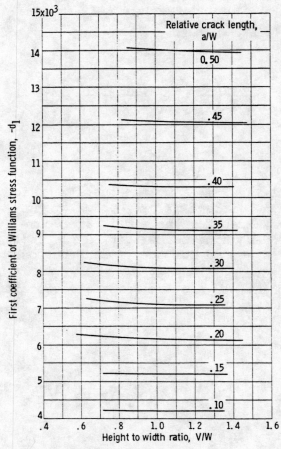

Figure 3. - Single-edge-notch specimen subjected to pure bending. Specimen width, 1 inch; applied bending moment, 200 inch-pounds; specimen thickness, 1/16 inch.

values computed for m equal to 11, 14, 17, 20, and 23. The computations were performed on a digital computer by using double-precision arithmetic (16 significant figures).

A value of K obtained by this method corresponds to a particular set of values of M, B, W, a, and V (fig. 1) selected for convenience of computation. For application, the results are better expressed more generally in terms of the dimensionless quantity $Y^2 = K^2 B^2 W^3/M^2$ (or in terms of Y), which depends only on the dimensionless ratios a/W and V/W. As will be shown, the effect of V/W on Y^2 is negligible for values of V/W greater than unity, so that for a specimen of adequate length, Y^2 is a function of a/W only. Consequently, a table or graph of Y^2 or Y against a/W is all that is needed for calculation of K for any given values of a, W, B, and M measured in a test.

RESULTS AND DISCUSSION

Pure Bending

The results of the computations for pure bending are summarized in figure 3. Each plotted point represents the average value of d_1 of five values computed for m equal to 11, 14, 17, 20, and 23. The plot shows the dependence of this average value of d_1 on V/W for each of nine values of a/W ranging from 0.1 to 0.5. The relative distance from the crack to the boundary at which a stress distribution corresponding to pure bending was imposed is represented by V/W (fig. 1). In practical terms, it corresponds roughly to the ratio of one-half the minor span to the depth of a four-point loaded beam. It is apparent that the dependence of d_1 on V/W is negligible if V/W exceeds unity. Essentially, the same conclusion was reached for single-edge-notch specimens loaded in uniform tension (ref. 1). Consequently, the K values were calculated from the uniform values of d_1 obtained when V/W was greater than unity. Strictly these K values should be considered to apply to four-point loaded bend specimens only when the minor span is 2W or greater.

The final results are given in table I in terms of Y^2 (i.e., $K^2 B^2 W^3/M^2$) as a function of a/W. The results obtained by a different analytical method (ref. 4) and those derived from experimental compliance measurements (ref. 5) are also tabulated for comparison. Experimental compliance measurements are used to derive values of the strain-energy-release rate with crack extension

TABLE I. - COMPARISON OF RESULTS OF PRESENT WORK WITH
CORRESPONDING RESULTS FROM REFERENCES 4 AND 5
IN TERMS OF DIMENSIONLESS QUANTITY
$K^2B^2W^3/M^2$ AS FUNCTION OF
RELATIVE CRACK LENGTH

Relative crack length, a/W	Results of -		
	Collocation boundary procedure	Reference 4	Reference 5
	$Y^2 = \dfrac{K^2B^2W^3}{M^2}$		
0.10	12.4	12.2	11.8
.15	18.5	-----	17.4
.20	25.3	25.2	24.2
.25	33.2	-----	32.15
.30	42.8	-----	41.9
.35	55.2	-----	53.9
.40	71.4	-----	68.6
.45	92.7	-----	88.9
.50	123.0	151.2	118.0

𝒢, rather than values of K directly. As discussed previously (ref. 2), the 𝒢 values were converted to K values according to the generalized plane stress equation $K^2 = E\mathcal{G}$, where E is Young's modulus.

In reference 4, results are given for only three values of a/W; of these, the two lower results are in excellent agreement with the present results, while the value corresponding to $a/W = 0.5$ is considerably higher than the present result. The agreement between the present results and those of reference 5 is good over the whole range. For the practical purpose of K_{Ic} measurement, the range of a/W between 0.15 and 0.25 is of most importance. In this range there is satisfactory agreement between all three sets of independent results.

The following empirical equation is a compact expression of the present results for values of a/W up to 0.35:

$$Y^2 = \frac{K^2B^2W^3}{M^2}$$

$$= 139\frac{a}{W} - 221\left(\frac{a}{W}\right)^2 + 783\left(\frac{a}{W}\right)^3$$

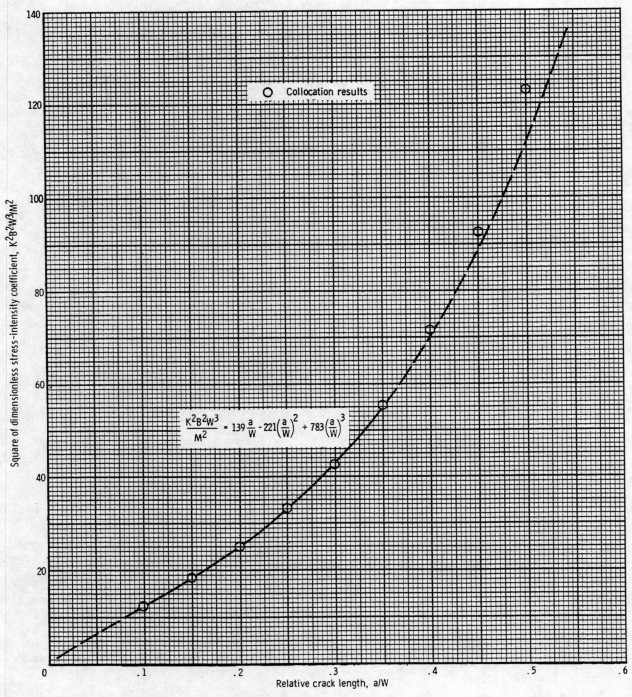

Figure 4. - Curve of least-squares best-fit cubic equation representing results for values of a/W from 0.1 to 0.35. Use of this equation outside this range is not recommended.

The equation was obtained by a least-square best-fit computer program for a cubic in a/W, incorporating the a priori condition that K should be zero when a/W is zero. Only the results for a/W up to 0.35 were used in fitting this equation because it is considered undesirable to use bend specimens having cracks deeper than about 0.35 W. Of the several quantities used to calculate a value of K_{Ic}, the one subject to the greatest uncertainty in measurement is the crack depth. Since the sensitivity of the value of K to a small variation in a/W increases with a/W, it is undesirable to use specimens having large a/W. On the other hand, the efficiency of bend specimens is low when a/W is less than 0.15, as shown in the next section. The optimum range of a/W for bend specimens appears to be between 0.15 and 0.25, and the usual value is 0.2. The curve representing the fitting equation is shown in figure 4, together with the collocation results. The curve is shown dashed in the ranges of a/W from 0 to 0.1 and from 0.35 upwards to emphasize that the equation is not intended to represent the collocation results outside the range of a/W between 0.1 and 0.35.

K_{Ic} Measurement Capacity in Pure Bending

One of the necessary conditions for a meaningful K_{Ic} test is that the specimen width must be sufficient to ensure that the stress field in the vicinity of the crack is sufficiently well represented by that of the assumed linear elastic fracture mechanics model. For reasons that are discussed in reference 3 it will be assumed here that the useful limit of applicability of the model to bend specimens will be reached if the nominal stress at the position of the crack tip reaches the yield strength of the material

$$\frac{6M}{B(W-a)^2} = \sigma_{YS}$$

where σ_{YS} is usually taken to be the 0.2 percent offset tensile yield strength. On this basis, a test result will be valid only if the abrupt crack extension ("pop-in") occurs at a value of M not exceeding $\sigma_{YS}B(W-a)^2/6$. The K value corresponding to this value of M may be defined as the K_{Ic} measurement capacity of a specimen of given dimensions and yield strength, denoted C_{IK}, providing also that the specimen thickness is sufficient as discussed in reference 3. Substituting in the equation

$$Y^2 = \frac{K^2 B^2 W^3}{M^2}$$

and transposing, we obtain

$$\frac{C_{IK}^2}{\sigma_{YS}^2 W} = \frac{Y^2 \left(1 - \frac{a}{w}\right)^4}{36}$$

This equation expresses the K_{Ic} measurement capacity of a bend specimen in terms of the dimensionless quantity $C_{IK}^2/\sigma_{YS}^2 W$, which is a function of a/W

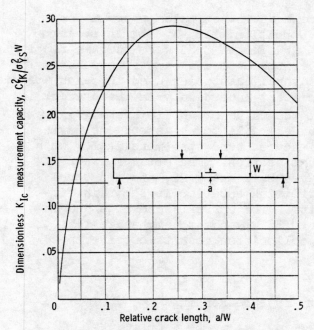

Figure 5. - Dependence of K_{Ic} measurement capacity C_{IK} on relative crack length for specimens in pure bending.

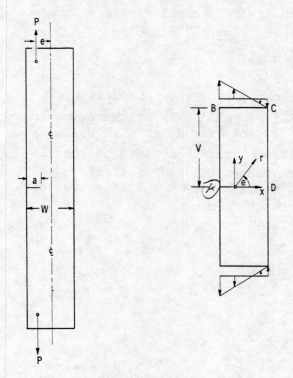

Figure 6. - Single-edge-notch specimen subjected to combined bending and tensile loads.

only since Y is a function of a/W only. For any given value of a/W, there is a unique value of $C_{IK}^2/\sigma_{YS}^2 W$, and hence C_{IK} is proportional to the yield strength of the material and to the square root of the specimen width.

Figure 5 is a plot of $C_{IK}^2/\sigma_{YS}^2 W$ against a/W, which shows that the range of a/W for greatest K_{Ic} measurement capacity is between 0.2 and 0.3. In this range, the C_{IK} for a specimen 3.5 inches deep and of adequate thickness is numerically about equal to the yield strength of the material.

Combined Bending and Tension

by Superposition

In the case of a single-edge-notch specimen loaded in tension through pins (fig. 6), the K_{Ic} measurement capacity might be expected to depend on the loading eccentricity parameter e/W as well as on a/W. It is therefore necessary to study this effect of e/W in connection with standardization of design of single-edge-notch tension specimens (ref. 3).

The load P acting through the off-center pins on the single-edge-notch specimen shown in figure 5 is statically equivalent to the combination of an equal load acting along the specimen centerline together with a couple of moment Pe, where e is the distance of the specimen centerline from the line through the pin centers. It should be noted that e is taken to be positive when the pin centers are on the same side of the centerline as the cracked edge and taken to be negative when they are on the other side. By the principle of superposition, the value of each stress component, and therefore the value of K, is equal to the sum of the values

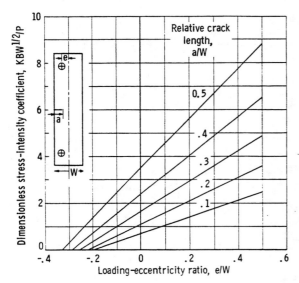

Figure 7. - Dimensionless stress-intensity coefficient as function of loading-eccentricity ratio for single-edge-notch specimens off-center pin-loaded in tension. (Ratio is positive when pin centers are on same side of centerline as cracked edge.)

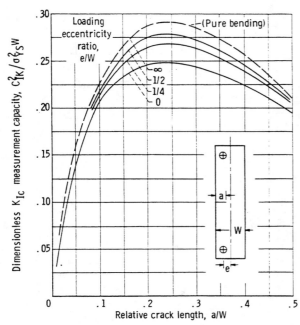

Figure 8. - Dependence of K_{Ic} measurement capacity C_{IK} of single-edge-notch specimens on relative crack length and loading-eccentricity ratio.

that would result individually from the action of P along the centerline and from the action of the couple Pe.

The component of K due to the action of the couple Pe is readily obtained from the present results for pure bending. A good approximation to the component of K due to the centerline tensile load P can be obtained from the results of reference 1. It should be appreciated that the results of reference 1 relate to a specimen loaded uniformly in tension normal to the ends which is not exactly equivalent to a specimen loaded through pins on the centerline. A pin-loaded single-edge-notch specimen bends slightly in proportion to the load, the net effect being that K/P is slightly less for a specimen pin-loaded along the centerline than for the same specimen uniformly loaded at the ends. The magnitude of this effect is discussed in reference 2 in which experimental compliance-measurement results for pin-loaded specimens are compared with the results of reference 1. For the present purpose the magnitude of the effect is negligible.

Figure 7 shows the dimensionless quantity $KBW^{1/2}/P$ against e/W for various values of a/W. By superposition, this quantity is equal to the sum of the component $(KBW^{1/2}/P)_t$, due to the uniform tensile load P, and the component $(KBW^{1/2}/P)_b$ due to the couple Pe. The tensile component was obtained from reference 1 (in that reference the symbol P denotes load per unit thickness, which is P/B in this report). The component $(KBW^{1/2}/P)_b$ is equal to Ye/W, because $M = Pe$, and therefore

$$Y = \left(\frac{K^2 B^2 W^3}{M^2}\right)^{1/2} = \frac{W}{e}\left(\frac{KBW^{1/2}}{P}\right)_b$$

For a given value of a/W, the values of

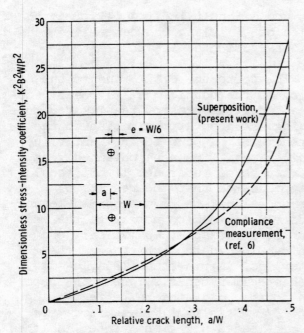

Figure 9. - Comparison of results of present work, for single-edge-notch tension specimens having loading eccentricity ratio of 1/6, with results of experimental measurements.

$(KBW^{1/2}/P)_t$ and of Y are unique, hence,

$$\frac{KBW^{1/2}}{P} = \left(\frac{KBW^{1/2}}{P}\right)_t + Y\frac{e}{W}$$

is a linear function of e/W, as shown in figure 7.

When the same criterion is used for this case as for pure bending, a test result will be valid only if abrupt crack extension occurs at a value of P not exceeding $\sigma_{YS}B(W-a)^2/(W+2a+6e)$. This condition was applied in the same manner as previously for pure bending, and values of $C_{IK}^2/\sigma_{YS}^2 W$ were calculated for combined bending and tension on the basis of the values of $KBW^{1/2}/P$ shown in figure 7. Figure 8 shows curves of $C_{IK}^2/\sigma_{YS}^2 W$ against a/W for values of e/W equal to 0, 0.25, and 0.5 and also shows the curve for pure bending from figure 5 for comparison. It is apparent that the measurement capacity is greatest in the range of a/W between 0.2 and 0.3 in all cases. The measurement capacity increases with e/W, but the magnitude of the effect is small, the difference between $C_{IK}/\sigma_{YS}W^{1/2}$ for $e/W = \infty$ (pure bending) and for $e/W = 0$ (uniform tension) being less than 10 percent. Uncertainty about the validity of the basis of comparison is at least of this order, so that for practical purposes it can be assumed that the measurement capacity of single-edge-notch specimens is independent of the manner of loading. For other reasons it is recommended elsewhere (ref. 3) that e/W should be zero for single-edge-notch specimens tested in tension.

A comparison of the results obtained by superposition for $e/W = 1/6$ with the experimental compliance-measurement results of reference 6 for a comparable pin-loaded specimen is shown in figure 9. The results of reference 6 cannot be considered very accurate for reasons discussed in reference 2; nevertheless, the agreement is fairly good for values of a/W up to about 0.3. The increasing discrepancy with increasing a/W beyond 0.3 is, in part, attributable to the fact that the bending, which occurs in the pin-loading of the compliance-measurement specimen, has the effect of slightly reducing the effective bending moment. This effect is neglected in the superposition calculations, as discussed earlier.

CONCLUSIONS

Stress-intensity factors computed by the boundary-collocation procedure for single-edge-notch specimens in pure bending were in good agreement with results derived from experimental compliance measurements. Because of this agreement between two entirely different methods, either result can be used with confidence.

The range of relative crack length a/W within which the K_{Ic} measurement capacity of a bend specimen is greatest between 0.2 and 0.3. This estimate results from the assumption that the nominal stress at the position of the crack tip should not exceed the yield strength in a valid K_{Ic} test.

Stress-intensity factors for single-edge-notched specimens loaded in combined bending and tension can be calculated by appropriate superposition of the available results for uniform tension and for pure bending. The K_{Ic} measurement capacity of single-edge-notch specimens that are loaded off-center in tension is only marginally influenced by the eccentricity of loading.

Lewis Research Center,
   National Aeronautics and Space Administration,
      Cleveland, Ohio, October 29, 1964.

APPENDIX - SYMBOLS

a	crack length or depth
B	specimen thickness
C_{IK}	estimate of maximum value of K_{Ic} that can be measured with specimen of given dimensions and yield strength
d_{2n}, d_{2n-1}	coefficients of Williams stress function
E	Young's modulus
e	distance of centerline of single-edge-notched tension specimen from line through loading pin centers
$\mathscr{G}$	strain energy release rate with crack extension per unit length of crack border, or crack extension force
K	stress-intensity factor of elastic stress field in vicinity of crack tip
K_{Ic}	critical value of K at point of instability of crack extension in first or open mode, a measure of plane strain crack toughness of material
L	length of bending moment arm
M	applied bending moment, $PL/2$

m	number of selected boundary stations used in collocation computation
P	total load applied to specimen
r	polar coordinate referred to crack tip
V	distance from crack to boundary at which stress distribution corresponding to pure bending was imposed
W	specimen width
x,y	Cartesian coordinates referred to crack tip
Y	dimensionless stress-intensity coefficient $KBW^{3/2}/M$ that is function of a/W only
θ	polar coordinate referred to crack tip
$\sigma_x, \sigma_y, \tau_{xy}$	stress components
σ_{YS}	0.2 percent offset tensile yield strength
χ	stress function

Subscripts:

b	bending moment component of Y
t	tensile component of Y

REFERENCES

1. Gross, Bernard, Srawley, John E., and Brown, William F., Jr.: Stress-Intensity Factors for a Single-Edge-Notch Tension Specimen by Boundary Collocation of a Stress Function. NASA TN D-2395, 1964.

2. Srawley, John E., Jones, Melvin H., and Gross, Bernard: Experimental Determination of the Dependence of Crack Extension Force on Crack Length for a Single-Edge-Notch Tension Specimen. NASA TN D-2396, 1964.

3. Srawley, John E., and Brown, William F., Jr.: Fracture Toughness Testing. Paper Presented at ASTM Meeting, Chicago (Ill.), June 22-26, 1964. (See also NASA TM X-52030, 1964.)

4. Bueckner, H. F.: Some Stress Singularities and Their Computation by Means of Integral Equations. Boundary Problems in Differential Equations, Langer, R. E., ed., Univ. of Wisconsin Press, 1960, pp. 215-230.

5. Lubahn, J. D.: Experimental Determination of Energy Release Rate for Notched Bending and Notched Tension. Proc. ASTM, vol. 59, 1959, pp. 885-913.

6. Sullivan, A. M.: New Specimen Design for Plane-Strain Fracture Toughness Tests. Materials Res. and Standards, vol. 4, no. 1, Jan. 1964, pp. 20-24.

7. Williams, M. L.: On the Stress Distribution at the Base of a Stationary Crack. Jour. Appl. Mech., vol. 24, no. 1, Mar. 1957, pp. 109-114.

8. Irwin, G. R.: Analysis of Stresses and Strains Near the End of a Crack Traversing a Plate. Jour. Appl. Mech., vol. 24, no. 3, Sept. 1957, pp. 361-364.

NASA Technical Note D-3092 (December 1965).

STRESS-INTENSITY FACTORS FOR THREE-POINT BEND SPECIMENS BY BOUNDARY COLLOCATION

by Bernard Gross and John E. Srawley

Lewis Research Center

SUMMARY

A boundary-value-collocation procedure was applied to the Williams stress function to determine values of the stress-intensity factor K for single-edge cracks in rectangular-section specimens subjected to three-point bending. The results are presented in terms of the dimensionless quantity $Y^2 = K^2 B^2 W^3/M^2$ where B and W are the specimen thickness and depth and M is the bending moment at midspan. The values of Y^2 as a function of relative crack depth a/W for three-point bending are appreciably lower than the corresponding values for pure bending (determined previously by the same method) and decrease as the ratio of support span to specimen depth S/W decreases. Plots of Y^2 against a/W are given for values of a/W up to 0.5 and S/W equal to 4 and 8.

The results were relatively insensitive to variations in the spread of the midspan load contact region, which was assumed to be related to the yield strength of the material. The results agreed fairly well with published results derived from experimental compliance measurements; one set gave higher values of Y^2 than the present method, and the other set gave lower values. The plane-strain fracture toughness measurement capacity of three-point bend specimens is somewhat lower than that of four-point bend specimens, but the difference is of negligible practical importance.

INTRODUCTION

Various types of crack-notch specimens are used for K_{Ic} plane-strain fracture toughness testing of materials (ref. 1). (Symbols are defined in appendix A.) The value of K_{Ic} determined in such a test is equal to that value of the crack-tip stress-intensity factor K at which the crack becomes unstable and extends abruptly in the opening mode. The stress-intensity factor K is proportional to the applied load and is a function of the

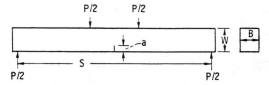

(a) Four-point bending (regarded as practically equivalent to pure bending in central test region near crack).

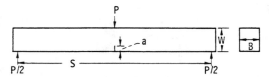

(b) Three-point bending (conventional force diagram).

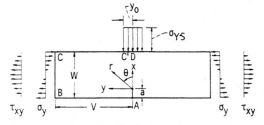

(c) Three-point bending (distributions of boundary stresses in central region extending distance V on either side of crack).

Figure 1. - Diagrammatic representation of beam loading.

specimen dimensions, particularly those of the crack, which depends upon the configuration of the specimen and manner of loading. The determination of an expression for K in terms of these factors for a particular type of specimen will be referred to as the K calibration for that specimen type. Various experimental and analytical methods of stress analysis for K calibration have been developed (ref. 2).

One important class of K_{Ic} specimens, which may be referred to briefly as single-edge-notch specimens, includes those of rectangular cross section having a single crack-notch extending from one edge. The different types of specimen within this class are distinguished primarily by the manner of loading. Previous reports by Gross, Srawley, and Brown dealt with the application of the boundary collocation method of stress analysis for K calibration to single-edge-notch specimens loaded in uniform tension (ref. 3) and to such specimens loaded either in pure bending or in combined bending and tension (ref. 4).

In practice, single-edge-notch specimens are often tested in three-point bending, whereas the results for pure bending given in reference 4 apply only to ideal four-point bending. In four-point bending (fig. 1(a)), providing that the loads are applied at positions sufficiently distant from the crack, the stress-intensity factor K depends upon four variables only, namely, the applied bending moment M, the crack length a, the specimen depth W, and the thickness B. The K calibration is conveniently expressed in terms of the dimensionless quantity $Y^2 = K^2 B^2 W^3 / M^2$, which is a function of a/W only.

In three-point bending (fig. 1(b)), however, there are two additional independent variables which might affect K. One of these variables is the support span S. For a given value of the maximum bending moment M = PS/4, there is a bending moment gradient and a shearing force, both inversely proportional to S. Thus, in terms of the dimensionless form of the K calibration, it is to be expected that Y^2 will be a function of S/W as well as a/W. The second additional variable concerns the distribution of contact pressure around the nominal position of the central loading point. The load is usually applied through a hard, cylindrical roller, and for the present purpose it is assumed that the specimen behaves in the contact region as a rigid-plastic solid with a well-defined yield strength σ_{YS}. The applied load P can then be regarded as evenly

TABLE I. – COMPARISON OF RESULTS OF PRESENT WORK WITH CORRESPONDING RESULTS FROM REFERENCES 4, 5, 6, AND 7 IN TERMS OF DIMENSIONLESS QUANTITY $K^2 B^2 W^3/M^2$ AS FUNCTION OF RELATIVE CRACK LENGTH

Relative crack length, a/W	Results of				
	Boundary collocation (a)	Reference 4 (b)	Reference 5 (a)	Reference 6 (c)	Reference 7
	$Y^2 = K^2 B^2 W^3/M^2$				
0.10	11.70	12.40	9.44	10.88	10.08
.15	17.30	18.50	13.92	17.28	-----
.20	23.47	25.30	19.20	24.48	22.29
.25	30.78	33.20	26.08	33.12	-----
.30	39.84	42.80	35.20	41.92	-----
.35	51.54	55.20	48.16	57.28	-----
.40	67.18	71.41	63.04	-----	-----
.45	88.78	92.70	83.36	-----	-----
.50	119.74	123.01	108.80	-----	140.48

[a]Ratio of support span to specimen depth, 8.
[b]Pure bending.
[c]Ratio of support span to specimen depth, 10.

distributed over a rectangular area of breadth B (the thickness of the beam) and length $2y_o$ (fig. 1(c)), such that $2y_o B$ is equal to P/σ_{YS}. Hence, the third variable which affects the value of Y^2 can be taken as $2y_o/W$.

Calibrations for K were computed for S/W values of 4, 6, 8, and 10 to cover what was considered to be a practical range. In the past, a ratio of 8 has been commonly employed, but there is an increasing tendency toward the use of smaller ratios in order to conserve test material. On the other hand, the accuracy of K calibrations for S/W less than 4 was considered dubious because of the increasing difficulty of representing the physical loading conditions accurately as this ratio becomes smaller.

The ratio $2y_o/W$ has an upper limit which is related to the condition for a valid K_{Ic} measurement, namely, that the nominal stress at the crack tip should not exceed the yield strength of the material (ref. 1). From this condition it can be shown that the upper limit of $2y_o/W$ is equal to $2(1 - a/W)^2 W/3S$. Calibrations for K were conducted in parallel by using this upper limit and taking zero to be the lower limit. The differences between the results for the two limits were so small that it was not necessary to consider intermediate values of $2y_o/W$.

Results of experimental compliance measurements on three-point bend specimens have been published by Irwin, Kies, and Smith (ref. 5) and by Kies, Smith, Romine, and Bernstein (ref. 6). Limited results of an analytical study by H. F. Bueckner have been published by Wundt (ref. 7). As shown in table I, there is sufficient lack of agreement between K calibration values derived from the results of these three references to warrant the undertaking of the present study. Furthermore, these references provide no information about the extent to which the K calibration depends on the parameters S/W and $2y_o/W$. In the interest of accurate measurement of K_{Ic}, it is important that the extent of the influence of these parameters should be known.

ANALYTICAL AND COMPUTATIONAL PROCEDURE

The method of analysis consists in finding a stress function χ that satisfies the biharmonic equation $\nabla^4 \chi = 0$ and also the boundary conditions at a finite number of stations along the boundary of a single-edge-notched specimen, such as shown in figure 1 (p. 2). The biharmonic equation and the boundary conditions along the crack are satisfied by the Williams stress function (ref. 8). Because of symmetry (fig. 1), the coefficient of the sine terms in the general stress function must be zero; hence,

$$\chi(r,\theta) = \sum_{n=1,2\ldots}^{\infty} \left\{ (-1)^{n-1} d_{2n-1} r^{n+(1/2)} \left[-\cos\left(n - \frac{3}{2}\right)\theta + \frac{2n-3}{2n+1} \cos\left(n + \frac{1}{2}\right)\theta \right] \right.$$

$$\left. + (-1)^n d_{2n} r^{n+1} \left[-\cos(n-1)\theta + \cos(n+1)\theta \right] \right\} \quad (1)$$

The stresses in terms of χ obtained by partial differentiation are as follows:

$$\left. \begin{aligned} \sigma_y &= \frac{\partial^2 \chi}{\partial x^2} = \frac{\partial^2 \chi}{\partial r^2} \cos^2\theta - 2 \frac{\partial^2 \chi}{\partial \theta\, \partial r} \frac{\sin\theta \cos\theta}{r} + \frac{\partial \chi}{\partial r} \frac{\sin^2\theta}{r} + 2 \frac{\partial \chi}{\partial \theta} \frac{\sin\theta \cos\theta}{r^2} + \frac{\partial^2 \chi}{\partial \theta^2} \frac{\sin^2\theta}{r^2} \\ \sigma_x &= \frac{\partial^2 \chi}{\partial y^2} = \frac{\partial^2 \chi}{\partial r^2} \sin^2\theta + 2 \frac{\partial^2 \chi}{\partial \theta\, \partial r} \frac{\sin\theta \cos\theta}{r} + \frac{\partial \chi}{\partial r} \frac{\cos^2\theta}{r} - 2 \frac{\partial \chi}{\partial \theta} \frac{\sin\theta \cos\theta}{r^2} + \frac{\partial^2 \chi}{\partial \theta^2} \frac{\cos^2\theta}{r^2} \\ -\tau_{xy} &= \frac{\partial^2 \chi}{\partial x\, \partial y} = \sin\theta \cos\theta \frac{\partial^2 \chi}{\partial r^2} + \frac{\cos 2\theta}{r} \frac{\partial^2 \chi}{\partial r\, \partial \theta} - \frac{\sin\theta \cos\theta}{r^2} \frac{\partial^2 \chi}{\partial \theta^2} - \frac{\sin\theta \cos\theta}{r} \frac{\partial \chi}{\partial r} - \frac{\cos 2\theta}{r^2} \frac{\partial \chi}{\partial \theta} \end{aligned} \right\} \quad (2)$$

The boundary collocation procedure consists in solving a set of $2m$ simultaneous algebraic equations which correspond to the known values of χ and either $\partial\chi/\partial x$ or $\partial\chi/\partial y$ at m selected stations along the boundary ABCC'D of figure 1(c); thus, values for the first $2m$ coefficients of the Williams stress function are obtained when the remaining terms are neglected. Only the value of the first coefficient d_1 is needed for the present purpose since the stress-intensity factor K is equal to $-\sqrt{2\pi d_1}$, as shown in reference 4.

The required values of χ and its first derivatives at the m selected boundary stations were obtained from distributions of bending moment, shear, and contact stresses (fig. 1(c)), equivalent to the concentrated loads of the conventional force diagram (fig. 1(b)). The equations for these boundary values, in dimensionless form, are as follows:

Along AB

$$\frac{\chi}{P} = 0; \quad W\frac{\partial(\chi/P)}{\partial x} = 0$$

Along BC

$$\frac{\chi}{P} = \frac{6}{BW^3}\left(\frac{S}{2} - V\right)\left[-\frac{x^3}{6} + \frac{x^2}{4}(W - 2a) + \frac{xa}{2}(W - a) + \frac{a^2}{2}\left(\frac{W}{2} - \frac{a}{3}\right)\right]$$

$$W\frac{\partial(\chi/P)}{\partial y} = \frac{-6}{BW^2}\left[\frac{-x^3}{6} + \frac{x^2}{4}(W - 2a) + \frac{xa}{2}(W - a) + \frac{a^2}{2}\left(\frac{W}{2} - \frac{a}{3}\right)\right]$$

Along CC'

$$\frac{\chi}{P} = \frac{S - 2y}{4B}; \quad W\frac{\partial(\chi/P)}{\partial x} = 0$$

Along C'D

$$\frac{\chi}{P} = \frac{1}{4B}\left[S - 2y - \frac{(y - y_0)^2}{y_0}\right]; \quad W\frac{\partial(\chi/P)}{\partial x} = 0$$

(3)

The distance V of the end boundary from the crack (fig. 1(c)), was chosen to be approximately 1.5 W, the exact value being different for different values of a/W as a matter of computational convenience. From physical considerations it is clear that the boundary should be chosen neither close to the crack nor close to the support point, because of the stress-field disturbances near these positions. Preliminary studies established that V/W equal to 1.5 was about optimum and that minor variations from 1.5 had a negligible effect on the K calibration. A more detailed discussion of the effect of the choice of V/W has been given for the case of pure bending in reference 4.

For each set of selected values of the primary variable a/W and the parameters S/W and $2y_0/W$, the collocation computation was carried out four times, using successively 15, 18, 21, and 24 boundary stations. In no case was the variation among the four d_1 values so obtained as great as 1 percent, and in most cases it was a small fraction of 1 percent. From this, together with the nature of the trends of the d_1 values, it was concluded that those values corresponding to 24 boundary stations were very close to the limit values for large numbers of boundary stations. These values were accordingly used to calculate the results reported herein.

The d_1 values so obtained were used to calculate values of the square of the dimensionless stress-intensity coefficient, namely $Y^2 = K^2 B^2 W^3/M^2$, where $M = PS/4$. In this general form the results are of more immediate utility than they would be in the form of values of K/M for a specimen of specific, arbitrary dimensions.

RESULTS AND DISCUSSION

Dependence of $K^2 B^2 W^3/M^2$ on S/W and $2y_0/W$

Figure 2 shows plots of Y^2 against S/W for constant a/W values of 0.1, 0.2, and 0.3. In each case the corresponding value of Y^2 for pure bending is shown as a horizontal dashed line for comparison. The values for pure bending were taken from reference 4 and, of course, are independent of S/W. It is clear that the value of Y^2 for three-point loading is always lower than that for pure bending, but that the difference decreases with increasing S/W. For a/W equal to 0.3, the ratio of the value of Y^2 for three-point bending to that for pure bending is 0.87 when S/W is equal to 4 and increases to 0.94 when S/W is equal to 10.

The trend of the curves in figure 2 indicates that Y^2 becomes increasingly sensitive to S/W as this ratio decreases. This is a consequence of the increasing complexity of the overall stress-field pattern with decreasing S/W and is the reason why computations were not conducted for S/W less than 4. It was considered that a K calibration

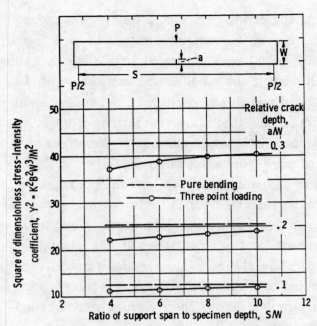

Figure 2. - Dependence of square of stress-intensity coefficient for three-point bending on ratio of support span to specimen depth.

for S/W substantially less than 4 would be of very dubious accuracy, in fact, more misleading than useful.

The values of Y^2 plotted in figure 2 are those for the upper limit values of the parameter $2y_o/W$, equal to $2(1 - a/W)^2 W/3S$. If the corresponding values for the lower limit of $2y_o/W$, equal to zero, had also been plotted, they would have been virtually indistinguishable from their companions. For S/W equal to 4, the values of Y^2 for the lower limit of $2y_o/W$ were about 1 percent greater than the plotted values. For S/W equal to 8, the values of Y^2 for the lower limit were less than 1/2 percent greater than the plotted values. For practical purposes, therefore, the effect of the parameter $2y_o/W$ on the K calibration can be considered negligible.

CALIBRATIONS OF K FOR S/W EQUAL TO 4 AND 8

Figure 3 shows plots of the computed values of Y^2 against a/W for S/W equal to 8 and 4, again for the upper limit values of the parameter $2y_o/W$.

The following empirical equations are compact expressions of the same results for values of a/W up to 0.35

$$Y_8^2 = 134\, a/W - 247(a/W)^2 + 813(a/W)^3 \tag{4}$$

$$Y_4^2 = 130\, a/W - 262(a/W)^2 + 820(a/W)^3 \tag{5}$$

where Y_8^2 refers to S/W equal to 8, and Y_4^2 to S/W equal to 4. These equations were obtained by least-squares-best-fit computer programs for cubics in a/W, incorporating the known condition that K should be zero when a/W is zero. Only the results for a/W up to 0.35 were used in fitting the equations since it is neither desirable nor

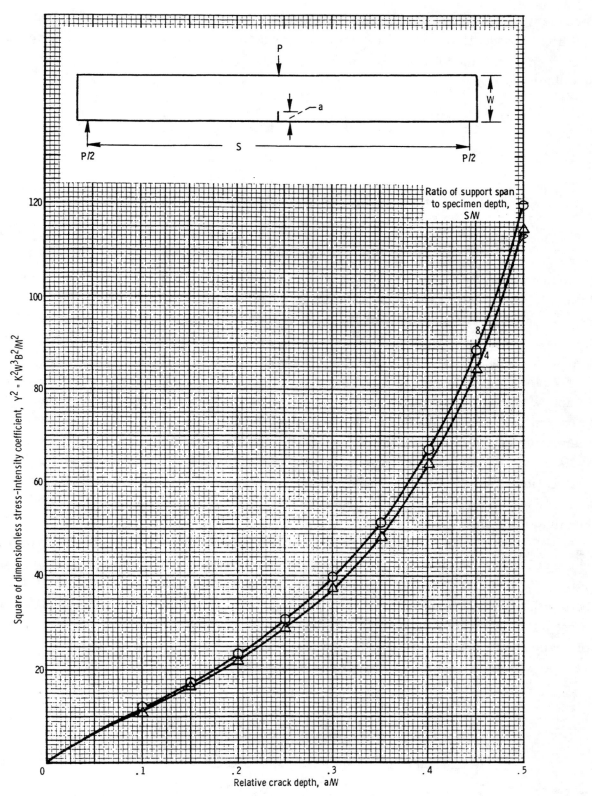

Figure 3. – Dependence of square of stress-intensity coefficient for three-point bending on relative crack depth. Bending moment, PS/4.

necessary to use bend specimens having cracks deeper than about 0.35 W. The results are fitted by equations (4) and (5) with an average deviation of less than 0.2 percent, the maximum deviation not exceeding 0.5 percent. The corresponding equation for pure bending (ref. 4) is

$$Y^2_{PB} = 139\, a/W - 221(a/W)^2 + 783(a/W)^3 \tag{6}$$

Comparison With Results of Independent Studies

The results of the present study for S/W equal to 8 are compared with the results of references 5, 6, and 7 in table I (p. 3). In general, the present results are in between the experimentally derived results of references 5 and 6, those of reference 5 being lower than the collocation values and those of reference 6 being higher. The results from reference 6 in table I are calculated from the fitting equation to the experimental data that is given in that reference. Values for a/W greater then 0.35 are omitted because the fitting equation obviously does not represent the experimental results in this range.

While the present results and those of reference 6 agree fairly well, the agreement is not as good as that demonstrated in reference 4 between the boundary collocation results for pure bending and experimental results for four-point bending. Furthermore, the agreement between the two sets of experimental results for three-point bending (refs. 5 and 6) is distinctly poorer than the agreement of either with the present boundary collocation results. These facts would appear to indicate that it is inherently more difficult to obtain an accurate K calibration for three-point bending than it is for four-point bending.

Reference 7 cites only three actual results from Bueckner's unpublished analysis of notched bars in bending. These results, converted to values of Y^2, are listed in table I. For a/W equal to 0.2, the agreement with the present result is quite good, in fact, better than the agreement with either set of experimentally derived results. The agreement for a/W values of 0.1 and 0.5, however, is no more than fair. This is a matter of academic rather than practical concern since the optimum range of a/W for K_{Ic} testing with bend specimens is between 0.15 and 0.35.

K_{Ic} Measurement Capacity in Relation to W

As discussed at some length in reference 1, the greatest value of K_{Ic} that can be measured accurately with a given bend specimen, or the K_{Ic} measurement capacity of the specimen, depends on both the depth W and the thickness B of the specimen. The

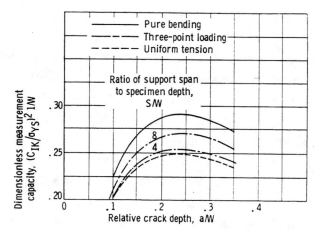

Figure 4. - Dependence of plane strain crack toughness measurement capacity on relative crack depth for single-edge-notch specimen subjected to various types of loading.

dependence on B is unrelated to the present work and will not be discussed here. Given a bend specimen of adequate thickness, the symbol C_{IK} is used to denote the maximum value of K_{Ic} that can be measured with acceptable accuracy by using that specimen. The currently accepted criterion for evaluating C_{IK} is that the nominal stress at the position of the crack tip should not exceed the yield strength of the material in a valid K_{Ic} test; that is, $6M/B(W - a)^2$ should not exceed σ_{YS}. Substituting in the expression $Y^2 = K^2 B^2 W^3/M^2$, and transposing give

$$C_{IK}^2/\sigma_{YS}^2 W = Y^2(1 - a/W)^4/36 \qquad (7)$$

The dimensionless quantity $C_{IK}^2/\sigma_{YS}^2 W$ is a K_{Ic} measurement efficiency factor which can be similarly evaluated for other types of specimens, as in reference 1. In all cases it is a function of a/W since Y^2 is a function of a/W. In the special case of three-point bending it is also somewhat dependent on S/W, to the same extent that Y^2 depends upon S/W. For any given values of a/W and S/W, there is a unique value of $C_{IK}^2/\sigma_{YS}^2 W$, and therefore, C_{IK} is proportional to the yield strength of the material and to the square root of the specimen depth W.

Plots of $C_{IK}^2/\sigma_{YS}^2 W$ against a/W for three-point bending with S/W equal to 4 and 8 are shown in figure 4. Also shown for comparison are similar plots for pure bending and uniform tension, taken from references 4 and 3, respectively. The measurement efficiency is greatest when a/W is about 0.25 in all cases. Furthermore, over the range of a/W from 0.15 to 0.35 the measurement efficiency is within 10 percent of the maximum value in each case. The differences in K_{Ic} measurement capacity according to the different methods of loading shown in figure 4 are insufficient to be of much practical importance. The degree of uncertainty about the criterion used to calculate the measurement efficiency factors precludes drawing any fine distinctions in this respect. For practical purposes, therefore, it is reasonable to assume that the measurement capacity of a single-edge-notch specimen will be independent of the manner in which it is loaded. A convenient working rule for all single-edge-notch specimens is that C_{IK} is about $\sigma_{YS} W^{1/2}/2$ (for a/W in the range 0.15 to 0.35).

CONCLUDING REMARKS

In this report the results obtained by the boundary collocation procedure are expressed in general form in terms of the square of the dimensionless stress-intensity coefficient $Y^2 = K^2 B^2 W^3/M^2$. In the case of pure bending, Y^2 is a function of the relative crack length a/W only. In the case of three-point bending, the computed values of Y^2 were appreciably lower than the corresponding values for pure bending (computed previously by the same method), the more so the smaller the ratio of support-span to specimen depth S/W. Thus, a different K calibration plot of Y^2 against a/W is needed for each different value of S/W that is used in plane-strain crack toughness testing with three-point bend specimens. Accurate plots of Y^2 against a/W and fitting equations are given for S/W equal to 8 and 4.

For S/W equal to 8 the maximum deviation of Y^2 for three-point loading from that for pure bending was about 7 percent. For S/W equal to 4 the maximum deviation was about 13 percent. Since Y^2 is proportional to K^2, the corresponding deviations for K are approximately half as great. It is considered that results for S/W substantially less than 4 would be of dubious accuracy and probably more misleading than useful.

The K calibration relation of Y^2 to a/W is also slightly affected by the spread of the contact pressure region around the center loading point. This spread depends upon the yield strength and toughness of the material and the size of the specimen tested. Over the practical range for acceptable plane-strain crack toughness measurements, the effect of this factor was at most 1 percent and can therefore be considered negligible.

The results agree fairly well with published results derived from experimental compliance measurements; one set gave higher values of Y^2 than the present method, and the other set gave lower values.

The plane-strain crack toughness measurement capacity as related to the depth of three-point bend specimens was estimated to be somewhat lower than that of four-point bend specimens, but for practical purposes this difference is probably of little importance. For all single-edge-notch specimens, whether tested in tension or bending, the measurement capacity is greatest in the range of a/W between 0.15 and 0.35.

Lewis Research Center,
    National Aeronautics and Space Administration,
        Cleveland, Ohio, August 20, 1965.

APPENDIX A

SYMBOLS

a	crack depth
B	specimen thickness
C_{IK}	estimated maximum value of K_{Ic} that can be measured with specimen of given dimensions and yield strength
d_{2n}, d_{2n-1}	coefficients of Williams stress functions
K	stress-intensity factor of elastic-stress field in vicinity of crack tip
K_{Ic}	critical value of K, at point of instability of crack extension in first or open mode, a measure of plane-strain crack toughness of material
M	applied bending moment
m	number of selected boundary stations used in collocation computation
P	total load applied to specimen
r	polar coordinate referred to crack tip
S	span
V	distance from crack to boundary selected for collocation analysis
W	specimen depth
x, y	Cartesian coordinates referred to crack tip
Y	dimensionless stress-intensity coefficient, $KBW^{3/2}/M$
$2y_o$	length over which applied load at center was assumed to be distributed
θ	polar coordinate referred to crack tip
σ_x	stress component in x-direction
σ_y	stress component in y-direction
σ_{YS}	0.2 percent offset tensile yield strength
τ_{xy}	shearing stress component
χ	stress function

REFERENCES

1. Srawley, John E.; and Brown, William F., Jr.: Fracture Toughness Testing Methods. Fracture Toughness Testing and Its Applications, STP No. 381, ASTM, 1965, pp. 133-198.

2. Paris, P. C.; and Sih, G. C.: Stress Analysis of Cracks. Fracture Toughness Testing and Its Applications, STP No. 381, ASTM, 1965, pp. 30-82.

3. Gross, Bernard; Srawley, John E.; and Brown, William F., Jr.: Stress-Intensity Factors for a Single-Edge-Notch Tension Specimen by Boundary Collocation of a Stress Function. NASA TN D-2395, 1964.

4. Gross, Bernard; and Srawley, John E.: Stress-Intensity Factors for Single-Edge-Notch Specimens in Bending or Combined Bending and Tension by Boundary Collocation of a Stress Function. NASA TN D-2603, 1965.

5. Irwin, G. R.; Kies, J. A.; and Smith, H. L.: Fracture Strengths Relative to Onset and Arrest of Crack Propagation. Proc. ASTM vol. 58, 1958, pp. 640-660.

6. Kies, J. A.; Smith, H. L.; Romine, H. E.; and Bernstein, H.: Fracture Testing of Weldments. Fracture Toughness Testing and Its Applications, STP No. 381, ASTM, 1965, pp. 328-356.

7. Wundt, B. M.: A Unified Interpretation of Room-Temperature Strength of Notched Specimens as Influenced by their Size. Paper No. 59-MET-9, ASME, 1959.

8. Williams, M. L.: On the Stress Distribution at the Base of a Stationary Crack. J. Appl. Mach., vol. 24, no. 1, Mar. 1957, pp. 109-114.

Reprinted from *International Journal of Fracture Mechanics*, Vol. 4(3), pp. 267-276
(September 1968), and errata Vol. 6, p. 87 (1970).

ELASTIC DISPLACEMENTS FOR VARIOUS EDGE–CRACKED PLATE SPECIMENS

Bernard Gross, Ernest Roberts, Jr., and John E. Srawley

(Lewis Research Center, National Aeronautics and Space Administration, Cleveland, Ohio)

ABSTRACT

The relative displacement per unit load of two conjugate points is used as a quantitative indicator of crack extension in plane strain fracture toughness K_{Ic} measurements. The necessary displacement data are presented here in dimensionless form for five types of single–edge–crack specimens: three–point bending, pure bending, remote axial tension, and eccentric tension of compact rectangular and tapered varieties. The results were obtained by a boundary collocation method of elastic analysis and are highly precise.

INTRODUCTION

In plane strain fracture toughness K_{Ic} testing, the relative displacement per unit load v_y/P of two conjugate points is used as a quantitative indicator of crack extension during the test[1-3]. The relation between v_y/P and relative crack length a/W depends on the type of specimen employed. The relations for various specimens are determined by linear elastic strain analysis, or by direct experimental measurements in which slots are used to simulate cracks. This report presents v_y/P data obtained by a boundary collocation method of analysis for several types of single–edge–crack plate specimens. The use of these specimens in K_{Ic} testing has been explained by Brown and Srawley[1,2].

The following types of loading were investigated (figs. 1 and 2): three–point bending having a 4/1 span to width ratio, pure bending, remote axial tension, and eccentric tension (compact rectangular and tapered varieties of specimens). The results are given in the form of tables of a dimensionless displacement coefficient as a function of the major variable, relative crack length a/W, and of the secondary variables x/W and y/W which represent the gage point locations in relation to the crack tip.

Displacement measurement for detection of crack extension in K_{Ic} test should not be confused with measurements of specimen compliance (reciprocal stiffness) for determination of energy release rates[1]. The intent of a compliance experiment is to determine the work done by the loading forces, and the displacement measured must be chosen appropriately. For evaluation of crack extension, any convenient gage points can be used for the displacement measurement. The most sensitive positions are those close to the crack, which are precisely the least suitable for compliance when the specimen is remotely loaded. However, for the eccentric tension specimens discussed in this paper, the gage point positions that are convenient for crack extension detection are also suitable for compliance determinations.

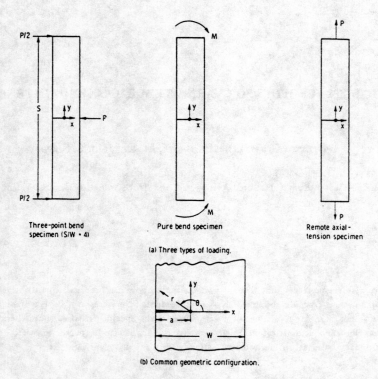

Figure 1: Bending— and axial—tension plate specimens.

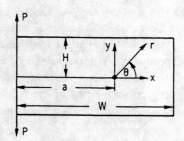

(a) Compact rectangular.

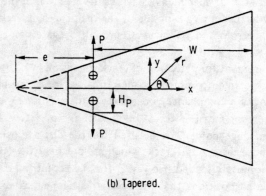

(b) Tapered.

Figure 2: Eccentric tension plate specimens.

SYMBOLS

Units may be in any consistent system since only dimensionless combinations are used in this report.

a	crack length
B	specimen thickness
d_{2n}, d_{2n-1}	coefficients of Williams stress function
E	Young's modulus
e	distance from wedge tip to line of load application
H	uniform depth of nontapered split arm
H_p	depth of tapered split arm at load line
K	stress intensity factor of crack tip elastic stress field
K_{Ic}	plane strain fracture toughness, measured in terms of opening mode stress intensity factor K_I in units of (stress) $\times$ (length)$^{1/2}$ (see refs. 1 and 2)
M	bending moment
P	load
r	polar coordinates referred to crack tip
S	span
U	displacement in radial direction
u_x	displacement in x–direction
V	displacement in tangential direction
v_y	displacement in y–direction between ±y locations
W	specimen width
x, y	Cartesian coordinate system referred to crack tip
θ	polar coordinate referred to crack tip
ν	Poisson's ratio
φ	harmonic function in displacement equation
χ	Airy stress function

ANALYSIS

The method of analysis is described in some detail by Gross and co–authors[4-8]. Its earliest use in solid mechanics is by Barta[9]. Considerable detail is given by Green[10] and Howland and Knight[11]. The method is called variously 'boundary collocation' and 'point matching'. Briefly, it consists of truncating a series solution to the appropriate partial differential equation, and making use of the boundary values at a finite number of points to evaluate its coefficients. It can be shown[12] that the biharmonic equation in terms of an Airy stress function properly describes the plane elastic problem. It is convenient to use the Williams stress function for our analysis[13,14]. It is an Airy stress function, it identically satisfies the biharmonic equation, and it identically satisfies the boundary conditions along the crack surface.

The equation to be solved is

$$\Delta^4 \chi = 0$$

where the origin of the polar coordinate system is the crack tip (fig. 3). The solution given by Williams is

$$\chi(r,\theta) = \sum_{n=1,2,3...}^{\infty} \left\{ (-1)^{n-1} d_{2n-1} r^{n+\frac{1}{2}} \left[-\cos\left(n-\frac{3}{2}\right)\theta + \frac{2n-3}{2n+1} \cos\left(n+\frac{1}{2}\right)\theta \right] \right.$$
$$\left. + (-1)^n d_{2n} r^{n+1} \left[-\cos(n-1)\theta + \cos(n+1)\theta \right] \right\}$$

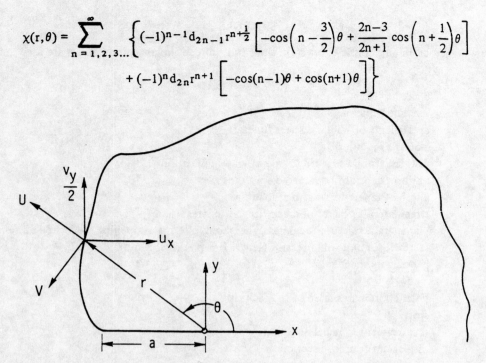

Figure 3: Geometry defining displacements in x- and y-directions.

The displacements in terms of the stress function and a harmonic function $\varphi(r,\theta)$ are given by Williams[13,14] and Coker and Filon[15]. In infinite series form they become

$$\varphi(r,\theta) = \sum_{n=1,2,3...}^{\infty} (-1)^{n+1} r^{n-1} 4 \left[\frac{-d_{2n-1}}{r^{1/2}} \frac{\sin\left(n-\frac{3}{2}\right)\theta}{n-\frac{3}{2}} + \frac{d_{2n}}{n-1} \sin(n-1)\theta \right]$$

$$2\mu V(r,\theta) = -\frac{1}{r} \frac{\partial \chi}{\partial r} + (1-\sigma) r^2 \frac{\partial \varphi}{\partial \theta}$$

$$2\mu U(r,\theta) = -\frac{\partial \chi}{\partial r} + (1-\sigma) r \frac{\partial \varphi}{\partial \theta}$$

where for plane stress

$$\sigma = \frac{\nu}{(1+\nu)}$$

for plane strain

$$\sigma = \nu$$

and for either plane stress or plane strain

$$\mu = \frac{E}{2(1+\nu)}$$

The resulting displacement equations are

$$V(r,\theta) = \frac{1}{2\mu} \sum_{n=1,2,3...}^{\infty} \left\{ (-1)^n d_{2n-1} r^{n-\frac{1}{2}} \left[\left(\frac{5}{2} + n - 4\sigma \right) \sin\left(n - \frac{3}{2}\right)\theta \right. \right.$$
$$\left. - \left(\frac{2n-3}{2} \right) \sin\left(n + \frac{1}{2}\right)\theta \right] + (-1)^n d_{2n} r^n \left[(n-\sigma) \sin(n-1)\theta \right.$$
$$\left. \left. - (n+1) \sin(n+1)\theta \right] \right\}$$

$$U(r,\theta) = \frac{1}{2\mu} \sum_{n=1,2,3...}^{\infty} \left\{ (-1)^n d_{2n-1} r^{n-\frac{1}{2}} \left[\left(\frac{7}{2} - n - 4\sigma \right) \cos\left(n - \frac{3}{2}\right)\theta \right. \right.$$
$$\left. + \left(n - \frac{3}{2} \right) \cos\left(n + \frac{1}{2}\right)\theta \right] + (-1)^{n+1} d_{2n} \left[(3-n-4\sigma) \cos(n-1)\theta \right.$$
$$\left. \left. + (n+1) \cos(n+1)\theta \right] \right\}$$

For the special case of $\theta = \pi$, it can be deduced from these equations that the displacements for plane stress are independent of Poisson's ratio ν, and that the displacements for plane strain are equal to $(1 - \nu^2)$ times those for plane stress. Since this simplification applies to nearly all the results obtained, it was convenient to compute the results for plane stress.

The displacements in the x— and y—directions (fig. 3) are

$$u_x = U \cos\theta - V \sin\theta$$
$$v_y = 2(U \sin\theta + V \cos\theta)$$

It is to be noted that v_y, as defined above, is the displacement indicated by a gage mounted across the crack, that is, the relative displacement of a pair of conjugate points.

The number of terms in the series, and hence the number of boundary points satisfied, was progressively increased until negligible changes occurred in the values of the displacement. The relationship between calculated displacement and the number of terms in the series is shown graphically in fig. 4 for one specimen configuration.

All quantities used in the analysis were made dimensionless. Hence, all crack lengths are relative to specimen width a/W, all coordinates are relative to specimen width x/W and y/W, and all displacements are relative to elastic modulus, specimen thickness, and either load or moment per unit depth W, $v_y BE/P$ and $v_y EBW/6M$.

RESULTS AND DISCUSSION

Dimensionless displacement coefficients for plane stress are tabulated for values of the relative crack length a/W up to 0.7 in Tables I to VI. The results for three—point bend specimens S/W = 4 (Tables I and II) are more detailed than the others because these specimens are of particular current interest for standardized K_{Ic} measurement[2]. Tables I and II show the displacements for different gage point locations, respectively; along the crack, and along the specimen edge normal to the crack. The displacement is considerably more sensitive to position along the crack than along the specimen edge normal to the crack. For this reason, it is desirable in practice to locate the gage on the edge of the specimen. However, in K_{Ic} measurement, the factor needed is the derivative of the logarithm

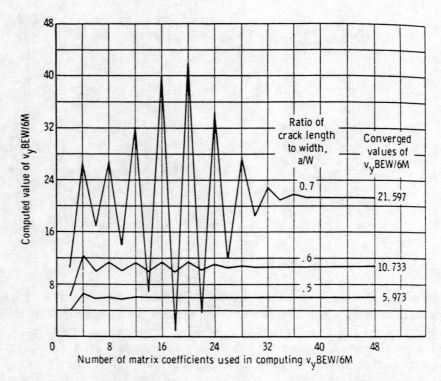

Figure 4: Typical variation of dimensionless displacement coefficient $v_y BEW/6M$ as function of number of coefficients used in computation. Displacements are for three–point bend specimens computed at specimen crack edge.

TABLE I

Dimensionless displacements along crack for three–point bend specimens ($S/W = 4$) (plane stress)

Gage location		a/W					
		0.3		0.5		0.7	
x/W	y/W	$EBv_yW/6M$	$EBu_xW/6M$	$EBv_yW/6M$	$EBu_xW/6M$	$EBv_yW/6M$	$EBu_xW/6M$
−a/32W	±0	0.316	0.0012	0.722	−0.0024	1.948	−0.0230
−a/16W	↓	.450	.0025	1.041	− .0045	2.876	− .0436
−a/8W		.644	.0052	1.525	− .0080	4.391	− .0789
−a/4W		.933	.0107	2.298	− .0120	7.050	− .1303
−a/2W		1.377	.0218	3.611	-----	12.019	− .1802
−a/W	↓	2.099	-----	5.972	-----	21.597	-----

of the displacement coefficient with respect to the logarithm of the relative crack length[1]*, and this factor is less sensitive to gage location than is the displacement coefficient itself. Therefore, the results obtained in K_{Ic} tests will show low sensitivity to small variations in gage location.

As a matter of interest, the displacements parallel to the crack u_x are given in Table I as well as the displacements normal to the crack. These lengthwise displacements are comparatively small

---

* For example, for remote axial tension, the factor is $[d \log(v_y EB/P)/d \log(a/W)]$ which is equal to $[(a/W)/(v_y EB/P)] [d(v_y EB/P)/d(a/W)]$ as in fig. 40 of ref. 1.

TABLE II

Dimensionless displacements at edge for three—point bend specimens (plane stress) with comparative experimental results (S/W = 4)

Gage location		a/W						Source
x/W	y/W	0.2	0.3	0.4	0.5	0.6	0.7	
		$EBv_yW/6M$						
−a/W	±0	1.184	2.091	3.520	5.973	10.733	21.597	Collocation
	± .10	1.217	2.109	3.529	5.987	10.751	21.625	Collocation
	± .20	1.453	2.293	3.696	6.154	10.940	21.865	Collocation
	± .50	2.385	3.082	4.388	6.817	11.632	22.656	Collocation
	± .10	----	1.94	3.46	5.60	9.74	19.9	Fisher, et al. (ref. 3)

and are not of any practical concern.

The last line in Table II lists experimental results by Fisher[3] for comparison. The experimental results are somewhat lower than the computed results for plane stress, but not as much as by the factor $(1 - \nu^2)$ for $\nu = 0.3$. Thus, the experimental results are bracketed by the computed results for plane stress and those for plane strain (to a close approximation). This would be expected since the region near the crack tip in the actual specimen approaches a state of plain strain, whereas regions remote from the crack tip are in a state of plane stress. The two—dimensional analysis is not capable of producing closer agreement with the experimental results. Experimental results are also compared with computed results in Table III for pure bending, and the differences are similar

TABLE III

Dimensionless displacements at edge for pure bend specimens (plane stress) with comparative experimental results

Gage location		a/W						Source
x/W	y/W	0.2	0.3	0.4	0.5	0.6	0.7	
		$EBv_yW/6M$						
−a/W	±0	1.260	2.241	3.763	6.368	11.354	22.575	Collocation
	±10	1.20	2.17	3.69	6.00	10.80	21.40	Fisher, et al. (ref. 3)

to those for three—point bending. The displacements are greater for pure bending than for three—point bending because of the different bending moment distributions.

Table IV gives results for single—edge—cracked specimens under remote axial tension on the assumption of uniform stress distribution at a distance not less than 0.8 W from the crack[4]. These results are intended to apply to pin—loaded specimens with pin centers not less than 3 W apart.

Tables V and VI give results for various compact tension specimens which are eccentrically loaded at positions close to the crackline and to the specimen edge[8]. Table V includes a comparison with experimental results from Bush and Wilson[16], which are in satisfactory agreement with the computed results in spite of the considerable difference of gage point location. It would

TABLE IV

Dimensionless displacements at edge for remote axial tension specimens (plane stress)

Gage location		a/W					
x/W	y/W	0.2	0.3	0.4	0.5	0.6	0.7
		EBv_y/P					
−a/W	±0	1.440	2.806	5.217	9.881	19.900	44.100

TABLE V

Dimensionless displacements at load line for eccentrically loaded compact rectangular tension specimens (plane stress)

Gage location		W/H	a/W					Source
x/W	y/W		0.3	0.4	0.5	0.6	0.7	
			EBv_y/P					
−a/W	±0	5/4	11.813	19.277	32.140	57.335	116.031	------
		5/3	14.266	22.968	37.015	63.320	122.050	------
		20/11	15.125	25.058	39.877	66.926	126.699	------
		25/12	18.075	29.473	46.229	75.234	136.966	------

x/W	y/W	W/H	a/W					Source
			0.333	0.389	0.444	0.500	0.556	
−a/W	±0	9/4	23.392	30.480	40.057	51.145	65.964	Collocation Bush, Wilson (ref. 16)
	±.875	9/4	24.96	30.96	38.52	49.20	63.36	

TABLE VI

Dimensionless displacements at load line for eccentrically loaded tapered tension specimens (plane stress)

Gage location		H_p/e	W/e	a/W					
x/W	y/W			0.2	0.3	0.4	0.5	0.6	0.7
				EBv_y/P					
−a/W	±0	0.2	3.2	162.8	312.7	474.2	641.9	810.0	985.4
		.3	4.2	97.3	170.8	242.4	313.6	387.5	481.0
		.4	5.0	67.0	107.8	148.3	187.3	234.9	306.3

be expected, however, that this type of specimen would be particularly insensitive to gage location along the specimen edge.

CONCLUSIONS

Boundary collocation is a satisfactory procedure for computing elastic displacements (per unit load) for single—edge—crack specimens. The results are highly precise and inexpensive of computer time, so that variations of specimen shape and gage location can be explored at little cost. To do the same experimentally would be quite expensive. The accuracy with which the results apply to actual specimens depends, of course, on the extent to which the assumed boundary conditions are equivalent to the actual load distributions.

Where the present results could be compared with existing experimental results the agreement is reasonably good (Tables II, III and V), and is adequate for K_{Ic} test purposes[2]. The results confirm the expected low sensitivity to small variations in gage point location when these points are on the specimen edge. The sensitivity to change of position of gage point location along the crack is greater.

Received July 20, 1967.

REFERENCES

1. W.F. Brown, Jr.; J.E. Srawley — *Plane Strain Crack Toughness Testing of High Strength Metallic Materials*, Spec. Tech. Publ. No. 410, ASTM (1966).

2. J.E. Srawley; M.H. Jones; W.F. Brown, Jr. — Mat. Res. & Standards, *7*, 6, 262—266 (June 1967).

3. D.M. Fisher; R.T. Bubsey; J.E. Srawley — 'Design and Use of Displacement Gage for Crack Extension Measurements', NASA TN D—3724 (1966).

4. B. Gross; J.E. Srawley; W.F. Brown, Jr. — 'Stress—Intensity Factors for a Single—Edge—Notch Tension Specimen By Boundary Collocation of a Stress Function', NASA TN D—2395 (1964).

5. B. Gross; J.E. Srawley — 'Stress—Intensity Factors for Single—Edge—Notch Specimens in Bending or Combined Bending and Tension by Boundary Collocation of a Stress Function', NASA TN D—2603 (1965).

6. B. Gross; J.E. Srawley — 'Stress—Intensity Factors for Three—Point Bend Specimens by Boundary Collocation', NASA TN D—3092 (1965).

7. B. Gross; J.E. Srawley — 'Stress—Intensity Factors by Boundary Collocation for Single—Edge—Notch Specimens Subject to Splitting Forces', NASA TN D—3295 (1966).

8. J.E. Srawley; B. Gross — Mat. Res. & Standards, *7*, 4, 155—162 (April 1967); see also NASA TN D—3820 (1967).

9. J. Barta — Z. Angew. Math. Mech., *17*, 3, 184—185 (June 1937).

10. A.E. Green — Proc. Roy. Soc. London, *A176*, 964, 121—139 (Aug. 28, 1940).

11. R.C.J. Howland; R.C. Knight — Phil. Trans. Roy. Soc. London, *A238*, 357—392 (1940).

12. I.S. Sokolnikoff — *Mathematical Theory of Elasticity*, McGraw—Hill Book Co., Inc., 2nd ed., (1956).

13. M.L. Williams — J. Appl. Mech., *24*, 1, 109—114 (March 1957).

14. M.L. Williams — J. Appl. Mech., *28*, 1, 78—82 (March 1961).

15. E.G. Coker; L.N.G. Filon — *A Treatise on Photo—elasticity*, Cambridge University Press, 2nd ed. (1957).

16. A.J. Bush; W.K. Wilson — 'Determination of Energy Release Rate for Biaxial Brittle Fracture Specimen', Westinghouse Electric Corp. Report No. WERL–8844–2 (August 1964). (Available from DDC as AD–611659).

RÉSUMÉ — On utilise le déplacement relatif par unité de change de deux points homologues comme indication quantitative de l'extension d'une fissure dans les mesures de la ténacité à la rupture K_{Ic} en état plan de déformation.

Les données sur ces déplacements sont présentées sous une forme adimensionnelle pour cinq types d'éprouvettes a fissure latérale simple: éprouvettes de flexion en trois points, de flexion pure, de traction axiale et de traction excentrique à section rectangulaire constante et variable.

Les résultats — extrêmement précis — ont été obtenus par une méthode d'analyse élastique de la correspondance des surfaces de rupture.

ZUSAMMENFASSUNG — Die relative Verdrängung von zwei konjugierten Punkten pro Belastungseinheit wird als quantitativer Indikator der Rissausdehnung bei planaren Bruchausdehnungsmassen K_{Ic} gebraucht. Die nötigen Verdrängungswerte sind in dieser Abhandlung in dimensionsloser Form für fünf Kategorien einzelachsiger Rissproben vertreten: Drei–punktbiegungen, reine Biegungen, fernachsige Spannung und exzentrische Belastung von kompakten, rechteckigen und zugespitzten Varietäten. Die Ergebnisse werden mittels einer Grenzkollokationsmethode erzielt und sind sehr genau.

Errata

Relating to the paper "Elastic Displacements For Various Edge-Cracked Specimens" by Gross, Roberts and Srawley, *International Journal of Fracture Mechanics*, Volume 4, No. 3, September 1968), Mr. D. V. Ramsamooj, Research Assistant, Ohio State University, Columbus, Ohio, brought the following corrections to the authors' attention. (See also M. L. Williams, "The Bending Stress Distribution at the Base of a Stationary Crack", Appendix, *J. Appl. Mech.*, March 1961).

The equation on page 270 regarding the V displacement should be

$$2\mu V(r,\theta) = -\frac{1}{r}\frac{\partial \chi}{\partial \theta} + (1-\sigma)r^2 \frac{\partial \psi}{\partial r}$$

and the displacement equations page 271 for U and V should be

$$V(r,\theta) = \frac{1}{2\mu}\sum_{n=1,2,3}^{\infty}\left\{(-1)^n d_{2n-1} r^{n-\frac{1}{2}}\left[(\tfrac{5}{2}+n-4\sigma)\sin(n-\tfrac{3}{2})\theta - \left(\frac{2n-3}{2}\right)\sin(n+\tfrac{1}{2})\theta\right] \right.$$
$$\left. +(-1)^n d_{2n} r^n [-(3+n-4\sigma)\sin(n-1)\theta + (n+1)\sin(n+1)\theta]\right\}$$

and

$$U(r,\theta) = \frac{1}{2\mu}\sum_{n=1,2,3}^{\infty}\left\{(-1)^n d_{2n-1} r^{n-\frac{1}{2}}\left[(\tfrac{7}{2}-n-4\sigma)\cos(n-\tfrac{3}{2})\theta + (n-\tfrac{3}{2})\cos(n+\tfrac{1}{2})\theta\right] \right.$$
$$\left. +(-1)^{n+1} d_{2n} r^n[(3-n-4\sigma)\cos(n-1)\theta + (n+1)\cos(n+1)\theta]\right\}$$

Only Table II is modified and is

Gage location		a/W							Source
x/W	y/W	0.2	0.3	0.4	0.5	0.6	0.7	0.8	
		$EBV_y W/(6M)$							
$-a/W$	± 0	1.159	2.075	3.497	5.942	10.672	21.446	51.370	Collocation
	$\pm .10$	1.166	2.077	3.497	5.943	10.672	21.445	51.370	Collocation
	$\pm .20$	1.188	2.081	3.497	5.941	10.671	21.442	51.454	Collocation
	$\pm .50$	1.465	2.261	3.622	6.043	10.768	21.544	51.626	Collocation
	$\pm .10$		1.94	3.46	5.60	9.74	19.9		Fisher et al. Ref. (3)

These results are based on solving a set of simultaneous equations in a least squares sense in that a greater number of boundary stations than unknowns in the stress function series were used.

WIDE RANGE STRESS INTENSITY FACTOR EXPRESSIONS FOR ASTM E 399 STANDARD FRACTURE TOUGHNESS SPECIMENS

John E. Srawley
Lewis Research Center, NASA
Cleveland, Ohio 44135 USA
tel: 216/433-4000

For each of the two types of specimens, bend and compact, described in the ASTM Standard Method of test for Plane Strain Fracture Toughness of Materials, E 399, a polynomial expression is given for calculation of the stress intensity factor K from the applied force P and the specimen dimensions. It is explicitly stated in the standard, however, that these expressions should not be used outside the range of relative crack length a/W from 0.45 to 0.55. While this range is sufficient for the purpose of E 399, the same specimen types are often used for other purposes over a much wider range of a/W; for example, in the study of fatigue crack growth. It is the purpose of this report to present new expressions which are at least as accurate as those in E 399-74, and which cover much wider ranges of a/W: for the three-point bend specimen from 0 to 1; for the compact specimen from 0.2 to 1. The range has to be restricted for the compact specimen because of the proximity of the loading pin holes to the crackline, which causes the stress intensity factor to be sensitive to small variations in dimensions when a/W is small. This is a penalty inherently associated with the compactness of the specimen.

The proposed expression *for the three-point bend specimen* is

$$(KB\sqrt{W})/P = \frac{3(S/W)\sqrt{\alpha}\,[1.99-\alpha(1-\alpha)(2.15-3.93\alpha+2.7\alpha^2)]}{2(1+2\alpha)(1-\alpha)^{3/2}} \tag{1}$$

for $0 \leq \alpha = a/W \leq 1$, and where B = thickness, W = width or depth, a = average crack length, and S = support span (= 4W ± 0.01W).

Using the same notation, the proposed expression *for the compact specimen* is

$$(KB\sqrt{W})/P = \frac{(2+\alpha)(0.886+4.64\alpha-13.32\alpha^2+14.72\alpha^3-5.6\alpha^4)}{(1-\alpha)^{3/2}} \tag{2}$$

for $0.2 \leq \alpha \leq 1$

Eqns (1) and (2) were devised to fit those available analytical results considered likely to be most accurate and reliable [1,2,3]. To examine the fidelity of these interpolation expressions it is convenient to express them in terms of dimensionless quantities which have finite, nonzero values throughout the range of a/W from 0 to 1, namely, for bend specimens, with S/W = 4, $F_B = (1-\alpha)^{3/2} KBW/P\sqrt{a}$; for compact specimens, $F_C = (1-\alpha)^{3/2}(KB\sqrt{W})/P$. Table 1 shows the comparison in these terms of the values F_{B1} and F_{C2}, obtained from (1) and (2) respectively, with the primary results, F_{B0} and F_{C0}, obtained from the sources indicated by the appended references. The accuracy of these primary results is not expected to be better than ± 0.25 percent, but the values are given to a higher degree of precision for the purpose of calculation of relative deviations.

From Table 1 it is apparent that (1) for the bend specimen agrees with the primary results within $\pm$ 0.2 percent, and that (2) for the compact specimen agrees with the primary results within $\pm$ 0.4 percent. Also, in both cases, the deviations are unsystematic and have very small average values. The polynomial interpolation expression in [2] also provides a good fit to the collocation analysis results, but was considered unsatisfactory because it has a finite value at a/W = 1 instead of the limit value, $3.978(1-a/W)^{3/2}$.

REFERENCES

[1] J. E. Srawley and B. Gross, *Engineering Fracture Mechanics* 4 (1972) 587-589.

[2] J. C. Newman, Jr., *Fracture Analysis*, STP 560, American Society for Testing and Materials, Philadelphia (1974) 105-121.

[3] J. T. Bentham and W. T. Koiter, *Mechanics of Fracture I - Methods of Analysis and Solution of Crack Problems*, Noordhoff International Publications, Leyden (1973) 159-162.

13 February 1976

Table 1
Comparison of Interpolation Expression Values with Primary Results

$\alpha = a/W$	BEND SPECIMENS (S/W = 4)			COMPACT SPECIMENS		
	F_{B0}	F_{B1}	$F_{B1}/F_{B0}-1$	F_{C0}	F_{C2}	$F_{C2}/F_{C0}-1$
0	11.932 [3]	11.940	0.0007	indefinite	1.772	---
0.1	---	9.147	---	---	2.585	---
0.2	7.513 [1]	7.519	0.0008	3.070 [2]	3.058	-0.0038
0.3	6.518 [1]	6.506	-0.0018	3.297 [2]	3.292	-0.0016
0.4	5.834 [1]	5.825	-0.0016	3.379 [2]	3.383	0.0012
0.5	5.317 [1]	5.325	0.0014	3.405 [2]	3.415	0.0030
0.6	4.919 [1]	4.927	0.0017	3.451 [2]	3.454	0.0010
0.7	4.602 [1]	4.596	-0.0012	3.544 [2]	3.541	-0.0008
0.8	---	4.321	---	3.679 [2]	3.685	0.0017
0.9	---	4.110	---	---	3.856	---
1.0	3.978 [3]	3.980	0.0005	3.978 [3]	3.978	0

STRESS INTENSITY FACTORS FOR DEEP CRACKS IN BENDING AND COMPACT TENSION SPECIMENS

INTRODUCTION

LARGE quantities of slow crack growth data under various combinations of cyclic and sustained loading are obtained and analyzed in terms of the history of the crack tip stress intensity factor[1, 2]. In some cases these data are obtained at crack lengths greater than those for which stress intensity calibrations are now available[3, 4]. Under these conditions a correct interpretation of the data is very difficult. In this note an accurate expression for the stress intensity factors for the single edge-cracked bend specimen[3], Fig. 1, and the compact tension specimen[3–5], Fig. 2, are given for deep cracks.

These expressions are based on results obtained from a boundary collocation analysis of the two test specimens. By collocating on the Williams stress function[6], accurate stress intensities were determined for crack depths up to $a/W = 0.8$. Beyond this depth the accuracy of the collocation results begins to decrease due to the proximity of the back free surface. But by expressing the valid results in the form of an appropriate dimensionless parameter, an extrapolation to $a/W = 1.0$ can be made with a minimum of error.

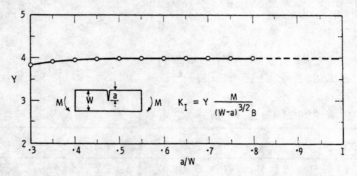

Fig. 1. Stress intensity factors for edge-cracked bend specimen. Points through which solid curve is drawn represent values obtained by boundary collocation. Dashed line is the extrapolation to $a/W = 1.0$.

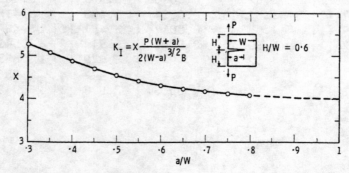

Fig. 2. Stress intensity factors for compact tension specimen. Points through which solid curve is drawn represent values obtained by boundary collocation. Dashed line is the extrapolation to $a/W = 1.0$.

SINGLE EDGE-CRACKED BEND SPECIMEN

The boundary collocation method was applied to the geometry and loading condition corresponding to a single edge-cracked specimen subject to pure bending (4 point loading). The boundary collocation was carried out on the Williams stress function and its normal derivative. In Fig. 1, the points which represent stress intensities calculated for a number of specific crack length between $a/W = 0.3$ and 0.8 are shown. Each point represents a stable value over a wide range of collocation point numbers. The points shown are in excellent agreement with results previously presented[4] over a smaller range of a/W. The values shown in Ref.[4] were also obtained by means of boundary collocation on the same stress function.

The stress intensities shown in Fig. 1 are presented here in the dimensionless form

$$Y = \frac{K_I(W-a)^{3/2}B}{M}.$$

In the limit this parameter must asymptotically approach a finite non-zero limit as a/W approaches 1.0. A limiting value of 3.99 is reached for values of a/W greater than 0.6. For values of a/W greater than 0.8, little error will be involved if Y is set equal to 4.00 as shown by the dashed line. In fact, $Y = 4.00$ should be accurate to within 1.0 per cent over the range of a/W from 0.5 to 1.0.

COMPACT TENSION SPECIMEN

In Fig. 2, the results obtained from a boundary collocation analysis of the compact tension specimen ($H/W = 0.6$) are shown for various a/W values between 0.3 and 0.8. These results are in excellent agreement with those previously presented[4, 9] over a more limited range of a/W. The stress intensities are presented here in the dimensionless form

$$X = \frac{K_I(W-a)^{3/2}B}{[P(a+W)/2]}.$$

This form is similar to that of the parameter Y in that the denominator represents the bending moment transferred through the uncracked ligament (W-a). As before, in the limit as a/W approaches one, this parameter must approach a finite non-zero limit. More specifically, this parameter should approach the same limiting value (3.99) approached by the Y parameter in the above bending analysis. As can be seen the values of X do appear to be approaching this limit as a/W approaches 1.0.

Very accurate estimates of X for values of a/W between 0.8 and 1.0 can be obtained by extrapolating the curve which passes through the points determined by collocation for $a/W < 0.8$ to a value of $X = 3.99$ at $a/W = 1.0$. A simple analytic form for this extrapolation is

$$X = 4.0 + \frac{(W-a)}{(W+a)}.$$

The second term on the right hand side represents the influence of the net tension carried through the uncracked ligament. This expression should be accurate to within 1.0 per cent over the range of a/W from 0.6 to 1.0. The accuracy of the expression is nearly the same for the WOL T type specimen ($H/W = 0.48$) over this range.

CONCLUSION

Simple expressions for the stress intensity factors of deep cracks in the single edge-cracked bend specimen and the compact tension specimen have been given. These relations were obtained by expressing the stress intensities determined by boundary collocation in the form of proper dimensionless parameters and extrapolating with a minimum of error to the limiting case ($a/W = 1.0$). Now stress intensity calibrations for these two specimens are available for all a/W ratios of practical interest.

Analytical Mechanics,
Westinghouse Research Laboratories,
Pittsburgh,
Penna. 15235, U.S.A.

W. K. WILSON

REFERENCES

[1] W. G. Clark, Jr., Ultrasonic detection of crack extension in WOL type fracture toughness specimens. *Mater. Eval.* **25**, 8 (1967).
[2] T. W. Crooker and E. A. Lange, Fatigue crack propagation in a high-strength steel under constant cyclic load with variable mean loads. *NRL Rep.* No. 6805, *Naval Res. Lab., Wash., D.C.*
[3] ASTM Committee E-24, Proposed method of test for plane-strain fracture toughness of metallic materials. 1969 *Book of ASTM Standards*, Part 31.
[4] W. F. Brown, Jr. and J. E. Srawley, *Plane Strain Crack Toughness Testing of High Strength Metallic Materials. ASTM Spec. Tech. Publ.* No. 410 (1966).

[5] E. T. Wessel, State-of-the-art of the WOL specimen for K_{Ic} fracture toughness testing. *Engng Fracture Mech.* **1**, 77–104 (1968).
[6] M. L. Williams, On the stress distribution at the base of a stationary crack. *J. appl. Mech.* **24**, 109–114 (1957).
[7] H. Neuber, *Theory of Notch Stresses*, 2nd Edn (1958). United States Atomic Energy Commission Translation AEC-tr-4547.
[8] P. C. Paris and G. C. Sih, Stress analysis of cracks. *Fracture Toughness. Testing and Its Application. ASTM Spec. Tech. Publ.* No. 381, pp. 30–83 (1965).
[9] E. Roberts, Jr., Elastic crack-edge displacements for the compact tension specimen. *Mater. Res. Stand.* 27 (1969).

NASA Technical Note D-6376 (August 1971).

AN IMPROVED METHOD OF COLLOCATION FOR THE STRESS ANALYSIS OF CRACKED PLATES WITH VARIOUS SHAPED BOUNDARIES[*]

By J. C. Newman, Jr.
Langley Research Center

SUMMARY

An improved method of boundary collocation was developed and applied to the two-dimensional stress analysis of cracks emanating from, or in the vicinity of, holes or boundaries of various shapes. The solutions, presented in terms of the stress-intensity factor, were based on the complex variable method of Muskhelishvili and a modified boundary-collocation method. The complex-series stress functions developed for simply and multiply connected regions containing cracks were constructed so that the boundary conditions on the crack surfaces are satisfied exactly. The conditions on the other boundaries were satisfied approximately by the modified collocation method. This improved method gave more rapid numerical convergence than other collocation techniques investigated.

INTRODUCTION

In the life of a structure subjected to cyclic loads, cracks may initiate at and propagate from geometric discontinuities (holes, cut-outs, edges, or flaws). In designing structures for fatigue and fracture resistance the stresses around these sites must be calculated if unexpected failures are to be avoided. In recent years the use of high-strength (crack sensitive) materials for greater structural efficiency has resulted in a series of aircraft service failures due to the presence of small cracks. The rates at which these cracks propagate and the size of the crack that causes failure are strongly influenced by the shape of the discontinuity and the type of loads applied near it.

The crack-tip stress-intensity factor (restricted to small-scale yielding) can account for the influence of component configuration and loading on fatigue-crack growth

---

[*]The information presented herein is based in part upon a thesis entitled "Stress Analysis of Simply and Multiply Connected Regions Containing Cracks by the Method of Boundary Collocation," offered in partial fulfillment of the requirements for the degree of Master of Science, Virginia Polytechnic Institute, Blacksburg, Virginia, May 1969.

and static strength (refs. 1 to 4). Many investigators (refs. 5 to 12) have obtained theoretical stress-intensity factors for cracks growing near stress concentrations or boundaries. However, for cracks in complicated structures, stress-intensity factors cannot always be obtained by closed-form analytical methods. They are often obtained by a series solution (such as boundary collocation) which, to be useful, should converge rapidly to the proper answer.

The objective of this paper is to present an improved method of collocation for calculating the influence of various boundary shapes on the stress-intensity factor in two-dimensional linear elastic bodies and to apply this method to a number of boundary-value problems involving cracks.

Stress-intensity factors were calculated for cracks emanating from a circular hole in an infinite plate and cracks emanating from a circular hole in a finite plate, a crack near two circular holes in an infinite plate, a crack emanating from an elliptical hole in an infinite plate, and a crack in a finite plate. These configurations were subjected to either internally or externally applied loads.

SYMBOLS

A_n, B_n, C_n, D_n	coefficients in the complex stress functions
a	half crack length
b	half axis of the elliptical hole perpendicular to the plane of the crack
C_{jn}, D_{jn}	coefficients for jth pole in the complex stress functions
c	half axis of the elliptical hole, in the plane of the crack
d	distance from the centerline of the crack to the center of the circular hole
F	stress-intensity correction factor
F_x, F_y	resultant force per unit thickness acting in the x and y directions, respectively
F_{sn}, G_{sn}	sth influence function for nth coefficients
f, g	resultant forces and displacements
f_o, g_o	specified resultant forces or displacements on the boundary
H	height of rectangular plate
$i = \sqrt{-1}$	
K	stress-intensity factor
K_T	stress-concentration factor
M	number of points at which the error in the boundary conditions was evaluated

N	number of coefficients in each series stress function
P	concentrated force per unit thickness acting in the y direction
p	pressure
R	radius of the circular hole
r	minimum radius of curvature for the ellipse
S	stress on the external boundary
u,v	displacement in the x and y direction, respectively
W	width of rectangular plate
z	complex variable, $z = x + iy$
z_j	location of pole in x-y plane, $z_j = x_j + iy_j$
α	angle between the x-axis and the normal to a boundary
β	constant in equation (1)
ζ	coordinate along the contour of a boundary
κ	material constant
λ	ratio of applied stresses on separate portions of a boundary
μ	Lamé's constant (shear modulus)
ν	Poisson's ratio
σ_n	normal stress at a boundary
σ_x	normal stress in x direction
σ_y	normal stress in y direction
τ_{nt}	shear stress at a boundary
τ_{xy}	shear stress
Φ, Ψ, ϕ, Ω	complex stress functions

Subscripts:

j,m,n,q,s   indices

METHOD OF BOUNDARY COLLOCATION

Boundary collocation is a numerical method used to evaluate the unknown coefficients in a series stress function, such as those developed in appendix A. The method begins with a general series solution to the governing linear partial differential equation. Certain terms are eliminated from the series by conditions of symmetry. The series is then truncated to a specified number of terms. The coefficients are determined through satisfaction of prescribed boundary conditions. The series solution finally obtained satisfies the governing equation in the interior of the region exactly and one or more of the boundary conditions approximately.

Various techniques have been used to satisfy boundary conditions. Conway (ref. 13) determined the coefficients from the criterion that the boundary conditions be satisfied exactly at a specified number of points on the boundary. Hulbert (ref. 14) and Hooke (ref. 15) selected the coefficients so that the sum of the squares of the stress residuals was a minimum for a specified number of points on the boundary. These two techniques have been used to analyze the stress state around a crack in a rectangular plate (refs. 11 and 14). Bowie (ref. 12) used a modified mapping-collocation technique to calculate the stress-intensity factors for a crack in a circular disk.

The present approach combines the complex variable method of Muskhelishvili (ref. 16) with a modified boundary-collocation method. The modification requires that the resultant forces on the boundary be specified (in a least-square sense) in contrast to previous work in which the boundary stresses were specified. (See appendix B.)

The resultant forces and displacements are expressed in terms of the complex stress functions $\phi(z)$ and $\Omega(z)$ (see appendix A) as

$$\beta \int \phi(z)dz + \int \Omega(\bar{z})d\bar{z} + (z - \bar{z})\overline{\phi(z)} = f(x,y) + ig(x,y) \tag{1}$$

The bars denote the complex conjugates. For resultant forces ($\beta = 1$) acting over the arc $\zeta - \zeta_0$ on the boundary

$$F_y - iF_x = -\left[f(x,y) + ig(x,y)\right]\Big|_{\zeta_0}^{\zeta} \tag{2}$$

For convenience, the location of ζ_0 was selected as the intersection of the boundary with either the x- or y-axis. For displacements ($\beta = -\kappa$) at a point ζ on the boundary

$$2\mu(u + iv) = -\left[f(x,y) + ig(x,y)\right]_{z=\zeta} \tag{3}$$

For the case of plane strain $\kappa = 3 - 4\nu$ and for plane stress $\kappa = \dfrac{3 - \nu}{1 + \nu}$.

In general, the truncated complex stress functions can be expressed as

$$\phi(z) = \sum_{n=0}^{N} A_n \tilde{\phi}_n(z) \qquad \Omega(z) = \sum_{n=0}^{N} B_n \tilde{\Omega}_n(z) \tag{4}$$

where $\tilde{\phi}_n$ and $\tilde{\Omega}_n$ are power-series functions of z. From equations (1) and (4) the expressions for f and g are

$$f = \sum_{n=0}^{N} A_n F_{1n} + \sum_{n=0}^{N} B_n F_{2n}$$

$$g = \sum_{n=0}^{N} A_n G_{1n} + \sum_{n=0}^{N} B_n G_{2n}$$
(5)

where F_{sn} and G_{sn} (s = 1,2) are influence functions derived from $\tilde{\phi}_n$ and $\tilde{\Omega}_n$ for the nth coefficients.

Having developed the expressions for f and g in terms of the unknown coefficients A_n and B_n, it is necessary to select a method for evaluating the coefficients from the specified boundary conditions. The two methods developed for this purpose were (1) direct specification of boundary forces and/or displacements and (2) least-squares specification of boundary forces and/or displacements.

Direct Specification of Boundary Forces and/or Displacements

In specifying the resultant forces and/or displacements on the boundary, equations (5) were evaluated at N equally spaced points on the boundary and the resulting equations were solved on a computer for the unknown coefficients. Further details on the computer and matrix solutions are given in the section entitled "Application of the Boundary-Collocation Method."

Least-Squares Specification of Boundary Forces and/or Displacements

In general, the computed boundary values are in error at any given point ζ_m because the series stress functions were truncated. The square of this error is

$$e_m^2 = \left\{ f_0 - \sum_{n=0}^{N} A_n F_{1n} - \sum_{n=0}^{N} B_n F_{2n} \right\}_m^2 + \left\{ g_0 - \sum_{n=0}^{N} A_n G_{1n} - \sum_{n=0}^{N} B_n G_{2n} \right\}_m^2 \quad (6)$$

where f_0 and g_0 are specified boundary values at point ζ_m. The coefficients are then evaluated by minimizing the squares of the errors at a specified number of points M on the boundary:

$$\frac{\partial \sum_{m=1}^{M} e_m^2}{\partial A_q} = 0 \qquad \frac{\partial \sum_{m=1}^{M} e_m^2}{\partial B_q} = 0 \quad (7)$$

where $q = 1, 2, \ldots, N$. Equations (7) result in a set of $2N$ linear algebraic equations for the unknown coefficients A_n and B_n

$$\left.\begin{aligned}\sum_{n=0}^{N} \alpha_{nq} A_n + \sum_{n=0}^{N} \beta_{nq} B_n &= \delta_q \\ \sum_{n=0}^{N} \beta_{qn} A_n + \sum_{n=0}^{N} \gamma_{nq} B_n &= \eta_q\end{aligned}\right\} \quad (8)$$

where

$$\alpha_{nq} = \sum_{m=1}^{M} \left(F_{1n} F_{1q} + G_{1n} G_{1q}\right)_m$$

$$\beta_{nq} = \sum_{m=1}^{M} \left(F_{2n} F_{1q} + G_{2n} G_{1q}\right)_m$$

$$\gamma_{nq} = \sum_{m=1}^{M} \left(F_{2n} F_{2q} + G_{2n} G_{2q}\right)_m$$

$$\delta_q = \sum_{m=1}^{M} \left(f_o F_{1q} + g_o G_{1q}\right)_m$$

and

$$\eta_q = \sum_{m=1}^{M} \left(f_o F_{2q} + g_o G_{2q}\right)_m$$

The second partial derivative of the square of the error with respect to the unknown coefficients was positive, indicating a definite minimum.

APPLICATION OF THE BOUNDARY-COLLOCATION METHOD

The improved collocation method was used to analyze a number of crack problems. The configurations investigated were grouped into four categories: (1) cracks emanating

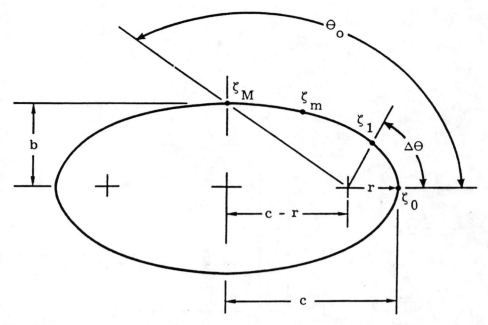

Figure 1.- Coordinate system used for describing the location of the collocation points on the circular and elliptic boundaries.

from a circular hole in an infinite plate, (2) a crack near two circular holes in an infinite plate, (3) cracks emanating from an elliptical hole in an infinite plate, and (4) a crack in a finite plate. In each category several boundary conditions were investigated.

In applying the method, it was necessary to specify the points on the boundary at which the error equation was to be evaluated. The locations on the circular and elliptic boundaries were specified by dividing the angle θ_0 into M equal increments ($\Delta\theta$) (see fig. 1). This procedure was selected because more points are concentrated on sections of the boundary which have smaller radii of curvature and, hence, larger stress gradients. For simplicity, this procedure was also employed on the external boundaries. In general, the value of M used in the solution of the various boundary-value problems was twice the total number of unknown coefficients in the series stress functions. For the case of cracks emanating from the elliptical hole in an infinite plate and from the circular hole in a finite plate, 160 coefficients were used in the stress functions. In all other cases 90 coefficients were used (except where noted).

A digital computer was used to solve the resulting equations using either single precision (14 significant digits) or double precision (29 significant digits). The equations were solved by a routine which employed Jordan's method (ref. 17). Using single precision, the computer time was 1 and 4 minutes for solving 90 and 180 equations, respectively. Double precision required approximately twice as much time as single precision. Double precision was used only in the cases involving the finite plate.

The results are presented in terms of a correction factor F which accounts for the influence of the various boundaries on the stress-intensity factor for a crack in an infinite plate. The correction factor is defined as the ratio of the stress-intensity factor for the particular case to that for a crack in an infinite plate subjected to the same loading. A table of numerical values of F and the stress-intensity equation are given in Tables I to VII.

TABLE I.— CRACKS EMANATING FROM A CIRCULAR HOLE IN AN INFINITE PLATE SUBJECTED TO BIAXIAL STRESS

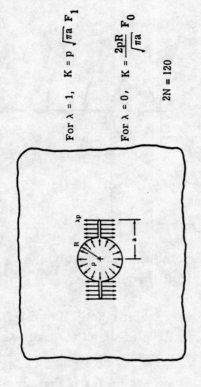

$$K = S\sqrt{\pi a}\, F$$

$2N = 120$

a/R	$F\,(\lambda=-1)$	$F_0\,(\lambda=0)$	$F_1\,(\lambda=1)$
1.01	0.3325	0.3256	0.2188
1.02	.5971	.4514	.3058
1.04	.7981	.6082	.4183
1.06	.9250	.7104	.4958
1.08	1.0135	.7843	.5551
1.10	1.0775	.8400	.6025
1.15	1.1746	.9322	.6898
1.20	1.2208	.9851	.7494
1.25	1.2405	1.0168	.7929
1.30	1.2457	1.0358	.8259
1.40	1.2350	1.0536	.8723
1.50	1.2134	1.0582	.9029
1.60	1.1899	1.0571	.9242
1.80	1.1476	1.0495	.9513
2.00	1.1149	1.0409	.9670
2.20	1.0904	1.0336	.9768
2.50	1.0649	1.0252	.9855
3.00	1.0395	1.0161	.9927
4.00	1.0178	1.0077	.9976

TABLE II.— CRACKS EMANATING FROM AN INTERNALLY PRESSURIZED HOLE IN AN INFINITE PLATE

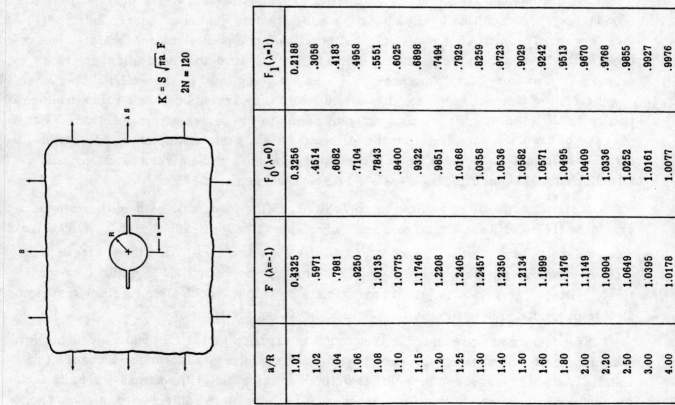

For $\lambda = 1$, $\quad K = p\sqrt{\pi a}\, F_1$

For $\lambda = 0$, $\quad K = \dfrac{2pR}{\sqrt{\pi a}}\, F_0$

$2N = 120$

a/R	$F_1\,(\lambda=1)$	$F_0\,(\lambda=0)$
1.01	0.2188	0.1725
1.02	.3058	.2319
1.04	.4183	.3334
1.06	.4958	.3979
1.08	.5551	.4485
1.10	.6025	.4897
1.15	.6898	.5688
1.20	.7494	.6262
1.25	.7929	.6701
1.30	.8259	.7053
1.40	.8723	.7585
1.50	.9029	.7971
1.60	.9242	.8264
1.80	.9513	.8677
2.00	.9670	.8957
2.20	.9768	.9154
2.50	.9855	.9358
3.00	.9927	.9566
4.00	.9976	.9764

TABLE III.- CRACK LOCATED BETWEEN TWO CIRCULAR HOLES IN AN INFINITE PLATE SUBJECTED TO UNIAXIAL STRESS

$K = S\sqrt{\pi a}\, F$

a/R	$F\left(\dfrac{d}{R}=2\right)$	$F\left(\dfrac{d}{R}=3\right)$	$F\left(\dfrac{d}{R}=4\right)$	$F\left(\dfrac{d}{R}=6\right)$	$F\left(\dfrac{d}{R}=10\right)$
0.01	0.1390	0.5234	0.7127	0.8661	0.9506
.25	.1728	.5355	.7172	.8671	-----
.50	.2683	.5698	.7303	.8700	.9512
.75	.4045	.6206	.7505	.8748	-----
1.00	.5502	.6799	.7758	.8810	.9528
1.50	.7809	.7958	.8324	.8969	.9553
2.00	.9058	.8819	.8847	.9150	.9585
2.50	.9640	.9351	.9250	.9328	.9622
3.00	.9902	.9654	.9527	.9486	.9663

TABLE IV.- CRACK APPROACHING TWO CIRCULAR HOLES IN AN INFINITE PLATE SUBJECTED TO UNIAXIAL STRESS

$K = S\sqrt{\pi a}\, F$

$\dfrac{a}{d-R}$	$F\left(\dfrac{d}{R}=1.15\right)$	$F\left(\dfrac{d}{R}=1.25\right)$	$F\left(\dfrac{d}{R}=1.5\right)$	$F\left(\dfrac{d}{R}=2\right)$	$F\left(\dfrac{d}{R}=4\right)$
0.01	4.3975	3.2630	2.1530	1.4687	1.0761
.1	4.4160	3.2739	2.1612	1.4738	1.0774
.2	4.4751	3.3143	2.1875	1.4898	1.0816
.3	4.5853	3.3856	2.2338	1.5183	1.0892
.4	4.7308	3.4946	2.3045	1.5624	1.1017
.5	4.9593	3.6538	2.4076	1.6279	1.1215
.6	5.2890	3.8850	2.5575	1.7249	1.1538
.7	5.7816	4.2271	2.7824	1.8740	1.2101
.8	-----	4.7825	3.1504	2.1230	1.3202
.9	-----	5.8028	3.8661	2.6350	1.5927

TABLE V.- CRACKS EMANATING FROM AN ELLIPTICAL HOLE IN AN INFINITE PLATE SUBJECTED TO UNIAXIAL STRESS

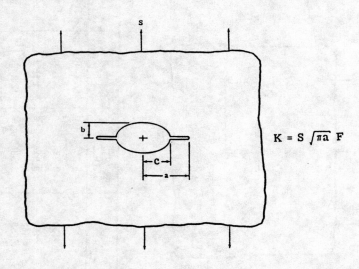

$K = S\sqrt{\pi a}\, F$

a/c	$F\left(\frac{b}{c}=0.25\right)$	$F\left(\frac{b}{c}=0.5\right)$	$F\left(\frac{b}{c}=1\right)$	$F\left(\frac{b}{c}=2\right)$	$F\left(\frac{b}{c}=4\right)$
1.02	0.9050	0.6757	0.4514	0.3068	0.2341
1.03	.9597	.7742	-----	-----	-----
1.04	.9865	.8398	.6082	.4297	-----
1.05	1.0013	.8861	-----	-----	.3622
1.06	1.0098	.9206	.7104	.5164	-----
1.08	1.0179	.9664	.7843	.5843	.4493
1.10	1.0206	.9925	.8400	.6401	.4960
1.15	1.0202	1.0258	.9322	.7475	.5901
1.20	1.0176	1.0357	.9851	.8241	-----
1.25	-----	-----	-----	-----	.7248
1.30	-----	1.0366	1.0358	.9255	-----
1.40	-----	1.0317	1.0536	.9866	.8494
1.50	-----	-----	1.0582	1.0246	-----
1.55	-----	-----	-----	-----	.9279
1.60	-----	-----	1.0571	1.0483	-----
1.80	-----	-----	1.0495	1.0714	1.0063
2.00	-----	-----	1.0409	1.0777	-----
2.10	-----	-----	-----	-----	1.0551
2.20	-----	-----	1.0336	1.0766	-----
2.40	-----	-----	1.0251	1.0722	1.0788

TABLE VII.- CRACKS EMANATING FROM A CIRCULAR HOLE IN A RECTANGULAR PLATE SUBJECTED TO UNIAXIAL STRESS

$K = S\sqrt{\pi a}\ F$

$\dfrac{H}{W} = 2$

2a/W	$F\left(\dfrac{2R}{W}=0\right)$
0	1.0000
.1	1.0061
.2	1.0249
.3	1.0583
.4	1.1102
.5	1.1876
.6	1.3043
.7	1.4891
.8	1.8161
.9	2.5482

2a/W	$F\left(\dfrac{2R}{W}=0.25\right)$
0.25	0
.26	.6593
.27	.8510
.28	.9605
.29	1.0304
.30	1.0776
.35	1.1783
.40	1.2156
.50	1.2853
.60	1.3965
.70	1.5797
.80	1.9044
.85	2.1806
.90	2.6248

2a/W	$F\left(\dfrac{2R}{W}=0.5\right)$
0.50	0
.51	.6527
.52	.8817
.525	.9630
.53	1.0315
.54	1.1426
.55	1.2301
.60	1.5026
.70	1.8247
.78	2.1070
.85	2.4775
.90	2.9077

TABLE VI.- WEDGE-FORCE LOADED CRACK IN A RECTANGULAR PLATE

$K = \dfrac{P}{\sqrt{\pi a}}\ F$

2a/W	$F\left(\dfrac{H}{W}=0.5\right)$	$F\left(\dfrac{H}{W}=0.75\right)$	$F\left(\dfrac{H}{W}=1.0\right)$	$F\left(\dfrac{H}{W}=2.0\right)$
0	1.0000	1.0000	1.0000	1.0000
.1	1.0921	1.0467	1.0279	1.0121
.2	1.3572	1.1863	1.1115	1.0497
.3	1.7721	1.4185	1.2499	1.1163
.4	2.3269	1.7431	1.4418	1.2191
.5	3.0554	2.1589	1.6866	1.3710
.6	4.0464	2.6587	1.9894	1.5958
.7	5.3985	3.2275	2.3772	1.9421
.8	7.0162	3.8858	2.9523	2.5309
.9	8.4078	4.9791	4.1665	3.7810

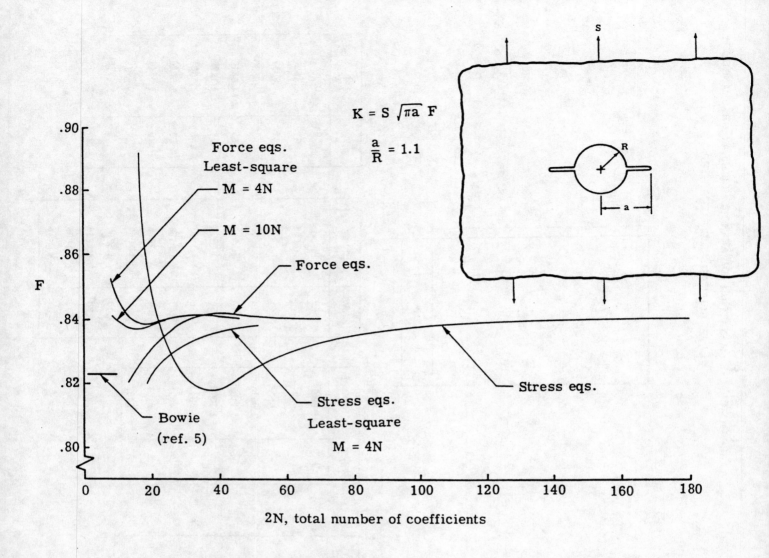

Figure 2.- Convergence curves for the problem of cracks emanating from a circular hole in an infinite plate subjected to uniaxial stress.

Cracks Emanating from a Circular Hole in an Infinite Plate

Remote biaxial stress.- For the case of cracks emanating from a circular hole, four collocation techniques were used to obtain the unknown coefficients in the stress functions. (See eqs. (A11) of appendix A.) The conditions at infinity were satisfied by adding to equations (A11) the $n = 0$ terms in equations (A10). The results of each of these techniques are compared in figure 2. The curves were obtained by specifying either stresses at discrete points on the boundary (stress equations) or resultant forces along the boundary (force equations), or by using a least-squares procedure with the stress equations or force equations. For the least-squares procedure employing resultant forces, two convergence curves are presented. One is for the case in which the number of points M was twice the total number of coefficients in the stress functions ($2N$ is the total number of coefficients in the stress functions); the other is for the case in which the number of points considered was five times the total number of coefficients. Convergence is seen to be rapid in both cases. For the least-squares procedure employing stress equations, the number of points considered was twice the total number

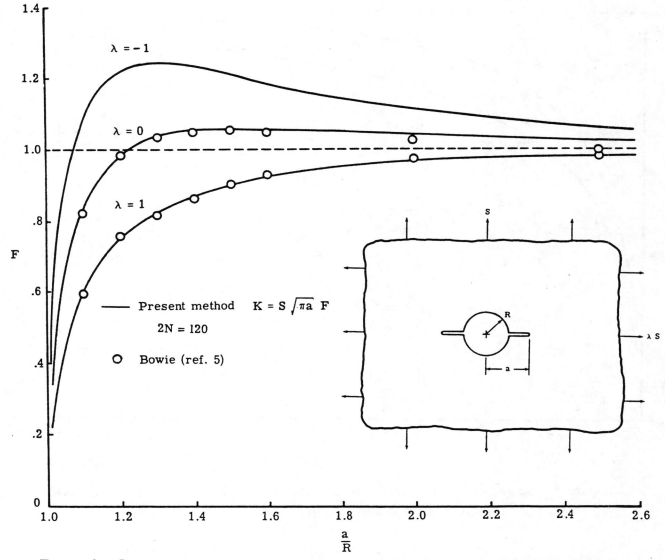

Figure 3.- Stress-intensity correction factors for cracks emanating from a circular hole in an infinite plate subjected to biaxial stress.

of coefficients. All techniques seemed to converge to the same value as the number of coefficients increased. The value of F was approximately 2 percent higher than Bowie's approximate solution (ref. 5) for the same configuration and loading. The boundary stresses also converged to their specified values as the number of coefficients increased. However, the analysis using the least-squares procedure with the force equations converged more rapidly than the other techniques.

The correction factors for three states of biaxial stress are shown in figure 3. The solid curves show the results obtained in the present investigation. The circles are Bowie's solution and were obtained from a table given in reference 18. The agreement is considered good.

A solution for other values of λ was constructed by superposition of the two cases, $\lambda = 0$ and $\lambda = 1$. The stress-intensity equation is

$$K = S\sqrt{\pi a}\left[(1 - \lambda)F_0 + \lambda F_1\right] \tag{9}$$

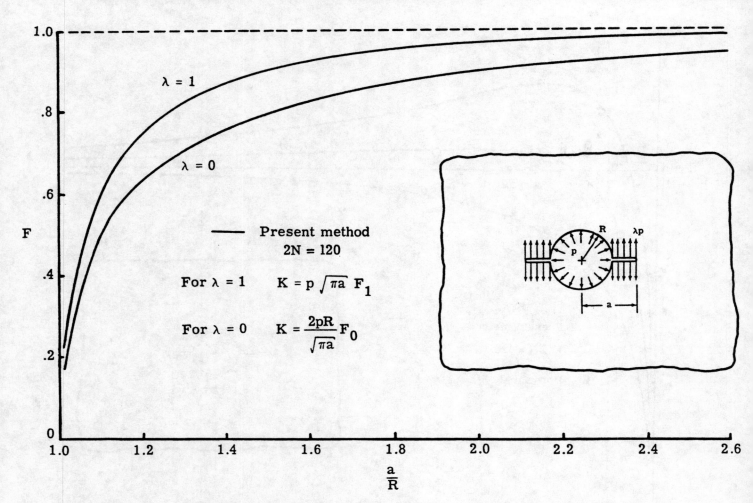

Figure 4.- Stress-intensity correction factors for cracks emanating from a circular hole in an infinite plate subjected to internal pressure.

where F_0 and F_1 are the correction factors for $\lambda = 0$ and $\lambda = 1$, respectively. The calculations for $\lambda = -1$ confirmed equation (9) to greater than eight significant digits. However, equation (9) is only valid for positive values of stress intensity. Negative stress intensities would indicate crack closure and the problem would become one of calculating contact stresses.

Internal pressure.- In a configuration like that in the preceding section, internal pressure p was applied to the hole boundary and λp to the crack surfaces. The stress functions for this case are given by equations (A11).

The correction factors for $\lambda = 0$ and $\lambda = 1$ are shown in figure 4. For the case in which no pressure is applied to the crack surfaces ($\lambda = 0$) the correction factor modifies the stress-intensity factor for wedge-force loading on the crack surfaces (ref. 18). Here the wedge-force loading 2pR is that produced by the internal pressure. The other case modifies the solution for a uniformly pressurized crack. A solution for other values was constructed by superposition of these two cases. The stress-intensity equation is

$$K = \frac{2pR}{\sqrt{\pi a}}(1 - \lambda)F_0 + \lambda p\sqrt{\pi a}\,F_1 \tag{10}$$

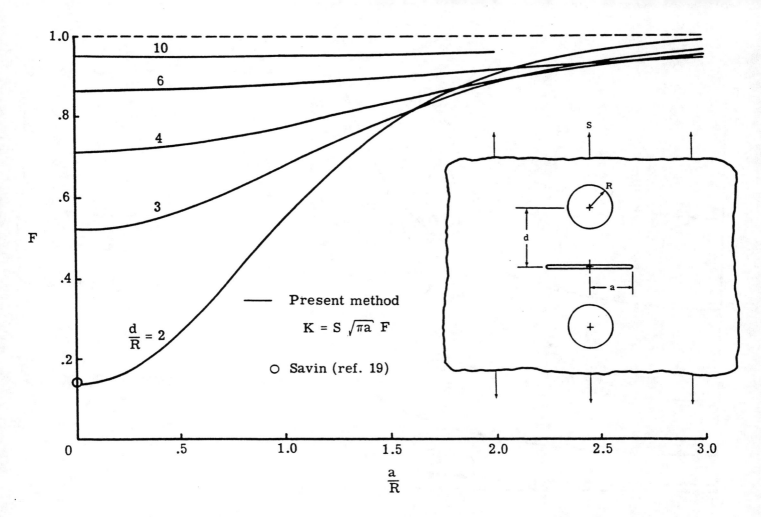

Figure 5.- Stress-intensity correction factors for a crack located between two circular holes in an infinite plate subjected to uniaxial stress.

where F_0 and F_1 are the correction factors for $\lambda = 0$ and $\lambda = 1$, respectively. Equation (10) is also valid only for positive values of stress intensity. The results indicate that the influence of the hole on the stress-intensity equation may be neglected for values of a/R greater than 2.5.

A Crack Near Two Circular Holes in an Infinite Plate

Crack located between the holes subjected to remote uniaxial stress.- In the case of two circular holes the stress functions, equations (A13), were used with poles located at $z_j = \pm id$ (see appendix A). The conditions at infinity were satisfied by adding to equations (A13) the $n = 0$ terms in equations (A10).

The correction factors are shown in figure 5 for several values of d/R. At small d/R values the holes had a pronounced effect on the correction factors. However, for d/R values greater than 10 or a/R values greater than 3 the influence of the holes may be neglected.

In the limit as the crack length approaches zero, the correction factor is equivalent to the local-stress-concentration factor σ_y/S at the origin. In terms of the remote stress and the local stress the stress-intensity factors are

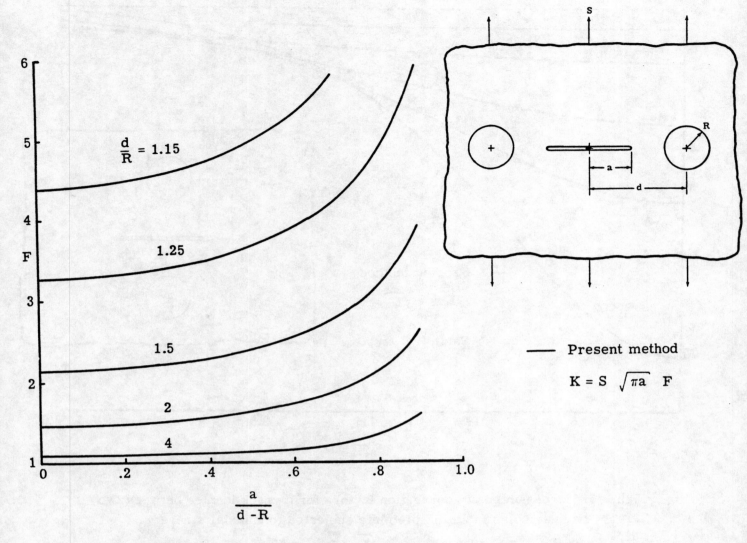

Figure 6.- Stress-intensity correction factors for a crack approaching two circular holes in an infinite plate subjected to uniaxial stress.

$$K = S\sqrt{\pi a}\, F = \sigma_y \sqrt{\pi a} \tag{11}$$

from which the correction factor is given as the ratio of σ_y to S. The circle plotted on the ordinate axis is the local-stress-concentration factor at the origin as given in reference 19 for a value of $\frac{d}{R} = 2$. The agreement is considered good.

Cracks approaching the holes subjected to remote uniaxial stress.- In this case the stress functions, equations (A13), were used with poles located at $z_j = \pm d$. Again, the conditions at infinity were satisfied by adding to equations (A13) the $n = 0$ terms in equations (A10).

The correction factors are shown in figure 6 for several values of d/R. The factors are plotted against the nondimensional crack length $\frac{a}{d-R}$. The correction factors increase from their initial value at $a = 0$ to large values as the crack approaches the edge of the hole. Again, the correction factor at $a = 0$ is equal to the local-stress-concentration factor at the point midway between the two holes.

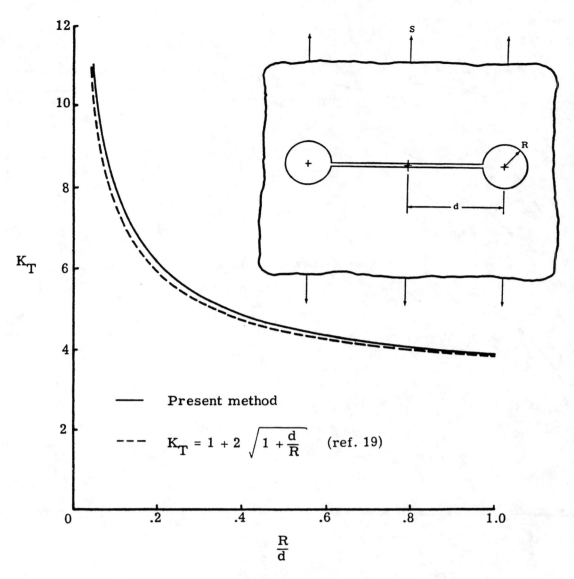

Figure 7.- Stress-concentration factor for two circular holes connected by a crack in an infinite plate subjected to uniaxial stress.

As the crack intersects the boundary of the holes, the concept of the stress-intensity factor K no longer exists. However, the stress-concentration factor K_T at the edge of the hole ($x = R + d$) does describe the severity of the notch. The stress-concentration factors are shown in figure 7 as a function of R/d ratios. The stress-concentration factors were compared with those calculated from an "equivalent" elliptical-hole solution as shown by the dashed line. The elliptical hole had a minimum radius of curvature equal to the radius of the circular holes and had an overall length equal to $2(R + d)$. The stress-concentration factors for the elliptical hole were only slightly lower than those calculated for the case of two circular holes connected by a crack. Thus, the elliptical-hole solution may be used to approximate this configuration.

Cracks Emanating from an Elliptical Hole in an Infinite Plate

<u>Remote uniaxial stress</u>.- For the case of cracks emanating from an elliptical hole (see fig. 8) the stress functions, equations (A13), contained multiple poles z_j on either

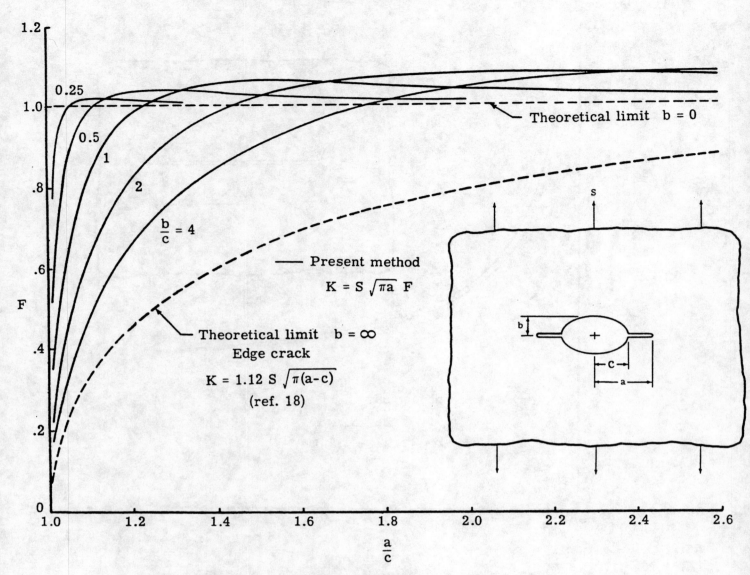

Figure 8.- Stress-intensity correction factors for cracks emanating from an elliptical hole in an infinite plate subjected to uniaxial stress.

the x- or y-axis. The $n = 0$ terms in equations (A10) were added to equations (A13) in order to satisfy the boundary conditions at infinity. For the case of $\frac{b}{c} = 0.5$ and 2, the value of J in equations (A13) was 4 and for $\frac{b}{c} = 0.25$ and 4 the value was 16. These values were determined by trial and error. The poles were always located on the major axis of the ellipse at the origin, at the center of the minimum curvature, and equally spaced between these points.

The correction factors are given in figure 8 for several values of b/c. The dashed curves show the theoretical limits expressed in terms of the correction factor F as the value of b approaches either zero or infinity. The stress-intensity factor in the limiting case ($b = \infty$) was obtained from reference 18 for the case of an edge crack in a semi-infinite plate. The other limit is where the elliptical hole reduces to a crack. The correction factors approached unity for all finite values of b/c as the crack length approached infinity. The edge-crack solution approached 1.12. For b/c ratios greater than 2 and large values of a/c the results indicated that the influence of the elliptical

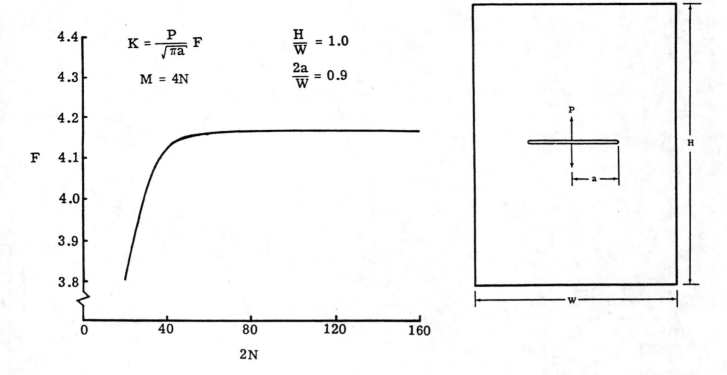

Figure 9.- Convergence curve for the problem of a wedge-force loaded crack in a rectangular plate.

hole may not be negligible. However, for small b/c ratios the correction factor approached unity very rapidly.

Crack in a Finite Plate

In the following section the stress intensity solutions for two cases of a crack in a rectangular plate are presented. Double precision was used in the computer solution of the resulting equations primarily because the least-squares procedure generated matrix elements whose values ranged over many orders of magnitude and the unknown coefficients may have been susceptible to round-off error. In fact, for some choices of the pertinent parameters the single-precision routine did not appear to converge. However, for the same parameters the double-precision routine did appear to converge as the number of coefficients increased.

Crack subjected to wedge-force loading.- For the case of a crack in a rectangular plate subjected to wedge-force loading on the crack surfaces (see fig. 9) the stress functions were taken to be

$$\left.\begin{array}{r}\phi(z) = \phi_0(z) + \dfrac{Pa}{2\pi z \sqrt{z^2 - a^2}} \\[2ex] \Omega(z) = \Omega_0(z) + \dfrac{Pa}{2\pi z \sqrt{z^2 - a^2}}\end{array}\right\} \quad (12)$$

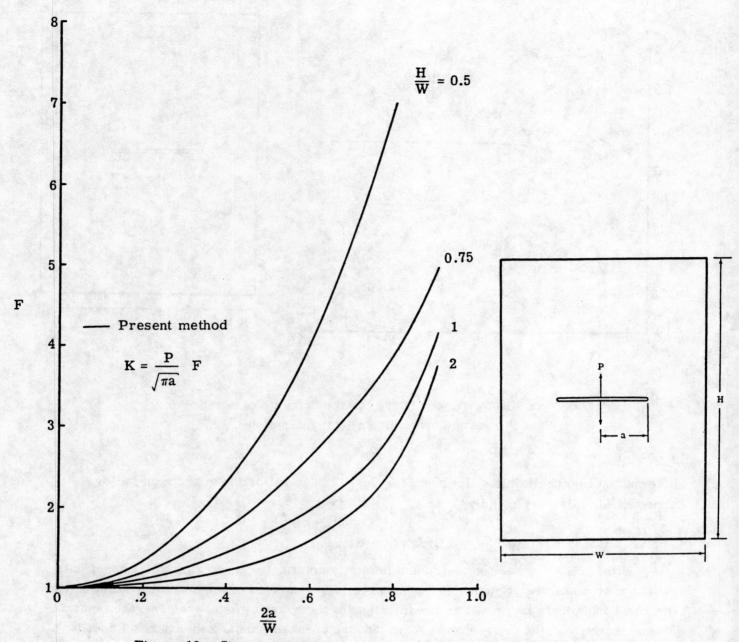

Figure 10.- Stress-intensity correction factors for a wedge-force loaded crack in a rectangular plate

where ϕ_0 and Ω_0 are given by equations (A10). The additional terms in equations (12) were added in order to account for the concentrated forces on the crack surfaces (ref. 9). The unknown coefficients from equations (A10) in equations (12) were evaluated by satisfying the conditions on the external boundary. To show convergence, an example problem with $\frac{2a}{W} = 0.9$ was solved for several values of $2N$ (see fig. 9). This example problem was selected because the close proximity of the crack tip to the boundary was expected to pose difficulties in achieving convergence. However, the convergence was rapid.

The correction factors for several values of H/W are shown in figure 10. For a given value of H/W the correction factors ranged from unity, at small crack length to plate width ratios, to very large values at large ratios.

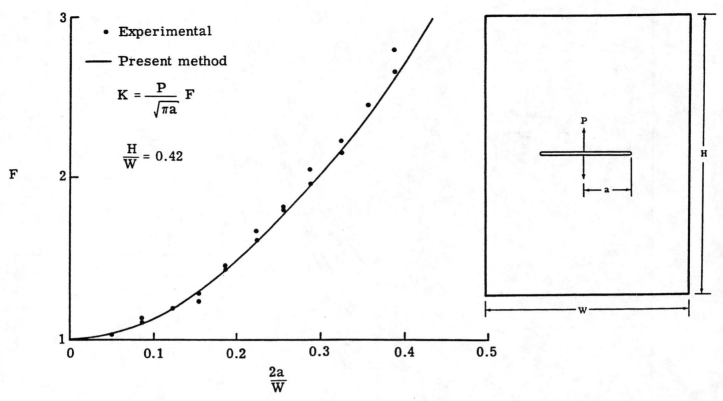

Figure 11.- Experimental and theoretical stress-intensity correction factors for a wedge-force loaded crack in a rectangular plate.

In figure 11, a comparison is made between experimental and theoretical stress-intensity correction factors for a wedge-force loaded panel. The material was 7075-T6 aluminum alloy cycled at constant amplitude with a load ratio (minimum load to maximum load) of 0.05. The experimental stress-intensity factors were obtained from measured crack growth rates used together with a curve of stress intensity as a function of crack growth rate (ref. 4). The agreement between the calculated and the experimental correction factors is considered good.

Cracks emanating from a circular hole subjected to uniaxial stress.- For the case of cracks growing from a circular hole in a rectangular plate subjected to a uniaxial stress (see fig. 12) the stress functions were taken to be the sum of equations (A10) and (A11) since both internal and external boundaries must be considered. The unknown coefficients were evaluated by satisfying the conditions on both boundaries.

The correction factors for several values of $2R/W$ and a plate aspect ratio (H/W) of 2 are shown in figure 12. As the hole radius approaches zero, the boundary-value problem reduces to that of a single crack in a rectangular plate. In this case all coefficients in equation (A11) are set equal to zero. The correction factors shown for $\frac{2R}{W} = 0$ agree almost identically with Isida's solution (ref. 20) for the case of a central crack in a finite width strip. The solid curve was slightly higher than Isida's solution for values of $\frac{2a}{W} > 0.7$. However, the difference may be due to the influence of the finite height.

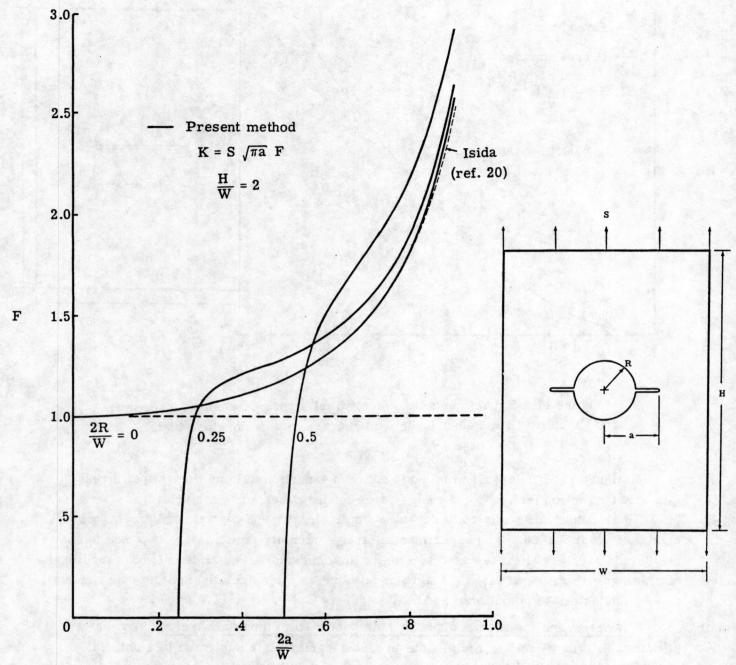

Figure 12.- Stress-intensity correction factors for cracks emanating from a circular hole in a rectangular plate subjected to uniaxial stress.

CONCLUDING REMARKS

An improved method of boundary collocation was presented and applied to the stress analysis of cracked plates having various boundary shapes and subjected to inplane loading. The solutions were based on the complex variable method of Muskhelishvili and a modification of the numerical method of boundary collocation. The complex series stress functions formulated for cracked plates automatically satisfy the boundary conditions on the crack surfaces. The conditions on the other boundaries were satisfied approximately by the series solution. Stress-intensity correction factors were presented for several configurations involving cracks in the presence of stress concentrations or boundaries. The least-squares boundary-collocation method used to minimize the resultant-force residuals on the boundary gave better numerical convergence in the boundary conditions than the other three collocation methods investigated. The other methods were (1) direct specification of boundary forces, (2) direct specification of boundary stresses at discrete points, and (3) least-squares specification of boundary stresses. The improved method was able to analyze more complex boundary-value problems than the other methods because of the more rapid convergence in the series solution.

The stress-intensity factors presented in this paper cover a moderate range of configurations and loadings. Most of the boundary-value problems solved do occur frequently in aircraft structures such as cracks growing from or near cutouts (windows, rivet holes, etc.) or from other structural discontinuities. Thus, the solutions presented here may be used to design a structure more efficient against fatigue and fracture or to help monitor structural damage due to fatigue loading.

Langley Research Center,
National Aeronautics and Space Administration,
Hampton, Va., June 15, 1971.

APPENDIX A

FORMULATION OF THE COMPLEX SERIES STRESS FUNCTIONS
FOR CRACKED BODIES

One of the major advances in the field of two-dimensional linear elasticity has been the complex-variable approach of Muskhelishvili (ref. 16). The representation of biharmonic functions by analytic functions has led to a general method of solving plane-strain and generalized plane-stress problems.

The following formulation for two-dimensional cracked bodies is based on the work of Muskhelishvili for an infinite plane region containing cracks and subjected to inplane loading (see fig. 13). The known surface tractions are applied to the boundaries of this region. The body forces are assumed to be zero and the material is assumed to be linear elastic, isotropic, and homogeneous. The equilibrium and compatibility equations are combined to form the biharmonic equation

$$\frac{\partial^4 U}{\partial x^4} + 2\frac{\partial^4 U}{\partial x^2 \partial y^2} + \frac{\partial^4 U}{\partial y^4} = 0 \qquad (A1)$$

where $U(x,y)$ is the Airy stress function.

The biharmonic function $U(x,y)$ can be expressed as

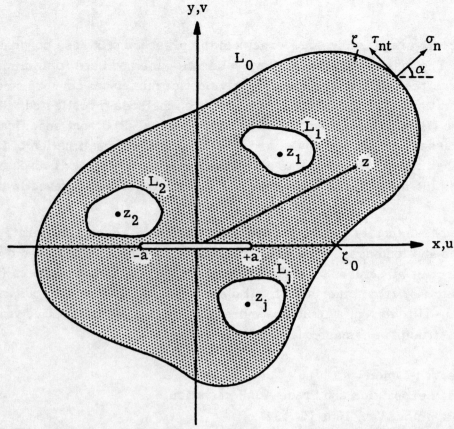

Figure 13.- Two-dimensional multiply-connected body containing a crack and subjected to surface tractions on the boundaries.

$$U(x,y) = \frac{1}{2} \text{Re}\left[\int \overline{\Psi}(z)\,dz + \int \Phi(z)\,dz + (\bar{z} - z)\Phi(z)\right] \tag{A2}$$

where $\Phi'(z) = \phi(z)$ and $\Psi'(z) = \Omega(z)$. The primes and bars denote differentiation and complex conjugates, respectively. Therefore, the generalized plane-stress and plane-strain problems are reduced to the determination of these functions subject to specified boundary conditions.

From reference 16, the stress functions in the neighborhood of a crack $(z = \pm a)$ can be expressed as

$$\left. \begin{aligned} \phi(z) &= \frac{Q_1(z)}{\sqrt{z^2 - a^2}} + Q_2(z) \\[1em] \Omega(z) &= \frac{Q_3(z)}{\sqrt{z^2 - a^2}} + Q_4(z) \end{aligned} \right\} \tag{A3}$$

where

$$Q_s(z) = \sum_{n=-\infty}^{\infty} A_{sn} z^n \tag{A4}$$

In the present investigation, the formulation of the complex series stress functions for simply and multiply connected regions containing cracks was restricted to the situation where the configuration and loading are symmetric about the x- and y-axis. The boundary conditions satisfied by ϕ and Ω for a stress-free crack surface are

$$
\begin{aligned}
&\text{(I)} \quad \sigma_y = \tau_{xy} = 0 && (|x| < a;\ y = 0) \\
&\text{(II)} \quad \tau_{xy} = v = 0 && (|x| \geq a;\ y = 0) \\
&\text{(III)} \quad \tau_{xy} = u = 0 && (|y| \geq 0;\ x = 0)
\end{aligned} \tag{A5}
$$

where the coordinate system used is shown in figure 13. The stresses expressed in terms of the stress functions are

$$
\begin{aligned}
\sigma_x + \sigma_y &= 2\left[\phi(z) + \overline{\phi(z)}\right] \\
\sigma_y - \sigma_x + 2i\tau_{xy} &= 2\left[(\bar{z} - z)\phi'(z) - \phi(z) + \bar{\Omega}(z)\right]
\end{aligned} \tag{A6}
$$

The displacements are given by equation (1).

The boundary conditions in equations (A5) define a unique relationship between the two analytic functions ϕ and Ω. The relationship for the series which contains the square-root term is $Q_1(z) = Q_3(z)$ and for the series which contains no square-root term, $Q_2(z) = -Q_4(z)$. These conditions insure the satisfaction of the boundary conditions (stress free) on the crack surfaces.

For a multiply connected region, in addition to the boundary conditions stated in equations (A5), the single-valuedness of displacements must be insured. This condition is stated as

$$\kappa \oint_{\Lambda_j} \phi(z)dz - \oint_{\Lambda_j} \Omega(\bar{z})d\bar{z} = 0 \tag{A7}$$

where Λ_j is the contour around each separate hole boundary L_j (see fig. 13).

The stress intensity factor K is obtained from the stress functions as

$$K = 2\sqrt{2\pi}\ \lim_{z \to a} \sqrt{z - a}\ \phi(z) \tag{A8}$$

As an example, consider a crack located along the x-axis in an infinite plate as shown in figure 14. The dashed lines L_0 and L_1 define the boundaries of the region (shaded area). The internal boundary L_1 has cracks growing from the edge of the hole to $x = \pm a$. The boundaries L_0 and L_1 may have any simple shape which is symmetric about the x- and y-axis and be subjected to any symmetric boundary conditions. The complex-series stress functions must be single-valued and analytic in the region between L_0 and L_1. This region does not include the crack which is represented by a branch

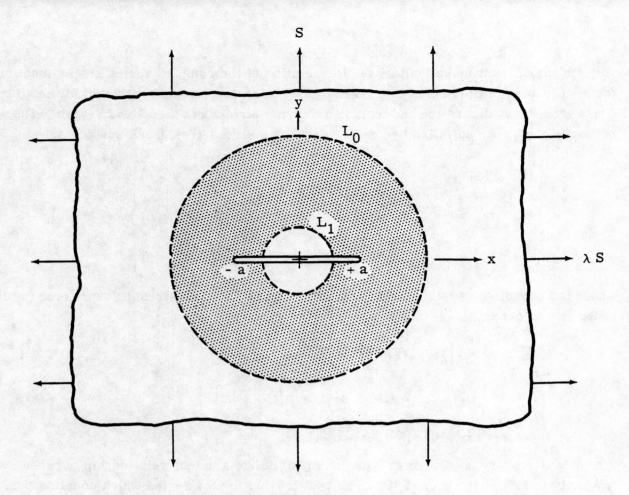

Figure 14.- Crack in an infinite plate subjected to biaxial stress.

cut. The stress functions are

$$\left.\begin{aligned}\phi(z) &= \phi_0(z) + \phi_1(z) \\ \Omega(z) &= \Omega_0(z) + \Omega_1(z)\end{aligned}\right\} \quad (A9)$$

The subscripts denote the functions which are used to satisfy the boundary conditions on boundaries L_0 and L_1, respectively. The stress functions used to satisfy the conditions on the external boundary L_0 are

$$\left.\begin{aligned}\phi_0(z) \\ \Omega_0(z)\end{aligned}\right\} = \frac{z}{\sqrt{z^2 - a^2}} \sum_{n=0}^{N} A_n z^{2n} \pm \sum_{n=0}^{N} B_n z^{2n} \quad (A10)$$

where the coefficients A_n and B_n are real. In the situation where the boundary L_0 is located at infinity $A_0 = \frac{S}{2}$, $B_0 = \frac{S}{4}(\lambda - 1)$ and the remaining coefficients for $n \geq 1$ are zero. For the internal boundary L_1 the stress functions are

$$\left.\begin{aligned}\phi_1(z) \\ \Omega_1(z)\end{aligned}\right\} = \frac{z}{\sqrt{z^2 - a^2}} \sum_{n=1}^{N} \frac{C_n}{z^{2n}} \pm \sum_{n=1}^{N} \frac{D_n}{z^{2n}} \quad (A11)$$

where the coefficients C_n and D_n are real. The stress functions, equations (A10) and (A11), automatically satisfy the boundary conditions on the crack surfaces. The conditions on the other boundaries $(L_0$ and $L_1)$ were approximated by the series solution. The stress-intensity factor K calculated from equations (A8), (A10), and (A11) is

$$K = \sqrt{\pi a} \left\{ \sum_{n=0}^{N} 2A_n a^{2n} + \sum_{n=1}^{N} \frac{2C_n}{a^{2n}} \right\} \tag{A12}$$

In the case of multiple circular holes or elliptical holes, the complex-series stress functions require the use of poles at various stations along the x- or y-axis. The stress functions are given by

$$\left.\begin{array}{c}\phi(z)\\ \Omega(z)\end{array}\right\} = \frac{z}{\sqrt{z^2 - a^2}} \sum_{j=1}^{J} \sum_{n=1}^{N} \frac{C_{jn}}{\left(z^2 - z_j^2\right)^n} \pm \sum_{j=1}^{J} \sum_{n=1}^{N} \frac{D_{jn}}{\left(z^2 - z_j^2\right)^n} \tag{A13}$$

where the coefficients C_{jn} and D_{jn} are real. In these stress functions the poles $\pm z_j$ must lie on either the x- or y-axis and be symmetrically placed about the other axis.

APPENDIX B

OTHER COLLOCATION METHODS INVESTIGATED

Direct Specification of Boundary Stresses

The boundary-collocation method treated in this section involves the specification of the normal- and shear-stress components at discrete points on the boundary. The boundary-value problem considered had cracks emanating from the edges of a circular hole in an infinite plate subjected to a biaxial stress (see fig. 15). The complex equation for the two stress components on the boundary is

$$\sigma_n - i\tau_{nt} = \phi(z) + \overline{\phi(z)} - \left[(\bar{z} - z)\phi'(z) - \phi(z) + \overline{\Omega(z)}\right]e^{2i\alpha} \tag{B1}$$

where the stress functions are

$$\left.\begin{array}{c}\phi(z)\\ \Omega(z)\end{array}\right\} = \frac{z}{\sqrt{z^2 - a^2}} \sum_{n=0}^{N} \frac{C_n}{z^{2n}} \pm \sum_{n=0}^{N} \frac{D_n}{z^{2n}} \tag{B2}$$

The coefficients C_0 and D_0 were determined from the stress conditions at infinity. The remaining coefficients were determined from the conditions that $\sigma_n = 0$ and $\tau_{nt} = 0$ at N equally spaced points on the boundary. The resulting equations were solved on a computer using single precision. The stress-intensity factor was calculated from equation (A8) as

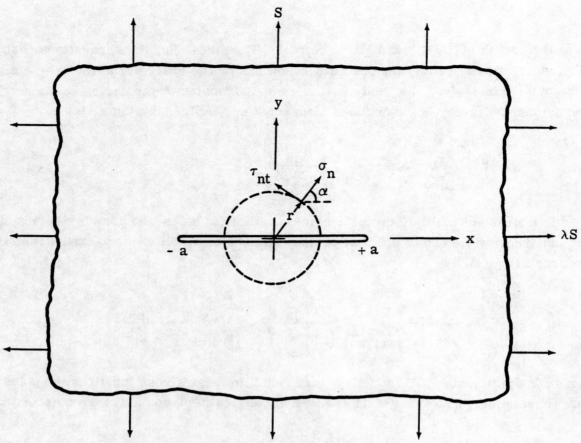

Figure 15.- Coordinate system used for the case of cracks emanating from a circular hole.

$$K = S\sqrt{\pi a}\left[1 + \sum_{n=1}^{N} \frac{2C_n}{a^{2n}}\right] \tag{B3}$$

where the term in the brackets is the correction factor F.

Least-Squares Specification of Boundary Stresses

The boundary-collocation method described in this section is similar to the one in the preceding section, except that here more equations are written in terms of the stress boundary conditions than there are unknown coefficients. The coefficients are then chosen so as to minimize the sum of the stress residuals on the boundary at a specified number of points M. The normal stress and shear stress components expressed in terms of the stress functions (eqs. (B2)) are

$$\left.\begin{array}{l}\sigma_n = \displaystyle\sum_{n=0}^{N} C_n F_{1n} + \sum_{n=0}^{N} D_n F_{2n} \\[2ex] \tau_{nt} = \displaystyle\sum_{n=0}^{N} C_n G_{1n} + \sum_{n=0}^{N} D_n G_{2n}\end{array}\right\} \tag{B4}$$

where F_{sn} and G_{sn} (s = 1,2) are the influence functions derived from the stress functions for the respective coefficient. Again, the coefficients C_0 and D_0 are determined from the stress conditions at infinity. Since the circular boundary is stress free, the square of the error in the boundary condition at point ζ_m is

$$e_m^2 = \left\{ \sum_{n=0}^{N} C_n F_{1n} + \sum_{n=0}^{N} D_n F_{2n} \right\}_m^2 + \left\{ \sum_{n=0}^{N} C_n G_{1n} + \sum_{n=0}^{N} D_n G_{2n} \right\}_m^2 \tag{B5}$$

The coefficients were then selected so that the sum of the squares of the errors at a number of points M on the boundary was a minimum

$$\left. \begin{array}{c} \dfrac{\partial \sum_{m=1}^{M} e_m^2}{\partial C_q} = 0 \\ \\ \dfrac{\partial \sum_{m=1}^{M} e_m^2}{\partial D_q} = 0 \end{array} \right\} \tag{B6}$$

where $q = 1,2,\ldots,N$. Equation (B6) results in a set of $2N$ linear algebraic equations for C_n and D_n

$$\left. \begin{array}{c} \sum_{n=0}^{N} \alpha_{nq} C_n + \sum_{n=0}^{N} \beta_{nq} D_n = 0 \\ \\ \sum_{n=0}^{N} \beta_{qn} C_n + \sum_{n=0}^{N} \gamma_{nq} D_n = 0 \end{array} \right\} \tag{B7}$$

where

$$\alpha_{nq} = \sum_{m=1}^{M} \left(F_{1n} F_{1q} + G_{1n} G_{1q} \right)_m$$

$$\beta_{nq} = \sum_{m=1}^{M} \left(F_{2n} F_{1q} + G_{2n} G_{1q} \right)_m$$

and

$$\gamma_{nq} = \sum_{m=1}^{M} \left(F_{2n}F_{2q} + G_{2n}G_{2q} \right)_m$$

These equations were solved on a computer using single precision. The resulting coefficients were used in equation (B3) to calculate the stress-intensity factor.

REFERENCES

1. Paris, Paul C.; Gomez, Mario D.; and Anderson, William E.: A Rational Analytic Theory of Fatigue. Trend Eng. (Univ. of Washington), vol. 13, no. 1, Jan. 1961, pp. 9-14.

2. Donaldson, D. R.; and Anderson, W. E.: Crack Propagation Behavior of Some Airframe Materials. D6-7888, Paper presented at Crack Propagation Symposium (Cranfield, England), Sept. 1961.

3. Anon.: Fracture Testing of High-Strength Sheet Materials: A Report of a Special ASTM Committee. ASTM Bull., no. 243, Jan 1960, pp. 29-40.

4. Figge, I. E.; and Newman, J. C., Jr.: Prediction of Fatigue-Crack-Propagation Behavior in Panels With Simulated Rivet Forces. NASA TN D-4702, 1968.

5. Bowie, O. L.: Analysis of an Infinite Plate Containing Radial Cracks Originating at the Boundary of an Internal Circular Hole. J. Math. Phys., vol. XXXV, no. 1, Apr. 1956, pp. 60-71.

6. Grebenkin, G. G.; and Kaminskii, A. A.: Propagation of Cracks Near an Arbitrary Curvilinear Hole. Sov. Mater. Sci., vol. 3, no. 4, July-Aug. 1967, pp. 340-344.

7. Fichter, W. B.: Stresses at the Tip of a Longitudinal Crack in a Plate Strip. NASA TR R-265, 1967.

8. Isida, M.: On the Determination of Stress Intensity Factors for Some Common Structural Problems. Eng. Fracture Mech., vol. 2, no. 1, July 1970, pp. 61-79.

9. Erdogan, Fazil: On the Stress Distribution in Plates With Collinear Cuts Under Arbitrary Loads. Proceedings of the Fourth U.S. National Congress of Applied Mechanics, Vol. One, Amer. Soc. Mech. Eng., 1962, pp. 547-553.

10. Hulbert, L. E.; Hahn, G. T.; Rosenfield, A. R.; and Kanninen, M. F.: An Elastic-Plastic Analysis of a Crack in a Plate of Finite Size. Applied Mechanics, M. Hetényi and W. G. Vincenti, eds., Springer-Verlag, 1969, pp. 221-235.

11. Kobayashi, A. S.; Cherepy, R. D.; and Kinsel, W. C.: A Numerical Procedure for Estimating the Stress Intensity Factor of a Crack in a Finite Plate. Trans. ASME, Ser. D.: J. Basic Eng., vol. 86, no. 4, Dec. 1964, pp. 681-684.

12. Bowie, O. L.; and Neal, D. M.: A Modified Mapping-Collocation Technique for Accurate Calculation of Stress Intensity Factors. Int. J. Fracture Mech., vol. 6, no. 2, June 1970, pp. 199-206.

13. Conway, H. D.: The Approximate Analysis of Certain Boundary-Value Problems. Trans. ASME, Ser. E: J. Appl. Mech., vol. 27, no. 2, June 1960, pp. 275-277.

14. Hulbert, Lewis Eugene: The Numerical Solution of Two-Dimensional Problems of the Theory of Elasticity. Bull. 198, Eng. Exp. Sta., Ohio State Univ.

15. Hooke, C. J.: Numerical Solution of Plane Elastostatic Problems By Point Matching. J. Strain Anal., vol. 3, no. 2, Apr. 1968, pp. 109-114.

16. Muskhelishvili, N. I. (J. R. M. Radok, transl.): Some Basic Problems of the Mathematical Theory of Elasticity. Third ed., P. Noordhoff, Ltd. (Groningen), 1953.

17. Fox, L.: An Introduction to Numerical Linear Algebra. Oxford Univ. Press, 1965.

18. Paris, Paul C.; and Sih, George C.: Stress Analysis of Cracks. Fracture Toughness Testing and Its Applications, Spec. Tech. Publ. No. 381, Amer. Soc. Testing Mater., c.1965, pp. 30-83.

19. Savin, G. N. (Eugene Gros, transl.): Stress Concentration Around Holes. Pergamon Press, Inc., 1961.

20. Isida, M.: On the Tension of a Strip With a Central Elliptical Hole. Trans. Japan Soc. Mech. Engrs., vol. 21, no. 107, 1955, p. 511.

ON THE FINITE ELEMENT METHOD IN LINEAR FRACTURE MECHANICS†

S. K. CHAN, I. S. TUBA and W. K. WILSON

Westinghouse Research Laboratories, Pittsburgh, Pa. 15235, U.S.A.

Abstract — The usefulness of the finite element method for the computation of crack tip stress intensity factors is established. Although ordinary finite element methods lack the ability to represent crack tip stress singularity, meaningful values for crack tip stress intensity factors can be obtained by a simple process. The results are compared not only to the results of other analytical solutions, but additional correlation is made of two different fracture test specimen types.

NOTATION

a measure of crack length. See Figs. 1, 3, 5 and 9.
B thickness of fracture test specimens
e_{ij} strain tensor components
E, G elastic and shear moduli
f_{ij} various functions of θ in crack tip equations
J integral defined by (6)
K_I stress intensity factor for mode I
K_{IC} plane strain critical stress intensity factor for mode I
K_I^* Fictitious stress intensity factor, evaluated on the basis of local results, near crack tip, by using crack tip equations. Its limiting value at the crack tip should equal K_I.
P applied load, and load at fracture
r, θ, z polar coordinates; usually the origin is placed at a crack tip
x, y, z orthogonal cartesian coordinates
u, v displacement components
λ, μ Lamme constants of elasticity
ν Poissons ratio
ρ mass density of material
σ_{ij} normal and shear stress components
ω angular velocity at fracture.

Several quantities are defined in the figures, tables or the text.

1. INTRODUCTION

IN THE engineering application of the concepts of linear fracture mechanics to the prediction of strength and life of cracked structures, a knowledge of the crack tip stress intensity factor as a function of applied load and geometry of the structure is necessary. This information, combined with the experimentally determined critical stress intensity factor and crack growth rates for the structural material make such predictions possible.

The concepts and results of linear fracture mechanics are already adopted by a great number of engineers. The theoretical and experimental results are summarized in several recent publications [1, 2], therefore no attempt will be made here for the same. It is only noted that from the practicing engineers point of view linear fracture mechanics works as follows. For a given flaw (mostly cracks are assumed), and principal mode of loading, a theoretical stress intensity factor is computed based on the laws and principles of elasticity. For the material which is used, critical values of stress intensity factors are determined by experiments, under the same principal mode of loading. The

†Presented at the National Symposium on Fracture Mechanics, Lehigh University, Bethlehem, Pa., June 17–19, 1968.

conditions, such as temperature and irradiation, should be the same during the experiment as in the actual application. If the computed theoretical stress intensity factor is less than the critical value by a safe margin, then the flaw is acceptable. Otherwise some sort of corrective measure must be taken to avoid failure, excessive crack growth or hazardous operation.

It has been established that linear fracture mechanics, as described briefly above, works very well. Of course there are still many unresolved questions which must be investigated. In this paper only a few of these problems will be considered.

A rigorous determination of the crack tip stress intensity factor requires an exact solution of the elasticity problem formulated for the cracked structure. In most cases exact solutions to the actual problem are very difficult or nearly impossible to obtain.

Existing theoretical stress intensity factors are the product of highly sophisticated mathematical analysis for idealized model configurations and loading conditions. Considerable number of these are available in tabulated form[1, 2], which are, of course, very valuable for the practicing engineers. These results can be applied to a wide variety of problems, with more or less success, depending on how well the real geometry and loading condition agree with those used in the mathematical model.

Since in many real situations it is not possible to find suitable model representation for which an exact solution is available, the need for a relatively straight forward numerical method is apparent, in order to estimate stress intensity factors.

One such method will now be presented for plane structures. The extension of this method to other types of cracked structures will also be discussed.

Faith must be gained for the validity of results by approximate methods. The approach must be applied at first to configurations where exact or well established solutions exist. Suitable procedure has to be developed for the computation of the stress intensity factor and accuracy of the solution evaluated. When satisfactory results are obtained then the method can be applied to problems where exact solutions are impossible to obtain. Although the solution will be approximate, the magnitude of error should be estimated.

The finite element method is suggested as the best candidate at the present time for obtaining approximate stress intensity factors, whenever exact solutions are not available.

Finite element methods are rapidly adopted in structural analysis. The reason is that the method is conceptually simple, easily adoptable to high speed electronic computations, applicable to large classes of geometries, materials and loading conditions, and can be made quite accurate.

The technical details of the finite element method are described in the literature[3] and will not be repeated in this paper.

The method given here, for the determination of crack tip stress intensity factors, of course, contains all of the inherent advantages of the finite element stiffness method. Various forms of mixed boundary conditions can be analyzed. Structures formed by dissimilar metals can be handled. Body forces, such as those in rotating apparatus and structures subjected to shock can be analyzed. In addition to thermal stresses, the effects of plastic flow in highly stressed regions away from the vicinity of the crack tip can be included in the analysis. Reinforced plane sheets, which are subject to plane deformations only, can also be handled, provided that the crack tip is not in the reinforcement.

While the method as presented here is oriented toward plane structures, the extension of the approach to other classes of structures is obvious. For example the stress intensity factors for circumferential cracks in axisymmetrically stressed solids of revolution can be obtained by using appropriate axisymmetric ring elements. Extensions to mode *III* conditions of longitudinally cracked prismatic bars in torsion [4], circumferentially cracked circular shafts of variable diameter subject to torsion, and to antiplane strain problems is also possible.

2. FINITE ELEMENT PROGRAM

The program used in this study is based on the displacement method. First order displacement functions are assumed, that is the displacements vary linearly over the element which results in constant strains and stresses on the element. The program is general purpose; plane-stress and plane-strain cases can be handled. Displacements and stresses can be determined in arbitrary plane shapes by replacing the actual geometry with an assemblage of triangular elements. The program accepts a variety of boundary conditions and loadings, such as, prescribed displacements, boundary tractions, body forces and temperature variations. The program is written in the Fortran IV language and is operated on a CDC6600 Computer.

3. FINITE ELEMENT GENERATORS

Since the finite element program is general purpose, for particular geometries additional programs are written to generate the necessary input information.

Element generators were created for four basic shapes. The configuration for the study of mesh size effects and for comparisons with results by the collocation method is shown on Fig. 1. Three different geometries have been used, as indicated in the figure, by varying the crack length. The corresponding element generator is such that all dimensions can be varied, and the element sizes away from the crack tip can be

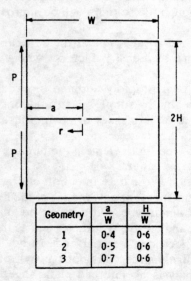

Geometry	$\frac{a}{W}$	$\frac{H}{W}$
1	0·4	0·6
2	0·5	0·6
3	0·7	0·6

Fig. 1. Configuration for the study of mesh size effects and for comparisons with results by the collocation method.

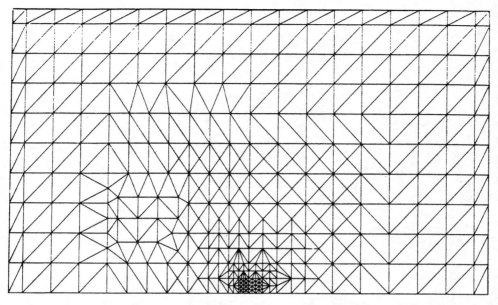

Fig. 2. A finite element representation of the simplified compact tension specimen. (Some nonuniformities are present because this representation is obtained by modifications of the mesh shown on Fig. 4.)

made larger or smaller. A corresponding typical mesh layout is shown in Fig. 2. In the crack tip area the possibility of independent refinement in mesh size is incorporated. On Fig. 2, around the crack tip two families of geometrically similar element groups are superimposed, one on top of the other, but smaller in size. Any number of element size reductions following this pattern can be carried out automatically. All necessary numbering of the elements and of the nodal points are carried out also automatically. Results of production plotting routines indicate all of these numbers, but they are omitted from Fig. 2, for the sake of clarity. For the same reason only one element size reduction is shown. In the actual computations the ratio of the smallest area used to the smallest area shown was $\frac{1}{64}$. The smallest elements used are difficult to see with the naked eye.

The second basic shape corresponds to the compact tension fracture test specimen, as is shown on Fig. 3. A corresponding typical element outline is shown on Fig. 4. The automatic mesh size reduction feature is included not only in these element generators but in all others with cracks.

The third basic shape is the case of a crack of length $2a$ in an infinitely large flat plate. No separate figures are included here for this problem, because the insert in Fig. 9 is descriptive enough.

The final basic shape is shown on Fig. 5, for the rotating fracture test specimens and a finite element setup is given on Fig. 6.

4. COMPUTATIONAL METHODS FOR STRESS INTENSITY FACTORS

Once the numerical solution has been obtained for a particular finite element representation, crack tip stress intensity factors can be estimated by the use of established crack tip relations. There are many possible methods which can be used. The

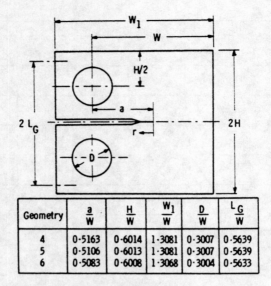

Fig. 3. Compact tension specimens used in analysis. Dimensional ratios are taken from actual test specimens.

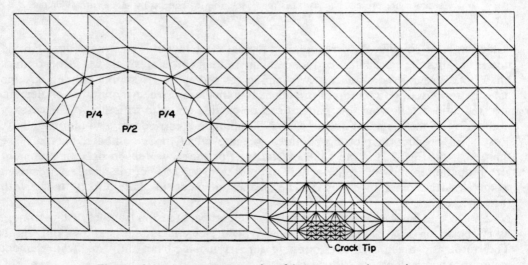

Fig. 4. A finite element representation of the compact tension specimen.

three specific methods considered here are: (1) displacement method, (2) stress method, and (3) line integral (energy) method. The major emphasis has been placed on the displacement method due to its relative simplicity and ease of interpretation. All three methods are conceptually straight forward and will be described below.

The displacement method involves a correlation of the finite element nodal point displacements with the well known crack tip displacement equations. The plane strain displacement equations are:

$$u_i = \frac{K_I}{G}[r/2\pi]^{1/2} f_i(\theta,\nu) \tag{1}$$

where

$$u_1 = u \quad u_2 = v$$
$$f_1(\theta,\nu) = \cos(\theta/2)[1 - 2\nu + \sin^2(\theta/2)]$$
$$f_2(\theta,\nu) = \sin(\theta/2)[2 - 2\nu - \cos^2(\theta/2)]$$

and the crack tip stress intensity K_I is a function of the geometry of the body containing the crack and of the applied loading conditions. A corresponding set of equations can be written for plane stress conditions. By substituting a nodal point displacement u_i^* at some point (r,θ) near the crack tip into (1) a quantity K_I^* can be calculated.

$$K_I^* = [2\pi/r]^{1/2} G u_i^*/[f_i(\theta,\nu)]. \tag{2}$$

From plots of K_I^* as a function of r for fixed values of θ and a particular displacement component, estimates of K_I can be made. If the substituted displacements were the exact theoretical values then the value of the K_I^* obtained as r approaches zero would be the exact value of K_I. Since the finite element displacements are rather inaccurate at an infinitesimal distance from the crack tip, this limiting process is not useable. Instead a tangent extrapolation of the K_I^* curve is used to estimate K_I. With suitable refinement of element size the K_I^* curves obtained from element displacements

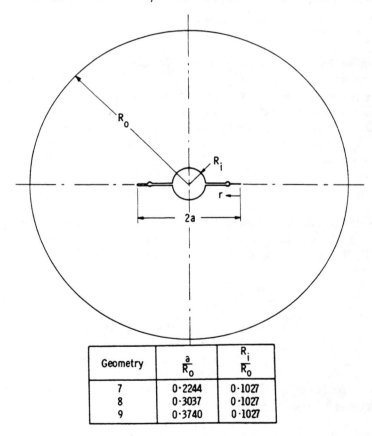

Geometry	$\frac{a}{R_o}$	$\frac{R_i}{R_o}$
7	0·2244	0·1027
8	0·3037	0·1027
9	0·3740	0·1027

Fig. 5. Rotating fracture test specimens.

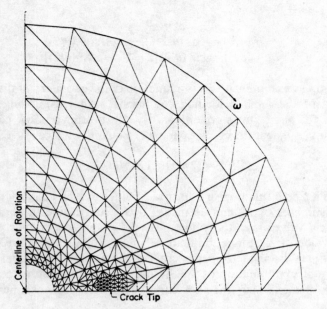

Fig. 6. A finite element representation of the rotating test specimen.

rapidly approach a constant slope with increasing distance (r) from the crack tip. The intercept of the tangent to the constant slope portion of the curve with the K_I^* axis is used as the K_I estimate. As will be shown when specific examples are considered this method gives very good estimates of K_I and as the exact displacements are approached by element displacements the exact value of K_I is approached. The most accurate estimates are obtained from the K_I^* curve corresponding to the v displacement on the crack surface (v_c).

$$K_I^* = \frac{(2\pi)^{1/2} E}{4(1-\nu^2)} \frac{v_c}{r^{1/2}}. \tag{3}$$

The stress method for determining crack tip stress intensity factors is similar to the displacement method. Now the nodal point stresses are correlated with the well known crack tip stress equations. For plane strain conditions the equations are

$$\sigma_{ij} = \frac{K_I}{(2\pi r)^{1/2}} f_{ij}(\theta) \tag{4}$$

where

$$f_{xx}(\theta) = \cos(\theta/2)[1 - \sin(\theta/2)\sin(3\theta/2)]$$
$$f_{yy}(\theta) = \cos(\theta/2)[1 + \sin(\theta/2)\sin(3\theta/2)]$$
$$f_{xy}(\theta) = f_{yx} = \sin(\theta/2)\cos(\theta/2)\cos(3\theta/2).$$

Nodal point stresses σ_{ij}^* in the vicinity of the crack tip can be substituted into (4) and values of K_I^* can be calculated

$$K_I^* = \frac{(2\pi r)^{1/2}}{f_{ij}(\theta)} \sigma_{ij}(r,\theta). \tag{5}$$

From plots of K_I^* as a function of r for a fixed θ and a particular stress component, estimates of K_I can be made. If the exact theoretical stresses were substituted into (5) then the intercept of the curve with the K_I^* axis at $r = 0$ would be the exact value of K_I. Due to the inability of the finite element method to represent the stress singularity conditions at the crack tip, the K_I^* curve for r greater than zero must again be extrapolated back to $r = 0$. The extrapolated value of K_I^* at $r = 0$ is the estimated K_I. Good K_I estimates by the stress method are obtained from the K_I^* curve corresponding to the σ_y stress on the $\theta = 0$ plane.

Rice has shown[5] that the value of the line integral

$$J = \int_\Gamma \left(W dy - T \cdot \frac{\partial \mathbf{u}}{\partial x} ds \right) \tag{6}$$

where Γ is an arbitrary contour surrounding the crack tip as shown in Fig. 7, is proportional to the square of the crack tip stress intensity factor. For plane strain conditions the following relation is given by Rice:

$$K_I = \left[\frac{JE}{(1-\nu^2)} \right]^{1/2}. \tag{7}$$

In (6) W is defined as the strain energy density, $\mathbf{T}$ is the traction vector defined according to the outward normal along Γ, $T_i = \sigma_{ij} n_j$, and $\mathbf{u}$ is the displacement vector. The line integral is evaluated in a counterclockwise sense starting from the lower crack

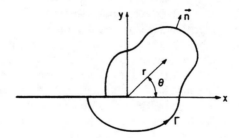

Fig. 7. Crack tip coordinates and typical contour Γ.

surface and continuing along the path Γ to the upper flat surface and ds is an element of arc length along Γ. Rice's path independent integral is applicable to linear and non-linear elastic materials. For linear elastic conditions under consideration

$$W(e) = \tfrac{1}{2}\lambda(e_{ii})^2 + \mu e_{ij} e_{ij}. \tag{8}$$

By numerically evaluating the integral of (6) for the finite element solution over a path surrounding the crack tip, an estimate of the crack tip stress intensity factor can be made by use of (7).

In the methods just presented and in the specific applications to follow, only mode I crack tip loading conditions are considered. For the plane problems under consideration the displacement and the stress methods as described can be extended to obtain mode *II* stress intensity factors, K_{II}, or combinations of K_I and K_{II}. For a mixed mode condition K_I and K_{II} estimates are made in the manner described above. To uncouple the mixed mode conditions, K_I estimates are made from the K_I^* curve constructed from the v displacement on the crack surface ($\theta = \pi$) by the displacement method, or from σ_y on the $\theta = 0$ plane by the stress method. Similarly the K_{II} component can most effectively be obtained from K_{II}^* curves constructed from the **u** displacement on the crack surface by the displacement method:

$$K_{II}^* = \frac{(2\pi)^{1/2}E}{4(1-\nu^2)} \frac{u_c}{(r)^{1/2}} \qquad (9)$$

or from τ_{xy} on the $\theta = 0$ plane by the stress method:

$$K_{II}^* = (2\pi r)^{1/2} \tau_{xy}. \qquad (10)$$

The value of the path independent line integral J (6) is related to the sum of the squares of the stress intensity components:

$$J = \frac{(1-\nu^2)}{E}[K_I^2 + K_{II}^2]. \qquad (11)$$

5. INFLUENCE OF ELEMENT SIZE

The influence of element size on estimated K_I appears to be approximately the same for all three methods considered. Since the displacement method is the simplest to work with and interpret it will be the method considered in detail here.

The effect of relative element size on the K_I^* curve as calculated by the displacement method is shown in Fig. 8. Here K_I^* calculated from the v displacement on the crack surface is shown as a function of r for a number of different mesh reductions for a specific proportion ($a/W = 0.5$ of the simplified CT geometry (Fig. 1). The finite element curves are compared with the K_I^* curve calculated from displacements obtained by a boundary collocation solution for the same geometry and loading conditions. The Williams' stress function[6–8] was used in the collocation process. The collocation curve shown in Fig. 8 is considered to be accurate within 0·50 per cent and is used as the basis for convergence considerations of the finite element method. Two independent phases of element size reduction are considered. The first phase is concerned with the size of elements very close to the crack tip. The size of elements in this area will be referred to as the inner element size. The second phase of reduction is concerned with the elements away from the crack tip and near the outer bounds of the geometry. Over this area the term outer element size will be used. Element size will be considered in terms of element area. The inner and outer element sizes for the cases shown in Fig. 8 are listed in Table 1. As indicated, cases 1–4 have the same outer element sizes but significantly different reductions in crack tip element sizes. The relative size of elements for case 1 is shown in Fig. 2. The inner element sizes are decreased in going from case 1 to case 4 by a factor of 256. Cases 4 and 5 have the same inner element size but the outer elements for case 5 are approximately one fourth of those of case 4.

From a study of Fig. 8 the following observations can be made. All of the finite element curves approach a constant slope as r/W increases. The higher the degree of inner element reduction, the more rapid the curve approaches a constant slope. For

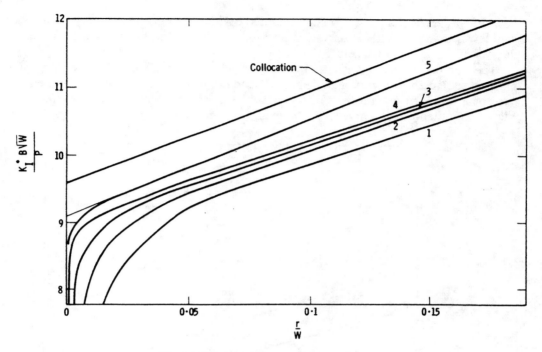

Fig. 8. Effect of mesh size on K_I^* for geometry 2.

Table 1. Element sizes for cases shown in Fig. 8

Case	Element size†-A/a^2	
	Inner (10^{-6})	Outer (10^{-2})
1	312	2
2	78	2
3	20	2
4	1·20	2
5	1·20	1

†A — Element area, a — Crack-length.

fixed outer element size, the sequential reduction in inner element size results in a convergence of the curves to a curve which is a straight line over a major part of its length with a sudden drop near $r/W = 0$. For fixed inner element size a reduction in the outer element size moves the constant slope portion of the curve closer to and more parallel with the collocation curve. It can be concluded, that the finer the outer element size is, the closer the actual curve is to the exact near but not at the crack tip. By sequential reduction of the inner element size the shape of the curve near the tip will deviate less and less from the curve of the exact solution which is a straight line.

The best estimate of the stress intensity factor which can be obtained from any finite element K_I^* curve is obtained by extrapolating the straight portion of the curve back to the vertical axis. The result obtained by the extrapolation of curve 5 is 4 per cent lower than the collocation value. On Fig. 8 since only part of the scale is used, the deviations are greatly exaggerated.

The curves of Fig. 8 apply to a case in which the uncracked section of the crack plane is subjected to a very high degree of bending. It is also beneficial to consider the case of a crack in a uniform stress field which represents the opposite extreme. In Fig. 9 the theoretical curve K_I^* obtained from Westergaards[9] exact solution is

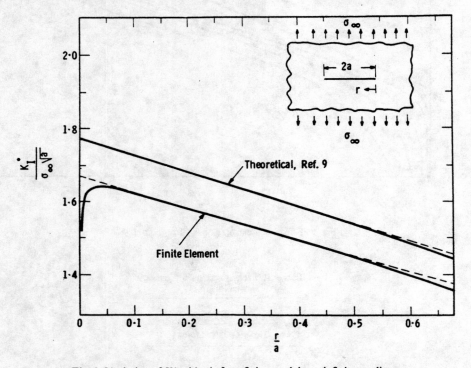

Fig. 9. Variation of K_I^* with r/a for a finite crack in an infinite medium.

compared to a finite element curve for a refined element representation. The comparison is essentially similar to that of Fig. 8, except that the slopes of the two curves are negative. The estimated stress intensity factor obtained from a linear back extrapolation of the straight portion of the finite element curve results in a value of 5·5 per cent lower than the exact value.

A plot of K_I^* values obtained from the stress method for the same geometry and loading condition is shown on Fig. 10. The element refinement is the same as that for case 5 in Table 1. The points indicate values calculated from nodal point stresses σ_y on the $\theta = 0$ plane. Whereas a smooth curve could be drawn through the K_I^*-displacement calculated points as shown in Fig. 8, such is not the case for the nodal stress calculated values. There is much more scatter with these points than with the corresponding displacement points. The courser the element size, the greater the scatter that is observed for the stress calculated K_I^* values. Therefore, the stress method K_I is estimated by fitting the points for small values of r/W with a straight line and taking the vertical axis intercept as K_I. Such an operation applied to the points of Fig. 10

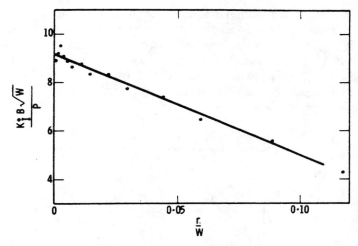

Fig. 10. K_I^* plot for stesss method (case 4).

results in an estimate which is 4·5 per cent lower than the collocation value. For courser element sizes the increase in scatter reduces confidence in this method. The reason for the scatter is of course that in the method constant stresses are computed for individual elements, and nodal point stresses are obtained by an averaging process.

The line integral method was also applied to this configuration. The contour integration path was taken over the outer boundary of the geometry and the integral was evaluated numerically. The strain energy densities were calculated from nodal point stresses and the nodal point forces were used as the surface tractions. The stress intensity factor estimated by this method for element size reduction of case 5 from Table 1 is 3·5 per cent less than the collocation value. The advantage of this method is that no extrapolations are required, only one number is calculated. The disadvantage is that for a single element representation it is difficult to estimate the degree of error in the estimated value, whereas for the other two methods a study of the K_I^* plot gives some indication of the degree of accuracy.

As these specific cases demonstrate, estimates of K_I values within engineering accuracy can be made by this approach. The higher the degree of accuracy required, the larger the number of elements needed. The primary limitation on the number of elements used is the available computer storage capacity. The stability of the total stiffness matrix appears to be relatively independent of the number of elements used or of the variation in element size over the structure.

6. RESULTS FOR COMPACT TENSION SPECIMENS

Compact tension fracture test specimens are the latest in an evolution series, in which they were preceded by the Manjoine and the WOL (wedge opening load) variety. Since the development of these specimens is well documented[10, 11], discussion here will be limited to the comparison of some recent experimental data to corresponding solutions by the finite element method. Greenberg, Wessel, and Pryle [12] have tested a Cr–Mo–V turbine rotor steel. The actual dimensional ratios from three of their test specimens are shown on Fig. 3. These same dimensions were used also in the analysis. A finite element representation is shown on Fig. 4. The ratio of the smallest element used to the smallest element shown is $\frac{1}{256}$. The results from the finite element method for the stress intensity factors are compared to collocation results on Fig. 11. Actual dimensions and fracture loads were used in the calculations. During the test procedure, the change ($2v_y$) in a gage length ($2L_G$ on Fig. 3) is plotted as a function of the applied load. The actual values at fracture of the ratio v_y/P, often referred to as compliance, are compared to the corresponding computed values in Table 2. The agreement is gratifying.

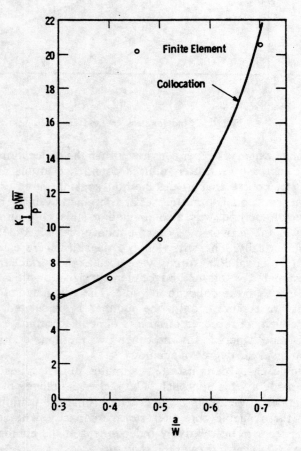

Fig. 11. Comparison of results of collocation and finite element solutions.

7. RESULTS FOR ROTATING TEST SPECIMENS

Rotating tests are also used to evaluate fracture properties of materials. Specimen preparation, test procedures and results can be found in [13]–[16] and they will not be reviewed here. The geometry considered is shown on Fig. 5. Once the test is run, the present practice is to evaluate K_{IC} from the following expression

$$K_{IC} = \sigma_0 \sqrt{\pi a} \tag{12}$$

where

$$\sigma_0 = \frac{3-2\nu}{8(1-\nu)}\rho\omega^2 R_0^2 \tag{13}$$

which is the stress at the center of a solid rotor of radius R_0, and the other quantities are defined in the nomenclature.

A triangular finite element representation of the rotating test specimen is shown on Fig. 6. In the actual computation at the crack tip much smaller elements were used than those shown on the figure. The ratio of the area of the smallest element used to the smallest element shown was $\frac{1}{256}$.

For the prediction of K_I the displacement method described in part 4 was used. For geometry 8 the variation of K_I^* with r/R_0 is shown on Fig. 12. Of course, from the

Table 2. Compact tension specimen results

Geometry	Experimental†			Finite element	
	Fracture load P (lb)	Crack opening $2v_Y$ (10^{-2} in.)	Compliance v_Y/P (10^{-4} in./lb)	Compliance v_Y/P (10^{-4} in./lb)	Stress intensity K_{IC} (ksi $\sqrt{\text{in.}}$)
5	5820	1·27	10·9	10·04	41·3
6	5630	1·07	9·5	9·79	39·1
7	5940	1·16	9·8	9·69	40·1

†Ref. [12].

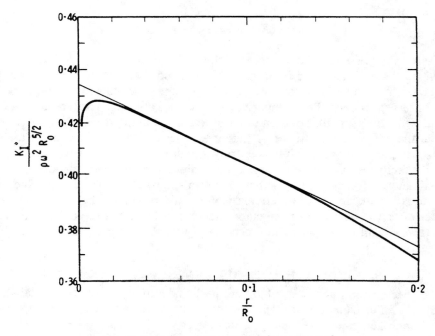

Fig. 12. Variation of K_I^* with r/R_0 for geometry 8.

nature of the approximations inherent to the method, it is known that the exact value of the stress intensity factor is 4–5 per cent higher than those predicted here.

The resulting factor for various crack lengths are given in Fig. 13. On the same figure two different comparisons are made. The empirical formula, (12), used by Sankey[16] crosses over the line predicted by the finite element solution. Another approximate method has been investigated recently by Williams and Isherwood[17] for plane stress problems with finite geometry. Modifying their method to plane strain conditions, the corresponding results are also shown.

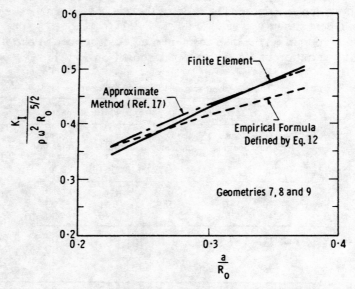

Fig. 13. Stress intensity factors for various a/R_0.

The rotating test specimens were machined from the same forging as the compact tension specimens discussed in the previous section. It is possible, therefore, to compare stress intensity factors for the Cr–Mo–V steel as obtained by the two different types of test specimens. The comparison is made in Table 3. It is interesting to note that regardless of the computational method used for obtaining the critical stress intensity factors, the values are consistently higher for the rotating test specimen, than for the compact tension specimens. If K_{IC} is truly a material property, then the critical stress intensity factors should be the same. The values for compact tension specimen are approximately 15 per cent lower than those for the rotating case. For engineering purposes, this is satisfactory. The comparison of this type must be carried out on large number of test results before one should seek reasons for discrepancies either in the basic theory or in the experimental process. That, of course, is beyond the objectives of this study.

8. POTENTIAL IMPROVEMENTS AND EXTENSIONS

The specific approach presented here is considered to be only the initial step in the use of the finite element methods to obtain crack tip stress intensity factors. This general approach can be refined in a number of ways. Higher order elements such as linear strain elements can be used in place of the constant strain elements considered

Table 3. Comparison of test results for a Cr–Mo–V turbine rotor steel

		Fracture specimen type			
		Compact tension (at 0°F)			Rotating (at −11°F)
	Geometry	4	5	6	8
K_{IC} (ksi $\sqrt{\text{in.}}$)	Load at fracture	5820 lb†	5630 lb†	5940 lb†	10,800 rev/min‡
	Finite element	41·3	39·1	40·1	47·0
	Collocation	42·6	40·35	41·4	
	Equation (12)				45·4
	Williams–Isherwood method				47·8

†Ref. [12].
‡Ref. [16].

here. Also combinations of elements of various orders with the higher order elements closer to the crack tip is possible. The use of a special crack tip element which includes the crack tip singularity condition as its assumed displacement pattern is also very promising. The potential of these variations is now under consideration by the writers.

9. CONCLUSIONS

It has been demonstrated in this paper that the finite element method, supplemented with special computational procedures, can be used to find crack tip stress intensity factors in various shapes under different types of loading conditions. The accuracy of the prediction is satisfactory without the use of excessive computer time, and of course, can be improved, within the limitations of the computer. Although the method is most significant for geometries and loading conditions where no exact solutions exist, it seems desirable to analyze both the structures containing the cracks, and the test specimens used to evaluate the material properties by the same method. The inherent approximations this way would become compensated for automatically.

Acknowledgements—The writers are greatly indebted to Mr. S. E. Gabrielse and Dr. C. Visser for the development of the basic computer program, to Mrs. V. Ho and Mr. D. Wei for programming the element generators, and to Mr. G. O. Sankey and Mr. E. T. Wessel for freely providing the results of their experimental work. All named persons are with the Westinghouse Research and Development Center.

REFERENCES

[1] C. F. Tiffany and J. N. Masters, Fracture toughness testing and its applications. *ASTM Spec. Tech. Publ.* No. 381, pp. 249–277 (1965).
[2] E. T. Wessel, W. G. Clark and W. K. Wilson, Engineering methods for the design and selection of materials against fracture. *U.S. Army Tank and Automotive Center Rep.*, Contract No. DA-30-069-AMC602(T) (1966).
[3] O. C. Zienkiewicz and Y. K. Cheung, *The Finite Element Method in Structural and Continuum Mechanics*. McGraw-Hill, New York (1967).
[4] G. Sih, Strength of stress singularities at crack tips for flexural and torsional problems. *Trans. ASME J. appl. Mech.* (1963).
[5] J. R. Rice, A path independent integral and the approximate analysis of strain concentration by notches and cracks. *J. appl. Mech.* 35, 379 (1968).

[6] M. L. Williams, On the stress distribution at the base of a stationary crack. *J. appl. Mech.* **24**, 109–114 (1957).
[7] W. K. Wilson, Analytic determination of stress intensity factors for the manjoine brittle fracture test specimen. *AEC Res. Devel. Rep.* No. WERL-0029-3 (1965).
[8] B. Gross, E. Roberts, Jr. and J. E. Srawley, Elastic displacement for various edge-cracked plate specimens. *NASA Tech. Note* No. D-4232 (1967).
[9] H. M. Westergaard, Bearing pressures and cracks. *Trans. ASME J. appl. Mech.* (1939).
[10] M. J. Manjoine, Biaxial brittle fracture tests. *Trans. ASME* 293–298 (1965).
[11] E. T. Wessel, State of the art of the WOL specimen for K_{IC} fracture toughness testing. *Engng Fracture Mech.* **1**, 77–103 (1968).
[12] H. D. Greenberg, E. T. Wessel and W. H. Pryle, Fracture toughness of turbine generator rotor forgings. Presented at *2nd National Symp. Fracture Mech., Lehigh University* (1968).
[13] D. H. Winne and B. N. Wundt, Application of the Griffith–Irwin theory of crack propagation to the bursting behavior of disks, including analytical and experimental studies. *Trans. ASME* **80** (1958).
[14] G. O. Sankey, Spin fracture tests of *NiMoV* rotor sheets in the brittle fracture range. *Proc. Am. Soc. Test. Mater.* **60**, 721–732 (1960).
[15] G. D. Cooper, G. O. Sankey and E. T. Wessel, Discussion of paper results of bursting tests of alloy steel disks and their application to design against brittle fracture. *Symp. Large Rotor Forgings, ASTM A. Mtg, Purdue University* (1965).
[16] G. O. Sankey, Spin tests to determine brittle fracture under plane strain. Presented at *SESA Spring Mtg, Albany* (1968).
[17] J. G. Williams and D. P. Isherwood, Calculation of the strain-energy release rates of cracked plates by an approximate method. *J. Strain Analysis* **3**, 17–22 (1968).

(*Received* 29 *May* 1968)

Résumé – On établit l'utilité de la méthode de l'élément fini pour les calculs des facteurs d'intensité de la tension de rupture en bout. Bien qu'habituellement les méthodes d'éléments finis ne peuvent pas représenter la singularité de la force de rupture en bout, il est possible d'obtenir par un procédé simple des valeurs significatives de ces facteurs d'intensité. Les résultats sont comparés à ceux des autres solutions analytiques, et une corrélation supplémentaire est effectuée avec types spécimens différents d'essai de rupture.

Zusammenfassang – Die Zweckmässigkeit der Methode der endlichen Elemente für die Errechnung von Spannungsfaktoren an Rissspitzen wird dargelegt. Obzwar die gewöhnlichen Methoden der endlichen Elemente nicht imstande sind die Singularität der Rissspitzenspannung vorzustellen, können mittels eines einfachen Verfahrens doch sinnvolle Werte für Spannungsfaktoren an den Rissspitzen erhalten werden. Die Ergebnisse werden nicht nur mit den Ergebnissen anderer analytischer Lösungen verglichen, sondern es wird auch eine zusätzliche Korrelation von zwei verschiedenen Typen von Bruchprobekörpern vorgenommen.

A NOTE ON THE FINITE ELEMENT METHOD IN LINEAR FRACTURE MECHANICS

THE POTENTIAL usefulness of the finite element method of stress analysis in linear fracture mechanics has been demonstrated in a recent paper by Chan, Wilson and Tuba[1]. These authors determined crack tip stress intensity factors for several standard laboratory test geometries by three methods, all of which involved using the finite element calculated stress and/or displacement fields close to the crack tip with established crack tip relations. In order to achieve satisfactory accuracy with these methods, it was necessary to incorporate in the finite element structural idealization an extremely fine grid near the crack tip. The present author has carried out a limited study with a finite element computer program[2], utilizing a method different from those explored in [1]; namely, the compliance or strain energy release rate method. The purpose of this note is to present these results, since it appears that this method offers the advantage of not having to employ such a fine grid in the structural idealization.

The configuration considered was the single-edge crack (SEC) specimen subjected to uniform, uniaxial tension (Fig. 1). The overall dimensions and applied loading duplicate those utilized by Srawley et al.[3], for experimental compliance measurements.

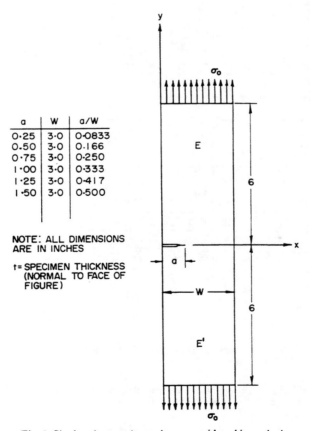

Fig. 1. Single-edge-crack specimen considered in analysis.

The single symmetry of the problem permitted consideration of one half of the specimen in the analysis, as shown in Fig. 2. Solutions were obtained by specifying the following force and displacement conditions for the boundary nodes:

Along $AB - P_x = 0, P_y = 0$
Along $BCD - u_y = 0$ (excluding C), $P_x = 0$
At Point $C - u_y = 0, u_x = 0$
Along $DE - P_x = 0, P_y = 0$
Along $EF - P_x = 0, \Sigma P_y = \sigma_0 Wt$
Along $FA - P_x = 0, P_y = 0,$

where u_x and u_y are nodal displacements in the x- and y-directions and P_x and P_y are nodal forces in the x- and y-directions. Other symbols are defined in Fig. 1.

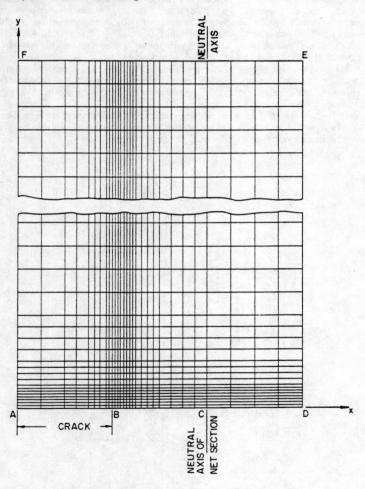

Fig. 2. Structural idealization of single-edge-crack specimen
($a = 1 \cdot 0$ in.).

For the specimen configuration and loading considered, it is known that the strain energy release rate, G, is related to the specimen compliance per unit thickness, c, by

$$G = \frac{P^2}{2} \frac{dc}{da}, \tag{1}$$

where

$$c = \frac{e}{P}, \quad P = \sigma_0 W t. \tag{2}$$

and e is the displacement over some gage length across the crack (i.e. points EE' in Fig. 1). The relationship expressed in (1) can be put in the following dimensionless form[3]:

$$\frac{EWG}{P^2} = \frac{E}{2} dc/d\left(\frac{a}{w}\right). \tag{3}$$

Also, it is known[4] that G and the stress intensity factor K are related by the following formulae:

$$G = \frac{K^2(1-\nu^2)}{E}, \text{ (plane strain)}, \tag{4}$$

$$G = \frac{K^2}{E}, \text{ (plane stress)}, \tag{5}$$

where E is Young's modulus and ν, Poisson's ratio. To develop an expression for G and K through the compliance method, it is necessary to calculate c for a sufficient number of values of a/W to determine $dc/d(a/W)$. Thus, a different idealization of the type shown in Fig. 2 was generated for each crack length studied. The crack length-to-width ratios (a/W) considered included 0·0833, 0·166, 0·250, 0·333, 0·417, and 0·500. The rather coarse grid size surrounding the crack tip in Fig. 2 was maintained constant for each value of a/W. Exclusive use was made of rectangular (linear strain) elements.

Calculations using an elastic modulus of $10·3 \times 10^6$ psi were made, and compliance was determined for gage points 8·0 in. apart at $x/W = 0·5$. These conditions duplicate those of tests reported in [3], so that the experimental and calculated compliance data can be compared directly. The experimental results are estimated to be accurate within $\pm \frac{1}{2}$ per cent in the range of a/W from 0·25 to 0·40[1]. Several experimental and calculated values of c are compared in Table 1 for nearly identical values of a/W. Although the experimental values of c are slightly greater than the calculated values, the agreement can be considered excellent.

Table 1. Comparison of calculated and experimentally determined values of compliance

Finite element results		Experimental data[3]	
a/W	$Ec/2$	a/W	$Ec/2$
0·166	1·395	0·170	1·424
0·250	1·498	0·251	1·532
0·333	1·730	0·331	1·729
0·417	2·084	0·412	2·088

Accurate determinations of K vs. a/W for the SEC specimen have also been obtained by Gross et al.[5], using boundary-collocation techniques. These results, expressed in a least squares, best fit polynomial form, are recommended by ASTM[6] for use with the SEC specimen. In order to obtain a comparison with these

Table 2. Comparison of calculated and ASTM-recommended K calibration for SEC specimen

	$KtW/Pa^{1/2}$		
a/W	ASTM[6]	Finite element	Difference in $KtW/Pa^{1/2}\%$
0·15	2·25	2·33	3·5
0·20	2·44	2·44	0
0·25	2·67	2·64	1·1
0·30	2·95	2·92	1·0
0·35	3·30	3·29	0·3
0·40	3·74	3·75	0·3
0·45	4·30	4·29	0·2
0·50	5·02	4·92	2·0

results, the calculated values of $Ec/2$ were first fit in the form of a polynomial in a/W. The polynomial was, in turn, differentiated to obtain an expression for G in accordance with (3), and then G was related to K by (5). Selected values of $KtW/Pa^{1/2}$ determined in this fashion are compared in Table 2 with ASTM recommended values. The agreement is good; the greatest difference in the two sets of values in the range of a/W from 0·15 to 0·5 is 3·5 per cent.

This limited study has shown that good results can be obtained with the strain energy release rate method without excessive grid size refinement in the vicinity of the crack tip. There appears to be two reasons for this. First, the fact that the finite element calculational method will underestimate stresses and displacements close to the crack tip is minimized by examining the strain energy of the entire body through compliance determinations. Second, although the calculated compliance across a given gage section may be underestimated, the magnitude of the underestimate should be relatively independent of crack length. Hence, the slope of the compliance vs. crack length curve, which is used for computing G and K, should be very close to the true curve.

General Electric Company, D. F. MOWBRAY†
Knolls Atomic Power Laboratory,
Schenectady,
N.Y. 12302, U.S.A.

REFERENCES

[1] S. K. Chan, I. S. Tuba and W. K. Wilson, On the finite element method in linear fracture mechanics. Presented at the *2nd Natn. Symp. Fracture Mech., Lehigh University* (1968).
[2] W. B. Jordan, A program to solve elastic structures by the use of finite elements. *General Electric Rep.* No. KAPL-M-6582 (WBJ-4) (1966).
[3] J. E. Srawley, M. H. Jones and B. Gross, Experimental determination of the dependence of crack extension force on crack length for a single-edge-notch tension specimen. *NASA Tech. Note* No. D-2396 (1964).
[4] G. R. Irwin, Elasticity and plasticity. *Encyclopedia of Physics.* Vol. 6, pp. 551–590. Springer, Berlin (1958).
[5] B. Gross, J. E. Srawley and W. F. Brown, Jr., Stress-intensity factors for a single-edge-notch tension specimen by boundary collocation of a stress function. *NASA Tech. Note* No. D-2395 (1964).
[6] W. F. Brown, Jr. and J. E. Srawley, Plain strain crack toughness testing of high strength metallic materials. *ASTM Spec. Tech. Publ.* No. 410, p. 12 (1966).

(*Received* 21 *October* 1968)

†Present address: Materials and Processes Laboratory, General Electric Co., Schenectady, N.Y. 12305, U.S.A.

FINITE ELEMENTS FOR DETERMINATION OF CRACK TIP ELASTIC STRESS INTENSITY FACTORS

DENNIS M. TRACEY[†]

Division of Engineering, Brown University, Providence, R. I. 02912, U.S.A.

Abstract — A new type of finite element is introduced which embodies the inverse square root singularity present near a crack in an elastic medium. Using this element near the tip in two typical cracked configurations, stress intensity factors within 5 per cent of accepted values were obtained with meshes having as few as 250° of freedom.

INTRODUCTION

It is well known that the elastic stress distribution near a crack tip in any two-dimensional deformation field exhibits an inverse square root singularity [1, 2]. The near-tip stress and displacement distributions for a symmetric "mode I" or tensile deformation are

$$\begin{Bmatrix} \sigma_{xx} \\ \tau_{xy} \\ \sigma_{yy} \end{Bmatrix} = K_I (2\pi r)^{-1/2} \cos(\theta/2) \begin{Bmatrix} 1 - \sin(\theta/2)\sin(3\theta/2) \\ \sin(\theta/2)\cos(3\theta/2) \\ 1 + \sin(\theta/2)\sin(3\theta/2) \end{Bmatrix}$$

$$\begin{Bmatrix} u_x \\ u_y \end{Bmatrix} = K_I (r/2\pi)^{1/2}/2G \begin{Bmatrix} \cos(\theta/2)[\kappa - 1 + 2\sin^2(\theta/2)] \\ \sin(\theta/2)[\kappa + 1 - 2\cos^2(\theta/2)] \end{Bmatrix} \quad (1)$$

where G is the elastic shear modulus and $\kappa = (3-4\nu)$ for plane strain and $\kappa = (3-\nu)/(1+\nu)$ for plane stress, ν being Poisson's ratio. The tension is applied along the y direction in this notation with the crack lying along $x \leq 0$. The polar coordinates (r, θ) are centered at the crack tip with the crack surfaces along $\theta = \pm\pi$. The strength of the singularity, K_I, is the prime unknown in any elastic crack analysis. The functional dependence of K_I upon applied load and geometric dimensions of a typical cracked configuration is generally tractable only through some method of numerical solution of the relevant elasticity equations.

A new type of finite element is introduced which when used near the crack tip results in a solution in excellent agreement with (1) allowing a very accurate determination of K_I. Determination of stress intensity factors by employing conventional types of elements has not been satisfactory due to the inability of these elements to represent the singular near tip deformation. Chan et al.[3] did an extensive finite element study of crack problems using constant stress triangular elements. An extrapolation of the solution away from the tip was used to estimate K_I since the near tip solutions were not reasonable. They report K_I of an edge-cracked plate under eccentric loads, gained from extrapolating a solution involving about 2000° of freedom, within 5 per cent of a collocation solution they obtained. The near tip solutions of problems reported in this paper yield K_I values within 5 per cent of accepted values when allowing as few as 250° of freedom.

[†]Research Assistant.

The triangular "singularity" element used in the present work has a stiffness which embodies the fields (1). These near tip elements were joined with quadrilateral isoparametric elements[4]. The two primary solutions reported are the doubly edge-cracked plane strain rectangle and the round circumferentially cracked bar under tension.

FORMULATION

The displacement functions of the two elements employed are expressed in terms of the elements' natural coordinates, call them (ξ, η). These coordinates are defined so that each edge of an element has one coordinate constant while the other varies linearly along it. The functional relationship between (ξ, η) and the physical coordinates (x, y) can be gained by considering a mapping of the physical element into a rectangle in a (ξ, η) parameter plane. For instance, the mapping

$$\mathbf{X} = \mathbf{X}_A (1-\xi)\eta + \mathbf{X}_B (1-\xi)(1-\eta) + \mathbf{X}_C \xi (1-\eta) + \mathbf{X}_D \xi\eta. \tag{2}$$

will give either of the two transofrmations of Fig. 1. $\mathbf{X}_A$ is meant to represent the position vector of node A in the physical plane. Notice that in Fig. 1b the edge along $\xi = 0$ with two distinct nodes A and B maps onto one vertex of the physical triangle so that in this case $\mathbf{X}_A = \mathbf{X}_B$ in (2). The inverse map for this triangular element in the notation of Fig. 2 is given by

$$\left.\begin{array}{l} \xi = s/s^* \\ \\ \eta = \tfrac{1}{2}(\tan\theta/\tan\alpha + 1) \end{array}\right\} \tag{3}$$

where (s, t) are local Cartesian coordinates and (r, θ) local polar coordinates with respect to (x, y).

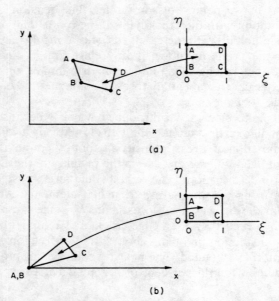

Fig. 1. (a) A typical quadrilateral element mapped into a square. (b) A typical near tip triangle mapped into a square.

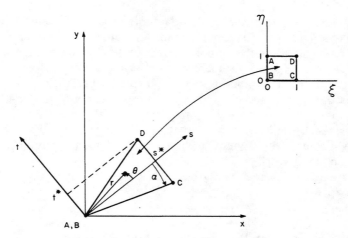

Fig. 2. A typical near tip triangle — square map.

If a crack were located on $x \leq 0$, a displacement function which embodies the singular field of (1) throughout the triangle of Fig. 2 is

$$\mathbf{u}(\xi, \eta) = \mathbf{b}_1 + \mathbf{b}_2 \sqrt{\xi} + \mathbf{b}_3 \sqrt{\xi \eta}. \tag{4}$$

Here $\mathbf{u}(\xi, \eta)$ is the physical cartesian displacement vector of a point within the element. The vectors $\mathbf{b}_1, \mathbf{b}_2, \mathbf{b}_3$ are the undetermined generalized displacements obtained from the three nodal point displacements $\mathbf{u}_{AB}, \mathbf{u}_C, \mathbf{u}_D$. The subscript AB is used to emphasize that although A is thought of as being distinct from B in the mapping, physically there is only one location (the crack tip) and one independent displacement vector at this location. If the crack tip serves as the center of a whole ring of triangular elements each with the displacement function (4) and if these are joined in the radial direction with quadrilateral isoparametric elements with displacement function

$$\mathbf{u}(\xi, \eta) = \mathbf{b}_1 + \mathbf{b}_2 \xi + \mathbf{b}_3 \eta + \mathbf{b}_4 \xi \eta. \tag{5}$$

the expected singularity has been completely embedded while displacement compatibility on inter-element boundaries is assured.

An alternate method of obtaining the displacement variation (4) within the near tip elements, the method used in the present study, is to consider each triangle as a four node quadrilateral with two nodes coincident at the crack tip but distinguishable by their "angular orientation" as in the map, Fig. 2. Then using the displacement function

$$\mathbf{u}(\xi, \eta) = \mathbf{b}_1 + \mathbf{b}_2 \sqrt{\xi} + \mathbf{b}_3 \sqrt{\xi \eta} + \mathbf{b}_4 \eta \tag{6}$$

along with the auxiliary constraint

$$\mathbf{u}_A = \mathbf{u}_B \tag{7}$$

will result in the displacement variation (4). This procedure was followed since the solutions reported here are the initial elastic solutions for complete elastic–plastic studies being undertaken. In the non-hardening plastic situation, a displacement variation is expected as the tip is approached along different angular rays[5]. To accommodate this behavior after the initial elastic solution without altering the mesh, the constraint (7) is relaxed and a displacement function appropriate for plastic deformation is substituted for (6). For non-hardening plasticity the isoparametric displacement function (5) is the appropriate choice[5]. The programming of this general procedure is clearly very compact. The auxiliary constraints alter the master stiffness matrix in the manner described by Hibbitt and Marcal[6]. The capability of specifying such constraints is available in the general purpose finite element program of P. V. Marcal which was used in this work.

Both the singularity element and isoparametric element have interpolation functions of the form

$$\mathbf{u} = \mathbf{N}(\xi, \eta)\boldsymbol{\delta} \tag{8}$$

where $\boldsymbol{\delta}$ contains the nodal displacement vectors of an element. From (8) is obtained the strain distribution

$$\boldsymbol{\epsilon} = \mathbf{B}(\xi, \eta)\boldsymbol{\delta}. \tag{9}$$

Without explaining the theory of the derivation which is adequately discussed in the literature[4], the element stiffness matrix for the plane strain problem is

$$\mathbf{K} = \iint \mathbf{B}^T \mathbf{D} \mathbf{B}\, dx dy \tag{10}$$

where $\mathbf{D}$ is the stress-strain matrix. $\mathbf{B}$ being dependent explicitly on the natural coordinates motivates a change of variables which results in

$$\mathbf{K} = \int_0^1 \int_0^1 \mathbf{B}^T \mathbf{D} \mathbf{B} |\mathbf{J}| d\xi f\eta \tag{11}$$

where $\mathbf{J}$ is the Jacobian of the transformation (2). For the arbitrary quadrilateral element the integrand of (11) is too complicated for exact integration. The procedure used evaluated the integrand at the centers of four equal area subregions of the element; these results were then summed and multiplied by one-fourth the total area. The integrand of (11) for the singularity element involves ξ only through terms of the form function $(\eta)/\xi$. The determinant of the Jacobian of the transformed element pictured in Fig. 2 is equal to $2\xi s^* t^*$. Hence numerical integration is necessary only in the variable η for this element. A two-point η integration rule was used in most of the problems solved.

The axi-symmetric problem solved involved mapping cross sections of ring elements from (ξ, η) to a $\phi = $ const. plane of a (ρ, z, ϕ) cylindrical coordinate system. The element stiffness was

$$\mathbf{K} = \int_0^1 \int_0^1 2\pi\rho(\xi, \eta) \mathbf{B}^T \mathbf{D} \mathbf{B} |\mathbf{J}| d\xi d\eta \tag{12}$$

A four point integration rule was used over the area in this case.

NUMERICAL RESULTS

To test the singularity element, a semi-infinite crack in an infinite plane was analyzed by considering a finite circular region centered at the crack tip, Fig. 3. Displacement

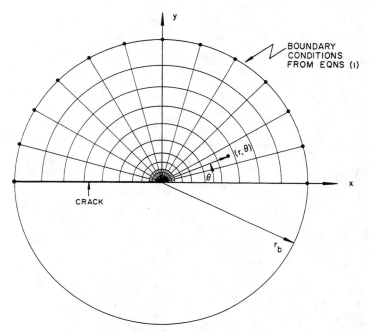

Fig. 3. Finite element treatment of isolated crack.

boundary conditions obtained from the displacement field of (1) were applied to the nodes at the boundary r_b. From symmetry only $y > 0$ was considered. Since the displacement at each point of this domain is expected to depend upon the square root of distance from the crack tip, the displacement function (6) was used for all of the elements in this problem. Of course the constraint (7) was applied only to crack tip elements. The irrational terms of (6) required that adjoining elements be mapped without translation in the radial direction to preserve the displacement character of (1). The mesh used consisted of 12 rings of trapezoidal shaped elements mapped as in Fig. 4 without translation (and for programming convenience, without stretch) in the s direction. The nodes described arcs of radius

$$[0, 0{\cdot}5, 1, 1{\cdot}625, (1{\cdot}5)^2, (2)^2, \ldots, (5)^2, (5{\cdot}5)^2].$$

The first four rings contained 24 elements at 7·5° while the remainder had 12 elements at 15° intervals. The nodes on $r = 2{\cdot}25$ not common to the adjacent 7·5° and 15° elements were constrained to move in a compatible manner[6]. There were a total of 192 elements and 229 nodes with 410° of freedom after the constraints (7) were applied. The integration of the element stiffnesses was accomplished by evaluating the integrand at the element midpoint and multiplying by the area.

The exact solution to this problem is of course given by (1); K_I having been specified in the boundary conditions. Excepting those points where the exact and computer solutions involved vanishingly small numbers when compared to significant values present in the solution, a percentage error was calculated at each node and stress point. As expected the solution increased in accuracy with distance from the crack. The nodes closest to the tip, $r = 0{\cdot}5$, had a vertical displacement u_y within $+1$ per cent

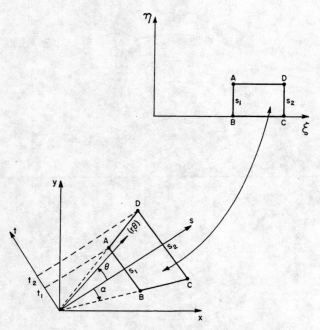

Fig. 4. An (x, y) isosceles trapezoid $\leftrightarrow$ (ξ, η) rectangle map.

of the exact value. The values of u_x for nodes in the range $0° \leq \theta \leq 158°$ were within $+2$ per cent of being correct while the node at $165°$ was $+5$ per cent in error. The node on the crack line, which should not displace horizontally according to (1), had a u_x which was 2 per cent of the value of the node at $0°$ which is a good approximation to zero. The stresses σ_x, σ_y, τ averaged $+2$ per cent error in their significant ranges with a maximum error of $+4$ per cent.

From this solution it can be anticipated that the singularity element gives very good near tip information for the general crack problem.

The next problem solved was the plane strain rectangular bar with symmetric edge cracks, Fig. 5. The geometry considered was $L/b = 4$, $b/a = 2$. From symmetry only one quadrant had to be considered, the symmetry lines were specified free of shear and fixed in the normal direction. The singularity element, (6) and (7), was employed in the first of fifteen rings of trapezoidal elements centered about the crack tip; all others were considered as isoparametric elements. The far field consisted of large rectangular elements. The trapezoids' nodes described arcs of radius

$$\frac{a}{36}[0, 0.5, 1, 1.625, (1.5)^2, (2)^2, \ldots, (5.5)^2, 37.27, 50.91, 72]. \tag{13}$$

Like the previous problem the first four rings contained $7.5°$ elements; thereafter, the elements covered $15°$. Where the arcs (13) cut the external boundary, arbitrary quadrilaterals were used. In all 252 elements and 300 nodes with $548°$ of freedom were used. It should be emphasized that the shape of the element has no effect on the logic of the programming in any way, this is the virtue of the isoparametric element. Yet the shape of the element completely determines the form of the assumed displacement distribution, that is why the singularity element is necessary.

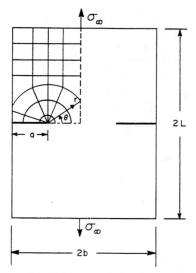

Fig. 5. Geometry and mesh design for the cracked rectangle and round bar.

Both the displacement and stress solution data near the crack tip can be used to estimate K_I; the estimate obtained from the displacements, of course, is the more fundamental and hence more reliable. For example, the last of (1) provides the formula

$$K_I = (2G\,(2\pi/r)^{1/2}/f_y(\theta))\,u_y(r,\theta) \tag{14}$$

where $f_y(\theta)$ stands for the function of θ in the equation.

The literature[7] has tended to express K_I for this particular type of problem in the form

$$K_I = \sigma_\infty (\pi a)^{1/2} \left(\frac{2b}{\pi a} \tan \frac{\pi a}{2b}\right) . C \tag{15}$$

where C is a correction factor multiplying the solution of a periodic array of cracks in a sheet under remote uniform stress. The values of u_y for the nodes closest to the crack tip (at $r/a = 1/72$) yielded values of C between 1·05 and 1·09 with an average of 1·07. The horizontal displacements predict C between 0·98 and 1·10 with an average value of 1·05. Hence a simple averaging of the near tip displacement solution gives C as 1·06. Bowie[8] solved this problem using approximate conformal mapping techniques and obtained a value of 1·02, a difference of 4 per cent from the present estimate. Whereas he expressed confidence that his result is within 1 per cent of being exact and since the previous problem was within 2 per cent, it is safe to say that the present solution is good to within 4 per cent.

The stresses at the midpoints of the near tip singularity elements were used to obtain a simple average value of 1·02 for C. The averaging excluded those stress data which were an order of magnitude lower than the peak stress at $r/a = 1/44$. Figures 6-8 present the raw stress data. The curves are the stresses which would result at $r/a = 1/44$ if K_I was given by (15) with $C = 1·02$. The stress solution amplifies the errors in displacement as expected; the data range -13 per cent to $+7$ per cent about the curves.

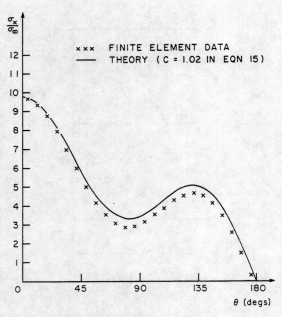

Fig. 6. σ_x at the near tip element centers.

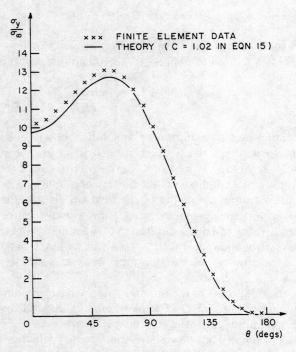

Fig. 7. σ_y at the near tip element centers.

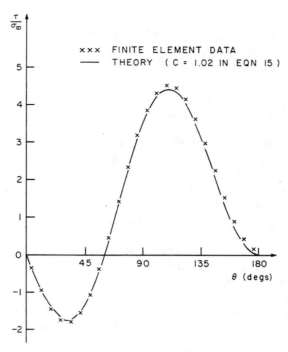

Fig. 8. τ at the near tip element centers.

The stresses in the range $0 \leq \theta \leq 45°$ are within 4 per cent of the curves. The $\sigma_x(\theta)$ result is poorest, reflecting the spread in the $u_x(\theta)$ data.

The last problem solved was the round tension bar with a circumferential edge crack. The geometry and mesh were the same as the previous problem when the cross-section of Fig. 4 is considered as that obtained by an axial cut. The outside diameter/inside diameter ratio (D/d) was thus equal to two. Paris and Sih[7] present stress intensity factors for this problem in the form

$$K_I = \sigma_{\text{net}} (\pi D)^{1/2} F(d/D) \qquad (16)$$

They provide a rough estimate of $F(1/2)$ as being equal to 0·227. Buechner in the discussion section of[7] reports a rigorous analysis of the problem involving singular integral equations resulting in $F(1/2) = 0·240$ which is expected to be no more than 1 per cent in error.

The near tip, $r/a = 1/72$, axial displacement values of the current solution predicted F to be 0·250 on an average with a spread between 0·245 and 0·259. The radial displacements yielded an average F equal to 0·238 with a range of 0·221–0·252. Thus, the total displacement average predicts $F = 0·244$ which is 2 per cent from Buechner's estimate. The stress solution was similar in accuracy to that of the plane strain rectangle.

DISCUSSION

The finite element procedure of dividing cracked configuration into triangular shaped singularity elements around the crack tip with adjoining trapezoidal shaped isoparametric elements was shown to yield very accurate stress intensity factors. It is noteworthy that good accuracy results from relatively coarse meshes. The edge cracked rectangle was considered as having 248° of freedom. The near tip displacements predicted K_I 5 per cent from Bowie[8]. The 548° of freedom formulation described above was 4 per cent from Bowie. Both these meshes are coarse compared to the 2000° of freedom problems solved by Chan *et al.*[3]. Computer core size is the essential consideration in determining the mesh refinement. The largest problem in this work required 3 C.P.U. min. on the IBM 360/67 at Brown University while the smallest required 1·5 minutes.

The numerical procedure outlined in this paper is similar in philosophy to the work of Wilson[9] who considered the circumferentially cracked round bar under torsion. The continuum was divided into triangular, linear displacement, ring elements joined to one "singularity" ring element of circular cross-section centered about the crack tip. The leading term of the Williams' eigenfunction expansion about a mode III crack tip [2] was embedded in the singularity element stiffness by an exact integration of the leading singular field over the circular cross-section of the element. The stress intensity factor K_{III} and the displacements of the nodes of the triangles not lying on the tip element boundary were the undetermined parameters of the problem. Using 1500° of freedom K_{III} values were predicted within 5 per cent of estimates gained from interpolating between exact shallow and deep crack solutions. The size of the near tip element was about the same as that used in the present work, $a/80$.

While Wilson evaluated the stiffness of the tip element by an exact integration of the expected singular field, an incompatibility results on the boundary of the circular element by joining straight-edged triangular elements there. The displacement function (4) used in the present work only approximates the smooth angular distribution of (1), but there are no incompatibilities introduced along inter-element boundaries in the process.

The fact that both approaches use only the leading term in the Williams expansion makes the choice of the tip element size very important. In both approaches it is most important to not have the tip element too large since the inverse square root stress variation is the only possibility allowed. Wilson[10] has used some collocation results to argue that the leading term is a good approximation to the actual stress state within a radius of about $a/80$. If the tip element is too small the constant stress elements used by Wilson would not be capable of adequately describing the large gradients present. The trapezoidal isoparametric elements used in the present work, on the other hand, have a large stress variation admissible within the element as well as a constant stress state. This results from the fact that one of the element's natural coordinates is linear in $\tan \theta$ (Fig. 5) which has a $1/r$ derivative with respect to x and y. The use of this element in conjunction with the singularity element is perhaps the reason why the present work obtained results as accurate as Wilson's with a mesh involving a small fraction of the number of degrees of freedom.

Acknowledgements — The helpful discussions of the material in this report with Prof. J. R. Rice are appreciated as is the help rendered by Prof. P. V. Marcal in the use of his finite element program. The computing was sponsored by the National Aeronautics and Space Administration under Grant NGL-40-002-080. The author undertook this project while a NASA Trainee at Brown University and completed it while employed at the U.S. Army Materials and Mechanics Research Center, Watertown, Mass.

REFERENCES

[1] G. R. Irwin, Analysis of stresses and strains near the end of a crack traversing a plate. *J. appl. Mech.* **24**, 361–364 (1957).

[2] M. L. Williams, On the stress distribution at the base of a stationary crack. *J. appl. Mech.* **24**, 109 (1957).
[3] S. K. Chan, I. S. Tuba and W. K. Wilson, On the finite element method in linear fracture mechanics. *Engng Fracture Mech.* **2**, 1, 1–17 (1970).
[4] O. C. Zienkiewicz and Y. K. Cheung, *The Finite Element Method in Structural and Continuum Mechanics*, p. 67, McGraw-Hill, London (1967).
[5] N. Levy, P. V. Marcal, W. J. Ostergren and J. R. Rice, Small scale yielding near a crack in plane strain: a finite element analysis. To appear in *Int. 1 J. Frac. Mech.* (1970–71).
[6] H. D. Hibbitt and P. V. Marcal, Hybrid finite element analysis with particular reference to axisymmetric structures. AIAA paper No. 70–137, presented at the AIAA 8th Aerospace Sciences meeting.
[7] P. C. Paris and G. C. Sih, Stress analysis of cracks. Fracture Toughness Testing and Its Applications, *ASTM Spec. Tech. Publ.* No. 381, 43–51 (1965).
[8] O. L. Bowie, Rectangular tensile sheet with symmetric edge cracks. *J. appl. Mech.* **31**, 208–212 (1964).
[9] W. K. Wilson, On combined mode fracture mechanics. Ph.D. Dissertation, University of Pittsburgh (1969).
[10] W. K. Wilson, in *Plane Strain Crack Toughness Testing of High Strength Metallic Materials* (Edited by W. F. Brown, Jr., and J. E. Srawley) *A.S.T.M. Spec. Tech. Publ.* No. 410, p. 75 (1967).

(*Received* 12 *October* 1970)

3-D ELASTIC SINGULARITY ELEMENT FOR EVALUATION OF K ALONG AN ARBITRARY CRACK FRONT

D. M. Tracey
Engineering Mechanics Development Department
Combustion Engineering, Incorporated
Windsor, Connecticut 06095, USA
tel: 203/688-1911

A three-dimensional finite element is introduced which embeds the inverse square root singularity along an arbitrarily curved crack front. The displacement interpolation function was designed to allow the type of singular deformation appropriate for a combined Mode I-Mode II crack front state. Test results demonstrate the element capable of rendering very accurate estimates of the stress intensity factor(s) along the front.

The modelling begins with a rectilinear approximation to the crack front. Each segment of the front then serves as the common edge of a group of wedge-shaped elements which encircle it. These are the singularity selments. They have six nodes, one at each vertex, with the triangular faces having a node at the ends of the front segment. Figure 1 is a schematic of the model. The front segment is taken as the z-axis of the local (x, y, z) and (r, θ, z) coordinate systems of the elements. The local strains e_{xx}, e_{xy}, and e_{yy} vary as $\sqrt{r}$ throughout the element and the strains e_{zz}, e_{zx}, and e_{zy} are non-singular.

The element interpolation function is written in terms of its natural coordinates (ζ, η, γ). The element edge along the crack front is taken as the face $\zeta=0$, the opposite quadrilateral face is $\zeta=1$, the focused quads determine $\eta=0,1$, and the base triangles define $\gamma=0,1$. The coordinate ζ is a linear function of r for this arbitrary wedge element and thus to embed the singular Mode I-Mode II deformation form the normal components of displacement (u_x, u_y) were chosen as functions of $\sqrt{\zeta}$. Using superscript notation abc to denote the (ζ, η, γ) position (a, b, c) of a node and a subscript n to denote the displacement component normal to the local z-axis, the interpolation function is

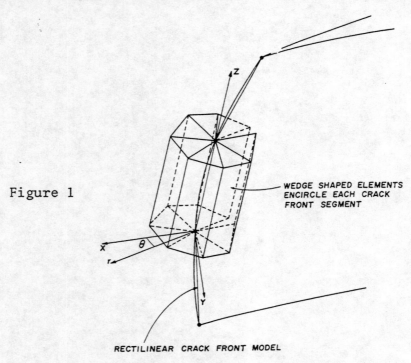

Figure 1

WEDGE SHAPED ELEMENTS ENCIRCLE EACH CRACK FRONT SEGMENT

RECTILINEAR CRACK FRONT MODEL

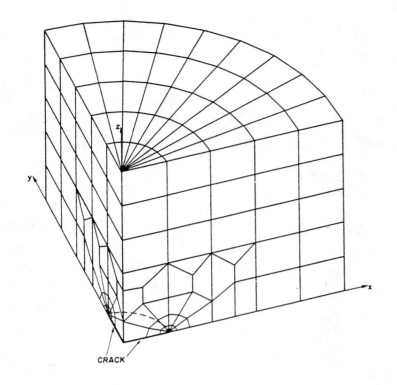

Figure 2

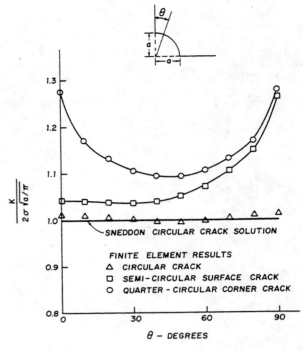

Figure 3

$$u_n = (1-\gamma) [u_n^{000} (1-\sqrt{\zeta}) + u_n^{100} (1-\eta)\sqrt{\zeta} + u_n^{110}\eta\sqrt{\zeta}]$$

$$+ \gamma [u_n^{001} (1-\sqrt{\zeta}) + u_n^{101} (1-\eta)\sqrt{\zeta} + u_n^{111}\eta\sqrt{\zeta}]$$

$$u_z = (1-\gamma) [u_z^{000} (1-\zeta) + u_z^{100} (1-\eta)\zeta + u_z^{110}\eta\zeta]$$

$$+ \gamma [u_z^{001} (1-\zeta) + u_z^{101} (1-\eta)\zeta + u_z^{111}\eta\zeta]$$

There is complete interelement compatibility when these elements encircle each crack front segment and eight-node isoparametric elements continue the mesh from $\zeta=1$. This element is the 3-D analogue of the 2-D element suggested by the author [1].

Three test problems were the isolated circular crack, the semi-circular edge crack, and the quarter-circular corner crack in an elastic material under remote tension. Figure 2 shows the periphery of the mesh used for these problems. The complete mesh was constructed from 10 planes of nodes oriented at 9° intervals about the global Z direction. Each plane had 58 independent nodes and 7 nodes along the Z-axis are common to all planes. There were then a total of 587 nodes and 486 elements. The radial and axial extent of the mesh was five times the crack radius a. Uniform tension $\sigma_{zz}=\sigma$ was applied on the face Z=5a. The singularity elements extended a distance a/20 from the front. The planes of symmetry for the circular crack case were Z=0, X=0, and Y=0; for the semi-circular crack they were Z=0, X=0, and only Z=0 was a symmetry plane for the corner crack case.

The K distribution was determined from the Z-displacement of the singularity element nodes on the crack face using the plane strain asymptotic relation $K = E\sqrt{(2\pi/r)}u_z/3.64$ for Poisson ratio equal to 0.3.

Figure 3 is a plot of K, normalized by the Sneddon [2] circular crack result $2\sigma\sqrt{(a/\pi)}$, versus position along the front for the three cases. The finite element circular crack result is within

1% of Sneddon. The surface crack result has the normalized K at the point of deepest penetration equal to 1.04 and it increases to 1.26 at the surface. This agrees very well with the numerical solution of Smith et al.[3]. The corner crack has a value of 1.10 at the middle of the front and this increases to 1.28 at the surfaces.

REFERENCES

[1] D. M. Tracey, *Engineering Fracture Mechanics* 3 (1971) 255-265.

[2] I. N. Sneddon, *Proceedings of the Royal Society*, London, Series A, 187 (1946) 229-260.

[3] F. W. Smith, A. F. Emery, and A. S. Kobayashi, *Journal of Applied Mechanics* 34 (1967) 953-959.

27 March 1973

THE COMPUTATION OF STRESS INTENSITY FACTORS BY A SPECIAL FINITE ELEMENT TECHNIQUE

P. F. WALSH[†]

Division of Building Research, CSIRO, Melbourne, Australia

Abstract—A special finite element method, for the computation of stress intensity factors, is presented in this paper. The special finite element consists of two regions. The stress and displacement distribution in the inner region is defined in terms of the singular stress field associated with the notch tip. The outer region of the special element contains conventional finite elements that are constrained to satisfy certain equilibrium and compatibility conditions on the interface between the two regions. The method is quite efficient and should allow the solution of problems outside the scope of present techniques. The validity of the procedure is confirmed by comparison with published solutions for some simple plane stress situations.

NOTATION

All matrices and column matrices are represented by a bold character. The transpose of a matrix $\mathbf{A}$ is denoted by $\mathbf{A}^T$. As a supplement to the symbols defined in the text the following notation has been adopted.

- **A** transformation matrix
- **B** transformation matrix from generalized to nodal displacements
- **P** force column matrix (various subscripts as discussed in text)
- **U** displacement column matrix (various subscripts as discussed in text)
- U_x x-displacement
- U_y y-displacement
- r distance from crack tip
- θ angle from x axis

1. INTRODUCTION

FRACTURE mechanics is concerned with the phenomenon of structural failure by catastrophic crack propagation at average stresses well below the yield strength. It has been shown by Leicester [5], that this problem arises not only in the sophisticated alloys used in aerospace structures, but also in simple notched timber beams.

One approach to the prediction, and hence prevention of such failures, is based on stress intensity factors which define the magnitude of the singularities in the stress field which occur in a linear elastic analysis of a structural component with an infinitely sharp notch. The currently available procedures, such as the collocation method, for the computation of stress intensity factors are restricted to problems involving uniform thickness and elastic properties. The finite element method of analysis has none of these restrictions, and moreover it is equally applicable to various notch angles. On the other hand, an infinitely sharp notch cannot be represented by a finite element mesh. This condition can be approached by using extremely small elements in the vicinity of the notch root but only at the expense of computational efficiency.

[†] Research Scientist.

In this paper, a method will be presented which has all the advantages of the finite element procedure in representing the structure. In the immediate vicinity of the notch, a special element will be developed that incorporates the theoretically exact stress patterns around the top of the notch. This region is then surrounded by a transition region to a conventional finite element mesh. The validity of the approach is confirmed by comparison with published solutions for some simple plane stress situations.

2. THEORY

The proposed special element consists of two regions as shown in Fig. 1. The stress distribution in the inner region can be defined by the stress intensity factors and their associated singular stress fields. The outer region consists of a conventional finite element mesh that is constrained to satisfy certain compatibility and equilibrium conditions on the interface between the two regions. The entire special element forms part of a larger finite element mesh that is analysed in the conventional manner.

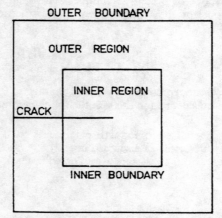

FIG. 1. General arrangement of special element.

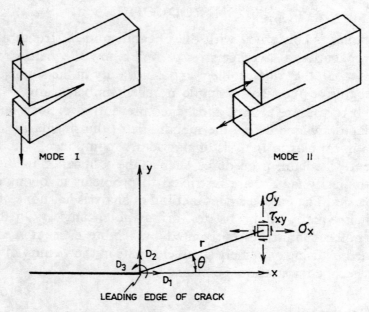

FIG. 2. Notation for plane strain singularities.

The expressions for the stiffness matrix of the hybrid element will be derived with particular reference to a sharp crack in an isotropic material which is subjected to plane strain conditions. In the immediate vicinity of the crack tip, the stresses and strains can then be defined by two stress intensity factors K_I and K_{II}, and their corresponding singular stress fields. With the notation shown in Fig. 2, these stress and displacement fields are given by Paris and Sih [6], as,

Mode 1

$$\sigma_x = \frac{K_I}{(2\pi r)^{\frac{1}{2}}} \cos\frac{\theta}{2}\left(1 - \sin\frac{\theta}{2}\sin\frac{3\theta}{2}\right)$$

$$\sigma_y = \frac{K_I}{(2\pi r)^{\frac{1}{2}}} \cos\frac{\theta}{2}\left(1 + \sin\frac{\theta}{2}\sin\frac{3\theta}{2}\right)$$

$$\tau_{xy} = \frac{K_I}{(2\pi r)^{\frac{1}{2}}} \sin\frac{\theta}{2}\cos\frac{\theta}{2}\cos\frac{3\theta}{2}$$

$$\sigma_z = \nu(\sigma_x + \sigma_y) \qquad (1)$$

$$U_x = \frac{K_I}{G}\left(\frac{r}{2\pi}\right)^{\frac{1}{2}} \cos\frac{\theta}{2}\left(1 - 2\nu + \sin^2\frac{\theta}{2}\right)$$

$$U_y = \frac{K}{G}\left(\frac{r}{2\pi}\right)^{\frac{1}{2}} \sin\frac{\theta}{2}\left(2 - 2\nu - \cos^2\frac{\theta}{2}\right).$$

Mode 2

$$\sigma_x = -\frac{K_{II}}{(2\pi r)^{\frac{1}{2}}} \sin\frac{\theta}{2}\left(2 + \cos\frac{\theta}{2}\cos\frac{3\theta}{2}\right)$$

$$\sigma_y = \frac{K_{II}}{(2\pi r)^{\frac{1}{2}}} \sin\frac{\theta}{2}\cos\frac{\theta}{2}\cos\frac{3\theta}{2}$$

$$\tau_{xy} = \frac{K_{II}}{(2\pi r)^{\frac{1}{2}}} \cos\frac{\theta}{2}\left(1 - \sin\frac{\theta}{2}\sin\frac{3\theta}{2}\right)$$

$$\sigma_z = \nu(\sigma_x + \sigma_y) \qquad (2)$$

$$U_x = \frac{K_{II}}{2G}\left(\frac{r}{2\pi}\right)^{\frac{1}{2}} \sin\frac{\theta}{2}\left(2 - 2\nu + \cos^2\frac{\theta}{2}\right)$$

$$U_y = \frac{K_{II}}{2G}\left(\frac{r}{2\pi}\right)^{\frac{1}{2}} \cos\frac{\theta}{2}\left(-1 + 2\nu + \sin^2\frac{\theta}{2}\right).$$

Although equations (1) and (2) are given for plane strain, the corresponding equations for plane stress can be obtained by suitably modifying the elastic constants.

Now the displacement of any point in the inner region can be expressed as a function of its position, the two stress intensity factors, K_I and K_{II} and the three components of rigid body displacement, D_1, D_2 and D_3 (see Fig. 2)

$$\begin{bmatrix} U_x \\ U_y \end{bmatrix} = \begin{bmatrix} U_{xI} & U_{xII} & 1 & 0 & -y \\ U_{yI} & U_{yII} & 0 & 1 & x \end{bmatrix} \begin{bmatrix} K_I \\ K_{II} \\ D_1 \\ D_2 \\ D_3 \end{bmatrix} \quad (3)$$

or in matrix form

$$\mathbf{U}_{xy} = \mathbf{M}\mathbf{U}_s.$$

In equation (3) U_{xI} and U_{yI} are the displacements of the point for a unit value of K_I as given by equation (1). The column matrix $\mathbf{U}_s$ contains the generalized "displacement" quantities $K_I, K_{II}, D_1, D_2, D_3$. It is convenient to introduce a column matrix, $\mathbf{P}'_s$, of generalized forces that correspond to the generalized displacements $\mathbf{U}_s$. A form of stiffness matrix relating these two quantities can be derived in the following manner. Consider the stresses produced by a unit value of the jth component of the $\mathbf{U}_s$ matrix and evaluate the work done as these stresses move through the displacement pattern produced by a unit value of the ith component of $\mathbf{U}_s$. By adopting this result as the (i,j) element of a matrix $\mathbf{S}_s$, then $\mathbf{P}'_s$ is defined by,

$$\mathbf{P}'_s = \mathbf{S}_s \mathbf{U}_s. \quad (4)$$

Since several of the components of $\mathbf{U}_s$ are simply rigid body displacements, many of the terms in $\mathbf{S}_s$ will be simply zero. The evaluation of the non-zero terms in $\mathbf{S}_s$ can be carried out by explicit or numerical integration along the boundary of the inner region.

For the outer region, the individual element stiffness matrices of the finite elements may be summed to give an equation which relates nodal forces P_m and nodal displacements U_m at discrete points on the inner and outer boundaries of the outer region, i.e.

$$\mathbf{S}_m \mathbf{U}_m = \mathbf{P}_m. \quad (5)$$

If the finite element mesh includes nodes that are not on either boundaries then the displacements of such nodes can be eliminated by partial Gauss–Jordan elimination to give equation (5).

In order to differentiate between the displacements and forces on the inner and outer boundaries, the suffices i and o are introduced, i.e.

$$\mathbf{U}_m = \begin{bmatrix} \mathbf{U}_o \\ \mathbf{U}_i \end{bmatrix}$$

and

$$\mathbf{P}_m = \begin{bmatrix} \mathbf{P}_o \\ \mathbf{P}_i \end{bmatrix}.$$

For the displacements of the finite element mesh to be compatible, at least at the nodes, with the displacements of the inner region the nodal displacements $\mathbf{U}_i$ must be restricted to a system that can be defined in terms of $\mathbf{U}_s$. Compatibility on the inner boundary, between nodes, is only approximately satisfied as the variation of the displacement field on the edge of the inner region would be different from that along the edges of the outer, finite element, region. This approximation can be refined by increasing the number of nodes. Now if the coordinates of the nodes on the inner boundary are successively substituted into equation (3) then the following matrix equation can be obtained.

$$\mathbf{U}_i = \mathbf{B}\mathbf{U}_s. \tag{6}$$

The nodal forces $\mathbf{P}_i$ may also be related to a system of generalized forces $\mathbf{P}_s''$ due to deformations within the finite element mesh by the contragredient transformation

$$\mathbf{P}_s'' = \mathbf{B}^T \mathbf{P}_i. \tag{7}$$

Equation (5) may now be modified to,

$$\begin{bmatrix} \mathbf{P}_o \\ \mathbf{P}_s'' \end{bmatrix} = \mathbf{A}\mathbf{S}_m \mathbf{A}^T \begin{bmatrix} \mathbf{U}_o \\ \mathbf{U}_s \end{bmatrix} \tag{8}$$

where

$$\mathbf{A} = \begin{bmatrix} \mathbf{I} & \mathbf{0} \\ \mathbf{0} & \mathbf{B}^T \end{bmatrix}. \tag{9}$$

Equilibrium then requires that the sum of the generalized forces due to deformation of the inner region, $\mathbf{P}_s'$, and due to deformation of the outer region $\mathbf{P}_s''$ is equal to the applied load $\mathbf{P}_s$ (normally zero).

Thus,

$$\begin{bmatrix} \mathbf{P}_o \\ \mathbf{P}_s \end{bmatrix} = \mathbf{A}\mathbf{S}_m A^T \begin{bmatrix} \mathbf{U}_o \\ \mathbf{U}_s \end{bmatrix} + \begin{bmatrix} \mathbf{0} & \mathbf{0} \\ \mathbf{0} & \mathbf{S}_s \end{bmatrix} \begin{bmatrix} \mathbf{U}_o \\ \mathbf{U}_s \end{bmatrix}$$

or,

$$\begin{bmatrix} \mathbf{P}_o \\ \mathbf{P}_s \end{bmatrix} = \mathbf{S} \begin{bmatrix} \mathbf{U}_o \\ \mathbf{U}_s \end{bmatrix}. \tag{10}$$

Where $\mathbf{S}$ is the stiffness matrix of the special element.

In the solution process the special element forms just one of the elements in a finite element mesh. The degrees of freedom associated with the components of $\mathbf{U}_s$ can be conveniently treated as displacements of imaginary nodes within the hybrid element. The solution of the entire problem then results in "displacement" of all the nodes and thus includes the stress intensity factors.

3. VERIFICATION

The accuracy of the proposed method for computing stress intensity factors depends primarily on the number of nodes on the inner and outer boundaries. To a lesser extent, the refinement of the finite element mesh that surrounds the hybrid element also affects the accuracy of the result. By increasing the number of nodes in the special element and in the surrounding mesh any desired accuracy should be possible.

So that a comparison may be carried out between the results of this method and the results reported in the literature, three simple structural configurations were investigated:
    (i) a double edge-notched plate in tension;
    (ii) a single edge-notched plate in tension;
    (iii) a single edge-notched plate in bending.

The material was taken to be isotropic with $\mu = 0.33$, and plain stress conditions were adopted. The three cases are shown in Fig. 3. Full advantage was taken of symmetry and as a result only one quarter of the double edge-notched plate and one half of the single edge-notched plates were considered in the analysis. Several values of the ratio of notch depth to plate width were considered.

The form of the special element adopted was rather crude. Only five nodes on the inner and on the outer boundaries were employed. This arrangement is shown in Fig. 4. Due to the symmetry, only two terms in the U_s matrix of generalized displacements were needed. These were the rigid body displacement in the x direction and K_I the stress intensity factor for mode 1. The size of the total stiffness matrix for the special element was thus only 12 by 12. The special element was contained in the relatively coarse mesh shown in Fig. 5. This mesh consisted of standard finite elements with linear displacement fields along the element edges. The results for the computer analysis of the three cases for various values of the ratio of notch depth to plate width are presented in Figs. 6–8. In these figures the computed results are compared with the results reported in the literature by various

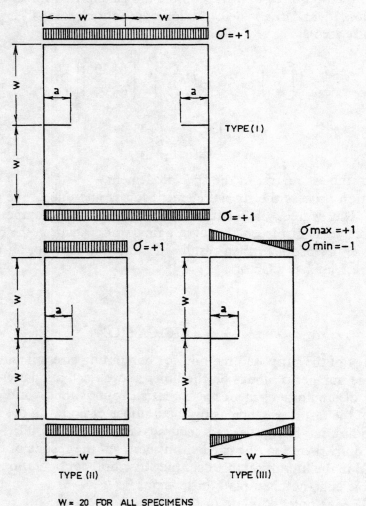

FIG. 3. Specimen configurations used for comparison between proposed method and published results.

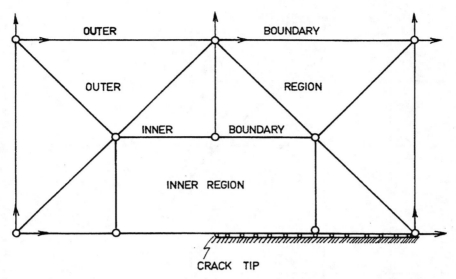

FIG. 4. Special element used for K_I computation symmetrical specimens.

authors (Beukner [1], Bowie [2], Gross and Srawley [3] and Gross et al. [4]). Despite the crude form of the special element the results are as accurate as is normally required. It might be noted that the computer execution time for each result was only 12 sec on a CDC 3600 computer.

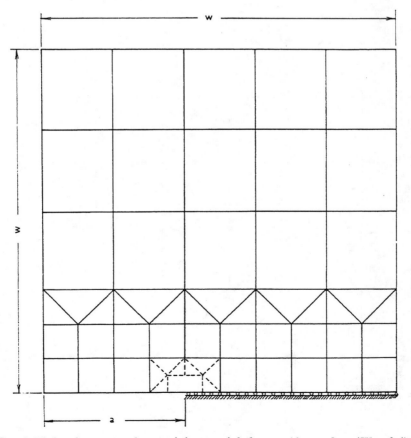

FIG. 5. Finite element mesh containing special element (drawn for $a/W = 0.4$).

357

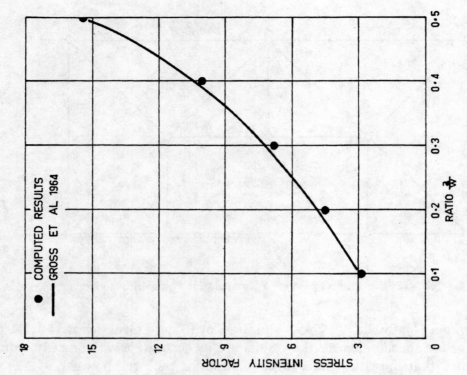

FIG. 7. Stress intensity factor vs. ratio of crack to specimen width for single edge notched specimen in tension.

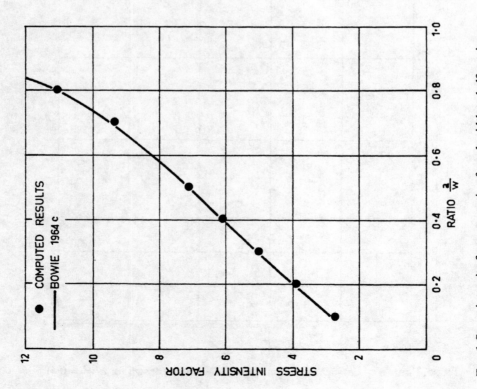

FIG. 6. Stress intensity factor vs. ratio of crack width to half specimen width, for a double edge notched specimen in tension.

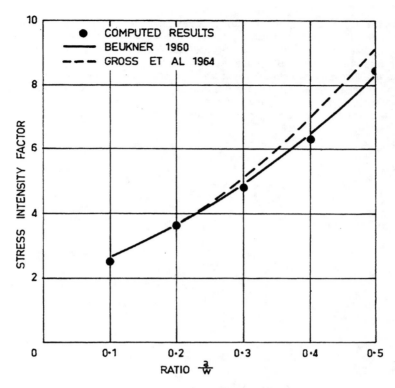

Fig. 8. Stress intensity factor vs. ratio of crack to specimen width for single edge notched specimen in bending.

4. CONCLUSION

The special element method presented in this paper is a practical and efficient procedure for determining stress intensity factors for a wide range of structural problems. Not only is the method computationally efficient but also it allows the solution of problems outside the scope of present techniques. Any form of notch singularity or notch angle can be considered, provided stress and displacements along the boundary of the notch can be defined numerically. Variations in thickness, point loads and many such complications may be included in this method without difficulty.

REFERENCES

[1] H. F. BEUKNER, Some Stress Singularities and their Computation by Means of Integral Equations, in *Boundary Value Problems in Differential Equations*, edited by R. E. LENGER. University of Wisconsin Press (1960).
[2] O. L. BOWIE, Rectangular tensile sheet with symmetric edge cracks. *J. appl. Mech.* (1964).
[3] B. GROSS and J. E. SRAWLEY, Stress-Intensity Factors for Single-Edge-Notch Specimens in Bending or Combined Bending and Tension by Boundary Collocation of a Stress Function, NASA TN D2603 (1964).
[4] B. GROSS, J. E. SRAWLEY and W. F. BROWN, Stress Intensity Factors for a Single-Edge-Notch Specimen by Boundary Collocation of a Stress Function, NASA TN D-2395 (1964).
[5] R. H. LEICESTER, The Size Effect of Notches, *Proceedings of the Second Australasian Conference on the Mechanics of Structures and Materials*, Adelaide, S.A. (1969).
[6] P. C. PARIS and G. C. M. SIH, Stress Analysis with Cracks, ASTM Special Technical Publication No. 381 (1964).

Абстракт—В работе даётся специальный метод конечного элемента, для расчета факторов интенсивности напряжений. Специаьный конеченый элемент состоит из двух областей. Определяются распределения напряжений и деформаций во внутренной области, в виде поля сингулярных напряений, связанного с вершиной надрезки. Внешняя область специального элемента залючает обыкновенные конечные элементы, которые приспособлены для удовлетворения некоторым условиям ровновесия и совместимости, на границе между двумя обастями. Этот метод вполне эффективный, благодаря чему можно получить решение задачи, вне рамок применяемых в настояще время способов. Важность процесса подтверждена путёи сравнения с опубликованными решениями для некоторых простых случаев плоских нарряжений.

ON THE USE OF ISOPARAMETRIC FINITE ELEMENTS IN LINEAR FRACTURE MECHANICS*

ROSHDY S. BARSOUM

Engineering Science Department, Power Systems Group, Combustion Engineering Inc., Windsor, Connecticut, U.S.A.

SUMMARY

Quadratic isoparametric elements which embody the inverse square root singularity are used in the calculation of stress intensity factors of elastic fracture mechanics. Examples of the plane eight noded isoparametric element show that it has the same singularity as other special crack tip elements, and still includes the constant strain and rigid body motion modes. Application to three-dimensional analysis is also explored. Stress intensity factors are calculated for mechanical and thermal loads for a number of plane strain and three-dimensional problems.

INTRODUCTION

The use of the finite element method in fracture mechanics has been quite extensive both in the elastic and elastic-plastic range.[1-7] A number of special crack tip finite elements have been developed,[8-10,20] where the displacement method has been used. Also the hybrid method has been used in developing special crack tip elements.[11] These special crack tip elements contain a singularity of the strain field at the crack tip, equal to the theoretical singularity.[17] One disadvantage of these special crack tip elements[9,10] is that they lack the constant strain and the rigid body motion modes. Therefore, they do not pass the patch test[16] and the necessary requirements for convergence[15] are not present.

From a practical point of view, the above considerations create a problem when these elements are used in thermal stress analyses. Completely erroneous results are obtained in calculating stress intensity factors for a thermal gradient, unless a special treatment is done in order to eliminate the effect of the constant thermal field across the singularity elements. One of the treatments devised was to perform two separate analyses, one model without a crack and with non-singular elements. This model is used to calculate the reactions along the crack due to the thermal gradient. These reactions are then applied in the opposite direction to the cracked model where the elements containing the singularity are used. The stress intensity factor is calculated from the final analysis, and the stresses from the superposition of the two solutions. It is clear that such an analysis is inconsistent, however, it gives reasonable results. The other treatment was to choose the thermal distribution such that the tip of the crack is at a zero average temperature. This is done by subtracting the average temperature at the tip of the crack from the temperatures elsewhere. Thus, there will be no thermal expansion across the singularity elements. This treatment, however, is not appropriate in the case of three-dimensional analysis where there is no average temperature that could be imposed on all the elements along the crack. Earlier thermal

---

* After submitting this paper, the author has learned of an independent development reporting a similar approach for 2-D problems, see R. D. Henshell, 'Crack tip finite elements are unnecessary', *Int. J. num. Meth. Engng*, **9**, 495–507 (1975).

Received 7 August 1974
Revised 8 October 1974

shock analyses have been performed using non-singular elements,[12] and therefore the above problems were avoided.

The idea proposed here is to use the 8-noded isoparametric element for plane strain, plane stress and axisymmetric analyses and the 20-noded isoparametric for three-dimensional crack tip analyses. The singularity in these elements is achieved by placing the mid-side node near the crack tip at the quarter point.[13,14]

It is well known that such elements in their non-singular formulation satisfy the essential convergence criteria,[15] namely, inter-element compatibility, constant strain modes, continuity of displacements, and rigid body motion modes. They also pass the patch test.[16] In this paper these various considerations are discussed for the singular case. The ease of using these singularity elements; since they exist in almost all general purpose programs, and their convergence characteristics, makes their application in linear fracture mechanics very tractable. The accuracy of the results in a reasonable mesh of a practical problem is very high as demonstrated by the example problems here.

SINGULAR QUADRATIC ISOPARAMETRIC ELEMENTS

The formulation of the isoparametric element stiffness is well documented.[15] Following the notation of Reference 15, the geometry of an 8-noded plane isoparametric element is mapped into the normalized square space $(\xi, \eta), (-1 \geqslant \xi \geqslant 1, -1 \geqslant \eta \geqslant 1)$ through the following transformations,

$$\left.\begin{aligned} x &= \sum_{i=1}^{8} N_i(\xi, \eta) x_i \\ y &= \sum_{i=1}^{8} N_i(\xi, \eta) y_i \end{aligned}\right\} \quad (1)$$

$$N_i = [(1+\xi\xi_i)(1+\eta\eta_i) - (1-\xi^2)(1+\eta\eta_i) - (1-\eta^2)(1+\xi\xi_i)]\xi_i^2 \eta_i^2/4 + (1-\xi^2)(1+\eta\eta_i)(1-\xi_i^2)\eta_i^2/2$$
$$+ (1-\eta^2)(1+\xi\xi_i)(1-\eta_i^2)\xi_i^2/2 \quad (1a)$$

where N_i are the shape functions corresponding to the node i, whose co-ordinates are (x_i, y_i) in the x–y system and (ξ_i, η_i) in the transformed ξ–η system. $(\xi_i, \eta_i = \pm 1)$ for corner points and zero for mid-side nodes. The displacements are interpolated by

$$\left.\begin{aligned} u &= \sum_{i=1}^{8} N_i(\xi, \eta) u_i \\ v &= \sum_{i=1}^{8} N_i(\xi, \eta) v_i \end{aligned}\right\} \quad (2)$$

The stiffness matrix is found as follows:

$$\{\varepsilon\} = [B] \begin{Bmatrix} u_i \\ v_i \end{Bmatrix} \quad (3)$$

$$[B] = \begin{bmatrix} \dfrac{\partial N_i}{\partial x} & 0 \\ 0 & \dfrac{\partial N_i}{\partial y} \\ \dfrac{\partial N_i}{\partial y} & \dfrac{\partial N_i}{\partial x} \end{bmatrix} \quad (4)$$

where

$$\begin{Bmatrix} \dfrac{\partial N_i}{\partial x} \\ \dfrac{\partial N_i}{\partial y} \end{Bmatrix} = [\mathbf{J}]^{-1} \begin{Bmatrix} \dfrac{\partial N_i}{\partial \xi} \\ \dfrac{\partial N_i}{\partial \eta} \end{Bmatrix} \quad (5)$$

where [**J**] is the Jacobian matrix and is given by

$$[\mathbf{J}] = \begin{bmatrix} \dfrac{\partial x}{\partial \xi} & \dfrac{\partial y}{\partial \xi} \\ \dfrac{\partial x}{\partial \eta} & \dfrac{\partial y}{\partial \eta} \end{bmatrix} = \begin{bmatrix} \cdots \dfrac{\partial N_i}{\partial \xi} \cdots \\ \cdots \dfrac{\partial N_i}{\partial \eta} \cdots \end{bmatrix} \begin{bmatrix} \vdots & \vdots \\ x_i & y_i \\ \vdots & \vdots \end{bmatrix} \quad (6)$$

The stress is given by

$$\{\sigma\} = [\mathbf{D}]\{\varepsilon\} \quad (7)$$

where [**D**] is the stress-strain matrix. The element stiffness [**K**] is then,

$$[\mathbf{K}] = \int_{-1}^{1} \int_{-1}^{1} [\mathbf{B}]^T [\mathbf{D}][\mathbf{B}] \det |\mathbf{J}| \, d\xi \, d\eta \quad (8)$$

In order to obtain a singular element to be used at the crack tip, the stress in equation (7) and the strain in equation (3) must be singular. This singularity is achieved by placing the mid-side node at the quarter points[13] of the sides.

Investigation of type of singularity

The form of $N_i(\xi, \eta)$ in all isoparametric elements (the Serendipity family[15]) are polynomials and hence, $\partial N_i/\partial \xi$, $\partial N_i/\partial \eta$ are non-singular. On the other hand, the strain in equation (3) can be written in a form by combining equations (3), (4), and (5) as

$$\{\varepsilon\} = [\mathbf{J}]^{-1} [\mathbf{B}'(\xi, \eta)] \begin{Bmatrix} u_i \\ v_i \end{Bmatrix} \quad (3a)$$

Therefore, the singularity could be achieved by requiring that the Jacobian [**J**] be singular at the crack tip. Or in other words, the determinant of the Jacobian det|**J**| to vanish at the crack tip, where,

$$\det |\mathbf{J}| = \frac{\partial(x, y)}{\partial(\xi, \eta)} \quad (9)$$

Case 1. Eight-noded quadrilateral with mid-side nodes of two sides at the quarter points, Figure 1.
For simplicity, the strength of the singularity will be found along the line 1–2 ($\eta = -1$), Figure 1. The shape functions evaluated along the line 1–2 are:

$$\left. \begin{aligned} N_1 &= -\tfrac{1}{2}\xi(1-\xi) \\ N_2 &= \tfrac{1}{2}\xi(1+\xi) \\ N_5 &= (1-\xi^2) \end{aligned} \right\} \quad (10)$$

From equation (1)

$$x = -\tfrac{1}{2}\xi(1-\xi)x_1 + \tfrac{1}{2}\xi(1+\xi)x_2 + (1-\xi^2)x_5 \quad (11)$$

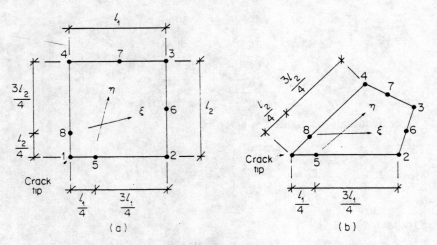

Figure 1. 2-D rectangular elements with mid-side nodes at the quarter points

Choosing $x_1 = 0$, $x_2 = L$, $x_5 = L/4$, then

$$x = \tfrac{1}{2}\xi(1+\xi)L + (1-\xi^2)\frac{L}{4} \tag{12}$$

therefore,

$$\xi = \left(-1 + 2\sqrt{\frac{x}{L}}\right) \tag{13}$$

In the Jacobian the term $\partial x/\partial \xi$ is given by

$$\frac{\partial x}{\partial \xi} = \frac{L}{2}(1+\xi) = \sqrt{\frac{x}{L}} \tag{14}$$

which makes the Jacobian singular at $(x = 0, \xi = -1)$. Considering only the displacements of points, 1, 2, and 5, the displacement u along the line 1–2 is given by

$$u = -\tfrac{1}{2}\xi(1-\xi)u_1 + \tfrac{1}{2}\xi(1+\xi)u_2 + (1-\xi^2)u_5 \tag{15}$$

And writing it in terms of x,

$$u = -\frac{1}{2}\left(-1+2\sqrt{\frac{x}{L}}\right)\left[2-2\sqrt{\frac{x}{L}}\right]u_1 + \frac{1}{2}\left(-1+2\sqrt{\frac{x}{L}}\right)\left[2\sqrt{\frac{x}{L}}\right]u_2 + \left(4\sqrt{\frac{x}{L}} - 4\frac{x}{L}\right)u_5 \tag{16}$$

The strain in the x-direction is then,

$$\varepsilon_x = \frac{\partial u}{\partial x} = \mathbf{J}^{-1}\frac{\partial u}{\partial \xi} = \frac{\partial \xi}{\partial x}\frac{\partial u}{\partial \xi} = -\frac{1}{2}\left[\frac{3}{\sqrt{(xL)}} - \frac{4}{L}\right]u_1 + \frac{1}{2}\left[-\frac{1}{\sqrt{(xL)}} + \frac{4}{L}\right]u_2 + \left[\frac{2}{\sqrt{(xL)}} - \frac{4}{L}\right]u_5 \tag{17}$$

The strain singularity along the line 1–2 is therefore, $1/\sqrt{r}$, which is the required singularity for elastic analysis.[17]

Case 2. Six-noded triangle with mid-side nodes at the quarter points Figure 2. This triangle is generated by collapsing the side 1–4 of the quadrilateral in Figure 1. In this case the singularity is investigated along the x-axis, $\eta = 0$.

$$x = -\tfrac{1}{4}(1+\xi)(1-\xi)l_1 - \tfrac{1}{4}(1+\xi)(1-\xi)l_1 + \tfrac{1}{2}(1-\xi^2)\tfrac{l_1}{4} + \tfrac{1}{2}(1+\xi)l_1 + \tfrac{1}{2}(1-\xi^2)\tfrac{l_1}{4} \quad (18)$$

$$x = (\xi^2 + 2\xi + 1)\tfrac{l_1}{4} \quad (19)$$

Therefore,

$$\xi = \left[-1 + 2\sqrt{\tfrac{x}{l_1}}\right] \quad (20)$$

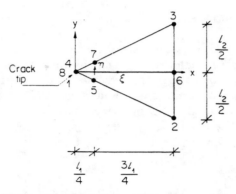

Figure 2. 2-D triangular element with mid-side nodes at the quarter points

which is the same as equation (13). And, therefore, in the Jacobian, the term $\partial x/\partial \xi$ is singular as before. The displacement and strain along the x-axis have terms similar to those in equations (16) and (17), respectively. The strain singularity along the x-axis is also $1/\sqrt{r}$.

Case 3. Three-dimensional twenty-noded cubic element with four mid-side nodes at the quarter points, Figure 3. The singularity in this case on the faces, 1234, 5678, is the same as for Case 1, which could be proved in a similar manner $1/\sqrt{r}$. It should be pointed out that the variation of the singularity along the ζ-direction is quadratic thus allowing a piecewise continuous change from one element to the next along the crack edge, thus allowing an accurate evaluation of the stresses from one element to the next that lies along the crack edge.

Case 4. Three-dimensional prism with four mid-side nodes at the quarter points (degenerate cube with one face collapsed) Figure 4. The singularity in this case on the faces 1234 and 5678 is also the same as for Case 2, $1/\sqrt{r}$. The variation in the ζ-direction is quadratic.

It was found that the elements described in Cases 2 and 4 are easier to use in generating a mesh and they give somewhat better results. This will be illustrated in the numerical results that follow.

Evaluation of the stiffness matrix

Although the strain (stress) in these elements is singular, their total strain energy should be finite, and hence their stiffness. This is only true, however, for the triangle and prism elements.

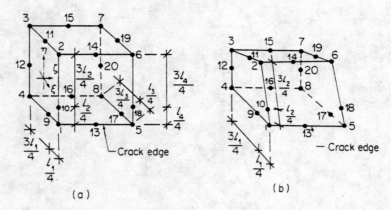

Figure 3. 3-D brick element with mid-side nodes at the quarter points

The stiffness of rectangular and brick elements was proved by Hibbitt[23] to possess a singularity at the crack tip. Using numerical integration eliminates this problem, but as will be shown later the numerical results of triangular and prism elements are much better than those of rectangular and brick elements. Gaussian quadrature was used in integrating the stiffness matrix. A 9-point Gaussian integration rule was used in 2-D problems, and 27 Gauss points in 3-D problems. The stresses were calculated at the integration points and their variation close to the crack tip was evaluated. This is also one of the advantages of using these elements.

Convergence of singular isoparametric elements

Although the isoparametric elements proposed here contain a singularity, they satisfy the necessary conditions for convergence.[15] Since the elements are isoparametric, i.e.

$$x^\alpha = N_i x_i^\alpha$$
$$u^\alpha = N_i' u_i^\alpha \qquad (21)$$

and

$$N_i = N_i' \qquad (22)$$

where $\alpha = 1, 2, 3$ corresponding to the co-ordinate system. From equation (22) it could be proved that they satisfy inter-element compatibility and continuity. Further, since $\sum_i N_i = 1$, then the elements satisfy the constant strain and rigid body motion conditions. Proofs of these theorems are given in Reference 15.

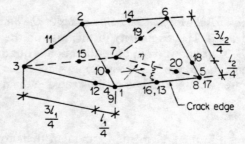

Figure 4. 3-D prism with mid-side nodes at the quarter points

In addition, a uniform temperature solution using these singularity elements, gives zero stress everywhere, and therefore the element satisfies the patch test.[16]

NUMERICAL RESULTS

In order to test the proposed element, the following tests were conducted. The MARC–CDC program[19] was used in performing these analyses.

Plane strain rectangular bar with symmetric edge cracks Figure 5

This problem was tested in Reference 9 for a mesh that contained 252 elements, 300 nodes, and 548 degrees-of-freedom. In this analysis 115 elements, 370 nodes, and 700 degrees-of-freedom are used. The first three rings contained 15 degrees elements, thereafter, the elements covered 30 degrees angle. A special tying of nodal displacements was used between the third and fourth ring to affect inter-element compatibility.

The stress intensity factor for this case is given by[17],

$$K_1 = \left[2G \left(\frac{2\pi}{r} \right)^{\frac{1}{2}} \bigg/ f_y(\theta) \right] u_y(r, \theta) \tag{23}$$

which is expressed for this type of problem in the form[18]

$$K_1 = \sigma_\infty (\pi a)^{\frac{1}{2}} \left(\frac{2b}{\pi a} \tan \frac{\pi a}{2b} \right) \cdot C \tag{24}$$

where C, the correction factor, used for comparison was $1 \cdot 02$.[18] The values of u_y for nodes closest to the crack tip in this analysis yielded C between $1 \cdot 03$ and $1 \cdot 04$ as compared with $1 \cdot 05$ and $1 \cdot 09$ given in Reference 9.

The error in the stress calculation, for the cases shown in Figures 6, 7, and 8 is within -9 per cent to $+4$ per cent as compared with -13 per cent to $+7$ per cent obtained in Reference 9.

Further tests of this problem were done in order to test the elements for use in three-dimensional analysis where fairly coarse meshes are required. Therefore, two plane-strain analyses were

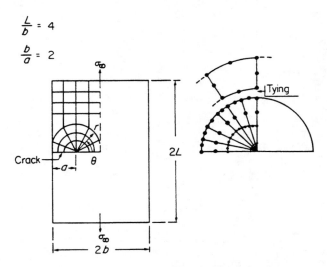

Figure 5. Plane strain rectangular bar with symmetric edge cracks

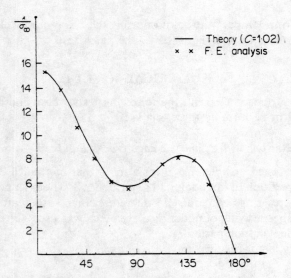

Figure 6. σ_x at the crack tip elements ($r/a = 0.0139$)

performed, Figures 9 and 10, with relatively small number of elements. The error in calculating K_I for the case in Figure 9 varied between $+2$ per cent and -23 per cent when calculated from the displacements at different points around the crack tip. The error in calculating K_I for the case in Figure 10 varied between $+10$ and -10 per cent for the same points around the crack tip. As discussed above, it is clear that the use of triangular shaped elements (Case 2) gives superior results. In the following examples the triangular elements and their counterpart, the prism, in three-dimensions will be used.

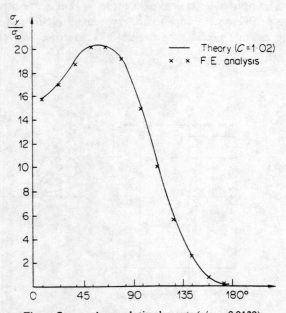

Figure 7. σ_y at the crack tip elements ($r/a = 0.0139$)

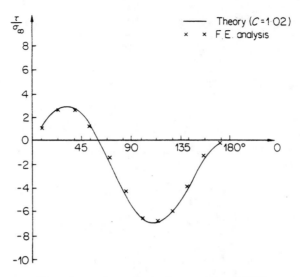

Figure 8. τ at the crack tip elements ($r/a = 0.0139$)

Thermal gradient across a rectangular bar with symmetric edge cracks in plane strain

The theoretical solution of this problem is found by calculating the thermal stresses in an uncracked continuum and then imposing the opposite stresses at the cracks on the cracked continuum. The solution of this problem is found by integrating the solution of the concentrated forces on the crack surface. Reference 18 gives the basic formulation for this approach.

The finite element solution of this problem, Figure 11, resulted in less than 10 per cent error in the calculation of the stress concentration factor.

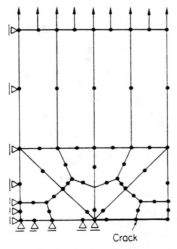

Figure 9. Coarse mesh—plane strain (rectangular elements)

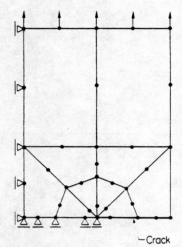

Figure 10. Coarse mesh—plane strain (triangular elements)

Three-dimensional solid analysis

Finally, in order to test the element in three-dimensional analysis, two analyses were studied using the prism crack element (Case 4). Figure 12 shows the model used in both analyses.

Plane-strain mechanical loads. As a test case the model in Figure 12 was used in evaluating the stress intensity factor of a non-symmetric crack a in a plane strain subject to a uniform tension at the ends. Theoretical value of the stress concentration factor K_I is found in Reference 18. The finite element model gave an error less than 5 per cent in calculating K_I, using the displacement closest to the crack. It should be noted that the stresses calculated at the integration points close to the crack edge ($r = 0.003a$), gave similar results.

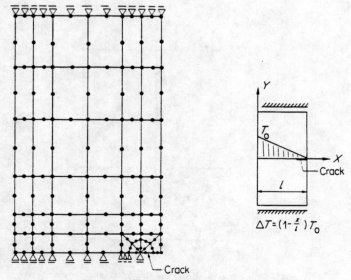

Figure 11. Thermal gradient—plane strain

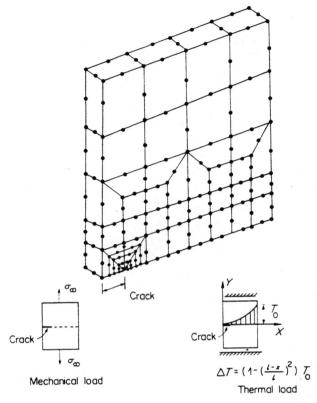

Figure 12. 3-D solid with crack, subject to mechanical and thermal loads

Plane-strain thermal loads. The thermal gradient in this problem was parabolic, the special crack tip elements had thus a gradient across them. The theoretical calculations were similar to those calculated in the second example. The finite element model gave, as in the previous case, less than 5 per cent error in calculating K_I.

Analysis of a compact tension specimen. The analysis of a compact specimen has become essential for the development of adequate strength theories. Since the specimen, Figure 13, is neither vanishingly thin (plane stress) nor the interior points are far from the outer surfaces (plane strain), therefore the analysis has to consider the three-dimensional character of the stress field. Due to the complexity of three-dimensional analysis, numerical methods have been used in such analyses.[21,22]

The finite element for this analysis is shown in Figure 13. Because of the double symmetry of the specimen and the loading only one quarter of the specimen was modelled. The model consisted of 150 elements and the nodes were located at the following distances from the crack front,

$r/a =$ 0·0125, 0·05, 0·075, 0·10, 0·15, 0·20, 0·25, 0·30, 0·35, 0·40, 0·55, 0·70, 0·85, 1·00

The crack front was divided into 12 equal spaces. A special tying was provided where the mesh changes from two elements to one element in order to ensure inter-element compatibility. The results were compared with those obtained by Tracey[22] using special singularity finite elements and 522 elements in the model. The stress intensity factor K was calculated from the displacement of the nodes closest to the crack front ($r/a = 0.0125$). At the centre, K was within 2 per cent of the

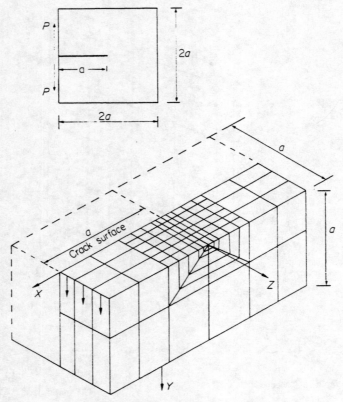

Figure 13. Compact tension specimen

plane strain solution ($K_{2D} = 7\cdot20\, p\sqrt{a}$). Reference 22 shows a variation of 1 per cent from K_{2D}. It is to be noted, however, that the mesh used in this analysis is much coarser than that used in Reference 22.

CONCLUSIONS

The use of quadratic isoparametric elements has been shown to provide an excellent crack tip element when the mid-side nodes closest to the crack are put at the quarter points. The element has been shown to contain $1/\sqrt{r}$ of singularity, thus providing a stress field which agrees with the theoretical stress singularity of linear fracture mechanics. It has also been shown that the element contains rigid body motion and constant strain modes, thus satisfying the necessary conditions for convergence. In addition, they satisfy the patch test.

Numerically it was demonstrated that triangular shaped elements give better results than quadrilateral elements whose stiffness is singular if integrated exactly.[23] It is therefore recommended to use only triangular and prism elements. The use of 9-point Gaussian integration for plane elements and 27 points for three-dimensional elements, provide good evaluation of the stiffness matrix and allow the calculation of the stresses very close to the crack tip, and hence, better evaluation of the K_I value from the stresses. The example problems show that the element gives excellent results even when very few elements are used. Calculation of the stress concentration due to thermal expansion poses no problem for these elements, as opposed to previous crack tip elements and gives excellent agreement with theoretical results.

ACKNOWLEDGEMENT

The numerical results were performed by R. W. Loomis and D. J. Ayres. The author would like to express further his thanks to Dr. D. J. Ayres for his critical review of the paper, his guidance and suggestions. Mr. C. H. Gilkey's encouragement is duly acknowledged.

REFERENCES

1. A. S. Kobayashi, O. E. Maiden, B. J. Simon and S. Iida, 'Application of the method of finite element analysis to two-dimensional problems in fracture mechanics', Univ. of Washington, Dept. of Mechanical Engng, *ONR Contract Nonr*–477(39), NR 064 478, TR No. 5 (1968).
2. G. P. Anderson, V. L. Ruggles and G. S. Stibor, 'Use of finite element computer programs in fracture mechanics', *Int. J. Fract. Mech.* 7, 63–76 (1971).
3. S. K. Chan, I. S. Tuba and W. K. Wilson, 'On the finite element method in linear fracture mechanics', *Engng Fract. Mech.* 2, 1–17 (1970).
4. N. Levy, P. V. Marcal, W. J. Ostergren and J. R. Rice, 'Small scale yielding near a crack in plane strain—a finite element analysis', Brown Univ., Div. of Engng, *TR NASA NIGL* 40–002–080/1 (1969).
5. 'The surface crack, physical problems and computational solutions', *ASME*, 1972 *Winter Annual Meeting*, Appl. Mech. Div. (Ed. J. Swedlow).
6. P. V. Marcal, P. M. Stuart and R. S. Bettes, 'Elastic-plastic behavior of a longitudinal semi-elliptic crack in a thick pressure vessel', Brown Univ., Div. of Engng, *HSSTP–TR–28* (1973).
7. R. H. Gallagher, 'Survey and evaluation of the finite element method in fracture mechanics analysis', *Proc. 1st Int. Conf. Struct. Mech. in Reactor Technology*, Berlin (1971).
8. E. Byskov, 'The calculation of stress intensity factors using the finite element method with cracked elements', *Int. J. Fract. Mech.* 6, 159–167 (1970).
9. D. M. Tracey, 'Finite elements for determination of crack tip elastic stress intensity factors', *Engng Fract. Mech.* 3, 255–265 (1971).
10. J. R. Rice and D. M. Tracey, 'Computational fracture mechanics', *Proc. Symp. Num. Meth. Struct. Mech.*, Urbana, Ill. (1971)..
11. T. H. H. Pian, P. Tong, and C. H. Luk, 'Elastic crack analysis by a finite element hybrid method', *Proc. Air Force 3rd Conf. Matrix Meth. in Struct. Mech.*, AFFDL–TR–71–160, Wright-Patterson A.F.B. (1971).
12. D. J. Ayres and W. F. Siddall, Jr., 'Finite element analysis of a reactor pressure vessel during emergency core cooling', *Petro. Mech. Eng. and Press. Vessel and Piping Conf.*, Denver, Colorado (1970), *ASME Paper No. 70–PVP–23*.
13. W. B. Jordan, 'The plane isoparametric structural element', *G. E. Co. Report KAPL–M–7112* (1970).
14. G. Steinmueller, 'Restrictions in the application of automatic mesh generation schemes by isoparametric coordinates', *Int. J. Num. Meth. Engng* 8, 289–294 (1974).
15. O. C. Zienkiewicz, *The Finite Element in Engineering Science*, McGraw-Hill, London, 1971.
16. B. M. Irons and A. Razzaque, 'Experiments with the patch test for convergence of finite elements', *Proc. Conf. Math. Foundations of Finite Element Method with Applications to Partial Differential Eqns*, Academic Press, 557–587 (1972).
17. G. R. Irwin, 'Analysis of stresses and strains near the end of a crack traversing a plate', *J. Appl. Mech.* 24, 361–364 (1957).
18. *Fracture Toughness Testing and Its Applications*, ASTM Spec. Tech. Publ. No. 381, 1965, 43–51.
19. MARC–CDC, Developed by MARC Analysis Research Corporation, Providence, R.I., 1974.
20. S. E. Benzley, 'Representation of singularities with isoparametric finite elements', *Int. J. num. Meth. Engng* 8, 537–545 (1974).
21. T. A. Cruse and W. Van Buren, 'Three-dimensional elastic stress analysis of a fracture specimen with an edge crack', *Int. J. Fract. Mech.* 7, 1–15 (1971).
22. D. M. Tracey, 'Finite elements for three-dimensional elastic crack analysis', *Nuclear Engng and Design* 26, 282–290 (1974).
23. H. D. Hibbitt, Private Communications, May 1973.

CRACK TIP FINITE ELEMENTS ARE UNNECESSARY*

R. D. HENSHELL AND K. G. SHAW

Mechanical Engineering Department, University Park, Nottingham, England

SUMMARY

It is shown that a singularity occurs in isoparametric finite elements if the mid-side nodes are moved sufficiently from their normal position. By choosing the mid-side node positions on standard isoparametric elements so that the singularity occurs exactly at the corner of an element it is possible to obtain quite accurate solutions to the problem of determining the stress intensity at the tip of a crack. The solutions compare favourably with those obtained using some types of special crack tip elements, but are not as accurate as those given by a crack tip element based on the hybrid principle. However, the hybrid elements are more difficult to use.

INTRODUCTION

Classical solutions for the stress and displacement fields around a crack tip are well known.[1-3] There is a singularity in the stress field and the stresses along any radial line from the tip of the crack (see Figure 1) are all proportional to $Kr^{-\frac{1}{2}}$ where K is known as the stress intensity factor. There is a value of K corresponding to each of the characteristic loading modes:

K_I—crack opening mode
K_{II}—shearing mode
K_{III}—twisting mode

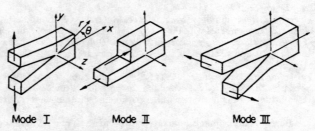

Figure 1. Crack tip nomenclature

To date most of the fracture mechanics work reported in the literature has been centred upon the first of these cases.

It is of considerable importance in many engineering situations to know the numerical value of the stress intensity factors since there are critical values of K which determine whether or not the crack will propagate.

---

* One of the journal editors and the referees of this paper had differing views on the title of this paper. The authors have left it in its original form to stimulate discussion.

Received 12 December 1973
Revised 12 June 1974

The finite element method is now established as a standard tool to use for the detailed determination of stresses in engineering structures and components. Since the method is usually based upon assumptions for displacements and/or stresses, which are defined in terms of polynomial functions over elements of finite size, it is not possible to obtain exact representation of the behaviour in the region of a singularity.

To overcome this difficulty many authors choose a finite element mesh in which there is a very substantial refinement around the tip of the crack.[4-6] It is expensive in computer time and data preparation effort to use these refined meshes and other authors have resorted to the inclusion of a special element in the region of the crack tip.[7-9] Naturally such elements include a singularity and in some cases this is the exact Westergaard solution. In one paper[10] an isoparametric four sided element was used at the crack tip but with one of the sides reduced to zero length to give a triangle with a singularity at one node. Further references to the use of finite element methods in solving fracture mechanics problems are given by Gallagher[11] and Wilson.[12]

In the present paper the determination of crack tip stress intensities is carried out using absolutely standard eight-noded isoparametric elements with 'mid-side' nodes displaced from their nominal position. This work is derived from the earlier work of one of the authors on permissible distortions in isoparametric finite elements[13-16] and from the other author's work on fracture mechanics.

THEORY

The standard eight-noded isoparametric element in an xy space which is transformed to a square in the $\xi\eta$ space with vertices at $(\pm 1, \pm 1)$ is considered. The behaviour of the mid-side nodes as they are moved away from their usual positions (Figure 2) is of interest, and to simplify the mathematics just one of the sides is considered as a one-dimensional element. This element, which is shown in Figure 3, has nodes at $\xi = -1, 0, +1$ and $r = 0, p, 2$. The undistorted element corresponds to $p = 1$. The assumptions for transformation and displacement take the forms:

$$r = a_1 + a_2\xi + a_3\xi^2 \tag{1}$$

$$u = b_1 + b_2\xi + b_3\xi^2 \tag{2}$$

where a_i and b_i are constants and $r = x/h$.

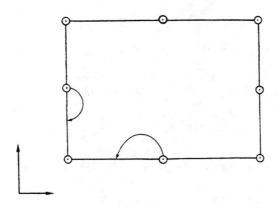

Figure 2. Distortion of an eight-noded element

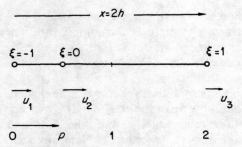

Figure 3. One-dimensional distorted element

Writing the equation (1) in terms of the nodal values of r to eliminate the a_i and then solving for ξ gives

$$\xi = \frac{-1 \pm \sqrt{(1-4p+4p^2+4(1-p)r)}}{2(1-p)} \tag{3}$$

Examination shows that the positive square root is the correct one.

Now the derivative of equation (3) is:

$$\frac{d\xi}{dr} = (1-4p+4p^2+4(1-p)r)^{-\frac{1}{2}} \tag{4}$$

This has a singularity when:

$$r = \frac{(1-2p)^2}{4(p-1)}$$

This singularity occurs at the $r = 0$ end of the element when $p = \frac{1}{2}$. This yields the following expressions for ξ, $d\xi/dr$ and for u

$$\xi = -1 + \sqrt{(2r)} \tag{5}$$

$$\frac{d\xi}{dr} = (2r)^{-\frac{1}{2}} \tag{6}$$

$$u = u_1 \frac{2-3\sqrt{(2r)}+2r}{2} + u_2(-2r+2\sqrt{(2r)}) + u_3 \frac{2r-\sqrt{(2r)}}{2} \tag{7}$$

where the constants b_i in equation (2) have been evaluated in terms of expression for the nodal values of u.

It is the stresses that are of interest and these are proportional to the strains. In the one-dimensional element the longitudinal strain is given by:

$$\frac{du}{dr} = u_1(1-\tfrac{3}{2}(2r)^{-\frac{1}{2}}) + u_2(-2+2(2r)^{-\frac{1}{2}}) + u_3(1-\tfrac{1}{2}(2r)^{-\frac{1}{2}}) \tag{8}$$

Equation (8) clearly shows that the singularity is of the order $r^{-\frac{1}{2}}$ as required by the Westergaard solutions. This suggests a simple rule; when elements with one 'mid-side' node are used at a crack tip the mid-side nodes should be moved from their usual position at the centre of each side to the '$\tfrac{1}{4}$' position as shown in Figure 4. Note that these elements at the crack tip are not in any

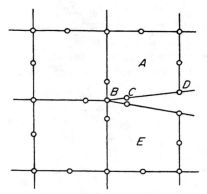

Figure 4. Distortion of elements in the region of a singularity

way special. An absolutely standard isoparametric element subroutine is used throughout the mesh.

A heuristic justification for distorting the elements in the region of a crack tip has now been provided. It is now interesting to consider some of the loads acting on an element which is adjacent to the free surface near the crack tip (element A in Figure 4). This element is subjected to a transverse force (say P) at only the node B and the basis of the finite element solution is that the transverse nodal loads at C and D are zero.

It can be shown[17] that a particular transverse pressure distribution $p(r)$ is equivalent on an energy basis to a set of loads at B, C and D. Conversely a particular set of nodal forces can be considered equivalent to a (non-unique) pressure distribution. A pressure distribution that can be written as a quadratic in ξ is chosen, since this is the simplest polynomial choice that can be made and since it happens to correspond to the form of the displacement polynomial. In particular it is necessary to find a pressure distribution that corresponds to loads of Q, O, O at B, C, D, respectively.

The procedure is to assume that $q(r)$, the load distribution is quadratic in r and then to find the work done along the boundary by the displacement $u(r)$. Following the methods of Reference 17 the nodal loads are found. Since the nodal loads are known, the constants in the $q(r)$ expression can be found and:

$$q(r) = \frac{dr}{dx} \cdot \frac{d\xi}{dr} \cdot Q \cdot \left(-\frac{3}{4} - \frac{3}{2}\xi + \frac{15}{4}\xi^2\right)$$
$$= \frac{Q}{h}\left[\frac{9}{2}(2r)^{-\frac{1}{2}} - 9 + \frac{15}{4}(2r)^{\frac{1}{2}}\right] \qquad (9)$$

The part of this expression within square brackets is plotted in Figure 5. It is clear that a varying pressure is indicated along the free surface of a crack which should be free from normal stress. In the next element away from the crack tip the pressure will be much closer to zero. Thus, the finite element formulation happens to imply a form of connection between the pair of elements A and E of Figure 4.

Therefore justification can be given for an important conclusion from numerical results which are described later: although the elements at the crack tip have reasonably accurate stiffnesses, the local values of stress and displacement in the element adjacent to the crack tips are poor. Thus the displacements on the elements adjacent to the crack tip can justifiably be ignored when calculating K. This is a further example of the 'modified Saint-Venant principle for finite elements':[13] the effects of inaccurate load modelling at some point dies out after one element or the usual 'Saint-Venant' distance, whichever is the greater.

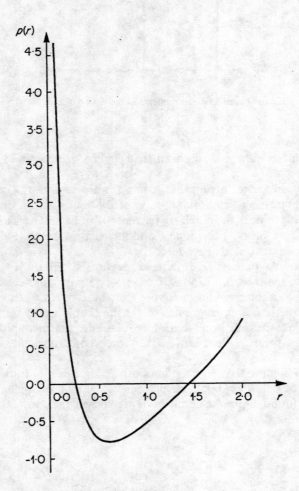

Figure 5. Load distribution on a free surface

DETERMINATION OF K VALUES

When the usual finite element analysis is complete the displacements and stresses everywhere can be found. Using the displacements and the Westergaard solutions a value of K may be calculated.[18] The predicted value of K is found to depend upon the point chosen for this purpose. It is found that reliable results are obtained by using the displacement at points along the free surface of the crack ($\theta = 180$ degrees).

Alternatively energy formulations are possible. For example the energy associated with crack opening may be written, for a plane strain case as

$$\frac{\delta U}{\delta a} = \frac{K_I^2}{(1-v^2)E} \tag{10}$$

where δU is the strain energy increase as the crack is extended δa.

To use equation (10) it is not necessary to solve the complete problem with two different lengths of crack. Normally the frontal elimination technique[19] is used and by choosing the crack tip elements to be last the numerical work is shortened. The elimination has reached a stage at which the remaining unknowns $\{X\}$ are those on the elements immediately around the crack.

Let

[S] be the reduced stiffness matrix of the remainder of the structure;

$\{F\}$ be the loads on the crack tip elements which are equivalent to the exterior loads on the structure and which are determined as the frontal process proceeds;

$[S_1]$ be the stiffness matrix of the group of crack tip elements (usually four in number).

$\{X^*\}$, $[S^*]$, $[U^*]$ are the corresponding values of $\{X\}$, $[S]$, $[U]$ after the crack is opened an amount δa.

All of the above matrices are small (e.g. of order 25 to 50)

After some manipulation we find

$$\delta U = U^* - U = \{X^*\}^T [S_1 - S_1^*]\{X\} \tag{11}$$

The computing procedure is therefore to carry out the frontal elimination until the point where the crack tip elements are reached and then to include these elements with and without an extended crack. The very last steps only are executed twice.

Difficulties arise with this energy technique if there are two modes present in the solution. The displacement comparison methods do not suffer from this drawback but tend to be less accurate.

SINGLE EDGE NOTCHED STRIP

Figure 6 shows the mesh used to solve this problem and Figure 7 shows the manner in which the calculated value of K_I varies at points along the line $\theta = 180$ degrees.

It is clear that the distorted elements are superior to the undistorted type when compared with the accurate solution due to Paris and Sih.[3] The intercept lines shown on the graphs represent linear extrapolations from the available data (ignoring the last element) to the crack tip.

DOUBLE EDGE NOTCHED STRIP IN TENSION

The idealization and problem are similar to the previous case. Estimations of K_I are shown in Figure 8. It is again apparent that distorted elements are best.

A comparison between the distorted standard isoparametric element and the special hybrid stress elements of Pian et al[20] has also been carried out. The idealization used was the same as that used by Pian et al, having three elements along the crack. Using this coarse mesh the value obtained for K_I, using the distorted isoparametric element, differed by 8·6 per cent from the value obtained by Bowie[21] using a boundary collocation technique. The value of K_I given by Pian et al for their best hybrid element differs from Bowie's solution by 0·4 per cent. However,

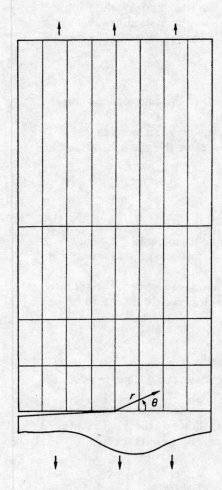

Figure 6. Mesh for a single edge notched strip

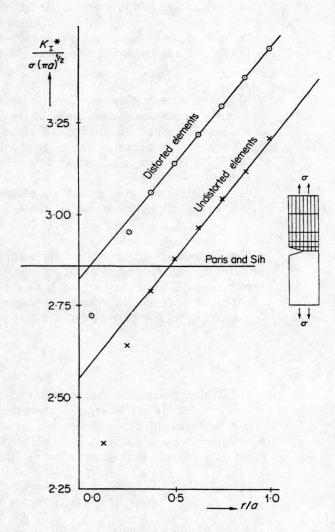

Figure 7. Single edge notched strip—K_I prediction

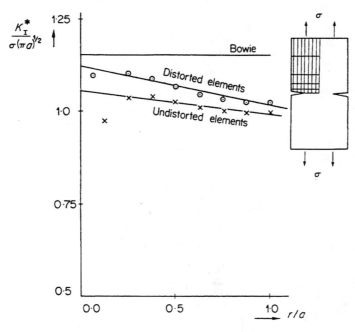

Figure 8. Double edge notched strip—K_I prediction

the present authors feel that the slight refinement of the isoparametric mesh necessary to obtain a required accuracy is worthwhile when compared with the disadvantages of using hybrid elements.

CENTRE CRACKED STRIP IN TENSION

The idealization and problem are similar to the previous cases. Estimates of K_I are shown in Figure 10. There is further justification for the distorted elements.

Jerram and Hellen[18] have also solved this problem using:
1. Linear displacement (constant stress) triangles.
2. Linear triangles, plus special linear elements at crack tip.
3. Quadratic displacement (linear stress) triangles.
4. Quadratic triangles plus special quadratic elements at crack tip. Some of their results are shown in Figure 9.

It can be seen that the distorted isoparametric element solution (147 nodes and 274 degrees-of-freedom) compares favourably with Jerram and Hellen's fourth solution (893 nodes and approximately 1,700 degrees-of-freedom).

MODE II CRACK

The example of a Mode II crack that was looked at is an edge notched plate subjected to antisymmetric shear loading.

The structure is shown in Figure 10 and was modelled with the idealization shown in Figure 11. The results obtained are shown in Figure 10 where they are compared with the solution obtained by Wilson.[22]

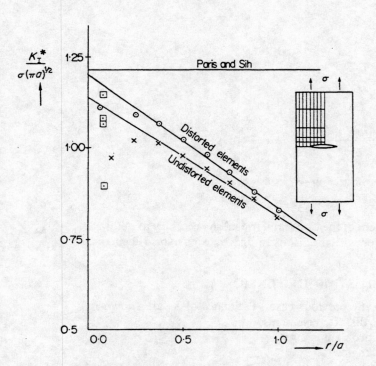

Figure 9. Centre cracked strip in tension—K_I prediction;
  (i) 414 constant stress triangles, 240 nodes[18];
  (ii) as (i) but with special crack tip elements[18];
  (iii) 414 quadratic triangles, 893 nodes[18];
  (iv) as (iii) but with special crack tip elements.[18]
  (i)–(iv) in order give improving accuracy and are shown as squared points

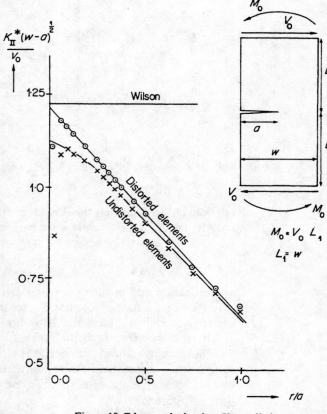

Figure 10. Edge notched strip—K_{II} prediction

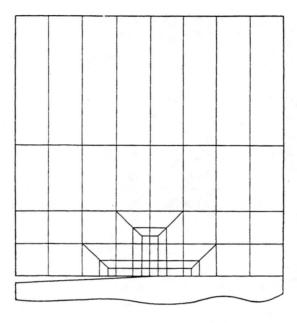

Figure 11. Mesh for K_{II} prediction

COMBINATION OF MODES I AND II

The example of a combined mode crack that was considered is the case of a plate subjected to tension and containing a central crack that is not perpendicular to the applied tension. The values of K_I and K_{II} depend upon the angle between the crack and the applied load (ϕ). Figure 13 shows the mesh used (228 elements, 740 nodes) and Figure 12 shows how K_I and K_{II} vary with ϕ. The accurate solution was obtained by superposition of three simple cases. Again it is apparent that the distorted elements give the better results.

PLATE BENDING PROBLEM

There are various plate bending problems in which a point transverse load causes a singularity. The simplest example is the case of the square simply supported plate with a central point load. An exact solution to the problem is available.[23] Finite element solutions were obtained using the eight-noded plate bending element described in Reference 14. Uniform meshes of elements were used in one quarter of the plate. Results obtained using distorted and undistorted elements are given in Table I.

At first sight the results for the distorted elements appear to be worse than those obtained using the undistorted mesh. In fact the element being used is non-conforming and therefore it can become too flexible. In the undistorted situation there happens to be a good balance between the extra stiffness introduced by the element and the non-conformity along the boundaries. The distorted elements are always a few per cent more flexible than the undistorted variety.

Table I. Displacement results for a plate bending problem. Square simply supported plate, Poisson's ratio $v = 0.3$

Mesh in $\frac{1}{4}$ of plate 'multiplication' etc.	Non-dimensionalized displacement and percentage errors			
	Undistorted $\frac{\delta E h^3}{Q c^2}$		Distorted $\frac{\delta E h^3}{Q c^2}$	
1×1	0.12664	(0.024)	0.12677	(0.079)
2×2	0.12699	(0.253)	0.12891	(1.768)
3×3	0.12683	(0.126)	0.12758	(0.718)

Exact result: $\delta E h^3 / Q c^2 = 0.12667$

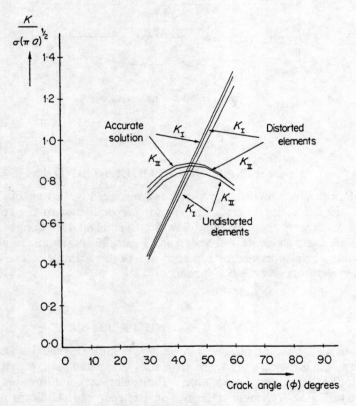

Figure 12. Combination of Modes I and II—K prediction

CONCLUSIONS

Special finite elements for crack tips are not necessary for plane stress/strain analysis. The whole structure can be analyzed using absolutely standard eight-noded elements. Those elements which are adjacent to the crack tip should be distorted to produce a singularity.

Figure 13. Mesh for combination of Modes I and II

ACKNOWLEDGEMENTS

The authors wish to acknowledge: P. Stanley for many suggestions during the preparation of this paper; The Science Research Council for providing financial support for one of the authors (KGS); The PAFEC group[17] for various subroutines.

APPENDIX

Notation

a = Length of crack
a, b = Arbitrary constants
c = Length of plate side
E = Young's modulus
$\{F\}$ = Force vector
h = Plate thickness
K_I, K_{II}, K_{III} = Stress intensity factors
p = r-co-ordinate of a mid-side node
q = Distributed load
Q = Point load
r = Distance from a singularity
$[S]$ = Stiffness matrix
u = Displacement in the Cartesian direction x or r
U = Strain energy
x, y = Cartesian axes
$\{X\}$ = List of displacements
ξ, η = Curvilinear or 'isoparametric' axis

v = Poisson's ratio
θ = Angular co-ordinate around crack tip
ϕ = Angle between crack and applied load

REFERENCES

1. H. M. Westergaard, 'Bearing pressures and cracks', *Trans. ASME, J. Appl. Mech.*, **6**, 49–53 (1939).
2. G. R. Irwin, 'Analysis of stresses and strains near the end of a crack traversing a plate', *Trans ASME, J. Appl. Mech.*, **24**, 361–364 (1957).
3. P. C. Paris and G. C. Sih, 'Stress analysis of cracks', *Am. Soc. Testing and Materials Special Tech. Publ.* 381.
4. S. K. Chan, I. S. Tuba and W. K. Wilson, 'On finite element method in linear fracture mechanics', *Engng Fract. Mech.* **2**, 1–17 (1970–71).
5. G. P. Anderson, V. C. Ruggles and G. S. Stibor, 'Use of finite element computer programs in fracture mechanics', *Int. J. Fract. Mech.* **7**, 63–75 (1971).
6. D. F. Mowbray, 'A note on the finite element method in linear fracture mechanics', *Engng Fract. Mech.* **3**, 173–176 (1970).
7. T. H. H. Pian and P. Tong and C. H. Luk, 'Elastic crack analysis by a finite element hybrid method', U.S. Airforce contract F44620–67–C–0019.
8. W. S. Blackburn, 'Calculation of stress intensity factors at crack tips using special finite elements', *The mathematics of finite elements and applications* (Ed. J. Whitman), 327–336, Brunel Univ. (1973).
9. D. M. Tracey, 'Finite elements for determination of crack tip elastic stress intensity factors', *Engng Fract. Mech.* **3**, 255–266 (1971).
10. N. Levy, P. V. Marcal, W. J. Ostergren and J. R. Rice, 'Small scale yielding near a crack in-plane strain. A finite element analysis', Brown Univ. Tech. Rep. for NASA, NGL–40–002–080/1.
11. R. H. Gallagher, 'Survey and evaluation of the finite element method in fracture mechanics analysis', *Proc. 1st Inst. Conf. Struct. Mech. Reactor Technology*, Compiled by T. A. Jaeger, Bundesanstalt für Materialprufung (BAM) Berlin, **6**, 637–653 (1971).
12. W. K. Wilson, 'Finite element methods for elastic bodies containing cracks', *Methods of Analysis and Solutions of Crack Problems* (Ed. G. C. Sih), Noordhoff, Netherlands, 1973, pp. 484–515.
13. T. J. Bond, J. H. Swannell, R. D. Henshell and G. B. Warburton, 'A comparison of some curved two-dimensional finite elements', *J. Strain Analysis*, **8**, 182–190 (1973).
14. R. D. Henshell, D. Walters and G. B. Warburton, 'A new family of curvilinear plate bending elements for vibration and stability', *J. Sound Vib.* **20**, 327–343 (1972).
15. R. D. Henshell, D. Walters and G. B. Warburton, 'On possible loss of accuracy in curved finite elements', *J. Sound Vib.* **23**, 510–513 (1972).
16. T. J. Bond, 'Some considerations of the finite element method in stress analysis', *Ph.D. Thesis*, Nottingham University, 1972.
17. R. D. Henshell (Ed.), 'PAFEC 70 +', Mechanical Engineering Dept, Nottingham University (1972).
18. K. Jerram and T. K. Hellen, 'The use of finite element techniques in fracture mechanics', C.E.G.B. Research Dept, Berkeley Nuclear Labs, KD/B/N2478 (1972).
19. B. M. Irons, 'A frontal solution program for finite element analysis', *Int. J. num. Meth. Engng*, **2**, 5–32 (1970).
20. T. H. H. Pian, P. Tong and C. H. Luk, 'Elastic crack analysis by a finite element hybrid method', *Proc. Conf.* Wright–Patterson A.F.B., Ohio (1971).
21. O. L. Bowie, 'Rectangular tensile sheet with symmetric edge cracks', *Trans. ASME*, **31**, 208–212 (1964).
22. W. K. Wilson, 'On combined mode fracture mechanics', Westinghouse Research Laboratories, Pittsburg, Penn., *Research Report* 69–1E7–FMECH–R1 (1969).
23. S. P. Timoshenko and S. P. Woinowsky-Krieger, *Theory of Plates and Shells*, McGraw-Hill, New York, 1959.

P. C. Paris,[1] *R. M. McMeeking,*[1] *and H. Tada*[2]

The Weight Function Method for Determining Stress Intensity Factors

REFERENCE: Paris, P. C., McMeeking, R. M., and Tada, H., "**The Weight Function Method for Determining Stress Intensity Factors,**" *Cracks and Fracture, ASTM STP 601,* American Society for Testing and Materials, 1976, pp. 471–489.

ABSTRACT: An alternate method to those of Bueckner and Rice is presented for the derivation of the two-dimensional weight function for determining crack tip stress intensity factors. The weight function has $r^{-1/2}$ type displacement singularity at the crack tip. It is shown that this singular field can be formulated using Westergaard stress functions in a manner similar to the method used for $r^{1/2}$ type fields in crack tip displacements.

A straight-forward approach for obtaining weight functions for cracked finite bodies is presented. This technique can be combined in a simple fashion with the finite element method. As an example, a weight function for the SEN strip is obtained in this manner. Moreover, closed form infinite body weight functions are also developed and used to derive some well-known stress intensity factor formulas.

KEY WORDS: crack propagation, fractures (materials), cracks, weighting functions, stress intensity factors

Bueckner [1][3] has devised a method of determining stress intensity factor solutions, which has also been discussed by Rice [2]. The method depends mainly on the reciprocal theorem and other energy-method like considerations. The method may also be extended to computations of displacements in a manner almost identical to that in Ref 3 (see Appendix B of Ref 3). Earlier similar "Green's Function" approaches are also found in Refs 4 through 6. For an elegant presentation of the method in all generality the reader is referred to Refs 1,2,7,8. However, in this section a more direct derivation of those results will be given. Nevertheless, a new approach to deriving weight functions will be presented in the

[1] Visiting professor and research assistant, respectively, Division of Engineering, Brown University, Providence, R. I. 02912.
[2] Staff scientist, Del Research Corporation, Hellertown, Pa.
[3] The italic numbers in brackets refer to the list of references appended to this paper.

second and third sections which do not require prior familiarity with weight functions. Therefore, the reader may prefer to simply note the results of this first section, Eqs 8 and 9, before proceeding directly into the later sections.

The analysis to follow will show that if the complete solution (for its stress intensity factor and displacements) to a crack problem for one loading system is known, then the solution (for K) for the same cracked configuration with any other loading may be obtained directly from the known solution. To show this consider a cracked body with loads, P_1, $P_2 \ldots P_N$, as the independently applied loads. From well-known results and definitions (for example, see pp. 1.10–1.12 and Appendixes A and B of Ref 3), the Griffith energy rate, $\mathscr{G}$, is

$$\mathscr{G} = \left.\frac{\partial U}{\partial a}\right|_P = \frac{1}{2} \sum_{i=1}^{N} \sum_{j=1}^{N} \frac{\partial C_{ij}(a)}{\partial a} P_i P_j = \frac{1}{2} \sum_{i=1}^{N} P_i \frac{\partial u_i}{\partial a} \quad (1)$$

where the displacements of loading points, u_i, can be written in terms of elastic compliance coefficients, $C_{ij}(a)$ as functions of crack length, a

$$u_i = \sum_{j=1}^{N} u_i^j = \sum_{j=1}^{N} C_{ij}(a) P_j \quad (2)$$

because of the reciprocal theorem $C_{ij} = C_{ji}$, which was used in writing the preceding equations.

Now on the other hand, the Griffith energy rate may be written in terms of stress intensity factors (see Ref 3, p. 1.12) as

$$\mathscr{G} = \frac{K^2}{E'} = \frac{1}{E'} \sum_{i=1}^{N} \sum_{j=1}^{N} k_i(a) k_j(a) P_i P_j \quad (3)$$

since the stress intensity factor, K, is linearly dependent on the loads, P_i, or

$$K = \sum_{i=1}^{N} K_i = \sum_{i=1}^{N} k_i(a) P_i \quad (4)$$

In this analysis E' is the effective modulus of elasticity; E, for plane stress or $E/1 - \nu^2$ for plane strain, and K_i the stress intensity factor for the ith load.

Now equating the double sums in both results for $\mathscr{G}$,[4] that is, Eqs 1 and 4, and noting that since this must be true for any values of the loads, P_1, $P_2 \ldots P_N$, the coefficients must be identical term by term, then

[4] As suggested by J. R. Rice, private communication, 1974.

$$\frac{k_i(a)k_j(a)}{E'} = \frac{1}{2}\frac{\partial C_{ij}(a)}{\partial a} \qquad (5)$$

Let a full solution be known for just one of the loads, say P_m. Then rearranging the latest result

$$k_i(a) = \frac{E'}{2}\frac{\partial C_{im}(a)}{\partial a}\frac{1}{k_m(a)}$$

or from Eq 4

$$k_i(a) = \frac{E'}{2}\frac{\partial C_{im}(a)}{\partial a}\frac{P_m}{K_m} \qquad (6)$$

By saying the solution is known for a load, P_m, it is meant that K_m is known and C_{im} is known since the displacements, u_i^m, for the load at P_m are presumed to be known and from Eq 2

$$C_{im} = \frac{u_i^m}{P_m}$$

Combining Eqs 4 and 6

$$K = \sum_{i=1}^{N} k_i(a)P_i = \frac{E'}{2}\frac{P_m}{K_m}\sum_{i=1}^{N}\frac{\partial C_{im}(a)}{\partial a}P_i \qquad (7)$$

Thus K can be found for loads, P_i, from results obtained from just one load, P_m. This is the desired result.

For arbitrary distributed tractions $T(s)$ over a surface, s, instead of discreet forces, P_i, the form of the result, Eq 7, becomes

$$K = \int h_m(s,a) T(s) \, ds \qquad (8)$$

where $h_m(s,a)$ is the "weight function" as determined entirely from the solution for a load (or loading system) characterized by m. For this result it is noted that

$$h_m(s,a) = \frac{E'}{2K_m(a)} \times \frac{\partial u^m(s,a)}{\partial a} \qquad (9)$$

where $u^m(s,a)$ are displacements at s in the direction of the tractions $T(s)$ but caused only by the loading system characterized by m.

A Special Method of Determining the Weight Function, Mode I

In a two-dimensional problem let loading State 1 be the known loading state (corresponding to m) where concentrated forces, P_1, are applied on the crack surface at a distance, c, from the crack tip. Let loading State 2 (for the same configuration) be one of arbitrary tractions for which it is desired to determine K_2. By the reciprocal theorem

$$P_1 u_2 = \int T_2 u_1 ds \qquad (10)$$

where the displacements u form reciprocal work products (for example, u_{22} is the displacement at the location of P_1 in the direction of P_1 but due to loading State 2) see Fig. 1.

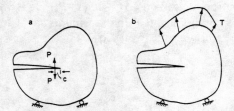

FIG. 1—*An elastic body with two states of loading:* (a) *State 1 and* (b) *State 2.*

Now, presume that the distance, c, from the crack tip to the loading forces in State 1 is very small (that is, approaching zero compared to other dimensions). Then the displacement u_2 will be within the crack tip stress field for State 2 or [3]

$$u_2 = 2v \left| \begin{matrix} r=c \\ \theta = \pi \end{matrix} \right. = \frac{4\sqrt{2}}{\sqrt{\pi} E'} K_{I2} \sqrt{c} \qquad (11)$$

Note that due to the symmetrical (with respect to the crack) force system selected for State 1 only Mode I fields contribute work-producing displacements, u_2, thus the K being computed here is only the Mode I component. Substituting Eq 11 and rearranging Eq 10 gives

$$K_{I2} = \frac{\sqrt{\pi} E'}{4\sqrt{2}} \frac{1}{P_1 \sqrt{c}} \int T_2 u_1 ds \qquad (12)$$

Now as c diminishes to zero let $P_1 \sqrt{c}$ remain a finite constant, choosing for later convenience

$$\frac{P_1 \sqrt{c}}{\pi} = B_I \qquad (13)$$

The Westergaard stress function for State **1** with the distances to boundaries very (infinitely) large compared to c is [3]

$$Z = \frac{P\sqrt{c}}{\pi(z+c)\sqrt{z}} \qquad (14)$$

which upon substituting Eq 13 and letting c approach zero is appropriate. This leads to a local crack tip field situation in State **1** of

$$Z_{I1}(z) = B_I/z^{3/2} \qquad (15)$$

since as c approaches zero, the effects of the boundaries may be neglected. Again, substituting results into Eq 12 gives

$$K_{I2} = \frac{E'}{4\sqrt{2\pi}\,B_I} \int T_2 u_1^1 ds \qquad (16)$$

where T_2 are the applied tractions and u_1^1 are now the corresponding displacements without applied loads, due to inserting a local Mode I singularity of $z^{-3/2}$ type [5] of strength B_I as in Eq 15 at the crack tip where K_2 is desired. The weight function is then

$$h_{Im}(a,s) = \frac{E' u_1^1 (a,s)}{4\sqrt{2\pi}\,B_I} \qquad (17)$$

Bueckner [1] obtained a similar result by a less direct approach.

It is easy to visualize inserting the local singularity described by Eq 15 into a finite element scheme to determine resulting displacements $u_1^1 (a,s)$, for all mesh points with **no** other loads present. This has distinct advantages over other finite element methods since the solution generated applies to *all* possible loadings. The advantage also applies when results are obtained using collocation or direct solutions employing this method.

This derivation has proceeded to consider Mode I only, and the resulting stress intensity factor in Eq 16 is of a Mode I type. However, it is possible to replace State **1** with its Mode II or Mode III counterparts and rederive results for K_{II} and K_{III}, as well as K_I, as follows.

[5] Let the $z^{-3/2}$ singularity be known as the "Bueckner Type" with its strength appropriately denoted as "B."

Mode II and Mode III Weight Functions

By repeating the derivation of the preceding section, that is, Eq 10 through Eq 17, but replacing State 1 on Fig. 1 with its Mode II or Mode III counterpart, as shown on Figs. 2a and b, then weight functions may be developed directly for Mode II and Mode III stress intensity factors. The results are shown in the following equations.

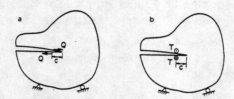

FIG. 2—*Alternate states of load for Modes II and III.*

Mode II

$$K_{II2} = \frac{E'}{4\sqrt{2\pi}\, B_{II}} \int T_2 u_1^{II} ds \tag{18}$$

where u_1^{II} is now caused by inserting a local field of the Bueckner type at the crack tip of interest, that is

$$Z_{II1}(z) = B_{II}/z^{3/2} \tag{19}$$

Mode III

$$K_{III2} = \frac{G}{2\sqrt{2\pi}\, B_{III}} \int T_2 u_1^{III} ds \tag{20}$$

where G is the shear modulus and u_1^{III} is caused by inserting the Bueckner singularity

$$Z_{III1}(z) = B_{III}/z^{3/2} \tag{21}$$

Therefore, it is equally easy to insert Mode II and Mode III singularities in problems to obtain the Mode II and Mode III weight functions

$$h_{IIm}(a,s) = \frac{E' u_1^{II}(a,s)}{4\sqrt{2\pi}\, B_{II}} \tag{22}$$

and

$$h_{IIIm}(a,s) = \frac{G\, u_1^{III}(a,s)}{2\sqrt{2\pi}\, B_{III}} \tag{23}$$

where now Eq 8 may be applied individually for each of the three modes of stress intensity factors.

For displacement boundary conditions see Appendix I.

Near Tip Bueckner Displacement Fields

The displacement fields may be computed for Bueckner type singularities, that is (see Eqs 15, 19, and 21)

$$Z(z) = B/z^{3/2}$$

for each mode. Making use of the usual r, θ coordinates measured from the crack tip, the results for plane strain displacement are as follows.

Mode I
$$u = \frac{B_I}{G\sqrt{r}} \cos \frac{\theta}{2} \left[2\nu - 1 + \sin \frac{\theta}{2} \sin \frac{3\theta}{2} \right]$$
$$v = \frac{B_I}{G\sqrt{r}} \sin \frac{\theta}{2} \left[2 - 2\nu - \cos \frac{\theta}{2} \cos \frac{3\theta}{2} \right] \quad (24)$$
$$w = 0$$

Mode II
$$u = \frac{B_{II}}{G\sqrt{r}} \sin \frac{\theta}{2} \left[2 - 2\nu + \cos \frac{\theta}{2} \cos \frac{3\theta}{2} \right]$$
$$v = \frac{B_{II}}{G\sqrt{r}} \cos \frac{\theta}{2} \left[1 - 2\nu + \sin \frac{\theta}{2} \sin \frac{3\theta}{2} \right] \quad (25)$$
$$w = 0$$

Mode III
$$u = 0$$
$$v = 0 \quad (26)$$
$$w = \frac{B_{III}}{G\sqrt{r}} 2 \sin \frac{\theta}{2}$$

For plane stress for Modes I and II replace ν by $\nu/(1 + \nu)$ and also note then $w \neq 0$.

For generating weight functions these fields should be inserted locally at the crack tip where K is desired. They are then the actual weight function displacements which should be used for loads near by the crack tip.

For a semi-infinite crack in an infinite body, they are the weight function displacements for the whole problem (two-dimensional problems). Some examples will follow making use of these results for very simple problems so as to illustrate the method. The power of the method is however only fully appreciated with more complicated problems when combined with numerical, finite element, and other procedures.

Closed-Form Weight Functions

Closed forms for Westergaard stress functions, Z and $\overline{Z}$, can be written to form weight function displacement fields throughout a body. The technique of finding such stress functions is much the same as for normal crack stress analysis problems except that the crack tip, for which the weight function is desired, will have a $z^{-3/2}$ (in Z) of strength B. Some examples are as follows (each applies to all three Modes I, II, and III)

$$\begin{Bmatrix} Z_I(z) \\ Z_{II}(z) \\ Z_{III}(z) \end{Bmatrix} = \begin{Bmatrix} B_I \\ B_{II} \\ B_{III} \end{Bmatrix} \frac{1}{z^{3/2}} \quad (27)$$

$$\begin{Bmatrix} \overline{Z}_I(z) \\ \overline{Z}_{II}(z) \\ \overline{Z}_{III}(z) \end{Bmatrix} = -\begin{Bmatrix} B_I \\ B_{II} \\ B_{III} \end{Bmatrix} \frac{2}{z^{1/2}} \quad (28)$$

Note, all Z and $\overline{Z}$ expressions apply to all three modes as indicated here.

$$Z(z) = \frac{iB}{z^{3/2}} \quad (29)$$

$$\overline{Z}(z) = -\frac{2iB}{z^{1/2}} \quad (30)$$

$$Z(z) = \frac{B\sqrt{2a}}{(z-a)^{3/2}(z+a)^{1/2}} \quad (31)$$

$$\overline{Z}(z) = -B\sqrt{\frac{2}{a}}\left(\frac{z+a}{z-a}\right)^{1/2} \quad (32)$$

$$Z(z) = \frac{2B\sqrt{2a}\,z}{(z^2 - a^2)^{3/2}} \quad (33)$$

$$\overline{Z}(z) = \frac{-2B\sqrt{2a}}{(z^2 - a^2)^{1/2}} \quad (34)$$

Symmetric about y-axis. Note, remove 2 to get K at one crack tip.

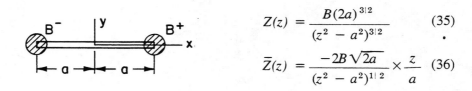

$$Z(z) = \frac{B(2a)^{3/2}}{(z^2 - a^2)^{3/2}} \quad (35)$$

$$\overline{Z}(z) = \frac{-2B\sqrt{2a}}{(z^2 - a^2)^{1/2}} \times \frac{z}{a} \quad (36)$$

Skew symmetric about y-axis. Note, remove 2 to get K at one crack tip.

$$Z(z) = \frac{B\sqrt{2a}}{(a - z)^{3/2}(a + z)^{1/2}} \quad (37)$$

$$\overline{Z}(z) = B\sqrt{\frac{2}{a}}\left(\frac{a + z}{a - z}\right)^{1/2} \quad (38)$$

$$Z(z) = \frac{2B\left(\dfrac{\pi}{w}\cos\dfrac{\pi a}{w}\right)^{3/2}\left(2\sin\dfrac{\pi a}{w}\right)^{1/2}\sin\dfrac{\pi z}{w}}{\left(\sin\dfrac{\pi z}{w} - \sin\dfrac{\pi a}{w}\right)^{3/2}\left(\sin\dfrac{\pi z}{w} + \sin\dfrac{\pi a}{w}\right)^{3/2}} \quad (39)$$

$$\overline{Z}(z) = \frac{-2B\left(\dfrac{2\pi}{w}\tan\dfrac{\pi a}{w}\right)^{1/2}\cos\dfrac{\pi z}{w}}{\left[\left(\sin\dfrac{\pi z}{w}\right)^2 - \left(\sin\dfrac{\pi a}{w}\right)^2\right]^{1/2}} \quad (40)$$

$$Z(z) = \frac{B\left(\dfrac{\pi}{w}\cos\dfrac{\pi a}{w}\right)^{3/2}\left(2\sin\dfrac{\pi a}{w}\right)^{1/2}}{\left(\sin\dfrac{\pi z}{w} - \sin\dfrac{\pi a}{w}\right)^{3/2}\left(\sin\dfrac{\pi z}{w} + \sin\dfrac{\pi a}{w}\right)^{1/2}} \quad (41)$$

$$\bar{Z}(z) = B\left\{\frac{-\left(\dfrac{2\pi}{w}\tan\dfrac{\pi a}{w}\right)^{1/2}\cos\dfrac{\pi z}{w}}{\left[\left(\sin\dfrac{\pi z}{w}\right)^2 - \left(\sin\dfrac{\pi a}{w}\right)^2\right]^{1/2}}\right.$$

$$\left. + i\cot\dfrac{\pi a}{w}\left(\dfrac{\pi}{w}\sin\dfrac{2\pi a}{w}\right)^{1/2}\Pi_c\left[\sin^{-1}\left(\dfrac{\sin\dfrac{\pi z}{w}}{\sin\dfrac{\pi a}{w}}\right), 1, \sin\dfrac{\pi a}{w}\right]\right\}$$

$$\Pi_c[\phi,n,k] = \int\frac{d\phi}{(1 - n\sin^2\phi)(1 - k^2\sin^2\phi)^{1/2}} \quad (42)$$

An Example of Finite Element Results

As mentioned earlier, weight function displacements can be obtained using finite element analysis by putting a small hole at the crack tip and inserting the Bueckner type field as boundary conditions on the hole. That is using Eqs 24 (or Eqs 25 or 26) as boundary conditions on the hole, with all other surfaces stress free in order to determine the State 1 displacements, u_1.

As an example, consider the single edge cracked strip for which a variety of loadings are available with previously tabulated results [9,3]. The particular strip selected here is (see Fig. 3), $a = b/2$, $2h = 6b$ with circular hole at the crack of radius, $r = 0.006944b$ and $\nu = 0.3$. A mesh of 352 elements and 398 nodes was used as shown on Fig. 3 (with the inner circles of elements removed and enlarged for clarity).

The table gives results obtained by simply inserting the displacement field, Eq 24, at 25 points on the upper half of the hole, and constraining

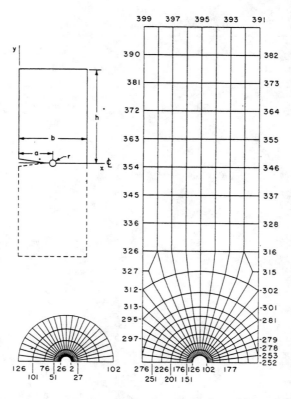

FIG. 3—*A single-edge cracked strip with a circular hole at the crack radius.*

points on the crack plane ahead of the crack to remain on that plane. In this manner only half of the problem needs to be treated by finite elements. The points on Table 1 are numbered as indicated on the mesh in Fig. 3 and located by coordinates x/b and y/b. The corresponding components of the displacements, u_1, for use in the integral to obtain K_{12} are

$$\bar{u}/b = \frac{E'(u_1)_x}{2\sqrt{2\pi}\, B_1 b} \tag{43}$$

$$\bar{v}/b = \frac{E'(u_1)_y}{2\sqrt{2\pi}\, B_1 b}$$

so that K_{12} may be obtained from components of tractions for one half of a problem (with respect to crack plane symmetry)

$$K_{12} \int (T_x \bar{u} + T_y \bar{v})\, ds \tag{44}$$

TABLE 1—*Weight function displacements for an edge cracked strip.*

$$\left(\frac{a}{b} = 0.5, \ \frac{h}{b} = 3, \nu = 0.3, \frac{r}{b} = 0.006944\right)$$

Point	x/b	y/b	$\bar{u}/b$	$\bar{v}/b$
2	0.5069	0	−2.736	0
27	0.5156	0	−1.767	0
102	0.5625	0	−0.595	0
177	0.75	0	0.285	0
26	0.4931	0	−0.000003	9.575
51	0.4844	0	−0.232	7.066
76	0.4722	0	−0.269	5.930
101	0.4566	0	−0.297	5.395
126	0.4375	0	−0.319	5.187
151	0.3888	0	−0.319	5.229
176	0.3264	0	−0.346	5.670
201	0.25	0	−0.368	6.374
226	0.1666	0	−0.385	7.214
251	0.0833	0	−0.393	8.084
276	0	0	−0.396	8.960
297	0	0.207	1.795	8.971
295	0	0.3837	3.719	8.956
313	0	0.5	4.988	8.933
312	0	0.6516	6.634	8.908
327	0	0.820	8.453	8.890
326	0	1.00	10.401	8.887
336	0	1.25	13.116	8.887
345	0	1.50	15.838	8.889
354	0	1.75	18.563	8.891
363	0	2.00	21.289	8.892
372	0	2.25	24.015	8.892
381	0	2.50	26.740	8.892
390	0	2.75	29.466	8.892
399	0	3.00	32.191	8.892
397	0.25	3.00	32.191	6.166
395	0.50	3.00	32.191	3.441
393	0.75	3.00	32.191	0.715
391	1.0	3.00	32.191	−2.010
382	1.0	2.75	29.466	−2.010
373	1.0	2.50	26.740	−2.010
364	1.0	2.25	24.015	−2.010
355	1.0	2.00	21.289	−2.010
346	1.0	1.75	18.564	−2.011
337	1.0	1.50	15.838	−2.012
328	1.0	1.25	13.109	−2.015
316	1.0	1.0	10.371	−2.017
315	1.0	0.820	8.386	−2.015
302	1.0	0.6516	6.515	−1.980
301	1.0	0.50	4.819	−1.912
281	1.0	0.3837	3.525	−1.744
279	1.0	0.207	1.774	−1.249
278	1.0	0.134	1.228	−0.865
253	1.0	0.0653	0.904	−0.437
252	1.0	0	0.805	0

NOTE—($\bar{u}$ and $\bar{v}$ values for all 398 mesh points are available from the authors).

Using this formula and Table 1, the following accuracies are observed in comparative results for various loadings:

(a) uniform tension applied at the ends ~ 2.9 percent (see Ref 3, p. 2.10),

(b) uniform pressure on the crack surface ~ < 4.9 percent,

(c) pure bending (moment applied anywhere beyond, $y/b = 1.5$) ~ 3.2 percent (see Ref 3, p. 2.13),

(d) four-point bending (with loads at $y/b = 1$ and 3) ~3.2 percent, and

(e) three-point bending (with a span of $4b$) ~3.6 percent (see Ref 3, p. 2.16).

This is only an example and better accuracies may be produced by finer finite element meshes. However, Table 1 allows the reader to calculate K_I results of comparable accuracies for any loading for this configuration. Its advantages and generality are thus obvious.

This same method applies, of course, to Mode II and Mode III simply using the proper fields, Eqs 25 and 26, as just implied. Moreover, it is obviously applicable to residual stress, thermal stress, and body force problems via simple superposition or direct application. The method is also open to three-dimensional finite element applications.

Closed Form Examples

Example 1

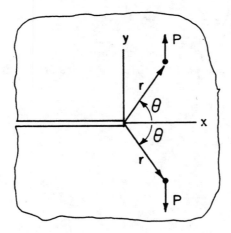

Consider an infinite sheet with a semi-infinite crack with a pair of equal and opposite forces (per unit thickness), P, as shown. The stress solution with a Bueckner type Mode I singularity (and no loading forces) is, as in Eq 27.

$$Z_{I1}(z) = \frac{B_I}{z^{3/2}} \text{ (any } z\text{)} \tag{45}$$

The stress intensity factor for the loads shown may be computed from

$$K_I = \frac{E'}{4\sqrt{2\pi}\, B_I} \int T_2 u_1^I \, ds \tag{46}$$

Now because of x-axis symmetry

$$\int T_2 u_1^I \, ds = 2P \times v_1\, (r,\theta) \tag{47}$$

The usual displacement relationships apply [3] or

$$2G v_1 = 2(1-\nu)\,\text{Im}\,\overline{Z}_{I1} - y\,\text{Re}\,Z_{I1} \tag{48}$$

where for plane stress ν can be replaced by $\nu/(1+\nu)$. Substituting Z_{I1} and taking $z = re^{i\theta}$, etc. (see also Eq 24)

$$v_1 = \frac{B_I}{G\sqrt{r}} \sin\frac{\theta}{2}\left[2(1-\nu) - \cos\frac{\theta}{2}\cos\frac{3\theta}{2}\right] \tag{49}$$

continuing the substitutions

$$K_I = \frac{P}{(1-\nu)\sqrt{2\pi r}} \sin\frac{\theta}{2}\left[2(1-\nu) - \cos\frac{\theta}{2}\cos\frac{3\theta}{2}\right] \tag{50}$$

For the special case $r = b$ and $\theta = \pi$, the result is as usual [3]

$$K_I = \frac{2P}{\sqrt{2\pi b}} \tag{51}$$

and for the special case $r = y_o$ and $\theta = \dfrac{\pi}{2}$ the result is easily obtained

$$K_I = \frac{P}{\sqrt{\pi y_o}}\left[\frac{5-4\nu}{4-4\nu}\right] \tag{52}$$

Therefore, it is seen that in this case getting the K_I through the weight function method gives direct algebraic access to a simpler form for some uses.

Note also from Eqs 25 and 26 that no Mode II or Mode III work is done by the forces, P, with these displacements, so $K_{II} = K_{III} = 0$.

Example 2

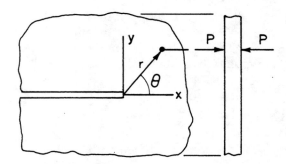

Consider a crack tip in a sheet which is subject to pinching forces, P, as shown. The (thinning) displacement per unit thickness is

$$u_1 = -\varepsilon_z = \frac{\nu}{E}(\sigma_x + \sigma_y) \tag{53}$$

which from the Westergaard stress function analysis becomes for

Mode I

$$u_1 = \frac{2\nu}{E} \, \text{Re} \, Z_I \tag{54}$$

Mode II

$$u_1 = -\frac{2\nu}{E} \, \text{Im} \, Z_{II} \tag{55}$$

for both modes

$$Z(z) = \frac{B}{z^{3/2}} = \frac{B}{r^{3/2}}\left(\cos\frac{3\theta}{2} - i\sin\frac{3\theta}{2}\right) \tag{56}$$

where substituting these results into Eqs 16 and 18 gives

$$K_I = \frac{\nu P}{2\sqrt{2\pi}\,R^{3/2}} \cos\frac{3\theta}{2}$$

$$K_{II} = \frac{-\nu P}{2\sqrt{2\pi}\,R^{3/2}} \sin\frac{3\theta}{2} \tag{57}$$

Acknowledgments

This work was principally accomplished at Brown University under support from the National Science Foundation—Materials Research Laboratory and the Atomic Energy Commission. The many suggestions of J. R. Rice are gratefully acknowledged as contributing much to this work.

APPENDIX I

Weight Functions for Displacement Boundary Condition

The previous derivations of weight functions assumed that all boundary conditions in State **2** are force loadings except for constraints just sufficient to eliminate rigid body modes. However, the reactions at those constraints are constituents of the forces T_2 applied to the weight function for computing K_2. The weight function for State **2** may be derived for boundary conditions of forces T_2 on the surface S_T and displacements u_2 on the surface S_u by imposing in State **1** boundary conditions of zero displacements on S_u. Then following in the same derivation as before Eq 16 becomes

$$K_{12} = \frac{E'}{4\sqrt{2\pi}\, B_\mathrm{I}} \left(\int_{S_T} T_2 u_1^\mathrm{I} ds - \int_{S_u} P_1^\mathrm{I} u_2 ds \right) \tag{58}$$

where u_1^I are the displacements in the altered State **1**, arising from the Bueckner type singularity of strength B_I at the crack tip, and P_1^I are the reactions in the altered State **1** arising at S_u due to the Bueckner type singularity. The Mode II and Mode III forms may be derived straight forwardly as before. Note that a problem with displacement boundary conditions on one part of the boundary requires a different weight function from a problem with displacement boundary conditions on another part of the boundary.

APPENDIX II

A Suggestion on Analytical Improvement of Numerical Solutions for Weight Functions

In the numerical weight function example given here and the preceding analysis, it was demonstrated that imposing only the Bueckner field of displacements (Eq 24, etc.) upon a hole at the crack tip gave reasonable results. However, the full weight function solution for a body of finite dimensions in general may be written as

$$Z = \frac{B + B_1 z + B_2 z^2 + B_3 z^3 + \cdots}{z^{1/2}} \tag{59}$$

where: $B_1, B_2, B_3, \ldots$, along with the addition of an appropriate state of uniform stress, are adjusted to suit the exterior boundary conditions. Thus, it is noted that using the Bueckner type singularity alone imposed inside a hole is imposing only the leading singularity. It thus appears that retaining all the singular terms, that is

$$Z = \frac{B + B_1 z}{z^{3/2}} \qquad (60)$$

would be a better approximation and lead to improvements in numerical analysis. The additional term is $B_1/z^{1/2}$ which is noted to be the usual type of crack tip stress singularity where

$$B_1 = \frac{K}{\sqrt{2\pi}} \qquad (61)$$

Therefore, selecting the appropriate value of B_1 for a given B is noted to require the determination of a crack tip stress intensity factor. The appropriate approach may be viewed as a superposition as shown in Fig. 4.

The actual solution desired is shown in Fig. 4a where both B and K are imposed through the implied field of Eqs 60 and 61 as displacement boundary conditions inside the hole. However, the appropriate value of K in terms of B (linear dependence) is not yet known. However, Fig. 4a can be regarded as the superposition of Fig. 4b and c. Now 4b is the infinite body solution with the Bueckner singularity alone imposed, whose exact solution is expressed by Eq 60 (with $B_1 = 0$). For the actual finite body in 4b we find tractions T_b on the boundary. Then in 4c we apply these tractions in a reversed sense, that is, $T_c = -T_b$ so that the superposition of 4b and c give the proper boundary conditions for 4a. The resulting K in 4c due to tractions T_c is the desired result.

Numerically the procedure is as follows:

1. Determine an initial (approximate) weight function imposing a Bueckner field, B, (only) within the hole in Fig. 4a. Note that the result is like the numerical example given in the paper.

2. Use the resulting initial weight function in Step 1 and the tractions T_c as determined from Fig. 4b, that is $-T_b$ to determine K.

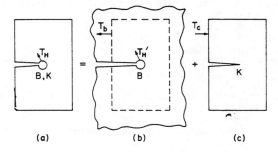

FIG. 4—*Superposition of loadings to determine and improve weight function.*

3. Impose the displacement field implied by K within the hole in Fig. 4a to get displacement corrections for the weight function in Step 1.

4. Combine the weight function displacements and corrections in Steps 1 and 3 to get the final weight function.

An additional improvement in computational ease can be gained in this procedure when using appropriate finite element programs for computation. The computation of K in Fig. 4c, that is, Step 2, from the initial weight function in Step 1 requires computation of the integral of tractions, $-T_b$, through weight function displacements, u_a, of the form

$$-\int u_a T_b ds \tag{62}$$

around the exterior boundary. However, by reciprocal theorem between Figs. 4a and b, it is found that

$$-\int u_a T_b ds = \int (T_H - T_H') u_H ds \tag{63}$$

where T_H are the resulting tractions at the hole due to imposed displacements in the hole in Step 1. The tractions, T_H, are normally available from the finite element results. T_H' are the tractions at the hole in Fig. 4b which are purely from the Bueckner type field which can easily be stored for computation, and u_H are the Bueckner field displacements of both Figs. 4a and b (already stored). Either form may be used, but it appears that for repeated computation on many configurations the form of the right-hand side of Eq 63 has advantages over the left.

Thus, it appears to be possible to make a very simple computation for K and a single repetition of the original weight function computation to obtain an improved weight function. Consequently, this suggestion is made for those interested in exploring the maximum possible computational accuracy obtainable through the methods in this work.

References

[1] Bueckner H. F., *Zeitschrift für Angewandte Mathemetik und Mechanik,* Band 46, 1970, Seite 529–545.
[2] Rice, J. R., *International Journal of Solids and Structures,* Vol. 8, 1972, pp. 751–758.
[3] Tada, H., Paris, P. C., and Irwin, G. R., *The Stress Analysis of Cracks Handbook,* Del Research Corporation, Hellertown, Pa. 1973.
[4] Paris, P. C., "The Mechanics of Fracture Propagation and Solutions to Fracture Arrester Problems," Document D2-2195, The Boeing Company, 1957.
[5] Barenblatt, G. I., in *Advances in Applied Mechanics,* Academic Press, New York, Vol. vii, 1962, p. 55.
[6] Sih, G. C., Paris, P. C., and Erdogan, F., "Crack Tip Stress Intensity Factors for Plane Extension and Plate Bending Problems," *Journal of Applied Mechanics, Transactions,* American Society of Mechanical Engineers, Vol. 29, 1962.
[7] Bueckner, H. F., *Zeitschrift für Angewandte Mathemetik und Mechanik,* Vol. 51, 1971, pp. 97–109.
[8] Bueckner, H. F., "Field Singularities and Related Integral Representations," *Methods of Analysis and Solutions of Crack Problems,* G. C. Sih, Ed. Noordhoff International Publishing, Chapter 3, 1972.
[9] Paris, P. C. and McMeeking, R. M., "Efficient Finite Element Methods for Stress Intensity Factors Using Weight Functions," *International Journal of Fracture Mechanics,* April 1975.

NUMERICAL EVALUATION OF ELASTIC STRESS INTENSITY FACTORS BY THE BOUNDARY-INTEGRAL EQUATION METHOD

Thomas A. Cruse
Mechanical Engineering Department
Carnegie-Mellon University
Pittsburgh, Pennsylvania

ABSTRACT

Numerical evaluations of elastic stress intensity factors are obtained by solution of certain integral equations, referred to as the boundary-integral equations, which relate the boundary tractions to the boundary displacements. The boundary-integral equation method has several advantages over other numerical methods such as finite elements for solving crack problems. These advantages are reduced problem size, reduced computer time requirements, and increased internal stress resolution.

The methods for evaluating stress intensity factors are developed through a study of a series of increasingly complex geometries. The first problem is a center-cracked planar body solved by the two dimensional version of the method. The second problem is the penny-shaped crack, solved by the three dimensional version of the method. Finally, the three dimensional problem of a semi-circular surface flaw in an essentially infinite body is studied. The numerical results for the elastic surface problem show a large rise in the stress intensity factor near the free surface. The surface flaw results raise certain questions as to the meaning of stress intensity factor for three dimensional problems.

The report summarizes the boundary-integral equation method, the crack modelling philosophy, and numerical results. The strengths and weaknesses of the boundary-integral equation method for studying crack problems are also discussed.

INTRODUCTION

The work reported herein forms the basis of a program of study of elastic surface flaw problems now underway. The goal of the present report is to discuss in detail the solution method, modelling philosophy, and means for evaluating elastic stress intensity factors (SIF). The numerical results reported are for a limited series of problems including a planar crack, an axisymmetric (penny-shaped) crack, and a semi-circular edge crack in an essentially infinite body. Future reports will discuss numerical results for surface cracks with non-constant curvature and for moderately thin plates. All problems being considered are of the mode I type with symmetric transverse loading.

Evaluation of stress intensity factors for the variety of surface flaw problems is apparently going to be in numerical rather than analytical terms owing to the complexity of the geometry. The boundary-integral equation (BIE) method of elastic stress analysis is a numerical method particularly suited to the stress analysis of cracked bodies. The BIE method is based on the numerical solution of a set of integral equation relations between boundary tractions and boundary displacements. The reduction to a boundary problem reduces problem dimensionality by one and has no requirement for discretization of the volume of the body. The reduced dimensionality of the problem significantly reduces computer core requirements. For a limited number of internal stress points such as needed for crack problems, the BIE method computer time requirements are much less than for equivalent finite element (FE) solutions. Further, the BIE method has superior interior stress resolution capability owing to the lack of an internal, or volume, discretization. The comparisons of size and time requirements are discussed more fully in [1].[1] The ability of the BIE method to adequately model two and three dimensional crack problems is established in this report.

The next section of this report reviews the BIE formulation, numerical solution strategy, and crack modelling schemes. The third section presents numerical data for the center-cracked plate solved in two dimensions and the penny-shaped crack solved in three dimensions. This data is used to verify the ability of the BIE method to model cracks and provides means for evaluating stress intensity factors. The fourth section presents a detailed discussion of the numerical results for the semi-circular elastic surface flaw problem. The section also contains a brief discussion of the difficulties in interpreting data for a truly three dimensional crack problem. The final section summarizes the results and reviews the strengths and weaknesses of the BIE method for the stress analysis of cracked bodies.

REVIEW OF THE SOLUTION METHOD

The boundary-integral equation for elasticity is an integral constraint equation between the boundary tractions and the boundary displacements. If the tractions (displacements) are everywhere specified then the BIE becomes an integral equation for the unknown boundary displacements (tractions). For a well-posed mixed boundary condition problem one obtains integral equations for all of the unspecified boundary data. Through the use of the BIE the dimensionality is reduced by one. Thus, a three dimensional *volume* problem becomes a two dimensional *boundary* problem. Once the unknown boundary data has been obtained by numerical solution of the BIE, the interior stresses are determined by integrals of the boundary data. The unknown stresses are computed only at specified interior and boundary points.

The mathematical basis of the BIE is the use of the reciprocal work theorem of elasticity together with the known fundamental solution to a point load in an infinite body. The reciprocal work theorem relates the known fundamental solution to the solution of the desired problem. Both the BIE and the interior stress identity are derived from the same reciprocal work relation.

Numerical solutions to the BIE are obtained in a very general, yet direct manner. The boundary of the body is modelled by a series of flat segments (straight lines for two dimensions; flat triangles for three dimensions). Over each boundary segment the components of the traction vector and the displacement

---

[1]Brackets enclose reference numbers; references are located at the end of this report.

vector are taken to be constant. The integral equations are thereby reduced to linear algebraic equations with the same number of equations and unknown boundary data. The algebraic equations are solved by standard matrix reduction algorithms for the unknown data. After the numerical BIE has been solved all boundary data is known, in a piecewise constant approximation. The interior stresses are then obtained directly by quadrature with no need for approximations to the interior stress data. The stresses may be calculated at any number of points at any locations.

The advantages of the BIE method are several. The method is very general and will handle arbitrary geometries and specifications of boundary data with equal ease. The boundary segment mesh is only generated for the boundary, requires few segments compared to finite element models, and has no bandwidth requirement. As discussed in [1] the system of algebraic equations is smaller than for comparable FE models and, for crack problems, computer time requirements are lower than for FE models. The time advantage occurs because of the ability to concentrate on a particular region for calculating interior stresses or using boundary displacements to obtain stress intensity factors in crack problems. It is also noteworthy that the interior stresses may be recalculated at a new set of interior points at any time by preserving the solution to the corresponding BIE.

The BIE method also has some limitations as a solution method. The method is not a general purpose solution tool with different element capabilities such as some FE programs. To date, the method is limited to elasticity although some current effort to include plasticity is meeting with success. The BIE method is not readily adaptable to solving multi-material problems such as layered material. Finally, its most serious drawback as discussed in [1] is the time required to generate full field solutions. That is, when the stresses are desired at *many* interior points the FE method shows a clear time advantage, particularly when the stress field is not too severe.

Thus it was concluded that the BIE method offers real potential for providing numerical solutions to elastic surface flaw problems. The solution to crack problems has been reported earlier [1]. The present work seeks to establish the validity of the modelling and data reduction methods for solving crack problems using the BIE method.

Field Equations and Numerical Solutions

The boundary-integral equation relates the boundary traction vector $t_i(Q)$[2] to the boundary displacement vector $u_i(Q)$ and, as derived in [2], is given by

$$u_j(P)/2 + \int_{\partial R} T_{ji}(P,Q)u_i(Q)ds(Q) = \int_{\partial R} U_{ji}(P,Q)t_i(Q)ds(Q) \quad (1)$$

In (1), points on the boundary ∂R are denoted by $P(\underline{x})$ and $Q(\underline{x})$ where $\underline{x}$ is a point with coordinates x_1, x_2, x_3.

The second order tensor kernels in (1) derive from the traction vector $t_i^*(Q)$ and the displacement vector $u_i^*(Q)$ for the fundamental solution of a unit load in an infinite medium with shear modulus μ and Poisson's ratio ν, as given by

---

[2] Standard indicial notation is used throughout this report with the range of the subscripts (1,2,3); summation is implied for repeated indices. The comma-index notation denotes partial differentiation. For clarity, some explicit summation is also used.

$$t_i^* = -[(1-2\nu)(\delta_{ji}r_{,k}+ \delta_{kj}r_{,i}-\delta_{ik}r_{,j}) + 3r_{,i}r_{,j}r_{,k}] e_j n_k /$$
$$[8\pi(1-\nu)r^2] = T_{ji}e_j \quad (2)$$

$$u_i^* = [(3-4\nu)\delta_{ij}+r_{,i}r_{,j}] e_j/[16\pi(1-\nu) \mu r] = U_{ji}e_j \quad (3)$$

In (2) and (3), the distance between the points $P(\underline{x})$, $Q(\underline{x})$ is denoted by $r(P,Q)$ where

$$r = r(P,Q) = \sqrt{[(x_{i|Q}- x_{i|P}) (x_{i|Q}-x_{i|P})]} \quad (4)$$

and the derivatives are denoted by

$$r_{,i} = \partial r/\partial x_{i|Q} = (x_{i|Q}-x_{i|P})/r(P,Q) \quad (5)$$

The unit load is applied in each of the three orthogonal directions denoted by e_j in (2) and (3). The unit *outward* normal vector at $Q(\underline{x})$ is denoted n_k in (2).

The stress tensor $\sigma_{ij}(p)$ at any interior point $p(\underline{x})$ is given in [2] as

$$\sigma_{ij}(p) = -\int_{\partial R} S_{kij}(p,Q)u_k(Q)ds(Q) + \int_{\partial R} D_{kij}(p,Q)t_k(Q)ds(Q) \quad (6)$$

The third order tensor kernels in (6) are reported in [2] and are combinations of the first derivatives of (2) and (3).

The solution strategy for a given elastic boundary value problem is first to solve (1) numerically for the unknown boundary data. Then, given the entirety of boundary tractions and displacements, the stresses at any interior point $p(\underline{x})$ are obtained by direct integration of the identity (6). The relation for the stress tensor at a boundary point $P(\underline{x})$ is given in [3]. Further discussion in this section is restricted to the assumption of a piecewise constant approximation to boundary data.

Let the boundary of the body ∂R be divided into N piecewise flat segments (taken herein to be triangles) denoted ∂R_n, n=1,$\cdots$,N. Over each segment assume the traction vector and the displacement vector are constant and denote these values $t_i(n)$ and $u_i(n)$. Application of the approximation to (1) results in the set of algebraic equations

$$u_i(m)/2 + \sum_{n=1}^{N} \Delta T_{ji}(m,n)u_j(n) = \sum_{n=1}^{N} \Delta U_{ji}(m,n)t_j(n) \quad (7)$$

where m=1,$\cdots$,N and

$$\Delta T_{ij}(m,n) = \int_{\partial R_n} T_{ji}(P_m,Q)ds(Q)$$
$$\Delta U_{ij}(m,n) = \int_{\partial R_n} U_{ji}(P_m,Q)ds(Q) \quad (8)$$

In (8) the points P_m are taken to the centroids of the mth boundary segment ∂R_m. The 3N algebraic equations in (7) may be written in matrix notation

$$[A]\{u\} = [B]\{t\} \quad (9)$$

For the displacem t-specified problem, (9) may be solved directly for the unknown traction vectors; for the traction problem, (9) may be solved for the unknown displacement vectors *after* fixing the body against rigid-body motions. The more usual boundary value problem is mixed-mixed. That is, some traction vector components are known for part of the boundary and some displacement vector components are known. At any given boundary segment three boundary data must be known and three boundary data unknown. For such problems (9) has 3N knowns and 3N unknowns; (9) may be solved, after rearrangement of the equations, by standard matrix reduction schemes to obtain the unknowns.

After (9) has been solved, the 3N boundary tractions and the 3N boundary displacements are known. Again replacing the boundary by N piecewise flat segments the interior stresses are calculated from (6) by the quadrature

$$\sigma_{ij}(p) = -\sum_{n=1}^{N} \Delta S_{kij}(p,Q_n) u_k(Q_n) + \sum_{n=1}^{N} \Delta D_{kij}(p,Q_n) t_k(Q_n) \qquad (10)$$

In (10) the kernels are calculated by integration of S_{kij} and D_{kij} as in (8). The kernels in (7) and (10) are calculated in closed form by specifying the location and orientation of the boundary segment.

The validity of the approximations in (7) and (10) has been verified by a number of simple problems [2], by the stress analysis of crack problems in three dimensions [1,3,4], by the stress analysis of some classical sphere problems [5], and in two dimensions by the stress analysis of a plate with an elliptical hole (Inglis problem)[6].

Modelling Crack Geometries

In finite element [7] and finite difference [8] models of cracked bodies the crack plane is usually a plane of symmetry which becomes a boundary for the discrete model. In both of these solution methods the volume is replaced by an internal mesh of finite elements or finite difference points. To model cracks the internal mesh must be highly refined in the crack tip vicinity to be able to adequately represent the stress singularity.

For two dimensional problems where the nature of the stress singularity is known it has been possible to use special elements at the crack tip [9] to reduce the crack modelling problem. For usual finite element solutions in two dimensions and all finite element solutions in three dimensions it is necessary to use special data handling techniques for obtaining values of the SIF, as discussed in [10,11].

Some of the same comments are applicable to BIE models of cracks. It is possible, for example to replace the plane containing the crack with a physical boundary. However, in adopting this model the continuous stress field ahead of the crack is replaced by a series of constant traction boundary segments. To obtain adequate resolution of the singular stresses ahead of the crack many boundary segments are required and, for three dimensions, the problem size becomes too large.

Another similar modelling problem is the need for element resolution near the crack tip. For the BIE method this means that the boundary segment size must be decreased near the crack tip in relation to the need for adequate solution resolution. Values of the SIF may be obtained from the crack opening displacements or transverse stresses ahead of the crack in the same ways as for finite elements [10,11]. These techniques are applied to crack data in the next two sections of this report.

To avoid the need for segments in the body on the crack plane it is possible, with fewer boundary segments, to model the entire boundary and solve for the basic symmetric part of the unknown boundary data. This is the method used most often in this report and has the further advantage of being able to model infinite bodies with uniform far-field tractions.

Assuming that the tractions and displacements vanish at infinity, ∂R in (1) is the remaining finite interior surface, with the normal pointing away from the body. Thus the loading on the finite boundary ∂R must be self-equilibrating. For bodies loaded at infinity the solution is obtained by two steps:

solving the infinite problem with no crack, and solving the infinite problem with uniform stresses applied to the crack; the answer is the sum of the two solutions.

Treating the infinite cracked body in this manner, a problem is encountered in evaluating the right hand side of (1). The tractions on the two sides of the crack at a given location must be equal and opposite, for each component of $t_i(Q)$. Let ∂R_1 be one side of the crack and ∂R_2 be the other side of the crack. Then the right hand side of (1) is given by

$$\int_{\partial R} U_{ji}(P,Q)t_i(Q)ds(Q) = \int_{\partial R_1} U_{ji}(P,Q)t_i(Q)ds(Q) + \int_{\partial R_2} U_{ji}(P,Q)t_i(Q)ds(Q) \tag{11}$$

The points $P(\underline{x})$ on ∂R_1 and ∂R_2 are the same points for an ideal crack; furthermore, $U_{ji}(P,Q)$ is independent of the local normal $n_i(Q)$ so that $U_{ji}(P,Q)$ is the same for both integrals on the right side of (11). Using these results plus the antisymmetry of the $t_i(Q)$, (11) can be seen to be zero

$$\int_{\partial R} U_{ji}(P,Q)t_i(Q)ds(Q) = 0 \tag{12}$$

Thus the information as to the boundary loading is lost and solution is not unique, if it exists. It is therefore necessary to come as close to modelling a perfect crack as possible without letting the crack surfaces coincide.

A model for the infinite body is to admit some finite but small separation between the crack faces. The limit on the separation is computer round-off error. The finite separation introduces a new modelling problem: how to join the two plane crack surfaces at the model crack tip? Since the boundary segments are piecewise flat, the segments joined at the crack tip will generally have an included angle less than 180° (See Fig. 1). The model for the boundary-integral equation method reported herein assumes the displacement vector is constant over these segments corresponding to a clamped-clamped boundary condition at the crack tip. From the eigensolution of such a problem [12] it is seen that the stress field near the crack tip has a modelling singularity which may be greater than the inverse-square root singularity for the real problem. This singularity has twice the square root singularity if the included angle ϕ in Fig. 1 is taken to be zero. Only as the included angle approaches 180° does the modelling singularity disappear.

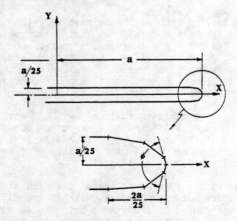

Fig. 1 Crack Tip Closure Model Detail

That the modelling singularity is important was discussed in [3] where the stress results did not achieve the expected inverse-square root behavior. A suitable boundary segment model for crack geometries is to join the two crack surfaces with ellipses, maintaining the included angle at the crack tip near 180°. This model has the disadvantage of introducing a finite stress concentration factor (SCF) at the model crack tip, rather than the theoretically infinite SCF. As the aspect ratio of the elliptical closure is increased (See Fig. 2) the SCF also increases. However, the size of the first segment at the model crack tip must decrease correspondingly to maintain the included angle near 180°. The model selected for the current study maintains a uniform crack face separation of 2/25th of the crack radius (or semi-length) with a 2:1 elliptical closure as shown in Fig. 1. Numerical results for BIE models of cracks are discussed in the next two sections.

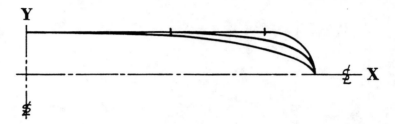

Fig. 2 Elliptical Crack Closure Schemes (Exaggerated Crack Face Separation)

EVALUATION OF THE STRESS INTENSITY FACTOR

In all numerical models of cracked bodies the key is the rapid and accurate extraction of information concerning the magnitude of the stress intensity factor for different geometries, loading and, in three dimensions, for different locations in the body. For two dimensional problems where the inverse-square root behavior is known to occur, special crack-tip-methods have evolved such as in [9] which allow for direct computation of the stress intensity factor. However, most numerical methods employed to date, including the BIE method, do not have a built-in crack tip behavior. Rather, the nature of the singularity must be extracted by some numerical scheme.

Several reports [10,11] have dealt with different schemes that have been developed for FE solutions. The basic schemes begin by seeking limits which, for the planar problem, come from [13]

$$\lim_{r \to 0} \sigma_y \sqrt{r} = K_I/\sqrt{2\pi}$$

$$\lim_{r \to 0} v/\sqrt{r} = 4(1-\nu^2)K_I/E\sqrt{2\pi}$$

(13)

where r is the distance from the crack tip, σ_y and v represent typical transverse stress and displacement components, and K_I represents the mode I stress intensity factor for a symmetrically loaded crack in tension. The art in finding K_I is the use of different curve-fits to the data $\sigma_y \sqrt{r}$ and $v/\sqrt{r}$ vs. r; the most popular is a least-squares fit to the straight line,

$$f(r) = aK_I + b\sqrt{r}$$

(14)

where b is sometimes taken to be zero, and then evaluating f(r) for r = 0.

The scheme used for the BIE method is the same and is discussed in greater detail below. The philosophy employed was to develop a standard scheme based on extensive investigation of the planar crack results, and to verify that scheme for the three dimensional model of the penny-shaped crack. The verified scheme is then employed in the next section to evaluate the SIF for a semi-circular surface flaw in a large body.

Two Dimensional Center-Cracked Tension Specimen

The two dimensional, planar crack problem was studied extensively to develop a consistent scheme for predicting the mode I stress intensity factor K_I for a symmetric, center-cracked, tension specimen as shown in Fig. 3(a). The two dimensional problem was particularly useful because of the small problem size and computer time[3] required to study different models, and because K_I is well-known for finite specimens [14].

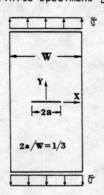
(a) Center-Crack Specimen

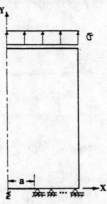
(b) Model 2D-3

Fig. 3 Planar Crack Geometry

Three representations of the specimen in Fig. 3(a) were investigated. Two are models based on the use of a finite crack face separation with an elliptical closure; the first (2D-1) uses ten segments to model the symmetric portion of the crack face and the second (2D-2) uses twenty-two. The third model (2D-3) is shown in Fig. 3(b) and uses the symmetry plane $x_2 = 0$ for an auxiliary surface. Ten segments are used to model the crack face and fifteen segments are used to model the auxiliary surface. In models 2D-1,2 the crack face separation is 2a/25; in model 2D-3 no separation was used.

The crack surface opening displacements are plotted in Fig. 4. It is seen that the displacements follow the predicted square root-distance relation predicted from the first term expansion given in [13] and shown in the second equation of (13). The straight line corresponding to the exact solution for 2a/w = 1/3 is taken from the value of K_I in [14]. The straight line fit to the 2D-1 model is high by about 14%; the fit to the 2D-2,3 models is high by about 6%.

The interior stresses computed for all three planar models are plotted in Fig. 5 together with the one term solutions taken from the displacement data in Fig. 4. Several conclusions may be drawn from Fig. 5. First, the stresses for 2D-1,2 do not correspond to the one term expansions and, further, the stresses do not fall on a straight line. The reason for the behavior of the stress data is in the model of the crack with finite separation. Model 2D-2 was found to have converged to a stable result and might be termed the exact answer for the 2D-1,2 model. Model 2D-3 had the same number of crack face segments as 2D-1 yet had the characteristic inverse square root stress behavior and a much better value of K_I than 2D-1. It is concluded that the stress data for a crack model with finite separation is not totally satisfactory for a prediction of K_I.

---

[3] A summary of some of the key problem sizes and solution times is given in the Appendix.

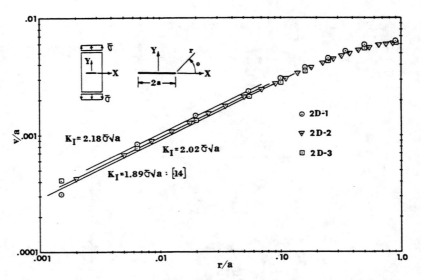

Fig. 4 Displacement Data for Planar Models ($\theta=0°$)

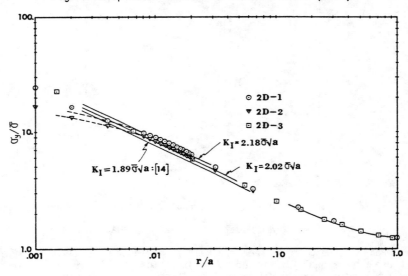

Fig. 5 Stress Data for Planar Models ($\theta=0°$)

The use of displacement data is not as simple for the three dimensional surface flaw BIE model and we therefore seek a relation between the data in Fig. 4 and Fig. 5. The stress data in Fig. 5 is seen to begin to diverge for $r/a < 0.008$, due to the finite SCF and the accuracy of the models 2D-1,2. The approach to predicting K_I from stress data expressed in (13) was used to investigate the stress data for small $r/a > 0.008$. The results for 2D-1,2 are shown in Fig. 6 together with the results from Fig. 5. It was found that the values of $\sigma_y\sqrt{r}$ reached a maximum value for $r/a > 0.013$, were constant out to $r/a = 0.016$ and then dropped steadily. It was also found for 2D-2 that the value of $\sigma_y\sqrt{r}$ dropped uniformly for $r/a < 0.008$ due to the presence of the finite SCF at the crack tip. It is to be expected that similar results will obtain for the equivalent three dimensional BIE models. A least-squares fit to the data for the range $0.008 \leq r/a \leq 0.013$ extrapolated to $r = 0$ gives good agreement between the stress and displacement data as shown in Fig. 6. It is concluded that to obtain a value of predicted K_I from the stress data, that agrees with the displacement data, the stresses for the range $0.008 \leq r/a \leq 0.013$ should be fit by a straight line according to (14) to obtain $\sigma\sqrt{r}$ at $r = 0$.

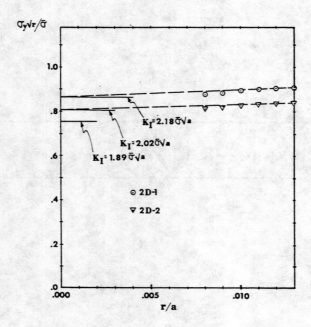

Fig. 6 K_I Extrapolation from Stress Data for Planar Models

Three Dimensional Model of the Penny-Shaped Crack

The problem of a penny-shaped crack in an infinite body was next modelled by the three dimensional BIE method to evaluate the ability of the BIE method to predict the stress intensity factor for three dimensional problems. For such a problem the finite crack face separation model must be used. A set of boundary segments equivalent to the 2D-1 model was selected and is shown, for a symmetrical octant of the surface, in Fig. 7. The separation distance was again $2a/25$ with an elliptical closure with aspect ratio 2:1. The problem size and computer time requirements for this problem are discussed in [1] and are summarized in the Appendix.

The crack opening displacement data is plotted in Fig. 8 as a function of the distance from the crack tip. As in the two dimensional data, the displacements for boundary segments beyond the first one follow the characteristic slope allowing for a direct calculation of K_I. The predicted K_I is about 17% high which corresponds quite closely with the two dimensional results.

The transverse stresses, σ_z, are plotted in Fig. 9 against the distance from the crack tip. Straight lines representing the first term of the stress

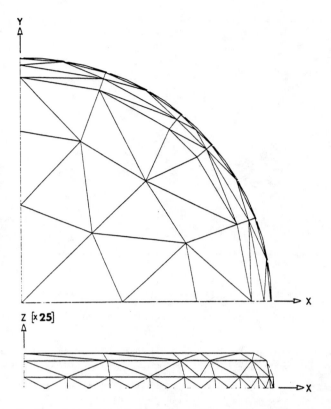

Fig. 7 Boundary Segments for Three Dimensional Crack Surface

expansion at small crack tip distance for two values of K_I are also plotted in Fig. 8. The lower line is for the exact solution taken from [15]

$$K_I = 2\bar{\sigma}\sqrt{\pi a}/\pi \tag{15}$$

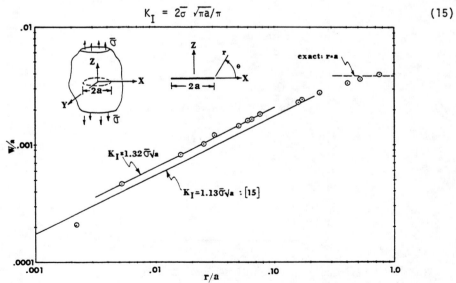

Fig. 8 Displacement Data for Penny-Shaped Crack Problem ($\theta=0°$)

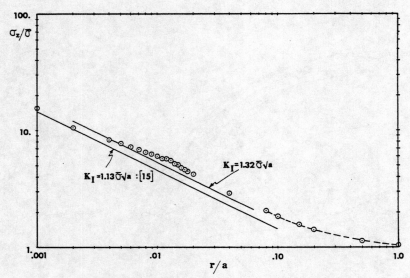

Fig. 9 Stress Data for Penny-Shaped Crack Problem ($\theta=0°$)

The upper line in Fig. 9 corresponds to a best fit to the displacement data in Fig. 8.

The stresses in Fig. 9 show the same behavior as for the two dimensional models 2D-1,2. That is, the stresses for $r/a < 0.008$ begin to flatten to a finite SCF (approx 8.3) for $r/a = 0$, but then rise for $r/a < 0.004$ due to the modelling singularity discussed above. It is also seen from the stress data that the transverse stress does not fall on a straight line of the characteristic inverse-square root slope.

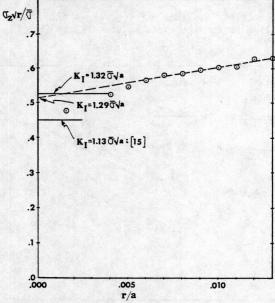

Fig. 10 K_I Extrapolation from Stress Data for Penny-Shaped Crack Problem

The transverse stresses were then plotted in Fig. 10 according to the scheme in (13). The two dimensional stress data in Fig. 6 showed that a straight line fit to the data for the range $0.008 \leq r/a \leq 0.013$ gave the value of K_I for $r/a = 0$ that best agrees with the value predicted from displacements. Fitting a straight line over this range, as shown in Fig. 9, gives a value of K_I that again compares very well with value from the displacements, in the case of the penny-shaped crack. It is therefore concluded that when displacement data cannot be conveniently used (as is the case for the surface flaw model discussed in the next section) the stress data over the range $0.008 \leq r/a \leq 0.013$ may be used to predict K_I for the three dimensional crack model.

SEMI-CIRCULAR SURFACE FLAW

The preceding section developed a basis for predicting the stress intensity factor from internal stresses or boundary displacements for a particular model of a circular crack (Fig. 7). The same model of the crack with separation is used to study the semi-circular surface flaw in an essentially infinite body. By using the same crack model it is possible to apply the same basis for predicting K_I, discussed in the preceding section. Surface flaw geometries other than the one reported herein are currently being studied using the same crack modelling philosophy.

The surface flaw was imbedded in a cube as shown in Fig. 11. The cube dimensions were chosen so as to isolate the crack from all but the front face (x=0). The adequacy of the cube size was verified by solving the additional problem of a penny-shaped crack in a cube of dimensions 6a:6a:6a. The stresses for the finite cube model were indistinguishable from the infinite body stress data discussed in the preceding section. The surface flaw data reported herein were obtained for uniform normal displacement applied to the faces $z = \pm H/2$.

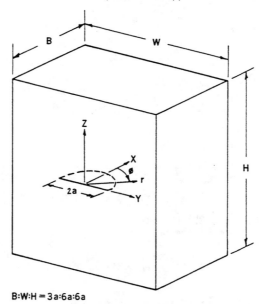

B:W:H = 3a:6a:6a

Fig. 11 Semi-Circular Surface Flaw Geometry

The displacement results for the stress free crack surface are no longer axisymmetric as they were for the penny-shaped crack. As the unknown boundary data are computed at the boundary segment centroids, the crack face displacements are not functions of radial location only. Thus it is not possible to obtain K_I from the displacements as was done for the penny-shaped crack model.

Future effort will be directed towards more adequate boundary segment models.

The stress intensity factor must therefore be computed from the internal transverse stresses on the plane of the crack. The scheme is to fit a straight line to the data $\sigma\sqrt{r}$ for the range of crack tip distance $0.008 \leq r/a \leq 0.013$ and taking the limiting value for $r/a = 0$. Sufficient internal stress points were selected for various radial and angular positions and are plotted in Fig. 12. The results for $0° \leq \phi \leq 60°$ follow the same slope as for the penny-shaped crack in Fig. 10. The value of K_I at $\phi=0°$ is about 3% higher than the penny-shaped crack, a result that agrees well with previous results [16]. The value of K_I increases by 16% at $\phi=60°$ relative to the $\phi=0°$ value. This increase is much more than obtained in [16]. A possible reason for this discrepancy is now noted.

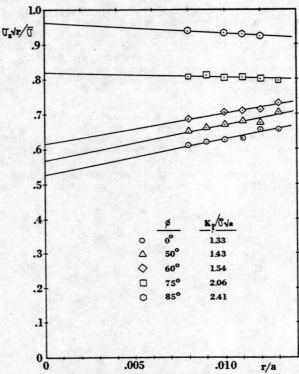

Fig. 12 K_I Extrapolations from Stress Data for Surface Flaw Problem

Between $\phi=60°$ and $\phi=75°$ the stress data in Fig. 12 change their character. Assuming that the slope of the data in Fig. 10 is a characteristic of the geometry used to model a crack, one concludes that the data shown for $\phi \geq 75°$ includes a singularity which is greater than the inverse-square root of the ideal crack. The data in Fig. 13 illustrate the change in stress behavior in another way. For $\phi < 75°$ the ratio of σ_z to its value for $\phi=0$ is the same for all three radial locations and is therefore the same as the variation in predicted K_I. The data for $\phi \geq 75°$ have a different radial dependence than for the $\phi=0°$ data.

The basis for the change in characteristic behavior of the stresses is shown in Fig. 14 where the values of the plane strain factor

$$\alpha = \sigma_\phi / [\nu(\sigma_r + \sigma_z)] \tag{16}$$

are plotted against tangential and radial locations. Values of $\alpha=1$ correspond to plane strain; $\alpha=0$, to plane stress. The contours of constant-α are uniform

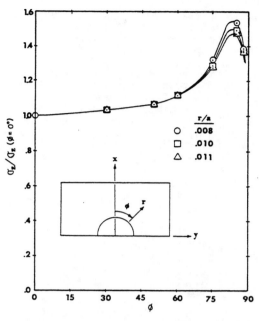

Fig. 13 Transverse Stresses at Different Radial Locations Ahead of Surface Crack

for r/a < 0.020 out to about $\phi=60°$. The contours for $\phi > 60°$ and r/a < 0.020 show a marked influence of the free surface-surface flaw intersection. It is the influence of the free surface which accounts for the change in the slopes of the data in Fig. 12.

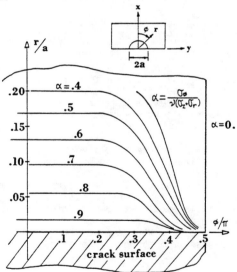

Fig. 14 Lateral Constraint Contours in Crack Plane for Surface Flaw Problem

The important question then becomes what are the values of K_I for $\phi > 60°$, assuming that the exact stress field due to the free-surface-crack intersection is still inverse-square root? The values of K_I reported in [1] were taken from the stresses at r/a = 0.010 and predicted over a 50% growth in K_I from $\phi=0°$ to the maximum K_I near $\phi=90°$. If the extrapolation in Fig. 12 is used to obtain K_I, the growth in K_I is more like 80%.

To obtain a different estimate of the size of K_I near the free surface the crack surface opening displacements were studied. Some of the displacement data are plotted in Fig. 15. The open data points are for segment centroids closest to $\phi=90°$; the closed data points are for centroids closest to $\phi=0°$. Plotted along with the data are two curves: one is an ellipse; the other is a parabola. Both curves have a finite value of curvature at R/a = 0 which is inversely proportional to the displacement predicted K_I, squared.

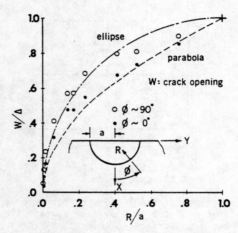

Fig. 15 Crack Opening Displacements for the Surface Flaw

The kinematics of crack opening require a displaced shape which has a finite root radius at the crack tip. Further, the crack profile for $\phi=90°$ must have zero slope at R/a = 1. The crack profile for $\phi=0°$ must only have the same displacement at R/a = 1 as the $\phi=90°$ profile. The simplest functions to satisfy these requirements are the ellipse ($\phi=90°$) and the parabola ($\phi=0°$). The displacement data in Fig. 15 are seen to behave much as the two curves. Since the curvature of the ellipse at R/a = 0 is one-half that of the parabola, the kinematic argument would predict a value of K_I at the surface about 40% higher than for $\phi=0°$. This result agrees most closely with the stress ratio data for r/a ≥ 0.010.

The discussion above points to two very important questions in the numerical study of surface flaws. The first is the need for an analytical investigation of the crack-free surface intersection in order to know the stress singularity for various values of ϕ. The second is the need to concentrate numerical results on the use of crack surface normal displacements to predict K_I. Not only does the displacement data behave more uniformly for the numerical studies reported herein, but its use eliminates the complications of the free surface effect on stresses. When the complications of including plasticity are included the stress problem becomes even more complex and a kinematical model may offer the only basis of a fracture criterion.

CONCLUSIONS

The data reported establishes a basis for the use of the BIE method to model three dimensional, elastic crack problems, including the surface flaw problem. The methods for estimating the stress intensity factor were developed on a two dimensional planar model, verified on the three dimensional model of the penny-shaped crack and applied to a semi-circular surface flaw. While the known errors for the first two problems were on the order of 14-17%, the use of a consistent model kept the error about constant in going from two dimensions to three dimensions. The stress data for the rise in stress intensity factor at the free surface of the surface flaw problem was found to be consistent with a simple kinematical model of crack opening.

On the other hand, the report also has shown some of the limitations of the BIE method. First, being a numerical technique, precise statements on singularities are not possible. Further, the BIE method does not allow for a practical boundary segment model of a perfect crack, but must model a flaw with a finite crack face separation. The number of boundary segments needed to obtain better accuracy for three dimensional problems is still prohibitive on today's computers.

Because of the finite separation model of the crack a special method had to be devised to use the internal stress data to predict the stress intensity factor. This method worked well enough for all problems except for the surface flaw, and then worked well except near the free surface. The problem of what is the stress intensity factor for a three dimensional problem was shown to be difficult and is usually not fully discussed in numerical studies. It was also found that the free crack surface displacements are probably the best source of stress intensity factor information. However, due to the current boundary segment model these displacements are not conveniently found.

Finally, the present BIE model is too crude. The boundary data model is not as accurate as a higher-order model (e.g. linear variation) and the solutions are limited to elasticity. Work is now underway to improve the model, add plasticity, and to study a broader range of surface crack geometries and loadings. This work will, of course, be reported as it is accomplished.

ACKNOWLEDGEMENTS

The author gratefully acknowledges the financial support of the U. S. Army Research Office — Durham under grant DA-ARO-D-31-124-72-G3; the advice, comments, and encouragement of Dr. J. L. Swedlow; the computer time supplied by Carnegie-Mellon University; and the careful preparation of the manuscript by Ms. Kathleen Sokol.

REFERENCES

[1] Cruse, T. A., "Application of the Boundary-Integral Equation Method to Three Dimensional Stress Analysis," *Journal of Computers and Structures* (to appear).

[2] Cruse, T. A., "Numerical Solutions in Three Dimensional Elastostatics," *International Journal of Solids and Structures*, Vol. 5, Dec. 1969, pp. 1259-1274.

[3] Cruse, T. A., and Van Buren, W., "Three-Dimensional Elastic Stress Analysis of a Fracture Specimen with an Edge Crack," *International Journal of Fracture Mechanics*, Vol. 7, No. 1, March 1971, pp. 1-15.

[4] Cruse, T. A., "Lateral Constraint in a Cracked, Three Dimensional Elastic Body," *International Journal of Fracture Mechanics*, Vol. 6, No. 3, Sept. 1970, pp. 326-328.

[5] Cruse, T. A. "Some Classical Elastic Sphere Problems Solved Numerically by Integral Equations," *Journal of Applied Mechanics*, Vol. 39, No. 1, March 1972, pp. 272-274.

[6] Cruse, T. A., and Swedlow, J. L., "Interactive Program for Analysis and Design Problems in Advanced Composites Technology," Technical Report AFML-TR-71-268, Dec. 1971.

[7] Swedlow, J. L. "Initial Comparisons Between Experiment and Theory of the Strain Fields in a Cracked Copper Plate," *International Journal of Fracture Mechanics*, Vol. 5, No. 1, March 1969, pp. 25-31.

[8] Ayres, D. J., "A Numerical Procedure for Calculating Stress and Deformation Near a Slit in a Three-Dimensional Elastic-Plastic Solid," *Engineering Fracture Mechanics*, Vol. 2, No. 2, Nov. 1970, pp. 87-106.

[9] Byskov, E., "The Calculation of Stress Intensity Factors Using the Finite Element Method with Cracked Elements," *International Journal of Fracture Mechanics*, Vol. 6, No. 2, June 1970, pp. 159-168.

[10] Chan, S. K., Tuba, I. S., and Wilson, W. K., "On the Finite Element Method in Linear Fracture Mechanics," *Engineering Fracture Mechanics*, Vol. 2, No. 1, July 1970, pp. 1-18.

[11] Oglesby, J. J., and Lomacky, O., "An Evaluation of Finite Element Methods for the Computation of Elastic Stress Intensity Factors," Report 3751, Naval Ship Research and Development Center, Bethesda, Maryland, Dec. 1971.

[12] Williams, M. L., "Stress Singularities Resulting From Various Boundary Conditions in Angular Corners of Plates in Extension," *Journal of Applied Mechanics*, Vol. 19, No. 4, Dec. 1952, pp. 526-528.

[13] Paris, P. C., and Sih, G. C., "Stress Analysis of Cracks," in *Fracture Toughness Testing and its Applications*, ASTM STP 381, American Society for Testing and Materials, Philadelphia, 1969.

[14] Brown, W. F., Jr., and Srawley, J. E., *Plane Strain Crack Toughness Testing of High Strength Metallic Materials*, ASTM STP 410, American Society for Testing and Materials, Philadelphia, 1966.

[15] Sneddon, I. N., and Lowengrub, M., *Crack Problems in the Classical Theory of Elasticity*, J. Wiley & Sons, New York, 1969.

[16] Smith, F. W., Emery, A. F., and Kobayashi, A. S., "Stress Intensity Factors for Semicircular Cracks, Part 2 — Semi-Infinite Solid," *Journal of Applied Mechanics*, Vol. 34, No. 4, Dec. 1967, pp. 953-959.

APPENDIX

Key Computer Usage Data

Model	Coeff. Matrix (Words)	No. Sym. Planes	No. Stress Points	1108 CPU Times (sec.) Boundary Solution	1108 CPU Times (sec.) Interior Solution
2D-1	1,444	2	21	3.9	7.1
2D-2	4,096	2	21	8.8	13.0
2D-3	4,900	1	11	6.1	0.
Penny	42,849	3	25	1070.	1058.
Surface	97,344	2	36	1330.	600.

Three-Dimensional Elastic Stress Analysis of a Fracture Specimen with an Edge Crack

T. A. CRUSE
Department of Mechanical Engineering, Carnegie-Mellon University, Pittsburgh, Pa. (U.S.A.)

W. VANBUREN
Mechanics Dept., Westinghouse Research Laboratory, Pittsburgh, Pa (U.S.A.)

(Received December 22, 1969 and in revised form January 26, 1970)

ABSTRACT
A numerical stress analysis of an elastic three-dimensional specimen similar to the compact tension specimen used in fracture investigations is presented. The numerical results are achieved using singular integral equations which are analogous to Green's boundary formula in potential theory. The analysis yields details of the stresses near the crack tip and clearly shows their three-dimensional character. Some results are also given to indicate the influence of thickness and Poisson's ratio on the stresses.

1. Introduction

This paper presents details of the stress analysis of a fracture specimen similar in geometry to specimens used in experimental work. The geometry and loading are indicated in Fig. 1. Two features of the specimen are significant. First, the specimen contains a sharp crack with a length of the same order of magnitude as the other dimensions of the specimen. Second, the thickness of the specimen is finite. Thus, the stress field in the specimen is highly three-dimensional and will show considerable dependence on the thickness variable, z.

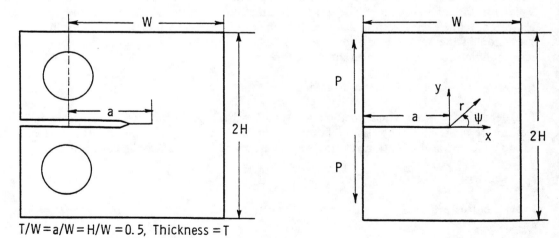

Figure 1. Comparison of fracture test specimen (left) with specimen used in the analysis (right).

The stresses arising from geometrical discontinuities have been extensively investigated. A reasonably current review [1] shows that much of the activity has been concerned with the three-dimensional character of the stress field. However, when the geometric discontinuity is near free boundaries the only problems that have been successfully attacked analytically are those with circular, cylindrical holes [2]. More recent attempts at analytic results for sharp cracks in finite bodies using eigenfunction expansions [3] and variational methods [4] yield only partial results. In these reports certain assumptions have been made on the nature of the thickness variation of the stresses and the results are fairly qualitative. Strictly numerical attempts [5–6] at the solution of the three-dimensional problem have also been largely unsuccessful. The compromises necessary to fit the problem into today's computers seriously limit the amount of detail available in the vicinity of the stress concentration.

It is no surprise then that most analytic results applied to the specimen in Fig. 1 are based on two-dimensional generalized plane stress or plane strain theories. A thorough summary of all the literature is not appropriate here since the results for many different geometries are in the literature [7]. Much of the two-dimensional work derives from the use of eigenfunction expansions [8,9], the complex variable formulation [10] and boundary point collocation [11, 12].

Since the specimen in Fig. 1 is not vanishingly thin and since no interior points are "far" from the unloaded surfaces, neither of the planar theories is appropriate. Proper interpretation of experimental results and the development of an adequate strength theory depend on obtaining details of the three-dimensional character of the stress field. It is the purpose of this paper to report on the application of a new analytical method to the solution of the three dimensional problem of a cracked, finite body. The analysis method has been applied extensively to the solution of two-dimensional problems[13–16] and recently has been extended to three-dimensional problems[17].

This new solution method, referred to here as the direct potential method, utilizes the numerical solution of linear integral equations which are analogous to Green's boundary formula in potential theory. In the direct potential method singular integral equations are written directly in terms of the physical boundary tractions and displacements. There is no need to introduce physically non-significant potentials or to use complex variables. The advantage to the numerical analyst using this procedure, aside from its generality, is the two-dimensional nature of the numerical analysis: all approximations and solutions are in terms of the boundary quantities. Thus the analyst need solve equations only on the surface of a body, *not throughout its volume*.

A review of the equations and the numerical procedures used in the direct potential method is given in the second section. The third section contains results of the stress analysis of the fracture specimen with comparisons to the other theories. The fourth section includes the discussion of the results and some conclusions on the three dimensional behavior. Some new analytic details are given in the Appendix.

2. Review of the Direct Potential Method

The details of the analytical method appear elsewhere [16], but are summarized in this section for reference*. The basic relationship utilized in the direct potential method is the boundary constraint equation relating surface tractions t_i to surface displacements u_i:

$$\tfrac{1}{2}u_i(P) + \int_{\partial R} u_j(Q) T_{ij}(Q, P) \mathrm{d}S(Q) = \int_{\partial R} t_j(Q) U_{ij}(Q, P) \mathrm{d}S(Q). \qquad (1)$$

In (1) the displacements u_i are those which satisfy Navier's equations in the absence of body forces:

$$\frac{1}{1-2\nu} u_{i,ij} + u_{j,ii} = 0 \qquad (2)$$

where ν is Poisson's ratio. The tractions t_i are related to the stresses σ_{ij} and the displacement gradients through Hooke's Law and the unit *outward* normal n_i as

$$t_i = \sigma_{ij} n_j = \left[\frac{2\mu\nu}{1-2\nu} \delta_{ij} u_{m,m} + \mu(u_{i,j} + u_{j,i}) \right] n_j \qquad (3)$$

* The notation used in this paper is the usual Cartesian tensor notation with implied summation on repeated indices. The range of the indices is 1, 2, 3. Partial differentiation with respect to the independent variables is denoted by the comma-index notation.

where μ is the shear modulus of the material.

The second order tensors, $U_{ij}(P, Q)$ and $T_{ij}(P, Q)$ are the jth components of the displacement and traction vectors at the point Q for Kelvin's problem of a unit point load in the x_i direction applied at the point P. They are given by

$$U_{ij}(Q, P) = \frac{1}{4\pi\mu}\left(\frac{1}{r}\right)\left[\frac{3-4\nu}{4(1-\nu)}\delta_{ij} + \frac{1}{4(1-\nu)}r_{,i}r_{,j}\right] \tag{4}$$

and

$$T_{ij}(Q, P) = -\frac{1-2\nu}{8\pi(1-\nu)}\left(\frac{1}{r^2}\right)\left[\frac{\partial r}{\partial n}\left(\delta_{ij} + \frac{3}{1-2\nu}r_{,i}r_{,j}\right) - n_j r_{,i} + n_i r_{,j}\right] \tag{5}$$

where the normal is evaluated at the point Q. The distance between the two points P, Q on the surface ∂R of the body R is given in terms of their coordinates x_{Qi} and x_{Pi} by

$$r = \sqrt{[(x_{Qi} - x_{Pi})(x_{Qi} - x_{Pi})]} \tag{6}$$

where P, Q are not summed. In (4) and (5) the differentiation is with respect to the point $Q(x)$, that is

$$r_{,i} = \frac{\partial r}{\partial x_{Qi}} = \frac{1}{r}(x_{Qi} - x_{Pi}). \tag{7}$$

The boundary conditions needed for a well-posed boundary value problem are given for the displacements and tractions respectively on ∂R as

$$\begin{aligned} u_i(Q) &= q_i & Q \in \partial R(u_i) \\ t_i(Q) &= \sigma_{ij}(Q) n_j(Q) = p_i, & Q \in \partial R(t_i) \end{aligned} \tag{8}$$

where $\partial R(u_i)$ and $\partial R(t_i)$ are the portions upon which the displacements and tractions are prescribed. When these boundary conditions are applied to (1) the unknown tractions and/or displacements may be calculated by a suitable numerical technique. Eq. (1) is strongly suggestive of Fredholm equations of the first and second type if displacements or tractions are everywhere specified. However, the equations are singular due to the existence of the term

$$\frac{1}{r^2}(n_j r_{,i} - n_i r_{,j}) \tag{9}$$

in (1). That is, it can be shown that

$$\lim_{Q \to P}[r^2(P, Q) T_{ij}(Q, P)] \neq 0 \tag{10}$$

whereas a limit of zero is required for Fredholm kernels. The first boundary integral in (1) therefore must be interpreted in the sense of the Cauchy Principal Value. It has been shown [18] however, that the integral equations (1) are regular for all admissible values of Poisson's ratio and that the index of the operator $K[u_j]$ in (1) is zero. Therefore, the Fredholm alternatives apply and all normal problems are soluble.

Once the solution for the surface quantities is completed the stresses may be generated at any desired interior point $p(x)$ by the following relation

$$\sigma_{ij}(p) = -\int_{\partial R} u_k(Q) S_{kij}(Q, p) dS(Q) + \int_{\partial R} t_k(Q) D_{kij}(Q, p) dS(Q).$$

The tensors S_{kij} and D_{kij} are given in [17]. Eq. (11) is determined from Somigliana's identity for the displacements and is discussed in greater detail in the Appendix. The determination of $\sigma_{ij}(P)$ where $P(x)$ is a boundary point is also given in the Appendix.

Analytic solutions to (1) are not available for problems with non-simple boundary shapes. The success in the use of the direct potential method is due to the fact that (1) may be replaced by a set of algebraic equations by assuming the boundary data can be replaced by a set of discrete values. That is, the boundary is divided into N triangles, ∂R_n, $n = 1, 2, ..., N$ and the displacements and tractions are assumed to be constant on each of these. Then (1) becomes

$$\tfrac{1}{2}u_j(P_m) + \sum_{n=1}^{N} u_i(Q_n)\Delta T_{ji}(P_m, Q_n) = \sum_{n=1}^{N} t_i(Q_n)\Delta U_{ji}(P_m, Q_n) \tag{12}$$

where

$$\Delta T_{ij}(P_m, Q_n) = \int_{\partial R_n} T_{ij}(P_m, Q)\,\mathrm{d}S(Q)$$
$$\Delta U_{ij}(P_m, Q_n) = \int_{\partial R_n} U_{ij}(P_m, Q)\,\mathrm{d}S(Q)\,. \tag{13}$$

Eq. (12) may be solved for the unknown values of the displacements and/or tractions by standard solution methods for algebraic equations. When these values are found the interior stresses are given by

$$\sigma_{ij}(p) = -\sum_{n=1}^{N} u_k(Q_n)\Delta S_{kij}(Q_n, p) + \sum_{n=1}^{N} t_k(Q_n)\Delta D_{kij}(Q_n, p)\,. \tag{14}$$

All integrations for ΔU, ΔT, ΔS and ΔD are known exactly in terms of the coordinates of the corners of the surface triangles, ∂R_n. Any number of interior solutions may be made once the boundary solution is obtained. Since the solution for $\sigma_{ij}(p)$ is performed at preselected points the analyst may concentrate on particular areas of interest and is not burdened with complete field solutions.

3. Stress Analysis Results

The solution method outlined in the previous section has been applied to the fracture specimen shown in Fig. 1. Although a variety of surface element refinements were investigated the results in this section are for two cases. The element arrays for Case I (128 triangles) and for Case II (488 triangles) are shown in Fig. 2. The influence of the element array will be noted when appropriate. The numerical results will be for Case II except where specifically noted.

Before detailing the results for the three-dimensional stress analysis it is appropriate to review the results from the two-dimensional theory. The stress field near the crack tip has been found for the two-dimensional theory to be singular [9]. That is, the stresses approach infinity in some fashion depending on geometry and loading as the crack tip is approached. While this result is not physically possible, the strength or intensity of the singularity has been shown to be of practical value in fracture mechanics. The results [7] for plane strain are given below:

$$\sigma_x = \frac{K_I}{\sqrt{(2\pi r)}} \cos\frac{\psi}{2}\left[1 - \sin\frac{\psi}{2}\sin\frac{3\psi}{2}\right] \tag{15a}$$

$$\sigma_y = \frac{K_I}{\sqrt{(2\pi r)}} \cos\frac{\psi}{2}\left[1 + \sin\frac{\psi}{2}\sin\frac{3\psi}{2}\right] \tag{15b}$$

$$\tau_{xy} = \frac{K_I}{\sqrt{(2\pi r)}} \sin\frac{\psi}{2}\cos\frac{\psi}{2}\cos\frac{3\psi}{2} \tag{15c}$$

$$\sigma_z = v(\sigma_x + \sigma_y) \text{ (plane strain)}; \; \sigma_z = 0 \text{ (plane stress)} \tag{15d}$$

$$u = \frac{K_I}{\mu}\sqrt{(r/2\pi)}\cos\frac{\psi}{2}\left(1 - 2v + \sin^2\frac{\psi}{2}\right) \tag{15e}$$

$$v = \frac{K_I}{\mu}\sqrt{(r/2\pi)}\sin\frac{\psi}{2}\left(2 - 2v - \cos^2\frac{\psi}{2}\right) \tag{15f}$$

$$w = 0 \text{ (plane strain)}; \; w \neq 0 \text{ (plane stress)}\,. \tag{15g}$$

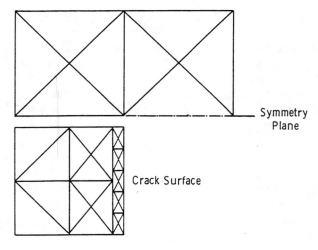

a. Case I - 128 Triangles

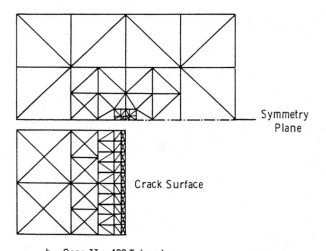

b. Case II - 488 Triangles

Figure 2. Surface triangle arrangements.

The stresses may be combined to obtain equations for the octahedral normal, or mean stress θ, and the octahedral shear stress τ_0:

$$\theta = \frac{1+v}{3} \frac{K_I}{\sqrt{(2\pi r)}} \cos \frac{\psi}{2} \tag{16a}$$

$$\tau_0 = \frac{\sqrt{2}}{3} \frac{K_I}{\sqrt{(2\pi r)}} \cos \frac{\psi}{2} \left[1 + 3 \sin^2 \frac{\psi}{2} - 4v(1-v)\right]^{1/2}. \tag{16b}$$

The stress intensity factor K_I appropriate to the type of loading and to the specimen geometry has been calculated for the two-dimensional problem using the boundary collocation method [11, 12]. The value of K_I for the geometry in Fig. 1 is 14.3 $(P\sqrt{a}/TW)$. In the data that follows in this section this value of K_I is used to depict the plane strain results. The results for the plane stress theory are obtained by replacing Poisson's ratio by $(v/1+v)$ in (15e) and (15f), setting Poisson's ratio to zero in (16) and using the same value of K_I.

The results in Fig. 3 show the calculated values of the crack opening displacement. The results show agreement with the slope that is predicted by (15f). The plane strain results are also shown for comparison and indicate that the predicted value of K_I is about 70% of that calculated for plane strain. The fact that the predicted value is low is compatible with other numerical results [19] obtained using a finite element simulation of the planar case.

The lower value of K_I may be attributed to discretization of the boundary data which stiffens the behavior of the model. This forces lower displacements near the crack tip as well

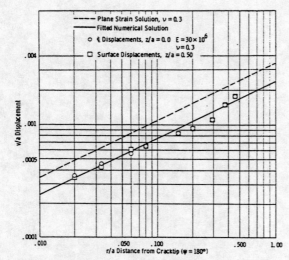

Figure 3. Crack surface displacements.

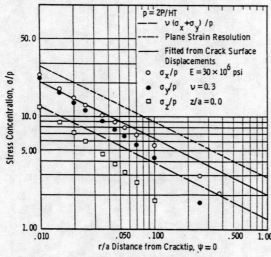

Figure 4. Interior stresses near cracktip.

as for the load application points. Comparison of the K_I for improved element refinements over the *external* surface of the specimen has shown significant increase in K_I. Refinement of the element size near the crack tip has the effect of extending the region of useful results to points closer to the crack tip.

The value of K_I obtained from the displacement data has been used to predict the stresses in the plane $\psi = 0$ ahead of the crack. Comparison of the calculated stresses at the center line with this prediction is shown in Fig. 4. This figure also contains the plane strain prediction of σ_x and σ_y. Fig. 4 shows that the stress σ_z falls below the plane strain values. This is shown by comparing σ_z to the predicted line for $v(\sigma_x+\sigma_y)$. The plane strain condition is approached as the crack tip is approached.

In finite element simulations for this type of problem [19] the stresses and displacements very near the crack tip show a marked decrease in the predicted value of K_I. The direct potential method, on the other hand, shows the opposite trend. This difference is to be expected when considering the theoretical predictions from the eigenfunction expansion [8] for the stresses. In order to have the two nearly adjoining surfaces at the crack tip displace rigidly the stress

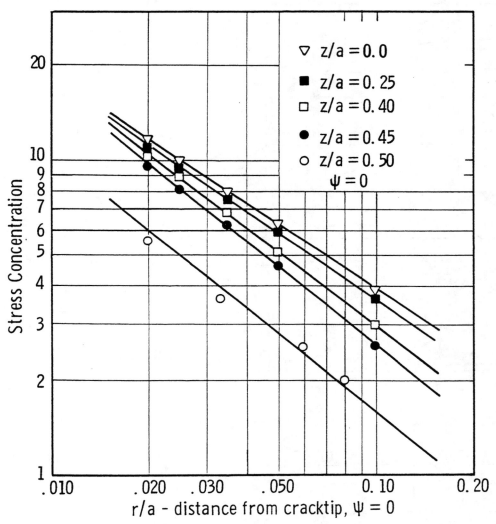

Figure 5. Mean stress variation with z/a

singularity may approach $1/r$ rather than $\sqrt{(1/r)}$. In the direct potential method this strongly singular behavior is forced due to the discretization of the surface displacements as each element has a constant displacement. For points taken closer than $r/a=0.010$ the results have shown a marked increase in slope. The most accurate means of identifying the order of the singularity is to use the displacement data near the crack tip rather than the stress data.

The centerline values of the mean stress in the plane of crack prolongation ($\psi=0$) for various values of z/a is shown in Fig. 5. The results show about a twenty percent growth in the slope of the data as the free surface is approached. Most of the change in slope occurs in the region $0.4 < z/a < 0.5$. The increase in slope indicates that a corner singularity may exist where the crack meets the free surface. The existence of a corner singularity would indicate that the width of the region of plane strain effects ($\sigma_z \cong \nu[\sigma_x + \sigma_y]$) increases as one approaches the crack tip.

The numerical results to date indicate that while there is probably not absolute accuracy in the magnitude of the stresses, the accuracy of the variations of the stresses is very good. This

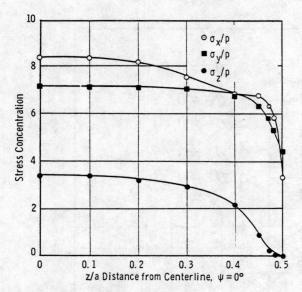

Figure 6. Stress variation through the z/a.

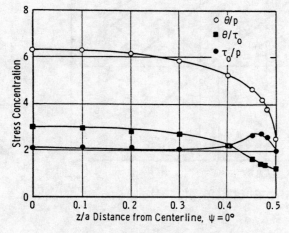

Figure 7. Octahedral stress variation through the thickness.

conclusion derives from the fact that major changes in surface element arrangement change the magnitudes of the stresses but not their relative variations. The remainder of the data emphasizes the variations of the stresses with various coordinates and parameters.

Fig. 6 shows the variation of the three normal stresses with respect to the thickness variable. The results are for $r/a = 0.050$ and are typical results. This value of r/a was chosen for the convenient use of calculated surface stresses at a boundary element. The values of the shear stress are not shown as they are an order of magnitude less than the normal stresses. The results clearly indicate a central region of uniform stress and a boundary layer where the stresses drop significantly to surface values. Thus, in the three-dimensional case, the strains ε_x and ε_y are probably fairly uniform and the decrease in the stresses σ_x and σ_y is probably due to the loss of σ_z. The variation of σ_z shows the correct zero value and zero slope at the free surface. Comparison of these results with photoelastic results [20] indicate strong qualitative agreement on the decrease of surface stresses. The sharpness of the drop is, however, related to the sharpness of the crack and to the distance from the crack.

Fig. 7 shows the same data as Fig. 6 except now calculated in terms of the mean stress, octahedral shear stress, and the ratio of the two. The interior region contains a high mean stress level relative to the octahedral shear stress. At the surface these two stresses are almost equal. Thus, there is a greater tendency to plastic flow near the surface and a greater tendency

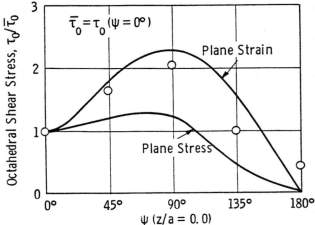

Figure 8. Octahedral shear stress distribution.

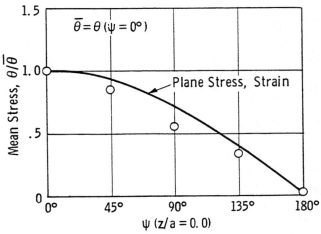

Figure 9. Mean stress distribution.

to dilatation and therefore brittle failure in the central core. This will become more pronounced closer to the crack. As shown in Fig. 10 the ratio of mean stress to octahedral shear stress increases with decreasing distance from the crack tip.

Fig. 8 and Fig. 9 show the variation of the octahedral shear stress and the mean stress with the angle, ψ. The results are compared with the plane strain and plane stress results of (16). The stresses are shown for the plane $z/a = 0$ and for $r/a = 0.020$. The data are in reasonable

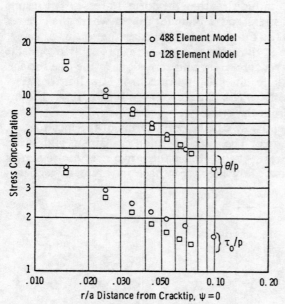

Figure 10. Element refinement comparisons.

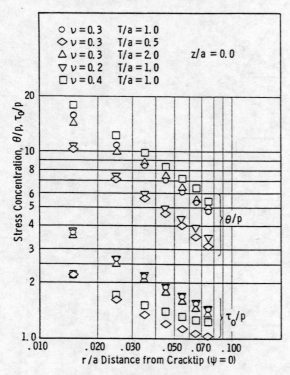

Figure 11. Stress variations with thickness and Poisson's ratio.

agreement with the plane strain results for the octahedral shear stress. The discrepancy at large angles is due to the fact that the one term singularity solution does not contain the remote stresses which are non-zero at $\psi = \pi$. This error is less significant for the mean stress variation. It is indicated from the data that the greatest tendency to flow extends on a plane roughly at $\psi = \pm \pi/2$ from the crack tip. In plane stress [19] this result occurs at $\psi = \cos^{-1} \frac{1}{3} \simeq \pm 70°$.

The last group of results were obtained to indicate the dependence of the stress intensity on the parameter T/a and Poisson's ratio. These results were obtained using the calculated values of mean stress and octahedral stress near the crack tip. Comparisons of the centerline values of the mean stress and the octahedral shear stress for the Case I and Case II models are shown in Fig. 10. The numerical results are in close agreement for small values of r/a. It was therefore justified to use the Case I model to generate the data reported in Fig. 11 through Fig. 13. Additional data to that in Fig. 10 was obtained for $T/a=\frac{1}{2}$ and $T/a=2$ at $\nu=0.3$ and for $\nu=0.2$ and $\nu=0.4$ at $T/a=1$. These results are shown in Fig. 11.

Fig. 12 and Fig. 13 were obtained from the values of the mean stress and the octahedral shear stress for $r/a=0.015$. These figures are normalized on the values of mean stress and octahedral shear stress for the case $T/a=1$ and $\nu=0.3$.

Fig. 12 shows the dependence of the centerline values of mean stress and octahedral shear

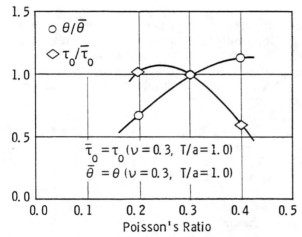

Figure 12. Stress variation with Poisson's ratio.

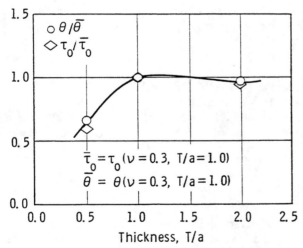

Figure 13. Stress variation with thickness.

stress for the location, $r/a=0.015$. In this case the dependence on Poisson's ratio is made complex as the two-dimensional theory shows explicit dependence on Poisson's ratio. If (16) is used to remove the explicit dependence on Poisson's ratio in both the mean stress and the octahedral shear stress the results show a clear dependence of K_I on Poisson's ratio. This dependence correlates for the two stress components to within 10%. Thus, two conclusions can be made. First, for a given value of Poisson's ratio, the explicit dependence on Poisson's ratio indicated in (16) is valid near the crack tip. However, for the three-dimensional case the stress intensity factor, K_I, appears to be dependent on Poisson's ratio, a result not expected from the planar cases.

Fig. 13 depicts the effect of the thickness parameter on both stresses. A number of conclusions is possible. For $T/a \geqq 1.0$ there appears to be no dependence of these stresses on T/a. For $T/a < 1.0$, however, the dependence on both is pronounced and is the same. This indicates that the variation may be in the stress intensity factor K_I. If the specimen is taken sufficiently thin ($T/a \to 0$) the stress intensity factor should increase, showing a return of the stress ratios in Fig. 13 to 1.0 as $T/a \to 0$. The stress intensity factor would then have to reach a minimum value for $0 < T/a < 0.5$.

4. Summary and Conclusions

The direct potential method has been used to analyze the stresses in a cracked specimen significant to the study of fracture mechanics. The results of this analysis have been summarized in this report. Some aspects of the two-dimensional analytical results have been observed, including the order of the stress singularity and the angular dependence of the stress field near the crack. The numerical results of the three-dimensional solution indicate a value lower than expected of the stress intensity factor.

Some results have been given to indicate the three-dimensional character of the stress field. It was found that the tendency to plastic flow increased in the vicinity of the free surface. It was also indicated that a corner singularity may exist where the crack meets the free surface, thereby reducing the thickness of the region influenced by the free surface. It was also found that the stress intensity factor at the center may be a function of the thickness parameter T/a and of Poisson's ratio.

A number of areas for further numerical studies are indicated. One of these is to analyze further the effect of the surface triangle size on the stress intensity factor. Another significant problem is that of the corner singularity that may exist due to the interaction of the crack and the free surface. Finally, the existence of a minimum value for stress intensity factor as a function of the parameters T/a and ν should yield some significant interpretations for fracture mechanics. These problems and others are currently under investigation.

Acknowledgements

This present work was supported in part by the NASA Research Grant NGR-39-002-023 and in part under a consulting agreement with Westinghouse Research Laboratories. Thanks are due to J. L. Swedlow and F. J. Rizzo for valuable discussions and to Carnegie-Mellon University Computation Center for providing some of the computer time used for this work.

Appendix

Boundary Identity for the Stress Tensor

The discussion in this section deals with the determination of the integral identity for the stress tensor σ_{ij} at points on the surface ∂R of a three-dimensional body R. The procedure is closely related to the determination of the boundary constraint equation (1). The reasons for special attention to the integral identity are threefold. First, the in-plane stress components at a surface are not known directly from the boundary conditions. They are, however, very important for the determination of the surface effect in cracked bodies of finite thickness [4]. Secondly, the numerical evaluation of the full stress tensor provides an error check by comparison to the known tractions at the surface. Thirdly, there are reasons to believe that analyses similar to that reported in this Appendix can be used to obtain closed-form, analytic results for the three-dimensional singularities near reentrant corners.

The method to be used begins with the formulation of Somigliana's identity for the displacement vector in the body due to surface tractions and displacement vector in the body due to surface tractions and displacements. This identity, as developed for the direct potential method, is given by

$$u_i(p) = - \int_{\partial R} u_j(Q) T_{ij}(Q, p) dS(Q) + \int_{\partial R} t_j(Q) U_{ij}(Q, p) dS(Q). \tag{A1}$$

The displacements $u_i(p)$ are evaluated at an interior point $p(x)$ while the integrals are carried out over the surface ∂R denoted by the surface point $Q(x)$. The second order tensors $U_{ij}(Q, p)$ and $T_{ij}(Q, p)$ are given in (4) and (5). Once the solution to (1) is determined then (A1) may be integrated directly to obtain $u_i(p)$.

Utilizing Hooke's law in terms of the displacement gradients, the stress tensor identity may be determined by appropriate differentiation of (A1):

$$\sigma_{ij}(p) = - \int_{\partial R} u_k(Q) S_{kij}(Q, p) dS(Q) + \int_{\partial R} t_k(Q) D_{kij}(Q, p) dS(Q). \tag{A2}$$

The third order tensors $S_{kij}(Q, p)$ and $D_{kij}(Q, p)$ are given in [17].

The purpose of this discussion is to report on the process whereby the point $p(x)$ in (A2) may be taken to the boundary point $P(x)$. The difficulty arises from the fact that $S_{kij}(Q, p)$ and $D_{kij}(Q, p)$ contain singularities of the order r^{-3} and r^{-2}, respectively, where $r(P, p)$ is the distance between the points $P(x)$ and $p(x)$.

Let $\bar{u}_i(Q, P)$ be defined as the rigid-body translation associated with the point $P(x)$ such that

$$u'_i(P) = u_i(P) - \bar{u}_i(P, P) \equiv 0. \tag{A3}$$

The stress tensor at $p(x)$ associated with the rigid-body displacement field is identically zero. Thus, the tractions are also zero. Then (A2) becomes

$$0 = \int_{\partial R} \bar{u}_k(Q, P) S_{kij}(Q, p) dS(Q). \tag{A4}$$

Fig. 14 shows a local coordinate system which has its origin at the point $P(x)$. The coor-

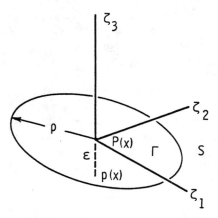

Figure 14. Local coordinate system at $P(x)$.

dinates ζ_1, ζ_2 are in the tangent plane and ζ_3 is in the direction of the exterior normal. The point $p(x)$ is located at a distance ε from $P(x)$ in the inward normal direction. Sufficiently near the origin the displacements $u_i(Q)$ can be represented by a Taylor series expansion in the surface coordinates

$$u_i(Q) = u_i(P) + \left.\frac{\partial u_i}{\partial \zeta_1}\right|_P d\zeta_1 + \left.\frac{\partial u_i}{\partial \zeta_2}\right|_P d\zeta_2 + \ldots \tag{A5}$$

when the surface has a unique tangent plane at $P(x)$. Then using (A5), (A3) becomes

$$u'_i(Q) = \left.\frac{\partial u_i}{\partial \zeta_1}\right|_P d\zeta_1 + \left.\frac{\partial u_i}{\partial \zeta_2}\right|_P d\zeta_2 + \ldots \tag{A6}$$

Now let the surface ∂R be divided into surfaces Γ and S where $\partial R = S + \Gamma$. Let Γ be the surface enclosed by a circle of radius ρ centered at $P(x)$ as shown in Fig. 14. Combining (A2) and (A3), and using the results in (A4), the stress identity becomes

$$\sigma_{ij}(p) = -\int_S u'_k(Q) S_{kij}(Q,p) dS(Q) + \int_S t_k(Q) D_{kij}(Q,p) dS(Q) +$$
$$- \int_\Gamma u'_k(Q) S_{kij}(Q,p) dS(Q) + \int_\Gamma t_k(Q) D_{kij}(Q,p) dS(Q). \tag{A7}$$

The first two integrals in (A7) are to be interpreted in the sense of the Cauchy Principal Value and are continuous as $p(x) \to P(x)$. If $\varepsilon \ll \rho$ then the last two integrals in (A7) may be integrated assuming that $t_k(Q)$ is continuous at $P(x)$. After some lengthy calculations the results are found for $p(x) \to P(x)$ to be

$$\sigma_{ij}(P) = -\int_S u'_k(Q) S_{kij}(Q,P) dS(Q) + \int_S t_k(Q) D_{kij}(Q,P) dS(Q) +$$
$$+ \tfrac{1}{2} t_k(P) A_{kij} - \mu \left.\frac{\partial u_k}{\partial \zeta_1}\right|_P B_{kij} - \mu \left.\frac{\partial u_k}{\partial \zeta_2}\right|_P C_{kij}. \tag{A8}$$

The tensors A_{kij}, B_{kij}, and C_{kij} are lengthy and contain material constants and transformation matrices between the local coordinates and the global coordinates. It can be shown that the last three terms in (A8) are identically equal to $\tfrac{1}{2}\sigma_{ij}(P)$. Thus,

$$\tfrac{1}{2}\sigma_{ij}(P) = -\int_S u'_k(Q) S_{kij}(Q,P) dS(Q) + \int_S t_k(Q) D_{kij}(Q,P) dS(Q) \tag{A9}$$

a formula closely analogous to the boundary identity (1).

Thus, there are really three possible methods for calculating $\sigma_{ij}(P)$. The first method is to use (A9). The second method and the one used in the report, is to calculate the in-plane displacement gradients by finite difference approximations and use the last three terms in (A8). The third method is to use (A8). The second method was chosen because it gives superior numerical results. The reason it does so is that it reflects the surface strains more directly than (A9) can because of the discretization of the boundary data.

It has also been found that the accuracy of generating the interior stresses near the surface is improved by subtracting out the rigid body motion as in (A3). One area of future concern is to utilize the same methods that were used in this section to obtain results near reentrant corners by including the possibility of two or more tangent planes at $P(x)$. This result would be of great significance in the interpretation of the results as outlined in this report.

REFERENCES

[1] E. Sternberg, Three-Dimensional Stress Concentrations in the Theory of Elasticity, *Applied Mechanics Reviews* 11, 1, (1968).
[2] J. B. Alblas, *Theory of the Three-Dimensional State of Stress in a Plate with a Hole*, H. J. Paris (Amsterdam((1957).
[3] R. J. Hartranft and G. C. Sih, The Use of Eigenfunction Expansions in the General Solution of Three-Dimensional Crack Problems, *J. Math. Mech.*, 19 (1969) 123.
[4] R. J. Hartranft and G. C. Sih, *An Approximate Three-Dimensional Theory of Plates with Application to Crack Problems*, Technical Report No. 7, Lehigh University Institute of Research, (May 1968).

[5] J. H. Argyris, Matrix Analysis of Three-Dimensional Elastic Media, Small and Large Displacements, *AIAA Jnl.* 3 (1965) 45.
[6] D. J. Ayres, *A Numerical Procedure for Calculating Stress and Deformation Near a Slit in a Three-Dimensional Elastic-Plastic Solid*, NASA Report TM X-52440 (1968).
[7] P. C. Paris and G. C. Sih, Stress Analysis of Cracks, in *Fracture Toughness Testing and Its Applications*, ASTM STP 381, *Amer. Soc. for Testing and Materials*, Philadelphia (1964).
[8] M. L. Williams, Stress Singularities Resulting from Various Boundary Conditions in Angular Corners of Plates in Extension, *J. Appl. Mech.*, 19 (1952) 526.
[9] M. L. Williams, On the Stress Distribution at the Base of a Stationary Crack, *J. Appl. Mech.*, 24 (1957) 109.
[10] H. M. Westergaard, Bearing Pressures and Cracks, *J. Appl. Mech.* 61 (1939) 49.
[11] B. Gross, J. E. Srawley and W. F. Brown, Jr., *Stress-Intensity Factors by Boundary Collocation for Single-Edge-Notch Specimens Subject to Splitting Forces*, NASA TN D-2395 (February 1966).
[12] L. E. Hulbert, *The Numerical Solution of Two-Dimensional Problems of the Theory of Elasticity*, Bulletin 198, Engineering Experiment Station, Ohio State University.
[13] F. J. Rizzo, An Integral Equation Approach to Boundary Value Problems of Classical Elastostatics, *Q. Appl. Math.* 25 (1967) 83.
[14] T. A. Cruse and F. J. Rizzo, A Direct Formulation and Numerical Solution of the General Transient Elastodynamic Problem-I, *J. Math. Anal. Appl.* 22 (1968) 244.
[15] T. A. Cruse, A Direct Formulation and Numerical Solution of the General Transient Elastodynamic Problem-II, *J. Math. Anal. Appl.*, 22 (1968) 341.
[16] F. J. Rizzo and D. J. Shippy, A Formulation and Solution Procedure for the General Non-Homogeneous Elastic Inclusion Problem, *Int. J. Solids Struct.*, 4 (1968) 1161.
[17] T. A. Cruse, Numerical Solutions in Three-Dimensional Elastostatics, *Int. J. Solids Struct.*, 5 (1969) 1259.
[18] S. G. Mikhlin, *Multidimensional Singular Integrals and Integral Equations*, Pergamon Press (1965).
[19] S. K. Chan, I. S. Tuba and W. K. Wilson, *On the Finite Element Method in Linear Fracture Mechanics*, Scientific Paper 68-1D7-FMPWR-P1, Westinghouse Research Laboratories, Pittsburgh, Pa., (April 1968).
[20] M. M. Leven, *Stress Distribution in the M4 Biaxial Fracture Specimen*, Research Report 65-1D7-STRSS-R1, Westinghouse Research Laboratories, Pittsburgh, Pa., (March 1965).

RÉSUMÉ

On présente une analyse numérique des contraintes dans une éprouvette à trois dimensions en sollicitation dans le domaine élastique, éprouvette similaire à l'éprouvette de traction utilisée dans les études de mécanique de rupture.

Les résultats numériques sont obtenus en recourant à des équations à intégrales singulières, qui sont analogues à l'équation aux limites de Green dans la théorie du potentiel.

L'analyse conduit à fournir la distribution complète des contraintes au voisinage de l'extrémité de la fissure et démontre clairement le caractère tridimensionnel de celle-ci.

L'influence de l'épaisseur et du module de Poisson sur les contraintes est également dégagée de certains résultats qui ont été obtenus.

ZUSAMMENFASSUNG

Es wird für eine dreidimensionale elastische Probe, ähnlich der kompakten Zugprobe für Bruchuntersuchungen, eine zahlenmässige Analyse der Spannungen dargelegt. Die numerischen Ergebnisse wurden über singuläre Integralgleichungen analog den Greenschen Formeln in der Potentialtheorie ermittelt. Die Untersuchung ergibt Einzelheiten über die Spannungen in der Umgebung der Rißspitze und zeigt klar deren dreidimensionalen Charakter. Es werden auch einige Resultate mitgeteilt die den Einfluß der Dicke und des Poissonschen Moduls auf die Spannungen herausstellen.

STRESS-INTENSITY FACTORS FOR A WIDE RANGE OF SEMI-ELLIPTICAL SURFACE CRACKS IN FINITE-THICKNESS PLATES

I. S. RAJU and J. C. NEWMAN, Jr.
NASA-Langley Research Center, Hampton, Virginia, U.S.A.

Abstract—Surface cracks are among the more common flaws in aircraft and pressure vessel components. Accurate stress analyses of surface-cracked components are needed for reliable prediction of their crack growth rates and fracture strengths. Several calculations of stress-intensity factors for semi-elliptical surface cracks subjected to tension have appeared in the literature. However, some of these solutions are in disagreement by 50–100%.

In this paper stress-intensity factors for shallow and deep semi-elliptical surface cracks in plates subjected to tension are presented. To verify the accuracy of the three-dimensional finite-element models employed, convergence was studied by varying the number of degrees of freedom in the models from 1500 to 6900. The 6900 degrees of freedom used here were more than twice the number used in previously reported solutions. Also, the stress-intensity variations in the boundary-layer region at the intersection of the crack with the free surface were investigated.

NOTATION

- a depth of surface crack
- b half-width of cracked plate
- c half-length of surface crack
- F stress-intensity boundary-correction factor
- h half-length of cracked plate
- K stress-intensity factor (Mode I)
- Q shape factor for an elliptical crack
- S applied uniform stress
- t plate thickness
- x, y, z Cartesian coordinates
- ν Poisson's ratio
- ϕ parametric angle of the ellipse

INTRODUCTION

SURFACE CRACKS are among the more common flaws in aircraft and pressure vessel components. Accurate stress analyses of these surface-cracked components are needed for reliable prediction of their crack growth rates and fracture strengths. Exact solutions to these difficult problems are not available; therefore, approximate methods must be used. For a semi-elliptical surface crack in a finite-thickness plate subjected to tension (Fig. 1), Browning and Smith[1], Kobayashi[2] and Smith and Sorensen[3] used the alternating method and Kathiresan[4] used the finite-element method to obtain the stress-intensity factor variations along the crack front for various crack shapes. For a deep semi-elliptical surface crack (with $a/t = 0.8$ and $a/c = 0.2$) subjected to tension, the stress-intensity factors obtained by Smith and Sorensen[3], Kobayashi[2] and Kathiresan[4] disagreed by 50–100%. The reasons for these discrepancies are not well understood.

This paper presents stress-intensity factors for shallow and deep semi-elliptical surface cracks in plates subjected to uniform tension. To test the validity of the present analysis, two crack configurations, both embedded in a large body subjected to uniform tension, were analyzed: (1) a circular (penny-shaped) crack and (2) an elliptical crack. These results are compared with exact solutions from the literature[5]. To verify the accuracy of the solutions for surface cracks in finite-thickness plates, convergence was studied by varying, from 1500 to 6900, the number of degrees of freedom in the finite-element models. The 6900 degrees of freedom used here were more than twice the largest number used previously[4]. These models were composed of singular elements around the crack front and isoparametric (linear strain) elements elsewhere. Mode I elastic stress-intensity factors were calculated by using a nodal-force method.

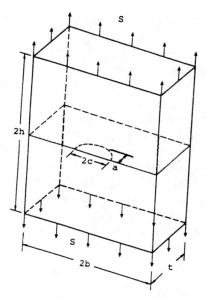

Fig. 1. Surface crack in a plate subjected to tension.

The stress-intensity variations in a boundary-layer region at the intersection of the crack with the free surface were also investigated for a typical semi-circular surface crack. Here the finite-element models used were highly refined in the boundary-layer region.

The computations reported herein were conducted on a unique computer called the STAR-100 at the NASA Langley Research Center.

THREE-DIMENSIONAL ANALYSIS

Several attempts [6–8] have been made to develop special three-dimensional finite-elements that account for the stress and strain singularities caused by a crack. These elements had assumed displacement or stress distributions that simulate the square-root singularity of the stresses and strains at the crack front. Such three-dimensional singularity elements [6, 9] were also used herein for the analysis of finite-thickness plates containing embedded elliptical or semi-elliptical surface cracks (see, for example, Fig. 1).

Finite-element idealization

Two types of elements (isoparametric and singular) were used in combination to model elastic bodies with embedded elliptical cracks or semi-elliptical surface cracks. Figure 2 shows a typical finite-element model for an embedded circular crack in a large body ($h/a = b/a = 5$).

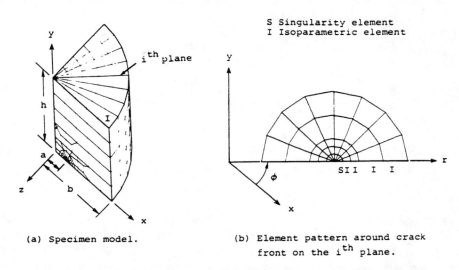

(a) Specimen model.

(b) Element pattern around crack front on the ith plane.

Fig. 2. Finite-element idealization for an embedded circular (penny-shaped) crack.

This model idealizes one eight of the body. Various numbers of wedges were used to form the desired configuration. Figure 2(a) shows a typical model with eight wedges. Each wedge is composed of elements that are identical in pattern to that shown in the $\phi =$ constant plane. The arrangement of the elements around the crack front is shown in Fig. 2(b). The isoparametric (linear strain) elements (denoted as I) were used everywhere except near the crack front. Around the crack front each wedge contained eight "singularity" elements (S) in the shape of pentahedrons. The "singularity" elements had square-root terms in their assumed displacement distribution and, therefore, produced a singular stress field at the crack front. Details of the formulation of these types of elements are given in Refs. [6] and [9] and are not repeated here.

The finite-element model for the embedded elliptical or semi-elliptical surface crack was obtained from the finite-element model for the circular crack by using an elliptic transformation. This transformation was needed because the stress-intensity factors must be evaluated from either crack-opening displacements or nodal forces along the normals to the crack front. If (x, y, z) are the Cartesian coordinates of a node in the circular-crack model and (x', y', z') are the coordinates of that same node in the elliptical-crack model, then the transformation is

$$\left. \begin{array}{l} x' = x \sqrt{\left(1 + \dfrac{c^2 - a^2}{x^2 + z^2}\right)}, \quad y' = y, \quad z' = z \quad \text{for} \quad \dfrac{a}{c} \leq 1 \\ \\ \text{and} \\ \\ x' = x, \quad y' = y, \quad z' = z \sqrt{\left(1 + \dfrac{a^2 - c^2}{x^2 + z^2}\right)} \quad \text{for} \quad \dfrac{a}{c} > 1 \end{array} \right\} \quad (1)$$

for x and z not at the origin. Figure 3 shows how circular arcs and radial lines in the x, z plane of the circular-crack model are transformed into ellipses and hyperbolas, respectively, in the x', z' plane of the elliptical-crack model using eqns (1). Because eqns (1) are not valid at the origin, a circle of very small radius, $a/1000$, was used near the origin in the circular crack model. The small circle maps onto an extremely narrow ellipse in the x', z' plane. The use of the small circle avoids ill-shaped elements near the origin in the elliptical-crack model. Figure 4 shows a typical finite-element model of a finite plate containing an elliptical crack. The transformation reduced the b/c ratio; therefore, in order to maintain $b/c \geq 4$, additional elements were added along the x'-axis to eliminate the influence of plate width.

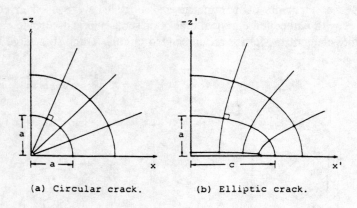

Fig. 3. Circle to ellipse transformation.

Stress-intensity factor

The stress-intensity factor is a measure of the magnitude of the stresses near the crack front. Under general loading, the stress-intensity factor depends on three basic modes of deformation (tension and in- and out-of-plane shear). But here only tension loading was considered and, therefore, only Mode I deformations occurred. The Mode I stress-intensity

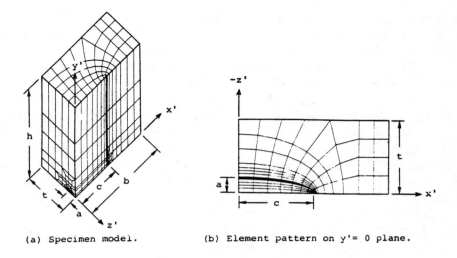

(a) Specimen model.   (b) Element pattern on $y' = 0$ plane.

Fig. 4. Finite-element idealization for a semi-elliptical surface crack.

factor, K, at any point along an elliptical or semi-elliptical crack in a finite plate is taken to be

$$K = S \sqrt{\left(\pi \frac{a}{Q}\right)} F\left(\frac{a}{t}, \frac{a}{c}, \phi\right) \qquad (2)$$

where S is the applied stress, a is the crack depth, Q is the shape factor for an ellipse and is given by the square of the complete elliptic integral of the second kind[5]. The boundary-correction factor, $F(a/t, a/c, \phi)$, is a function of crack depth, crack length, plate thickness and the parametric angle of the ellipse. The present paper gives values for F as a function of a/t and ϕ for $a/c = 0.2, 0.4, 0.6, 1.0$ and 2.0. The a/t values ranged from 0.2 to 0.8.

The stress-intensity factors from the finite-element models of embedded elliptical and semi-elliptical surface cracks were obtained by using a nodal-force method, details of which are given in the appendix. In this method, the nodal forces normal to the crack plane and ahead of the crack front are used to evaluate the stress-intensity factor. In contrast to the crack-opening displacement method[6], this method requires no prior assumption of either plane stress or plane strain. For a surface crack in a finite plate, the state of stress varies from plane strain in the interior of the plate to plane stress at the surface. Thus, the crack-opening displacement method could yield erroneous stress-intensity factors.

RESULTS AND DISCUSSION

In the following sections a circular (penny-shaped) crack and an elliptical crack completely embedded in a large body subjected to uniform tension were analyzed using the finite-element method. The calculated stress-intensity factors for these crack configurations are compared with the exact solutions to verify the validity of the present finite-element method.

For a semi-circular and semi-elliptical surface crack in a finite-thickness plate, convergence of the stress-intensity factors was studied while the number of degrees of freedom in the finite-element models ranged from about 1500 to 6900. The stress-intensity variations in a thin boundary-layer region at the intersection of the crack with the free surface were also investigated. The stress-intensity factor variations along the crack front for semi-circular ($a/c = 1$) and semi-elliptical ($a/c = 0.2, 0.4, 0.6, 1.0$ and 2.0) surface cracks were obtained as functions of a/t with $h/c = b/c \geq 4$. Whenever possible, these stress-intensity factors are compared with results from the literature.

Exact solutions

In this section a comparison is made between the stress-intensity factors calculated from the finite-element analysis and from the exact solution[5] for an embedded circular (penny-shaped) crack ($a/c = 1$) and an embedded eliptical crack ($a/c = 0.2$) in an infinite body. In the finite-element model, h and b were taken to be large enough that the free boundary would have a negligible effect on stress intensity. The boundary correction on stress intensity for a circular crack in a cylinder of radius b with $b/a = 5$ is about 1%[10]. Therefore, to simulate a large body the finite-element model was assigned the dimensions $h/a = b/a = 5$, along with 3078 degrees of freedom. The calculated stress-intensity factors along the crack front from the circular crack model were about 0.4% below the exact solution.

The embedded elliptical crack ($a/c = 0.2$) model, with $h/c = b/c = 4$ and $t/a = 5$, had 3348 degrees of freedom. The influence of finite boundaries on stress intensity was estimated to be about 1%. The finite-element analysis gave stress-intensity factors along the crack front generally within 1% of the exact solution, except in the region of sharpest curvature of the ellipse, where the calculated values were about 3% higher than the exact solution. Further refinement in the mesh size in this area gave more accurate stress-intensity factors.

Because the present method yielded stress-intensity factors for completely embedded circular and elliptical cracks generally within 0.4–1% of the exact solutions, the method was considered suitable for analyses of more complex configurations, provided that enough degrees of freedom were used to obtain good convergence.

Approximate solutions

To verify that the finite-element meshes used for the circular and elliptical crack models were sufficient to analyze cracked plates with free boundaries, through-the-thickness cracks with crack length-to-width ratios ranging from 0.2 to 0.8 were analyzed under plane strain assumptions. The meshes used here were exactly the same as those which occur on the $\phi = \pi/2$ plane of the circular and elliptical crack models. These meshes were then used to model the center-crack tension specimen. For crack length-to-width ratios (c/b) ranging from 0.2 to 0.6, the finite-element results were within 1.3% of the approximate solutions given in Ref. [10]. For $c/b = 0.8$, the finite-element result was 2% below the solution given in Ref. [10]. These results indicate that the mesh pattern along any plane has enough degrees of freedom to account for the influence of free boundaries on the stress-intensity factor.

Convergence

In the previous section the mesh pattern along any ϕ = constant plane was found to be sufficient to account for the influence of free boundaries even for very deep cracks. In this section, the mesh pattern in the angular direction, ϕ, is studied. Figures 5 and 6 show the results of a convergence study on the stress-intensity factors for a semi-circular surface crack and a semi-elliptical surface crack, respectively, in a finite-thickness plate. The larger numbers of degrees of freedom are associated with smaller wedges in the ϕ-direction and a more accurate representation of the crack shape.

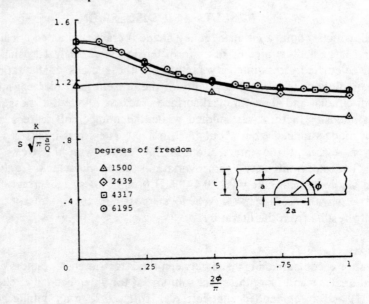

Fig. 5. Convergence of stress-intensity factors for a deep semi-circular surface crack ($Q = \pi^2/4$; $a/t = 0.8$).

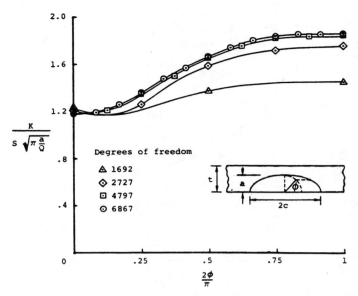

Fig. 6. Convergence of stress-intensity factors for a deep semi-elliptical surface crack ($Q = 1.104$; $a/t = 0.8$; $a/c = 0.2$).

For the semi-circular crack (Fig. 5), $h/a = b/a = 5$ and $a/t = 0.8$. This configuration was chosen because the close proximity of the back surface to the crack front was expected to cause difficulty in achieving convergence. The models were composed of either 2, 4, 8 or 12 equal wedges. The number of degrees of freedom ranged from 1500 to 6195. The two finest models (4317 and 6195 degrees of freedom) gave stress-intensity factors within about 1% of each other.

Figure 6 shows the convergence study for a semi-elliptical surface crack ($a/c = 0.2$). The h/c and b/c ratios were equal to four and, again, a/t was chosen as 0.8. The number of degrees of freedom ranged from 1692 to 6867. The eight-wedge model with 4797 degrees of freedom gave results within about 1% of those from the finest model.

The results shown in Figs. 5 and 6 demonstrate rapid convergence of the solution and indicate that the eight-wedge model may be used to obtain stress-intensity factors as a function of a/t and a/c. However, further consideration of the stress-intensity variations in the boundary-layer region at the intersection of the crack with the free surface must be given.

Boundary-layer effect

Hartranft and Sih[11] proposed that the stress-intensity factors in a very thin "boundary layer" near the free surface drop off rapidly and equal zero at the free surface. To investigate the boundary-layer effect, a semi-circular surface crack in a large body was considered. Three different finite-element models with 8, 10 and 14 wedges were analyzed. The eight wedge model, shown in Fig. 2(a), had eight equal wedges. The other models had nonuniform wedges and were obtained by refining the eight-wedge model near the free surface. The smallest wedge angle for the ten- and fourteen-wedge models were $\pi/32$ and $\pi/128$, respectively. The stress-intensity factors obtained from the three models are shown in Fig. 7. The maximum stress-intensity factor obtained from the eight-wedge model at the free surface was slightly less (1.4%) than the peak value obtained from the fourteen-wedge model. These results also show that the stress intensities near the free surface were affected by the refinements and drop off rapidly as proposed by Hartranft and Sih. However, the stress-intensity distributions in the interior ($\phi > \pi/16$) were unaffected by the refinements. These results strongly suggest that the stress intensities in the interior are insensitive to whatever is happening in the boundary layer. An estimate for the size of the boundary layer was obtained from Ref. [11], assuming that the crack was a through crack of length, $2a$, and this estimate is also shown in Fig. 7.

In view of the highly localized boundary-layer effect, the eight-wedge model was used subsequently to obtain stress-intensity factors as a function of a/c and a/t. The stress-intensity factor obtained at the free surface with the eight-wedge models were interpreted as an average stress-intensity factor *near* the free surface.

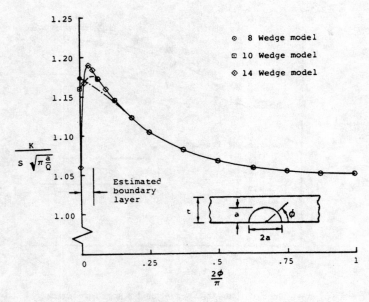

Fig. 7. Effects of mesh refinement near the free surface on the distribution of stress intensity for semi-circular surface crack ($Q = \pi^2/4$; $a/t = 0.2$).

Semi-circular surface crack in a finite-thickness plate

Figure 8 shows and Table 1 presents the stress-intensity factors for a semi-circular surface crack in a finite-thickness plate as a function of the parametric angle, ϕ, and the crack depth-to-plate thickness ratio, a/t. Near the intersection of the crack and the free surface, the stress-intensity factor increases more rapidly with a/t than at the deepest point ($\phi = \pi/2$). For each value of a/t, the stress-intensity factor calculated from the eight-wedge model is largest at the free surface.

For a semi-circular surface crack with $a/t = 0.55$, the stress-intensity factors calculated by Browning and Smith[1], who used the alternating method, were about 1–3% below the present results for various values of ϕ.

Semi-elliptical surface crack in a finite-thickness plate

The stress-intensity factors for a semi-elliptical surface crack ($a/c = 0.2, 0.4, 0.6$ or 2.0) in a finite-thickness plate as a function of the parametric angle, ϕ, and the crack depth-to-plate thickness ratio, a/t, are given in Table 1.

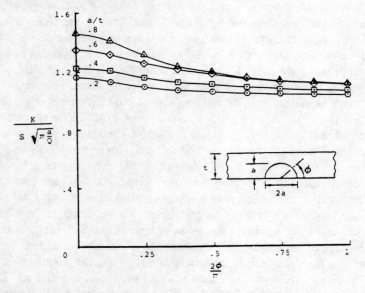

Fig. 8. Distribution of stress-intensity factors along crack front for a semi-circular surface crack ($Q = \pi^2/4$).

Table 1. Boundary correction factors, F, for semi-elliptical surface cracks subjected to tension ($\nu = 0.3$; $F = K/S\sqrt{(\pi a/Q)}$)

a/c	$2\phi/\pi$	a/t 0.2	0.4	0.6	0.8
0.2	0	0.617	0.724	0.899	1.190
	0.125	0.650	0.775	0.953	1.217
	0.25	0.754	0.883	1.080	1.345
	0.375	0.882	1.009	1.237	1.504
	0.5	0.990	1.122	1.384	1.657
	0.625	1.072	1.222	1.501	1.759
	0.75	1.128	1.297	1.581	1.824
	0.875	1.161	1.344	1.627	1.846
	1.0	1.173	1.359	1.642	1.851
0.4	0	0.767	0.896	1.080	1.318
	0.125	0.781	0.902	1.075	1.285
	0.25	0.842	0.946	1.113	1.297
	0.375	0.923	1.010	1.179	1.327
	0.5	0.998	1.075	1.247	1.374
	0.625	1.058	1.136	1.302	1.408
	0.75	1.103	1.184	1.341	1.437
	0.875	1.129	1.214	1.363	1.446
	1.0	1.138	1.225	1.370	1.447
0.6	0	0.916	1.015	1.172	1.353
	0.125	0.919	1.004	1.149	1.304
	0.25	0.942	1.009	1.142	1.265
	0.375	0.982	1.033	1.160	1.240
	0.5	1.024	1.062	1.182	1.243
	0.625	1.059	1.093	1.202	1.245
	0.75	1.087	1.121	1.218	1.260
	0.875	1.104	1.139	1.227	1.264
	1.0	1.110	1.145	1.230	1.264
1.0	0	1.174	1.229	1.355	1.464
	0.125	1.145	1.206	1.321	1.410
	0.25	1.105	1.157	1.256	1.314
	0.375	1.082	1.126	1.214	1.234
	0.5	1.067	1.104	1.181	1.193
	0.625	1.058	1.088	1.153	1.150
	0.75	1.053	1.075	1.129	1.134
	0.875	1.050	1.066	1.113	1.118
	1.0	1.049	1.062	1.107	1.112
2.0	0	0.821	0.848	0.866	0.876
	0.125	0.794	0.818	0.833	0.839
	0.25	0.740	0.759	0.771	0.775
	0.375	0.692	0.708	0.716	0.717
	0.5	0.646	0.659	0.664	0.661
	0.625	0.599	0.609	0.610	0.607
	0.75	0.552	0.560	0.560	0.554
	0.875	0.512	0.519	0.519	0.513
	1.0	0.495	0.501	0.501	0.496

Figure 9 shows the stress-intensity factors for a semi-elliptical surface crack ($a/c = 0.2$) as a function of the parametric angle and the crack depth-to-plate thickness ratio. For each value of a/t, the maximum stress-intensity factor occurs at the deepest point ($\phi = \pi/2$). Also, the maximum stress-intensity factor is larger for larger values of a/t.

Figure 10 shows the stress-intensity distributions for various semi-elliptical surface cracks as a function of a/c with $a/t = 0.8$. For smaller values of a/c, the normalized stress-intensity factor at $\phi = \pi/2$ is larger. The maximum normalized stress-intensity factor occurred at $\phi = \pi/2$ for $a/c \leq 0.4$ but occurred at $\phi = 0$ for $a/c \geq 1.0$.

Figures 11 and 12 shows stress-intensity factors obtained by several investigators for a semi-elliptical surface crack in a finite-thickness plate. Figure 11 shows the results for a surface

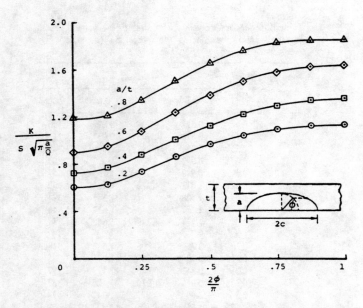

Fig. 9. Distribution of stress-intensity factors along crack front for a semi-elliptical surface crack ($Q = 1.104$; $a/c = 0.2$).

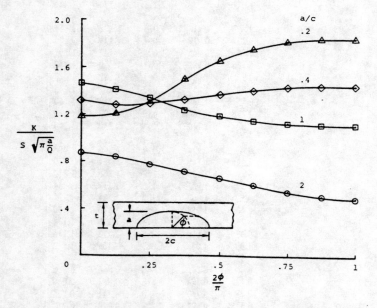

Fig. 10. Distribution of stress-intensity factors along crack front for semi-elliptical surface cracks ($a/t = 0.8$).

crack with $a/c = 0.2$ and $a/t = 0.8$. Smith and Sorenson[3] and Kobayashi[2] used the alternating method and Kathiresan[4] used the finite-element method to obtain stress-intensity factor variations along the crack front. These three solutions disagree by as much as 50–100%. The reasons for these discrepancies are not well understood. The present results, shown as solid symbols, are considerably higher than the previous solutions[2–4]. The results from Smith and Sorensen[3] are generally closer to the present results, though 10–25% lower.

Figure 12 shows a comparison of the maximum stress-intensity factors obtained by several investigators for a semi-elliptical surface crack as a function of a/t. The maximum stress-intensity factors occurred at $\phi = \pi/2$. The present results are shown as solid symbols. The open symbols show the results from Smith and Sorensen[3], Kobayashi[2,12] and Rice and Levy[13]. The results from Rice nad Levy, obtained from a line-spring model, are about 3.5% below the present results over an a/t range from 0.2 to 0.6. For $a/t > 0.6$, the Rice–Levy solution shows a reduction in stress intensity. The dash-dot curve shows the results of an approximate equation proposed by Newman[14] for a wide range of a/c and a/t ratios.

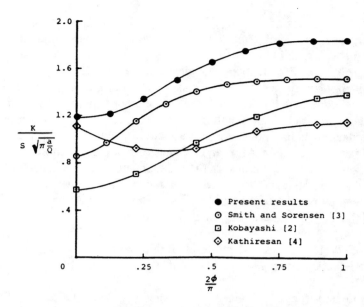

Fig. 11. Comparison of stress-intensity factors for a deep semi-elliptical surface crack ($Q = 1.104$; $a/t = 0.8$; $a/c = 0.2$).

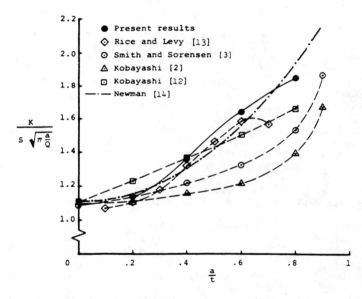

Fig. 12. Comparison of the maximum stress-intensity factor for a semi-elliptical surface crack as a function of a/t ($\phi = \pi/2$; $Q = 1.104$; $a/c = 0.2$).

Newman's equation is within ±5% of the present results over an a/t range from 0.2 to 0.8. Newman's equation gives a good engineering estimate for the maximum stress-intensity factor.

CONCLUDING REMARKS

A three-dimensional finite-element elastic stress analysis was used to calculate stress-intensity factor variations along the crack front for completely embedded elliptical cracks in large bodies and for semi-elliptical surface cracks (crack depth-to-half crack length ratios were 0.2 to 2.0) in finite-thickness plates. Three-dimensional singularity elements were used at the crack front. A nodal force method which requires no prior assumption of either plane stress or plane strain was used to evaluate the stress-intensity factors along the crack front.

Completely embedded circular and elliptical cracks were analyzed to verify the accuracy of the finite-element analysis. The stress-intensity factors for these cracks were generally about 0.4–1% below the exact solutions. However, for the elliptical crack the calculated stress-intensity factors in the region of sharpest curvature of the ellipse were about 3% higher than the exact solution. The numbers of degrees of freedom in the embedded crack models were about 3000. A convergence study on stress-intensity factors for semi-elliptical surface cracks in finite-thickness plates showed that convergence was achieved for both the semi-circular and the semi-elliptical surface crack with about 4500 degrees of freedom.

For the semi-circular surface crack the maximum stress-intensity factor occurred near the intersection of the crack with the free surface. On the other hand, for the semi-elliptical surface crack (crack depth-to-crack half length ratio of 0.2), the maximum stress-intensity factor occurred at the deepest point. For both the semi-circular and semi-elliptical surface crack the stress-intensity factors were larger for larger values of crack depth-to-plate thickness ratio.

For the semi-circular surface crack, the stress-intensity factors calculated by Browning and Smith[1] using the alternating method, agreed generally within about 3% with the present results. However, for semi-elliptical surface cracks (crack depth-to-plate thickness ratio of 0.2) Smith and Sorensen[3] using the alternating method gave stress-intensity factors in considerable disagreement (10–25%) with the present results. For semi-elliptical surface cracks the results from Rice and Levy[13] for crack depth-to-plate thickness ratios less than or equal to 0.6 and an approximate equation proposed by Newman[14] were in good agreement with the present results.

The stress-intensity factors obtained herein should be useful in correlating fatigue crack growth rates as well as fracture toughness calculations for the surface-crack configurations considered.

REFERENCES

[1] W. M. Browning and F. W. Smith, An analysis for complex three-dimensional crack problems. *Developments in Theoretical and Applied Mechanics*, Vol. 8, *Proc. 8th SECTAM Conf.* (1976).
[2] A. S. Kobayashi, Surface flaws in plates in bending. *Proc. 12th Annual Meeting of the Soc. of Engng Sci.*, Austin, Texas (1975).
[3] F. W. Smith and D. R. Sorensen, Mixed mode stress intensity factors for semi-elliptical surface cracks. *NASA CR-134684* (1974).
[4] K. Kathiresan, Three-dimensional linear elastic fracture mechanics analysis by a displacement hybrid finite element model. Ph.D. Thesis, Georgia Institute of Technology (1976).
[5] A. E. Green and I. N. Sneddon, The distribution of stress in the neighborhood of a flat elliptical crack in an elastic solid. *Proc. Cambridge Phil. Soc.* **46** (1959).
[6] D. M. Tracey, Finite element for three-dimensional elastic crack analysis. *Nucl. Engr. and Design* **26** (1974).
[7] R. S. Barsoum, On the use of isoparametric finite elements in linear fracture mechanics. *Int. J. Num. Meth. Engr.* **10**(1) 25–37 (1976).
[8] S. Atluri and K. Kathiresan, An assumed displacement hybrid finite element model for three-dimensional linear fracture mechanics analysis. *Proc. 12th Annual Meeting of the Soc. Engr. Science*, University of Texas, Austin, Texas (1975).
[9] I. S. Raju and J. C. Newman, Jr., Three-dimensional finite-element analysis of finite-thickness fracture specimens. *NASA TN D-8414* (1977).
[10] H. Tada, P. C. Paris and G. R. Irwin, *The Stress Analysis of Cracks Handbook*, p. 28.1. Del. Research Corp., C. (1973).
[11] R. J. Hartranft and G. C. Sih, An approximate three-dimensional theory of plates with application to crack problems. *Int. J. Eng. Sci.* **8**(8), 711–729 (1970).
[12] A. S. Kobayashi, Crack opening displacement in a surface flawed plate subjected to tension or plate bending. Presented at the 2nd Int. Conf. Mech. Behavior of Materials, Boston, MA (1976).
[13] J. R. Rice and N. Levy, The part-through surface crack in an elastic plate. *Trans. ASME J. Appl. Mech.* Paper No. 71-APM-20 (1970).
[14] J. C. Newman, Jr., Fracture analysis of surface- and through-cracked sheets and plates. *Engng Fracture Mech.* **4**, 667–689 (1973).

(*Received* 18 *July* 1978; *received for publication* 4 *September* 1978)

APPENDIX

Determination of stress-intensity factors using the nodal-force method

The nodal-force method, developed in Ref. [9] for evaluating stress-intensity factors is presented herein. In this method, the nodal forces normal to the crack plane and ahead of the crack front are used to evaluate the stress-intensity factor. In contrast to the crack-opening displacement method[6], this method requires no prior assumption of either plane stress or plane strain.

Consider an elliptic surface crack idealized by several wedges as shown in Fig. 13. Each wedge was composed of various number of elements. The nodal forces (F_i) normal to the $y' = 0$ plane along the junction (dashed curve) between wedges i and $i+1$ are assumed to be contributed from the normal stresses acting over one-half of the wedges on either side of the junction (shaded area). For clarity, the shaded area has been enlarged ($R \ll a$). The dashed curves are hyperbolas normal to the crack front. Nodal forces along these hyperbolas are needed to evaluate the stress-intensity factors.

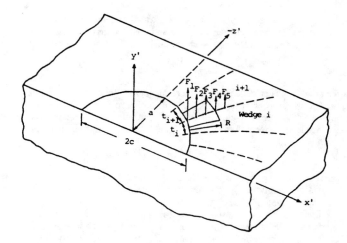

Fig. 13. Nodal forces on $y' = 0$ plane and at the junction between wedges i and $i + 1$.

The normal stress σ_y at any point in the shaded area is assumed to be given by

$$\sigma_y = \frac{K}{\sqrt{(2\pi r)}} + A_1' r^{1/2} + A_2' r^{3/2} + \ldots \qquad (3)$$

where r is the distance normal to the crack front. In general, K and A_i' are functions of the coordinate along the crack front. However, in the present analysis K and A_i' were assumed to be constant over the shaded area. This assumption is justified when the wedges are thin or the actual K variations across a wedge are small.

The total force acting in the y'-direction over the shaded area is given by

$$F_y = t_a \int_0^R \sigma_y \, dr \qquad (4)$$

where t_a is the average width of the shaded area ($t_a \approx (t_i + t_{i+1})/2$). Substituting eqn (3) into eqn (4) gives

$$F_y = t_a \left(\frac{K}{\sqrt{(2\pi)}} 2\sqrt{R} + A_1 R^{3/2} + A_2 R^{5/2} + \ldots \right) \qquad (5)$$

where A_i represents constants. Using only the first term on the righthand side of eqn (5), K is replaced by the apparent stress-intensity factor K_{ap} and is evaluated as

$$F_y = \frac{K_{ap}}{\sqrt{(2\pi)}} \sqrt{(R)} 2 t_a. \qquad (6)$$

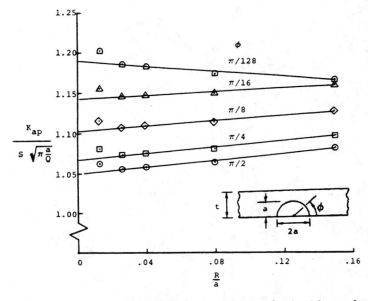

Fig. 14. Apparent stress-intensity factors for semi-circular surface crack ($a/t = 0.2$) as a function of R/a and the parametric angle, ϕ.

The force F_y was obtained from the finite-element analysis and was determined as follows. For example, let nodes 1–4 in Fig. 13 be the nodes lying on the junction between wedges i and $i+1$. Node 4 is a distance R from the crack front. The nodal forces F_j ($j = 1$–4) are obtained from the finite-element analysis. Thus, the total force acting over a distance R is

$$F_y = F_1 + F_2 + F_3 + F'_4 \tag{7}$$

where F'_4, the nodal force at node 4, is calculated using the elements that are connected to node 4 and are inside the shaded area. For various values of R, the number of nodal forces contributing to F_y would also vary.

The total force F_y obtained from the finite-element analysis is substituted into eqn (6), and the apparent stress-intensity factor is evaluated. The procedure is repeated for various values of R. Equating eqns (5) and (6) gives

$$K_{ap} = K + B_1 R + B_2 R^2 + \ldots \tag{8}$$

where B_i represents constants. From eqn (8), a plot of K_{ap} vs R is linear for small values of R and the intercept at $R = 0$, at the crack front, gives the stress-intensity factor at a particular location along the crack front. The procedure is repeated at various locations along the crack front.

Figure 14 shows the apparent stress-intensity factor for a typical semi-circular surface crack obtained by the nodal-force method plotted against the distance R/a for various values of the parametric angle, ϕ. If the K_{ap} value nearest the crack front ($R = 0$) is neglected, the relationship between K_{ap} and R/a are nearly linear, as expected. Therefore, the K_{ap} value closest to the crack front was neglected. For each value of ϕ, a straight line was fitted (by the method of least squares) to the K_{ap} results to evaluate the stress-intensity factor. The intercept of these lines at $R/a = 0$ gives the stress-intensity factor.

AN EMPIRICAL STRESS-INTENSITY FACTOR EQUATION FOR THE SURFACE CRACK

J. C. NEWMAN, JR.[†] and I. S. RAJU[‡]
NASA Langley Research Center, Hampton, VA 23665, U.S.A.

Abstract—This paper presents an empirical stress-intensity factor equation for a surface crack as a function of parametric angle, crack depth, crack length, plate thickness and plate width for tension and bending loads. The stress-intensity factors used to develop the equation were obtained from a previous three-dimensional, finite-element analysis of semielliptical surface cracks in finite elastic plates subjected to tension or bending loads. A wide range of configuration parameters was included in the equation. The ratios of crack length to plate thickness and the ratios of crack depth to crack length ranged from 0 to 1.0. The effects of plate width on stress-intensity variations along the crack front were also included.

The equation was used to predict patterns of surface-crack growth under tension or bending fatigue loads. The equation was also used to correlate surface-crack fracture data for a brittle epoxy material within ±10 percent for a wide range of crack shapes and crack sizes.

NOTATION

- a depth of surface crack, mm
- b half-width of cracked plate, mm
- C_A, C_B crack-growth coefficients (see eqns (A1) and (A2))
- c half-length of surface crack, mm
- F stress-intensity boundary-correction factor
- h half-length of cracked plate, mm
- K_I mode I stress-intensity factor, kN/m$^{3/2}$
- K_{cr} elastic fracture toughness, kN/m$^{3/2}$
- M applied bending moment, N·m
- N number of cycles
- n exponent in equation for crack-growth rate
- Q shape factor for elliptical crack
- S_b remote bending stress on outer fiber, $3M/bt^2$, Pa
- S_t remote uniform-tension stress, Pa
- t plate thickness, mm
- ΔK stress-intensity factor range, kN/m$^{3/2}$
- ϕ parametric angle of the ellipse, deg.

INTRODUCTION

SURFACE cracks are common flaws in many structural components. Accurate stress analyses of these surface-cracked components are needed for reliable prediction of their crack-growth rates and fracture strengths. However, because of the complexities of such problems, exact solutions are not available. Investigators have used experimental or approximate analytical methods to obtain stress-intensity factors for surface cracks under tension or bending loads. For a semielliptical surface crack in a plate of finite thickness (Fig. 1), Smith and Alavi[1], Smith and Sorensen[2] and Kobayashi et al.[3] used the alternating method to obtain the stress-intensity factor variations along the crack front for various crack shapes. Kathiresan[4] and Raju and Newman[5, 6] used the finite-element method to obtain the same information. These results were presented in the form of curves or tables. However, for ease of computation, results in the form of an equation are preferable.

The present paper presents an empirical equation for the stress-intensity factors for a surface crack as a function of parametric angle, crack depth, crack length, plate thickness and plate width for tension and bending loads. The equation was based on the stress-intensity factors obtained from a three-dimensional, finite-element analysis of semielliptical surface cracks in finite elastic plates subjected to tension or bending loads[7]. The present empirical equation covers a wide range of configuration parameters. The ratios of crack depth to plate thickness and the ratios of crack depth to crack length ranged from 0 to 1.0. The effects of plate width on stress-intensity factor variations along the crack front were also included.

[†]Research Engineer, NASA Langley Research Center, Hampton, VA 23665, U.S.A.
[‡]Assistant Research Professor, The George Washington University, Joint Institute for Advancement of Flight Sciences, NASA Langley Research Center, Hampton, VA 23665, U.S.A.

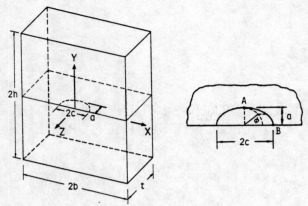

Fig. 1. Surface crack in a finite plate.

In Appendix A, the equation was used to predict surface-crack-growth patterns under tension or bending fatigue loads. These predicted patterns are also compared with measurements reported in the literature for steel, titanium alloy and aluminum alloy materials. In Appendix B, the maximum stress-intensity values from the equation were also used to correlate surface-crack fracture data reported in the literature for a brittle epoxy material.

STRESS-INTENSITY FACTOR EQUATION FOR THE SURFACE CRACK

An empirical equation for the stress-intensity factors for a surface crack in a finite plate subjected to tension and bending loads has been fitted to the finite-element results from Raju and Newman[5–7] for a/c values from 0.2 to 1.0. To account for the limiting behavior as a/c approaches zero, the results of Gross and Srawley[8] for a single-edge crack have also been used. Two types of loads were applied to the surface-cracked plate: remote uniform tension and remote bending. The remote uniform-tension stress is S_t in Fig. 2(a); the remote outer-fiber bending stress S_b in Fig. 2(b) is calculated from the applied bending moment M. The stress-intensity factor equation for combined tension and bending loads is

$$K_I = (S_t + H S_b)\sqrt{\pi \frac{a}{Q}} F\left(\frac{a}{t}, \frac{a}{c}, \frac{c}{b}, \phi\right) \tag{1}$$

for $0 < a/c \leqq 1.0$, $0 \leqq a/t < 1.0$, $c/b < 0.5$ and $0 \leqq \phi \leqq \pi$. A useful approximation for Q,

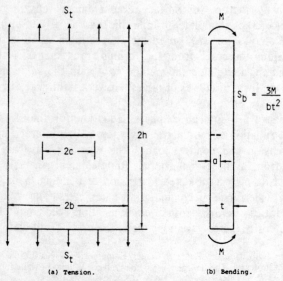

Fig. 2. Surface-cracked plate subjected to tension or bending loads.

developed by Rawe and used in Ref. [9], is

$$Q = 1 + 1.464 \left(\frac{a}{c}\right)^{1.65} \quad \left(\frac{a}{c} \leqq 1\right). \tag{2}$$

The functions F and H are defined so that the boundary-correction factor for tension is equal to F and the boundary-correction factor for bending is equal to the product of H and F. The function F was obtained from a systematic curve-fitting procedure by using double-series polynomials in terms of a/c, a/t, and angular functions of ϕ. The choice of functions was based on engineering judgment. The function F was taken to be

$$F = \left[M_1 + M_2\left(\frac{a}{t}\right)^2 + M_3\left(\frac{a}{t}\right)^4\right] f_\phi g f_w \tag{3}$$

where

$$M_1 = 1.13 - 0.09\left(\frac{a}{c}\right) \tag{4}$$

$$M_2 = -0.54 + \frac{0.89}{0.2 + (a/c)} \tag{5}$$

$$M_3 = 0.5 - \frac{1.0}{0.65 + (a/c)} + 14\left(1.0 - \frac{a}{c}\right)^{24} \tag{6}$$

$$g = 1 + \left[0.1 + 0.35\left(\frac{a}{t}\right)^2\right](1 - \sin\phi)^2. \tag{7}$$

The function f_ϕ, an angular function from the embedded elliptical-crack solution [10], is

$$f_\phi = \left[\left(\frac{a}{c}\right)^2 \cos^2\phi + \sin^2\phi\right]^{1/4}. \tag{8}$$

The function f_w, a finite-width correction from Ref. [11], is

$$f_w = \left[\sec\left(\frac{\pi c}{2b}\sqrt{\frac{a}{t}}\right)\right]^{1/2}. \tag{9}$$

The function H, developed herein also by curve fitting and engineering judgment, has the form

$$H = H_1 + (H_2 - H_1)\sin^p\phi \tag{10}$$

where

$$p = 0.2 + \frac{a}{c} + 0.6\frac{a}{t} \tag{11}$$

$$H_1 = 1 - 0.34\frac{a}{t} - 0.11\frac{a}{c}\left(\frac{a}{t}\right) \tag{12}$$

$$H_2 = 1 + G_1\left(\frac{a}{t}\right) + G_2\left(\frac{a}{t}\right)^2. \tag{13}$$

In this equation for H_2,

$$G_1 = -1.22 - 0.12\frac{a}{c} \tag{14}$$

$$G_2 = 0.55 - 1.05\left(\frac{a}{c}\right)^{0.75} + 0.47\left(\frac{a}{c}\right)^{1.5}. \tag{15}$$

For all combinations of parameters investigated and $a/t \leq 0.8$, eqn (1) was within ±5% of the finite-element results and the single-edge crack solution. (Herein, "percent error" is defined as the difference between eqn (1) and the finite-element results normalized by the maximum value for that particular case. This definition is necessary, especially for the case of bending, for which the stress-intensity factor ranges from positive to negative along the crack front.) For $a/t > 0.8$, the accuracy of eqn (1) has not been established. However, its use in that range appears to be supported by estimates based on the concept of an equivalent through crack. Results from eqn (1) for tension and bending are shown in Figs. 3 and 4, respectively, with the stress-intensity factor plotted as a function of ϕ for several combinations of a/c and a/t to illustrate the characteristics of the equation. The negative stress-intensity factors shown in Fig. 4 are of course, meaningful only in the presence of sufficient tensile loading to prevent contact between the crack surfaces.

Equation (1) was used in Appendix A to predict the growth patterns of surface cracks under tension and bending fatigue loads. The predicted growth patterns are in good agreement with previously published experimental measurements made on steel, titanium alloy, and aluminum alloy material. The maximum stress-intensity values from eqn (1) were also used in Appendix B to correlate surface-crack fracture data from the literature for a brittle epoxy material under tension loads. In these data, the ratios of crack depth to plate thickness ranged from 0.15 to 1.0 and the ratios of crack depth to crack length ranged from 0.3 to 0.84. The equation correlated 95% of the data analyzed to within ±10% of the calculated failure stress.

CONCLUDING REMARKS

Previously obtained stress-intensity factors from three-dimensional finite-element analyses were used to develop an empirical equation for the stress-intensity factor for both tension and bending loads. The equation applies, for any parametric angle, ratios of crack depth to crack length ranging from 0 to 1.0, ratios of crack depth to plate thickness ranging from 0 to 1.0, and ratios of crack length to plate width less than 0.5. For all configurations for which ratios of crack depth to plate thickness do not exceed 0.8, the equation is within ±5% of the finite-element results and the single-edge crack solution. For ratios greater than 0.8, no solutions are

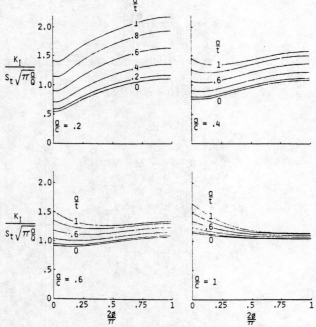

Fig. 3. Typical results from the stress-intensity factor equation (eqn 1) for a semielliptical surface crack in a plate under tension. ($c/b = 0$.)

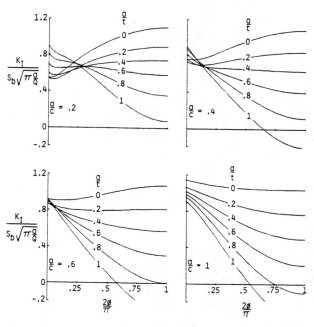

Fig. 4. Typical results from the stress-intensity factor equation (eqn 1) for a semielliptical surface crack in a plate under bending. ($c/b = 0$.)

available for direct comparison; however, the equation appears reasonable on the basis of engineering estimates.

The wide-range equation was used in Appendix A to predict the growth patterns of surface cracks under tension and bending fatigue loads. The predicted growth patterns were in good agreement with previously published experimental measurements made on steel, titanium alloy and aluminum alloy material. The equation was also used in Appendix B to correlate surface-crack fracture data from the literature on a brittle epoxy material. In these data, the ratios of crack depth to plate thickness ranged from 0.15 to 1.0 and the ratios of crack depth to crack length ranged from 0.3 to 0.84. The equation correlated 95% of the data analyzed to within ±10% of the calculated failure stress.

The stress-intensity factor equations presented herein should be useful for correlating fatigue-crack-growth rates as well as in computing fracture toughness of surface-cracked plates.

REFERENCES

[1] F. W. Smith and M. J. Alavi, Stress intensity factors for a penny shaped crack in a half space. *Engng Fracture Mech.* **3**(3), 241–254 (1971).
[2] F. W. Smith and D. R. Sorensen, Mixed mode stress intensity factors for semi-elliptical surface cracks. *NASA CR*-134684 (1974).
[3] A. S. Kobayashi, N. Polvanich, A. F. Emery and W. J. Love, Surface flaws in a plate in bending. *Proc. 12th Annual Meeting Soc Engng Sci.*, Austin, Texas (1975).
[4] K. Kathiresan, Three-dimensional linear elastic fracture mechanics analysis by a displacement hybrid finite element model. Ph.D. Thesis. Georgia Institute of Technology (1976).
[5] I. S. Raju and J. C. Newman, Jr., Improved stress-intensity factors for semi-elliptical surface cracks in finite-thickness plates. *NASA TM* X-72825 (1977).
[6] I. S. Raju and J. C. Newman, Jr., Stress-intensity factors for a wide range of semi-elliptical surface cracks in finite-thickness plates. *Engng Fracture Mech.* **11**(4), 817–829 (1979).
[7] J. C. Newman, Jr. and I. S. Raju, Analyses of surface cracks in finite plates under tension or bending loads. *NASA TP*-1578 (1979).
[8] Bernard Gross and John E. Srawley, Stress-intensity factors for single-edge-notch specimens in bending or combined bending and tension by boundary collocation of a stress function. *NASA TN* D-2603 (1965).
[9] J. G. Merkle, A review of some of the existing stress-intensity factor solutions for part-through surface cracks. U.S. Atomic Energy Commission *ORNL-TM*-3983 (1973).
[10] A. E. Green and I. N. Sneddon, The distribution of stress in the neighborhood of a flat elliptical crack in an elastic solid. *Proc. Cambridge Phil. Soc.* Vol. 46 (1950).
[11] J. C. Newman, Jr., Predicting failure of specimens with either surface cracks or corner cracks at holes. *NASA TN* D-8244 (1976).

[12] Paul C. Paris, The fracture mechanics approach to fatigue. *Fatigue—An Interdisciplinary Approach. Proc.* 10th *Sagamore Army Mat. Res. Conf.*, p. 107. Syracuse University Press (1964).
[13] D. L. Corn, A study of cracking techniques for obtaining partial thickness cracks of pre-selected depths and shapes. *Engng Fracture Mech.* 3(1), 45–52 (1971).
[14] C. S. Yen and S. L. Pendleberry, Technique for making shallow cracks in sheet metals. *Matls Res. Standards* 2(11), 913–916 (1962).
[15] David W. Hoeppner, Donald E. Pettit, Charles E. Feddersen, and Walter S. Hyler, Determination of flaw growth characteristics of Ti-6A1-4V sheet in the solution-treated and aged condition. *NASA CR*-65811 (1968).
[16] Kunio Nishioka, Kenji Hirakawa, and Ikushi Kitaura, Fatigue crack propagation behaviors of various steels. *The Sumitomo Search No.* 17, 39–55 (1977).
[17] F. W. Smith, The elastic analysis of the part-circular surface flaw problem by the alternating method. *The Surface Crack: Physical Problems and Computational Solutions*, (Ed. J. L. Swedlow), pp. 125–152 (1972).

APPENDIX A

Fatigue-crack-growth patterns of surface cracks

The stress-intensity factor equation (eqn 1) developed for surface cracks is used herein to predict fatigue-crack-growth patterns under tension and bending fatigue loads. The predicted growth patterns are compared with experimental data obtained from the literature for steels, titanium alloys, and aluminum alloys.

Procedure. The surface-crack configuraiton considered is shown in Fig. 1. Although eqn (1) gives the stress-intensity factor at any location along the crack front, only the values at the maximum-depth point A and at the front surface B were used to predict the crack-growth patterns. (See insert in Fig. 1.) The cracks were always assumed to be semielliptical with semiaxes a and c.

The crack-growth rates were calculated by assuming that the Paris relationship[12] between crack-growth rate and stress-intensity factor range is obeyed independently at points A and B at the crack front. Thus,

$$\frac{\mathrm{d}a}{\mathrm{d}N} = C_A \Delta K_A^n \tag{A1}$$

$$\frac{\mathrm{d}c}{\mathrm{d}N} = C_B \Delta K_B^n \tag{A2}$$

where ΔK is the stress-intensity factor range at point A or B, n is an exponent to be specified, and C_A and C_B are the crack-growth coefficients for points A and B, respectively. In this paper, n is assumed to be 4, a value which has been found to be applicable to a wide range of materials. Normally, C_A and C_B are assumed to be equal; however, experimental results[13, 14] for surface cracks under tension and bending fatigue loads show that small semicircular cracks tend to grow semicircular for low a/t ratios. Because the stress-intensity factor solution for the small semicircular crack shows that the stress intensity at point B is about 10% higher than the value at point A, the coefficient C_B was assumed to be

$$C_B = 0.9^n C_A \tag{A3}$$

so that a small semicircular crack would be predicted to initially retain its shape. Accordingly, eqn (A3) was used for all crack configurations considered. One reason C_A is not equal to C_B may be the changing relationship between the stress-intensity factor and the crack-growth rate as the stress state changes from plane stress on the front surface to plane strain at the maximum-depth point.

The number of stress cycles required for propagation of a surface crack from an initial half-length c_0 to a desired half-length c_f was obtained by a numerical integration of eqn (A2). This was accomplished by dividing the crack extension $(c_f - c_0)$ into a large number of equal increments Δc and assuming that each increment was created at a constant crack-growth rate. The constant growth rate for each increment was determined from eqn (A2) by using the crack configuration which existed at the start of that growth increment. For each increment of crack advance Δc at the surface, a new increment of crack depth Δa was computed from

$$\Delta a = \frac{C_A}{C_B}\left(\frac{\Delta K_A}{\Delta K_B}\right)^n \Delta c = \left(\frac{\Delta K_A}{0.9 \Delta K_B}\right)^n \Delta c \tag{A4}$$

This defined the crack configuration for the next growth increment, and the process was repeated until the crack depth reached the plate thickness.

Tension. Figure 5 shows the experimental and predicted fatigue-crack-growth patterns for surface cracks subjected to tension. The figure shows the a/c ratio plotted vs the a/t ratio for Ti-6A1-4V titanium alloy[15], 9% nickel steel[16] and 2219-T87 aluminum alloy.

The experimental procedure for the aluminum and titanium alloys was as follows. An electric-discharge machined (EDM) notch was used as a crack starter. The vertical bar in Fig. 5 denotes the range of EDM notch shapes (a/c) for the nine aluminum-alloy specimens. Each specimen was subjected to constant-amplitude cyclic loading for various numbers of cycles and then statically pulled to failure. The data points indicate the final fatigue-crack shapes and sizes measured from the broken specimens. Hence, a separate specimen was necessary to obtain each data point.

The experimental data for the nickel steel were obtained from one specimen which was subjected to two-level variable-amplitude loading and then pulled to failure. The amplitude change caused "marker bands" to be formed on the crack plane. The marker bands, in turn, were used to define the crack shape and size.

The dashed curves in Fig. 5 are the predicted fatigue-crack-growth patterns from eqn (1) and eqns (A1)–(A4). (Note that n is assumed to be equal to 4 and that the growth patterns are independent of the magnitude of C_A and C_B. See eqn A4.)

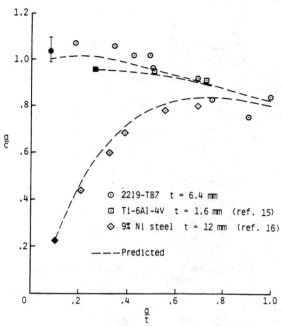

Fig. 5. Experimental and predicted fatigue-crack growth patterns for a surface crack in a plate under tension. (Solid symbols denote initial conditions.)

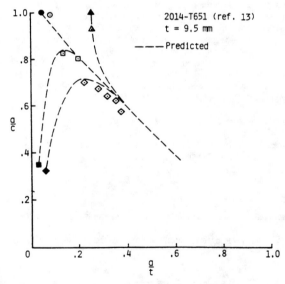

Fig. 6. Experimental and predicted fatigue-crack growth patterns for a surface crack in an aluminum alloy cantilever plate under bending. (Solid symbols denote initial conditions.)

The predicted growth patterns are in good agreement with the measurements made on the three materials. The solid symbols denote initial crack size and shape. For all initial crack shapes considered, the predicted a/c ratio was about 0.8 when the crack depth became equal to the plate thickness.

Bending. Figure 6 shows the experimental and predicted fatigue-crack-growth patterns for surface cracks in plates subjected to cantilever bending. Although the stress-intensity factor equation (eqn 1) used herein was developed for pure bending, the differences between crack-growth patterns for cantilever and pure bending are not expected to be large. The a/c ratio is plotted as a function of the a/t ratio for aluminum-alloy specimens. All data points for the 2014-T651 aluminum alloy[13] were obtained from separate specimens. The specimens were cycled under constant-amplitude loading and then statically pulled to failure. Again, the solid symbols denote the initial crack dimensions.

In Fig. 6, the predicted fatigue-crack-growth patterns (dashed curves) are in good agreement with the experimental data. The predicted results show that the cracks tended to approach a common propagation pattern, as pointed out by Corn[13].

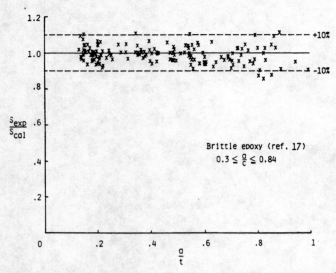

Fig. 7. Correlation of experimental failure stress S_{exp} to calculated failure stress S_{cal} for a brittle epoxy material as a function of a/t.

APPENDIX B

Fracture of surface-cracked brittle materials

The application of linear-elastic fracture mechanics to surface-cracked specimens is complicated because the stress-intensity factor solution is a function of the parametric angle ϕ. For surface cracks with $a/c < 0.6$, the maximum stress-intensity factor occurs at the maximum-depth point, $\phi = \pi/2$; for surface cracks with $a/c \geq 0.6$, the maximum stress-intensity factor occurs near the front surface $\phi = 0$. In fracture analyses, most investigators have used the stress-intensity factor at the maximum-depth point[9]. However, for a/c ratios greater than about 0.6, some investigators have used the value at the front surface because the stress-intensity factor is maximum there. In this paper, the maximum stress-intensity factors from eqn (1) were used. The maximum value occurred at either $\phi = 0$ or $\pi/2$.

Equation (1) was used to analyze data obtained by Smith[17] in a large number of fracture tests on surface-cracked tension specimens made of a brittle epoxy material. (Plane-strain plastic-zone size, based on the largest computed fracture toughness, was two orders of magnitude below minimum specimen thickness and, hence, consistent with brittle fracture behavior.) Thicknesses ranged from 2.5 to 9.5 mm, with $0.15 \leq a/t \leq 1.0$ and $0.3 \leq a/c \leq 0.84$. All specimens were 25-mm wide.

In this analysis, the specimens were arranged into five groups according to their date of manufacture. Specimen thicknesses were constant within each group. Fracture of all specimens was assumed to occur at the same value of stress-intensity factor (denoted herein by K_{cr}). The elastic fracture toughness K_{cr} for each group of specimens was obtained by averaging the calculated stress-intensity factor at failure K_{Ie} as

$$K_{cr} = \frac{1}{m} \sum_{i=1}^{m} (K_{Ie})_i \tag{B1}$$

where m is the number of specimens in a group. The K_{cr} values for the five groups of specimens were calculated as 675, 676, 713, 720 and 725 kN/m$^{3/2}$.

After K_{cr} was determined, eqn (1) was used to calculate failure stresses. The gross failure stresses S_{cal} were calculated from

$$S_{cal} = \frac{K_{cr}}{\sqrt{\pi \frac{a}{Q} F_{max}}} \tag{B2}$$

where F_{max} denotes the maximum value of F for a given a/c, a/t and c/b. Figure 7 shows the ratio of experimental failure stress S_{exp} to calculated failure stress S_{cal} plotted as a function of the a/t ratio. The solid line at unity denotes perfect agreement and the dashed lines denote ±10% scatter. The proposed equation correlated 95% of the data analyzed within ±10% for a wide range of a/t and a/c ratios.

Section Three
Experimental Methods

THE DYNAMIC STRESS DISTRIBUTION SURROUNDING A RUNNING CRACK - A PHOTOELASTIC ANALYSIS

A. A. WELLS*, British Welding Research Association, Abington, Cambridge, England,
and
D. POST**, University of Illinois, Urbana, Ill.

ABSTRACT

A photoelastic analysis is described for the dynamic stress distribution surrounding a running crack started from one edge of a plate of CR-39 in tension. By using the photoelastic fringe multiplication method, it was possible to obtain adequate numbers of isochromatic fringes with models $\frac{1}{8}$ in. thick, without the blurring at the root of the crack which results from using thicker models. A multiple spark technique of the type originated by Schardin was employed, and full frames of successive events were recorded on a stationary photographic plate. In addition, individual principal stress distributions were determined for running cracks in similar models by means of an optical interferometer and spark illumination.

It was found that the dynamic stress distributions in the vicinity of the crack approximated to static distributions in models extended at their ends by a fixed displacement. At greater distances from the crack the distributions approached those for constant load during fracture. For the crack unaffected by the presence of external plate edges, the surrounding zone of stress and strain disturbance was found to grow in all directions proportionally to crack length, as distinct from that surrounding a moving crack of constant length, where the disturbed field remains constant in area and is merely translated with the crack. In the models of limited width the average tensile stresses on remaining uncracked cross sections were found by individual principal stress measurements to increase steadily with length of crack to a maximum of about $1\frac{1}{2}$ times the original value for the almost completely cracked condition. Measurements of terminal crack velocity compared well with calculated values, but observed rates of acceleration were smaller than calculated.

INTRODUCTION

In connection with the study of brittle fractures in steel and other materials, it is of interest to observe the stress and strain distribution surrounding a running crack in an essentially elastic material. In particular, it is desired to know:
1. The manner in which this distribution differs from that surrounding a static crack under comparable loading conditions, and
2. The loading conditions in each equivalent static case which are appropriate to take into account any change of load during fracture.

Crack velocity measurements on numbers of brittle materials, including glasses, plastics and steel, indicate that values much above

---

Presented at the Spring Meeting of the Society for Experimental Stress Analysis in Boston, Mass., May, 1957.

* Visiting scientist at the U.S. Naval Research Laboratory during performance of this investigation.

** Formerly at the U.S. Naval Research Laboratory.

one-third the speed of longitudinal elastic waves in the material are seldom observed[1-7]*. From this observation it follows that points on a tensile specimen which are further away than about three crack lengths from the tip of a running crack can suffer no change of stress or load during fracture. Conversely, there will certainly be changes of stress in the path of the advancing crack, and there need be no equality of tensile load between the center and ends of a long specimen during fracture.

A theoretical comparison between the static and dynamic strain distributions for an internal crack of constant length in a large plate has been made by Yoffe[8]. In this instance, the plane of the crack was considered to be perpendicular to the direction of tension, with the crack moving along its own plane at given velocity. Thus, one end of the crack opens as the other closes. Yoffe established by means of this solution that the dynamic strain distribution at crack velocities up to one-third the velocity of longitudinal waves differed negligibly from the equivalent static distribution. The more usual case of fracture differs from Yoffe's case, in that the crack grows rather than is translated during propagation. Apparently no calculated dynamic strain distribution for such conditions is available in the literature. A most important difference in such a case where, for instance, the crack grows equally at both ends, or in from a free plate edge, is that the region away from the crack tip, bounded by any given value of particle velocity, may be predicted by similitude to increase in area as the square of the crack length, for constant crack velocity. Mott[9] has shown that the growing kinetic energy associated with this velocity field expansion, if fed only from released internal elastic strain energy during fracture, gives rise to a limiting crack velocity. This limit velocity has been calculated[10] on the basis that the dynamic strain distribution surrounding the crack is identical with the static distribution for equal load and extension of the infinite plate containing the crack. A value of 0.38 times the velocity of longitudinal waves was so obtained.

---

* Superiors in brackets pertain to references listed at the end of the paper.

In the study of fracture it is also of interest to examine the relationship between the strain energy release rate associated with an advancing crack and the demand for surface energy to create the crack surface. The well-known Griffith proposition[11], which has recently been applied to cases of brittle fracture in steel and aluminum alloys[12,13] as well as to glass and plastics, treats the surface energy as a material property. When the crack is just beginning to accelerate, under conditions where the strain energy release rate and surface energy just balance, it is unnecessary to examine the strain distribution dynamically, or to account for changes of load as the crack progresses. At an advanced stage of crack propagation, however, these two factors must be accounted for. In particular, it is of interest to examine the assumption that the equivalent static strain field for strain energy release rate determination is that of the so-called "fixed grips", where the ends of the specimen remain fixed as the slot or crack is lengthened.

The photoelastic analysis described below was designed with the use of new and existing techniques so that these questions might be elucidated, with respect to the fracture from one edge produced in a tensile plate of finite width.

PART I
RELATIVE RETARDATION MEASUREMENTS

EXPERIMENTAL METHODS

Dynamic Measurements - Optical System.

The passage of stress waves through plates of transparent plastic material has been examined by Senior and Wells[14] using condenser discharge tubes as an instantaneous light source, and by Christie[15] using multiple spark equipment as developed originally by Cranz and Schardin[16]. Photoelastic observations of dynamic crack propagation have recently been made by Schardin[17] using methyl methacrylate models. This material has low optical sensitivity, and Schardin's records unfortunately do not provide sufficient fringe orders for detailed analyses. Furthermore, the models were $\frac{5}{16}$ in. thick, and there was complete extinction of light within an appreciable region surrounding the crack tip. This was presumably caused by distortions of the plate surfaces near the crack, causing refraction of light away from optical paths reaching the camera lens; this distortion, or dimpling effect, becomes more pronounced for both thicker and tougher model materials.

It was therefore decided that $\frac{1}{8}$ in. thick transparent CR-39 would be employed in the present experimental work. This material offers a reasonably high stress- and strain-optical sensitivity, and exhibits a stress-strain relationship which is characteristic of highly brittle materials. The Schardin spark system was used for photographing dynamic events, with the optical system modified for the method of photoelastic fringe multiplication[18].

The optical system is shown schematically in Fig. 1. Four spark discharges were arranged to fire in sequence at predetermined intervals. Light from each spark was collimated in the model space and converged to enter only one camera lens. Partial mirrors of approximately 60 percent reflectance, which straddled the model as shown, were inclined sufficiently to direct the rays that traversed the model three times into a second row of four camera lenses. Thus, four successive dynamic events were recorded on a stationary photographic plate, and for each event, isochromatic patterns of 1X and 3X fringe multiplication were obtained. A representative sequence for a running crack is displayed in Fig. 3, with patterns of three times multiplication in the upper row and no modification in the lower row. Aside from the obvious virtue of providing greater experimental sensitivity, the availability of both 1X and 3X patterns expedited unique evaluation of fringe orders in the patterns, for one could deduce whether fringe order was increasing or decreasing in a given region by inspection of the two patterns.

Light field circular polarization was employed throughout. The field lenses were 10 in. diameter, with focal lengths of 45 in. and 60 in. for the source side and camera side, respectively. The camera was equipped with eight lenses of the type used in photographic enlargers, $7\frac{1}{2}$ in. focal length, stopped down to f-11.

Models and Method of Loading.

Models were rectangular plates sawed from $\frac{1}{8}$ in. CR-39 sheets. Observations were made for two types of specimens: approximately 5 in. x 15 in. and 3 in. x 19 in. An edge crack was made in each specimen before loading so

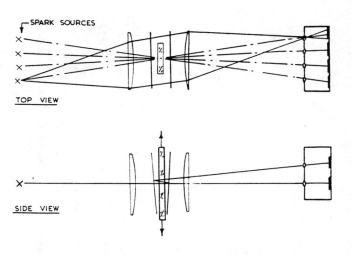

FIG. 1. OPTICAL SYSTEM EMPLOYED FOR DYNAMIC RELATIVE RETARDATION MEASUREMENTS.

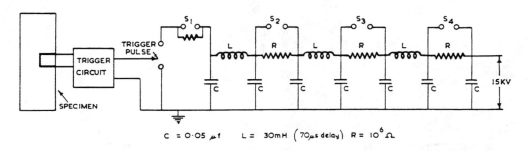

FIG. 2. SCHARDIN TYPE MULTIPLE SPARK UNIT.

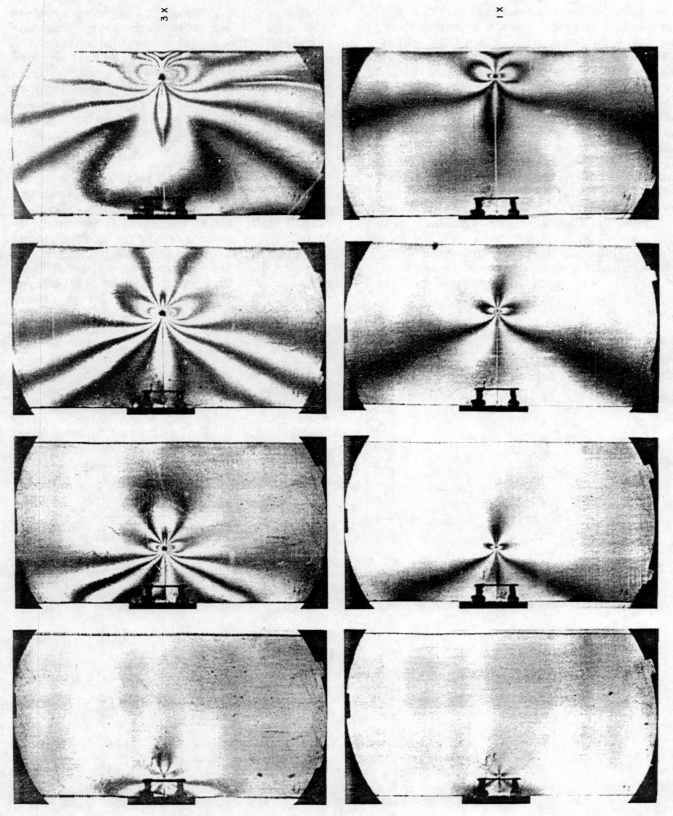

FIG. 3. ISOCHROMATIC DISTRIBUTION DURING CRACK PROPAGATION FOR 15 IN. x 5 IN. SPECIMEN.

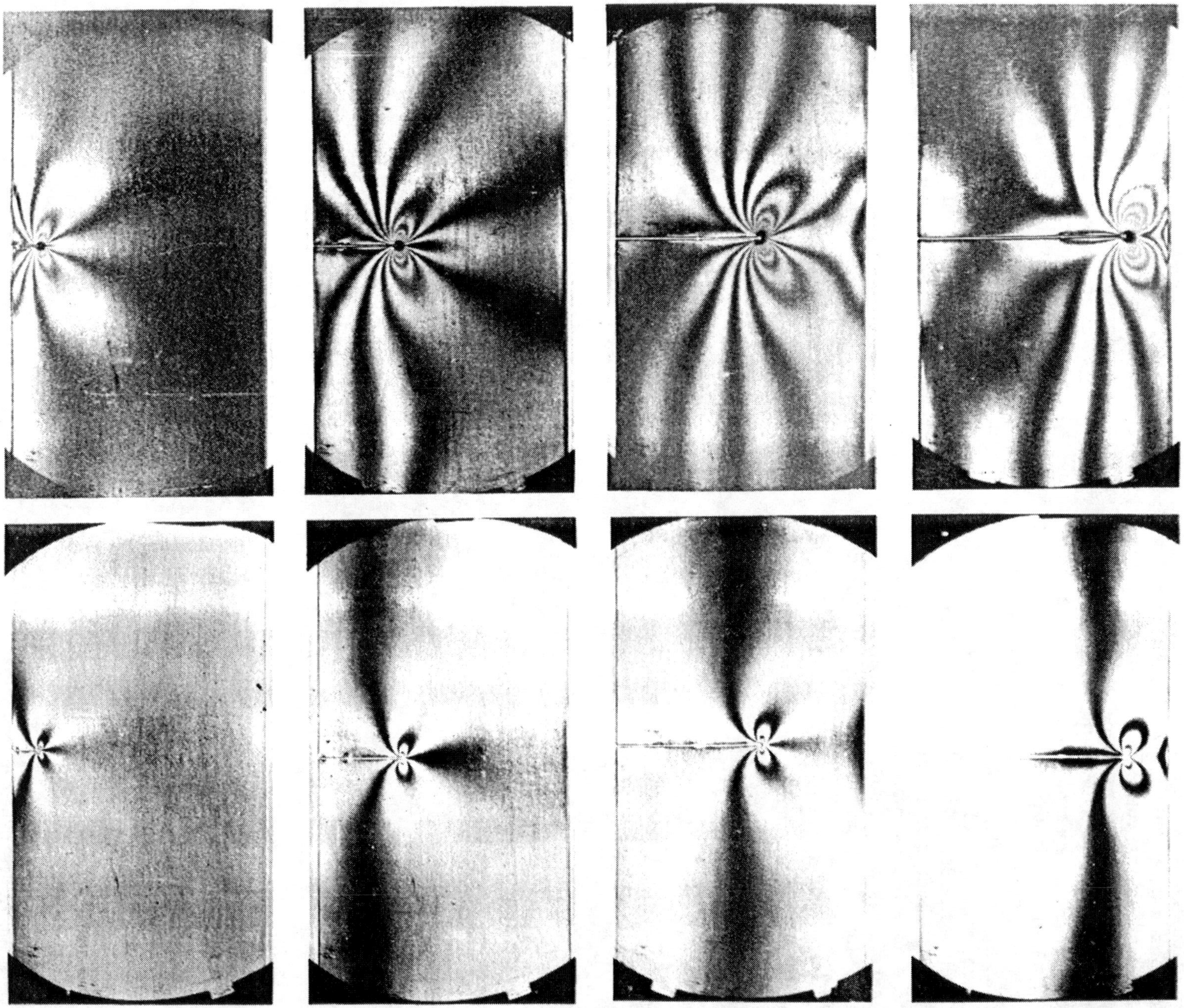

3 X
1 X

FIG. 4. ISOCHROMATIC DISTRIBUTIONS FOR STATICALLY LOADED SLOTTED SPECIMEN SIMULATING PATTERNS OF FIG. 3.

as to produce fracture at the required position from subsequent loading. This was produced as an extension to a fine saw cut by means of a blow on a razor blade placed in contact with the bottom of the cut. Each of the natural cracks so produced was made to a depth of approximately $\frac{1}{8}$ in. below the edge of the specimen. Each specimen was then clamped into pin-jointed, inherently rigid aluminum alloy grips, so that it could be subjected to an increasing, centrally applied tension, until fracture occurred. In earlier experiments, the load was built up to the fracture load in a period of about one minute. For the later work, a tensile blow was applied, but no load measurements were made. The quick application of load, estimated as 0.1 sec. loading time, was used in order to minimize birefringence creep, which had been observed to occur with the slower loading.

The Photographic Method.

In view of its simplicity and convenience, Schardin's [16] spark system was used, with four sparks having intermediate delay times of about 70 microseconds for earlier experiments, and 40 microseconds for later experiments. The arrangement is shown diagrammatically in Fig. 2. The first spark was set off by means of a trigger circuit coupled to a line of silver bearing paint applied to the specimen. When this conducting paint line was severed by the advancing crack, the trigger circuit transmitted a high voltage pulse to

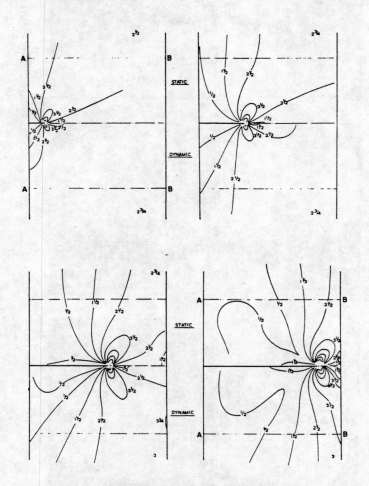

(a) FRINGE ORDER DISTRIBUTIONS.

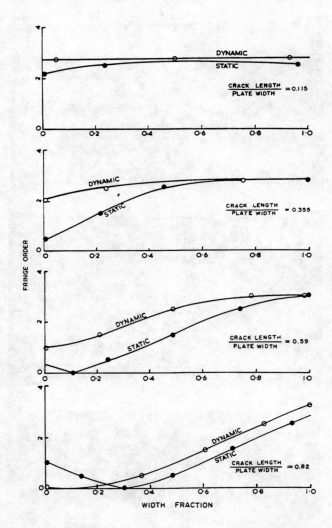

(b) FRINGE ORDER ALONG PLANE AB, ONE-HALF SPECIMEN WIDTH FROM CRACK PLANE. WIDTH FRACTION IS DISTANCE ALONG SPECIMEN DIVIDED BY WIDTH OF SPECIMEN.

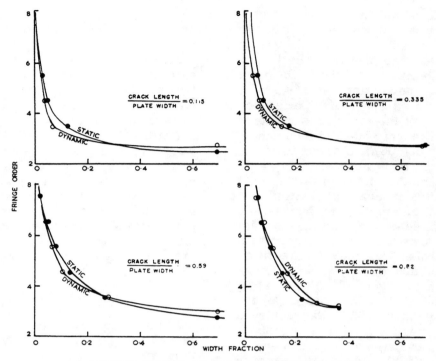

(c) FRINGE ORDERS ON PLANE OF MAXIMUM SHEAR
(60° TO CRACK PLANE, FROM CRACK ROOT).

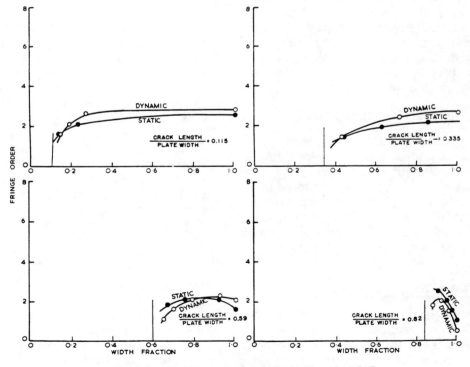

(d) FRINGE ORDERS ALONG UNCRACKED LIGAMENT.

FIG. 5. COMPARISON BETWEEN ISOCHROMATIC DISTRIBUTIONS IN DYNAMIC
AND STATIC CASES; SPECIMENS OF FIGS. 3 AND 4.

the third electrode of an auxiliary spark gap. This pulse discharged across the small gap to ground potential, thereby ionizing the air in part of the auxiliary gap, and reduced the gap resistance sufficiently to discharge the first condenser. Thus, the system was set into operation, with delayed spark discharges occurring in sequence at intervals prescribed by the magnitudes of capacitance and inductance.

With condensers of $0.05\mu f$, charged to 15 kv, discharging through gaps approximately $\frac{1}{8}$ in. across, it was found that sufficient light could be produced to register spark photographs of the one and three times multiplied fringes. An optimum compromise between fringe contrast and photographic sensitivity was obtained with the use of a Corning No. 3387 yellow filter passing radiation above 4200 Å, and Kodak spectroscopic plates, type 103-O, accepting radiation below 5000 Å. By this means, an optical bandwidth of about 800 Å was achieved with a probable mean wavelength of 4600 Å. Greater monochromacity would have been desired but was obtainable only at the expense of light intensity. This lack of monochromatic purity explains the reduction in contrast of higher order fringes in the 3X multiplied isochromatic patterns. From an examination of the closely spaced fringes near the tip of the moving crack, it was estimated that the effective spark duration was of the order of 1 microsecond, which was adequately short.

Time Interval Measurements.

For this purpose, measurements were made in some cases by means of two conducting paint lines intercepted successively by the crack, with the conducting lines coupled to a microsecond counter; for other specimens the time intervals between successive sparks were measured by means of a photocell pickup connected to an oscilloscope having time markers.

Static Measurements.

The isochromatic fringes obtained in the dynamic tests were compared to the results of similar static tests in which fine saw-cuts were used to simulate an advancing crack. Various lengths of saw-cuts were made, but these were always terminated by a small drilled hole which prevented fracture at low loads. The condition of "fixed grips" was chosen for loading in order to obtain a suitable comparison; the ends of each specimen were clamped in a relatively rigid fixture, and an extension of 0.022 in.* was applied uniformly across the width of the specimen. Fringe photographs for this static fixed grip condition were obtained with the same spark illumination and filters, and hence with the same effective wavelength, as the dynamic photographs; these patterns are shown in Fig. 4. In addition, measurements of static load and eccentricity necessary to achieve the plate extension for each slot length were made and were compared with similar values in a plate with a $\frac{1}{8}$ in. slot length, that is, in the condition of the cracked plate just prior to propagation (Figs. 12 and 13).

Material Calibration.

A discussion of material calibration is deferred to Appendix I, in which calibration for relative and absolute retardation measurements are dealt with simultaneously.

RESULTS

In Fig. 3 the photoelastic fringe photographs for one and three times multiplication are shown for a single specimen of 5 in. width, as viewed in the 10 in. diameter polariscope. It will be noted that there is slight loss of detail in the photographs for three times multiplication, due to the oblique incidence of several degrees, but that these indicate much more satisfactorily the birefringence distribution away from the crack. Conversely, the local fringe distribution near the root of the crack is shown more satisfactorily in the unmultiplied photographs. For the first photograph the crack has traveled only a fraction of an inch, and the strain disturbance covers a very small portion of the specimen. All the photographs show an extinction spot near the tip of the crack, and it was confirmed by a change of camera location in a static experiment, that this effect was due to deformation of the specimen in this region, causing a diverging lens effect. Thus the smaller magnitude of the spot in the unmultiplied fringe photographs is readily explained.

---

* See INTERPRETATION.

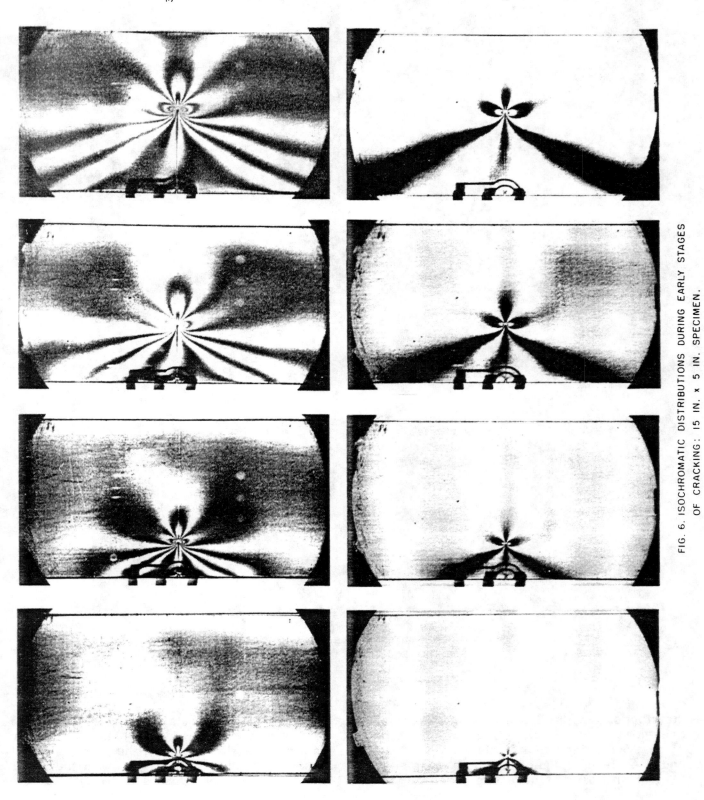

FIG. 6. ISOCHROMATIC DISTRIBUTIONS DURING EARLY STAGES OF CRACKING: 15 IN. x 5 IN. SPECIMEN.

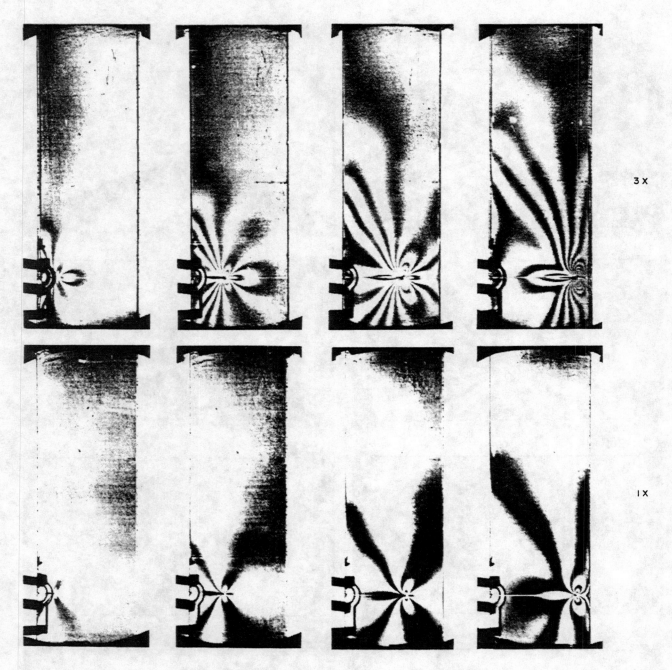

FIG. 7. ISOCHROMATIC DISTRIBUTIONS DURING CRACKING FOR ONE END OF 19 IN. × 3 IN. SPECIMEN.

Detailed comparisons of the isochromatic distributions in the dynamic and static cases are shown in Fig. 5 (a), (b), (c) and (d). Those of Fig. 5(a), translated directly from Figs. 3 and 4, are supplemented by plots along single planes, with all distances expressed as fractions of plate width. Thus, the results for plane AB [see Fig. 5(a)], which lies one-half plate width from the crack plane, are shown in Fig. 5(b). In Fig. 5(c), the plane is at 60° to the crack plane, converging with the latter at the crack root; this is approximately the

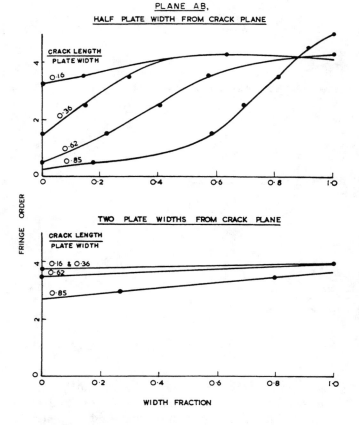

FIG. 8. TRANSVERSE ISOCHROMATIC DISTRIBUTIONS FOR SPECIMEN OF FIG. 7.

plane of maximum principal stress difference. Plots along the crack plane are shown for the remaining uncracked section, hereafter called the unbroken ligament, or ligament, in Fig. 5(d). For regions not too far from the crack, these distributions are in close agreement, excepting a consistent tendency for the dynamic fringe orders to be lower near the root of the running crack than for the corresponding position on the static specimen. This is discussed below. On the other hand, the divergence between the dynamic and static fringe patterns, in regions away from the crack, is already clearly evident for the plane AB in Fig. 5(b). In the dynamic case, there is markedly less disturbance of the original constant stress field as the cracking progresses, as compared with the static case.

The fringe photographs of Fig. 6 were obtained for a 5 in. wide plate in an identical manner to those of Fig. 3, except that time intervals between sparks were shorter, in order that events might be determined in greater detail for shorter cracks, where reflections from an opposite face would not be of significant account.

These results show clearly, without the need for further analysis, the growth of isochromatics in all directions proportional to crack length, which occurs dynamically under such conditions. Since each expanding isochromatic fringe represents a zone of constant principal stress difference, it defines in itself the extent of the disturbance generated by the crack.

In Fig. 7 fringe photographs are shown for a symmetrical 3 in. wide specimen, which is offset in the field of view in order to relate events remote from the crack with those near to the crack. The small degree of load relaxation on a plane two plate widths from the crack plane is shown for this specimen in Fig. 8.

Within regions closer to the crack, where a comparison may be made, the distributions of isochromatics in Fig. 7 are similar to those in Fig. 3, in spite of the difference of absolute specimen size. This is additionally seen in Fig. 8 for the plane one-half width from the crack, to be compared with Fig. 5(b). It is evident from the latter comparison that those differences which do exist between Figs. 7 and 3 primarily arise from different average stresses in the narrow and wide plates, respectively, at the onset of fracture in each case.

Velocity Measurements.

Crack velocities, averaged over short distances, were deduced by means of microsecond interval timers and by the interval between successive sparks of the Schardin system recorded by a cathode ray oscilloscope. The results are shown in Fig. 9. These velocities are expressed in terms of a velocity of longitudinal elastic waves in similar material, due to Clark[19]. The consistency of the measurements on different specimens is striking. Calculated crack velocities[10] are also plotted for comparison, using the relationship:

$$v = 0.38\sqrt{\frac{E}{\rho}}\left(1 - \frac{c_o}{c}\right), \qquad (1)$$

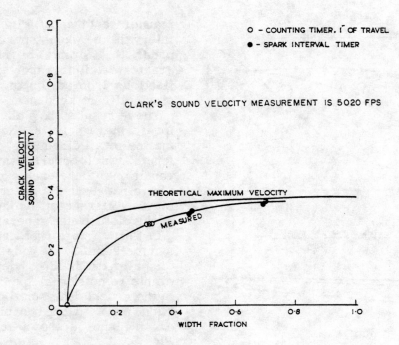

FIG. 9. CRACK VELOCITY MEASUREMENTS.

where E and ρ are Young's modulus and density, and c_o and c are the critical and subsequent crack lengths, respectively*. E/ρ is the longitudinal elastic wave velocity. With the model material and conditions of loading of these experiments, c_o was the original crack length of $\frac{1}{8}$ in. Although the terminal velocity is close to the calculated value, it appears that observed rates of acceleration were lower than those calculated.

PART II
ABSOLUTE RETARDATION MEASUREMENTS

EXPERIMENTAL METHOD

It was next desired to investigate the variation, with crack length, of total load carried by the unbroken ligament. The stress system along the ligament is strongly biaxial for cracks of appreciable length, and therefore total load cannot be determined from isochromatic measurements. The recently developed method of absolute retardation measurements[20], however, permits evaluation of individual principal stresses and appears to be suited for this phase of the dynamic analysis.

In this experimental method, a photoelastic model is observed in an optical interferometer; the two orthogonally polarized rays which propagate with different velocities through each point in the doubly refracting model give rise to two superimposed interference fringe patterns for the model. The two fringe shifts at any point ψ_p and ψ_q, caused by the application of load upon the model, are related to the principal stresses at that point, P and Q, by

$$P = \frac{1}{t}(k_1\psi_p - k_2\psi_q),$$
$$Q = \frac{1}{t}(k_1\psi_q - k_2\psi_p), \quad (2)$$

where t is the model thickness, and k_1, k_2 are calibration constants of the model material. Thus the individual principal stresses are directly evaluable from a single experiment.

The Naval Research Laboratory 9 in. diameter Series Interferometer was employed

---
* A correction incorporated in Eq. (1), with regard to the value given in Ref. [10], was kindly communicated by G. M. Boyd.

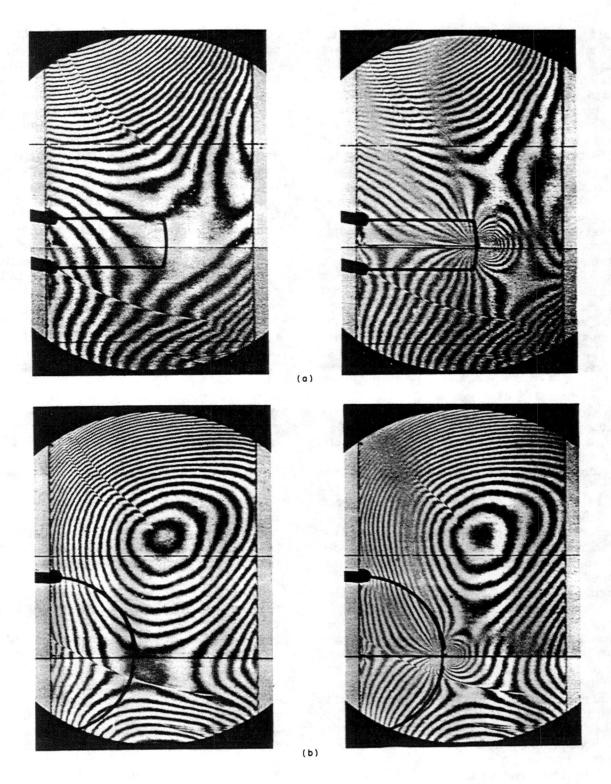

FIG. 10. REPRESENTATIVE INTERFERENCE FRINGE PATTERNS FOR NO-LOAD AND DYNAMIC CRACK PROPAGATION CONDITIONS FOR 15 IN. x 5 IN. SPECIMENS; CRACK LENGTH: (a) 3.10 IN., (b) 2.04 IN.

in this study. Although it was shown in Ref. [21] that observations at oblique incidence are possible, such observations proved impracticable in these experiments. With oblique incidence of collimated light on the interferometer, the "plane" of fringe localization was a severely warped surface in space, and the camera lens could not bring a large portion of the pattern into sharp focus at one setting. This condition prescribed illumination at normal incidence, and precluded the application of the Schardin system of high-speed photography. No other adequate high speed camera system was available, and therefore, single spark photography was employed. The optical system, then, consisted simply of a spark light source, the two 10 in. diameter field lenses described before, the interferometer, and a still camera.

The interferometer imposes requirements for monochromatic purity that are more stringent than before, and an interference filter was used to produce a spectral band of about 150 Å half-width, centered at 4200 Å. The light source consisted of a spark discharge from 15 kv to ground potential. The discharge was confined by an open ceramic tube of small bore, in order to increase intensity; this confinement, however, extended the effective exposure time of the light to about 3 microseconds, which was too long to yield clear fringes in the immediate vicinity of the crack tip - the region of greatest rate of change of fringe order. Photographs were recorded on Kodak type I-0 spectroscopic plates and developed in Kodak D-19 according to recommended practice.

Specimens of CR-39, nominally $\frac{1}{8}$ in. thick and 5 in. wide, were loaded to fracture as before, in approximately 0.1 sec. loading time. Representative interference patterns are shown in Fig. 10 for two different crack lengths. In each case a no-load pattern is required, for the model material is not optically flat, and the fringe order at every point in this pattern must be subtracted from the interference fringe orders at corresponding points in the subsequent dynamic pattern. The second illustration in each case shows the interference pattern an instant after the running crack severed the line of conducting paint used to trigger the spark. The cracks propagating through the plate are just visible above the horizontal line which runs across the entire optical field.

Since only one dynamic pattern was obtained for each specimen, a sequence of experiments was conducted to gather data for progressively increasing crack lengths. Dynamic patterns were photographed for crack lengths ranging from 0.15 to 0.80 of the plate width. While experimental conditions were similar for every specimen, corrections had to be applied to relate the results to a given constant breaking load (see INTERPRETATION). Breaking load was not measured experimentally but was derived from the interference patterns in a manner to be described later. These small differences in breaking load from specimen to specimen may be caused by variation in length and sharpness of the initial crack, and local variations in material toughness.

The two superimposed patterns are clearly indicated in the upper portions of the figure for the dynamic condition; in regions where a dark band of one interference fringe parameter falls upon a light band of the other parameter, hazy grey areas are formed, and in regions where the patterns are out of step by an integral number of fringes, clear black and white bands appear. As described in Ref. [20], the difference of fringe shift values, $\psi_p - \psi_q$, can be derived immediately from such patterns. The horizontal line in each figure is the upper edge of a Polaroid sheet (inserted in the optical path outside the interferometer) whose axis of polarization is parallel to the longitudinal axis of the specimen. Since the directions of principal stresses are all practically parallel and perpendicular to the direction of applied tension (except for a small region near the head of the advancing crack) this Polaroid transmits the light that is polarized in the direction of the p stress system, and absorbs the light polarized in the direction of the q stress system. Thus the fringe parameters are separated in the region covered by the Polaroid sheet, and the fringe shift ψ_p, at any point in this region, can be ascertained by subtracting the fringe order at the point for the no-load pattern from that for the subsequent stress pattern. (This fringe count is started at a point where the fringe shift is zero, here the free external corners on the crack axis). These values, ψ_p and $(\psi_p - \psi_q)$, provide complete fringe shift data. Alternatively, the individual p and q fringes can be traced across the superimposed pattern, and fringe separation by optical techniques can be dispensed with; this latter scheme was employed for some of the analyses.

Again, a discussion of material calibration constants relating to absolute retardation measurements will be deferred to Appendix I. Individual principal stresses were calculated by Eq. (2) from fringe shift measurements and appropriate calibration factors.

RESULTS

Distributions of principal stresses are shown for the unbroken ligament in Fig. 11, and for section AB, one-half plate width from the crack, in Fig. 12. Here, P and Q are larger and smaller principal stress, respectively. These curves are all corrected to represent a constant breaking load of 421 pounds; this small correction is based on the assumption that the magnitude of these elastic stresses are everywhere directly proportional to breaking load (see INTERPRETATION). Broken lines near the crack roots in Fig. 11 represent extrapolations beyond regions of clear information from the interference patterns. These extrapolations were made for load integration using an inverse square-root law relating to distance from the crack tip, in conformity with theory and other observations [22].

Loads that are instantaneously supported by section AB and by the unbroken ligament were obtained by integration of vertical stresses P along these sections. Load variations with crack length are given in Figs. 13 and 14.

With regard to stress distributions on plane AB (Fig. 12), the Q stresses were found to be zero, within limits of experimental accuracy, for each specimen with different crack widths. Thus, the stress system is very nearly uniaxial as close as one-half plate width from the crack, and the isochromatic fringes depend upon only longitudinal stress components for this plane. Accordingly, Figs. 12, 5(b) and 8, representing distribution for 5 in. x 15 in. and 3 in. x 19 in. specimens, may be mutually compared. The agreement between the three is satisfactory.

Conversely, transverse stress components along the ligaments (Fig. 11) are of relatively large magnitude. Existence of the high tensile Q stress components on the ligament shown in Fig. 11 illustrates the inadequacy of isochromatic information for this plane. In

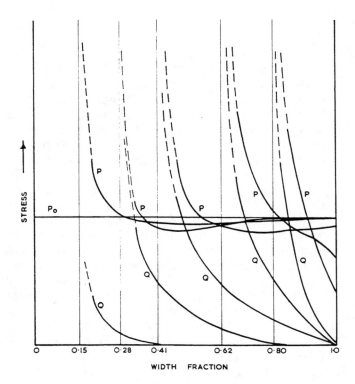

FIG. 11. STRESS DISTRIBUTION ON UNCRACKED LIGAMENT (FOR CRACK LENGTHS EQUAL TO WIDTH FRACTIONS SHOWN).

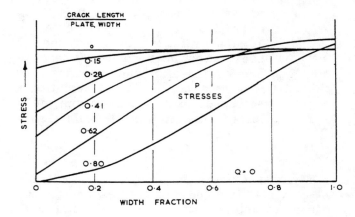

FIG. 12. STRESS DISTRIBUTION ON PLANE AB, ONE HALF PLATE WIDTH FROM CRACK PLANE.

Fig. 13, it is shown that longitudinal loads on ligaments are appreciably smaller than across those planes more remote from the crack. In fact, when the results are reduced to terms of average tensile stresses on a ligament, as in Fig. 14, it appears that no increase greater than 50 percent occurs at any stage of cracking, and movement of the line of load, depicted also in Fig. 14, shows only a limited amount of bending.

Dynamic calibration of the model material, CR-39, showed appreciable variation of stress-optical sensitivity with loading time. Accordingly, the best available estimate of effective loading time at each point was employed to determine calibration factors appropriate for stress determination at that point. This approach is discussed in Appendix I.

PART III
INTERPRETATION AND CONCLUSIONS

INTERPRETATION

A. Effects of Creep and Breaking Load Variation.

An attempt has been made to present the results with the minimum of ambiguity. Nevertheless, since CR-39 shows both creep and some lack of homogeneity with regard to fracture, it follows that certain corrections may be necessary in order to infer results for an ideal elastic material under plane stress or strain. With this in mind two principles were followed in the investigation, as follows:
1. The results have been limited to distributions of birefringence or stress at each crack length with respect to constant but not necessarily specified breaking load for each specimen,
2. Creep was treated as a variation of stress-optical sensitivity and modulus of elasticity with loading time.

Creep.

Mylonas[23], in showing for other transparent plastics that creep rates are proportional to applied stress, has established the general convenience of the latter treatment. The work of Clark[19], proceeding in parallel

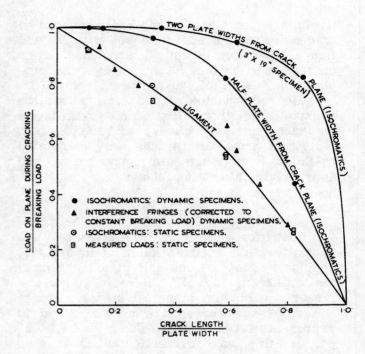

FIG. 13. LONGITUDINAL LOAD VARIATIONS WITH CRACK LENGTH FOR PLANES AT DIFFERENT DISTANCES FROM CRACK PLANE: 5 IN. AND 3 IN. SPECIMEN WIDTHS.

with this investigation, provided essential calibration information in this respect, by establishing the changes of each of the calibration factors within a wide range of loading time. Results of simple tests over the upper range of loading times for the material of this investigation were extrapolated to shorter times in accordance with Clark's observations for another batch of similar material, as reported in Fig. 15.

On account of (1) above, no difficulties arose with respect to results of Figs. 3, 6, and 7, since each group of photographs for different crack lengths was obtained for the same breaking specimen. On the other hand, since loading times on each specimen varied from 10^{-1} seconds away from the cracks to the order of 10^{-4} seconds nearer to them, it follows from Fig. 15 that the results cannot be interpreted directly in terms of principal stress differences, since the stress-optical coefficient changed by 20 percent over this range of loading times. Conversely, bearing in mind the fairly close proportionality at all

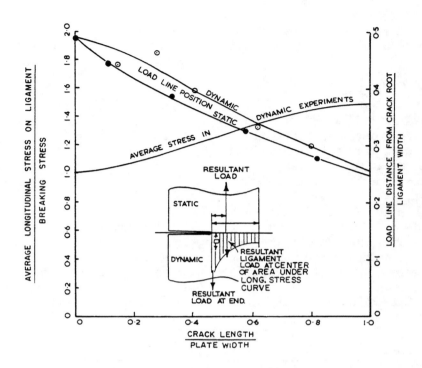

FIG. 14. RELATIVE MAGNITUDE OF AVERAGE STRESS ON LIGAMENT AND POSITION OF LOAD LINE DURING CRACKING.

loading times of stress-optical coefficient and modulus of elasticity, the results might be more easily expressed in terms of strains. In fact, the results remain in terms of birefringence in this instance; nevertheless, distributions shown in Figs. 5(a), (c) and (d) may be taken to represent stress differences, if it is kept in mind that values at points very near the running crack may be underestimated by amounts up to 20 percent if calibration for 10^{-1} sec. is used. In the case of stresses derived from interference measurements, variations in calibration constants were accounted for, and stress distributions are not limited by the above reservation.

Fixed Grips Extension.

All specimens were loaded quickly to fracture in order to minimize these creep effects, and direct and accurate breaking load measurements could not easily be made. Thus it was necessary to infer the required loading conditions in another way, for the static, slotted specimens of Fig. 4. Fundamentally, it was desired to examine the "fixed grips" situation, where each static specimen was extended at its ends by the same amount as the specimen of Fig. 3 at the instant of initiation. This extension may be calculated for such a running crack specimen; relevant data are as follows:

N, fringe order remote from crack, Fig. 3, for shortest crack length
= 2.75 at 3X multiplication,

c_σ, stress-optical coefficient for 10^{-1} sec. loading time, from Fig. 15
= 109 lbs/sq. in./fringe/inch thickness at 5460 Å wavelength.

E, Modulus of Elasticity, for 10^{-1} sec. loading time, from Fig. 15
= 390,000 lbs/sq. in.,

Mean wavelength of spark illumination
= 4600 Å,

Specimen,
length ℓ = 15 in.,
thickness $t = \frac{1}{8}$ in.

Thus the calculated extension, neglecting any transverse stress with regard to possible birefringence and strain effects when the crack length is very small ($\frac{1}{8}$ in.) becomes:

$$\frac{N}{t} C_\sigma \frac{1}{E} = \frac{2.75 \times 8}{3} \left(109 \times \frac{4600}{5460}\right) \frac{15}{390,000} = 0.026 \text{ in.}$$

Since the stress-optical coefficient and Young's modulus are closely proportional for a wide range of loading times, it follows that the loading time for the static specimen should be immaterial, provided that constant extension be maintained.

However, at the time of the experiment a constant extension of 0.022 in. for all the specimens of Fig. 4 was found to give the best correspondence between the dynamic and static fringe patterns, for regions in the vicinity of the unbroken ligament.

Since the calculated extension above is considered accurate, it follows that the discrepancy, as compared with the measured extension giving an equivalent strain pattern, may not be satisfactorily explained either in terms of creep or breaking load variations. Thus it represents a genuine difference between the dynamic case and the simple fixed grips case, and is further discussed below.

Correction for Random Breaking Loads - Interference Specimens.

The resultant loads transmitted by the unbroken ligament, or by section AB, as derived from the absolute retardation measurements, cannot be compared directly, since the cracks were propagating under the influence of somewhat different end loads for each absolute retardation specimen. A correction was em-

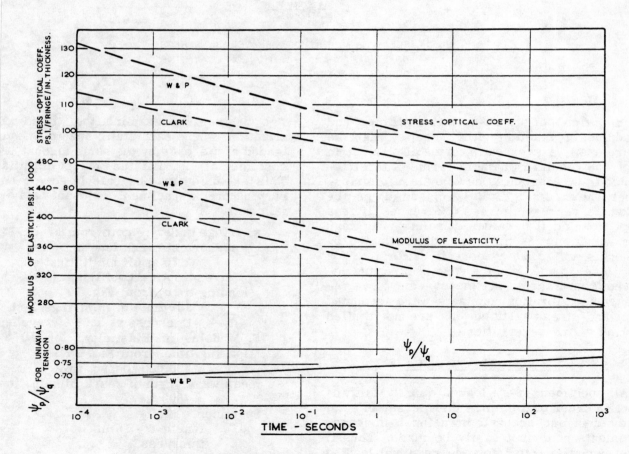

FIG. 15. CR-39 CALIBRATION: MATERIAL PROPERTIES vs. TIME AT WAVELENGTH 5460 Å. $\frac{1}{8}$ IN. CR-39 PLATES: WELLS AND POST; $\frac{1}{2}$ IN. CR-39 BARS: CLARK.

ployed in an effort to make the load values representative of the behavior of a single crack, propagating through a single specimen.

The actual breaking loads for these specimens are all approximately the same; correction for these small changes, based on the assumption that stress magnitudes are everywhere proportional to breaking load, is then justifiable.

On planes AB, one-half specimen width from the crack plane, transverse Q stresses are effectively zero. Also, the P stresses are oriented parallel to the direction of tensile load. The load on AB can therefore be evaluated from relative retardation and absolute retardation measurements as $\int_A^B (P-Q) da$ and $\int_A^B P da$, respectively.

The sequence of dynamic isochromatic patterns shown in Fig. 3, for a single propagating crack, was used as a datum for this correction. The loads represented by the first of the two integrals above were evaluated for this specimen, and they are given in Fig. 13 as a nondimensional plot against crack length. The second integral was determined for each absolute retardation specimen, and a correction factor $\int_A^B (P-Q) da / \int_A^B P da$ was applied to all stresses derived from the respective absolute retardation pattern. Thus, on the basis of proportionality between stress and initial breaking load, all results are reduced to the common datum established by Fig. 3.

The actual breaking load for the relative retardation specimen in Fig. 3 was calculated from the initial isochromatic fringe order, the material calibration factor, and dimensions of the specimen, as 421 pounds.

B. General.

If it is accepted from previous evidence[8,22] that dynamic stress distributions near running cracks are qualitatively similar to those in equivalent plates loaded under static conditions, it is now of interest to examine such similarities and differences as exist in the results of the investigation now described. Perhaps the most striking similarity lies in the manner of expansion of the disturbed strain field near the running crack, in all directions, as in Fig. 6, just as would be expected under static conditions. The geometrical similarity of the fringe patterns at each crack length is not distorted until the crack approaches the center of the specimen, as shown in Fig. 3. Here reflected stress waves play their part, in addition to the static boundary condition.

On the other hand, the presence of the stress waves impairs the static and dynamic similarity in an obvious way by introducing differences of load along the length of the specimen in the manner of Fig. 13, where the isochromatic and interference results are collected. It is seen that, whereas load falls off the ligament somewhat less than proportionally to crack travel, there is little drop of load at two plate widths distance from the crack until an advanced cracking stage. This is consistent with the finite velocity of the stress waves in communicating disturbances from the crack. With reference to that part of Fig. 14 derived from the interference specimens, it is seen that the average stress on the ligament increases by a maximum of 50 percent during cracking, and there is also an indication of some bending in the plane of the specimen, as shown in Fig. 12, and by the position of the load line, as in Fig. 14. Nevertheless, these are comparatively mild changes compared with those which would obtain if constant load and load line were maintained on the ligament during fracture. In fact, it is obvious that reasonably good agreement between the dynamic and static distributions for the ligament zone has been achieved, as shown in Fig. 5, using the approximate assumption of fixed grips, rather than fixed load during fracture. Three main sources of difference may be discussed.

Firstly, the calibration information of Fig. 15 suggests a loading time sensitivity which would affect the proportionality of stress to fringe order in the dynamic case for regions very near to the crack tip. In this region dynamic stresses would be underestimated by within 20 percent. Reference to Fig. 5(c) and (d), particularly for results at the middle two crack lengths, may illustrate this, as they show the fringe orders to be locally lower in the dynamic as opposed to the static case, for regions near the crack. If the loading time correction were to be applied in the dynamic case, the divergence in terms of stress in these regions, would be much reduced.

Secondly, the static specimens possessed slots which were not only wider than the crack, but were also terminated by drilled holes in order to prevent premature fracture. From the theoretical point of view, differences of stress distribution would be expected near the end of the slot or crack, for this reason alone. It is of interest to examine their theoretical magnitude. In order to secure a common elliptical coordinate system, the center of the drilled hole should be taken to be at a position one-half of its radius back from the end of the equivalent crack, and this was established in the models. Under such conditions, it is found for the crack plane, that the ratio between a longitudinal stress component at distance R from the root of the crack and that at a corresponding point in the slotted and drilled model, from the Inglis[24] elastic theory, is simply $2R/(2R+\rho)$, where ρ is the radius of the hole. This is so close to unity even at points near the hole in the static case, that it may be dismissed as a possibly significant source of variation from the unattainable ideal in the same case, of a static natural crack to match completely the running crack. Nevertheless, being less than unity, the ratio is also of the same sense as shown by observations in Fig. 5(c) and (d).

Thirdly, it should be borne in mind that the good dynamic and static comparison of Fig. 5 for regions near the ligament, depends upon a static plate extension 15 percent less than the true extension for fixed grips at the ends of the dynamic specimen. This may be explained in terms of the non-uniformity of load along the dynamic specimen. Let the length of the specimen be L, width B, crack length b at any time, and let σ be the average longitudinal stress near the grips at the onset of fracture. From Fig. 14, the average longitudinal stress on the ligament may be approximated as $\sigma' = \sigma[1+(b/2B)][1-(b/B)]$ for the dynamic case, by linearizing the average stress curve. In order to match the dynamic stress distribution near the ligament with that in a statically loaded plate, the latter should carry the lower average stress σ' along its entire length.

In the dynamic case, however, the average longitudinal stress near the grips is σ, and reduces gradually to σ' at the ligament. Hence, the total extension of the dynamic specimen lies between the limits $\sigma L/E$ and $\sigma' L/E$. From Fig. 13, curves of load, or average stress across the full plate width, can be plotted against distance from the plane of the ligament for each crack length. The area under these stress-length curves will be proportional to an extension from the grip to the cracking plane, averaged at the latter position over the whole plate width (both cracked and uncracked sections), and will thus be less than the original value when no crack existed.

This dynamic extension can be simulated by a schematic counterpart in which an unspecified length, kb, of the specimen is assumed to be unloaded by the crack to the lower average stress σ', across the full plate width; the remaining specimen length, L-kb, is then assumed to remain at the higher average stress level σ. This schematic behavior can be superimposed upon the average stress vs. length diagram as a stepped function, and by adjusting the length, kb/2, of the step to produce equal areas under the curves, the total extensions are made equal for the schematic and actual conditions. Such curves were plotted, and as a result, it appears that a representative value of $k = 2\frac{1}{2}$ would produce equal total extensions for the actual and schematic dynamic specimens.

Returning again to the correlation between static and dynamic extensions required to produce the best match of stress distributions near the ligament, the dynamic extension may be computed from the schematic distribution of average stress, whereas the static extension remains $\sigma'L/E$. Then, the difference, δ, between the static and dynamic extensions becomes

$$\delta = \frac{\sigma(L-kb)}{E} + \frac{\sigma'kb}{E} - \frac{\sigma'L}{E}.$$

When expressed as a percentage of the initial extension of the dynamic plate, $\Delta = \sigma L/E$, this reduces to

$$\frac{\delta}{\Delta} = \frac{b}{2B}\left(1 - \frac{kb}{L}\right)\left(1 + \frac{b}{B}\right) \times 100\%. \tag{3}$$

Using $k = 2\frac{1}{2}$, and $L/B = 3$ for the specimens of Figs. 3 and 4, the following values of δ/Δ were computed by Eq. (3):

b/B	0	0.2	0.5	0.8	1.0
δ/Δ, %	0	10	22	24	17

These values are of the order of the 15 percent reduction of extension found by trial, and thus this difference between static and dynamic fixed grip extensions is justified. The effect is considered important, even if the factor k may not be presently stated with precision. There is no reason, for instance, why it should be independent of the crack width ratio b/B, as has been assumed only for the sake of simplicity.

The inference from Eq. (3) is that dynamic stress conditions near the ligament are best matched by a static extension computed from the hypothetical schematic distribution of elongation with $k = 2\frac{1}{2}$. For short crack lengths, the possible error is small in any case. Nevertheless, neglect of this reservation in long specimens would be misleading.

At this stage of development of equivalent static conditions, additional data of Figs. 13 and 14 may be emphasized. Thus in Fig. 13 the progressive change of ligament load with crack length is shown for the static as well as for the dynamic specimens, and the agreement is excellent with regard both to external load and integrated fringe order measurements. These were all taken with the specimens of Fig. 4, and the fringe order estimate was made across the plane AB, one-half plate width from the crack plane, where transverse stresses could be neglected. Naturally, in the static as opposed to the dynamic case, the load could be taken constant along the specimen, and could be measured at any convenient cross section.

The static load line positions on the ligament, shown in Fig. 14, indicate somewhat more bending than in the dynamic case, and this effect also appears in Fig. 5(b). This effect could have been neutralized in the static, fixed grips case, if a thrust had been applied to the specimen along the crack plane in the direction of propagation, of small magnitude compared with the breaking load. Nevertheless, it would be undesirable to press the fixed grips analogy quantitatively to this extent; the point is mentioned primarily to illustrate the existence of the transverse reaction arising in the dynamic case, which is balanced purely by inertia forces from corresponding accelerations in the opposite direction, of parts of the plate near the crack plane.

It is interesting to note that the curves of measured and calculated crack velocity shown in Fig. 9 could be almost completely superimposed if the crack length scale were to be opened up about $2\frac{1}{2}$ times in the theoretical case. By inspection of Eq. (1) this would imply a rapid increase, after the start of cracking, of the critical crack length, hence surface energy, by the same amount. On the other hand, it is also possible that the true critical crack length could have been greater than the original value of $\frac{1}{8}$ in., irrespective of roughening, if an extension could have been produced in a creeping manner during load application.

CONCLUSIONS

1. The disturbance from a running crack in a brittle transparent material is shown by photoelastic observation to grow uniformly in all directions proportionally to crack length, when external boundaries are sufficiently distant to exert negligible influence of themselves.

2. In the absence of inelastic effects, the stress distribution in the vicinity of such a running crack approximates that in an equivalent statically loaded, slotted specimen, extended at its ends by a fixed displacement. This displacement is calculated on the basis of a schematic behavior in which a portion of the specimen length, kb, is assumed to carry the lower average longitudinal stress that exists in the crack plane, whereas the remaining length carries the full breaking stress. For relatively short cracks in a plate of given width, the length kb is not critical, but for representation of an advanced stage of cracking this length should be about $2\frac{1}{2}$ times the crack travel, in order to make allowance for the finite velocity of communication of the disturbance by means of elastic waves.

3. In regions more remote from the crack than those discussed in (2) above, there is little disturbance of stress arising from the crack propagation.

4. At an advanced stage of cracking the average stress on the uncracked ligament increases to a value estimated experimentally to be not greater than $\frac{3}{2}$ times the value at the onset of cracking. At this same advanced stage, a nominal amount of bending in the plane of the plate is also present on the ligament.

5. In the experiments described, measured crack velocities did not exceed 0.38 of the

longitudinal elastic wave velocity, in accordance with the theory quoted. Slower observed rates of crack acceleration, in comparison with theoretical values, are ascribed to uncertainties of critical crack length observation.

ACKNOWLEDGMENTS

The work described in this paper was conducted during 1954-1955 in the Mechanics Division of the Naval Research Laboratory, Washington, D. C., under the supervision of Dr. G. R. Irwin, Superintendent, as part of problem 62F01-03.

Gratitude is warmly expressed by the authors for the provision of excellent facilities, and for the arrangement by means of which one of the authors (Wells) was enabled during the course of the work to be employed as a visiting scientist on leave from the United Kingdom.

In addition to invaluable advice and encouragement received from Dr. Irwin, through his strong personal interest in furtherance of the work, much guidance and help was received from colleagues in the Mechanics Division, particularly from Dr. J. M. Krafft and Mr. A. B. J. Clark with regard to the dynamic properties of the photoelastic material, and from Mr. S. O. Bailey on the design and operation of the system of spark illumination.

Opinions expressed in the paper are the responsibility of the authors alone.

APPENDIX I

CALIBRATION CONSIDERATIONS

Mechanical and optical properties of the model material, CR-39, are significantly sensitive to rate of stress and time elapsed at fixed stress level. Accordingly, photoelastic measurements cannot be interpreted directly as stress magnitudes without reservations.

A dynamic calibration of CR-39 was conducted by Clark in cooperation with the present authors, and is reported separately[19]. Briefly, Clark impacted $\frac{1}{2}$ in. x $\frac{1}{2}$ in. CR-39 rods with high-velocity projectiles, and simultaneously measured mechanical strain (by electric strain gages) and birefringence (by phototube response to isochromatic fringes) as the longitudinal elastic wave passed the gage site on the bar. Clark's results for stress-optical coefficient and modulus of elasticity are presented in Fig. 15. To apply those results to the actual model material, samples of $\frac{1}{8}$ in. CR-39 plate were calibrated for stress-optical coefficient and modulus of elasticity for the longer range of loading time, and these data were extrapolated to shorter times in accordance with the trends established by Clark, as in Fig. 15. These curves are employed to represent all crack specimens of the experiment, for values measured statically matched within 2 percent for samples taken from each sheet of $\frac{1}{8}$ in. CR-39.

For the interference analyses, two calibration factors must be provided. In the case of uniaxial tensile stress, Eq. (2) reduces to

$$\sigma = \frac{1}{t}(k_1\psi_p - k_2\psi_q),$$
$$\theta = \frac{1}{t}(k_1\psi_q - k_2\psi_p), \quad (4)$$

where σ is the applied tensile stress. By virtue of the identity

$$2N = \psi_p - \psi_q,$$

where N is isochromatic fringe order, and the stress-optic relation

$$C_\sigma = \frac{(P-Q)t}{N},$$

where C_σ is stress-optical coefficient for relative retardations, Eq. (4) reduces to

$$k_2 = \frac{C_\sigma}{2\left(\frac{\psi_p}{\psi_q}\right)+1},$$
$$k_1 = \left(\frac{\psi_p}{\psi_q}\right)k_2. \quad (5)$$

Thus, calibration constants for absolute retardations are expressed in terms of the relative stress-optical coefficient, already established, and the ratio of fringe shifts developed at points subjected to uniaxial tension.

A static calibration (3-minute value) of ψ_p/ψ_q was ascertained by an additional calibration experiment utilizing the interferometer apparatus. A 10^{-1} sec. value was secured by observation of regions of uniaxial stress in the dynamic crack propagation experiments (Fig. 10). The sensitivity of this factor to loading time is minor, and a curve of ψ_p/ψ_q vs. time was drawn through these two points as a straight line in Fig. 15. This curve, together with the curve for stress-optical coefficient, was used to determine calibration factors of Eq. (2), for any loading time.

In the case of analysis of the isochromatic fringe patterns, the results were left in terms of birefringence. It was thought to be useful, however, to analyze the stresses from interference specimens, particularly those near the crack, taking loading time, hence creep, into account. The time applicable for a point that is remote from the crack and not yet affected by the presence of the crack is estimated as 10^{-1} sec. From Fig. 10 it may be estimated that a stress varies from its maximum value at the crack root to approximately that existing prior to the disturbance in about 1 or 2 in. on the crack plane. Thus at some point within 1 or 2 in. ahead of the crack traveling at about 2000 fps a time of the order of 10^{-4} sec. would be required to reach maximum stress. Since the stress at the crack tip is many times that in the undisturbed regions, the stress-optical coefficient for the tip of the crack might be assigned for an effective loading time of 10^{-4} sec. Where the change is of the order of magnitude of that caused by the initial breaking load, however, the over-all stress-optical coefficient should reflect the 10^{-1} sec. loading time. For example, a stress twice the magnitude of the undisturbed value might be computed with a coefficient midway between the 10^{-1} and 10^{-4} sec. values, and pro rata. This procedure was adopted and the estimated loading times for regions of major interest varied from 10^{-3} to 10^{-1} secs.

It is significant to observe here that application of variable calibration factors is peculiar to the problem at hand - in which the dynamic event occurs at a relatively long time interval after loading commences. For the more usual dynamic phenomena, the total effective loading time would not be likely to vary within a model by more than one order of magnitude, and for all practical purposes, constant calibration factors would be applicable.

REFERENCES

[1] Schardin, H. and Struth, W., Glastech Ber., Vol. 16 No. 7, p. 219, 1938.

[2] Edgerton, H. E. and Barstow, F. E., Jl. Amer. Ceramic Soc., Vol. 24, p. 131, 1941.

[3] Kennedy, H. E., Welding Res. Suppt., Vol. 10 No. 11, p. 597-s, 1945.

[4] Hudson, G. and Greenfield, M., Jl. App. Physics, Vol. 18, p. 405, 1947.

[5] Boodberg, A. and Collaborators, Welding Res. Suppt., Vol. 13 No. 4, p. 186-s, 1948.

[6] Smith, H. L. and Ferguson, W. J., Naval Res. Lab. Progress Rep., Apr. 1950.

[7] Robertson, T. S., Jl. Iron and Steel Inst., Vol. 175, p. 361, 1953.

[8] Yoffe, E. H., Phil. Mag., Vol. 42, p. 739, 1951.

[9] Mott, N. F., Proc. Conf. Brittle Fracture in Mild Steel Plates, B.I.S.R.A., p. 82, 1945. Also: Engineering, Vol. 165, p. 16, 1948.

[10] Roberts, D. K. and Wells, A. A., Engineering, Vol. 178, p. 820, 1954.

[11] Griffith, A. A., Phil. Trans. Roy. Soc., A., Vol. 221, p. 163, 1921. Proc. Int. Cong. App. Mech., Delft, p. 55, 1924.

[12] Orowan, E. and Felbeck, D. K., Welding Res. Suppt., Vol. 20, p. 570-s, 1955.

[13] Irwin, G. R., Naval Res. Lab., Report No. 4763.

[14] Senior, D. A. and Wells, A. A., Phil. Mag., Series 7, Vol. 37, p. 463, 1946.

[15] Christie, D. G., Trans. Soc. Glass Technology, Vol. 36, p. 74, 1952.

[16] Cranz, C. and Schardin, H., Zeit. Phys., Vol. 56, p. 147, 1929.

[17] Schardin, H., Kunststoff, Vol. 44 No. 2, p. 48, Feb. 1954.

[18] Post, D., Proc. Soc. Exp. Stress Analysis, Vol. XII No. 2, p. 143, 1955.

[19] Clark, A. B. J., Proc. Soc. Exp. Stress Analysis, Vol. XIV No. 1, p. 195, 1956.

[20] Post, D., Proc. Soc. Exp. Stress Analysis, Vol. XIII No. 2, p. 119, 1956.

[21] Post, D., Jl. Opt. Soc. Amer., Vol. 44 No. 3, p. 243, 1954.

[22] Post, D., Proc. Soc. Exp. Stress Analysis, Vol. XII No. 1, p. 99, 1954.

[23] Mylonas, C., Proc. 7th Int. Cong. App. Mech., London, Vol. 4, p. 165, 1948; see also Mylonas, C., Proc. Soc. Exp. Stress Analysis, Vol. XII No. 2, p. 129, 1955.

[24] Inglis, C. E., Trans. Inst. Naval Arch., Vol. 55, Pt. 1, p. 219, 1913.

DISCUSSION

by

G. R. IRWIN,

U. S. Naval Research Laboratory, Washington, D. C.,

of the Preceding Paper Entitled:
THE DYNAMIC STRESS DISTRIBUTION SURROUNDING
A RUNNING CRACK - A PHOTOELASTIC ANALYSIS

Assuming the stress field near the end of a running brittle crack can be essentially duplicated statically as shown in the paper, several analyses in extension of those furnished by the authors are of interest.

To obtain the simplest possible theoretical representation of the fringe patterns observed near the end of the crack, consider the stress equations which would be valid in the limit as the size of the crack tip region in question becomes very small compared to the crack length. In terms of coordinates as shown in Fig. 1 and neglecting non-linear or plastic strains, these stress equations are:

$$\sigma_y = K \frac{\cos \theta/2}{\sqrt{2r}} \left\{ 1 + \sin \frac{\theta}{2} \sin \frac{3\theta}{2} \right\}, \tag{1}$$

$$\sigma_x = K \frac{\cos \theta/2}{\sqrt{2r}} \left\{ 1 - \sin \frac{\theta}{2} \sin \frac{3\theta}{2} \right\} - \sigma_{ox}, \tag{2}$$

$$\tau_{xy} = K \frac{\cos \theta/2}{\sqrt{2r}} \sin \frac{\theta}{2} \cos \frac{3\theta}{2}. \tag{3}$$

These equations are characterized by two constants, K and σ_{ox}. A statement of the values of the two constants is therefore equivalent to a description of the stresses surrounding the end of the crack. As has been shown by the writer[1,2]* the stress intensity factor, K, is related to the crack extension force, $\mathscr{G}$, by the equation

$$K^2 = E \mathscr{G}/\pi, \tag{4}$$

where E is Young's modulus.

The maximum shearing stress in the plane of the plate is given by

$$\tau_m = \sqrt{\left(\frac{\sigma_y - \sigma_x}{2}\right)^2 + \tau_{xy}^2}.$$

Substituting into this relation from Eqs. (1), (2), and (3) one finds that the expected shape of isochromatic fringes can be predicted from the relation

$$(2\tau_m)^2 = \left\{ \frac{K}{\sqrt{2r}} \sin \theta + \sigma_{ox} \sin \frac{3\theta}{2} \right\}^2 + \left\{ \sigma_{ox} \cos \frac{3\theta}{2} \right\}^2. \tag{5}$$

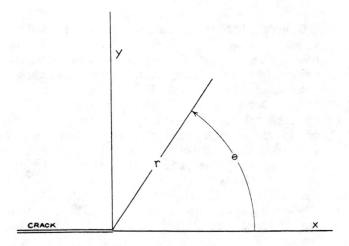

FIG. 1. COORDINATES x,y AND r,θ IN THE PLANE OF THE SHEET.

---

* Superiors in brackets pertain to references listed at the end of the discussion.

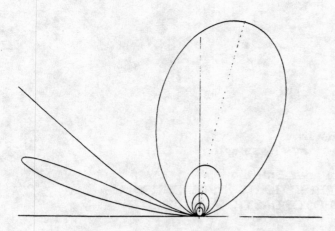

FIG. 2. THEORETICAL ISOCHROMATICS CLOSE TO THE END OF A CRACK. THESE CORRESPOND TO VALUES OF $2\tau_m/\sigma_{ox}$ OF 0.5, 1, 3, 5, 7, 9, AND 11 BEGINNING WITH THE CLOSED LOOP TO THE LEFT AND ENDING WITH THE SMALLEST LOOP.

Since the sum of the extensional stresses is an invariant,

$$\sigma_x + \sigma_y = \sigma_1 + \sigma_2,$$

where σ_1 and σ_2 are the principal extensional stresses in the plane of the plate. Thus, by adding Eqs. (1) and (2), one finds the fringe patterns for the interferometer experiments should correspond to the relation

$$\sigma_1 + \sigma_2 = 2\frac{\kappa}{\sqrt{2r}}\cos\frac{\theta}{2} - \sigma_{ox} \quad (6)$$

Graphical representations of lines of constant τ_m from Eq. (5) and lines of constant $(\sigma_1 + \sigma_2)$ from Eq. (6) compare well with the crack tip fringe patterns shown on the photographs by Wells and Post. Fig. 2, for example, shows a graph of selected theoretical isochromatics. By making two measurements on observed isochromatic loops close to the crack tip one can obtain experimental values of κ and σ_{ox}. The pair of measurements selected by the writer for trial were those of the radius value, r_m, and the angle value, θ_m, at the position on the isochromatic loop having the largest separation from the end of the crack.

In order to obtain expressions for κ and σ_{ox} in terms of the observed quantities r_m, θ_m, τ_m one may add to Eq. (5) a second relation expressing the fact that $\partial \tau_m / \partial \theta$ is equal to zero at the coordinate position r_m, θ_m. Through suitable algebraic rearrangement of terms these two equations then provide the following expressions for κ and σ_{ox}.

$$\kappa = \frac{2\tau_m\sqrt{2r_m}}{\sin\theta_m}\left\{1 + \frac{2\tan\frac{3\theta_m}{2}}{3\tan\theta_m}\right\}\left\{1 + \left(\frac{2}{3\tan\theta_m}\right)^2\right\}^{-\frac{1}{2}}, \quad (7)$$

and

$$\sigma_{ox} = \frac{2\tan\theta_m}{\cos\frac{3\theta_m}{2}}\left\{1 + \left(\frac{3\tan\theta_m}{2}\right)^2\right\}^{-\frac{1}{2}}. \quad (8)$$

Measurements of θ_m and r_m were made on seven of the running crack photographs of the paper. The isochromatic loop corresponding to $4\frac{1}{2}$ fringe order in the 3X multiplied pattern was most often used. The average value of σ_{ox} obtained was 2.7 in fringe order units corresponding to a θ_m of 78 degrees for the $4\frac{1}{2}$ fringe order isochromatic. The values obtained for κ^2 are shown in relation to the crack length in Fig. 3.

To obtain theoretical estimates of κ^2 for purposes of comparison one can use Westergaard's stress field equations for an infinite series of collinear equal-length cracks as explained in Ref. [2]. Employing this theoretical viewpoint one finds that

$$\kappa^2 = \frac{2\sigma^2 B \tan\frac{\pi b}{2B}}{\pi\left\{1 + \frac{8B}{\pi L}\log_e\left(\sec\frac{\pi b}{2B}\right)\right\}^2}. \quad (9)$$

In this equation b and B are the crack length and plate width, respectively, as used in the paper. L is the plate length assumed fixed during rapid crack extension, and σ is the value of σ_y remote from the crack at onset of rapid fracturing. Different curves for κ^2 as a function of b result from different assumptions regarding L.

To obtain agreement with the authors' measurements of average load on the unbroken ligament one may put L equal to $2B$. Values of κ^2 from this assumption are shown as the upper curve in Fig. 3. In some analyses of their work the authors considered that the equivalent fixed grip length, L, for their dy-

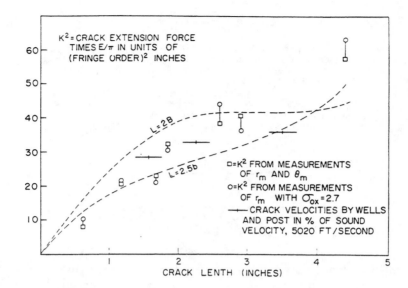

FIG. 3. EXPERIMENTAL DETERMINATIONS OF K^2 FROM ISOCHROMATICS USING PHOTOGRAPHS BY WELLS AND POST. THE DASHED CURVES ARE VALUES OF K^2 COMPUTED FROM EQ. (9) FOR TWO CHOICES OF THE EFFECTIVE "FIXED GRIP" SPECIMEN LENGTH L, AND ASSUMING A STRESS REMOTE FROM THE CRACK OF 5.18 IN FRINGE ORDER UNITS. ALSO SHOWN FOR COMPARISON ARE CRACK VELOCITY MEASUREMENTS BY WELLS AND POST.

namic experiments was 2.5 times the crack length. Although this assumption gives poor agreement with average load on the unbroken ligament, the corresponding curves for K^2, shown as the lower dashed line in Fig. 3, are in fair agreement with the experimental values of K^2. Actually the experimental points are well represented by a straight line through the origin. Considering the uncertain magnitude of the stress-optical coefficient and the expected increase of plate bending with crack length, the experimental values of K^2 are reasonable both in magnitude and trend. The fact that the crack extension force might be obtained from measurements on a single isochromatic loop close to the crack was unexpected at the outset of this study.

It is clear from Eq. (6) that experimental determinations of K and σ_{ox} might also be obtained by measurements of the interferometer fringe patterns. These measurements as well as others using isochromatic fringe patterns will be added to this study at a later time.

Shown also in Fig. 3 are average values of crack velocity at three crack lengths as reported by Wells and Post. The fact that these velocities are increasing with crack length less rapidly than the crack extension force suggests that a characteristic limiting upper velocity was approached under the selected experimental conditions. Experiments by Smith and Ferguson [3] have shown that crack velocity increases steadily with crack extension force until branching of the crack along multiple paths occurs. It would appear that for a crack extending under the action of a steadily increasing crack extension force, it is the roughening and branching of the crack which places a limit on velocity of crack extension. The fact that a calculation of limiting velocity can be made as referenced in the paper without considering development of multiple fracture paths is difficult to understand.

The results of the work reported by Wells and Post have been informally available to the writer and his associates during the past two years. Viewpoints based upon these results have been of great value in related fracture studies. We are indebted to the authors for their timely and skillful completion of a difficult task.

REFERENCES

[1] Irwin, G. R., "Onset of Fast Crack Propagation in High Strength Steel and Aluminum Alloys", Proc. of 1955 Sagamore Research Conference, Syracuse University Research Institute, Vol. 2, p. 289, Mar. 1956.

[2] Irwin, G. R., "Analysis of Stresses and Strains Near the End of a Crack Traversing a Plate", A.S.M.E., Applied Mechanics Div., Paper No. 57 - APM-22, June 1957.

[3] Smith, H. L. and Ferguson, W. J., "Report of NRL Progress", Apr. 1950.

AUTHORS' CLOSURE

Dr. Irwin's discussion adds much to the value of the paper and is significant from two viewpoints: (1) in revealing plane photoelasticity as an elegant method of determining crack extension forces* for different shapes of specimen and (2) in demonstrating a useful physical concept in which crack extension force is independent of strain energy terms. Since stress concentration factor is meaningless for a sharp crack, the alternative concept of variable crack extension force proportional to the span of any closed isochromatic loop is most valuable. Much research work, with which the paper was not concerned, has been addressed to establishment of the meaning of crack extension force in relation to failure by fast cracking in brittle materials.

The relation between crack extension force $\mathscr{G}$ and the isochromatic loop of span r_m, typified by Eqs. (4) and (7), is also dependent upon the inclination θ_m of the loop, and the latter is not without importance. Because of uncertainty

---
* Crack extension force as used by Irwin is synonymous with the term strain energy release rate used in the paper.

as to the exact position of the end of a crack, together with that of the other extreme end of a loop, measurement of θ_m may be subject to some error, say ± 2 degrees in any particular case. For a typical value of θ_m of 80 degrees, this would engender a possible error of ± 9 percent in the determination of κ and ± 18 percent for $\mathscr{G}$. The importance of correctly estimating the angle θ_m must therefore be emphasized. Nevertheless, the method promises to exceed in simplicity and directness most of the methods of crack extension force measurement now in use.

It is reassuring to find from the results of Fig. 3 that the crack extension forces measured with the method discussed above agree reasonably with values calculated from Eq. (9) with effective plate length for fixed grips taken as either 2 times plate width or 2.5 times crack length. Further understanding of these elastic wave effects, and the limiting speed of crack propagation as mentioned in the discussion, would be broadened if a theoretical elastic stress solution could now be produced for a fast extending crack.

An Investigation of Propagating Cracks by Dynamic Photoelasticity

A program was undertaken to provide experimentally an improved understanding of the elastic fields surrounding dynamic cracks and, in particular, stress fields surrounding arresting dynamic cracks

by W. B. Bradley and A. S. Kobayashi

ABSTRACT—A 16-spark gap, modified Schardin-type camera was constructed for use in dynamic photoelastic analysis of fracturing plastic plates. Using this camera system, dynamic photoelastic patterns in fracturing Homalite-100 plates, $3/8$ in. $\times$ 10 in. $\times$ 15 in. with an effective test area of 10 in. $\times$ 10 in., loaded under fixed grip condition were recorded. The loading conditions were adjusted such that crack acceleration, branching, constant velocity, deceleration and arrest were achieved.

The Homalite-100 material was calibrated for static and dynamic properties of modulus of elasticity, Poisson's ratio, and stress-optical coefficient. For dynamic calibration, a Hopkinson bar setup was used to record the material response under constant-strain-rate loading conditions.

The precise location of the dynamic isochromatic patterns in relation to the crack tip was determined by a scanning microdensitometer. This information was then used to determine dynamic stress-intensity factors which were compared with corresponding static stress-intensity factors determined by the numerical method of direct stiffness. Although the response of the dynamic stress-intensity factor to increasing crack length was similar to the static stress-intensity-factor response, the dynamic values were approximately 40 percent higher than the static values for constant-velocity cracks. For decelerating cracks, the peak values of dynamic stress-intensity factors were 40 percent higher than the corresponding static values.

Introduction

Theoretical analysis of fracture dynamics can be classified into the two categories of analysis by energy consideration and analysis by dynamic elasticity. The energy approach used by Mott,[1] Roberts and Wells,[2] and Berry[3] among others, yielded a relation for crack velocity as a function of material properties and crack length and provided an estimate of the terminal velocity of the crack. The method, however, cannot be used to explain crack deceleration and arrest. The approach of dynamic elasticity used by Yoffe,[4] Broberg,[5] Baker[6] among others, is limited to constant-velocity solutions and cannot relate crack acceleration and deceleration to geometric and loading conditions.

Experimental analysis of fracture dynamics follows closely the above two categories of theoretical analysis. One approach is to investigate the crack velocities in various materials without determining experimentally the stress field surrounding the crack tip as reported by Roberts and Wells,[2] Schardin,[7] and Clark and Irwin.[8] The other approach is to investigate the transient stress or strain field surrounding a propagating crack without an accurate correlation with crack velocities as reported by Wells and Post,[9] Stock and Pratt,[10] and Kobayashi, Bradley and Selby[11] among others.

The above brief review of the literature shows that there exists no unified theory on the general behavior of propagating cracks, and that available experimental information is not sufficient to test the available theories and their hypotheses and to resolve their conflicting predictions. Particularly lacking is an accurate characterization of the elastic stress field surrounding a crack tip which is propagating with a nonconstant velocity.

This program was undertaken to experimentally provide an improved understanding of the elastic fields surrounding dynamic cracks and, in particular, stress fields surrounding arresting dynamic cracks. To accomplish this a modified Schardin multiple-spark-gap, parallel-light transmission polariscope was developed. A brief description of the equipment, test specimen and experimental

---

W. B. Bradley was formerly a Research Associate at the University of Washington; he is presently Mechanical Engineer, Shell Development Co., Houston, Texas. A. S. Kobayashi is Professor, Department of Mechanical Engineering, University of Washington, Seattle, Wash.
Paper was presented at 1969 SESA Spring Meeting held in Philadelphia, Pa., on May 13–16.
The research reported in this technical report was made possible through support extended to the Department of Mechanical Engineering, University of Washington, by the Office of Naval Research under Contract Nonr-477 (39) NR 064 478. Reproduction in whole or in part is permitted for any purpose by the United States Government.

procedures is presented in the following.

Experimental Setup

The modified Schardin camera and the dynamic polariscope used in this investigation are similar to the system used by Dally and Riley.[12,13] The system is composed of sixteen triggerable spark-gap light sources, a 16-in.-diam parallel-light transmission polariscope, and a 16-lens, 11×14 in. view camera. The spark gaps are of enclosed design and have the ability of being individually triggered by their corresponding delay units, with delay capabilities continuously selectable over a range from 5 to 12,000 μsec. The spark gaps, arranged in a four-by-four array on 1.5-in. centers, are mounted together with their discharge capacitors and trigger transformers on a carriage traveling on overhead rails. The discharge of the 0.05-microfarad 30-kv capacitors across the 5-in. spark gaps is initiated by a 70-kv pulse from the trigger transformers which results in a spark duration of 1.25 μsec. The light from each spark gap is picked up by an EG&G Lite-Mike and recorded with Polaroid film on a Tektronix 547 oscilloscope, together with previously recorded timing marks, to provide a permanent record of the firing time of each spark.

The polariscope, mounted on two carriages traveling on the same rails as the light source, is composed of two field lenses, a polarizer, an analyzer and two quarter-wave plates arranged as a standard parallel-light, transmission polariscope. Also mounted on a movable carriage is the camera which was constructed from a 11×14-in. commercial view camera braced for improved rigidity, with 16 150-mm f 4.5 shutterless lenses mounted in a four-by-four array on 1.9-in. centers.

This modified Schardin system is capable of recording 16 2.1×2.1-in. images with each image representing a field of view at the model site of 10×10 in. together with a static resolution of 900 lines/in. Tri-X Ortho film which has sufficient film speed and adequately small grain size was used. The combination of this film, which has a spectral sensitivity to 5750 Å, together with a K-2 filter, which has a cutoff at 4800 Å, provides a 1000 Å bandwidth. This was found to be sufficiently monochromatic, while still providing enough light for adequate exposures.

The loading fixture shown in Fig. 1 is of fixed-grip design, utilizing a wedging system capable of hand-generated loads up to 2.5 tons. The displacement at the top and bottom edges of the specimen were measured by two LVDT's with their outputs recorded on an X-Y recorder to allow monitoring of the specimen displacements during the loading cycle. The signals from these two LVDT's were also fed to an oscilloscope to record the grip displacements during the dynamic fracturing portion of the test. The choice of this loading fixture provides versatility in specimen loading in that, besides uniform strain fields, increasing or decreasing strain fields along the width of the specimen may be selected.

Test Specimen

In selecting a material for the study of dynamic fracture, it was required that the material be of a

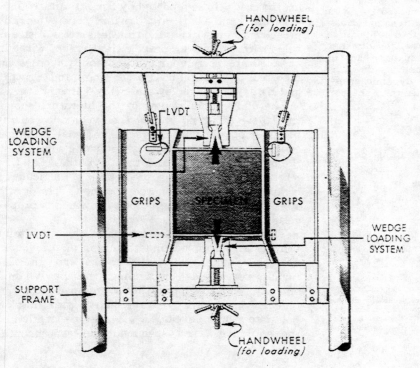

Fig. 1—Loading fixture

brittle nature, be suitably birefringent, and have material properties that are insensitive to changing strain rates. Clark and Sanford[14] reported on several materials which met the above qualifications, among which were Hysol 4290, Araldite 6020 with 45 pphr 901, and Homalite-100. Although Homalite-100 is not an ideal material for dynamic photoelastic work, the fact that its stress-optic coefficient is not strain-rate sensitive and that it is available in sheets of acceptable photoelastic quality dictated its use in this work. The dynamic and static properties of Homalite-100, along with the calibration methods used, are presented in a later section.

As an initially notched specimen, an edge crack was produced by sawing a 0.012-in.-wide crack to within approximately 0.1 in. of the desired crack length, halfway along the 15-in. side of the specimen, then splitting the test specimen with a razor blade to the desired crack length. Crack lengths from 0.06 to 0.58-in. long were produced in this manner in addition to two tests which used 0.012-in.-wide blunt cracks 0.52-in. long. The initiation of crack propagation was sensed by a crack wire painted in front of the crack tip with silver conducting paint and connected to the spark-delay-unit input.

Dynamic Calibration

The dynamic calibration of the Homalite-100 was accomplished using a Hopkinson-bar system similar to the one used by Clark and Sanford.[14] The test consisted of impacting the test bar with a steel projectile fired from an air gun and observing the effect of the impact at two stations along the bar. At the first station, the longitudinal compression-strain-time profile and photoelastic-fringe-time profile were simultaneously recorded and, at the second station, the compression-strain-time profile was again recorded in conjunction with the cross axis or Poisson strain-time profile. A $1/2$-in.-diam, 0.6-in.-long 4340 steel projectile with a 0.4-in.-long, 30-deg conical nose was used to impact a soft-aluminum cap placed in front of the test bar. With this arrangement and with a projectile velocity of approximately 1100 ips, a smooth ramp loading as shown in Fig. 2 was obtained. The comparison of the compression profile and the photoelastic profile at Station 1 yields the strain-fringe constant. The comparison of the compression profile and the Poisson profile at Station 2 provided the dynamic Poisson's ratio. The dynamic elastic modulus was obtained from the time of flight of the compression wave, as recorded by the difference in the time location of the Station 1 and Station 2 compression profiles, and the one-dimensional wave theory which provides the relation

$$\dot{C} = \sqrt{\frac{E}{\rho}} \quad (1)$$

The output from the photomultiplier tube and the three sets of strain gages were input to two Tektronix 555 dual-beam oscilloscopes and recorded with Polaroid type 107 film. The 555 scopes were both triggered by a signal from two magnetic pickups mounted on the gun barrel which sensed the passage of the projectile.

The results of the material calibration on the three sheets of Homalite-100 are presented, together with the results of Clark and Sanford,[14] in Fig. 3. The three sheets were from the same shipment and, although they show general agreement, sheet No. 2 appeared to differ somewhat from the other two sheets.

Test Results

Figures 4 and 5 show typical dynamic photoelastic patterns and locations of the propagating crack tip. Variations in crack velocities with crack lengths show that the combination of fixed grip loading and the 10 × 10-in. test-specimen geometry provided three different conditions for each test which affected the propagation of the crack. The test results are summarized in Table 1 where ϵ_{Top} and ϵ_{Bottom} represent the applied uniaxial strains at the two ends of the fixed grips of the test fixture. The first condition, approximating an

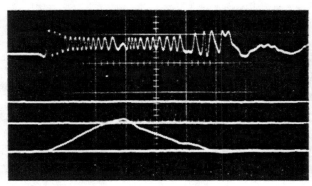

Time (50 μsec/cm)

Fig. 2(a)—Station 1: a typical photoelastic-fringe profile and compression strain profile for a 1.0-in.-long 30-deg conical-tip projectile

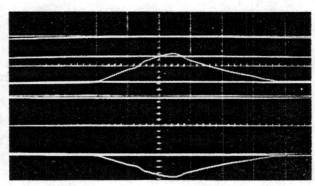

Time (50 μsec/cm)

Fig. 2(b)—Station 2: a typical compression strain profile and Poisson strain profile for a 1.0-in.-long 30-deg conical-tip projectile

infinite plate, controlled the crack propagation for the first 3 in. of crack extension, with the second condition, approximating an infinite strip under fixed grip loading, beginning to take over at this point.

The maximum velocity of crack propagation was found to be about 15,300 ips for single cracks and branching cracks for tests conducted with 0.375-in.-thick specimens. The 0.125-in.-thick specimens had a lower maximum velocity of approximately 14,900 ips with the exception of one test which had a velocity of 15,700 ips. Constant crack velocities lower than the maximum velocity were also produced with this loading arrangement, as is seen from tests No. 3 and 4 where velocities of 14,200 ips and 14,700 ips were produced by lowering the applied stress in the specimen. The effect of applied stress on the crack velocity was also observed in tests No. 12, 13, 14 and 15 where, with increasing initial stress, the rapid deceleration of the crack occurred at larger and larger crack lengths. The effect of fixed grip loading on the crack velocities raises the question as to whether the maximum steady-state crack velocity obtained in these tests was indeed the maximum crack velocity. Several test results show that the maxi-

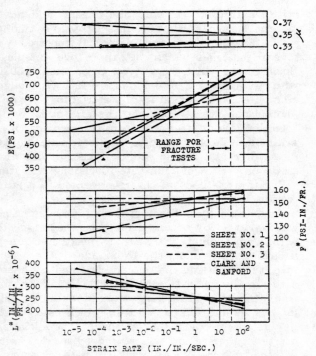

Fig. 3—Material properties of Homalite-100 as a function of strain rate

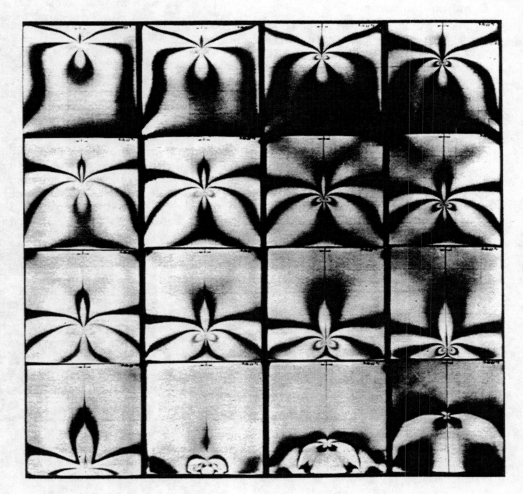

Fig. 4—Isochromatic patterns of dynamic crack propagation for test No. 1

Fig. 5—Isochromatic patterns of dynamic crack propagation for test No. 7

mum velocity was reached before entering the fixed grip region of influence; therefore, it can be concluded that the observed terminal velocity was indeed the maximum velocity of the crack. The results also agree with the maximum velocity of 15,000 ips reported by Beebe[15] for 26.5-in.-long by 13.5-in.-wide tension specimens of Homalite-100.

Dynamic Stress-intensity Factor

Once the isochromatic patterns characterizing the elastic fields associated with the dynamic crack are obtained, it is necessary to convert this pictorial information into quantities which characterize fracture dynamics of brittle materials. The stress-intensity factor, K_I, provides a convenient representation of the dynamic elastic fields associated with the propagating crack tip and, therefore, several methods were attempted for determining the dynamic stress-intensity factor from the isochromatic patterns. Irwin,[16] in a discussion of the work of Wells and Post,[9] developed a technique for the calculation of K_I from isochromatic patterns in which he modified the one-parameter representation of the static local-stress-distribution model. With this model, the maximum shear stress in terms of polar coordinates (r,θ) with the origin at the tip of the crack can be represented as

$$(2\tau)^2 = \left(\frac{K_I}{\sqrt{2\pi r}} \sin\theta + \sigma_{ox} \sin\frac{3\theta}{2}\right)^2 + \left(\sigma_{ox} \cos\frac{3\theta}{2}\right)^2 \quad (2)$$

where σ_{ox} is the pseudo boundary stress at $x = c$. Irwin also pointed out that a second equation could be obtained by noticing that the $\partial\tau/\partial\theta = 0$ at the tip of each isochromatic loop. With these two equations, the unknown quantities of K_I and σ_{ox} will be determined.

Applying this method to rapidly extending cracks in CR-39 from tests conducted by Wells and Post, Irwin obtained results which indicated a general correlation between the stress-intensity factor and the crack velocity in the initial acceleration phase of propagation, but found that the stress-intensity factor continued to increase while the velocity began to level off to a constant value. This method was again used by Pratt and Stock[10] and by Stock[17] where it was found that the procedure was applicable to $r_m{}^*$ values as large as

---
*r_m and θ_m represent in terms of the polar coordinate the location of the tip of the isochromatic where $\partial\tau/\partial\theta = 0$.

TABLE 1—SUMMARY OF EXPERIMENTAL RESULTS

Test No.	Sheet No. and Thickness (in.)	Initial Crack Length, in.	ϵ_{Top} ϵ_{Bottom}, in./in.	Terminal Crack Velocity, in./sec	K_c Dynamic Tests, psi $\sqrt{\text{in.}}$	Comment
1	No. 1-$^3/_8$	0.30	0.00119	15,200	606	
2	No. 3-$^3/_8$	0.30	0.00109 0.00109	15,300	530	
3	No. 3-$^3/_8$	0.40	0.000968 0.000988	14,200	504	
4	No. 3-$^3/_8$	0.58	0.00914 0.000870	14,700	527	
5	No. 2-$^3/_8$	0.53	0.00340 0.00344	15,400	1920	Blunt crack r = 0.004 in.
6	No. 2-$^3/_8$	0.53	0.00325 0.00325	15,200	1830	Blunt crack r = 0.004 in.
7	No. 3-$^3/_8$	0.20	0.00153 0.00156	15,300	655	One branch arrested
8	$^1/_8$	0.22	0.00172 0.00165	14,900	765	Second branching at 7.25 in.
9	$^1/_8$	0.17	0.00121 0.00124	14,900	498	
10	$^1/_8$	0.095	0.00194 0.00024	14,800	655	Decelerated
11	$^1/_8$	0.09	0.00296 0.000366	15,700	726	Decelerated, one branch arrested
12	No. 1-$^3/_8$	0.48	0.00104	14,800	573	Arrested gradually
13	No. 1-$^3/_8$	0.45	0.000848	13,900	461	Arrested gradually
14	No. 2-$^3/_8$	0.29	0.00127	15,100	607	Arrested gradually
15	No. 2-$^3/_8$	0.25	0.00124 6.2×10^{-6}	15,400	567	Arrested gradually
				Average	$\overline{579}$	

one-half of the length of the crack. Stock[17] pointed out that the method produced increasingly better results with the application of smaller and smaller values of r_m and developed an asymptotic-approach technique by utilizing information from several isochromatic loops. Wells and Post pointed out, in their closure to Irwin's discussion,[16] that the method was sensitive to the value of θ_m.* Using ± 2 deg as a reasonable estimate of the accuracy of the location of the tip, they found that a 9 percent uncertainty in the value of the stress-intensity factor was produced for a nominal value of $\theta_m = 80$ deg. In fact, the sensitivity of the method to the angular value increases with decrease in the magnitude of θ_m and approaches infinity at a value of 69.4 deg with the uncertainty in K_I varying from 3 percent at 90 deg to 70 percent at 72 deg.

In this particular program, the influence of the fixed grip boundaries caused isochromatic loops to exhibit a pronounced leaning with θ_m varying from 67 deg to 72 deg for r_m values as small as one sixth of the crack length. In order to reduce the sensitivity of the method to θ_m, the above pro-

cedure was modified as follows. By assuming

$$\sigma_{ox} = \sigma_\infty$$

or

$$\sigma_{ox} = \frac{K_I}{\sqrt{\pi a}} \quad (3)$$

thereby reducing Irwin's method to a one-parameter characterization. Then by substituting eq (3) into equation (2),

$$\tau = \frac{K_I}{2\sqrt{2\pi r}} f(\theta, r, a) \quad (4)$$

where

$$f(\theta, r, a) = \left(\sin^2\theta + 2\frac{\sqrt{2r}}{a} \sin\theta \sin\frac{3\theta}{2} + \frac{2r}{a} \right)^{1/2}$$

Selecting a point on each of two isochromatic loops, (r_1, θ) and (r_2, θ), the corresponding maximum shear stresses τ_1 and τ_2 can be subtracted to form:

$$\tau_2 - \tau_1 = \frac{K_I}{2\sqrt{2\pi}} \frac{f_2\sqrt{r_1} - f_1\sqrt{r_2}}{\sqrt{r_1 r_2}}$$

or solving for K_I

---
* See footnote on Page 110.

$$K_I = \frac{2\sqrt{2\pi}\,(\tau_2 - \tau_1)\sqrt{r_1 r_2}}{f_2\sqrt{r_1} - f_1\sqrt{r_2}} \quad (5)$$

This technique is considerably less sensitive to θ_m showing an uncertainty in K_I of less than 1 percent for 2-deg change. In addition, it does not exhibit the large increase in sensitivity with the reduction in the angular value. A comparison of the K_I values calculated by this method with the more precise method[18] shows good agreement, as can be seen in Fig. 6 where the values of the stress-intensity factor obtained from both methods are plotted as a function of time for test No. 13. Also plotted in this figure is the crack velocity which exhibits a lag in changes with respect to the changes in the dynamic stress-intensity factor.

Whereas this method yields satisfactory estimations of the stress-intensity factor for dynamic fracture, a plot of the isochromatics calculated by eq (2) using these values of K_I and σ_{ox} is low by approximately 6 to 8 percent. This error is caused by the prediction of σ_{ox} which is probably too low with respect to its time value. Although the effect of σ_{ox} on the calculation of τ is significant, its effect on the calculation of the stress-intensity factor is considerably diminished due to the partial cancelling of this error by the averaging of the two points in eq (5).

Evaluation of Data

All isochromatic data were obtained by scanning the dynamic photoelastic records with a Joyce-Loebel microdensitometer. From these data, the stress-intensity factors were computed by using eq (5).

Static and dynamic stress-intensity-factor curves of tests No. 1 are shown in Fig. 7 in which the stress-intensity factor increases to a crack length of approximately 4.5 in., then levels off to a constant value as predicted by static fixed-grip loading results, and then rises rapidly again as the free edge is approached. The general shape of this stress-intensity curve is influenced by the static boundary conditions rather than by the dynamic effect of the propagating crack, as can be seen by a comparison with the equivalent static analysis of Simon.[19] A similar comparison is also shown in Fig. 8 for test No. 13. In both cases, the static and dynamic stress-intensity factors coincide for crack velocities below 75 percent of the terminal velocities. It is also noted that the dynamic stress-intensity factors at terminal velocity and for decelerating crack are approximately 40 percent higher than the corresponding static values. Again, it can be seen in Fig. 7 that the changes in crack velocities lag slightly behind the changes in the dynamic stress-intensity factor.

The initiation of dynamic fracture under gradually increasing load produced an initial distortion wave which was observed in seven of the tests. The distortion wave produced two effects in the

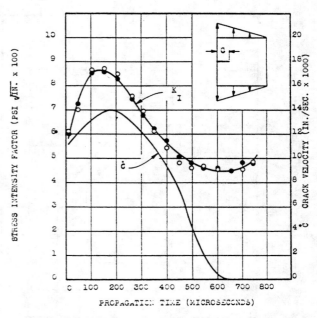

Fig. 6—A comparison of the dynamic stress-intensity factor calculated by the modified Irwin technique (solid circle) with the dynamic stress-intensity factor calculated by a least-squares fitting of Williams' stress function (open circle) to experimental isochromatic data for test No. 13. Also shown is the crack velocity

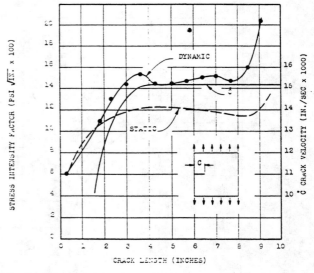

Fig. 7—Dynamic stress-intensity factor and crack velocity as a function of crack length calculated by the modified Irwin technique for test No. 1

tests, the first manifested as a slight increase in velocity at a crack length of about 3 in. corresponding to the reflected wave of equal sign returning from the grip boundaries, and the second a more noticeable reduction in velocity and stress-intensity factor at a length of approximately 5 in. The second effect corresponds to the distortion wave of opposite sign reflected from the free

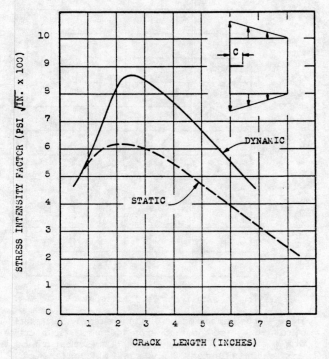

Fig. 8—A comparison of the dynamic stress-intensity factor with the static stress-intensity-factor solution of Simon for test No. 13

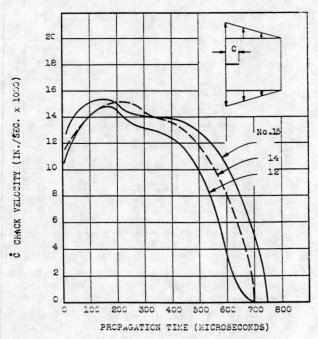

Fig. 9—Crack length and crack velocities as functions of propagation time for tests No. 12, 14 and 15

boundary of the specimen, and can clearly be seen in the stress-intensity-factor curve of test No. 1 (Fig. 7) and in the velocity curves of tests No. 12, 14 and 15, as shown in Fig. 9.

Conclusions

The following conclusions are reached based on the above results.*

1. A modification of Irwin's technique provides accurate stress-intensity factor results without restriction on the isochromatic-pattern inclination.

2. Static approximations to dynamic stress-intensity factors can be made for crack velocities below 75 percent of the terminal velocity in the region immediately following crack propagation.

3. The dynamic stress-intensity factors were approximately 40 percent higher than the corresponding static values at terminal crack velocity and decelerating cracks.

4. The change in the crack velocity was found to lag the change in the dynamic stress-intensity factor.

Acknowledgment

This investigation was supported by the Office of Naval Research Contract Nonr-477(39), NR 064 478. The writers wish to thank H. Liebowitz, formerly of ONR, N. Perrone and N. Basdekas of ONR for their patience and encouragement during the course of this investigation.

References

1. Mott, N. F., "Fracture of Metals, Some Theoretical Considerations," *Engineering*, 16–18 (January 2, 1948).
2. Roberts, D. K., and Wells, A. A., "The Velocity of Brittle Fracture," *Engineering*, 820–821 (December 24, 1954).
3. Berry, J. P., "Some Kinetic Considerations of the Griffith Criterion for Fracture—I and II, Equations of Motion at Constant Force," *Jnl. Mech. and Phys. of Solids*, 8, 194–206 and 207–216 (1960).
4. Yoffe, E. H., "The Moving Griffith Crack," *Phil. Mag.*, Ser. 7, 42, 739–750 (1951).
5. Broberg, B. K., "The Propagation of a Brittle Crack," *Arkiv for Phsik Band* 18 nr 10, 159–192 (1960).
6. Baker, B. R., "Dynamic Stresses Created by a Moving Crack," *Jnl. Appl. Mech.*, 449–458 (September 1962).
7. Schardin, H., "Velocity Effects in Fracture," *Fracture*, edited by B. L. Averbach, D. K. Felbeck, G. T. Hahn, and D. A. Thomas, John Wiley & Sons (1959).
8. Clark, A. B. J. and Irwin, G. R., "Crack Propagation Behaviors," EXPERIMENTAL MECHANICS, 6 (6), 321–330 (1966).
9. Wells, A. A. and Post, D., "The Dynamic Stress Distribution Surrounding a Running Crack—A Photoelastic Analysis," *Proc. SESA*, 16 (1), 69–92 (1958).
10. Pratt, P. L. and Stock, T. A. C., "The Distribution of Strain About a Running Crack," *Proc. Royal Society of London*, A. 285, 73–82 (1964).
11. Kobayashi, A. S., Bradley, W. B., and Selby, R. A., "Transient Analysis in a Fracturing Epoxy Plate with a Central Crack," *Proc. Intl. Conf. on Fracture, Sendai, Japan* (1965).
12. Dally, J. W. and Riley, W. F., "Stress Wave Propagation in a Half-Plane Due to a Transient Point Load," *Developments in Theoretical and Applied Mechanics*, 3, 357–377 Pergamon Press, New York (1967).
13. Riley, W. F. and Dally J. W., "Recording Dynamic Fringe Patterns with a Cranz-Schardin System," EXPERIMENTAL MECHANICS, 9(8)27N–33N (1969).
14. Clark, A. B. J. and Sanford, R. J., "A Comparison of Static and Dynamic Properties of Photoelastic Materials," *Proc. SESA XX* (2) 148–151 (1963).
15. Beebe, W. M., "An Experimental Investigation of Dynamic Crack Propagation in Plastics and Metals," *California Institute of Technology Tech. Report No. AFML-TR-66-249* (November 1966).
16. Irwin, G. R., Discussion and Author's Closure of the Paper, "The Dynamic Stress Distribution Surrounding a Running Crack—A Photoelastic Analysis," *Proc. SESA*, XVI (1), 93–96 (1958).
17. Stock, T. A. C., "Stress Field Intensity Factors for Propagating Brittle Cracks," *Intl. Jnl. of Fracture*, 3 (2), 121–130 (June 1967).
18. Bradley, W. B. and Kobayashi, A. S., "Fracture Dynamics—A Photoelastic Investigation," submitted for publication in *Jnl. Engr. Fracture Mech.*
19. Simon, B. J., "An Analysis of 2-Dimensional Fracture Mechanics Problems Using the Finite Element Method," *MS Thesis, University of Washington* (March, 1969).

---

* *Detailed discussion of all test results will appear in Ref. 18.*

STRESS DISTRIBUTION IN A TENSION SPECIMEN NOTCHED ON ONE EDGE

J. R. DIXON *National Engineering Laboratory, East Kilbride, Glasgow*
J. S. STRANNIGAN *National Engineering Laboratory, East Kilbride, Glasgow*
J. McGREGOR *National Engineering Laboratory, East Kilbride, Glasgow*

The stress distribution in a tension specimen notched on one edge was obtained photoelastically for several ratios of notch depth to specimen width. The stress-concentration factors agreed well with corresponding values derived from Neuber's theory of notch stresses. It was also shown that the stress-intensity factor for a tension specimen with a single crack on one edge, obtained by the collocation method, agreed well with that deduced from Neuber's theory.

NOTATION

a	Notch depth.
B	Specimen thickness.
b	$W - a$.
K	Stress-intensity factor.
P	$\sigma W B$.
r, θ	Polar co-ordinates.
V	Distance of section AA from notch section.
W	Specimen width.
Y	$KBW/Pa^{1/2}$.
α	(with subscripts 1 and 2) Stress-concentration factor for doubly externally deep-notched bars.
α	(with subscripts a, b, s, and d) Stress-concentration factor (Neuber's theory).
ρ	Notch-root radius.
σ	Applied stress.
σ	(with subscripts a, b, s, and d) Stress at root of notch (Neuber's theory).
σ_{max}	Maximum stress at root of notch.
σ_n	Nominal stress across notched (cracked section).
σ_x, σ_y	Normal stresses.

INTRODUCTION

It has been suggested (*1*)* that a tension specimen notched on one edge, of the type shown in Fig. 1, is particularly suitable for the determination of plane-strain fracture toughness values. This specimen appears to be more efficient than symmetrically cracked specimens with respect both to the material and required loading capacity.

The magnitude of the elastic-stress field in the immediate vicinity of a crack tip may be characterized by a parameter K called the stress-intensity factor; thus the distribution of the principal tensile stress (say σ_1) near the crack tip can be written

$$\sigma_1 = \frac{K}{(2\pi r)^{1/2}} f(\theta)$$

where r and θ are polar co-ordinates with their origins at the crack tip. The expression for K depends on the geometry of the specimen and loading conditions; for example, in an infinite plate containing a central crack of total length $2a$ subjected to an applied stress σ acting at right-angles to the axis of the crack, $K = \sigma(\pi a)^{1/2}$.

The value of K at the point of crack instability is a measure of the fracture toughness of the material and can be determined from tests on specimens containing sharp notches and cracks, provided that suitable expressions are available giving the stress-intensity factor K in terms of the specimen geometry and applied loads. Gross, Srawley, and Brown (*2*) obtained an analytical solution for the stress distribution in a tension specimen notched on one edge by means of a boundary-value collocation method. This is an approximate method of analysis for two-dimensional problems and is particularly useful when

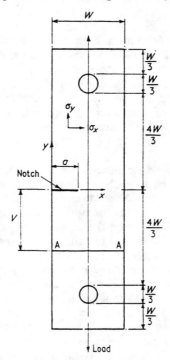

Fig. 1. Specimen notched on one edge

The MS. of this paper was received at the Institution of Mechanical Engineers on 21st February 1968 and accepted for publication on 18th July 1968. 33
* References are given in the Appendix.

exact solutions are not available. In the present case it consisted in finding a stress function which satisfied the biharmonic equation and the boundary conditions along the crack but only satisfied the remaining boundary conditions at the edge of the specimen at a finite number of stations. A uniform stress was assumed to act across the specimen at a distance $y = \pm V$ from the plane of the crack (Fig. 1) and the stress distribution was computed for several ratios of crack length to specimen width, a/W, between 0·04 and 0·5 (later extended to 0·6) and for values of V/W between 0·5 and 1·5. It was found that the stress-intensity factor was independent of V for V/W greater than 0·8.

In view of the current interest in the use of the tension specimen notched on one edge for fracture toughness testing and the need to have an accurate analytical expression for K, a photoelastic analysis was carried out of a single-edge-crack model having the geometry given in Fig. 1. The nature of the stress distribution between the crack and loading pins was determined to check whether a uniform stress existed across the specimen as assumed in the collocation analysis. In the model the crack was simulated by a slit having a finite end radius, and the variation of the stress-concentration factor σ_{max}/σ with a/W was compared with results derived from Neuber's theory of notch stresses. Values of the stress-intensity factor for a crack were also derived from Neuber's theory and compared with those obtained by the collocation method.

EXPERIMENTAL PROCEDURE AND RESULTS

A photoelastic model was machined from a sheet of Araldite CT200 $\frac{1}{8}$ in thick to the dimensions given in Fig. 1 with $W = 1\cdot5$ in. Load was applied through a lubricated sliding-fit pin at each end. The edge notch was in the form of a slit with an end radius of 0·031 in.

The model was loaded in tension before the slit was machined and it was found that there was an approximately uniform stress field between $Y = \pm W$. Slits of length a, giving a/W ratios of 0·2, 0·4 and 0·6, were machined and the stress distribution determined across section AA where $Y = V = 0\cdot9W$. From an inspection of the photoelastic fringe pattern for $a/W = 0\cdot2$, section AA was in the centre of a small length of the model subjected to approximately uniform stress conditions. The stress-concentration factor, stress at root of notch σ_{max}/nominal stress across gross section σ, was also determined for the above three a/W ratios and in addition for $a/W = 0\cdot66$ and 0·8.

The photoelastic analysis was carried out at room temperature by normal two-dimensional techniques. The interference fringes near the end of the slit were measured with a travelling microscope focused on an enlarged image. Readings along the x axis were taken as near as possible to the end of the slit and the boundary fringe value, giving σ_{max}/σ, obtained by graphical extrapolation. As the photoelastic fringe patterns give, in general, only the principal stress differences and principal directions, the distribution of the separate normal stresses σ_x and σ_y across section AA was obtained from a numerical integration of equilibrium equations using photoelastic data with the shear-difference method (3). A typical isochromatic fringe pattern for $a/W = 0\cdot4$ is shown in Fig. 2.

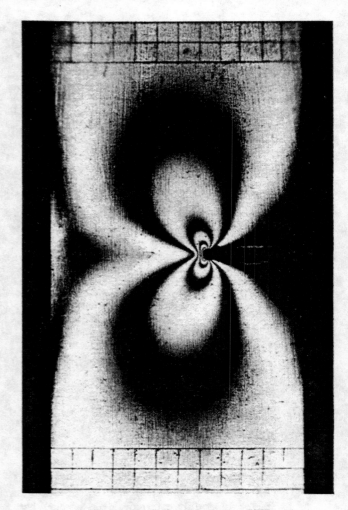

Fig. 2. Isochromatic fringe pattern: $a/W = 0\cdot4$

The distribution of σ_y across section AA, for $a/W = 0\cdot2$, 0·4, and 0·6, is plotted in Fig. 3; σ_x was negligibly small in all three cases. The distribution of $\sigma_y - \sigma_x$ across the notched section is shown in Fig. 4 for $a/W = 0\cdot2$, 0·4, and 0·6 and values of σ_{max}/σ are listed in Table 1 and plotted against a/W in Fig. 5. As expected, increased bending occurred across the notched section as a/W increased and, for a/W approximately greater than 0·3, σ_y became compressive near the edge $x = W$.

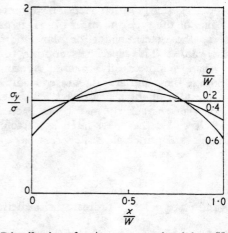

Fig. 3. Distribution of σ_y/σ across section AA at $V = 0\cdot9W$: photoelastic results

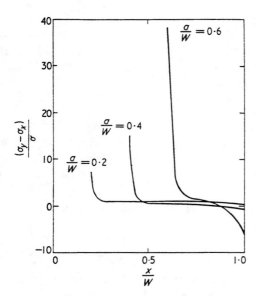

Fig. 4. Distribution of $(\sigma_y - \sigma_x)/\sigma$ across notched section: photoelastic results

Table 1. Stress-concentration factors

$\dfrac{a}{W}$	$\dfrac{\sigma_{max}}{\sigma}$	
	Neuber	Photoelastic
0	1·0	—
0·1	4·9	—
0·2	7·7	7·1
0·3	11	—
0·4	17	15
0·5	25	—
0·6	39	38
0·66	—	47
0·7	64	—
0·8	124	126

σ_{max} = maximum stress at root of notch.
σ = nominal stress across gross section.

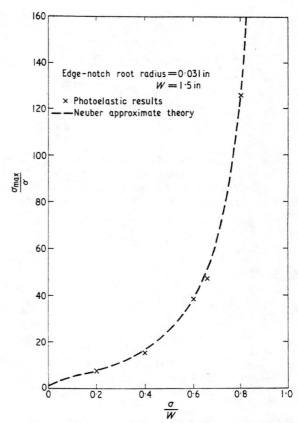

Fig. 5. Variation of stress-concentration factor σ_{max}/σ with a/W

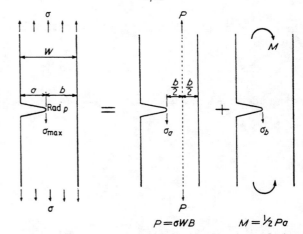

Fig. 6. Mathematical model for Neuber-type solution

NEUBER'S THEORETICAL SOLUTION

The stress-concentration factor σ_{max}/σ for a plate with a single edge notch, with uniform tension applied at infinity perpendicular to the centre-line of the notch, can be obtained from Neuber's theory of notch stresses (4). Neuber's notches are different in shape from the type of slit considered in this paper but it is assumed that his results apply. As sketched in Fig. 6 the problem can be divided into two parts: the uniform applied stress σ at infinity is represented as a combination of (a) a load $P = \sigma WB$ acting axially along a line distance $\tfrac{1}{2}(W-a)$ from the un-notched edge and (b) a bending moment $M = \tfrac{1}{2}Pa$. Neuber gives stress-concentration factors for loads (a) and (b) for shallow (small a/W) and deep (large a/W) notches; an empirical equation covers the intermediate (general a/W) case.

In the following analysis subscripts a and b are used with σ and α to denote loading cases (a) and (b); subscripts s and d are used to denote shallow and deep notches. Thus σ_{bd} denotes the stress at the root of a deep notch under bending load (b). Neuber's expressions are as follows.

For case (a):

$$\alpha_{as} = \frac{\sigma_{as}}{\sigma_n} = 3\left(\frac{a}{2\rho}\right)^{1/2} - 1 + \frac{4}{2+\left(\dfrac{a}{2\rho}\right)^{1/2}}$$

$$\alpha_{ad} = \frac{\sigma_{ad}}{\sigma_n} = \frac{\alpha_1 - 2C}{1 - \dfrac{C}{\left(\dfrac{b}{\rho}+1\right)^{1/2}}}$$

where

$$C = \frac{\alpha_1 - \left(\dfrac{b}{\rho}+1\right)^{1/2}}{\dfrac{4}{3\alpha_2}\left(\dfrac{b}{\rho}+1\right)^{1/2} - 1}$$

and
$$\alpha_1 = \frac{2\left(\frac{b}{\rho}+1\right)\left(\frac{b}{\rho}\right)^{1/2}}{\left(\frac{b}{\rho}+1\right)\tan^{-1}\left(\frac{b}{\rho}\right)^{1/2}+\left(\frac{b}{\rho}\right)^{1/2}}$$

$$\alpha_2 = \frac{4\left(\frac{b}{\rho}\right)^{3/2}}{3\left\{\left(\frac{b}{\rho}\right)^{1/2}+\left(\frac{b}{\rho}-1\right)\tan^{-1}\left(\frac{b}{\rho}\right)^{1/2}\right\}}$$

α_1 and α_2 are the stress-concentration factors for the two notches in a symmetrically and deeply notched specimen under tensile and bending loads respectively. For intermediate values of a/W the following empirical expression is used.

$$\alpha_a = \frac{\sigma_a}{\sigma_n} = 1 + \frac{(\alpha_{as}-1)(\alpha_{ad}-1)}{\{(\alpha_{as}-1)^2+(\alpha_{ad}-1)^2\}^{1/2}}$$

For case (b):

$$\alpha_{bs} = \frac{\sigma_{bs}}{\sigma_n} = 3\frac{a}{b}\left\{3\left(\frac{a}{2\rho}\right)^{1/2}-1+\frac{4}{2+\left(\frac{a}{2\rho}\right)^{1/2}}\right\}$$

$$\alpha_{bd} = \frac{\sigma_{bd}}{\sigma_n} = 3\frac{a}{b}\left\{\frac{2\left(\frac{b}{\rho}+1\right)-\alpha_1\left(\frac{b}{\rho}+1\right)^{1/2}}{\frac{4}{\alpha_2}\left(\frac{b}{\rho}+1\right)-3\alpha_1}\right\}$$

and for intermediate values of a/W,

$$\alpha_b = \frac{\sigma_b}{\sigma_n} = 1 + \frac{(\alpha_{bs}-1)(\alpha_{bd}-1)}{\{(\alpha_{bs}-1)^2+(\alpha_{bd}-1)^2\}^{1/2}}$$

The stress-concentration factor for the combined loading is given by

$$\frac{\sigma_{max}}{\sigma} = (\alpha_a+\alpha_b)\frac{\sigma_n}{\sigma} = (\alpha_a+\alpha_b)\left(1+\frac{a}{b}\right) \quad (1)$$

For an edge crack, ρ tends to zero, therefore

$$\alpha_{as} = 3\left(\frac{a}{2\rho}\right)^{1/2}$$

$$\alpha_{ad} = 8\left(\frac{\pi-3}{\pi^2-8}\right)\left(\frac{b}{\rho}\right)^{1/2}$$

$$C = \frac{2}{\pi}\left(\frac{4-\pi}{\pi-2}\right)\left(\frac{b}{\rho}\right)^{1/2}$$

$$\alpha_1 = \frac{4}{\pi}\left(\frac{b}{\rho}\right)^{1/2}$$

$$\alpha_2 = \frac{8}{3\pi}\left(\frac{b}{\rho}\right)^{1/2}$$

$$\alpha_{bs} = 9\frac{a}{b}\left(\frac{a}{2\rho}\right)^{1/2}$$

$$\alpha_{bd} = 4\frac{a}{b}\left(\frac{\pi-2}{\pi^2-8}\right)\left(\frac{b}{\rho}\right)^{1/2}$$

hence

$$\alpha_a = 24\left(\frac{a}{\rho}\right)^{1/2}\left(\frac{\pi-3}{\pi^2-8}\right)\left\{128\left(\frac{\pi-3}{\pi^2-8}\right)^2+9\frac{a}{b}\right\}^{-1/2}$$

$$\alpha_b = 36\frac{a}{b}\left(\frac{a}{\rho}\right)^{1/2}\left(\frac{\pi-2}{\pi^2-8}\right)\left\{32\left(\frac{\pi-2}{\pi^2-8}\right)^2+81\frac{a}{b}\right\}^{-1/2}$$

giving

$$\frac{\sigma_{max}}{\sigma_n} = \frac{12}{(\pi^2-8)}\left(\frac{a}{\rho}\right)^{1/2}\left[2(\pi-3)\left\{128\left(\frac{\pi-3}{\pi^2-8}\right)^2+9\frac{a}{b}\right\}^{-1/2}\right.$$
$$\left.+3(\pi-2)\frac{a}{b}\left\{32\left(\frac{\pi-2}{\pi^2-8}\right)^2+81\frac{a}{b}\right\}^{-1/2}\right] \quad (2)$$

as $\rho \to 0$.

The stress-intensity factor K can be derived from the corresponding stress-concentration factor through the relation (5)

$$K = \lim_{\rho \to 0} \tfrac{1}{2}\sigma_n(\pi\rho)^{1/2}\frac{\sigma_{max}}{\sigma_n} \quad . \quad . \quad (3)$$

It is convenient (1) to use the non-dimensional parameter Y, where

$$Y = \frac{KBW}{Pa^{1/2}} \quad . \quad . \quad . \quad (4)$$

Thus, for the specimen notched on one edge, $P = \sigma_n bB$, and equations (2)–(4) give

$$Y = \frac{6\pi^{1/2}}{(\pi^2-8)}\left(1+\frac{a}{b}\right)\left[2(\pi-3)\left\{128\left(\frac{\pi-3}{\pi^2-8}\right)^2+9\frac{a}{b}\right\}^{-1/2}\right.$$
$$\left.+3(\pi-2)\frac{a}{b}\left\{32\left(\frac{\pi-2}{\pi^2-8}\right)^2+81\frac{a}{b}\right\}^{-1/2}\right] \quad (5)$$

Values of σ_{max}/σ from equation (1) and Y from equation (5) are listed for various values of a/W in Tables 1 and 2.

DISCUSSION

Experimental values of σ_{max}/σ have been compared with the Neuber values from equation (1) in Fig. 5; they show good agreement even though Neuber's theory relates to a plate of infinite length with a uniform applied stress whereas the photoelastic model was of finite length which generally gave a non-uniform stress distribution away from the notch, as shown in Fig. 3.

The expression for $Y = KBW/Pa^{1/2}$, where K is the stress-intensity factor obtained by the collocation method for the tension specimen notched on one edge, is (1)

$$Y = 1 \cdot 99 - 0 \cdot 41\left(\frac{a}{W}\right) + 18 \cdot 70\left(\frac{a}{W}\right)^2$$
$$-38 \cdot 48\left(\frac{a}{W}\right)^3 + 53 \cdot 85\left(\frac{a}{W}\right)^4 \quad (6)$$

for $a/W \leqslant 0.6$. Values of Y for various a/W ratios are

Table 2. Stress-intensity factors

$\frac{a}{W}$	Y	
	Collocation	Neuber
0	2·0	1·9
0·1	2·1	1·9
0·2	2·4	2·2
0·3	2·9	2·8
0·4	3·7	3·7
0·5	5·0	5·1
0·6	7·1	7·4
0·7	—	12
0·8	—	23

$Y = \frac{KBW}{Pa^{1/2}}$.

K = stress-intensity factor.

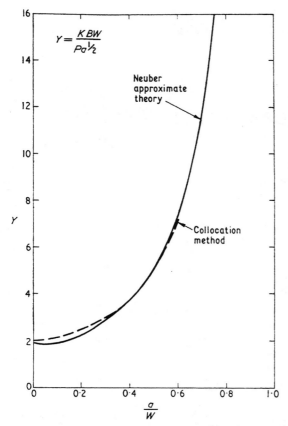

Fig. 7. Variation of Y (stress-intensity factor) with a/W

listed in Table 2. Values of Y from equation (6) are compared with those from Neuber's theory given by equation (5) in Fig. 7. The agreement is generally good, Neuber's values tending to be slightly lower for $a/W < 0.45$ and slightly higher for $a/W > 0.45$.

The agreement between Neuber's values for the stress-concentration factor σ_{max}/σ and stress-intensity factor K (or Y) with the photoelastic and collocation results respectively gives support to the use of the collocation method for calculating the stress-intensity factor. It seems that the non-uniform stress distribution at $V/W = 0.9$ in the photoelastic model, which would almost certainly occur in the tension-fracture test specimen notched on one edge, does not significantly affect the stress distribution (and hence the stress-intensity factor) in the immediate vicinity of the notch or crack. Thus the assumption of uniform stress at $V/W > 0.8$ in the collocation method is reasonable in practice. It should be mentioned, however, that errors in the value of K can occur if care is not taken to reduce pin friction to a minimum at the ends of the specimen (6).

ACKNOWLEDGEMENTS

This paper is published by permission of the Director, National Engineering Laboratory, Ministry of Technology. It is Crown copyright reserved and is reproduced by permission of the Controller of H.M. Stationery Office.

APPENDIX

REFERENCES

(1) BROWN, W. F. and SRAWLEY, J. E. 'Plane strain crack toughness testing of high strength metallic materials', *A.S.T.M. Special Tech. Pub.* No. 410, 1967.

(2) GROSS, B., SRAWLEY, J. E. and BROWN, W. F. 'Stress intensity factors for a single-edge-notch tension specimen by boundary collocation of a stress function', *N.A.S.A. tech. Note* D-2395, 1964.

(3) FROCHT, M. M. and GUERNSEY, R. jun. 'A special investigation to develop a general method for three-dimensional photoelastic stress analysis', *Tech. Notes natn. advis. Comm. Aeronaut., Wash.* TN 2822, 1952.

(4) NEUBER, H. *Kerbspannungslehre* 1958 (Springer, Berlin).

(5) IRWIN, G. R. 'Fracture mechanics', *Proc. Symp. Naval Structural Mechanics* (ed. by Goodier, J. N. and Hoff, N. J.) 1960, 557–591 (Pergamon Press, Oxford).

(6) POOK, L. P. 'The effect of friction on pin-jointed single-edge-notch fracture test specimens', *Int. J. Fracture Mech*, (to be published).

Photoelastic Determination of Stress-intensity Factors

Paper describes several methods for calculating
the stress-intensity factor from the photoelastically determined
stress distributions near the tip of a notch

by R. H. Marloff, M. M. Leven, T. N. Ringler and R. L. Johnson

ABSTRACT—The increasing number of analytical and numerical solutions for the crack-tip stress-intensity factor has greatly widened the scope of application of linear elastic fracture-mechanics technology. Experimental verification of a particular solution by elastic stress analysis is often a necessary supplement to provide the criteria for proper application to actual design problems.

In this paper, it is shown that the photoelastic technique can be used to obtain rather good estimates of the stress-intensity factor for various specimen geometries and loading conditions. Treated are the following cases: wedge-opening load specimen, several notched rotating-disk configurations, and a notched pressure vessel. A sharp crack is simulated by a relatively narrow notch terminating in a root radius of 0.010 in. or less. Stress distributions along the section of symmetry ahead of the notch tip are obtained using three-dimensional frozen-stress photoelasticity. The results are used to determine the stress-intensity factor, K_I, by three methods. Two of these are based on Irwin's expressions for the elastic stress field at the tip of a crack, and the other is a result of Neuber's hyperbolic-notch analysis. Agreement with available analytical solutions is good.

List of Symbols

a = radius of hole, in.
b = outside radius of disk, in.
B = thickness of WOL specimen, in.
c = distance from hole center to notch tip, in.
d = minimum section ahead of notch tip, in.
D_i = inside diameter of notched pressure vessel, in.
h = length of notch, $h = c-a$, in.
H = half height of WOL specimen, in.
K_I = stress-intensity factor for Mode I type of displacement, psi-in.$^{1/2}$
N = rotational speed, rpm
p = pressure, psi
P = load, lb
R_n = notch-root radius, in.
t = thickness of disk, in.
T = wall thickness of pressure vessel, in.
x, y, z = Cartesian-coordinate system
r, θ, z = cylindrical-coordinate system
$\sigma_x, \sigma_y, \sigma_z$ = stress components in Cartesian system
$\sigma_r, \sigma_\theta, \sigma_z$ = stress components in cylindrical system
σ_n = nominal stress, psi
$\sigma_{ymax}, \sigma_{\theta max}$ = maximum stress at the notch tip, psi
μ = Poisson's ratio
ρ = mass density, lb sec^2/in.^{4}
ω = rotational speed, radians/sec

Introduction

In the field of linear elastic fracture mechanics, the stress-intensity approach has been widely accepted as a valid means for predicting the behavior of a material in the presence of a crack or flaw. The stress-intensity factor, K, provides a single parameter representation for the stress conditions at the tip of a crack. In general, for each mode of fracture, the magnitudes of the elastic stresses are governed by appropriate values of K.[1,2]

The use of the stress-intensity criterion requires knowledge of K as a function of specimen and crack geometry and applied loads. Solutions, determined by both analytical and experimental methods, are known for a limited number of geometries, and the number of available solutions is rapidly increasing. Most of these solutions are limited to the Mode I fracture condition,* although considerable attention has been given recently to other modes as well.

Analytical methods for determining stress-intensity factors include complex variable mapping, boundary-collocation procedures, the finite-difference method, and various "strength of materials" approaches. Often, because of the mathematical rigor involved, boundary conditions must be rather simple; thus the solutions may not truly represent actual conditions.

Irwin, et. al.,[1] proposed an experimental method of determining K from the measured compliance (deflection per unit load) of a test specimen. As a result of a photoelastic analysis of propagating cracks conducted by Wells and Post,[3] Irwin[4] showed that the

---

R. H. Marloff, M. M. Leven and R. L. Johnson are associated with Westinghouse Research Laboratories, Pittsburgh, Pa. 15235. T. N. Ringler is associated with Westinghouse Electric, Bloomington, Ind. Paper was presented at 1971 SESA Spring Meeting held in Salt Lake City, Utah on May 18-21.
Original manuscript submitted: Jan. 5, 1971. Final manuscript received: April 8, 1971.

* In this mode, the opening mode, the crack surfaces move directly apart, perpendicular to the plane of the crack.

stress-intensity factor could be estimated from the isochromatic pattern at the tip of a sharp crack. Using this method, Smith and Smith[5] investigated mixed-mode stress-intensity factors for plates containing sharp edge cracks. They found, in general, that the true values of K were dependent on data taken very near the crack tip where resolution, geometric variations, and similar effects strongly influenced the results. Bradley and Kobayashi[6] also used the method in a photoelastic study of propagating cracks.

In this paper, several alternate methods are described for determining crack-tip stress-intensity factors from photoelastic data. A sharp crack is simulated by a slot or notch terminating in a small root radius. The methods are applied to the following cases: a wedge-opening load (Manjoine) specimen, notched rotating disks, and a notched pressure vessel.

The work reported herein summarizes a series of investigations in this area conducted at the Westinghouse Research Laboratories over an eight-year period.

Theory

The general form of the stress field at the tip of a crack is described in terms of the stress-intensity factor, K. For a through-the-thickness crack of zero root radius in a plate subjected to in-plane loads uniformly distributed through the thickness and symmetric about the plane of the crack, Irwin[2] gave the following general form for the elastic stress field near the tip:

$$\sigma_x = \frac{K_I}{\sqrt{2\pi r}} \cos\frac{\theta}{2}\left(1 - \sin\frac{\theta}{2}\sin\frac{3\theta}{2}\right) \quad (1)$$

$$\sigma_y = \frac{K_I}{\sqrt{2\pi r}} \cos\frac{\theta}{2}\left(1 + \sin\frac{\theta}{2}\sin\frac{3\theta}{2}\right) \quad (2)$$

$$\tau_{xy} = \frac{K_I}{\sqrt{2\pi r}} \cos\frac{\theta}{2}\left(\sin\frac{\theta}{2}\cos\frac{3\theta}{2}\right) \quad (3)$$

$$\sigma_z = \nu(\sigma_y + \sigma_x) \quad \text{(plane strain)} \quad (4)$$

or

$$\sigma_z = 0 \quad \text{(plane stress)} \quad (5)$$

where the subscript I in K_I designates the Mode I type fracture and the coordinate system is as shown in Fig. 1. Considering eq (2), K_I can be defined in terms of the distribution of stress, σ_y, along the plane of the crack ($\theta = 0$) as

$$K_I = \lim_{\substack{r \to 0 \\ \theta = 0}} \sigma_y \sqrt{2\pi r} \quad (6)$$

where σ_y is a function of r.

For a crack with a small but finite root radius R_n, the stresses will be finite at $r = 0$ and thus will not satisfy eqs (1-3); however, slightly away from the crack tip, the stress field can be closely approximated by these equations. Bowie and Neal[7] investigated the elastic-strain-energy release rate (which can be related to K_I) for an edge notch having a small root radius. In general, their results show that for notch-length-to-root-radius ratios greater than 2.0, the presence of the radius introduces at most a 2-percent

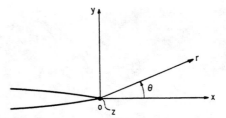

Fig. 1—Coordinates for stress field at the tip of a crack. Z axis is perpendicular to the plane of the paper

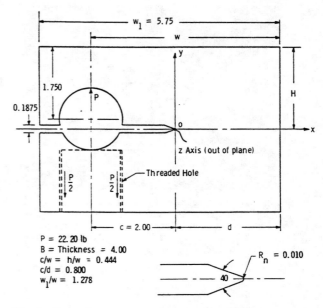

Fig. 2—Photoelastic model of WOL (wedge-opening load) specimen

error in the determination of K_I. Thus, eq (6) may be assumed to hold for small root radii.

Equation (2) forms the basis for two methods of determining K_I. Writing it in the form (for $\theta = 0$)

$$\sigma_y = \frac{K_I}{\sqrt{2\pi}} r^{-1/2} \quad (7)$$

we see that if the stress distribution has this form, a plot of σ_y as a function of $r^{-1/2}$ in the vicinity of the notch tip should be a straight line having the slope $K_I/\sqrt{2\pi}$.

The second method is based on Neuber's plastic-particle concept,[8] in which he suggested that the stress-concentration factor for a sharp notch might be calculated on the basis of an average elastic stress at an elementary particle of width ϵ at the notch tip. Irwin, et. al.,[1] suggested that this concept might be applicable to the calculation of the stress-intensity factor as well. Thus,

$$\sigma_{y\text{ave}} = \frac{1}{\epsilon}\int_0^\epsilon \sigma_y \, dr \quad (8)$$

Integrating eq (2) with $\theta = 0$ gives

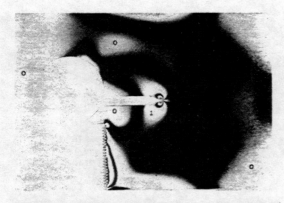

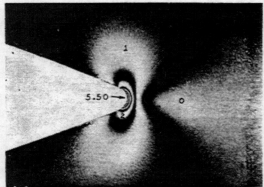

Fig. 3—Stress pattern in center slice taken from WOL specimen showing entire slice and enlargement near notch tip. Slice thicknesses are 0.125 in. (upper) and 0.0405 in. (lower). Notch-root radius is 0.010 in.

$$\sigma_{y\text{ave}} = K_I \sqrt{\frac{2}{\pi r}} \qquad (9)$$

or,

$$K_I = \sqrt{\frac{\pi}{2r}} \int_0^r \sigma_y \, dr \qquad (10)$$

Here r is a small distance from the notch tip.

Again based on Neuber's analysis of notch stresses, it has been shown[2,9] that the product of the maximum stress at the notch tip and the square root of the notch-root radius approaches a finite non-zero value as R_n approaches zero. This limit is related to K_I by

$$K_I = \lim_{R_n \to 0} (1/2 \sqrt{\pi R_n}) \, \sigma_{y\max} \qquad (11)$$

Note in this relationship, if $\sigma_{y\max}$ is inversely proportional to $\sqrt{R_n}$, K_I will become independent of the notch-root radius. As suggested in Ref. 22, an investigation to determine at what values of R_n a relatively constant value of K_I would result was made by considering the stress-concentration factors for a flat bar containing notches of various radii. Examining the limiting cases of a shallow elliptical notch and a deep hyperbolic notch under tension or bending,[10] for root-radius-to-maximum-section ratios, r/D (see Figs. 17 and 33, Ref. 10) less than 0.004, the quantity $\sqrt{R_n} \sigma_{y\max}$ in each case deviates from a constant value by less than 3 percent. It was assumed, then, that the same would be true for intermediate notch depths. Thus for small root radii compared to notch length or minimum section, a good approximation of K_I could be given by

$$K_I = \frac{1}{2} \sqrt{\pi R_n} \, \sigma_{y\max} \qquad (12)$$

Experimental Program

Photoelastic models representing several types of specimen geometries and loading conditions were made. A crack was simulated by a relatively narrow notch terminating in a small root radius of 0.010 in. or less. In all cases, root-radius-to-maximum-section ratios of less than 0.004 were provided to allow the approximation given by eq (12). Only the Mode I or crack-opening type of loading was considered.

Analyses were performed using standard three-dimensional stress-freezing and slicing methods. Maximum fringe orders at the notch tip were obtained by Tardy compensation and, in some cases, extrapolation, using very thin slices (less than 0.030 in.) and magnifications of 20X or more. As section O-x (see Fig. 1) is a plane of symmetry for all the specimens, the individual principal stresses were determined along this section by a combination of several methods:

1. Graphical integration of Filon's transformation of the Lame-Maxwell equilibrium equations.[11]
2. Subslicing in the x-z plane.
3. Slope equilibrium.[11]

Stress-intensity factors were calculated from the experimental stress distributions for the specimens using eqs (7), (10) and (12). For simplicity, we shall refer to these equations as Methods I, II and III, respectively. For the test-specimen geometries, convenient forms of these equations are

Method I:

$$\frac{K_I}{\sigma_n} = \sqrt{2\pi d} \left(\frac{x}{d}\right)^{1/2} \frac{\sigma_y}{\sigma_n} \qquad (13)$$

Method II:

$$\frac{K_I}{\sigma_n} = \sqrt{\pi d/2} \left(\frac{x}{d}\right)^{-1/2} \int_0^{x/d} \frac{\sigma_y}{\sigma_n} \, d\left(\frac{x}{d}\right) \qquad (14)$$

Method III:

$$\frac{K_I}{\sigma_n} = \frac{1}{2} \sqrt{\pi R_n} \frac{\sigma_{y\max}}{\sigma_n} \qquad (15)$$

where σ_n is an appropriate nominal stress, and x/d is a dimensionless distance from the notch tip.

Results

WOL Specimen

The WOL (wedge-opening load) specimen is the result of a considerable effort toward the development of a relatively small, efficient test specimen for valid fracture-toughness measurements. It is frequently referred to as the Manjoine specimen.[12] The photoelastic model is shown in Fig. 2. A notch-tip radius, R_n, of 0.010 in. was used and the specimen thickness, B, was 4.00 in. Load P (22.20 lb) was applied through a threaded nylon stud and a pin-and-clevis arrangement. After stress freezing, 8 slices approximately ⅛-in. thick were cut at various locations z through the thickness (z being measured from the center line of the model). These were thinned accordingly to allow determination of the peak stress at the notch tip. The upper stress pattern of Fig. 3 shows the entire center slice while the lower pattern is an enlargement near the notch tip. The variation of σ_y/σ_n through the thickness at the notch tip is shown in Fig. 4. The stress σ_n is the nominal direct plus bending stress on the section O-x and is given by

$$\sigma_n = \frac{P}{Bd}\left(4 + \frac{6c}{d}\right) = 19.75 \text{ psi} \quad (16)$$

Separation of the individual principal-stress components along the section O-x ahead of the notch tip was performed for the various slices as described above. Figure 5 shows the stress distributions for the center slice as a function of the dimensionless distance x/d, where d is the unnotched section length. These curves are typical of those for all of the slices. The stress-intensity factor was determined from this distribution by Methods I-III. A dimensionless form, $K_I \sqrt{c}\, B/P$ is used which can be obtained from the definition of σ_n as given by eq (16) and multiplication by the square root of the notch length, c. The resulting forms of eqs (13-15) for the particular WOL model geometry are, respectively

$$\frac{K_I\sqrt{c}\,B}{P} = 19.72 \left(\frac{x}{d}\right)^{1/2}\left(\frac{\sigma_y}{\sigma_n}\right) \quad (17)$$

$$\frac{K_I\sqrt{c}\,B}{P} = 9.86 \left(\frac{d}{x}\right)^{1/2}\int_0^{x/d}\left(\frac{\sigma_y}{\sigma_n}\right)d\left(\frac{x}{d}\right) \quad (18)$$

$$\frac{K_I\sqrt{c}\,B}{P} = 0.44 \left(\frac{\sigma_y}{\sigma_n}\right)_{\max} \quad (19)$$

Figure 6 shows graphically the application of these equations. Method III gives a constant value of 7.25. The curve representing eq (18), reaches a relatively constant value between approximately $x/d = 0.01$ and 0.025, or in the vicinity of 4 root radii from the notch tip. An average value over this range results in $K_I\sqrt{c}\,B/P = 6.54$ for Method II. The curve for Method I encompasses the range $0.001 \leq x/d \leq 0.025$ which represents the region best approximated by a linear relation. This was established by first plotting σ_y/σ_n as a function of x/d on log-log paper and then selecting the most linear region of the curve. An average slope representing a least-squares fit is

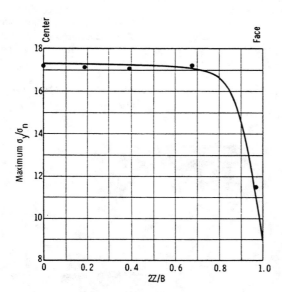

Fig. 4—Through-the-thickness variation of maximum stress at the notch tip for WOL specimen

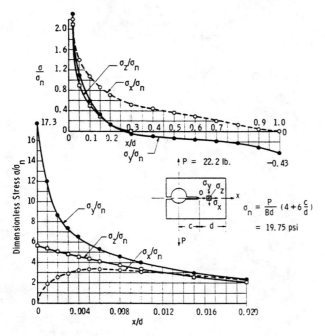

Fig. 5—Distribution of principal stresses along the section of symmetry o-x for the center section of WOL specimen

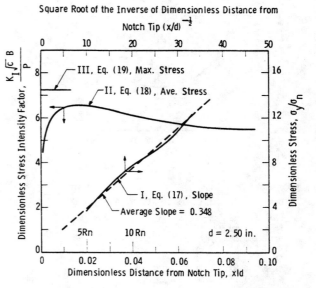

Fig. 6—Determination of the stress-intensity factor, $K_I\sqrt{c}\,B/P$ for the WOL specimen using methods I-III

TABLE 1—DIMENSIONLESS STRESS-INTENSITY FACTORS FOR WOL SPECIMEN

Method	$K_I \sqrt{c}B/P$
I (slope)	6.88
II (average stress)	6.54
III (maximum stress)	7.25
Analytical	6.30

shown by the dashed line, giving a value of $K_I \sqrt{c}\, B/P = 6.88$.

The variation of the maximum stress at the notch tip as shown in Fig. 4, is representative of the stress distribution near the tip as well. Consequently, the stress-intensity factor will also vary through the thickness of the specimen. This variation is due in part to the change from nearly plane-strain conditions in the center region to plane-stress conditions at the surfaces, and also because of the nonuniform loading produced by the centrally located loading stud. The average stress through the thickness is approximately 4.6 percent lower than that at the center. Using a boundary-collocation procedure, an analytical value of $K_I \sqrt{c}B/P = 6.30$ was obtained for a specimen with a zero notch-root radius, uniformly loaded through the thickness.[13] The stress-intensity factors as calculated above are based on the average stress to allow a more meaningful comparison.

Table 1 summarizes the various values of $K_I\sqrt{c}B/P$ for the WOL specimen. The maximum deviation from the analytical value is approximately 15 percent, for Method III.

Center-notched-disk Specimens

Test specimens having the configuration of a center-notched disk or rotor are used in the investigation of the fracture mechanics of alloy-steel turbine and generator-rotor forgings.[14,15] The configuration of the photoelastic models is shown in Fig. 7. Both single- and double-notch geometries (designated by SN and DN, respectively) were considered for various relative notch lengths, h/a. Specimen dimensions and test speeds are given in Fig. 7. All specimens were relatively thin disks, except for Model DN-1 which had a 4.00-in. thickness. The models were rotated by a steel drive plate attached to a drive shaft which passed through a central hole. Two steel pins transmitted rotation through oversize drive holes.

After stress freezing, center slices were removed

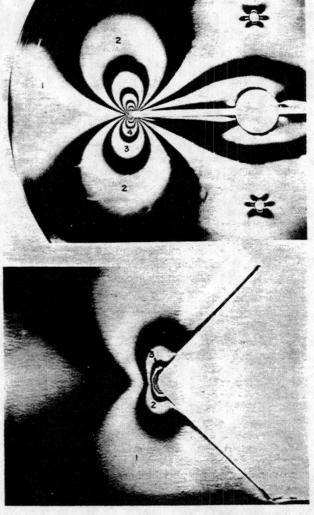

Fig. 8—Full-thickness stress pattern for doubly notched disk, model DN-4, for 0.336-in. slice thickness and enlargement near notch tip after reduction in thickness to 0.032 in. Notch-root radius is 0.010 in.

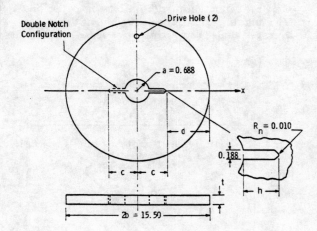

Model No.	h	h/a	t	C	C/b	N (rpm)	$\sigma_n = \frac{3+\mu}{8}\rho\omega^2 b^2$
DN-1	0.688	1.00	4.00	1.375	0.177	725	17.4 psi
DN-2	0.688	1.00	.374	1.376	0.177	650	14.2
DN-3	1.350	1.96	.375	2.040	0.263	725	17.4
DN-4	3.280	4.77	.375	3.970	0.512	595	11.7
SN-1	0.344	0.50	.363	1.032	0.133	700	16.3
SN-2	0.688	1.00	.360	1.376	0.177	699	16.2
SN-3	1.375	2.00	.364	2.063	0.266	654	14.9

Fig. 7—Configuration and dimensions of center-notched disk specimens

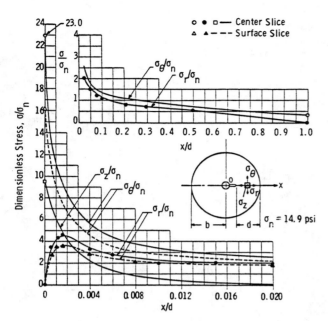

Fig. 9—Distribution of principal stresses on section of symmetry o-x for notched-disk model SN-3

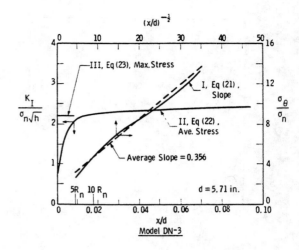

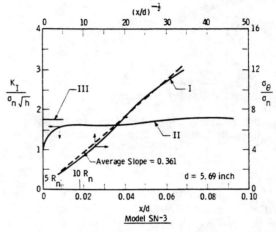

Fig. 10—Determination of dimensionless stress-intensity factor $K_I/\sigma_n \sqrt{h}$ for notched-disk models DN-3 and SN-3

from the specimens and were reduced in thickness as required to obtain the fringe orders at the notch tip. In addition, slices at the disk surfaces were taken to check the axial distribution of the stresses. Figure 8 shows the full-thickness stress pattern for Model DN-4 and an enlargement near the notch tip.

The individual principal-stress components σ_r and σ_θ were determined along the section of symmetry O-x for all the specimens; the axial stress σ_z was obtained for the single-notched specimens as well. Typical stress distributions are shown in Fig. 9 for Model SN-3. Other results are given in Refs. 16 and 17. The shape of the distributions is quite similar for all specimens, the differences being mainly within a distance of one- or two-root radii of the notch tip. The stresses are presented in units of a nominal stress, σ_n, at the center of a thin solid disk of the same diameter and material rotating at the same speed. Thus,

$$\sigma_n = \frac{3+\nu}{8} \rho \omega^2 b^2 \quad (20)$$

As in the WOL specimen, a rather high axial-stress component, σ_z, was observed near the notch tip, indicating a condition approximating plane strain. Thus, the presence of a small-radius notch produces a local triaxial-stress state in an otherwise plane-stress disk. An axial variation in the tangential- and axial-stress components was observed for the single-notched specimens; it would be expected that the double-notched specimens should exhibit this variation also. At the notch tip, the tangential stress was approximately 10 percent higher at the center than the average stress through the thickness.

A dimensionless stress-intensity factor, $K_I/\sigma_n \sqrt{h}$, where h is the notch length as shown in Fig. 7, was determined from the center-slice stress data for each of the disk specimens. The appropriate forms of eqs (13-15) for the disk geometries are

$$\frac{K_I}{\sigma_n \sqrt{h}} = 2.50 \left(\frac{d}{h}\right)^{1/2} \left(\frac{\sigma_\theta}{\sigma_n}\right) \left(\frac{x}{d}\right)^{1/2} \quad (21)$$

$$\frac{K_I}{\sigma_n \sqrt{h}} = 1.25 \left(\frac{d}{h}\right)^{1/2} \left(\frac{d}{x}\right)^{1/2} \int_0^{x/d} \left(\frac{\sigma_\theta}{\sigma_n}\right) d\left(\frac{x}{d}\right) \quad (22)$$

$$\frac{K_I}{\sigma_n \sqrt{h}} = \frac{0.089}{\sqrt{h}} \left(\frac{\sigma_\theta}{\sigma_n}\right)_{max} \quad (23)$$

Figure 10 shows the results of the application of these equations for Models SN-3 and DN-3. The curves are quite similar to those for the WOL specimen (see Fig. 6) as are those for the other disk specimens. The distances from the notch tip corresponding to 5 and 10 times the root radius are shown for reference. For Method II, all of the curves remained relatively constant over the range $0.01 < x/d < 0.04$. Thus the average value over this range was chosen as being a reasonable estimate of $K_I/\sigma_n\sqrt{h}$ for this method. The region shown in the curves for Method I represents the range of x/d from 0.001 to 0.050 which was found to be the most nearly linear region for all of the specimens.

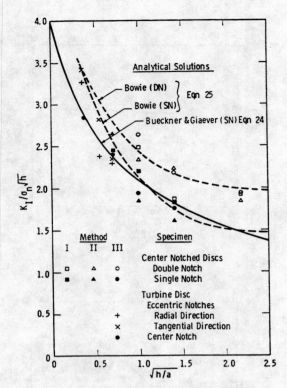

Fig. 11—Variation of dimensionless stress-intensity factor with relative notch length for notched rotating-disk specimens

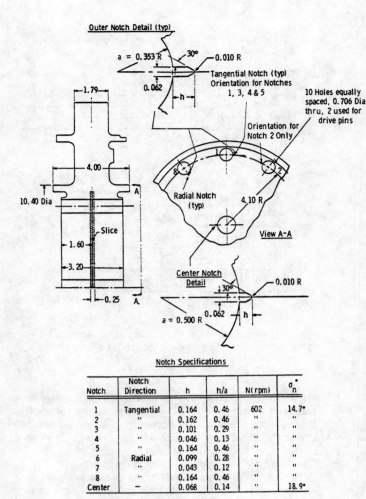

Fig. 12—Photoelastic model of a gas-turbine disk singly notched at center and through bolt holes

The results for the disk specimens are summarized in Fig. 11 where $K_I/\sigma_n\sqrt{h}$ is plotted as a function of $\sqrt{h/a}$. The data points for the double-notch case at $h/a = 1.00$ represent the average values for Models DN-1 and DN-2. Although the former was considerably thicker, (4.00 in. compared to 0.374 in.) within the limits of experimental error the results for the two cases were essentially the same.

A comparison is made with several analytical solutions which are appropriate to the rotating-disk configuration. Bueckner and Giaever[18] obtained a numerical solution for the stresses in an infinite disk containing a single notch at the bore from which they derived the following expression for the stress-intensity factor.[17]

$$K_I/\sigma_n = \pi\sqrt{\pi(c-1)/(c+1)^3}\,f(h/a) \quad (24)$$

where $c = (h + a)$ and $f(h/a)$ is a geometrical function. Bowie[19] analyzed the case of a very large rectangular plate under a uniform stress field and containing both single and double radial cracks emanating from the periphery of a bore hole. Winne and Wundt[15] adapted Bowie's analysis to the notched-disk problem and showed an equivalence of stress fields provided the overall crack length $2c$ is less than ¼ of the outer disk diameter. Bowie's expression for the stress-intensity factor has the form

$$K_I/\sigma_n = \sqrt{\pi a}/F(h/a) \quad (25)$$

where the function $F(h/a)$ is a different geometrical function for the single- and double-notch configurations. Results based on eqs (24) and (25) are shown in Fig. 11 for comparison with the experimental values.

Although a spread in the data is evident, the general trends in the results seem to agree reasonably well with that predicted by eqs (24) and (25). The largest deviation among the values for the various methods occurred for Model DN-4 ($\sqrt{h/a} = 1.40$). This is the result of what is believed to be a low value of $K_I/\sigma_n\sqrt{h}$ by Method I. It should be noted that Model DN-4 ($\sqrt{h/a} = 1.95$) violates the restriction of $c < 0.25b$ for eq (25). For this case, $c/b = 0.51$. For the range of notch lengths considered, eqs (24) and (25) for the single-notch case differ at most by only 7 percent. Thus, considering the spread in the data, the results are inconclusive insofar as verification of a particular solution is concerned.

Notched Turbine Disk

Figure 12 shows a photoelastic model of a gas-turbine disk. This model was utilized as a test specimen for non-centrally located notches. Through bolt holes, located on a 4.1-in. radius, had notches oriented in either the radial or tangential directions. Two adjacent tangential notches were cut facing one

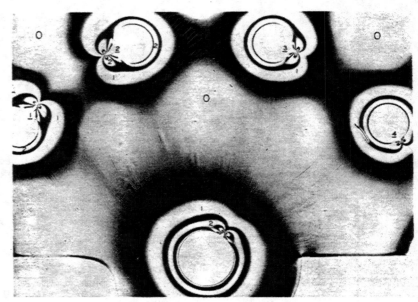

Fig. 13—Stress pattern in 0.25-in. slice taken from center of notched turbine disk. Notch numbers correspond to those in Fig. 12

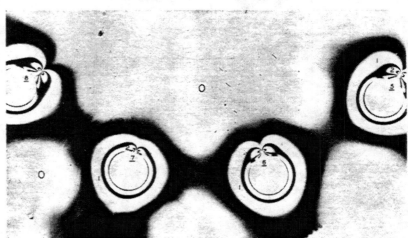

another to investigate possible interactions. Three relative notch lengths (h/a ratio) of approximately 0.5, 0.3 and 0.1 were considered. A relatively short, single notch at the bore was included to extend the range of the previous study. The table in Fig. 12 gives the notch specifications. Epoxy weights were cemented to the outside rim of the disk to simulate the effect of blade loading during rotation. The model was rotated in the fixture described above at a speed of 602 rpm.

Stress patterns for a ¼-in. center slice are shown in Fig. 13. Only the maximum stresses at the notch tips were determined in this test and these are given in Table 2. For consistency with the earlier results, the nominal stress σ_n is taken as the stress at the location of the hole centers in a solid disk. Here the actual disk configuration is used rather than a thin constant-thickness disk. The values of σ_n for the center and eccentric holes were determined from a disk computer program with both centrifugal and blade-loading effects considered. For the center notch, $\sigma_n = 18.90$ psi and for the eccentrically located notches, $\sigma_n = 14.69$ psi. Actually, at the 4.1-in. hole radius, the state of stress is not truly isotropic, as the tangential stress is about 10 percent higher than the radial stress. It is assumed that the average of the

TABLE 2—PEAK STRESSES FOR NOTCHED TURBINE-DISK MODEL

Notch and No. (see Fig. 12)		h/a	σ_n† (psi)	$\sigma_{\theta_{max}}$ (psi)	$\sigma_{\theta_{max}}/\sigma_n$
Center		0.14	18.9	158.2	8.4
Bolt Holes					
Radial	7	0.12	14.7	112.7	7.7
	6	0.28	14.7	124.4	8.5
	8	0.46	14.7	154.6	10.5
Tangential	4	0.13	14.7	128.8	8.8
	3	0.29	14.7	149.0	10.1
	5	0.46	14.7	154.5	10.5
	1	0.46*	14.7	151.3	10.3
	2	0.46*	14.7	156.7	10.7

* Adjacent notches facing one another.
† Calculated, disk program.

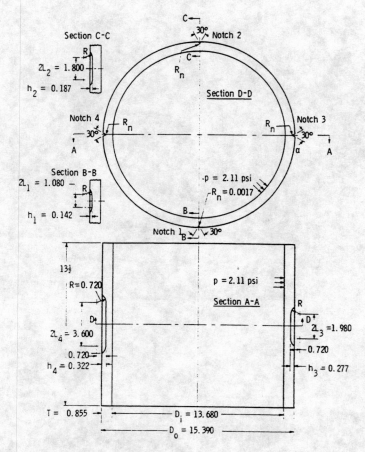

Fig. 14—Shape and dimensions of cylinder subjected to internal pressure and containing four longitudinal external notches. $D_i/T = 15.97$; $R_n/T = 0.00198$

two represents a nominal stress comparable to that for a central hole.

For the two shorter notch lengths, the radially oriented notches exhibited somewhat lower peak stress than did those in the tangential direction. However, for the longest notch, $h/a = 0.46$, this was not true, the magnitudes being about equal. Since there is no reason to assume that the former should not be true for this length and, furthermore, since the stresses for Notches 1, 2 and 5 were essentially equal (here it is assumed that no interaction occurs between Notches 1 and 2), it is believed that the peak stress for Notch 8 is perhaps in error.

Using Method III only, based on maximum stress [eq (23)], stress-intensity factors were computed for the three notch lengths as well as for the center notch. The results are given in Fig. 11 for comparison with the other notched-disk results. For the single center notch, good agreement with the Bueckner and Giaever solution was obtained, reinforcing the previous plane-disk results for larger h/a ratios, and verifying the validity of this relation for short notch lengths.

In view of the differences in geometry and loading conditions, it is probably not valid to compare the non-central notch results with the analytical solutions. As a first approximation, however, one might use these expressions to predict the stress-intensity factors for the cases.

Pressurized Cylinder with External Notches

A cylindrical vessel containing four external longitudinal notches was subjected to internal pressure. Figure 14 shows the general shape and dimensions of the photoelastic model. Each notch had a root radius of 0.0017 in. Vessel extensions having torospherical

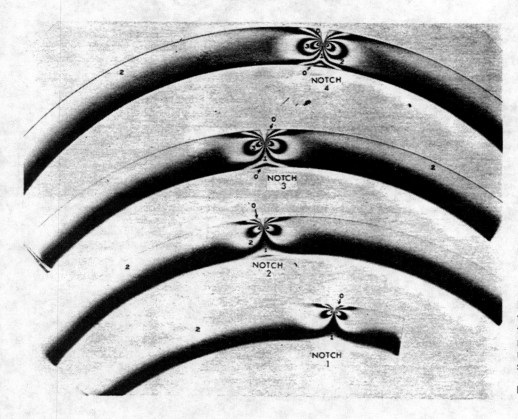

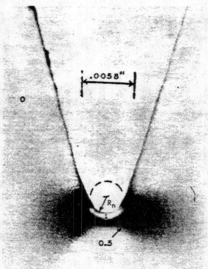

Fig. 15—(Left) Stress pattern in transverse slices taken through the center of the four notches for the longitudinally notched pressure vessel. Slice thicknesses are 0.25 in. Photograph at right shows enlargement near notch tip for Notch 1 at a slice thickness of 0.010 in. Notch-root radius R_n is 0.0017 in.

end caps were cemented to the ends of the section. An internal pressure of 2.11 psi was applied during the stress-freezing process. Transverse slices were cut through the center of the four notches and the stress patterns are shown at the top in Fig. 15.

A detailed stress analysis was carried out for Notch 1, having a relative notch length, h/T, of 0.166. An enlargement of the stress pattern near the notch tip for a slice thickness of 0.010 in. is shown in the lower half of Fig. 15. Stress separation was performed along the axis of symmetry. The stress distributions are shown as a function of dimensionless distance x/d from the notch tip in Fig. 16. The nominal stress, σ_n, represents the nominal hoop stress in the unnotched section of the cylinder and is given by

$$\sigma_n = \frac{p(D_i + T)}{2T} = 17.90 \text{ psi} \qquad (26)$$

A dimensionless stress-intensity factor, $K_I/\sigma_n\sqrt{h}$, was computed from the σ_θ/σ_n stress distribution by the three methods. For the particular geometry, eq (13)-(15) have the form

$$\frac{K_I}{\sigma_n\sqrt{h}} = 5.60 \left(\frac{\sigma_\theta}{\sigma_n}\right) \left(\frac{x}{d}\right)^{1/2} \qquad (27)$$

$$\frac{K_I}{\sigma_n\sqrt{h}} = 2.81 \left(\frac{d}{x}\right)^{1/2} \int_0^{x/d} \left(\frac{\sigma_\theta}{\sigma_n}\right) d\left(\frac{x}{d}\right) \qquad (28)$$

$$\frac{K_I}{\sigma_n\sqrt{h}} = 0.097 \left(\frac{\sigma_\theta}{\sigma_n}\right)_{max} \qquad (29)$$

These equations, when plotted, exhibit much the same behavior as those for the WOL and disk specimens (Figs. 6 and 10). The resulting values are given in Table 3. For Method I, the average slope was computed over the range $0.001 \leq x/d \leq 0.020$ and for Method II, the average value over the range $0.004 < x/d < 0.020$ was used.

If the effect of the longitudinal stress and the radial stress is neglected, the section containing the notch might be compared to a finite-width strip containing an edge notch and subjected to an eccentric tensile load. Gross and Srawley[20] investigated this case, and evaluated the stress-intensity factor as a function of relative notch-length-and-load eccentricity ratio. By choosing an eccentricity ratio such that a stress distribution similar to that in the vessel away from the notch is obtained, the value of $K_I/\sigma_n\sqrt{h}$ shown in Table 3 results. Method I agrees best with this value. Methods II and III, although in agreement with each other within 10 percent, show a considerable deviation from Method I. Thus, the spread of results for the notched vessel is considerably more than for the other specimens.

Discussion

The accuracy of the photoelastically determined stress-intensity factors depends first on the validity of the three methods described and, second, on the accuracy of the measured stress distributions near the notch tip.

Regarding the validity of the analytical methods, we first note that Methods I and II both depend on the assumed applicability of eq (1)-(3) to a crack

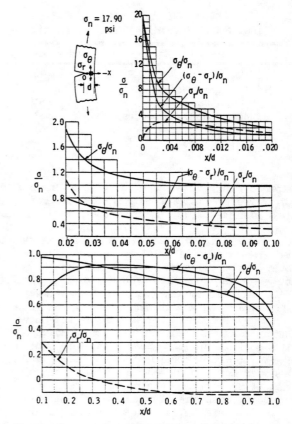

Fig. 16—Distribution of principal stresses along the section of symmetry o-x through center of Notch 1

having a non-zero root radius. Based on the results of several analytical investigations, Weiss and Yukawa[9] show that the stress distributions for cracks having sharp and finite root radii differ only at distances from the notch tip of less than one-fourth of the root radius. Creager[21] suggested that the finite root radius can be taken into account by using the following expression for the σ_y stress along the plane of the crack

$$\sigma_y = \frac{K_I}{\sqrt{2\pi r}}\left(1 + \frac{R_n}{2r}\right) \qquad (30)$$

instead of eq (2). Here r is measured from a point halfway between the notch tip and the center of the root radius. Evaluation of this equation shows that at a distance of one-fourth the root radius from the notch tip, the stress differs by only 4 percent over that given by eq (2). Thus, in light of these considerations, the effect of the finite root radius is assumed to be negligible. Another possible factor affecting accuracy is the influence of the notch-tip flank angle. In this investigation, the total angle varied between 30 deg and 60 deg for the various specimens. While the effect of this angle cannot be precisely determined, Fig. 36 of Ref. 10 indicates that the effect is probably small in the present tests.

The region in which the equations for Methods I and II apply probably depends on the length of the minimum section, d, ahead of the notch tip, since the

TABLE 3—STRESS-INTENSITY FACTORS FOR LONGITUDINALLY NOTCHED PRESSURE VESSEL

Method	$K_I/\sigma_n\sqrt{h}$
I (slope)	2.70
II (average stress)	2.15
III (maximum stress)	1.94
Single-edge-notch specimen (Ref. 20)	2.62

stress distribution will be influenced by the presence of a free or loaded boundary. For Method I, the best results for all the specimens were obtained within the range $0.001 < x/d < 0.02$. Rather than calculate K_I directly from the stress at a given location, it was felt that using an average value of slope over a small range of values would tend to average out errors due to small local disturbances in the stress field. For Method II, based on the average stress, the value of K_I was chosen where eq (14) remained relatively constant. For all of the specimens this occurred within the range $0.01 < x/d < 0.02$. Method II is perhaps the least valid of the three, because of the assumptions made in its derivation. First, it is assumed that a small amount of plastic straining occurs at the root of a sharp notch. Since this cannot occur in frozen-stress photoelasticity, this assumption is questionable. A second assumption is that this concept, although derived for the stress-concentration factor, is applicable to the determination of the stress-intensity factor as well.

Calculation of K_I based on the maximum stress at the notch tip. Method III, also represents an approximation. However, as discussed previously, for the root radii and notch lengths considered, the error in removing the limiting condition from eq (11) should be less than 3 percent.

Examining the accuracy of the experimental data, the greatest potential for error in the stress distributions occurs in the vicinity of the notch tip where a severe stress gradient exists for the σ_y or σ_θ stress. The accuracy here depends on the accuracy of both the measured principal stress difference and the stress-separation process. Values of K_I calculated by Methods I and II are severely affected by small changes in the stress distribution in this region. The resulting error is estimated to be less than ± 10 percent.

The accuracy in determination of the peak stress at the notch tip is governed to a large degree by the ability to resolve the notch-tip boundary. This resolution is greatly facilitated by a sufficient reduction in slice thickness. For the 0.010-in. root radius, the peak stresses should be better than 95 percent accurate. The smaller radius (0.0017 in.) used in the notched vessel will reduce this to perhaps 90 percent. Method III, in addition to requiring the peak stress, also requires a measurement of the notch-root radius. If the notch tip is of a constant radius, this can be measured directly from the final slice used to obtain the peak stress. In this investigation, the error in the measurement of the root radii is estimated to be ± 10 percent.

Although a considerable spread in the results of the three methods was evident for all of the specimens, Method I, in general, seemed to give the best results and is therefore the suggested one. However, Method III requires much less time, since only the peak stress is required. Thus, this method may be useful as a first approximation to the determination of K_I.

Acknowledgments

The authors wish to thank W. K. Wilson for his many helpful suggestions during the course of this investigation and, in particular, for the analytical effort connected with the WOL specimen. The assistance of J. Barcic in the machining and testing of the models is gratefully acknowledged.

References

1. Irwin, G. R., Kies, J. A. and Smith, H. L., "Fracture Strengths Relative to Onset and Arrest of Crack Propagation," Proc. ASTM, 58, 640-657 (1958).
2. Paris, P. C. and Sih, G., "Stress Analysis of Cracks, Fracture Toughness Testing and its Applications," ASTM Special Technical Publication No. 381.
3. Wells, A. A. and Post, D., "The Dynamic Stress Distribution Surrounding a Running Crack—A Photoelastic Analysis," Proc. SESA, 16 (1), 69-92 (1958).
4. Irwin, G. R., Discussion of Ref. 3 above, Proc. SESA, 16 (1), 93-96 (1958).
5. Smith, D. G. and Smith, C. W., "Photoelastic Determination of Mixed Mode Stress Intensity Factors," Presented at the Fourth National Symposium on Fracture Mechanics, Pittsburgh, Penna. (August 24-26, 1970).
6. Bradley, W. B. and Kobayashi, A. S., "An Investigation of Propagating Cracks by Dynamic Photoelasticity," EXPERIMENTAL MECHANICS, 10 (3), 106-113 (March 1970).
7. Bowie, O. L. and Neal, D. M., "The Effective Length of an Edge Slot in a Semi-Infinite Sheet Under Tension," Int. Jnl. Fracture Mech., 3 (2), 111-119 (June 1967).
8. Neuber, H., "Theory of Notch Stresses," Springer-Verlag, Berlin. Translation by United States Atomic Energy Commission (1958).
9. Weiss, V. and Yukawa, S., "Critical Appraisal of Fracture Mechanics," Fracture Toughness Testing and Its Applications, ASTM Special Technical Publication No. 381 (1965).
10. Peterson, R. E., "Stress Concentration Design Factors," John Wiley and Sons, Inc., New York, Chapter 2 (1953).
11. Frocht, M. M., "Photoelasticity," 1, John Wiley and Sons, Inc., (1941).
12. Manjoine, M. J., "Biaxial Brittle Fracture Tests," Trans. ASME, Jnl. Basic Engr., 87, 293-98 (June 1965).
13. Brown, W. F. and Srawley, J. E., "Plane Strain Crack Toughness Testing of High Strength Metallic Materials," Discussion by M. J. Manjoine, ASTM Special Technical Publication No. 410, 66-70 (1966).
14. Sankey, G. O., "Spin Fracture Tests of Nickel-Molybdenum-Vanadium Rotor Steels in the Brittle Fracture Range," Proc. ASTM, 60, 721-732 (1960).
15. Winne, D. H. and Wundt, B. M., "Application of the Griffith-Irwin Theory of Crack Propagation to the Bursting Behavior of Disks, Including Analytical and Experimental Studies," Trans. ASME, 80, 1643-1656 (Nov. 1958).
16. Marloff, R. H., "A Photoelastic Verification of Energy Release Rates for Effectively Notched Rotating Disks," M.S. Thesis, University of Pittsburgh (1964).
17. Ringler, T. N., "A Photoelastic Analysis of Singly Notched Spin Disk Specimens with Calculation of Stress Intensity Factors," M.S. Thesis, University of Pittsburgh (1970).
18. Beuckner, H. F. and Giaever, I., "The Stress Concentration in a Notched Rotor Subjected to Centrifugal Forces," Zeitschrift fur Angewante Mathematic und Mechanik, Band 46, Heft 5 (1966).
19. Bowie, O. L., "Analysis of an Infinite Plate Containing Radial Cracks Originating at the Boundary of an Internal Circular Hole," Jnl. Math. & Physics, 35, 60-71 (April 1956).
20. Gross, B. and Srawley, J. E., "Stress-Intensity Factors for Single-Edge-Notch Specimens in Bending or Combined Bending and Tension by Boundary Collocation of a Stress Function," NASA Technical Note D-2603 (Jan. 1965).
21. Creager, M., "The Elastic Stress Field Near the Tip of a Blunt Crack," M.S. Thesis, Lehigh University (1966).
22. Wahl, A. M., Leven, M. M. and Wilson, W. K., "Energy Release Rates for Biaxial Brittle Fracture Test Specimen," Research Report WERL-8844-1 Westinghouse Research Laboratories (Aug. 1964).

AN ASSESSMENT OF FACTORS INFLUENCING DATA OBTAINED BY THE PHOTOELASTIC STRESS FREEZING TECHNIQUE FOR STRESS FIELDS NEAR CRACK TIPS†

M. A. SCHROEDL, J. J. McGOWAN and C. W. SMITH

Department of Engineering Mechanics, College of Engineering,
Virginia Polytechnic Institute and State University, Blacksburg, Virginia 24061, U.S.A.

Abstract — Factors which influence crack tip stress field data are identified as:
(1) Non-linear zone near crack tip due to crack blunting.
(2) Normal stress parallel to crack surface.
(3) Location of region for data retrieval.

The Kolosoff–Inglis solution is used in order to assess effects of crack tip blunting. A set of stress freezing photoelastic experiments are conducted on plates containing through cracks and results are compared with the Westergaard solution in order to assess the effect of Item 2 using appropriate Item 3 locations. A data conditioning computer program is employed to yield accurate values of the stress intensity factor from photoelastic data.

NOMENCLATURE‡

$\sigma_{xx}, \sigma_{yy}, \tau_{xy}$	Stress components in plane normal to crack border
r, θ	Polar coordinates measured from crack tip
K_I	Mode I Stress Intensity Factor
τ_{max}, τ_m	Maximum shearing stress in plane perpendicular to crack border
τ_{m0}	Maximum remote shearing stress in plane perpendicular to crack border
n	Fringe order
f	Material fringe value psi-in./order
σ_{0x}	Normal stress parallel to crack plane
t	Model thickness
ρ	Crack root radius
δ	$\dfrac{\sigma_{0x}}{\sigma}$
σ	Remote uniform stress.

INTRODUCTION

PIONEER investigations into the use of the photoelastic technique for studying stress fields near crack tips are due to Post[1] and Wells and Post[2]. In discussing the latter paper, Irwin[3] pointed out that at least two parameters were necessary in order to completely describe the fringe pattern in the singular regions. He identified these parameters as the stress intensity factor and the normal stress parallel to the crack surface. The importance of the first parameter to fracture mechanics theory is well known. The need for the second parameter has been verified in studies by Smith and Smith[4–6], Kerley[7] and others. The adaptation of the stress freezing and slicing technique to photoelastic studies of crack tip stress fields has been made by Fessler and Mansell[8], Dixon and Strannigan[9], Liebowitz et al.[10], Smith and Smith[5], Marrs and Smith[11] and others. A rather extensive survey of results of calculating the stress intensity factor from photoelastic data has recently been published by Marloff et al.[12]. While these investigators generally recognize that the stress field bordering the crack

---

†Presented at the Symposium on Fracture and Fatigue, at the School of Engineering and Applied Science, George Washington University, Washington, D.C., May 3–5, 1972.

‡All stresses in pounds per square inch and all stress intensity factors in pounds per (in.)$^{3/2}$.

root is substantially different from the singular stresses of fracture mechanics due to a non-linear zone very near the crack tip, little attention has been given to the influence of Irwin's second parameter in these works except for approximate [13] and 'exact' [14] numerical techniques developed by Bradley and Kobayashi for use in determining the dynamic stress intensity factor quantitatively for two [13] and three [14] dimensional problems. A simplified approximate technique which yields accurate estimates of the stress intensity factor for the stress freezing problem with potential for application to three dimensional problems would be desirable. In order to achieve this, it appears that certain difficulties associated with measurement sensitivity must be overcome, together with an evaluation of the effects of the non-linear zone very near the crack tip and the normal stress parallel to the crack surface.

GENERAL CONSIDERATIONS

Irwin [3] described the stress field near the crack tip with the following equations:

$$\sigma_{xx} = \frac{K_I}{(2\pi r)^{1/2}} \cos\frac{\theta}{2}\left\{1 - \sin\frac{\theta}{2}\sin\frac{3\theta}{2}\right\} - \sigma_{0x}$$

$$\sigma_{yy} = \frac{K_I}{(2\pi r)^{1/2}} \cos\frac{\theta}{2}\left\{1 + \sin\frac{\theta}{2}\sin\frac{3\theta}{2}\right\} \tag{1}$$

$$\tau_{xy} = \frac{K_I}{(2\pi r)^{1/2}} \cos\frac{\theta}{2}\sin\frac{\theta}{2}\cos\frac{3\theta}{2}.$$

The notation is given in Fig. (1). Now while σ_{0x} has no influence upon the singular stress field itself, it does alter the fringe pattern (which is proportional to the maximum in-plane shear stress, τ_{max}) in the singular region. That is, from the stress-optic law,

$$\tau_{max} = \frac{nf}{2t} \tag{2}$$

and from plane stress theory,

$$\tau_{max} = \tfrac{1}{2}[(\sigma_{xx} - \sigma_{yy})^2 + 4\tau_{xy}^2]^{1/2}. \tag{3}$$

Now, substituting equations (1) into equation (3) there results:

$$\tau_{max}^2 = \frac{K_I}{(2\pi r)^{1/2}}\cos\frac{\theta}{2}\sin\frac{\theta}{2}\left[\frac{K_I}{(2\pi r)^{1/2}}\cos\frac{\theta}{2}\sin\frac{\theta}{2} + \sigma_{0x}\sin\frac{3\theta}{2}\right] + \frac{\sigma_{0x}^2}{4}. \tag{4}$$

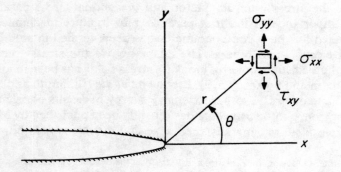

Fig. 1. Notation for equations (1).

Since σ_{0x} distorts the singular fringe loops, Irwin proposed measuring the distance from the crack tip to the farthest point on a given fringe loop together with its angle of inclination to the crack plane to completely characterize the pattern. He then substituted these coordinates into equation (4) and the equation $\partial \tau_{\text{mac}}/\partial \theta = 0$ and solved these equations simultaneously for K_I and σ_{0x} using equation (2) to define τ_{max}. In the closure to Irwin's discussion, Wells and Post pointed out that a high degree of accuracy was required in the two measurements in order to accurately compute K_I and σ_{0x}. (The authors have in fact found that this measurement accuracy cannot be consistently obtained.) Moreover, this method relies entirely upon a point measurement which does not take advantage of the whole field of data provided by the photoelastic technique. Therefore, in order to overcome the practical difficulties of Irwin's approach, the authors have developed a data conditioning computer program which accepts all data along an arbitrary line and utilizes statistical procedures in order to determine accurate values of K_I. The procedure thus takes advantage of more of the whole field of data and is applied in a region where K_I determination should be most accurate. The equations used are obtained by setting $\theta = \pi/2$ in equation (4), substituting equation (2) for τ_{max} and solving for K_I to yield:

$$K_I = (\pi r)^{1/2} \left[\left\{ 2\left(\frac{nf}{t}\right)^2 - \sigma_{0x}^2 \right\}^{1/2} - \sigma_{0x} \right]. \tag{5}$$

By applying equation (5) (neglecting the σ_{0x}^2 term) to all pairs of fringes (with n and r specified for each fringe), a set of values of K_I are generated with a known absolute deviation from the average. Rejecting all K_I values which have absolute deviations in excess of the average absolute deviation, a final average K_I is computed. Using this value of K_I an average τ_{0x} may be computed in an analogous manner but employing the exact form of equation (5).

The data conditioning program described above was developed to fit the type of experimental data obtained from fringe loops near crack tips, and the approximations utilized therein are directed towards the computation of accurate K_I values.

THE INVESTIGATION

(A) Non-linear zone near crack tip

A considerable amount of discussion is found in the literature ([7], [9] and [12]) regarding the feasibility of utilizing a round ended slot to simulate the crack in photoelastic work. This alteration in the geometry obviously alters the stress field just ahead of the crack tip. While it has been suggested by Weiss and Yukawa[15] that this effect is small and quite local ahead of the crack tip, Liebowitz et al.[10] have indicated a significant alteration of the local fringe pattern a distance of one root radius ahead of the slot tip for a root radius of 0·0035 in. The authors have obtained results similar to those of Liebowitz (Fig. 2). When the root radius is decreased by an order of magnitude to about 10^{-4} in. (as is found in a natural crack blunted by extension in a stress freezing cycle), fringe disturbances ahead of the crack tip are not detectable under twenty power magnification of the crack tip. However, there must still be a non-linear zone very near the crack tip due to finite rotations associated with crack blunting.

The fringe pattern associated with the singular stress field presents a readily measurable fringe gradient along a line passing through the crack tip but perpendicular to the crack surface. Moreover, the authors have observed the fringe pattern to be relatively

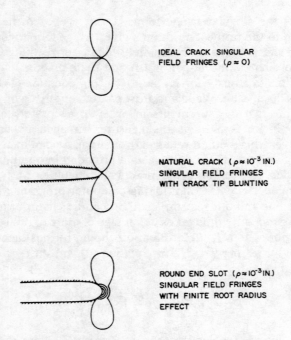

Fig. 2. Effect of finite root radius upon fringe pattern.

insensitive to small crack root radii in this region. This observation may be reinforced by the results of the elastic analysis shown in Fig. 3.

Using the Kolosoff-Inglis elastic solution for an elliptic hole in a plate under uniform biaxial tension, the maximum in-plane shear stress was computed for various ellipse root radii (ρ) and compared with the same quantity as computed from the singular stresses for a crack of the same length as the major diameter of the ellipse. The ratio of these values was plotted along a line perpendicular to the crack surface passing through the notch or crack tip. For a notch root radius $\rho/a = 0.01$, an error in τ_{max} of about two per cent results at an r/a of 0.04 but increases to about 7 per cent at an r/a of 0.02. For root radii less than $\rho/a = 0.001$, as used by the authors, virtually no effect upon τ_{max} results even for r/a of the order of 0.002.

By taking data only outside the affected zone near the crack tip, the influence of crack tip blunting or slot-radius may be virtually eliminated. It should be noted that in the experimental studies described in the sequel, the authors worked with 'natural' cracks, the root radii of which were not detectable under 600 power magnification in the unloaded state, but blunted during the stress freezing cycle to a root radius of 0.6×10^{-4} in. Moreover, the photoelastic material is linearly elastic throughout and the nonlinearity is associated only with finite rotations near the crack tip.

(B) *The experiments (effect of σ_{0x})*

A numerical study of equation (5) and its use in computing K_I and σ_{0x} revealed that while K_I could be computed quite accurately from the data conditioning program, when K_I was varied by less than 5 per cent the resulting σ_{0x} might vary by about 30 per cent. However, using the program described in the previous section, σ_{0x} need not be computed at all if only K_I is desired. In order to verify these conclusions, a set of photo-

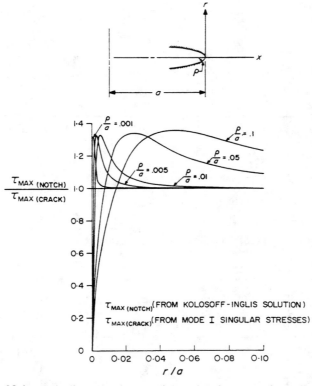

Fig. 3. Effect of finite root radius on in-plane maximum shearing stress along a line perpendicular to crack surface passing through the crack tip ($\theta = \pi/2$).

elastic experiments were performed. These tests were for a central through crack in a wide plate under uniaxial extension. This type of test was selected because a complete elastic solution was available from the Westergaard through crack solution[16] which allowed analytical determination of both K_I and σ_{0x} for comparison with experimental values (i.e. from the Westergaard solution $K_I = \bar{\sigma}(\pi a)^{1/2}$ and $\sigma_{0x} = \bar{\sigma}$).

As noted earlier, σ_{0x} tends to distort the fringe loops near the crack tip from those from the singular stresses alone. Figure 4 shows the nature of this distortion for a through crack in an infinite plate. For the Westergaard problem with uniform remote biaxial extension, the local fringe loops are symmetric about a line normal to and passing through the crack tip, and are represented by $\delta = 0$ in Fig. 4. For the case of uniaxial extension normal to the crack tip (i.e., $\sigma_{0x} = \sigma$) $\delta = 1\cdot 0$ and the fringe is extended normal to the crack surface and leans ahead of the crack. For negative σ_{0x}, the fringe is reduced in size and leans backwards (i.e. $\delta = -1\cdot 0$).

Two specimens were prepared from Hysol 4290 plates of nominally one half inch thickness. These plates were cut in half, edge slots were milled into the center of the cut halves, and edge cracks were tapped into each slot normal to the cut edges. The specimens were then glued back together forming a central crack. Typical specimen geometry is shown in Fig. 5. Specimens were then placed in a loading rig in an annealing oven and heated to 280°F, soaked for about 6 hr, loaded, and cooled under load at a rate not exceeding 5°F per hr. All loads were dead weights.

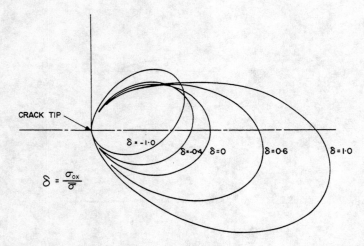

Fig. 4. Fringe loop distortion due to σ_{0x} for the Westergaard through crack solution (remote stress $\sigma_{xx} = \sigma_{yy} = \bar{\sigma}$).

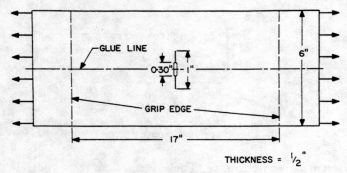

Fig. 5. Typical specimen geometry.

RESULTS AND DISCUSSION

A micro-photograph of the crack tip at the plate surface indicating the extent of crack tip blunting is shown in Fig. 6. The stress frozen models were immersed in a transparent walled tank containing a fluid of the same index of refraction as the model material and analyzed for fringes in the field of a circular polariscope. A uniform tension field was indicated well above and below the crack site. A typical fringe pattern from a tensile test is shown in Fig. 7. The leaning of the fringe loops towards the region ahead of the crack tip is due to the influence of σ_{0x}. It is important that the crack appears in the fringe photograph as a narrow band which comes to a point at the crack tip. When the crack appears broad and fuzzy and without a sharp tip, the crack was misaligned to the camera, the crack rotated out of its plane, or both. The resulting integrated optical effect is a reduction in fringe order very near the crack tip which leads to an increase in fringe gradient in the zone of K_I measurement. Since the computer program employed here computed $K(\text{Exp})$ from the fringe gradient, higher $K(\text{Exp})$ values would result.

A pilot test (Test 1) was run in order to perfect experimental techniques. The crack for this test rotated out of the plane perpendicular to the loads and so data from this test will be omitted. Data and results from the second test are shown in Table 1.

Fig. 6. Micro-photo of crack tip (2000X).

Fig. 7. Fringe pattern at crack tip (bright field) 6X.

Table 1. Test 2 Results

σ (psi)	K_I(Exp) (lb/in.$^{3/2}$)	K_I(Th) (lb/in.)	σ_{0x}(Exp) (psi)
5·03	6·79	6·42	6·58

The experimental K_I was obtained by averaging data from the two crack tips. Most of the difference of about 6 per cent between the experimental and theoretical values is believed to be due to experimental error. In Fig. 8, experimental data from the two crack tips in test 2 are shown. They were used to compute K(Exp) and σ_{0x} from which a curve of τ_m/τ_{m0} vs. r/a was developed. (The Westergaard solution indicates that the value of $\tau_{\max}$ computed from the singular stresses along $\theta = \pi/2$ differs from the exact solution at $r/a = 0\cdot30$ by less than 2 per cent.) The Fig. 8 curve is compared with a theoretical curve obtained from the Westergaard solution corrected for remote uniaxial extension in Fig. 9. The tendency of these curves to converge as $r/a \to 0$ verifies the agreement between K_I(Th) and K_I(Exp). The test specimen dimensions were gaged from standard test specimens designed for K_I measurement so it is not surprising that reasonable K_I correlation was obtained.

Perhaps a more interesting aspect of the results centers around values of σ_{0x}. Table 1 shows that values of σ_{0x} computed from the experiment were substantially different from the theoretical value $(\bar{\sigma})$. The difference in this case was about 27 per cent. As noted earlier, a computer study revealed that this much variation may simply be due to the 6 per cent difference in K_I values noted earlier. It should also be noted, in passing, however, that geometrical differences between the test specimen and the mathematical model might also contribute to this difference.

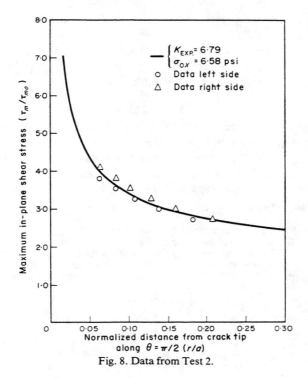

Fig. 8. Data from Test 2.

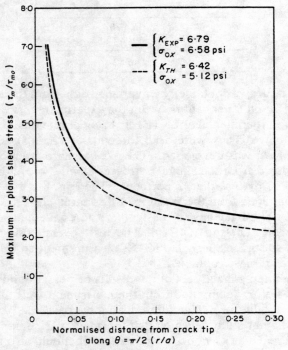

Fig. 9. Effect of σ_{0x} on $\tau_{\max}$.

It is interesting to note that, even though data were taken in the range $r/a = 0.04$ to 0.25 (where τ_m/τ_{m0} values differ by 15–20 per cent as seen in Fig. 9), the difference between $K_I(\text{Exp})$ and $K_I(\text{Th})$ was only 6 per cent.

Since K_I can be accurately computed without determining σ_{0x}, the way is open to study three dimensional crack problems of any geometry even though σ_{0x} is not known. Studies in this direction are well underway.

SUMMARY

Factors influencing the accuracy of the determination of the stress intensity factor using a stress-freezing photoelastic technique are discussed. A data conditioning program is developed and is coupled with the photoelastic technique for stress intensity factor determination. The degree of accuracy of the method is assessed by comparing with two-dimensional analytical results. It is noted that the method is currently being applied to three dimensional problems.

Acknowledgements—The authors are indebted to those whose pioneering efforts provided the basis for this study, especially G. R. Irwin and those authors noted in the introduction. They also wish to acknowledge the laboratory assistance of G. K. McCauley, the interest and suggestions of J. H. Underwood of Watervliet Arsenal and the encouragement of D. Frederick. This work was supported by the Department of Defense, Project Themis, Contract No. DAA-F07-69-C-0444 with Watervliet Arsenal, Watervliet, New York.

REFERENCES

[1] D. Post, Photoelastic Stress Analysis for an Edge Crack in a Tensile Field. *Proc. Soc. expl Stress Analysis* **12**, 99–116 (1954).
[2] A. A. Wells and D. Post, The Dynamic Stress Distribution Surrounding a Running Crack—A Photoelastic Analysis. *Proc. Soc. expl Stress Analysis* **16**, 69–92 (1958).
[3] G. R. Irwin, Discussion of Reference [2]. *Proc. Soc. expl Stress Analysis* **16**, 69–92 (1958).

[4] D. G. Smith and C. W. Smith, A Photoelastic Evaluation of the Influence of Closure and Other Effects upon the Local Bending Stresses in Cracked Plates. *Int. J. Frac. Mech.* **6**, 305–318 (1970).

[5] D. G. Smith and C. W. Smith, Influence of Precatastrophic Extension and Other Effects on Local Stresses in Cracked Plates under Bending Fields. *Expl Mech.* **11**, 394–401 (1971).

[6] D. G. Smith and C. W. Smith, Photoelastic Determination of Mixed Mode Stress Intensity Factors. *Engng Fracture Mech.* **4**, 357–366 (1972).

[7] B. Kerley, Photoelastic Investigation of Crack Tip Stress Distributions. GT-5 *Test Report Document* No. 685D 597, The General Electric Co. (1965).

[8] H. Fessler and D. O. Mansell, Photoelastic Study of Stresses Near Cracks in Thick Plates. *J. Mech. Engng Sci.* **4**, 213–225 (1962).

[9] J. R. Dixon and J. S. Strannigan, A Photoelastic Investigation of the Stress Distribution in Uniaxially Loaded Thick Plates Containing Slits. *NEL Report* No. 288, National Engineering Laboratory, Glasgow, Scotland (1967).

[10] H. Liebowitz, H. Vanderveldt and R. J. Sanford, Stress Concentrations Due to Sharp Notches. *Expl Mech.* **7**, 513–517 (1967).

[11] G. R. Marra and C. W. Smith, A Study of Local Stresses Near Surface Flaws in Bending Fields. *Proc. the Fifth National Symposium on Fracture Mech.* VPI-E-71-13, In press.

[12] R. H. Marloff, M. M. Leven, R. L. Johnson and T. N. Ringler, Photoelastic Determination of Stress elasticity. *J. expl Mech.* **10**, 106–113 (1970).

[13] W. D. Bradley and A. S. Kobayashi, An Investigation of Propagating Cracks by Dynamic Photoelasticity. *J. expl Mech.* **10**, 106–113 (1970).

[14] W. B. Bradley and A. S. Kobayashi, Fracture Dynamics—A Photoelastic Investigation *J. Engng Fracture Mech.* **3**, 317–332 (1971).

[15] V. Weiss and S. Yukawa, Critical Appraisal of Fracture Mechanics, *Fracture Toughness Testing and Its Applications.* ASTM STP 381 (1965).

[16] H. M. Westergaard, Bearing Pressures and Cracks. *Trans. ASME* Series E, *J. appl. Mech.* B9, A49-A53 (1939).

A GENERAL METHOD FOR DETERMINING MIXED-MODE STRESS INTENSITY FACTORS FROM ISOCHROMATIC FRINGE PATTERNS

ROBERT J. SANFORD
Ocean Technology Division, Naval Research Laboratory, Washington, DC 20375, U.S.A.

and

JAMES W. DALLY
Mechanical Engineering Department, University of Maryland, College Park, MD 20742, U.S.A.

Abstract—A general method is presented for determining mixed-mode stress intensity factors K_I and K_{II} from isochromatic fringes near the crack tip. The method accounts for the effects of the far-field, non-singular stress, σ_{ox}. A non-linear equation is developed which relates the stress field in terms of K_I, K_{II}, and σ_{ox} to the co-ordinates, r and θ, defining the location of a point on an isochromatic fringe of order N.

Four different approaches for the solution of the non-linear equation are given. These include: a selected line approach in which data analysis is limited to the line $\theta = \pi$ and the K-N relation can be linearized and simplified, the classical approach in which two data points at (r_m, θ_m) are selected where $\partial r_m/\partial\theta = 0$; a deterministic method where three arbitrarily located data points are used; and an over-deterministic approach where m (>3) arbitrarily located points are selected from the fringe field.

Except for the selected line approach, the method of solution involves an iterative numerical procedure based on the Newton–Raphson technique. For the over-deterministic approach, the method of least squares was employed to fit the K-N relation to the field data.

All four methods provide solutions to 0.1% providing that the input parameters r, θ, and N describing the isochromatic field are exact. Convergence of the iterative methods is rapid (3–5 iterations) and computer costs are nominal. When experimental errors in the measurements of r and θ are taken into consideration, the over-deterministic approach which utilizes the method of least squares has a significant advantage. The method is global in nature and the use of multiple-point data available from the full-field fringe patterns permits a significant improvement in accuracy of K_I, K_{II}, and σ_{ox} determinations.

INTRODUCTION

ONE OF THE most effective methods of experimentally determining the stress intensity factor for a body containing a crack is to analyze the isochromatic pattern obtained from a photoelastic model. Measurements of the fringe order N and position parameters r and θ, defined in Fig. 1, which locate one point or a number of points on a fringe loop, are sufficient to permit the determination of K_I, K_{II}, and σ_{ox}. Several investigators[1–10] have introduced methods for analyzing the isochromatic fringe data, but all of these methods are limited to determining only two (K_I and σ_{ox} or K_I and K_{II}) of the three quantities which affect the fringe pattern. Also, most of the methods developed to date employ isochromatic data from only one or two points in the fringe field and do not fully utilize the available data. The method presented here is more global in nature and the use of the full-field data permits a significant improvement in the accuracy of determining the stress intensity factors.

Wells and Post[1] and Irwin[2] first showed a method for determining the opening mode stress intensity factor K_I in the presence of a far-field stress, σ_{ox}. Irwin's method is only applicable if the shearing mode stress intensity factor $K_{II} = 0$. The method is based on selecting a point $r = r_m$ and $\theta = \theta_m$ on a given fringe loop (locus of constant τ_m) which is furthest removed from the crack tip, i.e.

$$\frac{\partial \tau_m}{\partial \theta} = 0 \tag{1}$$

where τ_m is the maximum in-plane shear stress.

Using eqn (1) and the governing equation for the isochromatic fringe pattern,

$$2\tau_m = Nf_\sigma/h \tag{2}$$

where f_σ is the material fringe value and h is the material thickness.

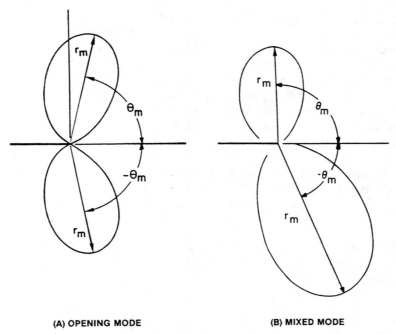

(A) OPENING MODE   (B) MIXED MODE

Fig. 1. Isochromatic fringe loops at a crack tip.

Irwin† then showed that

$$K_I = \frac{(Nf_\sigma/h)\sqrt{(2\pi r_m)}}{\sin \theta_m} \left[1 + \frac{2\tan(3\theta_m/2)}{3\tan \theta_m}\right]\left[1 + \left(\frac{2}{3\tan \theta_m}\right)^2\right]^{-0.5} \quad (3)$$

$$\sigma_{ox} = -\frac{2(Nf_\sigma/h)\cos \theta_m}{\cos(3\theta_m/2)[\cos^2 \theta_m + (9/4)\sin^2 \theta_m]^{0.5}}. \quad (4)$$

The accuracy which can be achieved with Irwin's approach depends upon θ_m [5] and the precision in locating the point r_m and θ_m. Errors of ± 3 degrees in measuring θ_m are common and can result in large errors in the determination of K_I and σ_{ox}.

Bradley and Kobayashi[3] have modified Irwin's approach and use data from two points (r_1, θ) and (r_2, θ) along a line intersecting two different fringe orders in a differencing technique to obtain

$$K_I = 2\sqrt{(2\pi)}(f_\sigma/h)(N_2 - N_1)\sqrt{(r_1 r_2)}/(f_2\sqrt{r_1} + f_1\sqrt{r_2}) \quad (5)$$

where f_1 and f_2 are functions of θ and σ_{ox}.

Schroedl and Smith[4] employ a differencing technique identical to that used by Bradley and Kobayashi except that θ is set equal to 90 degrees and the equation for K_I is simplified to obtain:

$$K_I = \sqrt{(2\pi)}(f_\sigma/h)(N_1 - N_2)\sqrt{(r_1 r_2)}/(\sqrt{r_2} - \sqrt{r_1}). \quad (6)$$

In many instances, these differencing techniques provide a more accurate method of analysis than Irwin's method[5]. However, these methods are all based on a two-parameter (K_I, σ_{ox}) analysis and errors[6] exceed $\pm 2\%$ if θ_m is outside of the range, $74 < \theta_m < 134$ degrees.

Etheridge and Dally[6] introduced a third parameter into the analysis by modifying the Westergaard stress function to more closely account for stress field variations near the crack tip. The three parameter method improves the accuracy over the entire range of θ_m and extends the range of analysis to $69 < \theta_m < 145$ degrees.

†Reference [2] contains a typographical error in the equation for determining σ_{ox}. The form presented in eqn (4) corrects this error.

Methods of analysis which include the shearing mode are much more limited. Chisolm and Jones[7] have treated the case for pure K_{II} and Smith and Smith[8] and Gdoutos and Theocaris[9] have developed a procedure for analyzing fringe loops due to the mixed mode condition where K_I and K_{II} occur together but the non-singular stress, σ_{ox} is not included.

Bradley and Kobayashi[10] introduced a multiple-point method of analysis to determine the coefficients in a Williams' stress function. The procedure used from 7 to 19 data points to determine 2–10 coefficients. Unfortunately, the convergence of the method was sensitive to errors in the data and the initial estimates of the coefficients.

In this paper a general method of analysis is introduced which relates the isochromatic fringe field (N, r, θ) to any arbitrary loading which produces K_I, K_{II} and σ_{ox}. Four different methods are introduced to solve the general relationship. The best of the four methods is based on a least squares approach where multiple data points are used to minimize the error in the K determination due to inaccuracies in measuring r and θ associated with a given fringe order N.

GENERAL RELATION BETWEEN THE ISOCHROMATICS AND THE K FIELD

For the purpose of this study, the stresses in the local neighborhood of a crack tip ($r/a \ll 1$) can be approximated by[11]:

$$\sigma_x = \frac{1}{\sqrt{(2\pi r)}} \left[K_I \cos\frac{\theta}{2}\left(1 - \sin\frac{\theta}{2}\sin\frac{3\theta}{2}\right) - K_{II} \sin\frac{\theta}{2}\left(2 + \cos\frac{\theta}{2}\cos\frac{3\theta}{2}\right) \right] - \sigma_{ox} \quad (7)$$

$$\sigma_y = \frac{1}{\sqrt{(2\pi r)}} \left[K_I \cos\frac{\theta}{2}\left(1 + \sin\frac{\theta}{2}\sin\frac{3\theta}{2}\right) + K_{II} \sin\frac{\theta}{2}\cos\frac{\theta}{2}\cos\frac{3\theta}{2} \right] \quad (8)$$

$$\tau_{xy} = \frac{1}{\sqrt{2\pi r}} \left[K_I \sin\frac{\theta}{2}\cos\frac{\theta}{2}\cos\frac{3\theta}{2} + K_{II} \cos\frac{\theta}{2}\left(1 - \sin\frac{\theta}{2}\sin\frac{3\theta}{2}\right) \right] \quad (9)$$

where r and θ are polar co-ordinates with the origin defined at the crack tip. The term σ_{ox} in eqn (7) has been included as a correction term to account in part for the effects of specimen geometry as intended originally by Irwin[2].

Recall the maximum in-plane shear stress τ_m is related to the cartesian components of stress by:

$$(2\tau_m)^2 = (\sigma_y - \sigma_x)^2 + (2\tau_{xy})^2. \quad (10)$$

Substituting eqns (7)–(9) and eqn (2) into eqn (10) gives the relation which defines the isochromatic fringe pattern in the local field near the crack tip as:

$$(Nf_\sigma/h)^2 = \frac{1}{2\pi r}[(K_I \sin\theta + 2K_{II} \cos\theta)^2 + (K_{II} \sin\theta)^2]$$

$$+ \frac{2\sigma_{ox}}{\sqrt{(2\pi r)}} \sin\frac{\theta}{2}[K_I \sin\theta(1 + 2\cos\theta) + K_{II}(1 + 2\cos^2\theta + \cos\theta)] + \sigma_{ox}^2. \quad (11)$$

SOLUTION OF THE N–K RELATIONSHIP

The N–K relation given in eqn (12) is non-linear in terms of the three unknown K_I, K_{II} and σ_{ox}. Several different approaches can be followed in the solution. The first approach involves the selection of a particular line in the field which reduces the complexity of eqn (12) permitting a direct solution. In the second and third approaches, the Newton–Raphson method[12] is employed to obtain an iterative solution in terms of the three unknowns. In the fourth approach, the Newton–Raphson method is coupled with a least-squares analysis to give an iterative solution based on the data taken at many points over the field. Each of these four approaches is described in subsequent sections.

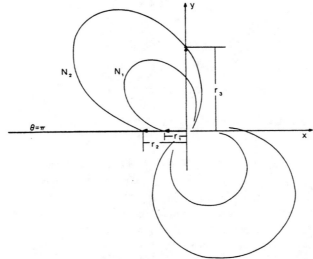

Fig. 2. Isochromatic fringes intersect the upper crack line ($\theta = \pi$) when $K_{II} < 0$.

SELECTED LINE APPROACH (TWO POINTS)

When $K_{II} < 0$, and $\sigma_{ox} \neq 0$, the isochromatic fringes do not form closed loops but intersect the upper crack line as illustrated in Fig. 2. Along this line, $\theta = \pi$ and eqn (11) reduces to:

$$(Nf_\sigma/h)^2 = \frac{4K_{II}^2}{2\pi r} + \frac{4K_{II}\sigma_{ox}}{\sqrt{(2\pi r)}} + \sigma_{ox}^2 \tag{12}$$

which is independent of K_I. Equation (12) can be written as

$$(Nf_\sigma/h) = \pm\left(\frac{2K_{II}}{\sqrt{(2\pi r)}} + \sigma_{ox}\right). \tag{13}$$

The choice of signs in eqn (13) is based on the fact that the isochromatic fringes intersect the $\theta = \pi$ line when $K_{II} < 0$ and intersect the $\theta = -\pi$ line when $K_{II} < 0$. Consider two different fringes $N_1 > N_2$ both of which intercept the upper edge of the crack ($\theta = \pi$) at positions r_1, r_2 respectively as shown in Fig. 2. Taking the negative sign option in eqn (13) and eliminating σ_{ox} by substitution leads to:

$$K_{II} = (f_\sigma/h)\sqrt{(\pi/2)}\frac{\sqrt{(r_1 r_2)}}{\sqrt{r_1} - \sqrt{r_2}}(N_1 - N_2). \tag{14}$$

Note that $K_{II} < 0$ as required since $N_1 < N_2$ and $r_2 < r_1$. Next, σ_{ox} is obtained from eqns (13) and (14) as

$$\sigma_{ox} = -(N_1 f_\sigma/h) - 2K_{II}/\sqrt{(2\pi r_1)}. \tag{15}$$

To determine K_I re-evaluate eqn (11) with $\theta = \pi/2$ to obtain

$$(N_3 f_\sigma/h)^2 = \frac{1}{2\pi r_3}[K_I^2 + K_{II}^2] + \frac{\sigma_{ox}}{\sqrt{(\pi r_3)}}[K_I + K_{II}] + \sigma_{ox}^2. \tag{16}$$

Equation (16) is quadratic in terms of K_I and thus

$$K_I = \frac{-b + \sqrt{(b^2 - 4ac)}}{2a} \tag{17}$$

where

$$a = 1/(2\pi r_3)$$
$$b = \sigma_{ox}/\sqrt{(\pi r_3)} \qquad (18)$$
$$c = (K_{II}^2/2\pi r_3) + K_{II}\sigma_{ox}/\sqrt{(\pi r_3)} + \sigma_{ox}^2 - (N_3 f_\sigma/h)^2.$$

The terms N_3 and r_3 are defined in Fig. 2. The plus sign is selected in front of the radical in eqn (17) because $K_I > 0$.

CLASSICAL APPROACH (TWO POINTS)

The classical approach is a selected point method of analysis and follows Irwin's[2] observation that eqn (1) is valid at the point r_m, θ_m on each fringe loop (see Fig. 1). This approach can also be used in the mixed mode analysis by employing information from two loops in the near field of the crack. Differentiating eqn (11) with respect to θ, setting $\theta = \theta_m$ and $r = r_m$ and using eqn (1) gives:

$$g(K_I, K_{II}, \sigma_{ox}) = \frac{1}{2\pi r_m}(K_I^2 \sin 2\theta_m + 4K_I K_{II} \cos 2\theta_m - 3K_{II}^2 \sin 2\theta_m)$$
$$+ \frac{2\sigma_{ox}}{\sqrt{(2\pi r_m)}}\left\{\sin\frac{\theta_m}{2}[K_I(\cos\theta_m + 2\cos 2\theta_m) - K_{II}(2\sin 2\theta_m + \sin\theta)]\right.$$
$$\left.+ \frac{1}{2}\cos\frac{\theta_m}{2}\left\{[K_I(\sin\theta_m + \sin 2\theta_m) + K_{II}(2 + \cos 2\theta_m + \cos\theta_m)]\right\}\right\} = 0. \qquad (19)$$

Next rewrite eqn (11) to obtain

$$f(K_I, K_{II}, \sigma_{ox}) = \frac{1}{2\pi r_m}\{[K_I \sin\theta_m + 2K_{II}\cos\theta_m]^2 + [K_{II}\sin\theta_m]^2\}$$
$$+ \frac{2\sigma_{ox}}{\sqrt{(2\pi r_m)}}\sin\frac{\theta_m}{2}[K_I \sin\theta_m(1 + 2\cos\theta_m) + K_{II}(1 + 2\cos^2\theta_m + \cos\theta_m)]$$
$$+ \sigma_{ox}^2 - (N_m f_\sigma/h)^2 = 0 \qquad (20)$$

where N_m is the fringe order corresponding to the point (r_m, θ_m).

Using the classical approach, the stress intensity factors, K_I and K_{II}, and the remote stress, σ_{ox}, are determined from isochromatic data taken from two points—one on each side of two fringes. In concept any two loops can be utilized; however, in practice more accurate results are obtained if one fringe is taken from the set of loops above the crack line and the other from the set of loops below the crack line.

Substituting the radii r_m and the angles θ_m from these two loops into a pair of equations of the form given in eqn (19) provides two independent relations in terms of the parameters of interest. The third equation is obtained by using eqn (20) with data from either the upper or lower loop. The three equations obtained in this manner are of the form

$$g_u(K_I, K_{II}, \sigma_{ox}) = 0$$
$$g_l(K_I, K_{II}, \sigma_{ox}) = 0 \qquad (21)$$
$$f_u(K_I, K_{II}, \sigma_{ox}) = 0$$

where the subscripts u, l refer to upper and lower loops respectively.

Although eqns (21) can be solved in closed form, the algebra becomes quite involved and a simpler approach using a numerical procedure based on the Newton–Raphson[12] method was employed. To review the Newton–Raphson method, consider an arbitrary function, h_k of the form

$$h_k(K_I, K_{II}, \sigma_{ox}) = 0 \qquad (22)$$

where $k = 1, 2$ or 3. If initial estimates are made for K_I, K_{II}, and σ_{ox}, and substituted into eqn (22), $h_k \neq 0$ since the initial estimates will usually be in error. To correct the estimates, a series of iterative equations based on a Taylor series expansion of h_k are written as:

$$(h_k)_{i+1} = (h_k)_i + \left(\frac{\partial h_k}{\partial K_I}\right)_i \Delta K_I + \left(\frac{\partial h_k}{\partial K_{II}}\right)_i \Delta K_{II} + \left(\frac{\partial h_k}{\partial \sigma_{ox}}\right)_i \Delta \sigma_{ox}, \qquad (23)$$

where the subscript i refers to the ith iteration step and ΔK_I, ΔK_{II}, and $\Delta \sigma_{ox}$ are corrections to the previous estimates. The corrections are determined so that $(h_k)_{i+1} = 0$, and thus, eqn (23) gives:

$$\left(\frac{\partial h_k}{\partial K_I}\right)_i \Delta K_I + \left(\frac{\partial h_k}{\partial K_{II}}\right)_i \Delta K_{II} + \left(\frac{\partial h_k}{\partial \sigma_{ox}}\right)_i \Delta \sigma_{ox} = -(h_k)_i. \qquad (24)$$

Applying eqns (24) to (21) and solving for the correction terms ΔK_I, ΔK_{II} and $\Delta \sigma_{ox}$ yields, in matrix notation,

$$\begin{bmatrix} \Delta K_I \\ \Delta K_{II} \\ \Delta \sigma_{ox} \end{bmatrix} = - \begin{bmatrix} \frac{\partial g_u}{\partial K_I} & \frac{\partial g_u}{\partial K_{II}} & \frac{\partial g_u}{\partial \sigma_{ox}} \\ \frac{\partial g_l}{\partial K_I} & \frac{\partial g_l}{\partial K_{II}} & \frac{\partial g_l}{\partial \sigma_{ox}} \\ \frac{\partial f_u}{\partial K_I} & \frac{\partial f_u}{\partial K_{II}} & \frac{\partial f_u}{\partial \sigma_{ox}} \end{bmatrix}_i^{-1} \begin{bmatrix} g_u \\ g_l \\ f_u \end{bmatrix}_i . \qquad (25)$$

The corrected values of K_I, K_{II}, and σ_{ox} are given by

$$(K_I)_{i+1} = (K_I)_i + \Delta K_I$$
$$(K_{II})_{i+1} = (K_{II})_i + \Delta K_{II} \qquad (26)$$
$$(\sigma_{ox})_{i+1} = (\sigma_{ox})_i + \Delta \sigma_{ox}.$$

The convergence of this method is rapid and usually four or five iterations are sufficient to obtain precise results for K_I, K_{II}, and σ_{ox}.

DETERMINISTIC APPROACH (THREE POINTS)

If data is selected from three arbitrary points (r_1, θ_1), (r_2, θ_2) and (r_3, θ_3), eqn (1) does not apply. Instead the Newton–Raphson method is applied to the solution of three simultaneous non-linear equations obtained from eqn (20) as:

$$f_k(K_I, K_{II}, \sigma_{ox}) = \frac{1}{2\pi r_k}\{[K_I \sin \theta_k + 2K_{II} \cos \theta_k]^2 + [K_{II} \sin \theta_k]^2\}$$

$$+ \frac{2\sigma_{ox}}{\sqrt{(2\pi r_k)}} \sin(\theta_k/2)[K_I \sin \theta_k (1 + 2 \cos \theta_k) + K_{II}(1 + 2 \cos^2 \theta_k + \cos \theta_k)]$$

$$+ \sigma_{ox}^2 - (N_k f_\sigma/h)^2 = 0 \qquad (27)$$

where $k = 1, 2$, or 3 and r_k, θ_k are coordinates defining a point on an isochromatic fringe of order N_k.

Following the iterative procedure outlined in the previous section, it is evident that the corrections on the initial estimates are given by

$$\begin{bmatrix} \Delta K_I \\ \Delta K_{II} \\ \Delta \sigma_{ox} \end{bmatrix} = - \begin{bmatrix} \frac{\partial f_1}{\partial K_I} & \frac{\partial f_1}{\partial K_{II}} & \frac{\partial f_1}{\partial \sigma_{ox}} \\ \frac{\partial f_2}{\partial K_I} & \frac{\partial f_2}{\partial K_{II}} & \frac{\partial f_2}{\partial \sigma_{ox}} \\ \frac{\partial f_3}{\partial K_I} & \frac{\partial f_3}{\partial K_{II}} & \frac{\partial f_3}{\partial \sigma_{ox}} \end{bmatrix}_i^{-1} \begin{bmatrix} f_1 \\ f_2 \\ f_3 \end{bmatrix}_i. \tag{28}$$

The convergence of this method is rapid and three or four iterations are sufficient for obtaining precise results for K_I, K_{II}, and σ_{ox}.

OVER DETERMINISTIC METHOD (MULTIPLE-POINTS)

The method of least squares involves the determination of K_I, K_{II}, and σ_{ox} so that eqn (11) is fitted to a large number of points over the isochromatic field. The fitting process involves both the Newton–Raphson method and the minimization process associated with the least squares method. Consider again the function of f_k given in eqn (27) except let $k = 1, 2, \ldots m$ where $m > 3$. The Taylor series expansions of f_k can be written as in eqn (23), and the iteration condition given in eqn (24) invoked to give an over-determined set of equations in terms of the corrections ΔK_I, ΔK_{II}, and $\Delta \sigma_{ox}$ of the form:

$$[f] = [a][\Delta K] \tag{29}$$

where the matrices are defined as

$$[f] = \begin{bmatrix} f_1 \\ - \\ - \\ - \\ - \\ f_m \end{bmatrix}_i ; [a] = - \begin{bmatrix} \frac{\partial f_1}{\partial K_I} & \frac{\partial f_1}{\partial K_{II}} & \frac{\partial f_1}{\partial \sigma_{ox}} \\ - & - & - \\ - & - & - \\ - & - & - \\ \frac{\partial f_m}{\partial K_I} & \frac{\partial f_m}{\partial K_{II}} & \frac{\partial f_m}{\partial \sigma_{ox}} \end{bmatrix}_i ; [\Delta K] = \begin{bmatrix} \Delta K_I \\ \Delta K_{II} \\ \Delta \sigma_{ox} \end{bmatrix}_i. \tag{30}$$

The least squares minimization process is accomplished by multiplying from the left both sides of eqn (29) by the transpose of matrix [a], to give

$$[a]^T [f] = [a]^T [a] [\Delta K] \tag{31}$$

or

$$[d] = [c][\Delta K]$$

where

$$[d] = [a]^T [f]$$
$$[c] = [a]^T [a].$$

Finally the correction terms are given by:

$$[\Delta K] = [c]^{-1}[d]. \tag{32}$$

The solution of eqn (32) gives ΔK_I, ΔK_{II}, and $\Delta \sigma_{ox}$ which are used to correct initial estimates of K_I, K_{II}, and σ_{ox} and obtain a better fit of the function f to m data points. Computer programs were written in BASIC to determine K_I, K_{II}, and σ_{ox} using data from either 10 or 20 points selected at arbitrary locations in the fringe field. A listing of the 10 point program is given in Ref. [13]. Again, convergence of the solution is quite rapid requiring only a few iterations to give precise estimates of K_I, K_{II}, and σ_{ox}.

EXAMPLE

A photoelastic model was machined from Homalite 100, to the dimensions shown in Fig. 3. The model was loaded with off-axis tensile forces, illustrated in Fig. 3, to produce a mixed-mode stress field with distortion present due to the non-singular stress, σ_{ox}. Photographs of the dark and light-field isochromatic fringe patterns in the neighborhood of one of the simulated crack tips are presented in Fig. 4.

Negatives of the fringe pattern were enlarged and θ, r, and N were measured for 20 data points. The 20 data points were carefully selected to minimize known sources of measurement error. The regions used were near the outside arc of the fringe loops where changes in r are small with respect to θ. Data was taken from the 2, 2.5, 3 and 3.5 fringe orders. To minimize errors due to the effect of gradients in σ_{ox} the maximum radius used was 0.243 in. (6.17 mm). To minimize measuring errors in determining the radius the minimum value used was 0.053 in. (1.35 mm).

Data from these 20 points was processed using the over-deterministic method and values of $K_I = 832$ psi $\sqrt{in}$ (0.915 MPa $\sqrt{m}$), $K_{II} = -602$ psi $\sqrt{in}$ (-0.662 MPa $\sqrt{m}$) and $\sigma_{ox} = -182$ psi (-1.26 MPa) were obtained. The accuracy of these estimates was assessed by comparing the experimental fringe patterns (Fig. 4) with a theoretical fringe pattern Fig. 5, plotted[14] using the computed values of K_I, K_{II}, and σ_{ox} and the measured value of f_σ (obtained from a disk specimen cut from the model after completion of the fracture mechanics analysis). For purposes of comparison with the experimental results, both the integer order (dark field) and half-order (light field) fringes are shown in the theoretical pattern. Also shown in Fig. 5 are the locations of the 20 data points used in the computations.

The fit of the 20 data points is in most instances quite close; however, discrepances are noted for points where $r > 0.2$ in. (5 mm). Differences are also larger above the crack line.

The differences are due to three different sources of error. First, the σ_{ox} stress field is not uniform over even the restricted region of data analysis. This is due to bending which introduces a gradient in σ_{ox} in the y direction. Second, the finite width of the crack, 0.048 in. (1.2 mm) and the geometry of the crack tip affect the shape of the isochromatic fringes used in the analysis (see the 2 and 2.5 order fringe loops on the upper side of the crack in Fig. 4). Finally measurement errors of about ± 0.005 in. (0.23 mm) in determining the radius occur due to difficulties in locating the crack tip and the fringe position.

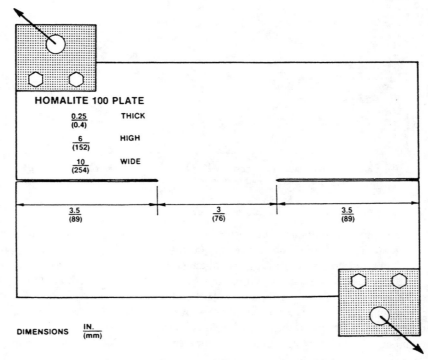

Fig. 3. Model Geometry and loading.

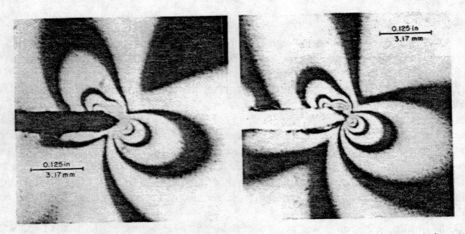

Fig. 4. Dark and light-field isochromatic fringes near the crack tip which were used in the analysis.

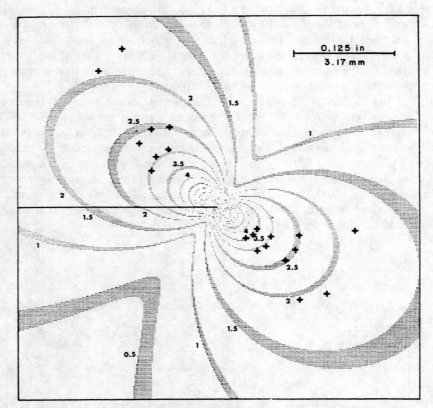

Fig. 5. Point plot of isochromatic fringe field for $K_I = 832$ psi $\sqrt{in}$, $K_{II} = -602$ psi $\sqrt{in}$, $\sigma_{ox} = -182$ psi. The 20 experimental data points used in the K determination are shown as crosses.

DISCUSSION

The four numerical methods previously described and the associated computer programs were verified and found to give precise results (accurate to 0.1%) providing the input data was exact. The data required for verification was obtained by substituting assumed values of K_I, K_{II}, and σ_{ox} into eqn (11). The resulting expression was then used to determine the fringe order at specified locations identified by arbitrary choices of r and θ.

Since all four methods give precise results with accurate input data, the selection of the most suitable method must be based on accuracy which can be achieved using experimental data containing measurement error. Measurement error arises due to difficulties in locating the crack tip and in identifying the fringe location.

In preparing photoelastic models, the crack is usually machined into the sheet of plastic. The crack has a finite width in the range of 0.020–0.050 in. (0.5–1 mm) and often the tip of the crack is not sharp. Location of the "correct" point at the blunt end of a simulated crack which represents the origin of the r, θ co-ordinate system is usually the largest source of measurement error. The circle of deviation is estimated to have a radius of 0.005 in. (0.13 mm). Efforts to manufacture models with very narrow slits and sharp tips are suggested in order to reduce the circle of deviation.

Isochromatic fringes have a finite width and location of the point associated with a specified fringe order requires that the position of the minimum intensity across the width of the fringe be precisely identified. Photographic techniques where long exposure times are used with high contrast film reduce the width of the fringe; however, errors in locating the point of minimum intensity may be ±2% of the radial position r.

Difficulties in locating the crack tip also produce serious errors in measuring θ. These errors depend upon the radius of the fringe loops being considered with the largest error occurring with the higher order fringes very close to the crack tip. Errors in θ of ±3 degrees can be expected for fringe loops with $r = 0.1$ in. (2.5 mm) and errors of ±8 degrees can be expected with $r = 0.035$ in. (0.9 mm).

Since the measurement errors are random, improvement can be made in the accuracy of the determination of K_I, K_{II}, and σ_{ox} by using the over deterministic approach. The use of say 20 data points gives 17 redundant points providing an averaging effect which markedly improves the accuracy of the determination. Experience (see Ref. [13]) has shown that error in determining K_I and K_{II} is reduced by a factor of 3–4 with the least squares fitting procedure.

The selected line approach is the most error prone of the four methods. The measurement of the radial position of the fringes intersecting the crack line is subject to two difficulties in addition to those previously listed. First, the fringes which intersect the crack approach the boundary at a rather shallow angle as illustrated in Fig. 2. Thus, the effect of a finite width crack produces large errors in the measurement of r_1 and r_2. Second, any machining stresses or time edge effects present in the model will influence the fringe crack boundary intersection point and affect measurement of r_1 and r_2.

The classical approach is also subject to significant errors as it is dependent upon the precise location of the (r_m, θ_m) point for eqn (1) to be valid. In a mixed mode field with σ_{ox}, fringe loops can be flattened[14] and determining θ_m can be difficult with significant measurement errors possible.

The deterministic approach where three arbitrary data points are selected usually will be more accurate than the classical or selected line approach because the three data points required can be taken from locations where measurement errors can be minimized. However, with the numerical methods available for an over-deterministic solution of eqn (11) it is not advisable to employ the deterministic approach which does not utilize more of the data provided by the photoelastic experiment.

CONCLUSIONS

A generalized method for determining mixed mode stress intensity factors K_I and K_{II} from isochromatic fringe patterns has been developed. The method is applicable when far-field stresses σ_{ox} occur and all three quantities K_I, K_{II}, and σ_{ox} are determined in the analysis

The K–N relation is non-linear and numerical methods proved to be quite useful in the solution of eqn (11). Indeed, the numerical methods provide a means for using the whole-field data available in a photoelastic experiment to improve the accuracy of the determination of the stress intensity factors.

The selected line, classical and deterministic approaches are not recommended. They were included here for completeness, to further the historical development of the topic and to preclude other investigators from attempting simplifications which lead to errors in determining the stress intensity factors.

The over-deterministic solution to the K–N relation is recommended. The least squares analysis which is incorporated into this method of solution tends to minimize the inaccuracies produced by the inevitable error in the measurements of r and θ. The computer costs involved in processing the whole field data is nominal since only 2–5 seconds of run time is required.

Computer generated plots of the isochromatic field such as the one shown in Fig. 5 are of great benefit because they provide a basis for comparison of the least-square fit of the data to the theoretical isochromatic field and should be used if facilities are available. Also, the method of stress intensity factor determination presented here is ideally suited to computerized analysis with image processing systems which provide an interactive method of analysis.

Acknowledgement—The research described herein was performed in the Structural Reliability Section, Ocean Technology Division of the Naval Research Laboratory (NRL) under the sponsorship of the Office of Naval Research, project RR 023-03-45. During the period when this research was performed J. W. Dally was on sabbatical leave at NRL under the terms of the Intergovernmental Personnel Act of 1970.

REFERENCES

[1] A. A. Wells and D. Post, The dynamic stress distribution surrounding a running crack—A photoelastic analysis. *Proc. of SESA* **16**, 69–93 (1958).
[2] G. R. Irwin, Discussion of Ref. [1]. *Proc. of SESA* **16**, 93–96 (1958).
[3] W. B. Bradley and A. S. Kobayashi, An investigation of propagating cracks by dynamic photoelasticity. *Experimental Mechanics* **10**(3), 106–113 (1970).
[4] M. A. Schroedl and C. W. Smith, Local stresses near deep surface flaws under cylindrical bending fields. *Progress in Flaw Growth and Fracture Toughness Testing, ASTM STP* **536**, pp. 45–63 (1973).
[5] J. M. Etheridge and J. W. Dally, A critical review of methods for determining stress-intensity factors from isochromatic fringes. *Experimental Mechanics* **17**(7), 248–254 (1977).
[6] J. M. Etheridge and J. W. Dally, A three parameter method for determining a stress intensity factors from isochromatic fringe loops. In *J. Strain Analysis.* **13**(2), 91–94 (1978).
[7] D. B. Chisholm and D. L. Jones, An analytical and experimental stress analysis of a practical Mode II fracture test specimen. *Experimental Mechanics* **17**(1) 7–13 (1977).
[8] D. G. Smith and C. W. Smith, Photoelastic determination of mixed mode stress intensity factors. *Engng Fracture Mech.* **4**(2), 357–366 (1972).
[9] E. E. Gdoutos and P. G. Theocaris, A photoelastic determination of mixed mode stress-intensity factors. *Experimental Mechanics* **18**(3) 87–97 (1978).
[10] W. B. Bradley and A. S. Kobayashi, Fracture Dynamics—a photoelastic investigation. *Engng Fracture Mech.* **3**, 317–332 (1971).
[11] P. C. Paris and G. C. Sih, Stress analysis of cracks. *ASTM STP* **381**, pp. 30–81 (1964).
[12] L. G. Kelley, *Handbook of Numerical Methods and Applications*, p. 99 Addison-Wesley, Reading, Mass. (1967).
[13] R. J. Sanford and J. W. Dally, Stress intensity factors in the TF-30 turbine engine 3rd stage fan disk. *NRL Rep.* 8202 (May 1978).
[14] J. W. Dally and R. J. Sanford, Classification of stress intensity factors from isochromatic fringe patterns. *Experimental Mechanics*, **18**(12), 441–448 (1978).

INVESTIGATION OF THE RUPTURE OF A PLEXIGLAS PLATE BY MEANS OF AN OPTICAL METHOD INVOLVING HIGH-SPEED FILMING OF THE SHADOWS ORIGINATING AROUND HOLES DRILLED IN THE PLATE

Peter Manogg*

ABSTRACT

The loading and unloading of a Plexiglas plate by a running fracture is investigated by means of a series of high-frequency cinematographic pictures of the shadows originating around drilled holes. Near the fracture tip, the stress distribution differs only slightly from that around a corresponding static crack. After the fracture tip, however, there are great deviations owing to the dynamic unloading of the plate. The velocity of the unloading process is considerably lower than the velocity of the transversal waves in the plate.

INTRODUCTION

The dynamics of the rupture of plates have been only partly explained [1-3]. The short duration of the process impedes the experimental determination of the dynamic stress distribution in the plate, knowledge of which is incomplete even for the "simple" case of the propagation of a rectilinear tensile fracture perpendicular to an external tensile stress [4,5]. In order to acquire more precise knowledge of the loading and unloading processes in a plate during a fracture, the shadows originating during the tensile fracture of a Plexiglas plate were filmed at high speed. Holes drilled in the plate at various measuring points enabled the direction and difference of the principal stresses at these points to be determined for various phases of the fracture process. The results obtained are compared with the stress distribution around a static crack corresponding to the fractures, and discussed. The latter stress distribution is known according to the theory of notch stresses [2,3] and can be expressed in the following approximation formulas for the vicinity of the crack tip [6]:

$$\sigma_x = p \sqrt{\frac{L}{2r}} \cos \frac{\theta}{2} \left(1 - \sin \frac{\theta}{2} \sin \frac{3\theta}{2}\right) - p$$

$$\sigma_y = p \sqrt{\frac{L}{2r}} \cos \frac{\theta}{2} \left(1 + \sin \frac{\theta}{2} \sin \frac{3\theta}{2}\right) \qquad (1)$$

$$\tau = p \sqrt{\frac{L}{2r}} \sin \frac{\theta}{2} \cos \frac{\theta}{2} \cos \frac{3\theta}{2}$$

in which r, θ are polar co-ordinates around the crack tip, L is the crack length, p is the external tensile stress perpendicular to the crack; σ_x, σ_y and τ are the normal and shear stresses related to a system of Cartesian co-ordinates x, y (Cf. fig. 1). The formulas apply to the area

---

* Ernst Mach Institute (Fraunhofer-Gesellschaft), Freiburg i. Br., West Germany. Now at the European School in Varese, Italy.

r≪L around the crack tip and to an infinitely large plate.

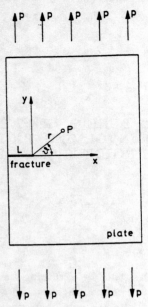

Fig.1. Stress distribution around a static crack.

THE METHOD[7,8] AND EXPERIMENTAL PROCEDURE

If a transparent plate is irradiated with parallel or slightly convergent light, the light will be refracted depending on the state of stress in the plate. In the case of materials such as Plexiglas, whose stress-induced birefringence in respect of the light refraction can be ignored, the light refraction is determined by the gradients of the principal stress sum at the relevant point in the plate. If a hole is drilled in an elastically-stressed plate the stress distribution around this circular notch is dependent on the state of stress that existed in the vicinity of this spot before the notch was made. The light refraction around such a drilled hole gives rise to a shadow in a reference plane parallel to the plate, the position and size of which shadow are characteristic of the undisturbed state of stress at the drilled hole. The axes of symmetry of the shadow image lie in the direction of the relevant principal stresses and the maximum diameter is a measure of the principal stress differences. Figure 2 shows the theoretical shadow image pertaining to a state of stress with the principal stresses p and q for the case p > q; the image pertaining to p < q should be envisaged turned $\pi/2$.

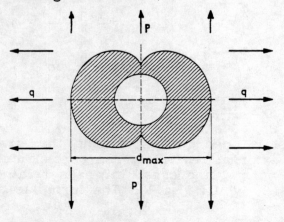

Fig.2. The shadow image of a circular notch in a plane state of stress with the principal stresses p and q (p > q). For p < q the figure is turned 90°. The shadow area is hatched.

The quantitative relationship of maximum shadow diameter d_{max} and principal stress difference in the case of a plate of thickness D is given, subject to a material-dependent normalizing factor f, by

$$|p-q| = f \frac{d_{max}^4}{Dz\,\mu^3 a_o^2} \qquad (2)$$

in which a_o is the radius of the drilled hole, z the distance between the plate and the reference plane, μ the relationship of the plate to the image in the reference plane (in the case of a parallel ray path $\mu = 1$).

For the investigation of the rupture process two Plexiglas plates were drilled at various points symmetrical to the fracture line. In the case of plate I (300 x 100 x 3.8 mm) the holes have a diameter of 8 mm and a pitch of 31 mm; in the case of plate II (300 x 100 x 2 mm) numerous holes with a diameter small (1 mm) in relation to their pitch ($\geq$ 10 mm) were drilled. The symmetrical arrangement of the measuring points was selected in order to increase and compare the accuracy of the results. The rupture process was initiated by the sudden application of a uniaxial external tensile stress at a point on the plate edge predetermined by a small crack notch, and the resultant shadows in a convergent ray path were filmed by means of a 24-flash high-speed cine camera apparatus according to Cranz-Schardin[9]. The picture frequency in both tests was 150,000 c/s; the remaining data are given in the captions to the relevant figures and picture series.

RESULTS AND DISCUSSION

Picture series I and II show the course of the rupture process in plates I and II. In each case the fracture runs from left to right perpendicular to the external tensile stress; the fracture tip is surrounded by an area of shadow. The shadow images around the drilled holes correspond in general to the theoretical picture; a slight displacement of the shadows in picture series I in relation to the non-refracted light falling through the hole is due to a slight wedging of plate I.

Comparison of the dynamic and static stress distributions

The values for the principal stress direction φ (related to the x and y axes (fig.1.)) derived from the shadow images around the holes in the case of picture series I and the principal stress difference $\Delta\sigma/\bar{p} = |p-q|/\bar{p}$ related to the rupture stress $\bar{p}$ ($=135$ kgf/cm²) are shown in figs. 3 and 4 as functions of the fracture length L (or the polar angle θ). For purposes of comparison the corresponding curves for a static crack according to formulae (1) are given. In the range 2 cm $\leq$ L $\leq$ 5 cm the measured values for the fracture agree largely with the stress pattern for a corresponding crack. The accuracy of measurement of $\Delta\sigma/\bar{p}$ is about $\pm$ 10%; the difference between the individually measured φ values and the common mean is generally less than 1.5 degrees (exception: shots 18 - 21 of picture series I, where the stress field in the vicinity of the holes varies greatly).

Deviations outside these error limits occur in the case of short fracture lengths for which the range of validity r « L of the approximation formulae (1) has been exceeded, as well as after the fracture, where the unloading of the plate becomes noticeable. If initially this dynamic effect is left out of account, it follows from figs. 3 and 4 that there is a large measure of similarity between the stress behavior near the fracture tip and that near the crack tip. This result is in agreement with investigations by Henschen[4] and Wells and Post[5], who have measured the dynamic stress along the extended fracture line (ligament) by interferometric and photoelastic methods. For the immediate vicinity of the fracture tip (r = 2.....7 mm) the similarity of the dynamic and the static stress distributions in the case of fractures and cracks also follows from the shape of the shadow image around the

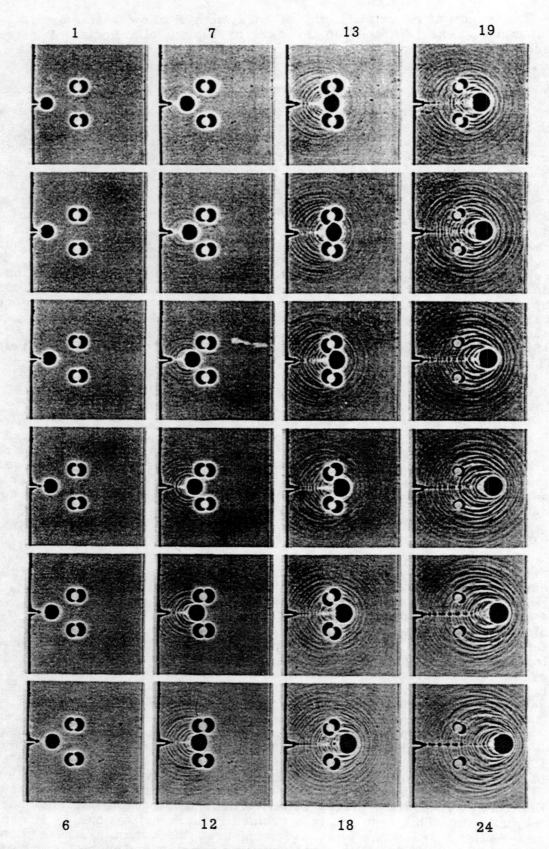

Picture Series I: High-speed cine-camera shots of the shadows occurring during the fracture of a Plexiglas plate. The tensile fracture runs between two drilled holes.
(Picture frequency 150,000 c/s, reference distance 200 cm, convergence of the ray path $\mu = 0.615$).

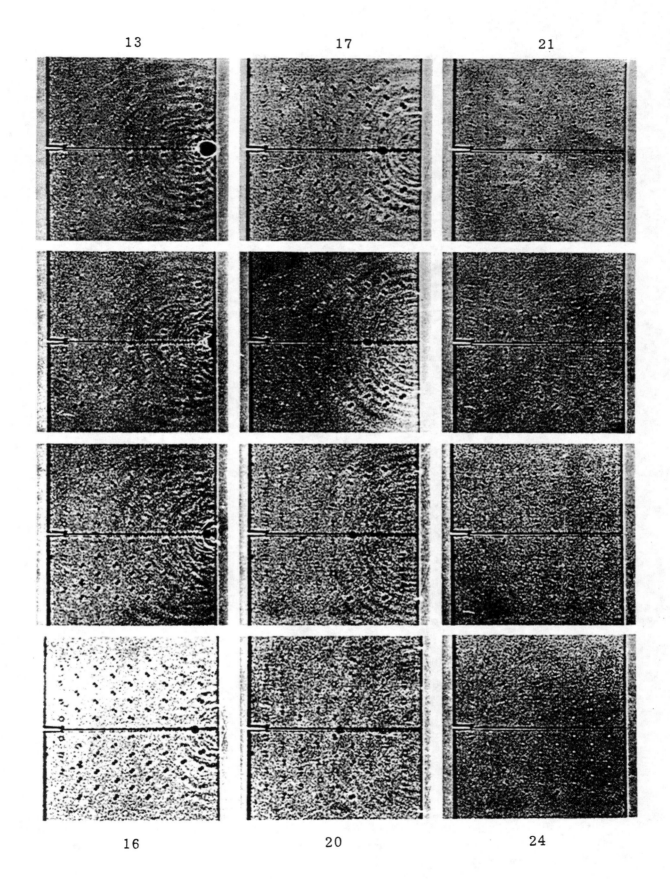

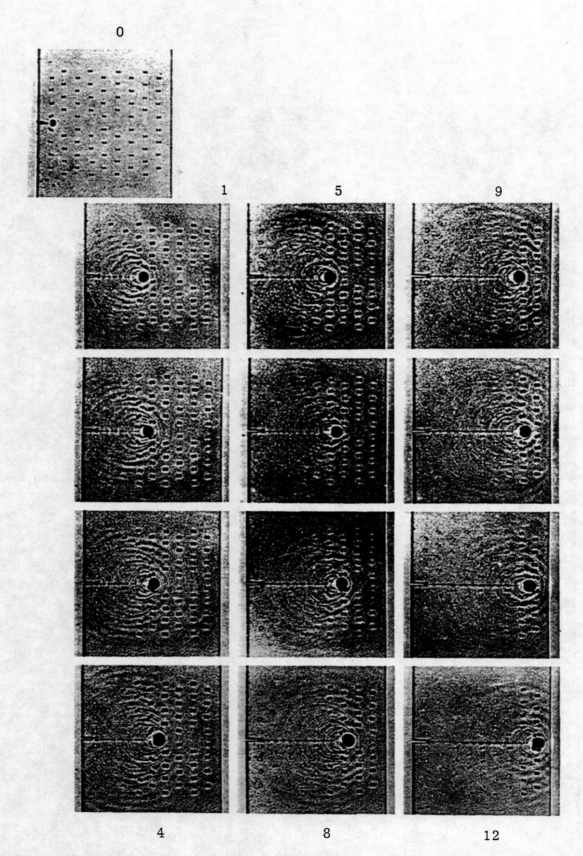

Pictures series II. High-speed cine-camera shots of the shadows occurring during the rupture of a Plexiglas plate. The plate has numerous holes drilled in it as measuring points. (Picture frequency 150.000 c/s, reference distance 100 cm, $\mu = 0.81$. The zero shot was obtained during a preliminary test.)

fracture tip, which can be derived by means of the stress formulae (1) for the static crack; analysis of this shadow image also makes it possible to determine the specific fracture energy and thus investigate the energy balance of the fracture process [7,10]

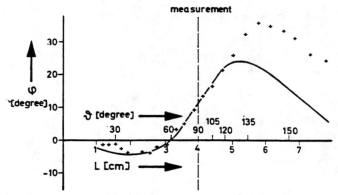

Fig. 3. The principal stress direction at the holes in plate I during the fracture process. The theoretical curve for a static crack of corresponding length (in accordance with (1)) has been inked in.

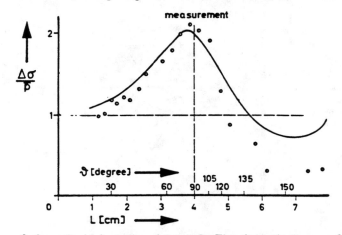

Fig. 4. The difference of the principal stresses in test I. The theoretical curve for a static crack in accordance with (1) has been inked in.

The unloading process in the plate

The shadow areas around the holes in test II (picture series II) are indicative of the state of stress for the individual phases of the rupture process in the entire plate. The axes of symmetry of the shadow images determine the directional field of the maximum and minimum principal stresses in the plate. The directional elements pertaining to the stress lines of the maximum principal stress are accentuated by strokes in the drawings of several shots of picture series II (fig. 5 a - f); in the case of some shadow areas the axis of symmetry of the maximum principal stress is discernible only by the light concentration of the caustic near the tips.

The course of the directional field of the stress lines and the size of the shadow images show clearly the unloading of the plate by the running fracture. A wedge-shaped unloaded zone behind the fracture tip travels with the fracture through the plate; the fracture velocity is constant within the accuracy of measurement and amounts to 635 ± 10 m/s. The velocity of the unloading v_e in the direction of the external tensile stress is ascertained by measurement of the wedge angle of the unloaded zone. The mean value of $47° \pm 2°$ for the first ten shots of the picture series corresponds at the fracture velocity indicated to an unloading velocity of 685 ± 80 m/s. The unloading of the plate perpendicular to the fracture line

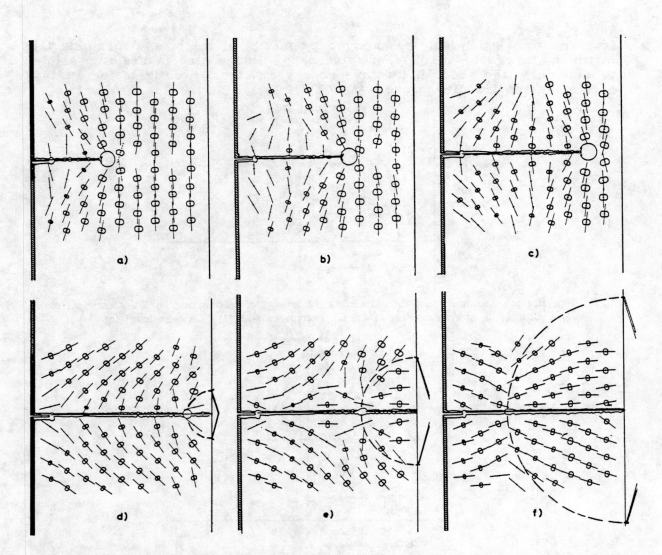

Fig. 5 a-f. The directional field of the stress distribution and the shadow outlines around the holes in test II (sketches of shots 1, 6, 10, 16, 18 and 22 of pictures series II).

therefore takes place at a velocity that is within the order of magnitude of the (maximum) fracture velocity and is considerably lower than the velocity of the transversal and longitudinal waves. The velocity of the edge wave visible after the rupture of the plate is about 1250 m/s (picture series II and fig. 5 d, e, f); the velocity of the transversal waves is approximately equal to, and that of the longitudinal waves is double, this figure.

Confirmation of the surprisingly low value of the unloading velocity is provided by figs. 3 and 4 (test I). The dynamic unloading of the circular holes, which is recognizable by the deviations from the "static behavior", according to formulae (1), starts at a polar angle of approximately $125°$, corresponding to a wedge angle of $55°$. Since the fracture velocity in the interval $\theta = 90° \ldots\ldots 125°$ averages 475 m/s - it rises from 250 m/s to 650 m/s between shots 1 and 24 - an unloading velocity $v_e = 475$ m/s. $\tan 55° = 680$ m/s follows, which tallies well with the value indicated above for test II.

The presence and recurrence of small shadow areas in the unloaded

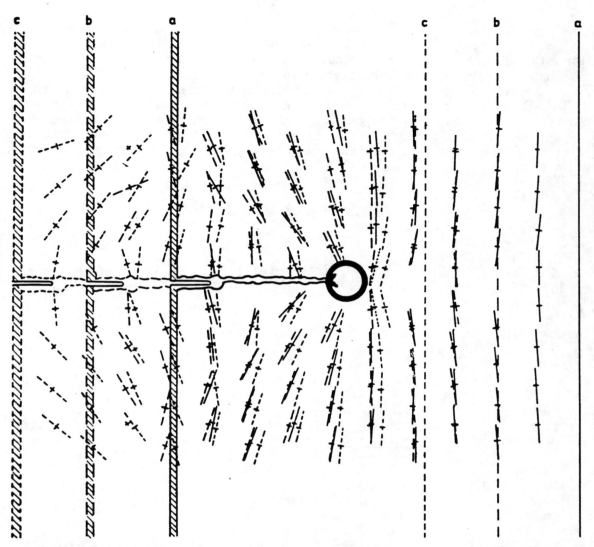

Fig. 6. Comparison of the stress distributions around the fracture tip for various phases of the fracture process according to fig. 5 a, b and c. For the sake of clarity only the directions of the maximum principal stresses at the measuring points are indicated.

zone shows that the unloading is incomplete and is in part cancelled out by waves reflected at the plate ends. The direction of the principal stresses is practically uniform shortly after the rupture of the plate (fig. 5 d) and indicates a shear stress parallel to the plate edges. This is partly broken down by the stress impulse occurring on rupture of the plate, which appears as an edge wave along the edge of the plate and the edge of the fracture and as a transversal wave impulse. The position of the transversal wave is determined by the edge waves along the plate edge (light concentration, air impulse wave) and along the fracture line (shadow area) and is easily discernible from the discontinuity of the principal stress directions in the plate; it is shown as a circular line in fig. 5 d - f. Unlike the actual edge waves, the stress impulse travelling along the fracture edges causes at the plate edge a concave deformation of the plate surface (shadow area!) and is damped and spread by the roughness of the fracture surface. The further stepwise reduction of the stresses in the plate is illustrated in fig. 5 e and f.

Various fracture phases

In fig. 6 pictures 5 a-c pertaining to various fracture phases of picture series II are drawn over each other in such a way that the fracture direction and the fracture tip coincide. For the sake of clarity only the directional elements of the maximum principal stresses at the holes are indicated. The directions of the principal stresses (and also the relationships of the shadow sizes) agree to a large extent in respect of the relevant fracture tip in all three fracture phases. The influence of the plate width on the course of the fracture is very slight. The dynamic tensile fracture therefore appears in this picture series as a quasistationary process that in the vicinity of the fracture tip can be closely described by the quasistatic crack model (1).

ACKNOWLEDGEMENTS

This investigation originated in the Ernst Mach Institute in Freiburg i. Br., West Germany. I should like to acknowledge the assistance of Prof. Dr. Ing. H. Schardin and Priv. Dozent Dr. habil. F. Kerkhof in suggesting and furthering this work.

Received June 16, 1966.

REFERENCES

1. H. Schardin — Z. Kunststoffe, 44, 2, 48–55 (1954).
2. F. Kerkhof — Kunststoffe, K. A. Wolf (ed.), Springer-Verlag, Berlin, 440–484 (1958).
3. G. R. Irwin — Handbuch der Physik, VI. S. Flugge (ed.), Springer-Verlag, Berlin, 551–590 (1958).
4. H. Henschen — Dissertation, Freiburg i Br. (1962).
5. A. A. Wells and D. Post — Proc. Soc. of Exp. Stress Anal., 16. 69–92 (1958).
6. I. N. Sneddon — Proc. Phys. Soc. (London), 187, 229–260 (1946).
7. P. Manogg — Dissertation, Freiburg i. Br. (1964).
8. P. Manogg — Glastechnische Berichte, 39, 323–329 (1966).
9. A. Stanzel — Kurzzeitphotographie, IV Int. Congress, Cologne, Helwich-Verlag, Darmstadt (1958).
10. P. Manogg — Phys. of Non-Crystalline Solids, Intl. Conf. Delft, North-Holland Publ. Co., Amsterdam (1964).

RÉSUMÉ – Les modifications de la charge dans une plaque de Plexiglas sous l'effet d'une fissure en cours de propagation ont été étudiées par diffraction de la lumière.

La technique consiste à filmer, à grande vitesse, les ombres projetées sur une plaque de référence par des trous forés dans la plaque; la position et les dimensions de ces ombres donnent une représentation quantitative de l'état de tension régnant autour de chaque trou.

On a pu, ainsi, montrer que la distribution des tensions, au voisinage du front de fissure, ne diffère que légèrement de celui régnant au voisinage d'une fissure statique. Par contre, de grandes déviations de l'état de tension ont été observées en arrière du font de fissure, en raison d'un déchargement dynamique de la plaque. La vitesse à laquelle se produit ce processus de déchargement est considérablement plus faible que la vitesse des ondes transversales dans le matériau considéré.

ZUSAMMENFASSUNG – Mit Hilfe einer schattenoptischen Anbohrmethode wird der Be- und Entlastungsvorgang durch einen laufenden Bruch in einer Plexiglasplatte anhand hochfrequenzkinematografischer Bildserien untersucht. In der Nähe der Bruchspitze erweist sich der Unterschied der Spannungsverteilung zu derjenigen um einen entsprechenden statischen Riß als gering. Hinter der Bruchspitze ergeben sich durch die dynamische Entlastung der Platte dagegen große Abweichungen. Die Geschwindigkeit des Entlastungsvorgangs ist wesentlich geringer als die Geschwindigkeit der Transversalwellen in der Platte.

ANALYSIS OF THE OPTICAL METHOD OF CAUSTICS FOR DYNAMIC CRACK PROPAGATION

ARES J. ROSAKIS

Division of Engineering, Brown University, Providence, RI 02912, U.S.A.

Abstract—In the interpretation of experimental data on dynamic crack propagation in solids obtained by means of the optical method of caustics, it has been customary to neglect the effect of material inertia on the stress distribution in the vicinity of the crack tip. In this paper, the elastodynamic crack tip stress field is used to establish the exact equations of the caustic envelope formed by the reflection of light rays from the surface of a planar solid near the tip of a propagating crack. These equations involve the instantaneous crack tip speed, the material parameters and the instantaneous dynamic stress intensity factor, and they can be used to determine the stress intensity factor for given material parameters and crack tip speed. The influence of inertial effects on stress intensity factor measurements for system parameters typical of experiments with PMMA specimens is considered. It is found that the stress intensity factor values inferred through a dynamic analysis may differ by as much as 30–40% from values based on a quasi-static analysis.

NOTATION

$C = \dfrac{dvz_0}{E}$

$C_{l,s}$ longitudinal, shear wave speeds
d thickness of specimen
D_t transverse diameter of caustic curve
E Young's modulus
$Z_{p'} = X + iY$
$J_p = x + iy$
K_I = mode I stress intensity factor for a stationary crack
K_{II} = mode II stress intensity factor for a stationary crack
$K_{st}(t)$ stress intensity factor for a running crack, evaluated using the static analysis
$K_d(t)$ stress intensity factor for a running crack, evaluated using the dynamic analysis
$K_{Id}(t)$ = mode I stress intensity factor for a running crack, evaluated using the dynamic analysis
$K_{IId}(t)$ = mode II stress intensity factor for a running crack, evaluated using the dynamic analysis

$K'(v,t) = \left[\dfrac{K_d(t)\cdot(1+\alpha_s^2)\cdot(\alpha_l^2+\alpha_s^2)}{4\alpha_l\cdot\alpha_s - (1+\alpha_s^2)^2}\right]$

$P(x,y)$ point on the actual specimen plane
$P'(X,Y)$ the image of point $P(x,y)$ on the plane of the screen
$r_l e^{i\theta_l} = x_l + y_l = x + i\alpha_l y$
$R^{(2)}$ specimen, plane
Δs optical path difference
u_z deformation of the surface of the specimen along the z direction
V_T terminal velocity of crack
$W = W_x + iW_y$, Light ray deviation vector
z_0 distance of the screen to the specimen

Greek symbols

$\alpha_{l,s} = \left[1 - \dfrac{V^2}{C_{l,s}^2}\right]^{1/2}$

$\lambda = \dfrac{C}{(2\pi)^{1/2}}$

$\mu = C\dfrac{K'(V,t)}{(2\pi)^{1/2}}$

v = Poisson's ratio

$\xi = \left[\dfrac{2\alpha_s}{1+\alpha_s^2}\cdot\dfrac{K_{IId}(t)}{K_{Id}(t)}\right]$

$\rho_0 = (\tfrac{3}{2}\mu\alpha_l)^{2/5}$ radius of the approximated initial curve in the (x_l, y_l) plane.
$\sigma_{1,2}$ principal stresses at a point
$\phi = \tan^{-1}\dfrac{K_{II}}{K_I}$

$\phi' = \tan^{-1} \zeta$
ω^- region inside the initial curve
ω^- region outside the initial curve
Ω^- region inside the caustic
Ω^- region outside the caustic

INTRODUCTION

AN OPTICAL method has been developed [1, 2–5] for the study of the stress singularity in the vicinity of a crack tip under conditions of plane stress. The method, known as the method of "caustics", has been extensively applied for the analysis of stress fields near the tip of stationary cracks [2–4]. Recently Kalthoff *et al.* [9, 10] and Theocaris *et al.* [6–8] presented work using the method of caustics for the study of propagating cracks. So far the method has been used under the basic assumption of a static solution for the stress–strain field near the crack tip. When dealing, however, with propagating cracks, an exact solution would require the introduction of the dynamic stress–strain distribution ahead of the crack tip, as well as the use of the dynamic elastic moduli E, v, for cases of dynamic loading conditions.

Theocaris was the first to measure and use the dynamic moduli [7, 8, 11], instead of the static ones, for the interpretation of caustic envelopes in cases of dynamic loading. The next step was that of Kalthoff *et al.* [9] who very recently introduced a correction factor to account for the error made by using the static analysis in cases of propagating cracks. The correction factor was obtained by using a series of simplifying assumptions based on numerical calculations.

In this paper the equations of the caustic envelope are obtained in a very simple form. Some approximations, based on analytical considerations are introduced without resorting to numerical arguments. A generalization of the results to include both opening and sliding mode propagating cracks is made in the Appendix.

1. CAUSTICS BY REFLECTION

An incident beam of parallel light rays is reflected from the near crack tip region of the specimen. Because of non-uniform contraction of the specimen in the thickness direction, the reflected rays deviate from parallelism and, under suitable conditions, generate a three-dimensional surface in space which separates an illuminated region from a dark region (see Fig. 1). That surface, composed of points of maximum luminosity or thermal effect, is called the "caustic" surface (Greek καυστικός, burning). The rays are tangent to the caustic surface, and the cross-sections of the surface can be observed as bright curves on a screen parallel to the specimen.

Let $u_z = u_z(x, y)$ be the normal displacement of the surface of the specimen due to the stresses at the vicinity of the crack tip. Consider a parallel beam of light illuminating the surface. If a screen is placed at a distance z_0 from the mid-surface of the specimen, a light ray impinging on point $P(x, y)$, will be reflected and recorded on the screen, giving an image point $P'(X, Y)$.

The deviation $W = W_x + iW_y$ of P' from the projection of P on the screen, will depend on the angle of reflection and on z_0. W will be a function of the slopes $\dfrac{\partial u_z}{\partial x}$, $\dfrac{\partial u_z}{\partial y}$ of the deformation surface, and of the distance z_0 of the screen to the specimen (see Fig. 2).

It has been shown [5] that

$$W = 2z_0 \operatorname{grad}_{x,y} [u_z(x, y)] \Rightarrow Z_{p'} = J_p + W = J_p + z_0 \operatorname{grad}_{x,y} [2u_z(x, y)] \qquad (1.1)$$

where $J_p = x + iy$ and $Z_{p'} = X + iY$. The quantity $2u_z(x, y) = \Delta S(x, y)$ represents the extra distance traveled by the ray because of the lateral deformation of the vicinity of the crack tip. ΔS is called the optical path difference.

Equation (1.1) is the governing equation of the mapping of points $P(x, y)$ of the specimen on to points $P'(X, Y)$ on the screen.

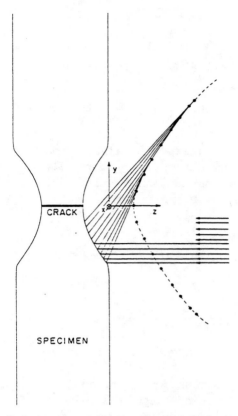

Fig. 1. Three dimensional caustic envelope formed by reflection (not in scale).

What appears on the screen is a completely dark area around the crack tip, surrounded by a bright region. The boundary between the bright and the dark regions is a highly luminous curve, the caustic curve. Equation (1.1) is a definite relation connecting points on the (X, Y) plane of the screen to generic points on the specimen plane (x, y). Experiment implies that there is a region in the (X, Y) plane on which no point of the (x, y) plane is mapped (shadow region). The whole (x, y) plane will therefore map in only part of the (X, Y) plane.

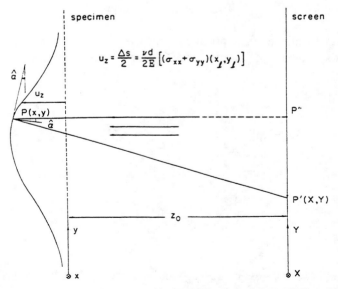

Fig. 2. The mapping of points $P(x, y)$ of the specimen, to points $P'(X, Y)$ on the screen.

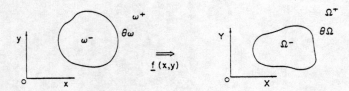

Fig. 3. Left: the initial curve; Right: the caustic curve.

It has been shown that there is a curve $\partial \omega$ in the (x, y) plane with the following properties: (i) every point outside $\partial \omega$ maps outside a specific curve $\partial \Omega$ in the (X, Y) plane. (ii) every point inside $\partial \omega$, maps again outside or on the same curve $\partial \Omega$. (iii) points on $\partial \omega$ will map on $\partial \Omega$ (see Fig. 3) or in standard mathematical notation:

$$\forall (x, y) \in \omega^- \upsilon \omega^+ \Rightarrow \underline{f}(x, y) \in \Omega^+ \upsilon \partial \Omega$$

$$\forall (x, y) \in \partial \omega \Rightarrow \underline{f}(\partial \omega) \subseteq \partial \Omega \qquad \text{(see Fig. 3)}$$

or equivalently:

$$\not\exists (x, y) \in R^{(2)} \ni \underline{f}(x, y) \in \Omega^- \Rightarrow f^{-1}(\Omega^-) = \emptyset.$$

The above statements imply that there are no points in the (x, y) plane (of the specimen) that are mapped into region Ω^-.

One can identify the Ω^- as the shadow region and Ω^+ as the bright one. The boundary $\partial \Omega$ between the bright and the shadow regions is, as we know from experiment, a highly luminous curve, the "caustic". High luminosity implies multiple mapping coming from both $\partial \omega$ and ω^-, and therefore the Jacobian of the transformation of points (x, y) on the specimen on to points (X, Y) on the image plane must vanish there. The condition $J = \dfrac{\partial (X, Y)}{\partial (x, y)} = 0$ provides the *locus* of points $\partial \omega$ which correspond to the luminous curve where multiple mapping occurs. $\partial \omega$ is called the "initial curve".

2. THE CAUSTIC CURVE OBTAINED BY A DYNAMIC FIELD AT THE VICINITY OF A PROPAGATING CRACK TIP

As we have seen in the previous section, the deviation from parallelism of a reflected light ray at a distance z_0 from the middle surface of the specimen, can be expressed by:

$$W = z_0 \operatorname{grad}_{x,y}[\Delta S(x, y)]$$

where ΔS is the difference in the optical path of the ray corresponding to a generic point $P(x, y)$. For the case of plane stress for caustics obtained by reflection, $\Delta S = 2u_z = \dfrac{\mathrm{d}v}{E}(\sigma_1 + \sigma_2)$. Substituting above:

$$W = z_0 \underline{\nabla}_{x,y}\left[(\sigma_1 + \sigma_2)\frac{\mathrm{d}v}{E}\right] = C\underline{\nabla}_{x,y}(\sigma_{xx} + \sigma_{yy}) \qquad (2.1)$$

where

$$C = \frac{\mathrm{d}v z_0}{E}.$$

Making use now of the expression for σ_{xx} and σ_{yy} established by Freund [12–14] which represent the stress components in the vicinity of the tip of a crack propagating with variable velocity $V(t)$, we have:

$$\sigma_{xx} = \frac{K_d(t)}{\sqrt{2\pi}} \cdot B_1(V) \left[(1 + 2\alpha_l^2 - \alpha_s^2) \frac{\cos(\theta_l/2)}{r_l^{1/2}} - \frac{4\alpha_l\alpha_s}{(1+\alpha_s^2)} \frac{\cos(\theta_s/2)}{r_s^{1/2}} \right] \quad (2.2)$$

$$\sigma_{yy} = \frac{K_d(t)}{\sqrt{2\pi}} \cdot B_1(V) \left[-(1+\alpha_s^2) \frac{\cos(\theta_l/2)}{r_l^{1/2}} + \frac{4\alpha_l\alpha_s}{(1+\alpha_s^2)} \frac{\cos(\theta_s/2)}{r_s^{1/2}} \right] \quad (2.3)$$

where:

$$r_l e^{i\theta_l} = x_l + iy_l = x + i\alpha_l y$$

$$\alpha_l = \left[1 - \frac{V^2}{C_l^2}\right]^{1/2}, \quad \alpha_s = \left[1 - \frac{V^2}{C_s^2}\right]^{1/2}.$$

$$B_1(V) = \frac{(1+\alpha_s^2)}{[4\alpha_l\alpha_s - (1+\alpha_s^2)^2]}$$

subscripts l and s referring to the longitudinal and shear way speeds, C_l and C_s, in the specimen material. It is worth noting that the above expressions were derived under the assumptions of elastic fracture behavior with V representing the instantaneous crack tip speed for non-uniform rates of crack growth. Adding (2.2) and (2.3) we obtain:

$$\sigma_{xx} + \sigma_{yy} = \frac{K_d(t)B_1(V)}{\sqrt{2\pi}} (2\alpha_l^2 - 2\alpha_s^2) \frac{\cos(\theta_l/2)}{r_l^{1/2}}$$

$$= \left[\frac{K_d(t)(1+\alpha_s^2)(\alpha_l^2 - \alpha_s^2)}{[4\alpha_l\alpha_s - (1+\alpha_s^2)^2]} \right] \sqrt{\frac{2}{\pi r_l}} \cdot \cos(\theta_l/2)$$

$$= K'(V,t) \cdot \sqrt{\frac{2}{\pi r_l}} \cdot \cos(\theta_l/2) \quad (2.4)$$

where $r_l e^{i\theta_l} = x_l + iy_l = x + i\alpha_l y$. Equation (2.4) gives the dynamic stress field $\sigma_{xx} + \sigma_{yy}$ at each point (x, y) as a function of the parameters (x_l, y_l) where $x_l = x$, $y_l = \alpha_l y$. It is worth observing that the dynamic expression (2.4) is of the same functional form as the static formula, $\sigma_{xx} + \sigma_{yy} = K_{st}\sqrt{\frac{2}{\pi r}} \cos(\theta/2)$ with two main differences: first, the existence of a multiplying factor

$$\frac{(1+\alpha_s^2)(\alpha_l^2 - \alpha_s^2)}{4\alpha_l\alpha_s - (1+\alpha_s^2)^2}$$

and second, the fact that the dynamic field has been scaled in the y direction only, by a factor α_l. One should note that (2.4) which gives the $\sigma_{xx} + \sigma_{yy}$ for a running crack reduces to the stationary for $V = 0$.

(i) *The Equations of the mapping*

For the case of a propagating crack the deviation W corresponding to the generic point (x, y) is given by:

$$W = C\nabla_{x,y}[(\sigma_{xx} + \sigma_{yy})(x_l, y_l)]$$

$$\Rightarrow W = C\frac{\partial}{\partial x_l}[(\sigma_{xx} + \sigma_{yy})(x_l, y_l)]\frac{\partial x_l}{\partial x} + iC\frac{\partial}{\partial y_l}[(\sigma_{xx} + \sigma_{yy})(x_l, y_l)]\frac{\partial y_l}{\partial y}$$

$$\Rightarrow W = C\frac{\partial}{\partial x_l}[(\sigma_{xx} + \sigma_{yy})(x_l, y_l)] + iC\frac{\partial}{\partial y_l}[(\sigma_{xx} + \sigma_{yy})(x_l, y_l)]\alpha_l.$$

The image $P'(X, Y)$ on the screen of the point $p(x, y)$ of the specimen will be given by:

$$\left.\begin{aligned}X &= x + C\frac{\partial}{\partial x_l}[(\sigma_{xx} + \sigma_{yy})(x_l, y_l)] \\ Y &= y + \alpha_l C\frac{\partial}{\partial y_l}[(\sigma_{xx} + \sigma_{yy})(x_l, y_l)]\end{aligned}\right\} \quad (2.5)$$

using now eqn (2.4) for the stresses and the fact that $x = r_l \cos\theta_l$ and $y = \dfrac{r_l \sin\theta_l}{\alpha_l}$, we get:

$$\left.\begin{aligned}X &= r_l \cos\theta_l + \mu r_r^{-3/2}\cos\frac{3\theta_l}{2} \\ Y &= \frac{r_l \sin\theta_l}{\alpha_l} + \alpha_l \mu r_r^{-3/2}\sin\frac{3\theta_l}{2}\end{aligned}\right\} \quad (2.6)$$

where

$$\mu = C\frac{K'(V, t)}{(2\pi)^{1/2}}, \quad 0 < \theta_l < 4\pi.$$

Equations (2.6) are the governing equations for the mapping of a generic point $P(x, y)$ of the specimen on to a point $P'(X, Y)$ of the screen. They are expressed with respect to the parameters (r_l, θ_l) which are connected to (x, y) by the relation $r_l e^{i\theta_l} = x + i\alpha_l y = x_l + iy_l$. It will become obvious from the following that by using (r_l, θ_l) as the parameters instead of (r, θ), the equations of the caustic will result in a very convenient form to work with.

It is worth noting that points $P(x, y)$ lying on a member of the family of ellipses $x^2 + \alpha_l^2 y^2 = \rho^2$ on the (x, y) plane, map on to points on the parameter plane (x_l, y_l) lying on the circle $x_l^2 + y_l^2 = \rho^2 = r_l^2$. Those points will in turn map through the transformation eqn (2.6) on to points $P'(X, Y)$ lying on the curve:

$$\left.\begin{aligned}X &= \rho\cos\theta_l + \mu\rho^{-3/2}\cos\frac{3\theta_l}{2} \\ Y &= \rho\frac{\sin\theta_l}{\alpha_l} + \alpha_l\mu\rho^{-3/2}\sin\frac{3\theta_l}{2}\end{aligned}\right\} \quad 0 < \theta_l \leq 4\pi. \quad (2.7)$$

Thus a member of the family of ellipses $x^2 + \alpha_l^2 y^2 = \rho^2$ defined on the actual plane of the specimen, will map on a member of the family of curves (2.7) defined on the plane of the specimen.

(ii) *The equations of the caustic*

The condition for the existence of a caustic curve surrounding the shadow region is the vanishing of the Jacobian of the transformation [3]. The condition $J = \dfrac{\partial(X, Y)}{\partial(x, y)} = 0$ in general will give a curve on the (x, y) plane. Points both inside and outside this curve will always map on or outside the caustic envelope. Points on this curve will map upon the caustic envelope. The curve obtained by setting $J = 0$, lying on the (x, y) plane is called the "*initial curve*".

$$J = \frac{\partial(X, Y)}{\partial(x, y)} = \alpha_l \frac{\partial(X, Y)}{\partial(x_l, y_l)} = 0 \Rightarrow \frac{\partial(X, Y)}{\partial(x_l, y_l)} = 0.$$

Using now eqn (2.6) and the polar form of the Jacobian we get:

$$\Rightarrow \frac{\partial(X[r_l, \theta_l] Y[r_l, \theta_l])}{\partial(r_l, \theta_l)} = 0$$

$$\Rightarrow \boxed{r_l^5 - \tfrac{9}{4}\mu^2 \alpha_l^2 + \tfrac{3}{2}\mu r_l^{5/2}(\alpha_l^2 - 1)\cos\frac{5\theta_l}{2} = 0} \ . \tag{2.8}$$

Since we have chosen to work with (r_l, θ_l) as parameters instead of the actual coordinates (r, θ), the condition $J = 0$ gives a curve with respect to (r_l, θ_l), which is the equation of the image of the "*initial curve*" in the (r_l, θ_l) plane. Equation (2.8) gives the region in the (r_l, θ_l) plane whose points map on to the caustic curve.

The *exact* equations of the caustic envelope are therefore given by:

$$\left.\begin{array}{l} X = r_l \cos\theta_l + \mu r_l^{-3/2} \cos\dfrac{3\theta_l}{2} \\[6pt] Y = \dfrac{1}{\alpha_l}\left[r_l \sin\theta_l + \alpha_l^2 \mu r_l^{-3/2} \sin\dfrac{3\theta_l}{2}\right] \end{array}\right\} \tag{2.9}$$

under the constraint:

$$r_l^5 - \tfrac{9}{4}\mu^2 \alpha_l^2 + \tfrac{3}{2}\mu r_l^{5/2}(\alpha_l^2 - 1)\cos\frac{5\theta_l}{2} = 0.$$

The curve described by eqn (2.8) has a very weak θ_l dependence, especially for low or medium values of V. Although the analytic expressions presented above are not unduly cumbersome, it is advantageous nonetheless to simplify them somewhat in order not to have to resort to numerical schemes. As it turns out, a very close approximation can be made regardless of the value of α_l.

The method of approximation adopted here is to find a set of circles each of which constitutes a best fit to (2.8) for each value of α_l. The procedure for finding the best fit circles is as follows: As shown above, a family of circles, on the (r_l, θ_l) plane, map on to a family of image curves given by eqn (2.6) (for different r_l) lying in the (X, Y) plane.

The circle which gives the closest fit to the curve represented by eqn (2.8), will be the one corresponding to that particular member of the (2.6) image family with the smallest enclosed area. This can be demonstrated as follows:

Let $r_l = \rho_0$ be the radius of the circle corresponding to the minimum area member of the (2.6) family (see Fig. 4). Let the dotted curve in the (x_l, y_l) plane correspond to curve (2.8), and the dotted curve in the (X, Y) plane to its image. Any point on $r_l = \rho_0$, whether inside or outside the dotted curve (2.8), will map outside the dotted image curve in the (X, Y) plane, except at those points where the curves touch (solid line).

Since the circle $r_l = \rho_0$ by hypothesis maps on to the minimum area member of the (2.6) family, every circle of smaller or larger radius than ρ_0 will map on to a member of the (2.6) family having a greater area than the curve corresponding to $r_l = \rho_0$. Consequently, the image of any other circle will lie further away from the dotted curve than the image of $r_l = \rho_0$

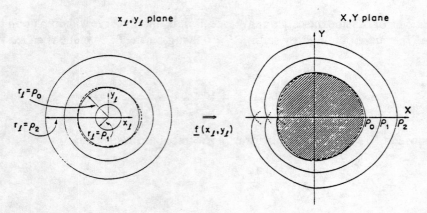

Fig. 4. Left: the image of the initial curve on the (x_l, y_l) plane (dotted), and its best circle approximation; Right: the caustic (dotted) and its best approximation curve.

does. The dotted curve in the (X, Y) plane is the caustic, and $r_l = \rho_0$ is the closest circle approximation to the (2.8) curve, since it gives an image which is the closest curve possible to the caustic.

3. EVALUATION OF ρ_0

In order to find ρ_0 let us consider the area of a random member of the family of images:

$$E = \frac{1}{2} \int_0^{4\pi} (XY'_{\theta_l} - YX'_{\theta_l}) \, d\theta_l.$$

Substituting the expressions for $X, X'_{\theta_l}, Y, Y'_{\theta_l}$ from (2.4) we get:

$$E = \frac{1}{2} \frac{4\pi}{\alpha_l} [r_l^2 + \tfrac{3}{2}\mu^2 \alpha_l r_l^{-3}].$$

The condition for an extremum gives:

$$\left.\frac{\partial E}{\partial r_l}\right|_{r_l = \rho 0} = 0 \Rightarrow \rho_0 = (\tfrac{3}{2}\mu\alpha_l)^{2/5} \tag{3.1}$$

which corresponds to a minimum, since $\left.\dfrac{\partial E^2}{\partial^2 r_l}\right|_{r_l = \rho 0} > 0$. Thus $r_l = \rho_0$ gives the minimum area member of the image (2.6) family.

Using the results of the above section, we conclude that $r_l = \rho_0 = (\tfrac{3}{2}\mu a_l)^{2/5}$ corresponds to the closest circle approximation to the curve represented by eqn (2.8). To find where the circle and the (2.8) curve touch, one can solve equations $r_l = \rho_0$ and eqn (2.8) simultaneously. One can thus see that the two curves touch at a total of 5 points. The coordinates of these 5 points in the (r_l, θ_l) plane are given by $r_l = \rho_0$ and by: $\cos\dfrac{5\theta_l}{2} = 0$. The latter provides 5 values of θ_l, namely, $\theta_l = \dfrac{\pi}{5}, \dfrac{3\pi}{5}, \pi, \dfrac{7\pi}{5}, \dfrac{9\pi}{5}$.

Since $r_l = \rho_0$ and the curve represented by eqn (2.8) touch in 5 points, their corresponding images in the (X, Y) plane will touch in 5 points as well.

4. (i) BEST APPROXIMATION OF THE INITIAL CURVE

Up to this point we have almost totally worked in the (r_l, θ_l) plane, investigating the best circle estimate to the (2.8) curve.

As we have seen above, the (2.8) curve is just the image of the initial curve on the parametric plane (x_l, y_l) and it has no physical meaning. Although it is easier to work in the

(r_l, θ_l) plane, it is instructive to investigate which is the best fit curve to the initial curve in the actual (x, y) specimen plane.

As shown above, the equation of the best circle approximation to the (2.8) curve (image of the initial curve in the (x_l, y_l) plane) is given by $r_l^2 = \rho_0^2 = x_l^2 + y_l^2 = (\tfrac{3}{2}\mu\alpha_l)^{4/5}$. In the actual (x, y) plane, the corresponding points (r, θ) lie on the ellipse $x^2 + \alpha_l^2 y^2 = (\tfrac{3}{2}\mu\alpha_l)^{4/5}$, since $x_l = x$, $y_l = \alpha_l y$. This ellipse is therefore the best approximation of the initial curve.

Thus we have seen that the initial curve in the (x, y) (specimen plane) for a propagating crack can be approximated by the ellipse:

$$x^2 + \alpha_l^2 y^2 = (\tfrac{3}{2}\mu\alpha_l)^{4/5} \qquad \text{for all values of } v \text{ and } \alpha_l, \alpha_s \qquad (4.1)$$

which reduces to:

$$x^2 + \alpha_l^2 y^2 = \left[\frac{3}{2} \left(\frac{(1+\alpha_s^2)(\alpha_l^2 - \alpha_s^2)}{4\alpha_l\alpha_s - (1+\alpha_s^2)^2} \right) \cdot \alpha_l \frac{K_d(t)\, dv z_0}{E(2\pi)^{1/2}} \right]^{4/5}$$

when the values of μ and C are substituted in (4.1).

(ii) *The behavior of the approximation curve as* $V \to 0$

As a check to the approximation curve (4.1) one could examine its behavior as $V \to 0$, $\alpha_l, \alpha_s \to 1$. Taking the limit as $\alpha_s, \alpha_l \to 1$, eqn (4.1) reduces to:

$$x^2 + y^2 = \left[\frac{3}{2} \frac{v}{E} \frac{dz_0 K_d(t)}{(2\pi)^{1/2}} \right]^{4/5} = \rho_{st}^2$$

which is the known equation of the initial curve obtained by using the stress field of a stationary crack [4]. Thus our approximation curve (4.1) reduced to the exact equation of the initial curve as $V \to 0$.

5. (i) THE APPROXIMATE EQUATIONS OF THE CAUSTIC CURVE

After the approximation is made, the eqn (2.9) of the caustic become:

$$\left. \begin{aligned} X &= r_l \cos\theta_l + \mu r_l^{-3/2} \cos\frac{3\theta_l}{2} \\ Y &= \frac{1}{\alpha_l}\left[r_l \sin\theta_l + \alpha_l^2 \mu r_l^{-3/2} \sin\frac{3\theta_l}{2} \right] \end{aligned} \right\} \qquad (5.1)$$

where

$$r_l = \rho_0 = (\tfrac{3}{2}\alpha_l\mu)^{2/5}.$$

Equations (5.1) can be expressed as follows:

$$\left. \begin{aligned} X &= \rho_0\left[\cos\theta_l + \tfrac{2}{3}\alpha_l^{-1}\cos\frac{3\theta_l}{2} \right] \\ Y &= \frac{\rho_0}{\alpha_l}\left[\sin\theta_l + \tfrac{2}{3}\alpha_l\sin\frac{3\theta_l}{2} \right] \end{aligned} \right\} \qquad (5.2)$$

where

$$0 < \theta_l \le 4\pi.$$

Equations (5.2) are the parametric equations of the caustic curve. As $V \to 0$, $\alpha_l, \alpha_s \to 1$, $\rho_0 \to \rho_{st}$ the above equations become the equations of a generalized epicycloid as predicted by the analysis of stationary cracks.

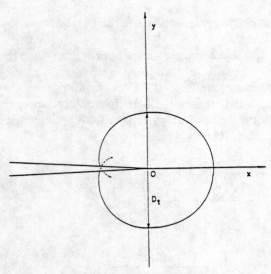

Fig. 5. Evaluation of the dynamic stress intensity factor $K_d(t)$ by measurement of D_t.

The parameter ρ_0 of the eqn (5.2) is a function of the elastic moduli E and v. For propagating cracks under dynamic loading conditions, the dynamic values of the moduli E and v must be calculated and used for the evaluation of ρ_0, as suggested by Theocaris et al. [7, 8, 11].

(ii) *Evaluation of $K_d(t)$*

By employing eqn (5.2) the geometric properties of the curve can be investigated (see Fig. 5). The transverse diameter D_t of the curve can be evaluated with respect to $K_d(t)$ at a given value of α_l. By setting $X = 0$ one obtains values of θ_l and hence values of the Y intercepts and D_t with respect to $K_d(t)$. Hence by measuring D_t, the stress intensity factor can be calculated.

(iii) *Concluding remarks*

Expressing the equations of the caustic envelope with respect to the parametric (r_l, θ_l) plane, one obtains more convenient and shorter forms than one would do expressing them with respect to the (r, θ) physical plane of the specimen.

The resulting equations are simplified further by using the approximation scheme described above. The approximations lead to an elliptical initial curve in the (r, θ) plane. This does not contradict intuition. It is known that for the cases of propagating cracks the $\sigma_{xx} + \sigma_{yy}$ stress field is scaled in the y direction, thus deforming the circular initial curve, corresponding to a static crack, to an ellipse as the velocity becomes finite.

The resulting equations are not limited to a certain range of velocities since the approximation curves depend each time on α_l and α_s. The method is therefore quite a general one, without the generality implying complications in the resulting equations. The simple form of the equations together with the lack of restrictions in the range of v justifies the use of the corrected formulae for cases of propagating cracks.

6. (i) CORRECTION DUE TO DYNAMIC EFFECTS

It is very instructive to calculate the error introduced in the evaluation of the dynamic stress intensity factor by using the static equations of the caustic curve, instead of the dynamic ones. Such a calculation will show if the use of the static analysis gives accurate enough results, and, if so, in which range of crack velocities.

For calculating the error, eqns (5.2) were used to evaluate the dynamic stress intensity factor for different velocities of a crack at a given material. The ratio of the calculated values over the ones corresponding to the same velocity, and obtained by the static analysis, were subsequently plotted against the velocity of the propagating crack.

The material for which the calculations were performed was PMMA (polymethylmethacrylate). PMMA was chosen since it behaves in close approximation to an ideally brittle material, that is, its behavior is well described by linear elastic fracture mechanics.

In addition to the above, PMMA has been used by a considerable number of investigators and therefore large numbers of experimental data are available in the literature concerning its material properties and fracture behavior (elastic moduli, limiting crack velocities, etc.).

The material parameters for PMMA in cases of experiments performed under conditions of low loading rates were taken to be the following:

$$\rho = 1200 \text{ kg m}^{-3} \quad \text{Density}$$
$$v = 0.34 \quad \text{Poisson's ratio}$$
$$E = 3200 \text{ MN m}^{-2} \quad \text{Young's modulus}$$

for which the values of the shear and longitudinal wave speeds were found to be equal to:

$$C_l = 2025 \text{ m s}^{-1} \quad \text{Longitudinal wave speed}$$
$$C_s = 997.5 \text{ m s}^{-1} \quad \text{Shear wave speed.}$$

The ratio $\dfrac{K_d(t)}{K_{st}(t)}$ was plotted v crack velocity (see Fig. 6i, ii).

(ii) *The error*

It can be seen from Fig. 6(i) that for low values of velocities, ranging between (0–200) m s^{-1}, the error is small, lying between 0 % to 2.0 %. For medium velocities in the range of (250–350) m s^{-1} it becomes significantly large, lying between 3.2 and 6.2 %.

Finally, for velocities in the range of 400–700 m s^{-1} (the terminal velocity of cracks in PMMA), the error becomes very large, ranging from 8.4 to 45.6 %.

The range of velocities considered in Fig. 6(i) was dictated by a theoretical estimate of the maximum terminal velocity of Mode I cracks in PMMA. The theoretical estimate of the terminal velocity was in agreement with experimental results obtained by a number of investigators.

(iii) *The terminal velocity*

A number of attempts, based on energy arguments, have been made to predict the maximum velocity of cracks propagating in brittle materials. In 1972 Bergvist [15] using the experimental results of Paxson and Lucas [16] for PMMA, arrived at an approximate expression for the terminal velocity in PMMA which was given by:

$$V_T = 0.69 C_s = 0.42 \left(\frac{E}{\rho}\right)^{1/2}. \tag{6.3}$$

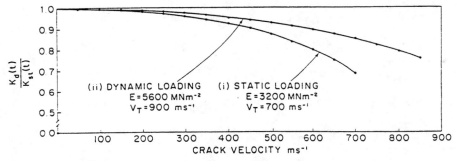

Fig. 6. The ratio $\dfrac{K_d(t)}{K_{st}(t)}$ v crack velocity in PMMA.

From (6.3) one can get a value for the maximum terminal velocity of the crack in PMMA. In our case the estimated velocity was found to be 686 m s^{-1}. The highest velocities in PMMA reported in the literature do not deviate much from this value. Schardin [17] reported the value of 655 m s^{-1}, Dulaney and Brace [18], 670 m s^{-1}, Cotterel [19] 670 m s^{-1}, Theocaris et al. [22] 670 m s^{-1} and Dahlberg [20] 700 m s^{-1}.

From the above one can see that for PMMA under conditions of static loading (low loading rates), the range of possible crack velocities is between 0 m s^{-1} and 700 m s^{-1}. The above estimated upper limit of crack tip velocities in PMMA, combined with the results of Fig. 6, show that there is a substantial range of velocities, between 350 m s^{-1} and 700 m s^{-1} corresponding to an error between 6.5 and 45.6%.

(ii) *Dynamic loading rates*

The results obtained above hold for cases of experiments performed under static loading conditions (low loading rates). Since PMMA is a viscoelastic material, the elastic moduli increase considerably at very high loading rates. To describe therefore the fracture behavior of PMMA for cases of high loading rates, one requires the knowledge of the elastic parameters corresponding to the loading rates used. In experimental investigations of the dynamic behavior of solids, a large number of devices have been used. These range from direct modification of conventional testing machines, resulting in higher rates of loading, to the use of impact testing, ultrasonic or high explosive techniques.

For cases of explosive loading, the dynamic value of Young's modulus was found to be almost double its static value. From experiments performed by Davies et al. [21] in PMMA, it was shown that the dynamic Young's modulus can reach a value as high as 6000 MN m^{-2}. The above value was obtained from experiments performed using a split Hopkinson bar, the specimen being loaded explosively. The compressive loading cycles were of 30 μs duration.

In relation to the method of caustics, Theocaris et al. calculated the dynamic values of the Young's modulus and Poisson's ratio by comparing the sizes of the caustics formed at the tips of stationary notches in PMMA under conditions of static or dynamic loading. His experiments showed that for loading rates of 0.35 s^{-1}, the Poisson's ratio of the material remained the same, whereas the Young's modulus increased to a value of $E = 4300$ Nm^2 [22]. He was the first to measure and introduce the dynamic moduli in the equations of the caustics used for the study of running cracks under conditions of impact loading [11].

Since the Young's modulus increases under conditions of dynamic loading, the Poisson's ratio remaining constant, the wave speeds must increase as well, resulting in a higher terminal velocity of the propagating crack. The error in cases of very high loading rates (higher values of $C_{l,s}$, lower values of $\alpha_{l,s}$) is expected to be less pronounced for small or moderate values of velocity, than the equivalent error obtained for cases of static loading. The range of obtainable velocities, however, will be increased (V_T going up with loading rate), still giving a considerable error at high velocities. To investigate the effect of loading rate on the difference between the static and dynamic analysis, values of $\dfrac{K_d(t)}{K_{st}(t)}$ were plotted vs crack tip velocity for case of experiments performed under conditions of impact loading. See Fig. 6(ii). The dynamic values of the parameters used were as follows:

$$\rho = 1200 \, \text{Kg m}^{-3} \quad \text{Density}$$
$$E = 5600 \, \text{MN m}^{-2} \quad \text{Dynamic Young's modulus}$$
$$\nu = 0.34 \quad \text{Poisson's ratio}$$

for which the shear and longitudinal wave speeds were found to be equal to

$$C_l = 2678.4 \, \text{m s}^{-1} \quad C_s = 1319 \, \text{m s}^{-1}.$$

For the case considered, the limiting velocity would be around 900 m s^{-1}. This value is verified by the experiments performed by Theocaris [11] who reported velocities up to 850 m s^{-1}.

(iii) *The error in cases of impact loading*

From comparison between curves (i)–(ii) in Fig. 6 one can see that the error is more pronounced for cases of low loading rates than for cases of higher rates. Figure 6(ii) shows that for low values of the crack velocity ranging between $(0-250)$ m s^{-1}, the error is negligible, lying between 0 and 1.5%.

For medium velocities in the range of $(300-500$ m s$^{-1})$ it is small, ranging between 2 and 7.8%. Finally, for velocities in the range of $550-900$ m s^{-1} (the terminal velocity for impact loading rates), the error becomes considerably larger, ranging from 9.6 to 38.0%.

7. THE EXACT EQUATIONS

The calculations described in the foregoing section were repeated using the exact dynamic equations of the caustic curve, given in (2.9). An IBM 360 computer was employed for the calculation. The results obtained were almost exactly coincident with the values shown in Fig. 6(i) and (ii). The values of $\dfrac{K_d(t)}{K_{st}(t)}$ were found to differ by less than 0.002 for every value of V.

The agreement in the results of the two cases is very close, indicating that the theoretically obtained approximation to eqn (2.9), described in Section 2(ii), is indeed justifiable.

CONCLUSIONS AND SUMMARY

The exact equations of the caustic curve for the case of a running crack were obtained in a simple form with respect to the (r_l, θ_l) parametric plane. These were further simplified by employing a theoretical argument. The error introduced by using the static analysis was then calculated by making use of *both* the exact and simplified dynamic formulae.

The numerical results obtained showed that the use of the approximate dynamic formulae is indeed justifiable. At the same time, it was shown that there is a considerably large range of velocities lying between $V_T/2$ and V_T corresponding to errors between 6.5 and 45.6%.

Acknowledgements—In discussing the paper and giving valuable criticism, the author is deeply indebted to Profs J. Duffy and L. B. Freund of the faculty of Mechanics of Solids and Structures, Brown University, for their guidance and support in the preparation of this report. Thanks are due also to Profs A. C. Pipkin and Visiting Assistant Prof G. Dassios of the faculty of Applied Mathematics, Brown University, for their kind advice on specific points. Lastly, the author would like to thank the office of Naval Research, Contract N00014-78-C-0051, and the Materials Research Laboratory, Brown University, for their support.

REFERENCES

[1] P. Manogg, *Anwendung der Schattenoptik zur Untersuchung des Zerreissvorganges von Platten* Dissertationsschrift an der Universität Freiburg (1964).
[2] P. S. Theocaris, Local yielding around a crack tip in plexiglass. *J. Appl. Mech.* **37**, 409–415 (1970).
[3] P. S. Theocaris, Reflected shadow method for the study of constrained zones in cracked plates. *Appl. Optics* **10**, 2240–2247 (1971).
[4] P. S. Theocaris and E. E. Gdoutos, An optical method for determining opening-mode and edge sliding-mode stress intensity factors. *J. Appl. Mech.* **39**, 91–97 (1972).
[5] P. S. Theocaris and E. E. Gdoutos, Surface topography by caustics. *Appl. Optics* **15**, 1629–1638 (1976).
[6] P. S. Theocaris, Dynamic propagation and arrest measurements by the method of caustics on overlapping skew-parallel cracks. *Int. J. Solids Structures* **14**, 639–653, (1978).
[7] D. D. Raftopoulos, D. Karapanos and P. S. Theocaris, Static and dynamic mechanical and optical behavior of high polymers. *J. Phys. D (Appl. Phys.)* **9**, 869–877 (1976).
[8] F. Katsamanis, D. Raftopoulos and P. S. Theocaris, Static and dynamic stress intensity factors by the Method of Transmitted Caustics. *J. Engin. Mat. and Tech.* **99**, 105–109 (1977).
[9] J. F. Kalthoff, J. Beinert and S. Winkler, Influence of dynamic effects on crack arrest. *Institut für Festkörpermechanik, Tech. Rep.* August (1978).
[10] J. F. Kalthoff, J. Beinert, S. Winkler and J. Blauel, On the Determination of the Crack Arrest-Toughness, *4th Int. Conf. Fracture*, Waterloo, Canada, **3**, 751 ff (1977).
[11] P. S. Theocaris and F. Katsamanis, Response of cracks to impact by caustics. *Engng Fracture Mech.* **10**, 197–210 (1978).
[12] L. B. Freund and R. J. Clifton, On the uniqueness of plane elastodynamic solutions for running cracks. *J. Elasticity* **4**, (4) 293–299 (1974).
[13] L. B. Freund, Dynamic crack propagation. *Mech. of Fracture* AMD **19**, 105–134 (1976).
[14] L. B. Freund, The analysis of elastodynamic crack tip stress fields. *Mech. Today* **3**, 55–91 (1976).
[15] H. Bergkvist, The motion of a brittle crack. *J. Mech. Phys. Solids* **21**, 299–339 (1973).
[16] T. L. Paxson and R. A. Lucas, *An Experimental Investigation of the Velocity Characteristics of a Fixed Boundary Fracture Model, Dynamic Crack Propagation* (Edited by G. C. Sih), Noordhoff, Leyden, p. 415 (1973).

[17] H. Schardin, Velocity effects in fracture. *Fracture*, p. 297. (Edited by B. L. Averback *et al.*) Wiley, New York (1959).
[18] E. N. Dulaney and W. F. Brace, Velocity Behaviour of a Growing Crack. *J. Appl. Phys.* 31, (12) 2233–2236 (1960).
[19] B. Cotterell, Velocity effects in fracture propagation. *Appl. Math. Res.* 4, 227–232 (1965).
[20] L. Dahlberg, Experimental studies of crack propagation in polymers. *R. Inst. Tech.*, Stockholm (1972).
[21] D. H. Davies and S. C. Hunter, The dynamic compression testing of solids by the method of the split Hopkinson pressure bar. *J. Mech. Phys. Solids* 11, 155–179 (1963).
[22] F. Katsamanis, D. Raftopoulos, P. S. Theocaris, The dependence of crack velocity on the critical stress in fracture. *Exp. Mech.* 17, (4) 128–132 (1977).
[23] L. B. Freund, Private Communication (1979).

(*Received* 23 *July* 1979; *received for publication* 23 *August* 1979)

APPENDIX

A1. (i) *Mixed mode propagating cracks*

As we have seen in Section 1 the deviation from parallelism of a reflected light ray at a distance z_0 from the middle surface of the specimen, can be expressed as:

$$W = C \, \text{grad}_{x,y} (\sigma_{xx} + \sigma_{yy}) \quad \text{where} \quad C = \frac{d \nu z_0}{E}.$$

When the crack tip deformation is a combination of the plane Modes I and II, the expressions for σ_{xx} and σ_{yy} developed by Freund [23] following the procedure of [12] are given by:

$$\sigma_{xx} = \frac{K_{Id}(t)}{\sqrt{2\pi}} B_I(V) \left[(1 + 2\alpha_l^2 - \alpha_s^2) \frac{\cos(\theta_l/2)}{r_l^{1/2}} - \frac{4\alpha_s \alpha_l \cos(\theta_s/2)}{(1 + \alpha_s^2) r_s^{1/2}} \right]$$

$$- \frac{K_{IId}(t)}{\sqrt{2\pi}} B_{II}(V) \left[(1 + 2\alpha^2 - \alpha_s^2) \frac{\sin(\theta_l/2)}{r_l^{1/2}} - (1 + \alpha_s^2) \frac{\sin(\theta_s/2)}{r_s^{1/2}} \right] \quad (A1.1)$$

and

$$\sigma_{yy} = \frac{K_{Id}(t)}{\sqrt{2\pi}} B_I(V) \left[-(1 + \alpha_s^2) \frac{\cos(\theta_l/2)}{r_l^{1/2}} + \frac{4\alpha_l \alpha_s}{(1 + \alpha_s^2)} \cdot \frac{\cos(\theta_s/2)}{r_s^{1/2}} \right]$$

$$- \frac{K_{IId}(t)}{\sqrt{2\pi}} B_{II}(V) \left[-\frac{\sin(\theta_l/2)}{r_l^{1/2}} (1 + \alpha_s^2) + \frac{\sin(\theta_s/2)}{r_s^{1/2}} (1 + \alpha_s^2) \right] \quad (A1.2)$$

where

$$B_I(V) = \frac{(1 + \alpha_s^2)}{[4\alpha_l \alpha_s - (1 + \alpha_s^2)^2]}$$

and

$$B_{II}(V) = \frac{2\alpha_s}{[4\alpha_l \alpha_s - (1 + \alpha_s^2)^2]}.$$

The above expressions were derived under the assumption of elastic fracture behaviour with V representing the instantaneous crack tip speed for nonuniform rates of crack growth. Adding (A1.1) and (A1.2) we obtain:

$$\sigma_{xx} + \sigma_{yy} = (\alpha_l^2 - \alpha_s^2) K_{Id}(t) B_I \sqrt{\frac{2}{\pi r_l}} \cos(\theta_l/2)$$

$$- (\alpha_l^2 - \alpha_s^2) K_{IId}(t) B_{II} \sqrt{\frac{2}{\pi r_l}} \sin(\theta_l/2).$$

Setting now

$$B_I K_{Id}(t)(\alpha_l^2 - \alpha_s^2) = K_I'$$

and

$$B_{II} K_{IId}(t)(\alpha_l^2 - \alpha_s^2) = K_{II}'$$

(A1.3)

we have:

$$\sigma_{xx} + \sigma_{yy} = K_I' \sqrt{\frac{2}{\pi r_l}} \cos(\theta_l/2) - K_{II}' \sqrt{\frac{2}{\pi r_l}} \sin(\theta_l/2). \quad (A1.4)$$

It is worth observing that the above expression describing the first stress invariant near the tip of the propagating crack is of the same functional form as the equivalent static one.

(ii) *The equations of the mapping.* The coordinates of a point $P'(X, Y)$ on the screen which is the image of a point $P(x, y)$ of the specimen are given by eqns (2.5) with respect to the first stress invariant, as follows:

$$X = x + C\frac{\partial}{\partial x_t}[(\sigma_{xx} + \sigma_{yy})(x_t, y_t)]$$
$$Y = y + \alpha_t C\frac{\partial}{\partial y_t}[(\sigma_{xx} + \sigma_{yy})(x_t, y_t)]$$

Substituting expression (A1.4) in the above equations, and using the fact that $x = r_t \cos \theta_t$ and $y_t = \frac{r_t \sin \theta_t}{\alpha_t}$, we get:

$$X = r_t \cos \theta_t + \lambda r_t^{-3/2}\left(K_I' \cos \frac{3\theta_t}{2} - K_{II}' \sin \frac{3\theta_t}{2}\right)$$
$$Y = \frac{1}{\alpha_t}\left(r_t \sin \theta_t + \alpha_t^2 \lambda r_t^{-3/2}\left(K_I' \sin \frac{3\theta_t}{2} + K_{II}' \cos \frac{3\theta_t}{2}\right)\right) \qquad (A1.5)$$

where $\lambda = \frac{C}{(2\pi)^{1/2}}$ and $0 < \theta_t < 4\pi$.

Equations (A1.5) are the governing equations for the mapping of a generic point $P(x, y)$ of the specimen on to a point $P'(X, Y)$ of the screen.

(iii) *The equations of the initial curve.* The condition for the existence of a caustic curve is the vanishing of the Jacobian of the transformation (A1.5).

Performing a calculation similar to the one described in Section 2(ii) for the case of an opening mode crack we get a curve with respect to the (r_t, θ_t) parameters which is the image of the "initial curve" in the (r_t, θ_t) plane. This curve is obtained by setting $J = \frac{\partial(X, Y)}{\partial(x, y)} = \alpha_t \frac{\partial(X, Y)}{\partial(x_t y_t)} = 0$, and in polar form it is given as:

$$r_t^5 - \tfrac{9}{4}\lambda^2 \alpha_t^2 (K_I'^2 + K_{II}'^2) + \tfrac{3}{2}\lambda r_t^{5/2}(\alpha_t^2 - 1)\left[K_I' \cos \frac{5\theta_t}{2} - K_{II}' \sin \frac{5\theta_t}{2}\right] = 0. \qquad (A1.6)$$

The above equations can be expressed in the following form:

$$r^5 - \tfrac{9}{4}\lambda^2 \alpha_t^2 (K_I'^2 + K_{II}'^2) + \tfrac{3}{2}\lambda r_t^{5/2}(\alpha_t^2 - 1)(K_I'^2 + K_{II}'^2)^{1/2} \cos \tfrac{5}{2}(\theta_t + \chi) = 0 \qquad (A1.7)$$

where

$$\chi = \tfrac{2}{5}\tan^{-1}\frac{K_{II}'}{K_I'} = \tfrac{2}{5}\tan^{-1}\left[\frac{K_{IId}(t)}{K_{Id}(t)} \cdot \frac{2\alpha_s}{(1 + \alpha_s^2)}\right]$$

for $K_{IId}(t) = 0$ the above reduces to eqn (2.8) in Section 2(ii). The exact equations of the caustic envelope are therefore given by:

$$X = r_t \cos \theta_t + \lambda r_t^{-3/2}\left(K_I' \cos \frac{3\theta_t}{2} - K_{II}' \sin \frac{3\theta_t}{2}\right)$$
$$Y = \frac{1}{\alpha_t}(r_t \sin \theta_t + \alpha_t^2 \lambda r_t^{-3/2}\left(K_I' \sin \frac{3\theta_t}{2} + K_{II}' \cos \frac{3\theta_t}{2}\right)) \qquad (A1.8)$$

under the constraint:

$$r_t^5 - \tfrac{9}{4}\lambda^2 \alpha_t^2 (K_I'^2 + K_{II}'^2) + \tfrac{3}{2}\lambda r_t^{5/2}(\alpha_t^2 - 1)\left[K_I' \cos \frac{5\theta_t}{2} - K_{II}' \sin \frac{5\theta_t}{2}\right] = 0$$

using the same approximation scheme as the one adopted in Section 3, we will try to find a set of circles each of which constitutes the best fit to (A1.6) for each value of α_t. To do so the area of a random member of the family of images (A1.5) is calculated with respect to r_t. The expression is subsequently differentiated, giving the critical radius $r_t = \rho_0$ corresponding to the minimum area image. The area of a member is given by:

$$E = \frac{1}{2}\int_0^{4\pi} (XY_{\theta_t}' - YX_{\theta_t}')\,d\theta_t.$$

Using expressions (A1.5) we can show that

$$E = \frac{1}{2}\frac{4\pi}{\alpha_t}[r_t^2 - \tfrac{3}{2}\lambda^2(K_I'^2 + K_{II}'^2)\alpha_t r_t^{-3}].$$

The conditions for the minimum gives

$$\left.\frac{\partial E}{\partial r_l}\right|_{r_l = \rho_0} = 0 \Rightarrow \rho_0 = [\tfrac{3}{2}\lambda\sqrt{K_I^{2'} + K_{II}^{'2}}\alpha_l]^{2/5} \tag{A1.9}$$

Thus $\rho_0 = [\tfrac{3}{2}\lambda\alpha_l\sqrt{K_I^{'2} + K_{II}^{'2}}]^{2/5}$ corresponds to the closest circle approximation to the curve represented by equation (A1.6). One can see that the two curves touch at a total of five points. The coordinates of these five points in the (r_l, θ_l) plane depend on the value of the ratio

$$\xi = \frac{K_{II}'}{K_I'} = \frac{2\alpha_s}{(1 + \alpha_s^2)} \cdot \frac{K_{IId}(t)}{K_{Id}(t)}$$

and will be given by substituting

$$r_l = \rho_0 = [\tfrac{3}{2}\lambda\alpha_l\sqrt{K_I^{'2} + K_{II}^{'2}}]^{2/5}$$

in (A1.6). The substitution gives:

$$\cot\left(\frac{5\theta_l}{2}\right) = \frac{K_{II}'}{K_I'} = \xi. \tag{A1.10}$$

For the special case of $K_{IId}(t) = 0 \Rightarrow K_{II}' = 0$, we have seen in Section 3, that θ_l takes the values of $\pi/5, 3\pi/4, \pi, 7\pi/5, 9\pi/5$ which are independent of the velocity of the crack.

For non-zero $K_{IId}(t)$, the coordinates of the five common points will be functions of both $K_{IId}(t)$ and α_s as shown in (A1.10).

Figure 7 shows the effect of a finite $K_{IId}(t)$. The axis of symmetry of the curve described by (A1.6) (dotted line) in the (x_l, y_l) plane, will be rotated through an angle $\chi = \tfrac{2}{5}\tan^{-1}\left[\frac{K_{IId}(t)}{K_{Id}(t)} \cdot \frac{2\alpha_s}{(1 + \alpha^2)}\right]$ with respect to the x_l axis, which coincides with the tangent to the crack tip.

(iv) *Best approximation of the initial curve in the* (x, y) *plane*. As shown above, the equation of the best circle approximation to the curve (A1.6) (image of the initial curve in the (x_l, y_l) plane) is given by:

$$x_l^2 + y_l^2 = [\tfrac{3}{2}\alpha_l\lambda\sqrt{K_I^{'2} + K_{II}^{'2}}]^{4/5}.$$

In the (x, y) specimen plane, the corresponding points (r, θ) lie on the ellipse

$$x^2 + \alpha_l^2 y^2 = [\tfrac{3}{2}\alpha_l\lambda\sqrt{K_I^{'2} + K_{II}^{'2}}]^{4/5} = \rho_0^2 \tag{A1.11}$$

which can be expressed as:

$$x^2 + \alpha_l^2 y^2 = \left[\frac{3}{2}\left(\frac{(1+\alpha_s^2)(\alpha_l^2 - \alpha_s^2)}{4\alpha_l\alpha_s - (1+\alpha_s^2)^2}\right)\alpha_l \frac{dvz_0}{E(2\pi)^{1/2}}K_{Id}(t)\right]^{4/5}(1+\xi^2)^{2/5} = \rho_0^2.$$

For $K_{IId}(t) = 0 \Rightarrow \xi = 0$, the above reduces to eqn (4.1) obtained for a mode I crack.

(v) *The behavior of the initial curve as* $V \to 0$. Letting $V \to 0$, both $\alpha_l, \alpha_s \to 1$ and $K_I' \to K_I, K_{II}' \to K_{II}$. Equation (A1.11) therefore reduces to

$$x^2 + y^2 = [\tfrac{3}{2}\lambda\sqrt{K_I^2 + K_{II}^2}]^{4/5} = \rho_{st}^2$$

which is the known equation of the initial curve obtained using the stress field of a stationary crack [4].

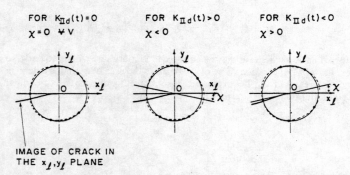

Fig. 7. The effect of a finite $K_{IId}(t)$: Dotted line: the image of the initial curve in the (x_l, y_l) plane: Solid line: its best circle approximation.

A2. (i) *The approximate equations of the caustic curve*

After the approximation is made, eqn (A1.8) become:

$$\begin{aligned} X &= r_t \cos\theta_t + \lambda r_t^{-3/2}\left(K_I' \cos\frac{3\theta_t}{2} - K_{II}' \sin\frac{3\theta_t}{2}\right) \\ Y &= \frac{1}{\alpha_t}(r_t \sin\theta_t + \alpha_t \lambda r_t^{-3/2}\left(K_I' \sin\frac{3\theta_t}{2} + K_{II}' \cos\frac{3\theta_t}{2}\right) \end{aligned} \quad (A2.1)$$

where

$$r_t = [\tfrac{3}{2}\lambda\alpha_t\sqrt{K_I'^2 + K_{II}'^2}]^{2/5} = \rho_0.$$

Equation (A2.1) can be expressed as follows:

$$\begin{aligned} X &= \rho_0\left[\cos\theta_t + \frac{2}{3}\frac{1}{(1+\xi^2)^{1/2}}\cdot\frac{1}{\alpha_t}\cos\frac{3\theta_t}{2} - \frac{2}{3}\frac{\xi}{(1+\xi^2)^{1/2}}\frac{1}{\alpha_t}\sin\frac{3\theta_t}{2}\right] \\ Y &= \frac{\rho_0}{\alpha_t}\left[\cos\theta_t + \frac{2}{3}\frac{1}{(1+\xi^2)^{1/2}}\alpha_t\sin\frac{3\theta_t}{2} + \frac{2}{3}\frac{\xi}{(1+\xi^2)^{1/2}}\alpha_t\cos\frac{3\theta_t}{2}\right] \end{aligned}$$

where

$$\xi = \frac{K_{II}'}{K_I'}.$$

Setting now $\xi = \tan\phi'$ the equations of the caustic curve can be finally expressed as:

$$\begin{aligned} X &= \rho_0\left[\cos\theta_t + \tfrac{2}{3}\alpha_t^{-1}\cos\left(\frac{3\theta_t}{2} + \phi'\right)\right] \\ Y &= \frac{\rho_0}{\alpha_t}\left[\sin\theta_t + \tfrac{2}{3}\alpha_t\sin\left(\frac{3\theta_t}{2} + \phi'\right)\right] \end{aligned} \quad (A2.2)$$

where

$$\phi' = \tan^{-1}\frac{K_{II}'}{K_I'} = \tan^{-1}\left[\frac{2\alpha_s}{1+\alpha_s^2}\frac{K_{IId}(t)}{K_{Id}(t)}\right].$$

The above equations are the parametric equations of the caustic curve for a case of a mixed mode propagating crack.

(ii) *Behavior as $V \to 0$*. As $V \to 0$ and $\alpha_{l,s} \to 1$, $\omega' \to \tan^{-1}\left(\frac{K_{II}}{K_I}\right) = \phi$, $\theta_t \to \theta$, $\rho_0 \to \rho_{st}$ and the above equations become:

$$\begin{aligned} X &= \rho_{st}\left[\cos\theta + \frac{2}{3}\cos\left(\frac{3\theta}{2} + \phi\right)\right] \\ Y &= \rho_{st}\left[\sin\theta + \frac{2}{3}\sin\left(\frac{3\theta}{2} + \phi\right)\right] \end{aligned} \quad (A2.3)$$

which, as expected, are the equations of the caustic curve for the case of a stationary mixed mode crack [4].

ON CRACK-TIP STRESS STATE: AN EXPERIMENTAL EVALUATION OF THREE-DIMENSIONAL EFFECTS

ARES J. ROSAKIS[†] and K. RAVI-CHANDAR[‡]

Graduate Aeronautical Laboratories, California Institute of Technology, Pasadena, CA 91125, U.S.A.

(Received 28 April 1984; in revised form 22 March 1985)

Abstract—The extent of the region of three-dimensionality of the crack-tip stress field is investigated using reflected and transmitted caustics. The range of the applicability of two-dimensional near tip solutions is thus established experimentally. The experiments are performed using Plexiglass and high-strength 4340-steel compact tension specimens. A wide spectrum of thicknesses is investigated. At each thickness, measurements are performed at a variety of distances r from the crack tip, ranging from $r/h = 0$ to $r/h = 2$, where h is the specimen thickness. The results indicate that plane-stress conditions prevail at distances from the crack tip greater than half the specimen thickness, while *no* significant plane-strain region is detected. The experimental results are also compared to the crack-tip boundary-layer solution of Yang and Freund[1], and the numerical results of Levy, Marcal and Rice[2]. Their solutions are consistent with the results of this work and approach the plane stress field at $r/h = 0.5$. In addition, and unlike what might be commonly expected, the analytical solution[1] exhibits no plane-strain behavior very near the crack tip. This behavior is in good agreement with the results of both the transmission and the reflection experiments.

1. INTRODUCTION

It has long been recognized that in plates of finite thickness, the stress field at the vicinity of stress raisers is three-dimensional in nature. The problem of a hole in an infinite plate has been successfully addressed in the works of Sternberg and Sadowsky[3] and Alblas[4]. The more complex problem of a crack in a thick plate has been attempted by a number of authors[5-8], but no complete solution has been obtained as yet.

The purpose of this paper is not to solve this complex problem in detail, but rather to explore regions where two-dimensional approximations would be acceptable. In particular, we attempt to identify the regions in which local experimental measurements based on two-dimensional theory can be performed with confidence. Such local measurements are based on the use of photoelasticity, interferometry or caustics (shadowgraphy), where data are usually obtained at some characteristic distance r from the tip. The constraint on the location of r, if a two-dimensional view point is adopted, is the existence of a K-dominant field (negligible high-order term effects and contained yielding). However, in plates of finite thickness, three-dimensional effects near the tip must assume a role at least as important as the above constraints. In the sequel such three-dimensional effects were investigated for Plexiglas in transmission and 4340 steel in reflection, using the optical method of caustics.

Recently, Yang and Freund[1], motivated by similar concerns regarding experimental measurements made in the vicinity of the crack tip, have explored the three-dimensional crack problem using a boundary-layer approach. Their analytical results have a close relationship to the present investigation, and will be compared in detail in Section 5. Also, the numerical results of the same problem presented by Levy, Marcal and Rice[2] will be compared with experiment.

[†] Assistant Professor, Graduate Aeronautical Laboratories, California Institute of Technology, Pasadena, CA 91125, U.S.A.

[‡] Formerly, Research Fellow, Caltech. Assistant Professor, Department of Mechanical Engineering, University of Houston, TX 77004, U.S.A.

2. THEORETICAL CONSIDERATIONS

Consider a semi-infinite mode-*I* crack in a homogeneous isotropic linear-elastic solid of uniform undeformed thickness h. The crack is assumed to be mathematically sharp with a straight crack front. If a local Cartesian coordinate system is introduced at the crack with its origin situated at the middle of the crack front (see Fig. 1), then the stress field at the vicinity of the crack tip can be expressed as a function of the dimensionless coordinates x_3/h, r/h, as well as the angular variation ϑ. In particular

$$\sigma = \sigma(x_3/h, r/h, \vartheta), \tag{2.1}$$

where $r^2 = x_1^2 + x_2^2$, $\vartheta = \tan^{-1}(x_2/x_1)$.

The choice of h as a normalization parameter is dictated by the observation that, for the problem considered, the thickness is the only relevant length scale. In addition, expression (2.1) must be such as to reduce to the familiar two-dimensional singular asymptotic stress field when conditions of either plane stress or plane strain are approached. More specifically, the value of the nondimensional variable r/h must reflect the changing nature the stress field as the crack front is approached from infinity. For $r/h \to \infty$, plane-stress–like conditions are expected to prevail—this being true for either very thin specimens and/or for large distances from the crack line. Also for $r/h \to 0$, and $x_3/h \neq \pm \frac{1}{2}$, a plane-strain type of field is expected to dominate†—this being true for either thick specimens and/or small distances from the crack line. Thus, in general,

$$\sigma = \sigma(x_3/h, r/h, \vartheta) \to \frac{K_{2D}}{\sqrt{2\pi r}} \cdot f(\vartheta) \begin{Bmatrix} K_{2D} & \text{for} & r/h \to \infty \\ K_E & \text{for} & r/h \to 0 \end{Bmatrix}. \tag{2.2}$$

where K_{2D} and K_E are the mode-*I* plane-stress and plane-strain stress-intensity factors, respectively, and $f(\vartheta)$ is a tensor function of ϑ, completely determined by the two-dimensional asymptotic near-tip analysis[9]. The region near the tip, where r/h is neither very large nor very small, is in general an area where three-dimensional effects are expected to be very strong. This region becomes particularly important in fracture mechanics for the following two reasons: First, from the theoretical point of view, accurate solutions describing the stress behavior there have been particularly elusive. Second, from the experimental point of view, it seems that despite its great three-

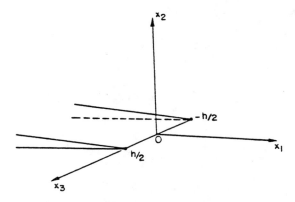

Fig. 1. Semi-infinite crack in plate of thickness h.

---

† As $r/h \to 0$ and $x_3/h = \pm \frac{1}{2}$ (points on specimen surface near the crack front) the nature of the singularity deviates from its two-dimensional nature due to the existence of a free corner there[5–7]. On the other hand, for $x_3/h \neq \pm \frac{1}{2}$ (points inside the specimen) conditions of plane strain are expected to dominate if the crack front is approached close enough.

dimensionality, this region has been selected in a number of experiments as a site of measurements, unfortunately interpreted on the basis of two-dimensional solutions.

In this work we will not attempt the solution of the very complicated three-dimensional problem. Instead, we have chosen to investigate the size of the annular zone of three-dimensional effects. In order to do so, we have introduced a couple of simplifying assumptions, which do not affect the three-dimensional nature of the field. Our assumptions do not predict the detailed character of the three-dimensional state, and have been introduced on the basis of qualitative preliminary experimental observations, which will be discussed later.

In particular we have assumed that the components of the stress field can be expressed as follows:

$$\sigma_{\alpha\beta} = K_1(x_3/h, r/h) \cdot \frac{1}{\sqrt{2\pi r}} \cdot \Sigma_{\alpha\beta}(\vartheta),$$

$$\sigma_{\alpha 3} = K_2(x_3/h, r/h) \cdot \frac{1}{\sqrt{2\pi r}} \cdot \Sigma_{\alpha 3}(\vartheta), \qquad (2.3)$$

$$\sigma_{33} = K_3(x_3/h, r/h) \cdot \frac{1}{\sqrt{2\pi r}} \cdot \Sigma_{33}(\vartheta).$$

where $K_i(x_3/h, r/h)$, $i = 1, 2, 3$ are unknown functions of x_3/h and r/h. In eqns (2.3), and in the remaining paper, Greek indices will have the range 1, 2.

Expressions (2.3) assume that only the ϑ dependence is separable, and do *not* provide any information concerning the nature of the r/h dependence of the stress field.

The separability of the ϑ dependence, as well as the requirement that expressions (2.3) reduce to eqns (2.2) for $r/h \to 0$ and $r/h \to \infty$, imply that

$$\Sigma_{11} = f_{11} = \cos\frac{\vartheta}{2}[1 - \sin\vartheta/2 \sin 3\vartheta/2],$$

$$\Sigma_{22} = f_{11} = \cos\frac{\vartheta}{2}[1 + \sin\vartheta/2 \sin 3\vartheta/2], \qquad \forall r/h, \qquad (2.4)$$

$$\Sigma_{12} = f_{12} = \sin\vartheta/2 \cos\vartheta/2 \cos 3\vartheta/2,$$

$$\Sigma_{33} = f_{33} = 2\nu \cos\vartheta/2.$$

Also for the plane-stress limit, as $r/h \to \infty$, $K_{1,3}(x_3/h, r/h) \to K_{2D}$ and $K_2(x_3/h, r/h) \to 0$. And that for the plane-strain limit, as $r/h \to 0$, $K_1(x_3/h, r/h) \to K_E$ and $K_{2,3}(x_3/h, r/h) \to 0$. Thus the governing equations can be written as

$$\sigma_{\alpha\beta} = K_{2D}\hat{K}_1(x_3/h, r/h) \frac{f_{\alpha\beta}(\vartheta)}{\sqrt{2\pi r}},$$

$$\sigma_{\alpha 3} = K_{2D}\hat{K}_2(x_3/h, r/h) \frac{f_{\alpha 3}(\vartheta)}{\sqrt{2\pi r}}, \qquad (2.5)$$

$$\sigma_{33} = K_{2D}\hat{K}_3(x_3/h, r/h) \frac{f_{33}(\vartheta)}{\sqrt{2\pi r}},$$

where

$$\hat{K}_i(x_3/h, r/h) = \frac{K_i(x_3/h, r/h)}{K_{2D}}.$$

3. THE METHOD OF CAUSTICS

Physical principle

Consider a family of light rays parallel to the x_3 axis, incident on a planar specimen whose midplane lies at $x_3 = 0$ (cf. Fig. 2). Upon reflection from an opaque specimen, or refraction through a transparent specimen, the rays undergo an optical path change dictated by the stress field, and in general will deviate from parallelism. Under certain stress gradients, the reflected or refracted rays will form an envelope in the form of a three-dimensional surface in space. This surface, which is called the *caustic surface*, is the locus of points of maximum luminosity in the reflected or transmitted light fields. The deflected rays are tangent to the caustic surface. If a screen is positioned parallel to the $x_3 = 0$ plane, and so that it intersects the caustic surface, then the cross-section of this surface can be observed as a bright curve (the so-called *caustic curve*) bordering a dark region (the *shadow spot*) on the screen. Suppose that the incident ray, which is reflected from or transmitted through point $p(x_1, x_2)$ on the specimen, intersects the screen at the image point $P(X_1, X_2)$ (cf. Fig. 2.). The X_1, X_2 coordinate system is identical to the x_1, x_2 system, except that the origin of the former has been translated the distance z_0 to the screen (z_0 can be either positive or negative). The position of the image-point P will depend on the gradient of the optical path change ΔS, introduced by the medium and on the distance z_0, and is given by [10]

$$X_i = x_i \pm z_0 \frac{\partial(\Delta S)}{\partial x_i}. \tag{3.1}$$

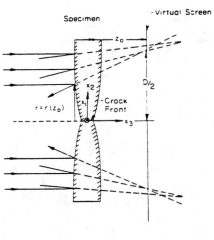

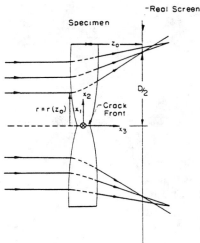

Fig. 2. Schematic of the formation of the caustics by (a) reflection and (b) transmission.

For the reflection case, $\Delta S = -2u_3(x_1, x_2, h/2)$ where u_3 is the displacement in the x_3 direction.

Equation (3.1) is a mapping of the points on the specimen onto points of the screen. If the screen intersects the caustic surface, then the resulting caustic curve on the screen is a locus of points of multiple reflection. That is, for points on the caustic curve, the mapping (3.1) is not invertible and the determinant of the Jacobian matrix of the transformation must vanish, i.e.

$$J(x_1, x_2; z_0) = \frac{\partial(X_1, X_2)}{\partial(x_1, x_2)} = 0. \tag{3.2}$$

Equation (3.2) is the necessary and sufficient condition for the existence of a caustic curve. The points on the specimen for which the Jacobian vanishes are the points from which the rays forming the caustic curve emerge. These rays are tangent to the caustic surface at its intersection with the plane of the screen. The locus of the points on the specimen mapped onto the caustic curve is called the *initial curve*. Equation (3.2) describes the initial curve and its strong parametric dependence on z_0. Rays emerging from within the initial curve map outside the caustic curve, and therefore only the rays emerging from the initial curve determine the geometry of the caustic curve. This characteristic of the optical mapping is vital to this work. By varying the position z_0 of the screen, the size of the initial curve (radius r) is varied. Thus the near-tip topology can be investigated at various distances r from the crack tip.

Caustics by reflection

For opaque specimens, caustics are formed by the reflection of light rays from the highly polished specimen surface. The shape of the caustic curve depends on the near-tip normal displacement u_3 of the plate face initially at $x_3 = h/2$, which is given by

$$u_3(x_1, x_2, h/2) = h \int_0^{1/2} \epsilon_{33}(x_1/h, x_2/h, x_3/h) \, d(x_3/h).$$

where ϵ_{33} is the direct strain in the thickness direction. For a linearly elastic solid,

$$\epsilon_{33} = \frac{1}{E}[\sigma_{33} - \nu(\sigma_{11} + \sigma_{22})], \tag{3.3}$$

and thus using (2.5),

$$u_3(r/h, \vartheta, h/2) = \frac{2h\nu K_{2D}}{E\sqrt{2\pi r}} \rho(r/h) \cos(\vartheta/2) \tag{3.4}$$

where

$$\rho(r/h) = \int_0^{1/2} [\hat{K}_3(x_3/h, r/h) - \hat{K}_1(x_3/h, r/h)] \, d(x_3/h) \equiv \frac{K^{\text{EXP}}(r/h)}{K_{2D}}.$$

We have chosen to denote the function $K_{2D}\rho(r/h)$ as K^{EXP}, since it may be possible to measure this quantity for different values of r/h. Appropriate variations of z_0 allow us to obtain initial curves at different distances r from the crack tip.

Caustic by transmission

For transparent specimens the optical path change ΔS depends on both local changes in thickness and on local changes of refractive index, and can be expressed in a manner similar to the two-dimensional case in [10] as

$$\Delta S(x_1, x_2) = 2(n - 1)h \int_0^{1/2} \epsilon_{33} \, d(x_3/h) + 2h \int_0^1 \Delta n \, d(x_3/h), \quad (3.5)$$

where the change in refractive index Δn is given by

$$\Delta n = A(\sigma_{11} + \sigma_{22} + \sigma_{33}),$$

and where A is the stress-optic constant[10]. Substituting the above together with (2.5) and (3.3) into (3.5) gives

$$\Delta s(r/h, \vartheta) = \frac{2hc_\sigma K_{2D}}{\sqrt{2\pi r}} \rho'(r/h) \cos \vartheta/2, \quad (3.6)$$

where

$$\rho'(r/h) = \int_0^{1/2} \left(\hat{K}_1(x_3/h, r/h) + \frac{\xi}{c_\sigma} \hat{K}_3(x_3/h, r/h) \right) d(x_3/h) \equiv \frac{K^{\text{EXP}}(r/h)}{K_{2D}},$$

and

$$c_\sigma = \left[A - (n - 1) \frac{v}{E} \right], \quad \text{plane-stress optical constant.}$$

$$\xi = \left[(n - 1) \frac{v}{E} + Av \right].$$

Evaluation of the mapping

Substituting eqn (3.4) or (3.6) into the equation of the mapping (3.1), one obtains

$$X_1 = r \cos \vartheta + \gamma K^{\text{EXP}} r^{-3/2} \cos 3\vartheta/2 + \gamma \frac{\partial K^{\text{EXP}}}{\partial r} r^{-1/2} \cos \vartheta/2 \cos \vartheta,$$

$$X_2 = r \sin \vartheta + \gamma K^{\text{EXP}} r^{-3/2} \sin 3\vartheta/2 + \frac{\gamma \partial K^{\text{EXP}}}{\partial r} r^{-1/2} \sin \vartheta/2 \sin \vartheta. \quad (3.7)$$

where

$$\gamma = \begin{cases} \dfrac{z_0 v h}{(2\pi)^{1/2} E}, & K^{\text{EXP}} = \rho(r/h) K_{2D} \quad \text{for reflection,} \\ \dfrac{z_0 c_\sigma h}{(2\pi)^{1/2}}, & K^{\text{EXP}} = \rho'(r/h) K_{2D} \quad \text{for transmission.} \end{cases}$$

The first two terms of the right-hand side are the traditional two-dimensional caustic terms.

The presence of the third term should alter the geometry of the caustic curve and therefore the relation between $K^{\text{EXP}}(r/h)$ and the dimensions of the caustic. However, extensive experimental observations of the caustic geometry (in particular, the aspect ratio) have demonstrated that the epicycloidal shape expected on the basis of the first two terms is retained (cf. Fig. 3). The photographs in Fig. 3 demonstrate that despite drastic changes in r/h, the caustic-curve shape remains unchanged. This suggests that the effect of the third term in expression (3.7) must be small compared to the second

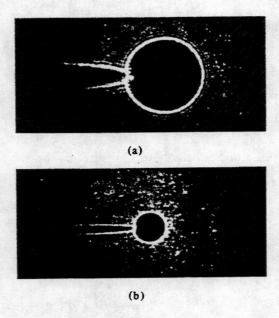

Fig. 3. Comparison of two caustic-curve shapes obtained at different distances from the crack tip. Both photographs correspond to the same experiment: (a) $r/h = 0.9$: (b) $r/h = 0.15$

term. This means that the ratio of the two terms is small. implying that

$$\frac{1}{K^{\text{EXP}}} \frac{\partial K^{\text{EXP}}}{\partial r} = o\left(\frac{1}{r}\right) \quad \text{as} \quad r \to 0. \tag{3.8}$$

Assuming the validity of eqn (3.8). one recovers from eqn (3.7) the traditional caustic equations. with the exception that the experimentally determined stress intensity K^{EXP} now varies with r/h. The caustic curve is then an epicycloid. and the "initial curve" is a circle of radius r. The radius of the initial curve r depends on the choice of z_0, and was varied during the experiment. By changing z_0. a number of optical measurements were performed at different distances from the crack tip.

The initial curve

As noted earlier in this section. the condition for the existence of a caustic curve on a screen at $x_3 = z_0$, is the vanishing of the Jacobian of the transformation (3.7). With reference to eqn (3.2), and assuming the validity of eqn (3.8), the condition that the determinant of the Jacobian matrix must vanish becomes

$$J = r - \left(\frac{3}{2} \gamma K^{\text{EXP}}\right)^{2/5} \tag{3.9}$$

Equation (3.9) describes the initial curve on the specimen surface. Then substitution of eqn (3.9) into eqn (3.7) yields the equation of the corresponding caustic curve in the X_1, X_2 plane, parametric in angle ϑ. For the case under consideration, the equation of the caustic curve becomes

$$X_1 = r[\cos \vartheta + 2/3 \cos 3\vartheta/2],$$
$$X_2 = r[\sin \vartheta + 2/3 \sin 3\vartheta/2], \tag{3.10}$$
$$-\pi < \vartheta < \pi,$$

where r is constrained by eqn (3.9). Equation (3.10) is the equation of an epicycloid, whose shape remains unchanged as r is varied. This result was confirmed (see Fig. 3) during the experiment, and thus assumption (3.8) was justified.

On the other hand, the absolute size of the caustic curve depends on the value of K^{EXP}, the bulk material properties, z_0, and the optical constants. In fact, eqn (3.10) is a relationship among all of these parameters. Thus if the values of the material, geometrical and optical parameters are known, then eqn (3.10) provides a relationship between the size of the caustic curve and the local value of K^{EXP}. Adopting the maximum transverse diameter $D/2 = \max(X_2)$ as the characteristic dimension of the caustic curve, the value of K^{EXP}, measured at a distance r (initial-curve radius) from the crack tip, can be expressed as

$$K^{EXP} = \frac{9.34 \times 10^{-2} D^{5/2}}{z_0 ch}, \qquad (3.11)$$

where

$$c = \begin{cases} c_\sigma & \text{for transmission.} \\ v/E & \text{for reflection.} \end{cases}$$

Also, from eqns (3.9) and (3.11), it can be demonstrated that the maximum transverse diameter of the caustic is related to the radius of the initial curve by the formula

$$D = 3.163r. \qquad (3.12)$$

4. EXPERIMENTAL CONSIDERATIONS

In order to undertake an experimental investigation of thickness effects, careful experiments have to be designed so as to isolate this effect from other geometry-related effects on the stress field. This requirement implies that the only length scale in the problem is the thickness, or, equivalently, that the in-plane dimensions be large in comparison to the thickness. However, cost considerations limit the in-plane dimensions, especially when large thicknesses are involved and a compromise between the more important mechanics considerations and cost essentially dictates the sizes of the specimens. The final choice of specimen geometry is illustrated in Fig. 4. Two materials were chosen for these experiments. Plexiglas was selected for investigation in transparent cases because of its easy availability in various thicknesses. For investigating opaque specimens, a 4340 martensitic steel, heat treated to 843°C and oil quenched, was selected because its high yield strength precluded plasticity effects that might hinder the thickness-effect investigation.

The plane stress-intensity factor for the geometry shown in Fig. 4 can be obtained from analysis, and is given by [11]

$$K_{2D} = \frac{P\sqrt{a}}{h} F_1\left(\frac{a}{b}, \frac{l}{b}, \frac{d}{l}\right). \qquad (4.1)$$

where F_1 is a function of crack length a, ligament length d, specimen height l, and the distance between loading points $2d$. Also, P is the applied load. This value of K_{2D} will be used in the sequel to compare with the experimentally determined values of the stress-intensity factor, K^{EXP}. The method of caustics is used to determine the value of K^{EXP}. The analytical basis for the use of the traditional caustics relations has been discussed in Section 3. Thus

$$K^{EXP} = \frac{9.34 \times 10^{-2}}{z_0 ch} D^{5/2}. \qquad (4.2)$$

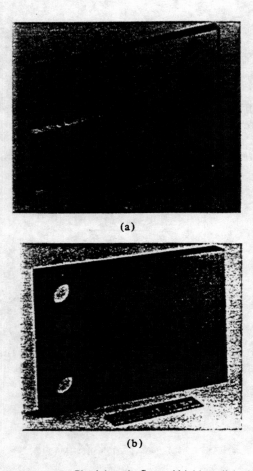

Fig. 4. The specimen geometry: (a) Plexiglas; (b) flat and highly polished Martensitic 4340 carbon steel.

where z_0 is the reference distance, h is the plate thickness, D is the maximum transverse diameter of the caustic, and c is the appropriate optical constant: c_σ for plane stress, c_ϵ for plane strain, or v/E for reflection. D is related to the initial-curve radius r by the formula $D = 3.163\ r$, see Section 3.

Two sets of experiments were performed: the first set with transparent Plexiglas, and the second set with martensitic 4340 steel. For the Plexiglas experiments, the plane-stress value of $c = c_\sigma = 1.08 \times 10^{-10}$ m^2/MN[10] was used in evaluating K^{EXP}. For the steel specimen, the value of $c = v/E$, where v is the Poisson's ratio, and E is the Young's modulus. In these two sets of experiments, there is a fundamental difference in the way that information is extracted experimentally. In Plexiglas, the light rays travel through the specimen and emerge with the integrated effect of the stress field in the specimen. Whereas in the opaque specimen, the strain ϵ_{33} accumulates through the thickness, thereby presenting a dimpled surface from which light rays reflect to form the caustic curve. In other words, the Plexiglas experiments provide an optically-averaged value of K^{EXP}, and the 4340-steel experiments give a materially-averaged value of K^{EXP}. However, since both materials have similar Poisson's effect, one might reasonably expect that the thickness effects in the two cases might present similarities.

The specimen preparation was a major part of the whole experiment. While the exterior geometry is easily machined, achieving proper crack-tip conditions is of primary importance. In order to compare the experimental results obtained for various thicknesses, the crack front must not possess any curvature. This was easily achieved in steel specimens where the crack was made using an EDM spark-cutter. The metallic specimen is shown in Fig. 4(a). For the Plexiglas specimens, after a lot of trial and error, it was found that the best way of producing straight crack fronts was to start with a machined Chevron notched crack front and grow the crack front under quasi-

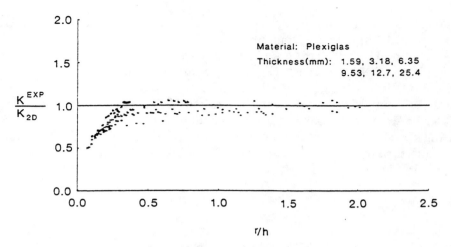

Fig. 5. Ratio of stress-intensity factor inferred from local shadow spot measurements (transmitted caustics) to analytical two-dimensional value, vs distance from crack tip divided by specimen thickness. Material: Plexiglas.

static conditions, rather than cyclic loading conditions. The resulting crack front from a specimen of 6.35-mm thickness is shown in Fig. 4(b).

In order to use the method of caustics in an opaque material, the undeformed surface of the specimen has to be a flat reflector. This was achieved by grinding, lapping and polishing the specimen to have a half-a-wavelength surface flatness. For Plexiglas, no special surface preparation was necessary. In the experiment, a collimated laser beam passes through the specimen thickness in the transparent case and reflects from the deformed surface in the opaque case. Due to the specimen deformation, the originally collimated light beam forms a caustic surface, as discussed in Section 3. A camera is focused on a plane at a distance z_0 from the specimen, and the caustic curve on that plane is photographed. From the diameter of the caustic, the stress-intensity factor is calculated using eqn (4.2). This corresponds to making a measurement at a distance r from the crack tip, where r is the initial-curve radius. Since there are no light rays inside the caustic surface (cf. Fig. 2), this implies that the light rays that pass through the specimen at radii less than r do not provide any information in this reference plane. Therefore, by varying the distance z_0, it is seen from Fig. 2 that measurements of K^{EXP} may be made at various distances r from the crack tip. This is easily done by changing the plane-of-focus of the camera. However, in order to retain the same magnification, the camera itself was moved to obtain different z_0 and r values.

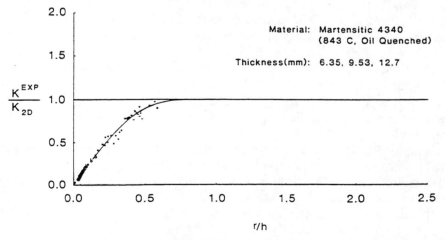

Fig. 6. Ratio of stress-intensity factor inferred from local shadow spot measurements (reflected caustics) to analytical two-dimensional value, vs distance from crack tip divided by specimen thickness. Material: Martensitic 4340 carbon steel.

Plexiglas specimens, of thicknesses 1.59–25.4 mm, were tested in an Instron loading machine at loads varying from 100 to 250 N. The z_0 distance was varied between 0.005 and 10 m. This resulted in a r/h variation from about 0.01 to 2. The results are plotted in normalized variables. The experimentally obtained K^{EXP} is normalized by K_{2D} and plotted as a function of r/h in Fig. 5 for Plexiglas. Three thicknesses (6.35, 9.53 and 12.7 mm) of the 4340 steel were tested in reflection at loads varying from 1 to 16 KN. The ratio of r/h was varied from about 0 to 0.6. In the reflection arrangement, increasing r/h above 0.6 proved to be increasingly difficult due to limitations in the experimental arrangement. The experimental results of K^{EXP}/K_{2D} are plotted as a function of r/h in Fig. 6.

5. DISCUSSION

The most striking feature of the experimental results displayed in Figs. 5 and 6 is the r/h similarity. In particular, the Plexiglas results contain data from 12 different tests using six thicknesses. Each of the tests covers different but overlapping range of r/h. In all cases, the K^{EXP}/K_{2D} exhibits the same r/h dependence. Also, in the case of the 4340 steel, r/h is the scaling parameter. This common r/h dependence will be referred to here as the "master curve", which, however, is different in transmission and reflection. Furthermore, it is clear from both the curves that for $r/h > 0.5$, K^{EXP} approaches K_{2D}. Since a plane-stress caustics analysis was used in evaluating K^{EXP}, and in view of the r/h similarity, it is apparent that *plane-stress conditions prevail at distances from the crack tip which are greater than half the specimen thickness.*

For $r/h < 0.5$, three-dimensional effects seem to take over and the validity of a two-dimensional analysis becomes questionable. Moreover, experimental measurements based on two-dimensional analyses, if performed at $r/h < 0.5$, are likely to contain large errors which increase as r/h decreases. The usual practice has been to make local measurements at very small distances from the tip, to ensure a K-dominant two-dimensional asymptotic field (implying negligible higher-order term effects). As demonstrated in Fig. 5, *the error due to the three-dimensional effects in the near-tip region ($r/h < 0.5$) seems to be more important than the neglect of higher-order terms in the far field* ($0.5 < r/h < 2.0$), where the plane-stress result seems to be valid.

The existence of a region with strong three-dimensional effects is not surprising, and was anticipated in Section 2. Also, as r/h decreases, a plane-strain regime is expected, particularly on the basis of plane-strain fracture toughness concepts. The results displayed in Fig. 5 and 6 ought to identify the extent of a plane-strain region. In transmission K^{EXP}/K_{2D} must tend to the ratio of $c_\epsilon/c_\sigma < 1$, since the plane-stress analysis of caustics was used. The master curve of Fig. 5 shows that K^{EXP}/K_{2D} is continuously decreasing below $c_\epsilon/c_\sigma = 0.7$, instead of reaching a constant value over a substantial range of r/h. This demonstrates that *the region where plane-strain conditions are likely to prevail is surprisingly small.* In reflection, K^{EXP}/K_{2D} must tend to zero under plane-strain conditions, since reflections from a flat surface will not produce a caustic ($\partial u_3/\partial r = 0$). In Fig. 6, it is seen that K^{EXP}/K_{2D} indeed goes to zero, but *not* over a significant range of r/h (i.e. it goes to zero only right at the crack tip). This is consistent with the results of transmission in indicating that the extent of the possible plane-strain region is *small*.

To graphically illustrate the consequences of the experimental result in the reflecting specimen, Fig. 6 was replotted in terms of the variation of $u_3(r/h, \vartheta = 0, x_3 = h/2)$ as a function of r/h, using eqn (3.4). This provides the profile of the deformed-specimen surface at the vicinity of the tip. This profile is plotted in Fig. 7. It demonstrates that deviations from the plane-stress solution occur for $r/h = 0.5$. This deviation can be expected on the grounds that the plane-stress solution predicts a physically unacceptable unbounded displacement u_3 as $r/h \to 0$.

As r/h decreases below 0.5, a minimum in the displacement profile is indicated at $r/h = 0.4$. For $r/h \to 0$, the present experimental results displayed in Fig. 7 indicate

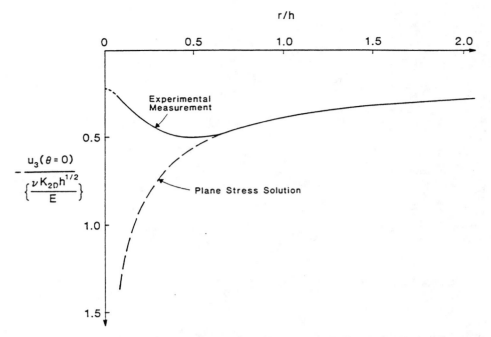

Fig. 7. Normalized out-of-plane displacement. $u_3(x_1, 0, h/2)$, vs distance ahead of the crack tip divided by specimen thickness (obtained by reflected caustics measurements).

that u_3 may not vanish at the crack tip. Further experimental work using interferometry is underway to investigate the details of this region.

The experimental results discussed above are in excellent agreement with the finite-element calculations of Levy, Marcal and Rice[2] and the recent boundary-layer solution of Yang and Freund[1]. Some of the results of [2] are displayed in Fig. 8, where it is demonstrated that σ_{33} approaches zero at $r/h = 0.5$, indicating the establishment of plane-stress conditions at this distance. In the work by Yang and Freund, the state of stress in an elastic plate containing through-cracks is investigated with a view toward assessing the influence of transverse shear on the crack-tip stress and deformation fields. A crack-tip boundary-layer solution is thus obtained, based on the assumption of uniform through the thickness extensional strain. Their results are displayed in Fig. 9 and 10.

In Fig. 9, the nondimensional out-of-plane displacement $u_3(x_1, 0, h/2)$, is plotted vs $(r/\epsilon)^{1/2}$, which is to be compared with our Fig. 7. (ϵ is a length parameter proportional to h.) In Fig. 10, the ratio of $u_3(x_1, 0, h/2)$ and the corresponding plane-stress value is plotted vs $(r/h)^{1/2}$. This ratio is proportional to K^{EXP}/K_{2D} and can be compared to Fig. 6.

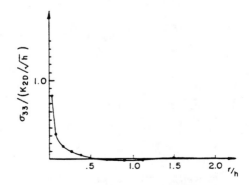

Fig. 8. Variation of midplane σ_{33} with normalized distance from the crack tip (from Levy, Marcal and Rice[2]).

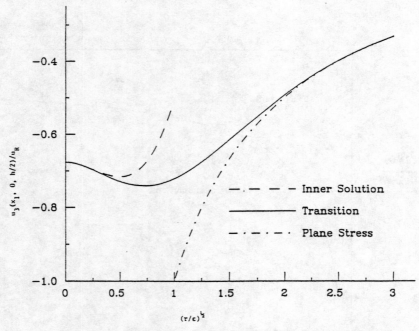

Fig. 9. Normalized out-of-plane displacement, $u_3(x_1, 0, h/2)$ showing inner boundary-layer transition and outer plane-stress regions vs normalized distance from the crack tip. Analytical result by Wang and Freund[1].

The analytical results are in good agreement with the present experiments in a number of issues. In particular, their solution is found to merge smoothly with the plane-stress result, at distances from the tip of one-half the plate thickness. Figure 9 also displays a minimum in the displacement profile, but the location of this minimum is closer to the crack tip than in the present experimental results. Finally, and as demonstrated in Fig. 8, the lateral contraction u_3 on the specimen surface does *not*

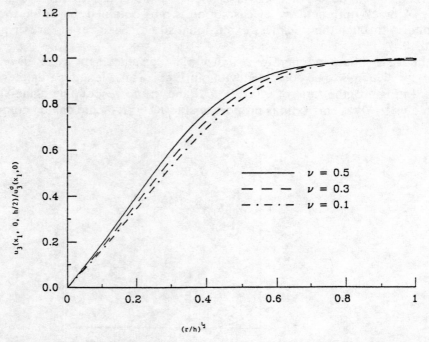

Fig. 10. Out-of-plane displacement derived by corresponding value $u_3^o(x_1, 0)$ or plane stress, vs normalized distances ahead of the crack tip for $\nu = 0.1, 0.3, 0.5$. Analytical results by Wang and Freund[1].

exhibit plane-strain behavior very near the crack tip, "in contrast to what is commonly assumed in problems of this type"[1].

6. SUMMARY AND CONCLUSIONS

1. The three-dimensional nature of the crack-tip field scales with thickness.
2. Plane-stress conditions prevail at distances from the crack tip greater than half the specimen thickness.
3. No significant plane-strain region is observed. Only right at the crack tip ($r/h = 0$), u_3 takes a constant value indicating conditions of generalized plane strain.
4. The r/h dependence of K^{EXP}/K_{2D} suggests that for $0 < r/h < 0.5$ the stress gradients are weaker than the corresponding two-dimensional gradients.
5. The experimental results are consistent with the analysis of Yang and Freund[1] and the numerical results of Levy, Marcal and Rice[2].

Acknowledgements—The authors would like to express their gratitude to Mr. W. Yang and Prof. L. B. Freund of Brown University, for providing them with a copy of their work, as well as the results displayed in Figs. 9 and 10, prior to publication of their work. The authors would also like to acknowledge many helpful discussions with Prof. J. Knowles, Prof. W. G. Knauss and Prof. C. D. Babcock of the California Institute of Technology. The research support of the National Science Foundation, through Grant No. MEA-83-07785, is gratefully acknowledged. One of us (K.R.C.) would also like to acknowledge the support of the National Science Foundation, through Grant No. MEA-8215438, during part of this work.

REFERENCES

1. W. Yang and L. B. Freund. Transverse Shear Effects for Through-Cracks in an Elastic Plate. Brown University Report (1984).
2. N. Levy, P. V. Marcal and J. R. Rice. Progress in three-dimensional elastic–plastic stress analysis for fracture mechanics. *Nucl. Eng. Des.* **17**, 64–75 (1971).
3. E. Sternberg and M. A. Sadowsky. Three-dimensional solution for the stress concentration around a circular hole in a plate of arbitrary thickness. *J. Appl. Mech.* **16**, Trans. ASME **71**, 27–38 (1949).
4. J. B. Alblas. Theorie van de driedimensionale spanningstoestand in een doorboovde plaat. Dissertation. Technische Hogeschool Delft. J. J. Paris, Amsterdam, The Netherlands (1957).
5. J. P. Benthem. State of stress at the vertex of a quarter infinite crack in a half space. *Int. J. Solids Structures* **13**, 479–492 (1977).
6. E. S. Folias. On the three-dimensional theory of cracked plates. *J. Appl. Mech. Trans. ASME* **42**, 663–674 (1975).
7. Z. P. Bažant and L. F. Estenssoro. General numerical method for three-dimensional singularities in cracked or notched elastic solids. Proceedings of the 4th International Conference on Fracture. University of Waterloo. *Fracture* **3**, 371–385 (1977).
8. R. J. Hartranft and G. C. Sih. An approximate three-dimensional theory of plates with application to crack problems. *Int. J. Eng. Sci.* **8**, 711–729 (1970).
9. M. L. Williams. On the stress distribution at the base of a stationary crack. *J. Appl. Mech.* **24**, 109–114 (1957).
10. J. Beinert and J. F. Kalthoff. Experimental determination of dynamic stress-intensity factors by the method of shadow patterns. In *Mechanics of Fracture*. (Edited by G. C. Sih), Vol. 7. Noordhoff, London, The Netherlands (1981).
11. H. Tada, P. Paris and G. Irwin. *The Handbook of Stress Intensity Factors*. Del Research Corporation (1973).

AN OPTICAL METHOD FOR DETERMINING THE CRACK-TIP STRESS INTENSITY FACTOR†

E. SOMMER‡

Department of Mechanics, Lehigh University, Bethlehem, Pa. 18015, U.S.A.

Abstract — A simple interference method has been used to measure the opening displacement of cracks in glass plates under eccentric tension load. The results have been compared with analytical predictions from the boundary collocation method and the Westergaard analysis. Since the crack opening displacement is related analytically to the stress intensity factor, K, a K-calibration curve for the crack-line-loaded, single-edge notched plate has been established.

NOTATION

K stress intensity factor
P load
a crack length
$\bar{a} = a - c$ reduced crack length — distance crack tip — center of load
W width of the plate
$\bar{W} = W - c$ reduced width
c distance from the center of the load to the nearest side of the plate
B thickness of the plate
L length of the plate
E Young's modulus
ν Poisson's ratio
v crack opening displacement } (see Fig. 4)
r distance from the crack tip
ϕ phase difference
λ_i wave length of monochromatic light in medium i
n_i index of refraction for medium i
$\alpha_\kappa, \beta_\kappa$ angles of incidence and refraction for interfaces κ, μ
h distance between fracture surfaces according to interference condition
E $7 \cdot 10^5 \text{ kg/cm}^2$
ν 0·25
n_g 1·52
n_l 1·33 (H_2O) and $= 1·74$ (CH_2I_2)
λ_{Na} 0·589 μ.

INTRODUCTION

THE GOAL of the following investigations was to experimentally determine the stress intensity factor K in a case, where due to complicated boundary conditions, an exact analysis becomes difficult. Glass plates were used for this investigation for several reasons: Because of its brittleness, an almost ideal elastic behavior corresponding to analytical predictions based on the assumption of linear elasticity can be expected. Because of its transparency, optical methods can be applied. In addition, it was of interest to carry out fracture experiments on rectangular glass plates under eccentric tension load in continuation of work which has been done on stress corrosion in glass by Irwin[1] and by Wiederhorn[2].

A simple arrangement has been used to produce and make in situ measurements of interference fringes between the opposing crack surfaces under load. Using interference fringes, the crack opening displacement will be measured in terms of half wave

†Presented at the National Symposium on Fracture Mechanics, Lehigh University, Bethlehem, Pa., June 17–19, 1968.
‡Presently at Ernst–Mach-Institut, Freiburg, Germany.

lengths of light. Since the crack opening displacement is related to the stress intensity factor K[3], a K-calibration (i.e. K as a function of crack length) can then be established.

ANALYTICAL PRELIMINARY REMARKS

If elastic conditions are assumed, the stress intensity factor K for eccentrically loaded tension specimens can be estimated as follows:

Based on the Westergaard analysis, the K-value for the case of one pair of forces $2P$ in the center of a slot of the length $2a$ in an infinite plate of thickness B (Fig. 1) is given by [3]:

$$K_\infty = \frac{2P}{B\sqrt{(\pi a)}}. \tag{1}$$

The stress intensity factor K_{fin} for a plate of finite width $2W$ may be obtained in the usual manner [3] by using a periodical solution (Fig. 2). The resulting K-value is:

$$K_{\text{fin}} = \frac{2P}{B\sqrt{\left(W \sin \frac{\pi a}{W}\right)}}. \tag{2}$$

For $a/W \ll 1$ (2), of course, reduces to (1).

The stress intensity factor for a single-edge-notched plate of finite width W, loaded in eccentric tension, may be obtained from (2) by considering only one-half of the plate (Fig. 3)

$$K = \frac{1 \cdot 3 \cdot 2P}{B\sqrt{\left(W \sin \frac{\pi a}{W}\right)}}. \tag{3}$$

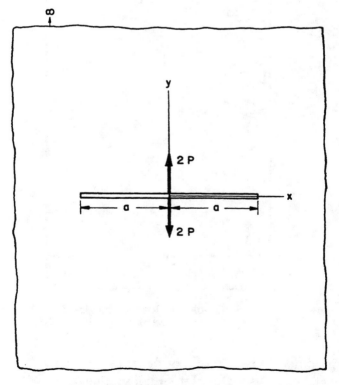

Fig. 1. Crack in an infinite plate with a pair of forces in the center.

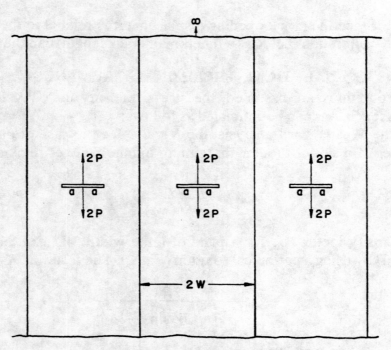

Fig. 2. An infinite array of colinear cracks in an infinite plate.

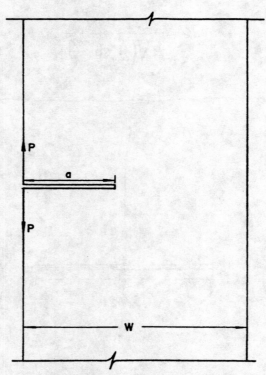

Fig. 3. Crack-line loaded, single-edge-notched plate.

The factor 1·3 is included here to correct for the boundary condition that the normal stresses along the cracked side of the plate must vanish[4].

For experimental reasons, the load will be applied in the notch at a short distance from the side of the plate (at $a = c$). If this loading point is considered as a new coordinate center, (3) becomes

$$K = \frac{1\cdot 3 \cdot 2P}{B\sqrt{\left(\bar{W}\sin\frac{\pi\bar{a}}{\bar{W}}\right)}} \quad \text{i.e.} \quad \frac{KB\bar{W}^{1/2}}{P} = \frac{1\cdot 3 \cdot 2}{\sqrt{\left(\sin\frac{\pi\bar{a}}{\bar{W}}\right)}} \tag{3a}$$

where the plate width W is replaced by $\bar{W} = W - c$, and the crack length a by $\bar{a} = a - c$.

Since the effect of the bending moment produced by the loads P is not included in (3a), the actual values of K can be expected to deviate from that given by (3a) for large values of $\bar{a}$. The effect of the bending moment is expected to raise and shift the minimum value of the K-calibration curve to a lower value of $\bar{a}$.

For comparison, a numerical solution for K given by Brown and Srawley[5] for the eccentrically loaded single-edge notched plates may be used.

$$\frac{KB\bar{W}^{1/2}}{P} = \frac{0\cdot 54\,(1 - \bar{a}/\bar{W}) + 2\cdot 17\,(1 + \bar{a}/\bar{W})}{(1 - \bar{a}/\bar{W})^{3/2}}. \tag{4}$$

This equation represents an adaption of the solution for a semi-infinite crack approaching the free edge of a half plane[3] and was found by Gross and Srawley (see [5]) to fit boundary collocation results fairly well for crack line loaded specimens when the crack approaches the far boundary. Because the compared boundary collocation values were based on computations for single edge notched plates of square dimensions that are somewhat different from those used in this investigation, exact agreement between (4) and the experimental results should not be expected. Equations (3a) and (4) are plotted in (Figs. 10 and 11).

The crack opening displacement v may be calculated using the Westergaard analysis [3]. For fracture mode I and plane strain conditions, and if $r \ll a$, the crack opening displacement is given by (Fig. 4):

$$2v = \frac{4K}{E}(1 - \nu^2)\sqrt{\left(\frac{2r}{\pi}\right)}. \tag{5}$$

For small distances r, a parabolic opening of the crack surfaces is predicted by (5), (Fig. 4).

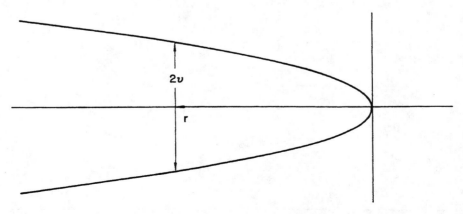

Fig. 4. Parabolic crack opening displacement.

To avoid the restriction $r \ll a$, the exact solution for the special case of a pair of forces in the center of a notch of an infinite plate may be considered:

$$2v = \frac{4P}{EB\pi}(1-\nu^2)\log\frac{1+\sqrt{\left(1-\left(\frac{a-r}{a}\right)^2\right)}}{1-\sqrt{\left(1-\left(\frac{a-r}{a}\right)^2\right)}}. \qquad (6)$$

If a series approximation is made and terms of higher order in r/a are neglected, (6) becomes

$$2v = \frac{4K_\infty}{E}(1-\nu^2)\sqrt{\left(\frac{2r}{\pi}\right)}\left(1+\frac{5}{12}\frac{r}{a}\right) \qquad (6a)$$

where K_∞ is given by (1).

By including a first-order approximation in r/a, crack opening displacement measurements at distances farther away from the crack-tip than that permitted by (5) may be utilized in the experimental determination of K. Although there may be some uncertainty in the precise value of the factor $\frac{5}{12}$ (which is not expected to be too different from $\frac{5}{12}$), this is more than offset by the improvement in accuracy in the determination of K engendered by the inclusion of larger numbers of displacement measurements.

These analytical results may also be compared with numerical data derived from the boundary collocation method. The displacements $2v$ for certain values of $\bar{a}/\bar{W}(c = 0.15$ in.$)$ and for two different plate widths ($W = 6$ in. and $W = 8$ in.), have been computed by Srawley and Gross[6], Table 1. The numbers in brackets in Table 1 were calculated from (6a) in a relative scale; i.e. $\sqrt{(r)}\left(1+\frac{5}{12}\frac{r}{a}\right)$ times a factor which is normalized to the computed values for $r = 0.2$ in. In the range which is of experimental interest, the deviation of the compared values is less than 4 per cent maximal. This shows that this deviation is within the measuring error and supports the previous contention that the value $\frac{5}{12}$ is quite reasonable.

Table 1

(W/in.)	$\bar{a}/(W-c)$	$r = 0.1$ in.	$r = 0.2$ in.	$r = 0.3$ in.	$r = 0.4$ in.	$r = 0.5$ in.	$r = \bar{a}$
				$EB(2v)/P$			
6	0.2	1.74	2.49	3.08	3.60	4.0(6)	6.7
		(1.70)	[2.49]	(3.14)	(3.75)		
	0.3	2.22	3.18	3.95	4.62	5.24	11.5
		(2.20)	[3.18]	(3.99)	(4.71)		
	0.4	2.89	4.16	5.17	6.06	6.88	18.8
		(2.90)	[4.16]	(5.19)	(6.09)		
	0.5	3.94	5.68	7.09	8.33	9.47	31.6
	0.6	5.75	8.32	10.43	12.30	14.05	56.7
8	0.2	1.68	2.4(0)	2.9(8)	3.4(8)	3.9(5)	8.0
		(1.66)	[2.40]	(3.01)	(3.56)		
	0.3	2.08	2.98	3.70	4.34	4.92	13.5
		(2.08)	[2.98]	(3.72)	(4.36)		
	0.4	2.63	3.78	4.69	5.50	6.24	21.7
		(2.64)	[3.78]	(4.70)	(5.49)		
	0.5	3.49	5.02	6.24	7.32	8.32	35.4
	0.6	4.98	7.18	8.96	10.54	11.99	61.4

INTERFERENCE METHOD

The experimental setup used to measure the crack opening displacement by interferometrical means can be seen in (Fig. 5). Parallel monochromatic light is produced. Due to the oblique incidence of the light, interference fringes between the crack surfaces in the glass plate are produced only if the crack is filled with a liquid; otherwise total reflection at the first crack surface (reflection dense/rare medium) would result. The distance of these interference fringes from the leading edge of the crack is measured with a traveling microscope. A more detailed consideration is given in the following paragraphs.

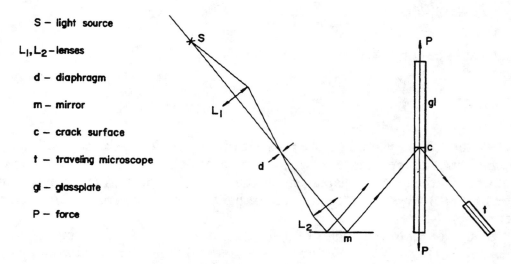

Fig. 5. Schematic diagram of the optical setup.

To get the interference condition for this case, an idealized slot filled with a liquid may be considered (Fig. 6). If the wave length of the monochromatic light and the index of refraction of the liquid in the slot and those of the surrounding medium (glass) are λ_l and n_l and λ_g and n_g, respectively, the phase difference between the two interfering rays a and b is given, according to (Fig. 6) by

$$\phi = \frac{2\pi}{\lambda_l}(\bar{AB}+\bar{BC}) - \frac{2\pi}{\lambda_g}\bar{AD} = \frac{2\pi}{\lambda} \cdot 2h \cdot n_l \cdot \cos\beta_{II}. \tag{7}$$

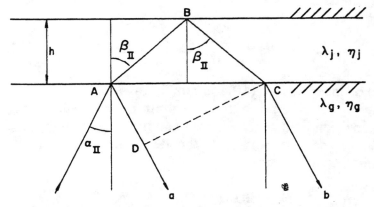

Fig. 6. Schematic diagram of light reflections at the crack surfaces for establishing the phase condition for light interference.

The last part of this equation results from geometrical considerations and the use of the law of refraction ($\sin \alpha_{II}/\sin \beta_{II} = n_l/n_g$. $\lambda = n_i \lambda_i$ is the wave length in vacuum.

An additional phase change of π takes place for the reflection at the interface of a dense to rare medium. Extinction of light occurs if the complete phase change is an odd multiple of π. This interference condition is:

$$\phi = m\pi = \frac{2\pi}{\lambda} \cdot 2h \cdot n_l \cdot \cos \beta_{II} + \pi \quad (m = 1, 3, 5, ..) \tag{8}$$

or

$$2q\pi = (m-1)\pi = \frac{2\pi}{\lambda} 2hn_l \cdot \cos \beta_{II} \left(q = \frac{m-1}{2} = 0, 1, 2, 3, ..\right). \tag{9}$$

This interference condition may be rewritten as

$$h = \frac{\lambda \cdot q}{2n_l \cos \beta_{II}} \quad (q = 0, 1, 2, ..). \tag{10}$$

It is clear then, if λ, n_l, and β_{II} are prescribed, the change in the separation distance h is given in terms of the change in q, the number of light interference extinction. For the case of a crack, the opening displacement at various distances from the crack tip may be obtained by (10) where q is the number of light-interference fringes counted from the crack tip.

To relate $\cos \beta_{II}$ to the angles of light incidence (α_I^i) and refraction (α_I^e), which can be measured, (Fig. 7), the following geometrical consideration would be helpful. Because it cannot be assumed that the crack surface is always perpendicular to the plate surface, a small additional angle $\pm \epsilon$ will be included.

If one calculates the angles β_I^i and β_I^e from the angles of incidence α_I^i and refraction α_I^e using the law of refraction; (i.e. $\sin \alpha_I/\sin \beta_I = n_g$) and by using the following relations (Fig. 7).

$$\begin{aligned}(\beta_I^i - \epsilon) + \alpha_{II} &= 90° \\ (\beta_I^e + \epsilon) + \alpha_{II} &= 90°\end{aligned} \tag{11}$$

one obtains

$$\sin \alpha_{II} = \cos \tfrac{1}{2} (\beta_I^i + \beta_I^e).$$

The angle α_{II} again is related to β_{II} by the law of refraction; (i.e. $\sin \alpha_{II}/\sin \beta_{II} = n_l/n_g$) as follows:

$$\cos \beta_{II} = \sqrt{\left(1 - \left(\frac{n_g}{n_l}\right)^2 \sin^2 \alpha_{II}\right)}. \tag{12}$$

This result is to be inserted into (10) for calculating h.

In the foregoing consideration, the assumption was made that the direction of crack propagation is perpendicular to the optical axis. Although this condition was not exactly fulfilled for each experiment, the deviation can be neglected because it involves only a small cosine error.

EXPERIMENTS AND RESULTS

The fracture tests have been conducted on glass plates (Plateglass, Pittsburgh Plate Glass Company) of the sizes ($L \times W \times B$: $10 \times 6 \times 0.25$ in. and $10 \times 8 \times 0.25$ in.).

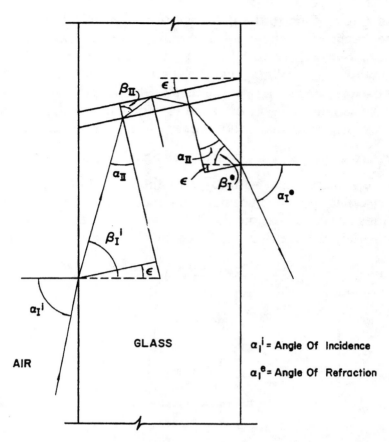

Fig. 7. Crack position in the plate for getting geometrical angle relations.

The material was not especially annealed. A notch was cut at the center of the 10 in. side of each plate, and a crack was initiated at the notch root with the aid of thermal stresses. The level of crack-line loads (applied by means of high strength steel wires) were chosen to produce a very slow rate of crack propagation (approximately 10^{-6} to 10^{-4} in./sec) to ensure good measuring accuracy.

The position of the interference fringes produced by monochromatic Na-light ($\lambda = 0\cdot 589\,m\mu$) as well as the crack length were measured with a traveling microscope on the center line of the crack surfaces. The resolution of the microscope is 10^{-4} in.

The crack was filled with either purified water (H_2O; $n_l = 1\cdot 33$) or diiodomethane (CH_2I_2; $n_l = 1\cdot 74$). Because of the higher index of refraction of diiodomethane, more fringes per unit length can be evaluated (see (10)) and greater accuracy can be achieved. The improved accuracy can be attained only if the index of refraction would remain constant over the time period necessary to perform one experiment. The latter condition is better fulfilled with water since the indices of refraction of aqueous solutions of fairly high concentration are nearly equal to that of pure water. On the other hand, water has a very strong influence on crack growth in glass. For the same load, the crack velocity is increased by several orders of magnitude due to stress corrosion effects [1, 2].

Since the goal is to calculate the stress intensity factor K from the crack opening displacement measurements, it is necessary to prove first that the experimental behavior of the crack opening agrees with (5) or (6a) and the numerical values calculated by Srawley and Gross. Due to the influence of real material properties as well as the simplicity of the idealized crack model, a deviation could be expected.

Two representative experimental curves are shown in (Figs. 8 and 9). The square of the displacement, $2v$, is plotted vs. the distance from the crack tip r. Although it is possible to measure the crack opening displacement accurately down to the order of one-half the wave length of light, the smallest corresponding position from the crack tip r is still three orders of magnitude larger. The slope of each (solid) curve which represents the experimental data can be reduced by the correction factor $\left(1+\frac{5}{6}\frac{r}{a}\right)$ according to (6a). In agreement with (5), straight lines (dotted), then, can be constructed. In comparison to the displacement calculated by Srawley and Gross (see +), however, the measured displacements of the second example—and of the remaining experiments—are larger than that predicted by this computation. If one corrects this

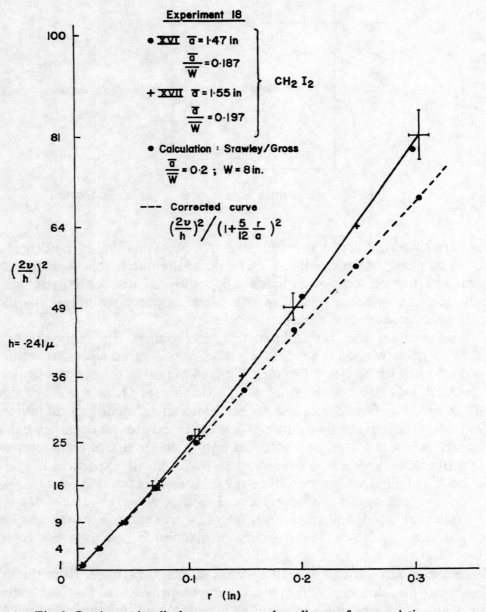

Fig. 8. Crack opening displacement squared vs. distance from crack tip.

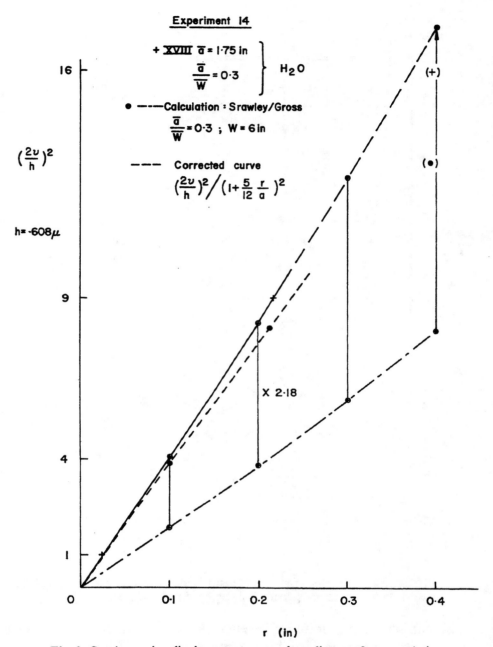

Fig. 9. Crack opening displacement squared vs. distance from crack tip.

deviation by a proportionality factor (for a detailed consideration of this correction, see the end of this part), the calculated values may be brought into a fit with the experimental curve. For distances r, that are large in comparison to the plate thickness, two principal errors would be introduced:
  (a) The model used for correction is too simple, (6a).
  (b) Although the experimental data are taken along the center line of the crack, the plane strain condition assumed for all calculations is not valid anymore.

Therefore, only small values of r, i.e. the positions of the first few interference fringes, and the corresponding displacements, (10–12) have been used for the K-calculation. Figure 10 shows a dimensionless plot of $KB\bar{W}^{1/2}/P$ vs. the reduced crack length $\bar{a}/\bar{W}$. The data of six experiments are compared with the analytical curves,

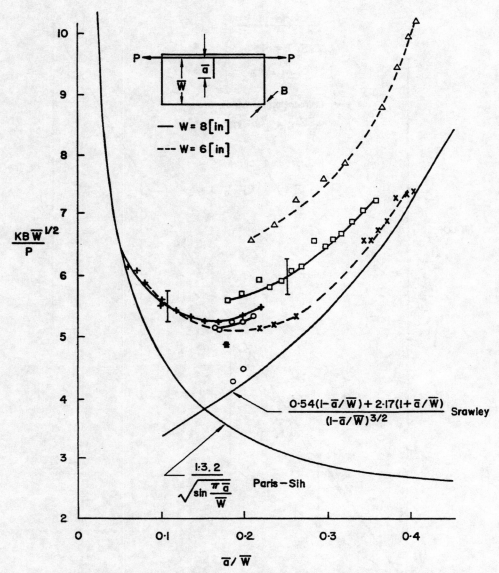

Fig. 10. Dimensionless plot of stress intensity factor K vs. crack length $\bar{a}$ for the eccentrically loaded, single-edge-notched plate.

(3 and 4). Each experimental point represents the mean value of up to ten single values averaged over a small range of different crack lengths. Although the experimental curves seem to be consistent in themselves, there are large deviations from each other.

An attempt to explain this behavior may be seen in the following discussion: As mentioned earlier, the investigated plates are not annealed. Therefore, residual stresses may exist and influence the results.

Based on the assumption that the internal stress, σ, (in the center layer of the plate) is uniformly distributed over the plate, an additional K-value

$$K_2 = 1\cdot 3 \sigma \sqrt{\pi a} \tag{13}$$

must exist. Due to this effect, the measured crack opening would be larger than expected from (6a) and may be described as

$$2v \propto k\sqrt{(r)} \propto (K_1 + K_2)\sqrt{(r)} \propto K_1\left(1 + \frac{K_2}{K_1}\right)\sqrt{(r)}. \tag{14}$$

If the proportionality factor F between the displacement calculated by the boundary collocation method and the measured crack opening is considered to be $(1 + K_2/K_1)$, a rough estimate of the internal stress σ may be obtained for certain points $\bar{a}/\bar{W}$. The results are given in Table 2. Inserting these σ-values into (13) and subtracting the resulting K_2-values from the original K, one obtains the curves plotted in (Fig. 11). Although this correction is based on a rough estimation, reasonable agreement with the experimental results was achieved.

Table 2

Exper.	22		21		19	18	14		13
$\bar{a}/\bar{w}$	0·202	0·299	0·203	0·202	0·20	0·191	0·301	0·405	0·402
σ (kg/cm²)	(2·9)	2·3	1·8	2·3	0·0	(5·4)	3·6	4·2	0·8

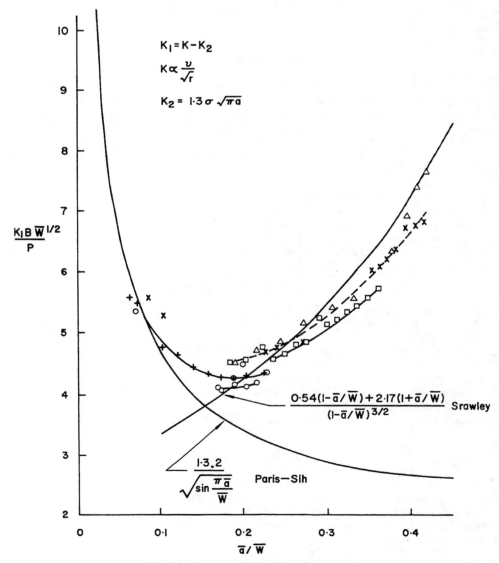

Fig. 11. Corrected K-calibration curve.

The maximal experimental errors are estimated as follows:

For the displacement $\quad \dfrac{\Delta h}{h} = \dfrac{\Delta 2v}{2v} = \pm 3\cdot 5$ per cent.

For the distance r $\quad \dfrac{\Delta r}{r} \leq \pm 5$ per cent.

For the K-value $\quad \dfrac{\Delta K}{K} \leq \pm 6$ per cent.

The error of the averaged K-values plotted in (Fig. 11) is probably less than the stated error of $\Delta K/K$ of ± 6 per cent. The error could be greatly reduced if a perpendicular, instead of oblique, incidence of light would be applied, and if the method would be used for other materials, for example, transparent plastics which would have a larger crack opening displacement due to a lower Young's modulus. On the other hand, it would be necessary to take into account the plastic and viscoelastic behavior of these materials which could be neglected in glass because of the very small zone of inelastic strain at the crack border.

SUMMARY

A simple interference method has been used to measure the opening displacement of cracks in glass plates under eccentric tension load. The results have been compared with analytical predictions from the boundary collocation method and the Westergaard analysis. Since the crack opening displacement is related analytically to the stress intensity factor, K, a K-calibration curve for the crack-line-loaded, single-edge notched plate has been established.

Acknowledgements—The author wishes to thank Professors G. R. Irwin and P. C. Paris for their many suggestions and discussions, and to express his appreciation to Dr. J. E. Srawley and Mr. B. Gross for the computation of crack opening displacement values.

This research was supported by the Advanced Research Projects Agency of the Department of Defense and was monitored by the Naval Research Laboratory under Contract No. NONR 610 (09).

REFERENCES

[1] G. R. Irwin, *Naval Research Lab. Rep.* No. 1678 (1966).
[2] S. M. Wiederhorn, *National Bureau of Stand. Rep.* No. 8618 (1965), *Rep.* No. 8901 (1965), and *Rep.* No. 9442 (1966).
[3] P. C. Paris and G. C. Sih, Symposium on fracture toughness testing and its applications. *ASTM Spec. Tech. Publ.* No. 381 (1965).
[4] J. R. Rice, Letter to E. J. Ripling, September (1966); Copy to P. C. Paris.
[5] W. F. Brown and J. E. Srawley, Plane strain crack toughness testing of high strength metallic materials. *ASTM Spec. Tech. Publ.* No. 410 (1966).
[6] J. E. Srawley and B. Gross, Letter to E. Sommer, January (1968).

(*Received* 1 *April* 1968)

Résumé—On employé une simple méthode d'interférences pour mesurer le déplacement de l'ouverture de fissures dans des plaques de verre sous une tension excentrique. Les résultats sont comparés à ceux des prédictions analytiques de la méthode de collocation limite et l'analyse de Westergaard. Purisque le déplacement de l'ouverture de la fissure est liée, du point de vue analytique, au facteur d'intensité de la force, K, on a établi une courbe de calibrage-K pour la plaque à entaille sur un côté et ligne de fissure chargée.

Zusammenfassung—Es wurde eine einfache Interferenzmethode angewandt, um die Rissöffnung von Brüchen in Glasplatten unter exzentrischer Belastung zu vermessen. Die Ergebnisse wurden mit Berechnungen nach einem Randwertverfahren und nach der Westergaardanalysis verglichen. Da ein analytischer Zusammenhang zwischen der Rissöffnung und dem Spannungsfaktor K besteht, wurde eine Eichkurve für K für den Fall der Belastung einer einseitig gekerbten Platte durch ein Kräftepaar angegeben, das parallel zum Plattenrand am Ende der Kerbe angreift.

AN OPTICAL-INTERFERENCE METHOD FOR EXPERIMENTAL STRESS ANALYSIS OF CRACKED STRUCTURES†

P. B. CROSLEY, S. MOSTOVOY and E. J. RIPLING

Materials Research Laboratory Inc., 1 Science Road, Glenwood, Illinois 60425, U.S.A.

Abstract—An optical interference technique can be used to measure the crack opening in a cracked transparent model of a structure under load. The measured crack opening as a function of distance from the crack tip can be used in conjunction with the opening mode crack tip stress field equations to determine the opening mode stress intensity factor. Application of this method to double cantilever beam specimens comprised of two glass arms bonded together with an epoxy adhesive gave results which were in close agreement with K-calibrations based on compliance measurements. The measuring techniques and the scheme of analyzing the data could be used to obtain K-calibrations of more complex structures.

INTRODUCTION

MANY fracture problems can be handled by linear elastic fracture mechanics and the crack tip stress field can be adequately described by a single parameter, the opening mode stress intensity factor K_i. The stress analysis of the cracked structure reduces to a problem of finding the constant of proportionality between the applied loads (or nominal stresses) and the stress intensity factor for a crack of a given size and location. This type of stress analysis, commonly referred to as K-calibration, can be approached by different analytic and experimental techniques, and results for a variety of structures are available in the literature. Analytic approaches depend on satisfying boundary conditions remote from the crack; for complex structures the procedures become increasingly difficult and/or inaccurate. The experimental method of compliance calibration is practical for structures which are loaded in a simple manner and in which the displacement of the loading points is large and sufficiently sensitive to crack size to permit an accurate definition of compliance as a function of crack size. For the K-calibration of complex structures it would be most desirable to make measurements in the immediate vicinity of the crack, i.e. in the actual region where the parameter K_i best describes the stress field. Measurements of this type have been made with strain gauges and by photoelastic techniques, but such techniques have not supplanted analytical ones as the main source of reliable K-calibration.

An alternative experimental method described in this paper is based on measurement of the crack opening as a function of the distance from the leading edge of the crack. A technique of this type had been previously described by Sommer[1]. The method uses a transparent model of the structure to be analyzed, and an optical interference method is used to measure the crack opening. With the model under load, monochromatic light directed on the crack plane is reflected separately from the opposing crack surfaces. Alternate bands of constructive and destructive interference give a definition of the crack opening as a function of distance from the crack tip and of position along the crack front. Comparison of the measured crack opening against the

†Presented at the Third National Symposium on Fracture Mechanics, Lehigh University, Bethlehem, Pa. August 25–27, 1969.

shape and magnitude of the crack opening predicted by the Mode I crack tip stress field is the basis for estimating the parameter. K_i.

The optical interference method has promise not only as an adjunct to current procedures for the analysis of simpler structures, but also as a means of treating complex structures for which reliable analyses are not available. Because the actual measurements are made only in the region of the crack tip, boundary conditions remote from the crack need not be treated explicitly, and, therefore, complicated shapes are basically no more difficult to analyze than simple ones. Moreover, the method should be applicable to cases where the stress intensity factor varies along the crack front, thus furnishing a means for handling three-dimensional problems.

The method requires that a model of the structure be made from a transparent material, that a crack of the desired size and shape be introduced, and that it be possible to direct light on the crack plane and observe the reflected light. The problems of fabricating a complex model might be simplified by glueing together transparent members. With this technique a flaw could be introduced into the adhesive joint during manufacturing. This would not only allow for starter cracks of almost any shape or size, but the crack surface would be reasonably flat and smooth so as to produce a distinct interference fringe pattern.

While the potential simplicity of the optical interference technique is obvious, its reliability must be established. For this purpose, the method was applied to the K-calibration of specimens for which reliable compliance calibrations were already available. Double cantilever beam specimens were made by cementing together two glass arms with an epoxy adhesive, and cracks were produced either in the epoxy itself or at the epoxy-glass interface. Both uniform and tapered specimens with and without side grooves were used.

BACKGROUND

Relationship between crack shape and stress intensity factor

The y-direction displacement (V) associated with the opening mode crack tip stress field is given by

$$V = \frac{2(1+\nu)}{E} K_i \left(\frac{r}{2\pi}\right)^{1/2} \sin\frac{\theta}{2} \left\{ 2 - 2\sigma - \cos^2\frac{\theta}{2} \right\} \tag{1}$$

where

E = Young's modulus
ν = Poisson's ratio
r, θ = polar coordinates referred to the crack tip (Fig. 1)
$\sigma = \nu$, for plane strain
$= \nu/(1+\nu)$, for plane stress.

The crack opening ($2V_\pi$) as a function of the distance d measured back from the crack tip is obtained by substituting $\theta = \pi$ in (1). This yields

$$2V_\pi = \frac{8(1-\nu^2)}{E} K_i \sqrt{\frac{d}{2\pi}} \tag{2a}$$

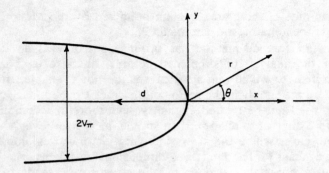

Fig. 1. Notations used in measuring crack profile.

for plane strain, and

$$2V_\pi = \frac{8}{E} K_i \sqrt{\frac{d}{2\pi}} \qquad (2b)$$

for plane stress. With the appropriate equation (2a or 2b), measurement of the crack opening ($2V_\pi$) vs. the distance d from the crack tip can be used to calculate the stress intensity factor K_i under the applied load at which the measurements are made.

For a linear elastic material (1) and (2) are exact in the limit as r (or d) approaches zero. As the distance from the crack tip increases, higher order terms contribute increasingly to the general stress function and hence to the shape of the crack opening. The nature and magnitude of the deviation from (2) must depend on the configuration of the cracked structure, but it should be independent of the load level. Thus, in terms of purely linear elastic considerations, it would appear best to measure the crack opening as close as possible to the crack tip, and at the highest possible loads to produce the largest values of $2V_\pi$.

In a real material, high stresses close to the crack tip lead to non-linear behavior, and a corresponding inexactness of (2) at small values of d. The extent of the non-linear behavior depends obviously on the material from which the structure is fabricated, and it increases as the load level (or value of K_i) is increased. To minimize the influence of the non-linear behavior at the crack tip it would appear best to make crack opening measurements under low levels of applied load and at values of d which are large relative to the anticipated size of the crack plastic zone.

The above considerations indicate that the relationship between K_i and $2V_\pi$ given by (2) should be most closely met at some intermediate range of distances d; i.e. close enough to the crack tip so that the inverse square-root singularity adequately describes the stress field, yet far enough to avoid serious errors resulting from non-linear behavior. The limitation is, of course, not unique to this method of K calibration. It reflects the basic assumption involved in describing the crack tip stress field by the parameter K_i, and the basic approximation behind applying linear elastic fracture mechanics to the evaluation of real materials. The practical problem for K-calibration is to find a range of d values and load levels at which K_i can be accurately determined from measurements of $2V_\pi$.

Measurement of crack opening

Optical interference fringes can be used to measure the crack opening ($2V_\pi$) in transparent materials. Monochromatic light directed perpendicular to the plane of the crack is reflected from the top and bottom surfaces of the crack. Destructive interference of the reflected waves occurs when the path length difference is an odd number of half-wavelengths, i.e. when

$$2V_\pi = \tfrac{1}{2}(n-\tfrac{1}{2})\lambda \quad n = 1, 2, 3, \ldots$$

and constructive interference occurs when

$$2V_\pi = \tfrac{1}{2}n\lambda \quad n = 1, 2, 3, \ldots$$

where λ is the wavelength.

Thus the fringe order (n) gives the crack opening ($2V_\pi$) in terms of the wavelength of the light used, and measurement of the distance of the fringe from the crack front gives the value of d.

Specimen characteristics

Measurements of the crack opening profile were made on double cantilever beam specimens of the type shown in Fig. 2. The specimens were made from two glass arms joined with a 0·005 in. thick layer of epoxy adhesive. Accurate values of the strain energy release rate, G_i, were available from compliance calibrations[2]. For the uni-

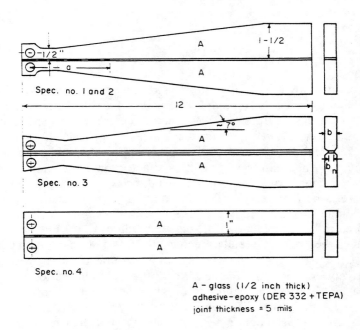

Fig. 2. Specimen configurations.

form beam specimens

$$G_i = \frac{P^2}{2b_n} \frac{8}{Eb} \left\{ \frac{3(a+0.6h)^2}{h^3} + \frac{1}{h} \right\} \tag{3a}$$

and for the tapered beam specimens

$$G_i = \frac{P^2}{2b_n} \frac{8}{E_b} m' \tag{3b}$$

where

P = applied load
b_n = crack width
b = specimen width
a = crack length
h = beam height at the distance a from the line of action of the load
m' = an experimental calibration constant equal to 90 in.$^{-1}$ for the specimens used in this experiment.

The relation between G_i and K_i is

$$G_i = \frac{1-v^2}{E} K_i^2 \tag{4a}$$

for plane strain, and

$$G_i = \frac{1}{E} K_i^2 \tag{4b}$$

for plane stress.

These equations obviously apply only to structure which everywhere have the same value of E. Hence the use of (4) imply the use of monolithic specimens. During the course of this work several attempts were made to prepare such specimens from glass, plexiglass and epoxy. It was not possible however, with any of these materials, to produce cracks which were planar over a sufficient area to yield a distinct interference fringe. On the other hand, introduction of satisfactory cracks in adhesive specimens of the type shown in Fig. 2 could be done without difficulty.

While there is no ambiguity in defining a strain-energy-release-rate (G_i) for these adhesive specimens, representation of the crack tip stress field by the parameter K_i could be influenced by the fact that the elastic modulus of the epoxy adhesive is much less than that of the glass adherends. Because the 0·005 in. thickness of the epoxy layer is small compared to the range of d values over which crack opening was to be measured, it was tentatively assumed that the crack opening would not differ markedly from that which would be observed in a monolithic glass specimen. Hence, the analysis was carried out on the basis that (4) **could** be used to define K_i from the experimental G-calibration. Were the assumption wrong that the specimen could be analyzed as a monolithic glass specimen, one would expect that the crack opening profile would not describe the parabolic shape predicted by (2) and/or that the values of K_i determined from crack opening measurements would not agree with the values derived from compliance calibration.

EXPERIMENTAL PROCEDURE

The equipment arrangement used for this study is shown schematically in Fig. 3. Light from a sodium lamp ($\lambda = 5890 \, \text{Å} = 23 \cdot 2 \, \mu\text{-in.}$) was reflected from a partial mirror through the top half of the specimen onto the crack plane. The light reflected from the crack surfaces passed back through the partial mirror and was photographed through a low power microscope at a magnification of 5–10X. The actual magnification was determined by photographing a scale through the top half of a broken specimen.

The specimens were loaded with dead weights ranging from 0–15 lb in most cases, but from 0–20 lb in one test. In most of the tests, the load was increased from zero to the maximum load in 1 lb steps and a photograph was taken at each load level. General experience indicated, however, that except at very low load levels (5 lb or less) the interference fringe pattern was not influenced by whether or not the previous applied load was higher or lower than the current one. Moreover, cracks within the epoxy layer, introduced in an increasing load test, gave the same results as cracks at the epoxy-glass interface, formed by subcritical crack extension.

A series of photographs obtained on one specimen is shown in Fig. 4. In each photograph the crack front is quite distinct, and the interference fringes are on the whole parallel to the crack front and equally spaced across the specimen thickness. There are local variations which generally appear in the same location on all of the photographs. Also, the specimens are not stress free at zero applied load. The K calibration, however, depends not on the actual crack opening, but rather on the change in the

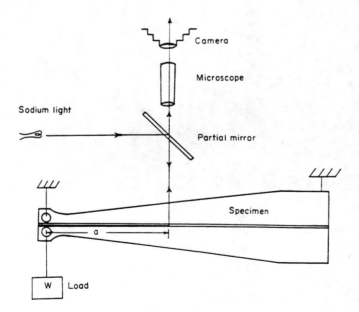

Fig. 3. Schematic drawing of test set-up.

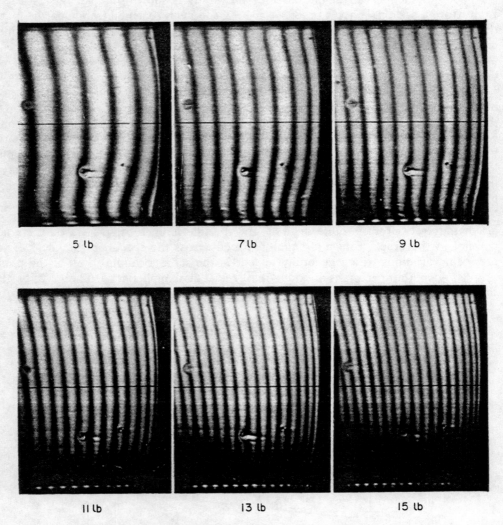

Fig. 4. Interference fringe patterns at different load levels for specimen No. 4.

crack opening with the application of load; therefore, residual effects or local irregularities should not affect the calibration procedure.

Measurements were made on the photographs along a reference line drawn perpendicular to the crack front and approximately at the mid-thickness of the specimen. To eliminate the effect of local irregularities, the line was located in the same place on all photographs with respect to flaws or bubbles in the joint line. The distance from the crack tip at which each destructive interference fringe (black line) crossed the reference line was measured to the nearest 0·01 in. on the photograph. Measurements at selected load levels of the photographs in Fig. 4 gave the crack opening shapes shown in Fig. 5. Each point is based on the location of a single destructive interference fringe. The crack opening is plotted in terms of sodium wavelengths and the distance from the crack front is the distance measured on the film. Actual dimensions, indicated on the figure, show that the crack opening is exaggerated approximately 1000:1 with respect to the distance d. The opening profiles appear to be roughly parabolic and equal load increments give equal increases in opening at a given value of d even though the results do not extrapolate to zero opening for zero applied load.

RESULTS AND DISCUSSION

Each destructive interference fringe gives a value of d corresponding to a crack opening $2V_\pi$. From these numbers an apparent value of the stress intensity factor, K_i^*, can be calculated by rewriting (2b) as

$$K_i^* = \frac{1}{4}\sqrt{\frac{\pi}{2}} E \frac{2V_\pi}{\sqrt{d}} \qquad (5)$$

This equation corresponds to the plane stress state of deformation. Ideally, all of the K_i^* values corresponding to a given load level should be equal. In fact, there was a

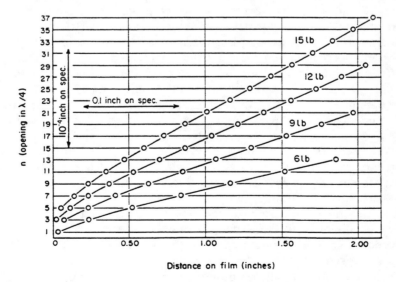

Fig. 5. Crack opening shape for specimen No. 4.

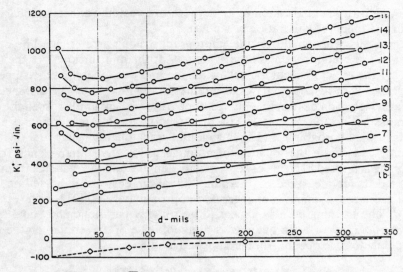

Fig. 6. $K_i^* = \tfrac{1}{4}\sqrt{\pi/2}\, E\, 2V_\pi/\sqrt{d}$ as a function of applied load and distance from the crack front for specimen No. 4.

systematic dependence of K_i^* on the distance d from the crack tip as can be seen in Fig. 6. Two features stand out: there is a gradual, nearly linear, increase in K_i^* with d, and superimposed on this is a tendency for the K_i^* values to rise rather sharply at very low values of d especially at the higher load levels. These features reflect the two limitations discussed earlier, viz., increasing influence of non-singular terms as d is increased, and non-linear deformation very close to the crack tip. Another effect, which was already pointed out with reference to Fig. 5 is the fact that the crack opening is not identically zero at zero applied load. The analysis necessary for a K calibration involves systematically eliminating these various effects which are not directly linked to the inverse square root singularity which defines the stress intensity factor.

The first step in reducing the data was to express the residual (zero load) crack opening as an effective increment, K, of the stress intensity factor. This was done by plotting K_i^* vs. load at selected values of d, as shown in Fig. 7. It can be seen that K_i^* increases linearly with load, but extrapolation does not give $K_i^* = 0$ at zero load. The intercepts in Fig. 7 define values of ΔK_o which are plotted as the dashed curve in Fig. 6. The effective value, ΔK_o, at zero load defined by this procedure was negative, indicating a residual stress state which forced the crack surfaces together. In some specimens the zero load correction was negative at low values of d but became positive at larger distances from the crack tip. However, once the ΔK_o are subtracted from the K_i^* values, the results reflect only elastic loading increments and the shape of the opening is no longer a factor in the analysis.

After subtracting the zero load increment (ΔK_o) from the K_i^*, the data were normalized by dividing by the applied load. The result was a single curve of K_i/P vs. d incorporating all of the data points in Fig. 6. This plot, which is shown in Fig. 8, has the same general features as the individual curves in Fig. 6; viz., a systematic increase with increasing d, superimposed on which is a sharp upturn very close to the crack tip. Ignoring the non-linear effect near the crack tip, the remainder of the curve is well represented by a straight line which intersects the ordinate at 54.0 in.$^{-3/2}$.

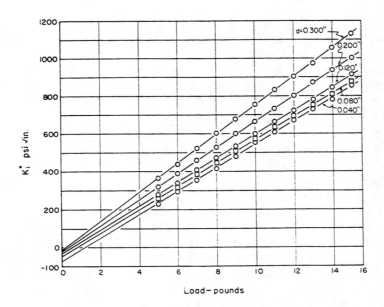

Fig. 7. Extrapolation to determine ΔK_0 at different values of d.

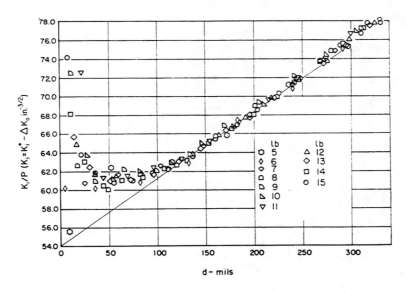

Fig. 8. K_i/P as a function of d for specimen No. 4.

The contribution of non-singular terms to the crack opening, and hence apparent value of K_i, should vanish at the mathematical position $d = 0$. Thus, if curves of K_i/P vs. d have a sufficiently simple shape to permit an unambiguous extrapolation, the intersection of the extrapolated curve with the $d = 0$ axis, should yield the proper value of K_i/P. In the case treated here, the curve was linear at values of d greater than 0·1 in., and the straight line extrapolation gave 54·0 in.$^{-3/2}$ for K_i/P. For this specimen the value of K_i/P based on compliance calibration was 53.1 in.$^{-3/2}$ using the plane stress conversion for G_i to K_i. The agreement is within the accuracy which might be expected from either method individually.

In carrying out the extrapolation, it was assumed that the shape of the curve near the crack tip reflected non-linear behavior which, for the purpose of the analysis, could be ignored. It is possible to assess the contribution of the non-linearity by treating its effect as an apparent shift in the crack front. For this purpose the influence of a 0·005 in. increase in apparent crack length was assessed by using the quantity $(d+0.005)$ in place of d in (5). The resulting plot is shown in Fig. 9. The effect of the crack length correction was virtually to eliminate the upturn in the curve of K_i/P at low values of d so as to produce a linear curve over the entire range of d values. The magnitude K_i/P obtained by extrapolation was, however, not greatly affected. It was reduced to approximately 52 in.$^{-3/2}$ as compared to 54 in.$^{-3/2}$ without the correction. Thus, it was felt that as a general procedure, the additional complication of an explicit accounting for crack tip non-linearity was not justified in view of its modest effect on the final result.

Measurements made on other specimens were analyzed in the same way. Curves of K_i/P for these specimens are shown in Figs. 10–12. The curves represent uniform and tapered specimens with and without side grooves as summarized in Table 1. All of the curves are generally comparable to the one already discussed, and values of K_i/P, uncorrected for crack tip nonlinearity, were obtained by extrapolation of the linear portion of the curves. These results compared to the values obtained from compliance calibration are summarized in Table 1.

The results in Table 1 are based on assumed plane stress deformation both for the description of the crack opening (2b) and for the conversion from G_i to K_i (4b). The use of the plane stress equations in preference to plane strain seems reasonable because a state of plane strain should be approached only very near to the crack tip

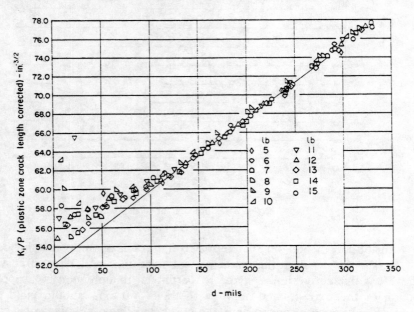

Fig. 9. Effect of 0·005 in. crack length correction on K_i/P for specimen No. 4.

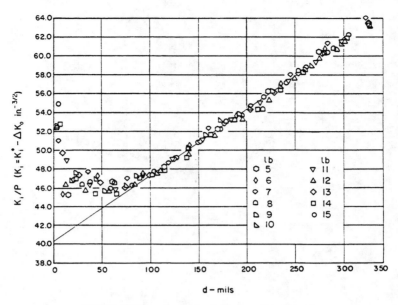

Fig. 10. K_i/P as a function of d for specimen No. 1.

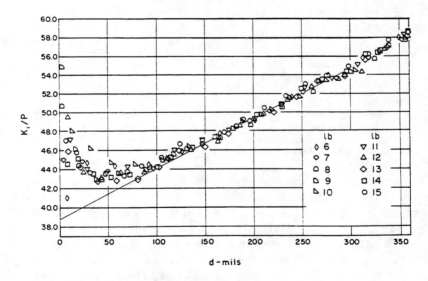

Fig. 11. K_i/P as a function of d for specimen No. 2.

where the stresses change most rapidly with r. At the positions where the crack opening measurements were made there should be little constraint to through-the-thickness contraction. With the assumption of plane stress deformation the values of K_i/P calculated from the crack opening were consistently higher than compliance calibration values. An assumption of plane strain would have produced a larger error in the same direction. The fact that the crack opening method gave consistently lower values is

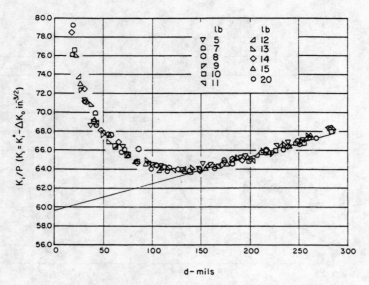

Fig. 12. K_i/P as a function of d for specimen No. 3.

Table 1. Comparison of K_i/P obtained by optical interference method with K_i/P obtained by compliance calibration

Specimen	Type	K_i/P optical (psi-$\sqrt{\text{in.}}$/lb.)	K_i/P, compliance§ (psi-$\sqrt{\text{in.}}$/lb.)
4 (705)	Uniform. $b_n/b = 1.0$, CoB†	54.0	53.1
1 (418)	Tapered. $b_n/b = 1.0$, CoB	40.1	37.9
2 (452)	Tapered. $b_n/b = 1.0$, IF‡	38.8	37.9
3 (μ)	Tapered. $b_n/b = 0.5$, CoB	59.7	53.7

† CoB = Crack in epoxy layer (center of bond).
‡ IF = Crack at epoxy-glass surface.
§ Compliance calibration procedure from [3].

most likely attributable to the non-linear deformation at the crack tip. As was demonstrated, the use of a plastic zone correction tends to lower the value obtained for K_i/P. For the first three specimens represented in Table 1, the agreement between the optical and the compliance values were sufficiently close that a correction did not seem warranted. For the last specimen represented in the table, the agreement was not as good. On comparing Fig. 12 with the corresponding curves for the other specimens, it is seen that specimen No. 3 showed a greater degree of crack tip non-linearity than the other specimens. Consequently, it would seem logical to apply a larger plastic zone correction to these data than to the others, which would have improved the agreement. However, since the correction is somewhat arbitrary, the uncorrected results were felt to better represent the accuracy of the optical interference technique.

SUMMARY

Crack openings measured by an optical interference technique can serve to estimate values of the opening stress intensity factor to within a few per cent of the values obtained by other methods. The requirements for its application are that the structure be modeled with a transparent material (at least in the vicinity of the crack) and that it be possible to see light reflected from the crack surfaces. The results indicate that a composite specimen with a crack introduced in the glue-line can be analyzed as though it were a monolithic specimen. This feature should greatly simplify the extension of the optical interference technique to complex structures. Measurements are made only in the vicinity of the crack tip, so that remote boundary conditions need not be considered explicitly; and the analysis of the measurements is straightforward. The analysis can be made in such a manner that the effect of non-singular terms and of non-linear deformation at the crack tip can be readily assessed. Because measurements can be made as a function of position along the crack front, the method is not restricted to two-dimensional analyses.

Acknowledgements — The many helpful suggestions made by Professor G. R. Irwin are gratefully acknowledged. Portions of this work were carried out under the direction of C. F. Bersch, of NAVAIR, whose encouragement was also most appreciated.

REFERENCES

[1] E. Sommer, An optical method for determining the crack-tip stress intensity factor. *Engng Fracture Mech.* **1**, 4, 705 (1970).
[2] S. Mostovoy, P. B. Crosley and E. J. Ripling, Use of crack-line-loaded specimens for measuring plane-strain fracture toughness. *ASTM J. Mater.* **2**, 3, 661 (1967).
[3] S. Mostovoy and E. J. Ripling, Fracture toughness of an epoxy system. *J. appl. Polym. Sci.* **10**, 1351 (1966).

The Determination of Mode I Stress-intensity Factors by Holographic Interferometry

A technique for estimating K_I using PMMA models is described and applied to a problem involving a transverse crack emanating from a circular hole in sheets of differing widths

by T. D. Dudderar and H. J. Gorman

ABSTRACT—The use of linear elastic fracture mechanics generally depends upon the availability of suitable analytical or numerical solutions for the relevant crack-tip stress-intensity factor, K. Convenient experimental verification of such solutions is a valuable aid to their correct application and can provide a practical substitute in real design situations of great complexity.

A convenient, new experimental technique for estimating the Mode I stress-intensity factor using holographic interferometry and test pieces cut from thin sheets of commercially available polymethylmethacrylate is described and demonstrated. The test pieces can readily be prepared to model any desired Mode I geometry and boundary conditions. In addition, a prior self-calibration procedure can be employed to enhance both convenience and accuracy. Real-time interference-fringe data from the crack-tip region are easily reduced and plotted to yield a straight line whose slope provides a one-parameter evaluation of the effect of geometry on the stress-intensity factor. This information, together with the crack length and applied stress, completely defines K.

Introduction

The basis of the widely accepted theory of linear elastic fracture mechanics involves the concept of a stress-intensity factor, K, associated with the assumed presence of a very sharp flaw or crack.[1] Specifically, it is defined by the simple relationship

$$K = \gamma \sigma^\infty \sqrt{\pi c} \quad (1)$$

where σ^∞ is the applied stress, c is the half length of the crack or flaw, and γ is a term related to geometry. Brittle fracture occurs whenever K reaches a critical value, K_c (representative of the material), due to either an increase in the applied stress, σ^∞, or a growth of the flaw length, $2c$, or both. As such, K_c

is a material parameter of use to both the materials scientist and the designer or materials user because it provides a quantitative evaluation of the conditions of fracture for a given material which can be applied to any prototype component or test piece for which the stress-intensity factor, K, can be determined.

Since the use of K_c requires the determination of K and, consequently, γ for any given geometry, considerable effort has been expended in the development of suitable analytical solutions. For the simplest case, that of an infinite sheet under uniaxial tension applied normal to the crack line, it can easily be shown that the geometrical term is unity. For more realistic and, consequently, complicated geometries, the results may vary considerably. Therefore, boundary-collocation procedures, finite-difference methods, or more complex mapping technique from mathematical elasticity, must be employed to develop the needed solutions. Many of these can now be found in the literature.[2] Frequently, the prototype geometry and boundary conditions may be of such complexity that significant idealization is required for the analysis; hence, the solutions may not accurately describe the problem at hand. Consequently, experimental evaluation is often of considerable importance to the rigorous application of the techniques of fracture mechanics.

The original experiment method was proposed by Irwin and Kies[3] as a means of determining a related parameter, the strain-energy release rate G. This is done by measuring the compliance, $C = dh/dP$, of a given test piece for a range of different crack lengths and then computing the strain-energy release rate from the derived relationship

$$G = \frac{1}{2} P^2 \frac{\partial C}{\partial c} \quad (2)$$

where P is the applied load per unit thickness and h is the distance between loading points. The stress-intensity factor can then be determined from a simple relationship involving G and the elastic constants (such as $K = \sqrt{EG}$ for plane stress). This technique

---

T. D. Dudderar is a Member of Technical Staff, Bell Laboratories, Murray Hill, NJ 07074. H. J. Gorman is Engineer, U. S. Army Electronics Command, Fort Monmouth, NJ; was formerly with Bell Laboratories.
Paper was presented at 1972 SESA Fall Meeting held in Seattle, WA on Oct. 17-20.
Original manuscript submitted: Feb. 11, 1972. Revised version received: Nov. 2, 1972.

has been used successfully many times and, like all such experimental methods, is subject to potentially significant errors, should conditions give rise to rapidly changing second derivatives of the experimental data.

Another technique also proposed by Irwin[4] and employed by various investigators including Marloff et al.[5] uses photoelasticity to determine stress-intensity factors directly from the measured stress distribution at the root of a machined notch in a photoelastic model. According to the mathematical theory of elasticity, the stress distribution near the root of an ideally sharp crack is related to the desired stress-intensity factor by the following field equations common to all Mode I* loading conditions regardless of geometry:[2]

$$\sigma_x = \frac{K_I}{(2\pi\rho)^{1/2}} \cos\left(\frac{\theta}{2}\right)\left[1 - \sin\frac{\theta}{2}\sin\frac{3\theta}{2}\right] - \sigma^\infty$$

$$\sigma_y = \frac{K_I}{(2\pi\rho)^{1/2}} \cos\left(\frac{\theta}{2}\right)\left[1 + \sin\frac{\theta}{2}\sin\frac{3\theta}{2}\right] \quad (3)$$

$$\sigma_{xy} = \frac{K_I}{(2\pi\rho)^{1/2}} \sin\frac{\theta}{2}\cos\frac{\theta}{2}\cos\frac{3\theta}{2}.$$

Here ρ is the radial distance from the crack tip and θ the angle measured from the crack line. Unfortunately, at nonzero angles, the actual stress distribution encountered under experimental conditions is usually too sensitive to details of the crack-tip geometry to provide a reliable basis of comparison. Consequently, for most effective correlation, data should be taken along the crack line, $\theta = 0$. In this case, the field equations simplify to

$$\sigma_x = \frac{K_I}{(2\pi\rho)^{1/2}} - \sigma^\infty$$

$$\sigma_y = \frac{K_I}{(2\pi\rho)^{1/2}} \quad (4)$$

and $\sigma_{xy} = 0$.

In essence, the photoelastic technique measures the principal-stress difference[6] which by eqs (4) is given by

$$\sigma_1 - \sigma_2 = \sigma_y - \sigma_x = \sigma^\infty.$$

As a result, the photoelastic technique suffers from a relative insensitivity to the stress distribution in the area of maximum interest. However, the present technique, holographic interferometry, completely inverts this situation by providing a means of measuring the principal-stress sum, or

$$\sigma_1 + \sigma_2 = \sigma_x + \sigma_y = \frac{2K_I}{(2\pi\rho)^{1/2}} - \sigma^\infty. \quad (5)$$

Consequently, holographic interferometry is most sensitive in the region of maximum interest.

The general theory of holographic interferometry is well known and only certain particularly relevant features need be reviewed here. In earlier investigations,[7,8] it has been demonstrated that holographic interferometry is well suited to studies of the strain field associated with the tip of a sharp crack in a stressed body. Because of the need for precise image alignment at the crack tip, use of the real-time holographic-interferometry technique is the most desirable. Using this technique, the hologram provides a means of storing an optical replica of the light wavefront from the test piece which can be reconstructed at any time as a reference. Optical interference with light from the test piece generates a live fringe pattern which is related to changes in the state of stress or deformation of the test piece, if all other factors are retained unchanged.

For the present investigation, very thin sheets of transparent polymethylmethacrylate (PMMA) were used as test-piece material. As demonstrated in Ref. 7, they provide for predominantly two-dimensional elastic behavior over the region of interest. An essential advantage over the sheet resins used for photoelasticity is the fact that machined slots in PMMA sheets can readily be sharpened with a razor blade to produce the sharp-crack limiting case defined by the theory. This provides a practical and repeatable test piece which accurately reproduces the fracture conditions often encountered in real-world situations.

Theory

For the general case of real-time holographic interferometry with plane transparent specimens, the light-intensity distribution can be written as[9]

$$I = \frac{a^2}{2}\left\{1 + \cos\left[\left(\frac{2\pi}{\lambda}\right)C_p(\sigma_1 + \sigma_2)l\right]\right.$$
$$\left. \cdot \cos\left[\left(\frac{2\pi}{\lambda}\right)\cdot C_m(\sigma_1 - \sigma_2)l\right]\right\}^\dagger \quad (6)$$

where C_p and C_m are stress-optic constants, λ is the light wavelength, and l is the geometrical light-path length through the test piece and a is the light amplitude. For PMMA, $C_m \approx 0$ (little or no photoelastic sensitivity) and the intensity expression may be simplified to

$$I = \frac{a^2}{2}\left\{1 + \cos\left(\frac{2\pi}{\lambda}\right)C_p(\sigma_1 + \sigma_2)l\right\}. \quad (7)$$

Therefore, dark fringes or isopachics appear whenever $I = 0$ or

$$\frac{2\pi}{\lambda}C_p(\sigma_1 + \sigma_2)l = \pi, 3\pi, 5\pi, \ldots, (2N-1)\pi$$

$$N = 1, 2, 3 \text{ etc.}$$

Defining the fringe count as $m = \frac{2N-1}{2}$ then

$m = \frac{C_p}{\lambda}l(\sigma_1 + \sigma_2)$ with black fringes at $m = 1/2$, $3/2, 5/2 \ldots$ etc. Substituting from eq (5) yields

$$m = \frac{C_p}{\lambda}l\left(\frac{2K_I}{(2\pi\rho)^{1/2}} - \sigma^\infty\right) \quad (8)$$

---

* There are three modes of fracture loading which are usually identified. The present paper will be confined to Mode I which is the most commonly considered. In Mode I, the test piece is loaded so that the crack surfaces move apart perpendicular to the crack plane. Accordingly, K becomes K_I.

† This result was originally derived by Nisida and Saito[10] for a Mach-Zehnder interferometer with a stressed birefringent test piece.

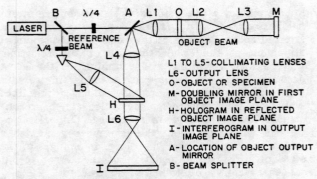

Fig. 1—Schematic diagram of the double-pass holographic interferometer

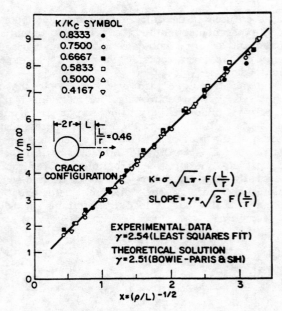

Fig. 2—Plot of reduced data for $L/r = 0.46$

as a description of the fringe pattern in the vicinity of the tip of any crack. Far away from the crack tip, where only the applied stress, σ^∞, is present, $\sigma_1 \equiv \sigma^\infty$ and $\sigma_2 \equiv 0$, so that the associated fringe order is simply $m^\infty = C_p l \sigma^\infty / \lambda$. This may be used to normalize eq (8). By eq (1), $K_I = \gamma \sigma^\infty \sqrt{\pi c}$ so that

$$\frac{m}{m^\infty} = \gamma \, (\rho/2c)^{-1/2} - 1 \qquad (9)$$

and a data plot of m/m^∞ vs. $(\rho/2c)^{-1/2}$ will yield a straight line with the desired geometrical term, γ, as its slope.

Experiment

In order to initially evaluate the proposed technique, a circular hole with one side crack in an infinite sheet in uniaxial tension was chosen as a suitable test geometry. Elastic solutions to the problem were published by Bowie[11] and the results were used by Paris and Sih[2] to derive a geometrical correction factor $F(L/r)$ where $L = 2c$, the crack length, and r is the radius of the hole. Following Paris and Sih,

$$K_I = F\left(\frac{L}{r}\right) \sigma^\infty \sqrt{\pi c}$$

so that $\gamma = \sqrt{2}\, F\left(\dfrac{L}{r}\right)$ provides the desired comparison of experiment and analysis.

To represent the infinite elastic sheet, a thin but wide (0.028 in. × 8 in. × 8 in.) strip of PMMA was prepared with a central flaw of a total length, $2a = 2r + L$, of 0.800 in., one tenth the sheet width ($b/a = 10$). Because it was necessary to use a very thin specimen in order to suppress the region of three-dimensionality at the crack tip,[7] it was also advantageous to use a double-pass focussed-image interferometer arrangement similar to the interferometer used by O'Regan and Dudderar.[9] The interferometer is shown schematically in Fig. 1. The output can be taken either from a partially reflecting mirror located near the spatial filter-beam spreader at A, or a full mirror at the same point with the lens L1 shifted so as to slightly skew the returning beam. The latter technique was adopted because it provides a four-fold increase in output efficiency with negligible loss of precision because the doubling mirror images the object back on itself. The resulting double-pass arrangement yields twice the sensitivity of a single-pass system and provided a maximum fringe count of 9 to 11 fringes across the region of interest before the test piece fractured. A dead-load test frame with equilibrated grip assemblies designed to minimize bending were used to stress the specimen. For greater flexibility, both the specimen and the hologram were provided with adjustable mounts. The hologram mount was used to clear the initial fringe field and maintain its orientation, while the specimen mount was used to compensate for any movement and maintain precise registration during loading.

The reference hologram was recorded at a minimum tare load necessary to hold the specimen in position and support the bottom grip, or about 1 kg (less than 2 percent of the smallest load analyzed). After processing and reregistration of the hologram, the resulting interferogram was completely cleared of fringes at tare, despite a rather irregular specimen surface. Therefore, all subsequent interference patterns evolved wholly from the developed stress fields as loads were applied. Interferograms were recorded for a dozen evenly spaced loads to fracture. Crack-line data taken from the six interferograms recorded at K/Kc between 0.417 and 0.8333 were reduced* and plotted as shown in Fig. 2. For the present specimen ($L/r = 0.46$), the analysis[2] predicts an $F(L/r)$ of 1.77 and a slope, γ, of 2.51, which compares very well with the experimental slope of 2.54 obtained by least-squares fitting data taken along the most linear portion of the data plot, or to within 4 to 5 percent of the half flaw length from the crack tip. This suppresses the effects of three dimensionality at the crack tip, again following the earlier work.[7] These

---

* Values of m^∞ were calculated from independent calibration data showing a fringe sensitivity of 30 psi/fringe/in., which compares favorably with the result obtained by Nisida and Saito[10].

Fig. 3—Interferograms taken from second PMMA specimen at successively reduced widths. The magnification is 1×

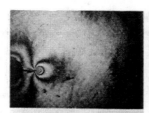
b/a = 9.80, σ = 499 psi

b/a = 4.90, σ = 1167 psi

b/a = 3.67, σ = 1111 psi

b/a = 2.30, σ = 1017 psi

b/a = 1.94, σ = 1320 psi

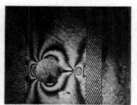
b/a = 1.47, σ = 1322 psi

data are generally less accurate, anyway, because of their greater sensitivity to details of the crack-tip geometry which cannot be controlled or even known with absolute precision.† The interferograms recorded at lower loads were ignored because they contained fewer fringes than could readily be used to compute slopes with confidence. On the other hand data taken from interferograms recorded at the two highest loads were omitted because of the occurrence of some irregular propagation of the crack tip approaching fracture.

The technique was then applied to the more realistic problem of the same hole, plus crack flaw centrally located in a strip of finite width and loaded in tension. To do this, another sheet specimen was first prepared with the same 8 in. × 8 in. geometry, but an L/r of 1.07, and the original experiment was repeated. However, this time the load was raised to only 60 percent to 70 percent of the estimated load for fracture. After testing, the specimen was unloaded, removed, machined to a reduced width, $2b$, of 4 in., replaced and retested using the original hologram. This procedure was repeated for four successively smaller widths, Figs. 3, to generate the data shown in Table 1. Because of the cyclic nature of this procedure some crack growth became apparent at the third loading which gave a new L/r of 1.22 and a slightly reduced $F(L/r)$ of 1.27. Therefore, a new reference hologram was made and successive load maxima were limited to about 40 percent of the estimated values required for fracture. For all SP2 tests, only those fringes lying within the linear portion of the reduced-data plot (between 5 percent and 50 percent of a half flaw length from the crack tip) were used for the analysis.

Discussion

The present technique provides a reasonably effective means of directly estimating Mode I stress-intensity factors. The data presented in Table 1 indicate that the finite-width test piece with L/r a little greater than 1 behaves like an infinite width sheet for width-to-flaw-length ratios above about 3.5, and that, by the time the b/a ratio is reduced to 1.5, the stress-intensity factor has risen to twice that which would apply for the infinite sheet. The slight drop-off seen in the results taken from the third loading ($b/a = 3.67$) was associated with some growth of the crack itself, as well as a certain loss of confidence due to an insufficiency of reliable fringe data.

Several areas of uncertainty were encountered which might also be expected to contribute to a small overall deviation in the results. For example, the actual location of the crack tip is ambiguous, as has already been mentioned. This is compounded by physical conditions which make it impossible to record interferograms which are simultaneously focussed with equal accuracy on both the fringe field and the crack. As a result it is often easier to measure relative fringe spacings with precision than to properly relate them to the distance from the effective crack tip as seen on the interferogram. However, any error resulting from a crack-tip misidentification will result in a loss of linearity at the high end of the plot of reduced data. That being the case, a small correction in the presumed location of the crack tip, within the range of uncertainty, can usually be made to compensate and render a more-linear plot. Further, for cracks with moderately irregular or "non-ideal" tips, this procedure might provide a means of defining an "effective" crack length in terms of the resulting stress field averaged through the thickness.

As mentioned previously, the three-dimensionality or "plane-strain" condition at the crack tip also creates some problems by suppressing the fringe count to an increasing degree as the crack tip is approached.[7] This necessitates the use of thin-sheet test pieces and relatively long cracks (large L/t ratios) in order to confine this phenomena to a restricted portion of the interferogram or test piece. Using the results of the earlier investigation, it can be seen that, for the present geometry, this error becomes relatively insignificant for distances from the crack tip beyond 4 to 5 percent of the half flaw

---

† *Because the crack front cannot ever be made absolutely straight and square to the surfaces, its effective position and, hence, the crack length, must always be ambiguous to a greater extent than resolution and repeatability of measurement might otherwise permit.*

TABLE 1—POLYMETHYLMETHACRYLATE TEST PIECES 1 AND 2

Test Piece No.	No. of Fringes Analyzed	Width 2b (in.)	Width/Flaw Length 2b/2a	Crack Length/ Circle Radius L/r	Analytic F_A (L/r) for b/a = ∞	Experimental F_E (L/r)	Ratio $\frac{F_E(L/r)}{F_A(L/r)}$
1*	54	8	10.00	0.46	1.775	1.769	1.01
2*	27	8	9.80	1.07	1.33	(1.33)*	(1.00)*
2	18	4	4.90	1.07	1.33	1.29	0.97
2	11	3	3.67	1.07	1.33	1.22	0.93
2	20	1.97	2.3	1.22	1.27	1.53	1.20
2	17	1.66	1.94	1.22	1.27	1.68	1.32
2	18	1.26	1.47	1.22	1.27	2.58	2.01

* Test piece 1 was made from independently calibrated material with a sensitivity of 30 psi/fringe/in. Test piece 2 was self-calibrated at b/a = 9.8 to match the analytic result obtained for b/a = ∞ which gave a sensitivity of 31.2 psi/fringe/in. Both values are reasonable for PMMA from different sources and probable experimental variations.

length. On the other hand, the local-stress-field solution used to derive eq (9) becomes less accurate as the distance from the crack tip is increased. This will be seen mainly as a loss of linearity at the low end of the reduced-data plot and indicates the maximum limit on the range of acceptable data for the proposed technique.

The need to provide an independent calibration with its inherent uncertainties can be avoided by a self-calibration procedure of both the sheet and crack. If, prior to machining the test piece to the desired geometry of the prototype, it is simply tested as a large sheet with a crack in it, a direct calibration can be made. For example, a thin PMMA sheet with a relatively short center-cut slot terminating in razor-sharpened ends might first be loaded, but not to fracture, in uniaxial tension. This would effectively model the special case of a crack in an infinite sheet. A real-time holographic interferogram of the region of maximum interest would then provide fringe-distribution data for this special case where γ is known to be unity. Subsequently, a plot of fringe order divided by applied stress, m/σ^∞, vs. $(\rho/2c)^{-1/2}$ would yield a straight line whose slope, s, represents a reduced calibration factor relating the fringe order to the stress, $m = s\sigma$. This calibrated unbroken sheet can then be machined to any desired prototype geometry (involving, if possible, the existing crack). It can then be retested using the original hologram to acquire new data for a plot of m/m^∞ vs. $(\rho/2c)^{-1/2}$ which would then provide a definitive γ for the evaluation of K in the usual manner.

In a similar manner, other geometries for which satisfactory solutions exist can be used to calibrate the material. In fact, this was done with the second specimen in the present study because it was made from a sheet of PMMA supplied by a different manufacturer for which no independent calibration had been made. Data taken from its first loading at maximum width was analyzed to provide a sensitivity about 4 percent higher than that of the first specimen. Consequently, for SP2, the initial values of slope and $F(L/r)$ coincide with the theoretical values (and are shown in parenthesis in Table 1) while all successive values are computed using the resulting, more appropriate, sensitivity.

Because the results depend only on the slope of the reduced-data plot, the occurrence of errors due to irregular shifts in the absolute-fringe order (which may result from air currents or unwanted long-term motions, or just erroneous counting) have little significant effect.

Conclusions

Holographic interferometry can be used to estimate linear-elastic stress-intensity factors. Thin-sheet polymethylmethacrylate models of prototype or real-world geometries can be used as test pieces and holographic interferograms recorded to measure their linear-elastic response to applied loadings. Reduced isopachic-fringe data along the crack line can be least-squares fitted to provide a straight line whose slope, γ, defines the dependence of the stress intensity on geometry and boundary conditions.

Acknowledgments

The authors wish to express their appreciation to L. Green for preparation of the PMMA test pieces and his assistance in the development of the experimental apparatus.

References

1. Irwin, G. R., "Fracture Mechanics," *Structural Mechanics*, Pergamon Press, New York, 557-92 (1960).
2. Paris, P. C. and Sih, G., "Stress Analysis of Cracks," *Fracture Toughness Testing and Its Applications*, ASTM Special Technical Publication 381, 30-81 (1965).
3. Irwin, G. R. and Kies, J. A., "Critical Energy Rate Analysis of Fracture Strength," *Welding Research Supplement*, 33, 193S-198S (April 1954).
4. Irwin, G. R., "Discussion (of paper by Wells & Post)," *Proceedings SESA*, 16 (1), 93-96 (1958).
5. Marloff, R. H., Leven, M. M., Ringler, T. N. and Johnson, R. L., "Photoelastic Determination of Stress-intensity Factors," EXPERIMENTAL MECHANICS, 11 (12), 529-539 (1971).
6. Frocht, M. M., "Photoelasticity," 1, 2, John Wiley & Sons, New York (1948).
7. Dudderar, T. D. and O'Regan, R., "Measurement of the Strain Field Near a Crack Tip in Polymethylmethacrylate by Holographic Interferometry," EXPERIMENTAL MECHANICS, 11 (2), 49-56 (1971).
8. Dudderar, T. D. and O'Regan, R., "Holographic Interferometry in Materials Research and Fracture Mechanics," *Int'l. J. of Nondestructive Testing*, in press.
9. O'Regan, R. and Dudderar, T. D., "A New Holographic Interferometer for Stress Analysis," EXPERIMENTAL MECHANICS, 11 (6), 241-247 (1971).
10. Nisida, M. and Saito, H., "A New Interferometric Method for Two-dimensional Stress Analysis," EXPERIMENTAL MECHANICS, 4 (12), 366-376 (1964).
11. Bowie, O. L., "Analysis of an Infinite Plate Containing Radial Cracks Originating at the Boundary of an Internal Circular Hole," *J. of Mech. and Phys.*, 35, 60-71 (1956).

The Determination of Mode I Stress-intensity Factors by Holographic Interferometry

Paper by T. D. Dudderar and H. J. Gorman was published in the April 1973 issue of EXPERIMENTAL MECHANICS, pages 145-149

Discussion

by M. E. Fourney

The authors are to be congratulated on a very interesting extension of earlier work by Dudderar and O'Regan.[1] The method proposed in the paper for determining the Mode I stress-intensity factor appears quite attractive; for that reason I feel that certain discrepancies should be discussed.

The stress-intensity factor is obtained using the relationship

$$\sigma_1 + \sigma_2 = \frac{2K_1}{(2\pi\rho)^{1/2}} - \sigma_\infty$$

from the work of Paris and Sih.[2] This expression is valid for plane stress and in the near field of the crack (normally taken as less than 5 percent of crack length). However the authors have determined the stress-intensity factor from the linear portion of their data that lies between 5 percent and 50 percent of the half crack length and specifically ignoring data within the near field. Also the condition of plane stress cannot be met at the crack tip as was shown earlier by one of the authors.[1]

Theocaris[3] has approached the same problem from a different point of view. Using the same material, he determined the stress-intensity factor by investigating the plastic zone surrounding the crack tip. The size of plastic zone was determined from the change in optical thickness, which is proportional to the sum of the principal stresses. He found that the plastic zone extended 12 percent to 35 percent of the crack length for loads up to the critical value for the onset of fracture. The value of the stress-intensity factor thus obtained agreed with the work of Kobayashi[4] and others. Theocaris thus concluded that, for cracked Plexiglas, the model introduced by Irwin[5] (i.e., the crack tip surrounded by a circular plastic zone) is the correct one. The only notable differences in these two tests were: Theocaris used an edge-notched specimen, while Dudderar and Gorman used a circular hole with a single side-crack specimen, and the thickness-to-crack-length ratio used by Theocaris was several times larger than in the present paper.

It would appear that Dudderar and Gorman have correlated their experimental data with a purely elastic analysis, however not in the near field of the crack tip as intended by the theory; yet partially within the plastic zone as determined experimentally by Theocaris. The fact that the method works is both its salvation and damnation. I would like to suggest that perhaps the test specimen geometry plays an important role. The presence of the circular hole would add to the effective crack length and would influence the stress distribution outside of the near-field crack.

References

1. Dudderar, T. D. and O'Regan, R., "Measurements of the Strain Field Near a Crack Tip in Polymethylmethacrylate by Holographic Interferometry," EXPERIMENTAL MECHANICS, 11 (2), 49-56 (1971).
2. Paris, P. C. and Sih, G., "Stress Analysis of Cracks," Fracture Toughness Testing and Its Applications, ASTM Special Tech. Pub. 381, 30-81 (1965).
3. Theocaris, P. S., "Local Yielding Around a Crack Tip in Plexiglas," J. of Applied Mechanics, 37 (2), 409-415 (June 1970).
4. Kobayashi, A. S., "Method of Collocation Applied to Edge-Notched Finite Strip Subjected to Uniaxial Tension and Pure Bending," Boeing Co., Seattle, WA, Document No. D 2-23551 (Aug. 1964).
5. Irwin, G. R., "Linear Fracture Mechanics, Fracture Transition, and Fracture Control," Engrg. Fracture Mechanics, 1 (1968).

M. E. Fourney is Associate Professor of Engineering and Applied Science, School of Engineering and Applied Science, University of California, Los Angeles, CA 90024.

Authors' Closure

The authors agree that it would be better to use stress data taken from within the near-field region for the determination of stress-intensity factors. However, because of the stress-triaxiality conditions studied in the earlier work by Dudderar and O'Regan,[1] it becomes necessary to compromise by admitting data from the region lying next beyond the near field and, for this reason, we claim only a technique for *estimating* K, albeit a rather effective one. So long as the reduced data plot of m/m^∞ vs. $(\rho/L)^{-1/2}$ is linear, it represents an extension of near-field approximation at the same slope. As $(\rho/L)^{-1/2}$ approaches zero, away from the crack tip, the plot becomes less linear and the associated data are omitted because they no longer approximate the leading term solution. We also agree that the hole itself contributes to the effective crack length, and probably does influence the entire linear portion of the reduced data plot.

The authors are unable to satisfactorily account for the findings of Theocaris[2] in relation to their own experience with polymethylmethacrylate. In this paper, he employed specimens of almost identical geometry and material as were used in the original holographic investigation by Dudderar and O'Regan.[1] However, at no time in the latter study were plastic zones observed at the crack tips, especially on the enormous scale, up to 110 percent of the crack length, reported by Theocaris.

In our experience, the material response was predominantly linear elastic to well within 1 percent of the crack tip for loads up to 95 percent of fracture initiation. This was supported by the absence of any circular fringe-field discontinuities in the at-load interferograms and disappearance of almost the entire fringe field upon removal of the load. We feel that the caustic observed in the work of Theocaris should not be interpreted as a plastic zone as it is normally defined in fracture mechanics or materials science, but rather represents an optical anomaly related to the development of large transverse elastic strain gradients and the characteristics of his unique interferometer. In any event, he too must compute K from data lying well beyond the near-field region at the crack tip.

References

1. Dudderar, T. D. and O'Regan, R., "*Measurement of the Strain Field Near a Crack Tip in Polymethylmethacrylate by Holographic Interferometry,*" EXPERIMENTAL MECHANICS, 11 (2), 49-56 (1971).
2. Theocaris, P. S., "*Local Yielding Around a Crack Tip in Plexiglas,*" J. Appl. Mechs., 37 (2), 409-415 (June 1970).

FRACTURE ANALYSIS BY USE OF ACOUSTIC EMISSION[†]

H. L. DUNEGAN,[‡] D. O. HARRIS[§] and C. A. TATRO[‡]

Lawrence Radiation Laboratory, University of California, Livermore, Calif. 94550

Abstract—This report contains results of acoustic emission studies on flawed and unflawed specimens of aluminum and beryllium. Acoustic emission from the flawed specimens is found to begin at stress levels far below the general yield stress. A theoretical model given here indicates that the total number of acoustic emission signals from a specimen containing a crack should be proportional to the fourth power of the stress intensity factor obtained from a sharp-crack fracture mechanics analysis. This is in disagreement with experimental data from single-edge-notched fracture toughness specimens of the two materials, which indicate that acoustic emission varies more like the sixth to eighth power of the stress intensity factor. An example is given to show how the acoustic emission data obtained on fracture toughness specimens can be used to nondestructively test the fracture strength of an engineering structure.

NOTATION

SEN	single-edge-notched
K	stress intensity factor
r_y	plastic zone size
ϵ_{ys}	yield strain
ϵ_u	plastic strain
N	acoustic emission total counts
$\dot{N}$	acoustic emission count rate
V_p	volume of plastic zone
B	specimen thickness
F_c	critical load
K_{ce}	elastic critical stress intensity factor (not plastically corrected)
K_c	critical stress intensity factor (plastically corrected)
K_{Ic}	critical stress intensity factor (plain strain)
G_{ce}	critical energy release rate (not plastically corrected)
σ_{ys}	yield stress

INTRODUCTION

THE ACCELERATED interest and new analytical methods that have become available in the past few years in the field of fracture mechanics have created a demand for better methods of nondestructively detecting flaws in critical structures. This demand results from the fact that the stress intensity factor (K), and therefore fracture, is dependent on a flaw size parameter. Most nondestructive testing efforts have been primarily directed toward determining the size or shape of a flaw, and not with methods of directly measuring the stress intensity. The purpose of this paper is to show that a characteristic relationship exists between acoustic emission and the stress intensity factor that holds great promise for a means of nondestructively determining the stress intensity factor and therefore the fracture characteristics of a structure.

When a metal is subjected to loads it emits low level sound, a phenomenon known as acoustic emission. It has been well established[1–6] that these sounds are associated

---

[†]Presented at the National Symposium on Fracture Mechanics, Lehigh University, Bethlehem, Pa., June 19–21, 1967. Work done under the auspices of the U.S. Atomic Energy Commission.
[‡]Materials Engineering Section of Support Division, Mechanical Engineering Department, Lawrence Radiation Laboratory.
[§]Now in the Division of Engineering Mechanics, Stanford University, Stanford, California.

with permanent deformation processes such as twinning, slip, microcrack formation, etc. Most investigations associated with this phenomenon have centered around the study of unflawed specimens, but recently attention has been turned to the investigation of sounds emitted by a notched or cracked body. The work on cracked bodies[7, 8] employed sensing transducers of relatively low sensitivity, and the investigation using more sensitive transducers[9] did not concern a realistic crack, but a sharply machined notch.

In this investigation, we were primarily interested in developing a nondestructive method of determining the fracture characteristics of flawed structures, so a program was initiated to determine the acoustic emission characteristics of parts containing realistic cracks. This was accomplished by using single-edge-notched (SEN) fracture toughness specimens in a testing machine designed specifically for acoustic emission tests. Experiments were also performed in this machine on unflawed specimens to determine the difference in the characteristic emission between the flawed and unflawed specimens.

A theoretical analysis is given which shows that the acoustic emission due to plastic deformation processes taking place at the tip of the crack in a stressed specimen should vary as the fourth power of the stress intensity factor. Experimental results show that the stress intensity factor calculated from specimens of beryllium and 7075 aluminum with different crack lengths is the normalizing factor for acoustic emission observed from these specimens, but that the exponent varies more like the sixth to eighth power. Although these results do not agree with the theoretical analysis, they have important advantages when applied to stress intensity analysis of flawed structures. This higher exponent will tend to weigh heavily the larger flaws in an actual structure so that we might expect the acoustic emission as a function of load on a structure undergoing test to closely resemble the acoustic emission as a function of stress intensity factor in a fracture toughness specimen. Examples are given of how the fracture load or flaw size of a structure can be determined from information of this type.

THEORETICAL CONSIDERATIONS

Previous investigations have shown that acoustic emission from metals is associated with plastic deformation processes. Unflawed tensile specimens have been observed to be very quiet until shortly before yielding occurs. Acoustic emission is high during and shortly after yielding, then decreases as further straining takes place. These findings indicate that if a structure or specimen is acoustically monitored during loading, and emission is observed prior to general yielding, then there must be a stress concentration that is raising the stress level above the yield point in some regions. A crack would provide such a stress concentration, and the presence of cracks would result in emission below the general yield point.

The stresses in the vicinity of an elastic crack tip are controlled by a single parameter, K, known as the stress intensity factor[10–13]. This parameter depends on the body geometry, the shape, size, and location of the flaw, and the type of loading that is present. It is known for a wide variety of geometries[10]. Since the acoustic emission in a flawed specimen stressed below general yield is dependent on the plastic strains in the vicinity of the crack tip, and these strains in turn depend on the stress intensity factor, it would be expected that acoustic emission characteristics should be closely related to the stress intensity factors for the flaws present. Hence, it would be expected that acoustic emission for a given series of specimens in which only the flaw size was changed would be dependent on the stress intensity factor for the flaw, rather than the load or nominal stress on the specimen.

A model which predicts the relationship between the observed total emission counts and the stress intensity factor is based on the following observations and assumptions:

1. It has been observed that a metal gives the highest acoustic emission rate (number of observed emission bursts per second) when strain levels are close to yielding. Emission is less at lower and higher strains, as shown schematically in Fig. 1.

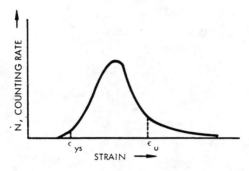

Fig. 1. Rate of acoustic emission as a function of strain.

2. The approximate size and shape of the zone of plastically deformed material adjacent to the crack tip can be calculated from the equations for the elastic stress field in the vicinity of the crack tip[14]. The strains (and stresses) are elastic outside this zone, near the yield point at the borders of the zone (elastic-plastic boundary), and plastic inside the zone. The approximate size and shape of the plastic zone is shown in Fig. 2. The value of r_y is approximately $1/2\pi(K/\sigma_{ys})^2$.

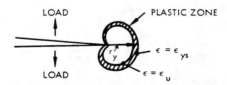

Fig. 2. Schematic of plastic zone at the tip of a crack. Material in the shaded zone is responsible for the majority of acoustic emission.

3. The strains are assumed to vary as $r^{-1/2}$ where r is the radial distance from the crack tip. This is the variation when the strain field is elastic, $\epsilon = c/\sqrt{r}$.

4. Since most of the emission from the material occurs when the strains are between ϵ_{ys} and ϵ_u (see Fig. 1), it seems reasonable to assume that the observed acoustic emission count rate, $\dot{N}$, is proportional to the rate of increase of the volume of the material which is strained between ϵ_{ys} and ϵ_u. Regions which are strained more than ϵ_u will not contribute significantly to the count rate, nor will regions strained below ϵ_{ys}. This assumption can be written as

$$\dot{N} \propto \dot{V}_p$$

where V_p is the volume of material which is strained between ϵ_{ys} and ϵ_u.

5. It is now necessary to estimate V_p. It is known that $\sigma_{ys}/E \approx \epsilon_{ys} = \epsilon$ at $r = r_y = 1/2\pi(K/\sigma_{ys})^2$. This can be used to determine the proportionality constant, c, in the relationship (from 3 above) as follows:

$$\epsilon = \frac{c}{\sqrt{r}} = \frac{\sigma_{ys}}{E} \approx \frac{c}{\sqrt{(1/2\pi(K/\sigma_{ys})^2)}}.$$

Solving for c,

$$c = \frac{\sigma_{ys}}{E}\sqrt{\left(\frac{1}{2\pi}(K/\sigma_{ys})^2\right)} = \frac{K/E}{\sqrt{(2\pi)}}.$$

The strain then varies according to

$$\epsilon = \frac{c}{\sqrt{r}} = \frac{K/E}{\sqrt{(2\pi r)}}.$$

This can be used to calculate the value of r at which $\epsilon = \epsilon_u$, which will be denoted as r_u.

$$r_u = \frac{1}{2\pi}\left(\frac{K}{E\epsilon_u}\right)^2$$

Now, taking the plastic zone to be approximately circular,

$$V_p \approx \pi(r_{ys}^2 - r_u^2)B = \pi B\left\{\left[\frac{1}{2\pi}\left(\frac{K}{E\epsilon_{ys}}\right)^2\right]^2 - \left[\frac{1}{2\pi}\left(\frac{K}{E\epsilon_u}\right)^2\right]^2\right\}$$

$$= \frac{\pi B}{4\pi^2 E^4}\left[\frac{1}{\epsilon_{ys}^4} - \frac{1}{\epsilon_u^4}\right]K^4 = \frac{B}{4\pi}\frac{\epsilon_u^4 - \epsilon_{ys}^4}{(E\epsilon_u\epsilon_{ys})^4}K^4$$

where B is the plate thickness. Thus

$$V_p \propto K^4.$$

Now, applying the assumption $\dot{N} \propto \dot{V}_p$ leads to

$$\dot{N} = \frac{dN}{dt} \propto \dot{V}_p = \frac{d}{dt}(V_p) = \frac{d}{dt}(K^4).$$

Integrating to give the total counts, N, as a function of the stress intensity factor, K, immediately gives

$$N \propto K^4.$$

This relationship predicts that if all the observed acoustic emission pulses are added up as the test proceeds, then at any time the total number of counts will be proportional to the fourth power of the stress intensity factor which is present for the flaw in the specimen at that time.

It is of interest to note that the assumptions leading to the fourth power exponent can be modified in several ways without changing the value of the exponent. For instance, the value of the exponent is not changed by assuming: (1) the total count is proportional to the volume of plastically deformed material; (2) the strains in the plastic zone vary as r^{-1}, rather than $r^{-1/2}$ as exists for mode III deformation; (3) the shape of the plastic zone is the elongated shape shown in [14], rather than circular. Thus, it is seen that the fourth power is not highly dependent on the assumed model.

EXPERIMENTAL PROCEDURE

Previous results by the authors[5] showed that the acoustic emission contains very broad band frequency components, which makes it convenient to operate at much higher frequencies than used by other investigators. This has two distinct advantages over earlier methods. (1) The need for soundproof chambers to eliminate extraneous noise no longer exists; a thin sheet of plywood is sufficient for most purposes, and this is only necessary to eliminate high frequency sources of noise such as keys jingling, etc. (2) Narrow banding techniques can be employed to increase the signal-to-noise ratio of the detection system; gains of 100 dB are easily accomplished by narrow banding around the resonance frequency of the transducer used for detecting the acoustic emission. These two features of the detection system used in this investigation have taken acoustic emission testing techniques out of the realm of the laboratory and into practical application.

Tensile tester

The frequency range used in our immediate investigation is between 30 and 150 kc. The primary consideration for choosing this frequency band is that it is high enough to eliminate most airborne noise, yet low enough that the data can be recorded on magnetic tape for further analysis. We found in our tests on tensile specimens that normal testing machines generate noise in this frequency band, so we designed a quiet tensile testing machine for conducting acoustic emission tests on tensile and fracture toughness specimens. The machine (Fig. 3) consists of a U-shaped frame with a flexure pivot to reduce its rigidity. There are no moving parts in the main frame, a factor that reduces extraneous noise during loading. The specimen is connected between the arms of the frame, which are pushed apart by the hydraulic ram at the open end, thereby loading the specimen.

Nylon pads were inserted between mating surfaces, when possible, to reduce friction and provide sound and electrical insulation. Further sound isolation was provided by the small-diameter wires which were used to connect the grips to the rest of the machine, as shown in the photograph.

The load was applied by pressurizing the hydraulic ram. The pressurization system consisted of a high-pressure accumulator (pressure reservoir) with a small-diameter capillary leading to the ram. The accumulator was charged to the desired pressure and the oil bled through the capillary to the ram, thus providing a nearly constant loading rate. A more complete description of the pressurization system is included in [5].

Specimen preparation

Two types of specimens were used in this investigation: smooth tensile specimens, and cracked fracture toughness specimens. Two materials were studied: Brush Beryllium Company designation N50A beryllium, and 7075-T6 aluminum. The beryllium is of relatively high purity ($\sim$99·3 per cent). It has a yield strength of 30 ksi and an ultimate strength of 48 ksi.

The specimen configuration chosen for the portion of the investigation dealing with flawed bodies was the single-edge-notched (SEN) plate in tension. This configuration was chosen because we wanted to obtain fracture toughness data from the cracked specimens, and the SEN plate in tension consumes a minimum of material[15] for fracture toughness specimens. This was of primary consideration, due to the high cost

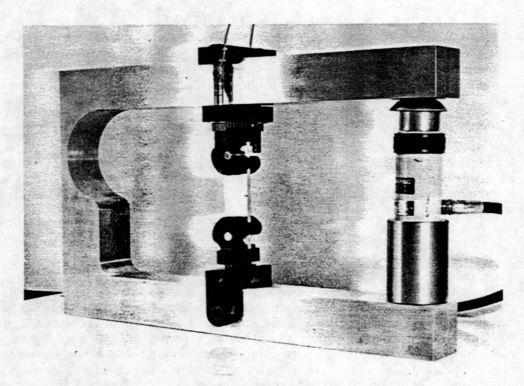

Fig. 3. Tensile tester.

of the N50A beryllium used in the majority of the tests. Another advantage of the SEN specimens is ease in fabrication and testing. All cracked specimens were $1\frac{1}{2}$ in. wide, 6 in. long, and proportioned according to the recommendations of Brown and Srawley[15]. The aluminum specimens were $\frac{1}{4}$ in. thick, and the beryllium 0·10 in. thick.

The aluminum fracture specimens were precracked by fatiguing. Due to the difficulty of producing controlled fatigue cracks in beryllium at room temperature, another method of precracking was devised. As shown in Fig. 4, a compression zone was produced on the faces of specimens, just below the machined notch root. A wedging force, which tended to open the machined notch, was then applied by pounding a chisel into the notch. The crack would run from the root of the notch and be arrested in the zone of compression. The length of the crack was easily controlled by the location of the compression zone, and good cracks were consistently obtained. A more complete coverage of the precracking procedure and other considerations concerned primarily with the fracture toughness testing aspects of the beryllium work is contained in [16].

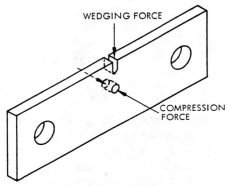

Fig. 4. Precracking setup.

The tensile specimens were $1\frac{1}{2}$ in. wide and 6 in. long with a centrally located reduced section.

A preloading procedure was devised to eliminate extraneous noise and acoustic emission from the highly stressed portion of the specimen around the pins. This procedure consisted of gluing the pins into the specimen with rubber cement, and then preloading the area around the pins, using the fixture shown in Fig. 5. The maximum preload was taken to be larger than the anticipated fracture load, and was applied in a conventional testing machine. This procedure was based on the observation that when a metal is loaded, it will not emit sounds until the load becomes larger than any previously applied load. Thus, any emission which would normally originate at the highly stressed region around the pins is eliminated during the preloading procedure, and any emission which is observed during the ensuing test therefore comes from the region around the crack or from the reduced section of the tensile specimen.

The hooklike grips seen in Fig. 3 were necessary in order to easily accommodate the specimen after the pins were glued in.

The four-arm bridges on the bolts were used to measure the total load by adding the output of the bridges. The difference in output was used for adjusting the nuts to provide symmetrical loading. A switching unit was constructed to easily change between sum and difference modes of operation.

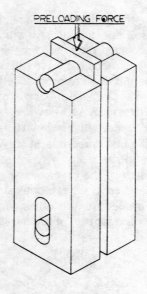

PRELOADING FIXTURE

Fig. 5. Preloading fixture.

Instrumentation

The electronic system for detection of the acoustic emission and recording of the data consists of the following components (a block diagram is provided in Fig. 6):

Sensing transducer. A PZT transducer with a fundamental thickness mode resonance of 400 kc. Two types were used, one with a 1½-in. dia. which gave a cross-coupling resonance at 70 kc and another with a diameter of ¾ in. which gave a cross-coupling resonance at 120 kc.

Preamplifier. A low-noise preamplifier with 80 dB gain. This unit also had a band

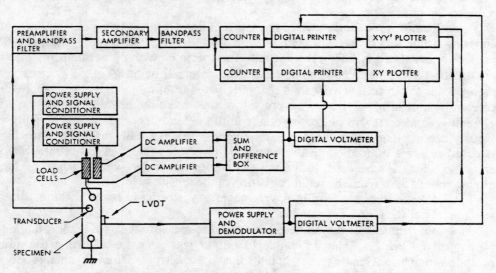

Fig. 6. Block diagram of electronics for acoustic emission system.

pass filter which was set at 30–100 kc for the first type transducer above and 100–300 kc for the second type.

Secondary amplifier. Another amplifier to further amplify the transducer output (gain 10 dB). This provided an overall system gain of 100 dB (10^5).

Variable band pass filter. Normally operated at 30–100 kc for first type transducer and 90–150 kc for second type.

Digital counters. Counters to detect the number of acoustic emission pulses detected by the transducer. One counter was set to measure incoming counts per second, the other was set to measure the total number of incoming counts as the test proceeded.

Digital printers. Printers to record, in digital form, the data from the counters as well as from the digital voltmeters used to measure crack-opening displacement and load.

XY plotters. Plotters to plot directly, in graphical form, various parameters of interest (as indicated in the block diagram).

Other support instrumentation, an Ampex Fr 1300 tape recorder, a Panoramic spectrum analyzer for determining the frequency response of transducers and the acoustic emission, and an oscilloscope for observing the acoustic emission pulses are not included in the block diagram.

RESULTS

Unflawed specimens

Figure 7 is an XYY′ plot of the acoustic emission and stress as a function of strain from an N-50A beryllium tensile specimen. This plot was obtained from the analog of the digital information of the counter set to read counts per second. As a result, it gives the rate of emission as a function of stress and strain in the specimen.

Figure 8 is an XYY′ plot of the summation of emission on the same specimen as Fig. 7. The steps in the acoustic emission curve represent time intervals of 2 sec.

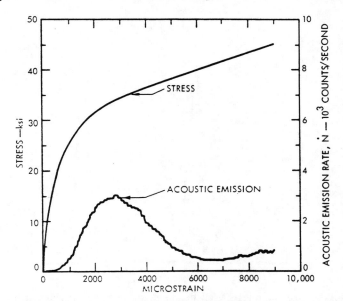

Fig. 7. Stress and acoustic emission rate as functions of strain, for a beryllium tensile specimen.

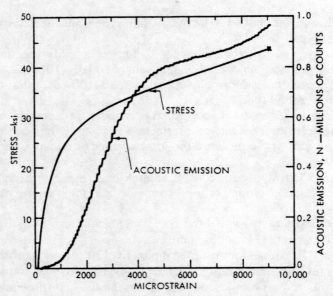

Fig. 8. Stress and total acoustic emission as functions of strain, for the beryllium tensile specimen of Fig. 7.

Figure 9 shows the acoustic emission rate as a function of stress for a beryllium and 7075-T6 aluminum specimen. It has been consistently observed in plots of this type that the acoustic emission is symmetrical when plotted as a function of stress. This symmetry has allowed us to perform a little black magic and predict where the specimen will fail. This is accomplished by observing the XY plotter as the test progresses until the peak of the emission curve becomes well defined. A mirror image of the initial part of the curve up to the peak is then folded around an axis bisecting the peak. The stress corresponding to the value where this mirror image extrapolates to zero emission is

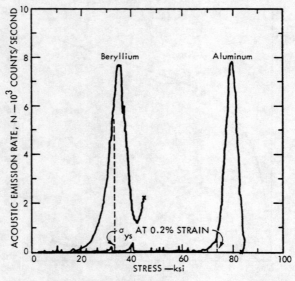

Fig. 9. Acoustic emission rate as a function of stress for beryllium and aluminum specimens.

taken as the fracture stress. This method is accurate to a few per cent, much of the time to within 1 per cent.

Flawed specimens

Figure 10 presents acoustic emission plots as a function of load for six SEN beryllium fracture toughness specimens. All of these specimens contained a sharp crack. The plots in this figure are seen to differ considerably from one another. This is an expected result, and is primarily due to variation in the fracture load resulting from differences in the crack length. However, the tests with similar crack lengths (1, 3, 4)

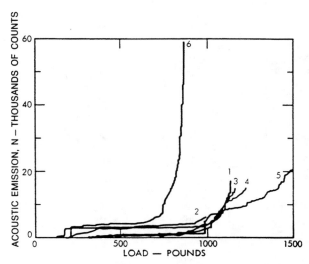

Fig. 10. Acoustic emission as a function of load for six beryllium fracture specimens.

show similar behavior. Figure 11 is the same summation acoustic emission data, only this time plotted as a function of stress intensity factor. These plots are seen to bear a greater resemblance to one another than the corresponding plots using load as the independent variable. This indicates that the emission behavior is more closely associated with stress intensity factor than with tensile load.

Figure 12 shows acoustic emission load plots of two SEN fracture toughness specimens of 7075-T6 aluminum with different crack lengths. Figure 13 is the same data plotted as a function of stress intensity factor. Note again that the stress intensity factor tends to normalize the data for the two different crack lengths.

Fracture toughness data

A summary of the fracture toughness values obtained from the acoustically monitored cracked specimens is presented in Table 1. Additional fracture toughness data on beryllium specimens which were not acoustically monitored is contained in [16]

Experimental values of exponent in N and K^s

The values of the exponent in the relation

$$N \propto K^s$$

were obtained from log–log plots of the information contained in Figs. 10 and 12 and are shown in Table 2.

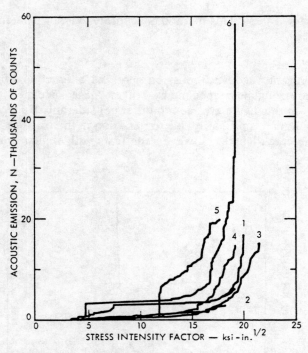

Fig. 11. Acoustic emission as a function of stress intensity factor, for the six beryllium fracture specimens of Fig. 10.

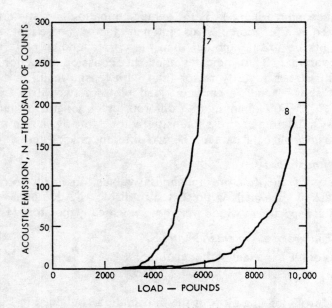

Fig. 12. Acoustic emission as a function of load for two aluminum specimens.

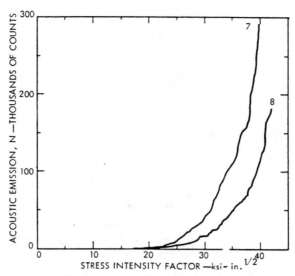

Fig. 13. Acoustic emission as a function of stress intensity factor, for the two aluminum specimens of Fig. 12.

Table 1. Summary of fracture toughness data

Test no.	Material	F_c (kips)	a_0 (in.)	K_{ce} (ksi(in.)$^{1/2}$)	G_{ce} (in.-#/in^2)	K_c† (ksi(in.)$^{1/2}$)	K_c‡ (ksi(in.)$^{1/2}$)
1	Be	1·14	0·56	19·6	8·46	28·9	21·1
2	Be	0·99	0·60	18·8	7·78	24·5	19·7
3	Be	1·18	0·58	21·4	10·1	§	22·9
4	Be	1·22	0·52	19·1	8·03	27·2	20·6
5	Be	1·51	0·41	17·7	6·90	24·2	18·7
6	Be	0·87	0·67	19·3	8·20	25·4	20·6
7	Al	6·16	0·54	40·5	164	46·8	—
8	Al	9·56	0·38	41·5	172	47·4	—

†Calculated using conventional plasticity correction: $r_y = 1/2\pi(K_c/\sigma_{ys})^2$.
‡Calculated using modified plasticity correction: $r_y = 1/2\pi(K_c/\sigma_{ult})^2$.
§Plasticity excessive.

Crack-opening displacement measurements

A linear variable differential transformer was used to measure the opening displacement at the open end of the crack. Due to the high modulus of elasticity and low toughness of the beryllium, the displacements involved are very small, and little information was obtained on this material. The displacements involved in the aluminum specimens were much larger, and more meaningful results were obtained which are shown in Fig. 14. A jump in the displacement was observed whenever a large burst of emission occurred.

DISCUSSION OF RESULTS

Note in Fig. 7 that the acoustic emission begins to reach a peak count rate at approximately 0·2 per cent strain, the value of strain normally used to define the engineering yield stress from a stress–strain curve. Figure 9 illustrates that yield strain defined by 0·2 per cent offset strain is not a realistic picture of yield in this material. According to the acoustic emission record, yield actually begins in individual grains at

Table 2. Experimentally determined values of the exponent, s, in the relationship $N \propto K^s$

Run no.	s†	s‡	s§
1	11·9	8·7	7·2
2	‖	—	—
3	8·6	5·3	4·6
4	10·5	7·6	5·3
5¶	7·0	5·9	4·6
6	7·5	6·5	5·7
7	6·0	—	—
8	6·9	—	—

†Calculated from elastic stress intensity factor (no plasticity correction).

‡Calculated from plastically corrected stress intensity factor using a stress of 40 ksi in the equation for the plastic zone size.

§Calculated from plastically corrected stress intensity factor using a stress of 30 ksi (yield stress) in the equation for the plastic zone size.

‖Not calculated, did not agree with any other runs.

¶Calculated after subtracting 'fictitious' early jump of 5000 counts.

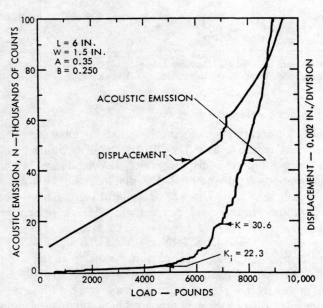

Fig. 14. Acoustic emission and crack-opening displacement as functions of load, for a single-edge-notched specimen of 7075-T6 aluminum.

more like 15,000 psi instead of the 30,000-psi yield strength ordinarily assigned to this material. The results of several tests have shown that the yield strength of this material can more consistently be defined by acoustic emission tests than by normal stress–strain curves due to the lack of a well-defined proportional limit in this material. The acoustic emission rate increases with increasing strain rate, thus the absolute value of the acoustic emission at the peak will vary with strain rate and, of course, with the gain used in the electronics. The characteristic shape of the curve remains the same, at least within the range of strain rates we have used thus far. It was observed by Schofield[3] as well as ourselves that two types of emission are present when a specimen is stressed: (1) a high-frequency, low-amplitude emission which we have called *healthy* emission, and (2) a higher amplitude, spike-type emission that occurs just prior to failure on an unflawed specimen, but is the main contributor to the emission from a specimen containing flaws.

Note in Fig. 8 that the acoustic emission is approximately proportional to strain during the most active region. Plots similar to Fig. 8 were obtained with specimens of varying gage length, with all other parameters such as gain, strain rate, etc. held constant. Indications were that the acoustic emission is proportional to the amount of volume undergoing plastic strain. This gave experimental credence to the assumptions in the theoretical discussion that the summation emission was proportional to the volume of material undergoing plastic strain.

It is of interest to note that the acoustic emission of the cracked specimens begins at stresses well below the yield stress of the net section, whereas the emission from the unflawed specimen began at stresses near yielding. Also, the emission observed from the flawed specimens consisted primarily of high amplitude spikes, instead of the *healthy* emission observed from the unflawed specimens. A comparison of Figs. 8 and 10 will illustrate the difference. The unflawed specimen shows a smoothly varying increase in the acoustic emission, whereas the flawed specimens show discontinuous jumps resulting from high-amplitude random emission. Also the total number of counts was an order of magnitude higher in the unflawed specimen due to the larger amount of material being strained. These differences between unflawed and flawed specimens suggest the applicability of acoustic emission to flaw detection, a result discussed more fully in the next section.

The calculations for the plastically corrected[14] values of the fracture toughness of the aluminum specimens in Table 1 are straightforward, and were carried out using the conventional plasticity correction for plain stress,

$$r_y = \frac{1}{2\pi}\left(\frac{K}{\sigma_{ys}}\right)^2.$$

However, the use of this plastic zone size in calculating the plastically corrected fracture toughness of the beryllium specimens leads to unreasonable results. The value of r_y calculated using the above equation was approximately 0·15 in. for the beryllium specimens. This is a fairly large value, and indicates that extensive plastic deformation would occur over a fairly large region in the vicinity of the crack tip. However, no effects of plastic deformation were observed on the broken portions of the beryllium specimens. All fracture surfaces were completely flat; no shear lip whatsoever was observed. Lateral contraction was also small. These factors indicate that the calculated plastic zone size of 0·15 in. was not really indicative of the extent of plastic flow near the crack tip. This discrepancy could be caused by the extensive strain hardening which occurs in this material. The presence of strain hardening would tend to reduce the size of the plastic zone, because higher stresses could be achieved within the zone. Thus, it may be more reasonable to use a stress higher than the yield stress in the equation for plastic zone size when estimating its size; therefore, the ultimate stress was used to calculate a set of fracture toughness values. Using the ultimate stress gives a lower limit of the plastic zone size, thus giving a lower limit of the plastically corrected fracture toughness. Use of the yield stress gives an upper limit. Thus, by using the two values of the stress, upper and lower bounds are obtained for the plastically corrected value of K_c. These upper and lower bounds, as well as the elastically calculated fracture toughness, are presented in Table 1. The value of the plastic zone calculated using the ultimate stress was approximately 0·03 in., a much more reasonable value than the 0·15 in. calculated using the yield stress.

As observed from the data in Table 1, the plasticity correction changes the fracture toughness of the beryllium more than it does that of the aluminum. This is because the ratio K_c/σ_{ys} is considerably smaller for the aluminum. The elastically calculated values of the stress intensity factor, K, are good approximations of the plastically corrected values for the aluminum; this is not so true for the beryllium.

The value of the exponent in Table 2 depends on the plasticity correction used. Since the plasticity correction does not greatly alter the stress intensity factor for the aluminum, values of the exponent, s, were calculated for this material only from elastic values.

Values of s for the beryllium were obtained from the elastic results, using the conventional plasticity correction, and from results using a modified plasticity correction in which an elevated stress was used in the equation for the plastic zone size. The value of the exponent is seen to decrease as the calculated plastic zone size becomes larger. However, it never decreases to the theoretical value of 4 but remains considerably larger than this.

The values of s for the aluminum are also consistently larger than 4. Thus, the proposed model does not accurately predict the experimental behavior. This model is a preliminary one, and is included for the sake of interest. At the present time, it would be difficult to rationalize a model which would agree better with the experimental value of the exponent [6–8]. If one assumes that the count rate is proportional to the square of the rate of change of volume of material which is strained between σ_{ys} and σ_u (rather than linearly proportional), an eighth power would result which agrees much better with experimental values. However, it is difficult to justify such an assumption.

It is of interest to note that the value of s for the aluminum agrees quite closely with the values for the beryllium which were calculated using the most realistic plasticity correction (that is, a stress of 40 ksi in the equation for r_y). These values typically range between 6 and 8, having an average value of 6·7.

Referring to Fig. 14, we observe that the displacement-load curve begins to become nonlinear at the point where the acoustic emission activity begins. This indicates that the emission in this range is due to plastic deformation occurring in the near vicinity of the crack. At 7000 lb, the displacement shows a series of abrupt changes that correspond to large acoustic bursts amounting to several thousand counts in some instances. This indicates that pop-in occurred at this point. The stress intensity calculated at this point was 30·6 ksi(in.)$^{1/2}$ which agrees well with results of other investigators [17, 18] for K_{Ic} fracture toughness of this material.

The acoustic emission from the N50A beryllium was found to begin in earnest at a value of K of approximately 15 ksi(in.)$^{1/2}$ (Fig. 11). This closely corresponds to the plane strain value of the fracture toughness (K_{Ic}) for this material[16], which further points out the applicability of acoustic emission to the determination of plane strain values of fracture toughness on materials where displacement measurement devices are too insensitive.

APPLICATIONS OF RESULTS

The results of this investigation have shown that the acoustic emission of flawed and unflawed specimens differs considerably, and that the emission characteristics of flawed specimens are highly dependent on the stress intensity factor of the flaw present. These results suggest that estimates of the flaw size and failure load of flawed structures could be determined by acoustic emission techniques. This may be of wide significance in the nondestructive testing of stressed structures.

Acoustic emission could be applied to a complex structure to estimate its fracture load and the sizes of flaws in it. This could be accomplished in the following manner:

1. The acoustic emission as a function of the stress intensity factor would have to be determined for the material under consideration. The results of several tests with different-size known flaws of the type anticipated in service (such as semi-elliptical surface cracks, single-edge cracks, etc.) would be conducted to obtain the typical N–K plot. The thickness of material to be used in the final structure would be most appropriate; the value of the critical stress intensity factor, K_c, could also be determined in these tests by taking the specimens all the way to failure.

2. The structure under consideration would be acoustically monitored during initial loading to obtain the value of the emission counts N_t (subscript t refers to the value determined in this test) corresponding to a given load, F_t.

3. The value of the stress intensity factor, K_t, corresponding to the given load, F_t, could be obtained from the typical N–K curve by using the value of N_t determined in step 2.

The fracture load, F_c, could then be determined from the relationship

$$\frac{F_t}{K_t} = \frac{F_c}{K_c}.$$

This relationship results from the linear dependence of the stress intensity factor on the applied load[10].

5. An estimate of the flaw size could be obtained, for an assumed flaw shape, by applying the stress analysis for the geometry under consideration.

The above series of tests could be used to predict the failure of structures of any geometry composed of materials susceptible to brittle failure.

If numerous flaws are present, the sixth to eighth power dependence of the emission on the stress intensity factor would tend to favor the larger flaws. Therefore, a structure containing several flaws would probably give a somewhat higher total number of counts, but the shape of the emission curve as a function of the loading should still be the same.

Obviously, the problem in applying this approach is to acoustically determine the stress intensity factor corresponding to the given load. The results shown in Figs. 11 and 13 are encouraging, but much additional work is required to see if the emission can be related more closely to the stress intensity factor.

CONCLUSIONS

Results of these studies of the acoustic emission characteristics of N50A beryllium and 7075 aluminum indicate that there is a marked difference between the acoustic emission from an unflawed tensile specimen and one containing a sharp crack. Emission is observed to begin at stresses far below general yield in the specimens containing sharp cracks. We have demonstrated that a sharp-crack fracture mechanics approach can be used to obtain correlation between the stress intensity at the crack tip and the acoustic emission characteristics. A theoretical model shows that the acoustic emission from a cracked specimen should vary as the fourth power of the stress intensity factor, whereas experimental results show a variation between the sixth and eighth power. We have no explanation for this discrepancy at present.

It appears that knowledge of the fracture toughness of a material and the acoustic emission characteristics as a function of stress intensity can be used to determine the fracture load of a structure from information obtained in an acoustic emission test. This technique has been illustrated by an example.

Acoustic emission appears to us to be a potentially valuable new tool for the nondestructive fracture analysis of engineering structures, and we believe that these results show this potential.

Acknowledgements — We wish to thank Dr. Alan Tetelman of Stanford University for his valuable suggestions and encouragement during the course of this investigation. We further wise to thank Albert Brown, Bernard Kuhn, Thomas Freeman and Paul Westling for their assistance in obtaining and reducing the experimental data.

REFERENCES

[1] J. Kaiser, Untersuchungen uber das Auftreten Gerauschen Beim Zugversuch. Ph. D. Thesis, Techn. Hochsch., München (1950); *Arch. Eisenhütt Wes.* **24**, 43–45 (1953).

[2] C. A. Tatro and R. Liptai, Acoustic emission from crystalline substances. *Proc. Symp. Physics and Nondestructive Testing*, pp. 145–158. Southwest Research Institute (1962).

[3] B. H. Schofield, Acoustic emission under applied stress. *Air Force Materials Lab., Tech. Rep.* No. ASD-TDR-63-509, Part I (1963).

[4] B. H. Schofield, Acoustic emission under applied stress. *Air Force Materials Lab., Tech. Rep.* No. ASD-TDR-63-509, Part II (1964).

[5] H. L. Dunegan, C. A. Tatro and D. O. Harris, Acoustic emission research. *Lawrence Radiation Lab., Livermore, Rep.* No. UCID-4868 Rev. 1 (1964).

[6] R. M. Fisher and J. S. Lally, Microplasticity detected by an acoustic technique. *Proc. Conf. Deformation of Crystalline Solids*, Ottawa, Canada, August 1966. To be published.

[7] A. T. Green, C. E. Hartbower and C. S. Lockman, Feasibility study of acoustic depressurization system. *Aerojet-General Corp. Rep.* No. NAS 7-310 (1965).

[8] M. H. Jones and W. F. Brown, Acoustic detection of crack initiation in sharply notched specimens. *Mater. Res. Stand.* **4**, 120–129 (1964).

[9] B. H. Schofield, Investigation of applicability of acoustic emission. *Air Force Materials Lab. Tech. Rep.* No. AFML-TR-65-106 (1965).

[10] P. C. Paris and G. Sih, Stress analysis of cracks. *Fracture Toughness Testing and Its Applications*, ASTM Spec. Tech. Publ. No. 381, pp. 30–81. American Society for Testing and Materials, Philadelphia, Penn. (1965).

[11] G. R. Irwin, Fracture mechanics. In *Structural Mechanics*, pp. 557–591. Pergamon Press, Oxford (1960).

[12] M. L. Williams, On the stress distribution at the base of a stationary crack. *J. appl. Mech* **24**, 109–114 (1957).

[13] G. R. Irwin, Analysis of stresses and strains near the end of a crack traversing a plate. *J. appl. Mech* **24**, 361–364 (1957).

[14] F. A. McClintock and G. R. Irwin, Plasticity aspects of fracture mechanics. *Fracture Toughness Testing and Its Applications*, ASTM Spec. Tech. Publ. No. 381, pp. 84–113. American Society for Testing and Materials, Philadelphia, Penn. (1965).

[15] W. F. Brown and J. E. Srawley, Fracture toughness testing methods. *Fracture Toughness Testing and Its Applications, ASTM Spec. Tech. Publ.* No. 381 pp. 133-244. American Society for Testing and Materials, Philadelphia, Penn. (1965).
[16] D. O. Harris and H. L. Dunegan, Fracture toughness of beryllium. *Lawrence Radiation Lab., Livermore, Rep.* No. UCRL-70255 (1967).
[17] J. G. Kaufman and H. Y. Hunsicker, Fracture toughness testing at Alcoa Research Laboratories. *Fracture Toughness Testing and Its Applications, ASTM Spec. Tech. Publ.* No. 381, pp. 290-308. American Society for Testing and Materials, Philadelphia, Penn. (1965).
[18] M. H. Jones and W. F. Brown, Acoustic detection of crack initiation in sharply notched specimens. *Mater. Res. Stand.* 4, 120-127 (1964).

(*Received* 6 *February* 1967)

Résumé — Cet exposé rend compte des résultats de travaux sur l'émission acoustique d'éprouvettes en aluminium et en béryllium, fêlées et non fêlées. On trouve que, en ce qui concerne les éprouvettes fêlées, l'émission acoustique commence à des niveaux de tension bien plus faibles que la tension limite normale. D'après un exemple théorique donné ici, le nombre total de signaux acoustiques provenant d'une éprouvette ayant une fêlure, devrait être proportionnel à la quatrième puissance du facteur d'intensité de la force de tension, obtenu à part d'une analyse de la mécanique de la rupture. Ceci est contraire aux donées expérimentales d'éprouvettes à encoche simple utilisées pour l'essai de résistance à la rupture, ce qui indique que la variation de l'émission acoustique se rapproche davantage de la sixième à la huitième puissance du facteur d'intensité de la force de tension. Un exemple donné montre comment les données sur l'émission acoustique obtenues à partir des éprouvettes pour l'essai de résistance à la rupture peuvent être utilisées pour un essai non destructif de la force de rupture d'une structure mécanique.

Zusammenfassung — Dieser Bericht enthält die Ergebnisse von Untersuchungen über akustische Emissionen von fehlerhaften und fehlerfreien Proben aus Aluminium und Beryllium. Es wurde festgestellt, dass die akustische Emission von den fehlerhaften Proben bei beträchtlich unterhalb der allgemeinen Streckgrenze gelegenen Spannungsniveaus beginnt. Ein hier angeführtes, theoretisches Modell deutet darauf hin, dass die Gesamtzahl der akustischen Signale, die durch eine einen Riss enthaltende Probe ausgesandt werden, zur vierten Potenz des der Spannungsfaktors, der sich aus der Bruch analysis für den scharfen Riß ergibt, proportional sein sollte. Das steht im Widerspruch zu den Versuchsdaten von einseitig gekeibten Proben zur Bestimmung der kerbzähigkeit der beiden Werkstoffe, die viel eher auf eine Änderung in der akustischen Emissionmit der sechsten bis achten Potenz des Spannungsfaktors hinweisen. Es wird an einem Beispiel gezeigt, wie die an Bruchzähigkeitsproben erhaltenen Emissionsdaten für eine zerstörungsfreie Prüfung der Bruchfestigkeit technischer Konstruktionen verwertet werden können.

Acousto-elastic measurement of stress and stress intensity factors around crack tips

A.V. CLARK, R.B. MIGNOGNA and R.J. SANFORD

Acousto-elasticity predicts that the phase velocity of sound waves in a material will be changed slightly by stress. For a slightly orthotropic plate in a state of plane stress, the shear stress σ_{xy} can be calculated from $\sigma_{xy} = (B \sin 2\phi)/2m$ once measurements of the acoustic birefringence B and the angle ϕ have been made. The birefringence is the difference in phase velocity between SH waves polarized along the 'fast' and 'slow' acoustic axes, ϕ is the angle between the acoustic axes in stressed and unstressed state, and m is an acousto-elastic constant for the material.

For symmetrical, two-dimensional crack-opening problems, σ_{xy} can be expressed as a series expansion of stress functions, each of which satisfies the equilibrium equation. The coefficients in the expansion allow the appropriate boundary conditions to be satisfied. The stress intensity factor K_1 is the coefficient of the leading term in the series.

Values of σ_{xy} and K_1 for an ASTM standard test specimen made of 2024 aluminium were acousto-elastically determined. These were compared with those obtained from a similar photoelastic specimen. Good agreement was obtained for both σ_{xy} and K_1.

KEYWORDS: ultrasonics, stress measurement, crack tips, acoustic birefringence, aluminium

Introduction

The use of ultrasound for stress measurements in opaque materials is an area of active interest. Measurement of small stress-induced changes in phase velocity has been investigated as a means of determining stress in materials[1-5] for over 20 years. However, until recently, it appeared that such measurements might be of limited practical value.

One reason for the resurgence of interest in ultrasonic stress measurements is the ability to rapidly obtain large amounts of ultrasonic data from computer-controlled scanning systems. Such a system has been developed by Kino and his coworkers.[6] They used a longitudinal-wave transducer in a liquid buffer to scan the stressed specimen. Variations in phase velocity were mapped out; these variations are proportional to the sum of the in-plane principal stresses (first stress invariant) for specimens in a state of plane stress. The scanning system was used to determine the first stress invariant in regions near

(a) a central, circular hole in a plate subjected to far-field tension,

(b) a notch in a double edge-notched panel which is subjected to far-field tension.

Good agreement between theory and experiment was reported.[7]

This system was also used for other interesting experiments:

(a) residual-stress mapping in a disc cut from a hydrostatically extruded rod,

(b) determination of stress intensity factor in a centre-cracked panel.

These results (and several others) are reported in Ref. 8. In the latter experiment it was noted that the experimentally obtained stress intensity factor differed by about 20% from the theoretical value.

A method for measurement of three-dimensional stress fields was proposed by Bennett, Husson, and Kino[9] where the phase difference for two coaxial transducers was compared. One transducer emitted a focused beam, while the other emitted an unfocused beam. The phase-shift difference between the beams results (approximately) from differences in stress in the focal plane. The phase difference for the stress field surrounding a slot was calculated; the slot was loaded by forcing its sides apart. Good agreement between calculated and measured phase difference was obtained.

Further impetus for using ultrasonic stress measurement is the need to determine applied axial stress in threaded fasteners. This problem has been successfully addressed by Heyman et al.[10-12] In his method a longitudinal wave was launched into a stressed specimen. The resonant frequencies of the threaded fastener were monitored and related to the applied stress. Changes in the resonant frequencies were observed using either cw techniques,[10] or a phase-locked loop system.[11,12]

A.V. Clark and R.B. Mignogna are in the Nondestructive Evaluation Section, Structural Integrity Branch, Naval Research Laboratory, Washington, DC, 20375, USA. R.J. Sanford is at the Department of Mechanical Engineering, University of Maryland, College Park, MD 20740, USA. Paper received 5 August 1982.

The use of shear waves has also been considered as a method to map out stress fields. For a homogeneous, isotropic material, the difference in phase velocity (acoustic birefringence) between two orthogonally polarized shear waves is given by

$$\frac{V_1 - V_2}{V_0} = A(\sigma_1 - \sigma_2).$$

where V_1 and V_2 are the phase velocities of waves polarized along the σ_1 and σ_2 principal stress directions respectively, V_0 equals $1/2\,(V_1 + V_2)$, and A is the acousto-elastic constant. The body is assumed to be in a state of plane stress, and the SH waves propagate in the anti-plane direction. Since longitudinal waves give only information about the sum of principal stresses for the plane-stress state, it is desirable to use shear-wave measurements to obtain the difference of principal stresses.

Unfortunately, most materials of interest are not isotropic and exhibit some anisotropy ('texture') due to the forming process. In particular, rolled plates of cubic materials (such as aluminium and mild steel) are orthotropic even in the absence of any applied stress.[13] It appears that texture affects the quality of stress measurements made with longitudinal waves much less than those made with shear waves. Consequently, the use of shear waves as a stress measurement tool has lagged behind that of longitudinal waves, due in part to lack of a theory which properly accounts for the effect of texture on acoustic birefringence.

An early attempt to account for texture was reported by Mahadevan,[14] who found that for aluminium, copper, stainless and mild steels, the birefringence was a function both of stress and of the transducer frequency. The birefringence due to applied stress was found to be independent of frequency; the frequency-dependent component was attributed to texture.

Iwashimizu and Kubomura calculated the birefringence for a stressed orthotropic material. The material was treated as a non-linear elastic solid. The effect of anisotropy was retained in the second-order elastic moduli, but not the third-order moduli.[15]

Okada used a novel approach to the texture problem.[16] Rather than treating the material as being non-linear elastic, he used an approach similar to that used in optics. He assumed a second-order index of refraction tensor N_{ij} whose eigenvalues were the velocities of acoustic waves propagating in the material. The incremental change ΔN_{ij} in the tensor was assumed to be a linear function of stress

$$\Delta N_{ij} = \alpha_{ijkl}\,\sigma_{kl}$$

For an orthotropic material, Okada assumed that there were nine independent coefficients α_{ijkl}, by analogy to the nine second-order elastic moduli for orthotropic elastic solids. The birefringence was obtained from the velocities of the orthogonally polarized shear waves.

Okada verified his theory for 5052 aluminium rolled plate in uniaxial tension and compression. Okada noted that his theory required the use of three acousto-elastic constants, which gave a better fit to his data than the theory of Iwashimizu and Kubomora. In a later paper[17] Okada verified his theory for 1100 aluminium rolled plate in tension. He also mapped out the stress field around a circular hole in a plate in far-field tension.

In Okada's theory, the birefringence is proportional to the shear stress, referenced to a coordinate system oriented along the directions of material symmetry in the orthotropic solid. He noted that once the shear stress was determined, it was (in principle) possible to obtain the normal stresses by integrating the stress equilibrium equations. Consequently, shear-wave stress measurements have the potential for characterizing the complete stress field for a body in a state of plane stress.

In this paper, we applied the theory of Okada to measurement of the stresses and stress intensity factor in the region around a crack tip in a plate. Our motivation here was twofold:

(a) To determine whether the stress intensity factor can be obtained with greater accuracy than was reported in Ref. 8 (where longitudinal waves were used as the ultrasonic probe).

(b) To attempt to map out the stress field near a crack tip using shear waves. The problem is complicated by the rather steep stress gradients around the crack tip.

Acousto-elastic theory for anisotropic solids

All tests were performed on specimens made from 2024-T351 aluminium rolled plate. Due to the rolling process, this material displays slight anisotropy, or 'texture'.

For specimens with no applied load, pure-mode SH waves could be propagated in the plate-thickness direction when a SH wave transducer was oriented either parallel or perpendicular to the rolling direction. The phase velocity of a SH wave polarized along the rolling direction was slightly larger than that of a SH wave polarized perpendicular to the rolling direction.

In the absence of stress, these phase velocities are the eigenvalues of the Christoffel equation for wave propagation through solids with negligible applied stress[18]

$$(C_{ijkl}\,n_j n_k - \rho_0 V^2 \delta_{il})\,v_l = 0 \qquad (1)$$

Here C_{ijkl} are the second-order elastic moduli, ρ_0 is the material density in the unstressed state, and δ_{il} is the Kronecker delta function. A monochromatic plane-wave with direction cosines n_j is assumed to propagate through the solid; the particle displacement vector has components v_l.

The eigenvectors of the Christoffel equation give the particle displacement directions for pure-mode waves, which for the 2024 aluminium plate are parallel and perpendicular to the rolling direction. A (local) coordinate system can be constructed whose axes are parallel to these eigenvectors; the axis parallel to the eigenvector having the larger phase velocity will be called the 'fast' acoustic axis, and that parallel to the eigenvector corresponding to the smaller phase velocity will be called the 'slow' axis. In the unstressed state, these axes are the Y_0 and X_0 axes, respectively, as shown in Fig. 1.

In the initial (unstressed) state, the acoustic birefringence B_0 is defined as

$$B_0 = \frac{V_{y0} - V_{x0}}{\langle V_0 \rangle} \qquad (2)$$

where V_{y0} and V_{x0} are velocities of waves polarized along the fast and slow axes respectively, and $\langle V_0 \rangle \equiv 1/2\,(V_{y0}+V_{x0})$.

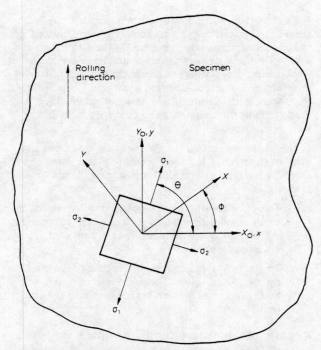

Fig. 1 Coordinate systems used in the analysis for SH wave propagation through rolled plate. X_0 and Y_0 denote the initial slow and fast acoustic axes, respectively; X and Y are the slow and fast axes in the stressed state; x and y are axes parallel to the initial acoustic axes and the principal stresses are labelled σ_1 and σ_2 ($\sigma_1 > \sigma_2$)

The presence of a stress field has two effects:

(1) the acoustic axes, in general, are no longer oriented along directions of material symmetry,

(2) the phase velocities are stress-dependent.

The form of the Christoffel equation, accounting for the effects of a superposed stress-field and material anisotropy, is quite complex. The general equation is given in tensor form in[19] for an anisotropic elastic solid with the non-linear stress-strain relation

$$\sigma_{ij} = C_{ijkl} E_{kl} + C_{ijklmn} E_{kl} E_{mn} \quad (3)$$

with C_{ijklmn} being the third-order elastic moduli, and E_{ij} are the Lagrangian strains.

The form of the generalized Christoffel equation for the special case of a slightly orthotropic solid was developed by Iwashimizu and Kubomura.[15] They retained the effect of anisotropy in the second-order moduli, but ignored it in the third-order moduli. They found that the acoustic axes in the stressed-solid rotated through an angle ϕ relative to the initial acoustic axes as shown in Fig. 1. To calculate ϕ they gave the formula

$$\tan 2\phi = \frac{M(\sigma_1 - \sigma_2) \sin 2\theta}{B_0 + M(\sigma_1 - \sigma_2) \cos 2\theta} \quad (4)$$

where σ_1 and σ_2 are the principal stresses and M is the acousto-elastic constant with dimensions of inverse stress. The principal stress axes are inclined at an angle θ to the initial acoustic axes. A state of plane stress has been assumed.

Iwashimizu and Kubomura derived a birefringence equation of the form[15]

$$B = \{ [B_0 + M(\sigma_1 - \sigma_2) \cos 2\theta]^2 +$$
$$+ M(\sigma_1 - \sigma_2) \sin 2\theta]^2 \}^{1/2} \quad (5)$$

where B is defined as

$$B \equiv (V_y - V_x)/\langle V \rangle \quad (6)$$

V_y and V_x are the phase velocities of SH waves polarized along the fast (Y) and slow (X) acoustic axes, respectively. Note that, in general, the acoustic axes and principal stress axes do not coincide.

A different approach to calculating the birefringence was taken by Okada.[16] He assumed the existence of an index of refraction tensor N_{ij} for the solid, by analogy with optical methods. He further assumed a linear variation of N_{ij} with stress

$$N_{ij} = N^°_{ij} + \alpha_{ijkl} \sigma_{kl} \quad (7)$$

where $N^°_{ij}$ is the tensor in the initial state and the α_{ijkl} are coefficients analogous to the second-order elastic moduli. In fact, Okada assumed the existence of nine independent α_{ijkl} for a slightly orthotropic material, just as there are nine independent second-order elastic moduli. The velocities of the SH waves are the eigenvalues of the N_{ij} tensor, and the acoustic axes are oriented parallel to the eigenvectors. Okada derived a birefringence equation of the form

$$B = \{ [B_0 + M_1(\sigma_1 + \sigma_2) + M_2(\sigma_1 - \sigma_2) \cos 2\theta]^2 + [M_3(\sigma_1 - \sigma_2) \sin 2\theta]^2 \}^{1/2} \quad (8)$$

where B_0 and B are as defined previously and M_1, M_2, M_3 are acousto-elastic constants. The angle ϕ between the initial and stressed acousto-axes is calculated from

$$\tan 2\phi = \frac{M_3(\sigma_1 - \sigma_2) \sin 2\theta}{B_0 + M_1(\sigma_1 + \sigma_2) + M_2(\sigma_1 - \sigma_2) \cos 2\theta} \quad (9)$$

Note that Okada's theory requires that three acousto-elastic constants be known. His theory reduces to that of Ref. 15 if $M_1 = 0, M_2 = M_3$.

For a state of plane stress, $(\sigma_1 - \sigma_2) \sin 2\theta = 2\sigma_{xy}$. Note that the theories of both Refs 15 and 16 predict that $\phi = 0$ for zero shear-stress. Consequently, birefringence measurements alone cannot discriminate between the cases of

(a) an isotropic material with normal stresses,

(b) an anisotropic material with zero normal stresses,

(c) an anisotropic material with normal stresses.

We chose to use the Okada theory, since it contains the theory of Iwashimizu and Kubomura as a special case. From (8) and (9), the shear stress is given by

$$\sigma_{xy} = \frac{B \sin 2\phi}{2M_3} \quad (10)$$

so that the birefringence measurements can be used to map out the shear-stress field.

Experimental technique for birefringence measurement

A modified version of the pulse-echo-overlap (PEO) method[20-22] was used to determine the orientation of acoustic axes and the acoustic birefringence. A block diagram of a system which uses this technique is shown in Fig. 2. Measurement of round-trip delay time with this system had an accuracy of 10 ppm. We used a 10 MHz ac-cut quartz shear-wave transducer having an active element 1.8 mm in diameter. A small transducer was necessary to achieve good spatial resolution when measuring stresses around the crack tip.

The transducer was mounted in a spring-loaded goniometer device. A viscous fluid acoustically coupled the transducer. The spring-loaded device, shown in Fig. 3, was designed to maintain a constant couplant thickness as the transducer was rotated. However, our estimate of the experimental errors in the birefringence measurement was approximately 5% and was attributed mainly to slight variations in couplant thickness.

The approximate orientation of the acoustic axes was determined by rotating the transducer and observing the arrival times of the echoes on an oscilloscope; maximum and minimum arrival times correspond to the slow and fast acoustic axes respectively. Once an acoustic axis was approximately located in this fashion, cycle-for-cycle overlap of the first two echoes was obtained with the PEO system. Using the PEO technique, the oscilloscope was triggered by a highly stable cw source at a frequency corresponding to the reciprocal of the round-trip time of SH waves in the stressed specimen (refer to Fig. 2).

If the transducer was oriented along an acoustic axis, the cycle-for-cycle overlap was maintained for subsequent echoes to be compared with the first echo; for example, first and second, first and third, first and fourth, etc. If the transducer was not oriented along the axis, the overlap was not maintained as the subsequent echoes were compared. This fact was used as a check to see whether correct orientation had been achieved for each axis. By this technique we were able to determine the angle ϕ to within $\pm 2°$.

Since SH waves polarized along both the fast and slow axes travel through the same material at each location, the birefringence is given by

$$B = \frac{f_f - f_s}{1/2\,(f_f + f_s)}$$

where f_f and f_s are the overlap frequencies associated with the fast and slow axes respectively. Once the acousto-elastic

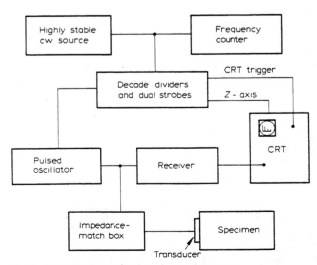

Fig. 2   Block diagram of pulse-echo-overlap system

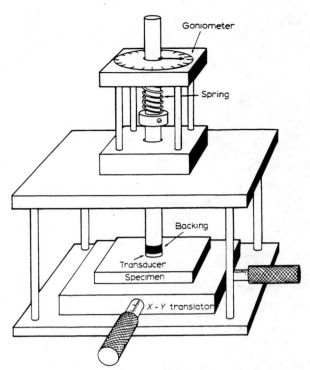

Fig. 3   Spring-loaded goniometer device incorporating an $X-Y$ translation stage used to position the specimen with respect to the 1.8 mm quartz transducer

constant M_3 is known, the shear stress σ_{xy} can be calculated from the measured values of B through (10). We obtained the acousto-elastic constants from uniaxial tension tests.

Uniaxial tension tests

For uniaxial tension σ, (8) becomes

$$B = [(B_0 + M_1\sigma + M_2\sigma\cos 2\theta)^2 \\ + (M_3\sigma\sin 2\theta)^2]^{1/2} \tag{11}$$

Recall that θ is the angle between the X_0-axis and the principal stress direction. Uniaxial tension specimens were cut from rolled 2024 aluminium plate at angles of $0°$, $45°$, $60°$, and $90°$ to the X_0-axis.

For $\theta = 0°$ and $90°$, the birefringence is proportional to stress

$$B = B_0 + (M_1 + M_2)\sigma \quad (\theta = 0°) \tag{12a}$$

$$B = B_0 + (M_1 - M_2)\sigma \quad (\theta = 90°) \tag{12b}$$

A straight-line fit was made to the experimentally obtained birefringence as a function of stress, σ, from which M_1 and M_2 were determined.

M_3 can be determined from the birefringence curve for $\theta = 45°$

$$B = [B_0 + 2M_1 B_0\,\sigma + (M_1^2 + M_3^2)\sigma^2]^{1/2} \tag{12c}$$

The $\theta = 60°$ specimen serves as a check case for internal consistency.

Figure 4 shows the results of measurements of B as a function of σ. Note the linear variation of birefringence for

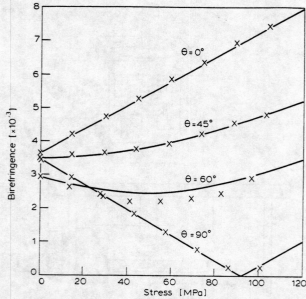

Fig. 4 Acoustic birefringence as a function of stress for uniaxial tensile specimens for $\theta = 0°$, $45°$, $60°$ and $90°$. —— theory; X data points

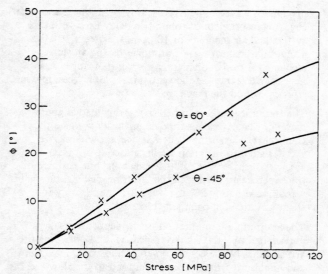

Fig. 5 Rotation of acoustic axes ϕ as a function of stress for uniaxial tensile specimens for $\theta = 45°$ and $60°$. —— theory; X data points

$\theta = 0°$ and $90°$, in accordance with theory. The theory gives a good fit to the data for $\theta = 45°$, and reasonable agreement to the data for $\theta = 60°$. The measured values of the acousto-elastic constants for 6.3 mm (0.25 in) thick 2024-T351 aluminium were: $M_1 = 0.01 \times 10^{-5}$ MPa^{-1}, $M_2 = 3.64 \times 10^{-5}$ MPa^{-1}, and $M_3 = 3.32 \times 10^{-5}$ MPa^{-1}.

It should be noted that the acousto-elastic constants are temperature-dependent. Consequently, all our acousto-elastic measurements were performed at a constant temperature of 20°C.

With known values of the acousto-elastic constants, it is possible to predict the variation of ϕ as a function of σ from (9). Values of ϕ were measured for all four uniaxial tension specimens. For the range of tensile stresses in our experiments, we found agreement with theory for all values of θ tested (0°, 45°, 60°, 90°). For reasons of clarity only the values of ϕ for $\theta = 45°$ and $\theta = 60°$ are shown in Fig. 5.

Measurements of stress in a fracture specimen

Since Okada's theory gave good results for the uniaxial tension specimens, we attempted to apply it to a case of more practical interest; namely, a plate with a simulated crack. We fabricated a modified compact tensile specimen, shown in Fig. 6. This is a well-characterized fracture specimen with stress intensity factor K_I given in Ref. 23. A 76.2 mm (3 in) long cut was made perpendicular to the rolling direction of the plate. A 25.4 mm (1 in) diameter hole was drilled through the plate, with the hole centred on the crack axis. A loading device was placed in the hole and drove the sides of the cut apart with a constant, measured force P.

Series representation of the stress field

This symmetric specimen with a symmetric load is in mode I (crack opening mode) deformation. The general solution for the stress field in mode I specimens with a single-ended crack has been derived by Sanford.[24] The stresses are given as series expansions of stress functions and derivatives of stress functions. The stress functions guarantee that the stress equilibrium equations will be satisfied; the coefficients in the series expansions allow the boundary conditions to be satisfied on the (finite) fracture specimen.

In particular, the solution for the shear stress is

$$\sigma_{xy} = \sum_{n=0}^{N} A_n r^{n-1/2} \sin \psi \cos(n-3/2)\psi + \\ \sum_{n=0}^{N} \beta_n r^n [\sin n\psi + n \sin \psi \cos(n-1)\psi] \quad (13)$$

which is valid outside the plastic zone around the crack tip.* Here r is the distance from the crack tip to a point (X_0, Y_0) and ψ is the polar angle $\tan^{-1}(Y_0/X_0)$. The stress intensity factor K_I is proportional to the coefficient of the leading term in the series

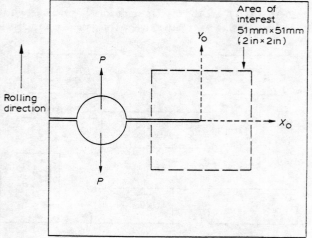

Fig. 6 Schematic diagram of the modified compact tensile specimen. The region where acoustic measurements were made is labelled 'area of interest'. Also shown are the initial acoustic axes, X_0 and Y_0, and the rolling direction

*For the maximum load P used in our tests, the plastic zone is estimated to be less than 3 mm in diameter

$$K_I = -(8\pi)^{1/2} A_0 \qquad (14)$$

Note that the leading term is dominant in the region very close to the crack tip.

The coefficients A_n and β_n can be determined in several ways; for example, boundary collocation, finite-element methods, etc. For the specimen geometry shown in Fig. 6, the coefficients have been determined by Sanford et al[25] using photoelasticity. In their method, the photoelastic fringes (isochromatics) are contours of constant maximum in-plane shear stress, τ_m, which can be expressed in terms of the Cartesian stress components as

$$\tau_m = \{[1/2(\sigma_x - \sigma_y)]^2 + \sigma_{xy}^2\}^{1/2} \qquad (15)$$

The functional forms of the normal stresses σ_x and σ_y are given in Ref. 24 as series expansions with coefficients A_n and β_n. Consequently, the isochromatic field contains information about the coefficients.

To obtain the coefficients, a match was made (in the least-squares sense) between the experimental fringe pattern and the functional form for the maximum shear-stress, with the coefficients A_n and β_n as the fitting parameters. To improve the accuracy, the sampling procedure of Sanford and Chona[26] was employed. The fitting procedure was performed 100 times using randomly selected sub-sets of 30 data points, and the resulting coefficients were then averaged. In the absence of residual stress, the stress field represented by these coefficients is the same as that in the aluminium fracture specimen.

Contours of constant σ_{xy} using coefficients obtained in Ref. 25 are shown in Fig. 7. Note that the shear stress vanishes along the crack axis due to symmetry of geometry and loading. This symmetry also requires that the shear stress be anti-symmetric about the crack axis, as shown.

Acousto-elastic measurement of stress field

The results of the photoelastic studies allowed us to predict the theoretical applied stress in the aluminium fracture specimen. To see whether the acousto-elastic stress measurements would reproduce this stress field, we measured the acoustic birefringence B and the angle ϕ at 66 sample points contained in a 51 mm × 51 mm (2 in × 2 in) square centred on the crack tip. Values of σ_{xy} were then determined from (10). We did not make a raster scan; rather, points were chosen to map out characteristic features of the stress field. For example, data were taken along the crack axis where σ_{xy} should vanish. We also sampled at 14 pairs of points which are reflections of each other in the crack axis, as a check for any asymmetry in the shear-stress field.

To obtain the coefficients in (13) from our acousto-elastic data, we used a procedure similar to that of Sanford and Chona for photoelasticity.[26] From our set of 66 values of σ_{xy}, we randomly chose a sub-set of 20 values and calculated the corresponding A_n and β_n by a linear, least-squares fit. This operation was performed 100 times, each time with a randomly chosen sub-set of 20 values of σ_{xy}.

We averaged the 100 sets of coefficients to obtain the values shown in Table 1; these are listed in descending order of importance in calculating the expansion (13). The value of β_0 is omitted since the term it multiplies in (13) vanishes identically. Note the good agreement with values obtained from the photoelastic experiment. Recall that the stress intensity factor K_I is related to A_0 by $-(8\pi)^{1/2}A_0$.

Contours of constant σ_{xy} were calculated using the acousto-elastically determined coefficients, and are shown in Fig. 8. Note the similarity between this figure and the contours of Fig. 7. In view of the good agreement between the photoelastically determined coefficients and those obtained acousto-elastically, this similarity is to be expected.

The anti-symmetric nature of the contours in Fig. 8 is due to the anti-symmetry of the functional forms in (13). We

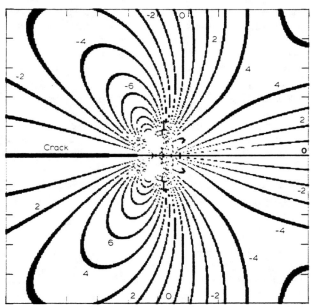

Fig. 7 Theoretical shear-stress contours within a 51 mm × 51 mm (2 in × 2 in) region of the modified compact tensile specimen centred at the crack tip. Theoretical contours were generated from the optical birefringence data of Ref. 25. Contours are plotted for $\sigma_{xy} = (N \pm 0.08)$ ksi, $N = 0, 1, 2 \ldots$; to obtain stress in MPa, multiply by 6.9

Table 1. Comparison of coefficients in the series expansion for σ_{xy}. Coefficients were obtained from photoelastic and acousto-elastic data

Data type	K_I [ksi in$^{1/2}$] (MPa m$^{1/2}$)]	A_1 [ksi in$^{-1/2}$] (MPa m$^{-1/2}$ × 10^2)]	A_2 [ksi in$^{-3/2}$] (MPa m$^{-3/2}$ × 10^3)]	β_1 [ksi in^{-1}] (MPa m^{-1} × 10^2)]	β_2 [ksi in^{-2}] (MPa m^{-2} × 10^3)]
Photoelastic	26.0 (28.6)	4.69 (2.03)	3.24 (5.52)	1.44 (3.91)	0.00 (0.00)
Acousto-elastic	25.7 (28.3)	4.75 (2.06)	3.43 (5.84)	1.52 (4.13)	0.13 (1.39)

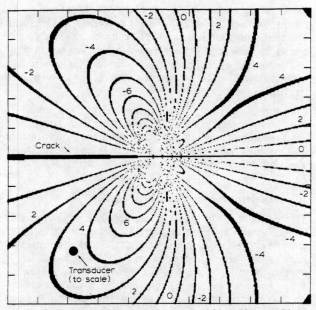

Fig. 8 Experimental shear-stress contours within a 51 mm x 51 mm (2 in x 2 in) region of the modified compact tensile specimen centred at the crack tip. Experimental contours were generated from the acousto-elastic data. The size of the transducer used in the experiment is also shown. Contours are plotted for $\sigma_{xy} = (N \pm 0.08)$ ksi; to obtain stress in MPa, multiply by 6.9

did note some (apparently random) asymmetry in the stresses measured at 14 pairs of sampling points located symmetrically about the crack axis. However, the average of the stresses at these points was approximately 1.04 MPa, which is less than our overall estimated error of ± 3.45 MPa. The average shear-stress measured along the crack axis was also about one-third of our experimental error.

Residual stress

The acousto-elastic shear-stress measurements map out the sum of the residual and applied shear-stresses. To make a valid comparison of the theoretical shear-stress field with the acousto-elastically determined shear-stress field, it is necessary that the latter be due entirely to applied shear-stress.

One strong indication that this criterion was met is the fact that the average shear stress measured acousto-elastically, 1.034 MPa (150 psi), is less than the experimental error. Since the applied stress field is known to be anti-symmetric, any dc shear-stress level must be due to residual stress; conversely, the absence of a dc level is a necessary condition for zero residual shear stress.

For an anisotropic single crystal, the initial (unstressed) acoustic axes are along directions of material symmetry. In both the theory of Refs 15 and 16, the rolled plate is assumed to behave as if it were an anisotropic single crystal. If this assumption is valid, then the initial acoustic axes for the fracture specimen should be parallel and perpendicular to the rolling direction.

Recall that (9) predicts that $\phi = 0$ when $\sigma_{xy} = 0$. Consequently, if the plate behaves as an anisotropic single crystal, a sufficient condition for zero residual shear-stress is that the acoustic axes, measured prior to loading the specimen, be parallel and perpendicular to the rolling direction.

Measurements of B and ϕ were made at 19 points in the 51 mm x 51 mm sampling region before the aluminium fracture specimen was loaded. The average birefringence was 3.24×10^{-3} with a standard deviation of 0.19×10^{-3}, and the average angle ϕ was less than $0.1°$ relative to a coordinate system oriented parallel and perpendicular to the rolling direction. The polarization direction of the quartz transducer was determined from its Laue x-ray diffraction pattern.

Since the average value of ϕ was essentially zero for no applied load, the residual shear-stress vanishes if the plate material behaves as an anisotropic single crystal. There may be residual normal stresses, as pointed out previously; however, we are only interested in comparing acousto-elastic measurements of applied shear-stress with the theoretical shear-stress, so residual normal stresses have no effect on this comparison.

Suppose that the apparent orthotropy of the plate as measured with acoustic methods is due to both material anisotropy (in the unstressed state) and residual stress (as opposed to anisotropy only). In this case, the true axes of material symmetry in the unstressed state may not be oriented parallel and perpendicular to the rolling direction.

Since the measured birefringence was essentially constant in the state of zero applied stress, it is reasonable to assume that any residual stresses which exist in the plate are also constant. The presence of a uniform residual stress field causes a uniform rotation of the acoustic axes from the initial (unstressed) axes (see (9)).

These residual stresses (if such exist) cannot be determined solely with the birefringence method used here, since the true orientation of the initial unstressed acoustic axes is indeterminate with acoustic methods. However, it is possible to use Okada's theory to show that the birefringence method measures only the applied stresses superposed on an undetermined, uniform residual stress field.

Recall that Okada assumes that the index of refraction tensor N_{ij} obeys the linear relationship

$$N_{ij} = N_{ij}^{\circ} + \alpha_{ijkl}\sigma_{kl} \qquad (15)$$

Let $\sigma_{kl} = \sigma_{kl}^a + \sigma_{kl}^r$ where σ_{kl}^a is the applied stress and σ_{kl}^r is the uniform residual stress; N_{ij}° is the index of refraction in the initial (unstressed) state. Since N_{ij}° and $\alpha_{ijkl}\sigma_{kl}^r$ are constants

$$N_{ij} = \hat{N}_{ij}^{\circ} + \alpha_{ijkl}\sigma_{kl}^a \qquad (16)$$

where the effect of the uniform residual stress-field has been absorbed into $\hat{N}_{ij}^{\circ} = N_{ij}^{\circ} + \alpha_{ijkl}\sigma_{kl}^r$. Now the eigenvalues and eigenvectors of N_{ij} give the velocities and polarization directions of SH waves in the presence of applied stress. The eigenvectors are rotated through an angle ϕ relative to the 'initial' acoustic axes; these 'initial' axes are pure-mode polarizations for a state of uniform residual stresses. The angle ϕ is given by (9), where all stresses in that equation are applied stresses. Consequently, even if a uniform residual stress field exists in our fracture specimen, what we have measured is the applied stress.

Summary

In these experiments, we have verified the applicability of Okada's theory to rolled 2024-T351 aluminium plate. We measured the stress field around the crack tip of a well-characterized fracture specimen, using a small transducer for good spatial resolution. From a relatively small number of data points, we were able to determine the entire stress field in the region of interest around the crack tip; we did this by taking advantage of the known functional form of the general solution for single-ended mode-I fracture specimens. Since the approach used here does not require *a priori* knowledge of the series coefficients, the method can be applied to any two-dimensional geometry containing an opening-mode crack. The shear-stress-field and the stress intensity factor obtained acousto-elastically were both in agreement with values obtained by optical methods.

Acknowledgement

The research reported in this paper was conducted in the Nondestructive Evaluation Section of the Naval Research Laboratory under the sponsorship of the Office of Naval Research. The authors are indebted to Mr J.P. Waskey for assistance in making specimens and loading fixtures for the acousto-elastic experiments.

References

1 **Benson, R.W., Raelson, V.J.** Acoustoelasticity, *Prod. Eng.* 30 (29) (1959) 56-59
2 **Crecraft, D.I.** The Measurement of Applied and Residual Stresses in Metals Using Ultrasonic Waves, *J. Sound Vib.* 5 (1) (1967) 173-192
3 **Hsu, N.N.** Acoustical Birefringence and Use of Ultrasonic Waves for Experimental Stress Analysis, *Exp. Mech.* 14 (5) (1974) 169-176
4 **Noronha, P.J., Wert, J.J.** An Ultrasonic Technique for the Measurement of Residual Stress, *J. Testing and Eval.* 3 (1975) 147-152
5 **Blinka, J., Sachse, W.** Application of Ultrasonic-pulse-spectroscopy Measurements to Experimental Stress Analysis, *Exp. Mech.* 16 (12) (1976) 448-453
6 **Ilic, D.B., Kino, G.S., Selfridge, A.R.** Computer-controlled System for Measuring Two-dimensional Velocity Fields, *Rev. Sci. Instrum.* 50 (12) (1979) 1527-1531
7 **Kino, G.S. et al,** Acoustoelastic Imaging of Stress Fields, *J. Appl. Phys.* 50 (4) (1979) 2607-2613
8 **Kino, G.S., et al,** Acoustic Measurements of Stress Fields and Microstructure, *J. Nondestructive Eval.* 1 (1) (1980) 67-77
9 **Bennett, S.D., Husson, D., Kino, G.S.** Measurement of Three-Dimensional stress Variation, 1981 Ult. Symp. Proc., IEEE Group on Sonics and Ultrasonics
10 **Heyman, J.S.** A cw Ultrasonic Bolt Strain Monitor, *Exp. Mech.* 17 (1977) 183-187
11 **Heyman, J.S., Chern, E.J.** Ultrasonic Measurements of Axial Stress, ASTM Symp. on Ult. Measurements of Stress, April 1981 (to be published)
12 **Heyman, J.S.** Pulsed Phase Locked Loop Strain Monitor, NASA Patent Disclosure LAR 12772-1 (1980)
13 **Papadakis, E.P.** Ultrasonic Study of Simulated Crystal Symmetries in Polycrystalline Aggregates, *IEEE Trans. Sonics and Ultrasonics,* SU-11 (1964) 19-29
14 **Mahadevan, P.** Effect of Frequency on Texture-Induced Ultrasonic Wave Birefringence in Metals, *Nature,* 211 (1966) 621-622
15 **Iwashimizu, Y., Kubomura, K.** Stress-Induced Rotation of Polarization Directions of Elastic Waves in Slightly Anisotropic Materials, *Int. J. Solids Structures,* 9 (1973) 99-114
16 **Okada, K.** Stress-Acoustic Relations for Stress Measurement by Ultrasonic Technique, *J. Acoust. Soc. Japan* (E) 1 (3) (1980) 193-200
17 **Okada, K.** Acoustoelastic Determination of Stress in Slightly Orthotropic Materials, *Exp. Mech.* 21 (1981) 461-466
18 **Auld, B.** Acoustic Fields and Waves in Solids, Sec. 6A, 8D, Wiley, New York (1973)
19 **Johnson, G.C.** Acoustoelastic Theory for Elastic-Plastic Material, *J. Acoust. Soc. Am.* 70 (2) (1981) 591-595
20 **Papadakis, E.P.** Ultrasonic Attenuation and Velocity in Three Transformation Products in Steel, *J. Appl. Phys.* 35 (1964) 1474-1482
21 **Papadakis, E.P.** Ultrasonic Phase Velocity by the Pulse-Echo-Overlap Method Incorporating Diffraction Phase Corrections, *J. Acoust. Soc. Am.* 42 (1967) 1045-1057
22 **Chung, D.H., Silversmith, D.J., Chick, B.B.** A Modified Ultrasonic Pulse-Echo-Overlap Method for Determining Sound Velocities and Attenuation of Solids, *Rev. Sci. Instrum.* 40 (1969) 718-720
23 ASTM Standard E561, 1978 Annual Book of ASTM Standards, Part 10, ASTM, Philadelphia (1978) 589-607
24 **Sanford, R.J.** A Critical Re-examination of the Westergaard Method for Solving Opening Mode Crack Problems, *Mech. Res. Comm.* 6 (1979) 289-294
25 **Sanford, R.J. et al,** A Photoelastic Study of the Influence of Non-Singular Stresses in Fracture Test Specimens, NUREG/CR-2179, University of Maryland, (August, 1981)
26 **Sanford, R.J., Chona, R.** An Analysis of Photoelastic Fracture Patterns With a Sampled Least-Squares Method, Proc. of the 1981 Spring Meeting, Soc. Exp. Stress Analysis (1981) 273-276

Strain Energy Release Rate for Radial Cracks Emanating from a Pin Loaded Hole

D. J. CARTWRIGHT
Department of Mechanical Engineering, University of Southampton, England

G. A. RATCLIFFE
British Aircraft Corporation, Filton, England

(Received December 30, 1970; in revised form May 21, 1971)

ABSTRACT
The strain energy release rate for two equal radial cracks emanating from a central hole in a finite width strip is determined by experimental compliance. The load is transmitted to the strip through a neatly fitting pin in the central hole. Comparisons are made with a recently available solution for the case where the loading is that of uniform tension remote from the hole.

List of Symbols

a	= Semi-crack length
a_n, b_n, c_n	= Polynomial coefficients
C	= Compliance
D	= Diameter of hole
E	= Young's modulus of elasticity for the strip
G	= Strain energy release rate
H	= Length of centrally cracked plate
K_I, K_{II}	= Stress intensity factors
K_t	= Stress concentration factor at intersection of hole boundary with y-axis based on gross stress
L	= Length of crack from circumference of hole
P	= Pin load
R	= Radius of hole
S	= Gross stress in strip remote from hole
t	= Thickness of strip
W	= Width of strip
λ	= Dimensionless crack length L/R

1. Introduction

An essential in the application of current fracture mechanics theory to studies of fatigue crack growth rate, residual static strength and stress corrosion, is a knowledge of the strain energy release rate or the stress intensity factor. It is the purpose here to present the strain energy release rate for equal radial cracks emanating from a central hole in a finite strip when the load is transmitted to the strip by a pin in the hole (Fig. 1). This represents a commonly occurring structural configuration which has not been reported heretofore in the literature.

Pin loading exhibits a special problem in that local stress around the hole and hence the stress intensity factor will depend on the particular way in which the pin transmits its load through the hole boundary to the strip. This is likely to be affected by:

(i) The modulus ratio of the pin/strip materials

(ii) The clearance between the pin and the hole
(iii) The ratio of the thickness of the strip to the pin diameter

In the uncracked situation the stress concentration at the point of minimum cross-section has been shown to be relatively insensitive to pin clearance [1]. (The main effect of pin clearance is to increase and shift the maximum stress in the direction of the reduced arc of contact). As the stress intensity factor depends for its value on the crack line stresses in the precracked field, pin clearance can be expected to have a small effect on K_I, providing the crack resides on the line of minimum section. In the present study the pin was chosen to be a neat fit in the hole having in all cases a clearance of 0.1%.

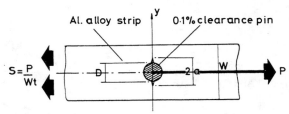

Figure 1. Pin loading.

To minimise the effect of pin bending, the strip thickness to pin diameter was kept in all cases to less than 0.4. This follows some photoelastic results of Meek [2] where it was found that, for uncracked situation, below a t/D ratio of 0.4 pin bending had negligible effect. The modulus ratio of the pin to strip material was 3, this being used to represent a hard pin in a hole. Having a low modulus strip gave the additional experimental advantage of larger displacements.

2. Method

The well-known method of compliance measurement [3] has been adapted for present purposes. In application this method demands that three variables be obtained, namely: load, extension and simulated crack length. Further to this, if the strain energy release rate is to be found, the numerical data must be differentiated. Loading was carried out in an Instron machine, the displacement control of which was found to be very suitable. The extensometer designed for the purpose incorporated a miniature linear differential transformer (L.V.D.T.) as the sensing device. This extensometer was comprised of two location clamps, one of which housed the L.V.D.T., a variable length sensing rod and a guide piece all mounted as shown in Figure 3. The gauge length over which the extension is measured must envelop the perturbation in the stress field created by the discontinuity. This minimum gauge length was determined photo-elastically for each specimen type. Output of the load and displacement signal to an $x-y$ plotter enabled the compliance at each crack length to be conveniently plotted several times so that accuracy could be improved by averaging. Six cycles were made at each crack length. The crack was simulated by a fine sawcut terminated by a small drilled hole ($\frac{1}{32}$ in and $\frac{1}{16}$ in) the effective crack length being taken as the length to the terminal hole boundary minus one half the terminal radius [4].

3. Theory

For a state of plane stress assuming that Mode I and Mode II deformations exist [5]

$$EG = K_I^2 + K_{II}^2 \tag{1}$$

where the strain energy release rate G may be written in terms of the compliance derivative as [3]

$$G = \frac{P^2}{2t} \frac{dC}{d(2L)}. \tag{2}$$

For the particular configuration under consideration a convenient form of (2) is

$$\frac{(EG)^{\frac{1}{2}}}{S(\pi a)^{\frac{1}{2}}} = \frac{W}{2R[\pi(1+\lambda)]^{\frac{1}{2}}} \left[\frac{d(ECt)}{d\lambda}\right]^{\frac{1}{2}}. \qquad (3)$$

A multiple regression computer programme was written to fit polynomials of the form

$$ECt = a_0 + \sum_{n=2}^{N} a_n \lambda^n \qquad (4)$$

to the experimental compliance measurements. Such a fit suppresses the coefficient a_1 to zero which ensures that $G=0$ at zero crack length [4].

Differentiating (4) and substituting into (3) leads to

$$\frac{(EG)^{\frac{1}{2}}}{S(\pi a)^{\frac{1}{2}}} = \frac{W}{2R}\left[\frac{\lambda}{\pi(1+\lambda)}\sum_{n=0}^{N-2} b_n \lambda^n\right]^{\frac{1}{2}}. \qquad (5)$$

The coefficients b_n and the degree $(N-2)$ in (5) are to be determined.

4. Test Case

The method of analysis can be tested with the work of Forman and Kobayashi [6] who provided a solution for the axial rigidity of a perforated strip in simple tension. When the perforation takes the form of a central crack the compliance expression becomes

$$C = \frac{H}{EtW\left[1 - \frac{2\pi W F_0}{H}\right]} \qquad (6)$$

where $F_0 = 0.25k^2 + 0.14871k^4 + 0.109687k^6 + 0.085314k^8 + 0.068697k^{10} + 0.06156k^{12}$ and $k = 2a/W$.

For this configuration $K_{II}=0$ and writing $(EG)^{\frac{1}{2}} = K_I$ equation (5) takes the form

$$\frac{K_I}{S(\pi a)^{\frac{1}{2}}} = \left[\frac{1}{\pi}\sum_{n=0}^{N-2} c_n \left(\frac{2a}{W}\right)^n\right]^{\frac{1}{2}}. \qquad (7)$$

Equation (6) enables values of ECt to be provided at various $2a/W$. The values of ECt so found provide accurate data for the multiple regression programme from which the c_n's in

$\frac{2a}{W}$	$K_I / S\sqrt{\pi a}$		
	eq^n 7	ref. 7	ref. 8
0.1	1.008	1.005	1.0061
0.2	1.025	1.021	1.0249
0.3	1.057	1.051	1.0583
0.4	1.110	1.100	1.1102
0.5	1.187	1.176	1.1876
0.6	1.301	1.292	1.3043
0.7	1.477	1.478	1.4891

Figure 2. Comparison of finite width correction factors for a centrally cracked strip.

equation (7) are determined. By this means values of $K_I/S(\pi a)^{\frac{1}{2}}$ appropriate to a centrally cracked finite width panel may be found. The values achieved in this manner from equation (7) are shown in Fig. 2 together with those of Koiter [7] and Newman [8] for the same case.

5. Results

As a preliminary to making measurements on the pin loaded configuration the stress intensity factors were found for the simpler case where the strip is loaded uniformly at each end remote from the hole (Fig. 3). At the time this work was performed no other solution for this configuration was available. Recently a collocation solution [8] has been completed and so a valuable comparison becomes possible.

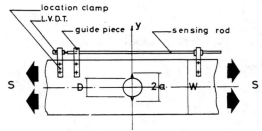

Figure 3. End loading.

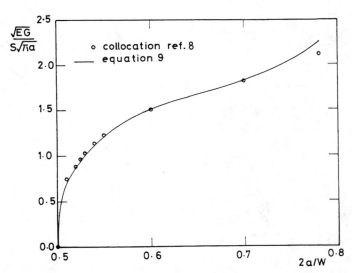

Figure 4. Comparison of results for fig. 3.

For this case $E = 72 \, GNm^{-2}$ (10.47×10^6 p.s.i.) and $t = 127$ mm (0.5 in) and the best fit polynomial becomes

$$ECt = 5.866 + 7.827\lambda^2 - 11.64\lambda^3 + 11.81\lambda^4 . \tag{8}$$

Putting this into equation (3) with $K_{II} = 0$ and $W/D = 2$ yields

$$\frac{K_I}{S(\pi a)^{\frac{1}{2}}} = 2\left[\frac{\lambda}{\pi(1+\lambda)} \{15.65 - 34.92\lambda + 47.24\lambda^2\}\right]^{\frac{1}{2}} . \tag{9}$$

This equation together with the results from the collocation solution appear in Fig. 4.

For the pin loaded hole Fig. 1, $E = 69.6 \, GNm^{-2}$ (10.11×10^6 p.s.i.) and $t = 122$ mm (0.48 in) the best fit polynomials obtained are

For $W/D = 2$, $0 < \lambda < 0.6$

$$\frac{d(ECt)}{d\lambda} = 17.71\lambda - 80.75\lambda^2 + 174.6\lambda^3 - 111.0\lambda^4 \tag{10}$$

For $W/D = 4$, $0 < \lambda < 1.8$

$$\frac{d(ECt)}{d\lambda} = 2.51\lambda - 4.115\lambda^2 + 2.791\lambda^3 - 0.563\lambda^4 \qquad (11)$$

For $W/D = 5$, $0 < \lambda < 2.5$

$$\frac{d(ECt)}{d\lambda} = 0.975\lambda - 0.66\lambda^2 + 0.18\lambda^3 . \qquad (12)$$

These are substituted into equation (3) together with the appropriate value of W/D. To demonstrate the difference, results for pin loading are plotted on Fig. 5 together with the collocation results for end loading at corresponding ratios of W/D.

6. Discussion of Results

Fig. 5. shows the strain energy release rate for the pin loaded and end loaded configurations and exhibits some rather interesting features. The curve for $W/D = 4$ rises more rapidly and to a higher value than its end-loaded counterpart. The curve then exhibits a slight decrease. At $W/D = 2$ the pin loaded case curve rises at about the same rate but does not achieve a higher value than with end-loading; also it does not decrease. This can be explained in the following way:

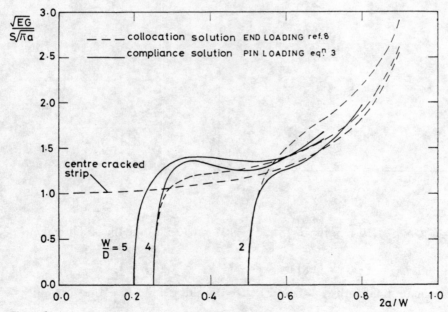

Figure 5.

Fig. 6 shows the stress concentration factor K_t for both pin and end-loading for various W/D ratios. As W/D increases from 2 to 5 the K_t's for the two types of loading diverge. From equilibrium considerations the higher K_t the more rapidly will the axial stress fall across the net section. Thus at $W/D = 4$ or 5 the higher K_t will initially cause the stress intensity factor for pin loading to increase more rapidly. This will not apply to the $W/D = 2$ case where the K_t's are more nearly the same. In addition to the stress concentration effect the outer boundaries of the strip will cause an increase in the stress intensity factor. For the $W/D = 2$ case this will occur almost immediately whilst at $W/D = 4$ or 5 the boundary is relatively remote and will not affect the crack tip so soon.

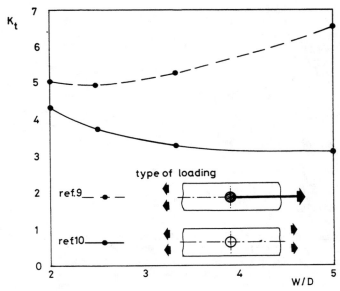

Figure 6. Comparison of stress concentration factors.

In short, the behaviour of the stress intensity factor will depend on the relative balance between the effect of, crack growth (proportional to $a^{\frac{1}{2}}$), a decreasing stress field and the proximity of boundary surfaces.

7. Effect of Asymmetry

With pin loading the distribution of stress is asymptotic about the line of minimum cross-section (y-axis). The effect of this asymmetry will be to create shear stresses along the crack line and hence a non zero K_{II}. The method of solution adopted here can only determine G and cannot uncouple K_I and K_{II}. A paper by Emery et al. [11] is useful in quantifying the effect of this asymmetry. This paper considers the stress intensity factor for a radially cracked pipe carrying a cryogenic liquid, the asymmetry here being created by filling the pipe to various depths. Using results from [8] the ratio of K_{II}/K_I for pin loading can be estimated to be at worst $\frac{1}{10}$, indicating as expected that the opening mode of deformation is dominant. Re-writing equation (1) as

$$(EG)^{\frac{1}{2}} = K_I \left[1 + \left\{\frac{K_{II}}{K_I}\right\}^2\right]^{\frac{1}{2}}$$

it is seen that the error in equating $(EG)^{\frac{1}{2}} = K_I$ for pin loading is less than 1%. Indeed, even if K_{II}/K_I where $\frac{1}{3}$, which is extremely unlikely, the error would be of order 5%.

Acknowledgements

One of us (D.J.C.) gratefully acknowledges the receipt of a grant from the Ministry of Aviation and Supply held through Structures Department, R.A.E., Farnborough, England for the period during which this work was carried out.

REFERENCES

[1] H. L. Cox and A. F. C. Brown, Stresses Round Pins in Holes, *The Aeronautical Quarterly*, Nov. (1964).
[2] R. M. G. Meek, *Effect of Pin Bending on the Stress Distribution in Thick Plates loaded Through Pins*, N.E.L. Report No. 311.
[3] G. R. Irwin and J. A. Kies, Critical Energy Rate Analysis of Fracture Strength, *Welding Research Supplement*, April (1954).

[4] J. E. Srawley, H. Jones and B. Gross, *Experimental determination of the dependence of crack extension force on the crack length for a single-edge-notch tension specimen.* N.A.S.A. TN D-2396 Aug. (1964).

[5] G. R. Irwin, *Analytical aspects of crack stress field problems*, T. and A.M. Report No. 213, Department of Theoretical and Applied Mechanics, University of Illinois, March (1962).

[6] R. G. Forman and A. S. Kobayashi, On the axial rigidity of a perforated strip and the strain energy release rate in a centrally notched strip subjected to uniaxial tension, *Journal of Basic Engineering* (1964).

[7] W. T. Koiter, *Note on the stress intensity factors for sheet strips with cracks under tensile loads.* Report 314 Laboratory of Engineering Mechanics, Technological University, Delft (1965).

[8] J. C. Newman, Jr., Private Communication Aug. (1970), To be published as a N.A.S.A. tech. note, Langley Research Centre, Hampton, Virginia.

[9] P. S. Theocaris, The stress distribution in a strip loaded in tension by means of a central pin, *Journal Applied Mechanics*, 78 Mar. (1956).

[10] R. C. J. Howland, On the stress distribution in the neighbourhood of a circular hole in a strip under tension, *Trans. Phil. Roy. Soc.*, 229 (1930).

[11] A. F. Emery, C. F. Barrett and A. S. Kobayashi, Temperature distributions and thermal stresses in a Partially Filled Annulus, *Experimental Mechanics*, Dec. (1966).

RÉSUMÉ

Par vérifications expérimentales, on détermine le taux de dissipation de l'énergie de déformation dans le cas de deux fissures identiques issues radialement d'un trou situé au centre d'une tôle mince de largeur finie.

La charge est appliquée à la tôle au moyen d'un axe étroitement ajusté dans le trou. On procède à une comparaison entre le cas traité et celui, récemment solutionné, où la charge est appliquée de manière uniforme à une certaine distance d'un trou.

ZUSAMMENFASSUNG

Der Dissipationsgrad der Verformungsenergie für zwei identische Risse, welche radial von einem in der Mitte eines Feinbleches endlicher Breite liegenden Loch ausgehen, wird experimentell bestimmt. Die Last wird durch einen genau in das Loch passenden Dorn auf das Blech übertragen.

Der hier behandelte Fall wird mit dem vor kurzem gelösten Problem, wo die Last in einer gewissen Entfernung des Loches gleichmäßig aufgebracht wird, verglichen.

Strain-Gage Methods for Measuring the Opening-Mode Stress-Intensity Factor, K_I

by J.W. Dally and R.J. Sanford

ABSTRACT—Measurements of strain near a crack tip with electrical-resistance strain gages do not usually provide a reliable measure of K_I because of local yielding, three-dimensional effects and limited regions for strain-gage placement. This paper develops expressions for the strains in a valid region removed from the crack tip, and indicates procedures for locating and orienting the gages to accurately determine K_I from one or more strain-gage readings.

Introduction

Although Irwin[1] first suggested the use of strain gages to determine the stress-intensity factor near the tip of a crack in 1957, little progress has been made in implementing this suggestion. The primary reason for the delay in the development of a suitable method involves the finite size of the strain gage. Questions arise regarding the effects of the strain gradients on the gage output, the magnitude of the strains to be measured if the gage is placed in close proximity to the crack tip, and the relative size of the gage compared to the size of the near-field region. A secondary reason for the delay is the availability of other experimental methods for determining the stress-intensity factor. Kobayashi[2] has described methods based on compliance measurements and photoelasticity. In addition Mannog[3] and Theocaris[4] have demonstrated the application of the method of caustics in a wide range of plane bodies containing cracks. Finally, Barker et al.[5] have shown an accurate numerical technique for determining the stress-intensity factor from full-field displacement data which can be obtained with either moiré or speckle photography.

This paper demonstrates that strain gages can be effectively employed to measure the stress-intensity factor. The application considered here is the determination of the opening-mode stress-intensity factor K_I in a plane body with a through crack.

The area adjacent to the crack tip is divided into three regions as shown in Fig. 1. The innermost region close to the crack tip, region I, is not a valid region for data acquisition because of nonlinearities caused by yielding or ambiguities concerning whether the stress state is plane stress or plane strain. The intermediate region, region II, is a valid area where the strain field can be represented within a specified accuracy by a multiparameter theory containing K_I and coefficients of higher order terms as unknowns. Region III, the outermost region, represents the far field where the truncated series describing the strain field is not sufficiently accurate. The boundary between region II and region III will depend upon the accuracy specified and the number of terms retained in the multiparameter theory. The area of region II is sufficiently large to accommodate common electrical-resistance strain gages.

The error due to strain gradient is first minimized by placing the strain gages sufficiently far from the crack tip, and is then eliminated by a simple integration procedure. The influence of the strain gradient in the θ direction is not treated because it is no larger than that normally encountered in strain-gage applications where the strain field varies as a trigonometric function of θ.

Multiparameter Representation of the Strain Field

Sanford[6] has shown that the Westergaard[7] equations should be generalized to solve fracture-mechanics problems where the stress field in the local neighborhood of the crack tip is influenced by the proximity of boundaries and points of load application. The stresses expressed in this generalized form are given by

$$\sigma_{xx} = ReZ - yImZ' - yImY' + 2ReY$$
$$\sigma_{yy} = ReZ + yImZ' + yImY' \tag{1}$$
$$\tau_{xy} = -yReZ' - yReY' - ImY$$

where, for a single-ended crack, the stress functions Z and Y can be represented by

$$Z(z) = \sum_{n=0}^{N} A_n z^{n-1/2} \tag{2}$$
$$Y(z) = \sum_{m=0}^{M} B_m z^m$$

J.W. Dally (SEM Fellow) and R.J. Sanford (SEM Fellow) are Professors, University of Maryland, Department of Mechanical Engineering, College Park, MD 20742.
Original manuscript submitted: February 6, 1986. Final manuscript received: March 11, 1987.

where

$$z = x + iy \tag{3}$$

with the coordinate system defined in Fig. 2. By substituting eqs (1) into the plane stress–strain relations:

$$\epsilon_{xx} = \frac{1}{E}[\sigma_{xx} - \nu\sigma_{yy}]$$

$$\epsilon_{yy} = \frac{1}{E}[\sigma_{yy} - \nu\sigma_{xx}] \tag{4}$$

$$\gamma_{xy} = \frac{1}{\mu}\tau_{xy}$$

one obtains the following generalized equations for the strain field.

$$E\epsilon_{xx} = (1-\nu)ReZ - (1+\nu)yImZ' - (1+\nu)yImY'$$
$$\qquad + 2ReY$$

$$E\epsilon_{yy} = (1-\nu)ReZ + (1+\nu)yImZ' + (1+\nu)yImY'$$
$$\qquad - 2\nu ReY \tag{5}$$

$$\mu\gamma_{xy} = -yReZ' - yReY' - ImY$$

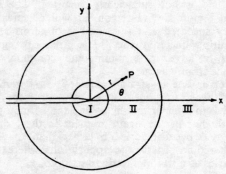

Fig. 1—Schematic illustration of the three regions associated with the near field for a compact-tension specimen at $a/W = 0.5$

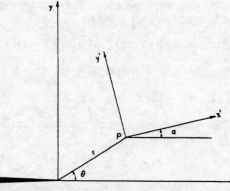

Fig. 2—Definition of coordinate systems Oxy and Px'y'

The strain field can be expressed exactly by using the infinite series representation of the stress functions Z and Y given in eq (2). This exact approach cannot, however, be utilized in practice due to the infinite number of unknown coefficients A_n and B_m. It is necessary to truncate the series and to accept a specified error in the representation of the strain field. In this paper, a four-term representation is described where the first two terms of each of the series for Z and Y have been retained. Of course, a higher-order theory could be developed, but the number of strain gages required to determine A_n and B_m would become prohibitively large.

Setting $n = 0, 1$ and $m = 0, 1$ gives

$$Z = A_0 r^{-1/2}[\cos(\theta/2) - i\sin(\theta/2)] +$$
$$\qquad A_1 r^{1/2}[\cos(\theta/2) + i\sin(\theta/2)] \tag{6}$$

$$Y = B_0 + B_1 r[\cos\theta + i\sin\theta]$$

Substituting eqs (6) into eqs (5) gives

$$E\epsilon_{xx} = A_0 r^{-1/2}\cos(\theta/2)[(1-\nu) -$$
$$(1+\nu)\sin(\theta/2)\sin(3\theta/2)] + 2B_0 + A_1 r^{1/2}\cos(\theta/2)[(1-\nu)$$
$$+ (1+\nu)\sin^2(\theta/2)] + 2B_1 r\cos\theta$$

$$E\epsilon_{yy} = A_0 r^{-1/2}\cos(\theta/2)[(1-\nu) +$$
$$(1+\nu)\sin(\theta/2)\sin(3\theta/2)] - 2\nu B_0 +$$
$$A_1 r^{1/2}\cos(\theta/2)[(1-\nu) - (1+\nu)\sin^2(\theta/2)] - 2\nu B_1 r\cos\theta$$

$$\mu\gamma_{xy} = (A_0/2)r^{-1/2}\sin\theta\cos(3\theta/2) -$$
$$(A_1/2)r^{1/2}\sin\theta\cos(\theta/2) - 2B_1 r\sin\theta \tag{7}$$

This representation of the strain field could be employed with data from strain gages positioned at arbitrary points $[P_1(r_1,\theta_1), P_2(r_2,\theta_2), P_3(r_3,\theta_3)$ and $P_4(r_4,\theta_4)]$ all located in region II and oriented in either the x or y directions to obtain a solution for A_0 which is related to K_I by

$$K_I = \sqrt{2\pi}A_0 \tag{8}$$

However, it is advantageous to consider orientation of the strain gages at an arbitrary angle α, as illustrated in Fig. 2, to explore the possibility of eliminating some of the terms in the strain-field representation by gage positioning and orientation.

Four-Parameter Strain Field Relative to a Rotated Coordinate System

The strains relative to a rotated coordinate system (x', y') with its origin at an arbitrary point $P(r,\theta)$ as defined in Fig. 2 are determined from the first invariant of strain,

$$\epsilon_{x'x'} + \epsilon_{y'y'} = \epsilon_{xx} + \epsilon_{yy} \tag{9}$$

and the complex form of the strain-transformation equations,

$$\epsilon_{y'y'} - \epsilon_{x'x'} + i\gamma_{x'y'} = (\epsilon_{yy} - \epsilon_{xx} + i\gamma_{xy})e^{2i\alpha} \tag{10}$$

Substituting eq (5) into eq (10) leads to

$$2\mu\epsilon_{x'x'} = \frac{1-\nu}{1+\nu}(ReZ + ReY) -$$
$$(yImZ' + yImY' - ReY)\cos 2\alpha -$$
$$(yReZ' + ReY' + ImY)\sin 2\alpha$$

$$2\mu\epsilon_{y'y'} = \frac{1-\nu}{1+\nu}(ReZ + ReY) +$$
$$(yImZ' + yImY' - ReY)\cos 2\alpha +$$
$$(yReZ' + yReY' + ImY)\sin 2\alpha \quad (11)$$

Again letting $n = 0,1$ and $m = 0,1$ in eq (2) and substituting the results for the truncated series into eq (11) yields

$$2\mu\epsilon_{x'x'} = A_0 r^{-1/2}[k\cos(\theta/2) - (1/2)\sin\theta\sin(3\theta/2)\cos 2\alpha$$
$$+ (1/2)\sin\theta\cos(3\theta/2)\sin 2\alpha] + B_0(k + \cos 2\alpha) +$$
$$A_1 r^{1/2}\cos(\theta/2)[k + \sin^2(\theta/2)\cos 2\alpha - (1/2)\sin\theta\sin 2\alpha]$$
$$+ B_1 r[(k + \cos 2\alpha)\cos\theta - 2\sin\theta\sin 2\alpha] \quad (12)$$

$$\mu(\epsilon_{y'y'} - \epsilon_{x'x'}) = (A_0/2)r^{-1/2}\sin\theta[\sin(3\theta/2)\cos 2\alpha -$$
$$\cos(3\theta/2)\sin 2\alpha] -$$
$$B_0\cos 2\alpha + (A_1/2)r^{1/2}\sin\theta[\cos(\theta/2)\sin 2\alpha -$$
$$\sin(\theta/2)\cos 2\alpha] + B_1 r[2\sin\theta\sin 2\alpha - \cos\theta\cos 2\alpha] \quad (13)$$

where

$$k = (1-\nu)/(1+\nu) \quad (14)$$

Equation (12) gives the relations between K_I and the strain $\epsilon_{x'x'}$ measured with a single-element strain gage oriented at an angle α with respect to the $P(x'y')$ coordinate system. Equation (13) gives the relation between K_I and the output of a rectangular rosette with the two elements connected to adjacent arms of a Wheatstone bridge. There are many options regarding gage positioning and orientation to simplify eqs (7), (12) and (13). However, before considering these options it is necessary to more closely examine region II and to determine its bounds, since it is essential that the gage position $P(r,\theta)$ be located within this region to avoid excessive errors.

Crack-Tip Regions

The size and shape of region II can be determined for say $\epsilon_{x'x'}$ by comparing the results of eq (12) with an exact solution of a representative fracture-mechanics problem. To illustrate this procedure, consider the compact-tension specimen with $a/W = 0.5$ subjected to pin loading. Chona et al.[8] have determined the coefficients A_0, A_1, A_2, B_0, B_1 and B_2 to obtain a six-parameter solution for the stress field which is considered to represent an exact solution over a region of some extent around the crack tip. Approximate values of the strains $\epsilon_{x'x'}$ for $\alpha = 60$ deg were determined from eq (12) over the field by using one or more of these coefficients. For example, a one-parameter representation of $\epsilon_{x'x'}$ utilizes A_0 with

$A_1 = A_2 = B_0 = B_1 = B_2 = 0$; a two-parameter solution utilizes A_0 and B_0; a three-parameter solution utilizes A_0, B_0, and A_1; and finally a four-parameter solution contains A_0, B_0, A_1, and B_1.

A point by point comparison was made over the field to determine the difference between the exact (six coefficient) and approximate values of $\epsilon_{x'x'}$. A plotting routine was then used to provide maps of the area around the crack tip where the differences were within ± 2, ± 5, and ± 10 percent. The results obtained for the compact-tension specimen over a ± 0.25 W sized zone about the crack tip are shown in Fig. 3. An examination of these results shows two very clear trends. Firstly, the size of region II increases markedly as additional terms are added to the series representation of $\epsilon_{x'x'}$. Note that the second term (i.e., B_0) does not affect the size or shape of region II because of the choice of $\alpha = 60$ deg (for $\nu = 1/3$) which eliminates this term in eq (12) for all values of r and θ. Secondly, the size of region II depends strongly on the accuracy required. However, in all cases considered, ± 2, ± 5 and ± 10-percent accuracies, the size of region II is large enough to accommodate several strain gages.

Region II has been divided into two parts in Fig. 3, namely IIa and IIb. Region IIb is a valid region from an accuracy viewpoint; however, the strain $\epsilon_{x'x'}$ in IIb will be quite low and is not a suitable area for strain-gage placement. The magnitude of the strain in IIa is much larger; gage placement should be restricted to this area.

The inner boundary of region II is determined by a circle of radius:

$$r = h/2 \quad (15)$$

where h is the thickness of the plane body. Rosakis and Ravi-Chandra[9] have shown experimentally that the state of stress at the crack tip is three dimensional in region I and is not represented by either plane stress or plane strain. Plane-stress conditions exist only when $r > h/2$.

Strain-Gage Position and Orientation

There are many possible approaches to determining K_I by employing eqs (7), (12) and (13). Only three approaches will be described here to illustrate some of the procedures to be followed in reducing the theory to practice. These approaches include: (1) the single gage—three-parameter solution, (2) the two gage-four-parameter solution and (3) the rectangular rosette—two- or three-parameter solution with temperature compensation. Each of these three approaches is covered as individual cases below.

Case I: Single Gage—Three-Parameter Solution

Consider first eq (12) for $\epsilon_{x'x'}$ and note that the B_0 term can be eliminated if

$$\cos 2\alpha = -k = -(1-\nu)/(1+\nu) \quad (16)$$

Next, set the coefficient of the A_1 term to zero.

$$k + \sin^2(\theta/2)\cos 2\alpha - (1/2)\sin\theta\sin 2\alpha = 0$$

which can be satisfied if

$$\tan(\theta/2) = -\cot 2\alpha \quad (17)$$

These results show that a single-element gage can be used to provide the data necessary for a three-parameter solution for K_I providing the angles α and θ are selected to satisfy eqs (16) and (17). These angles depend only on a Poisson's ratio, ν, of the specimen material as indicated in Table 1.

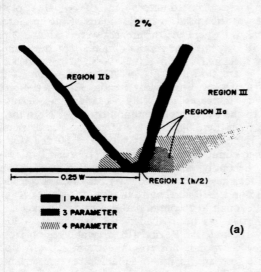

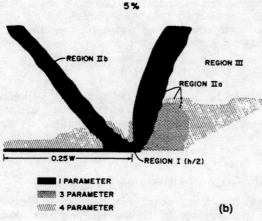

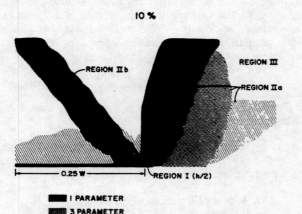

Fig. 3—Size of the valid region for $\epsilon_{x'x'}$ with $\alpha = 60$ deg and $\nu = 1/3$ as a function of the number of terms used in the series representation for a compact-tension specimen with $h = 0.250$ in. (6.3 mm). (a) Two-percent accuracy, (b) five-percent accuracy, (c) ten-percent accuracy

Take, for example, an aluminum specimen with $\nu = 1/3$. Then $\alpha = \theta = 60$ deg, and the gage is placed at any point in region II along a 60-deg radial line drawn from the crack tip. For this example (i.e., $\nu = 1/3$) eq (12) reduces to

$$2\mu\epsilon_{x'x'} = A_0 r^{-1/2} 3\sqrt{3}/8 \qquad (18)$$

Substituting eq (8) into eq (18) and solving for K_I gives

$$K_I = E\sqrt{(8/3)\pi r}\, \epsilon_{x'x'} \qquad (19)$$

Measuring the gage position r and recording the strain $\epsilon_{x'x'}$ from the single gage gives the data necessary to determine K_I with an accuracy consistent with the differences associated with region II.

Case II: Two Gages—Four-Parameter Solution

It is possible to obtain the data necessary to determine K_I from a four-parameter representation by using two strain gages, providing these gages are both placed in region II with the same value of θ and α as specified in Table 1. It is clear then that eq (12) subjected to the restrictions of eqs (16) and (17) reduces to

$$2\mu\epsilon_{x'x'} = A_0 r^{-1/2}[k\cos(\theta/2) + (k/2)\sin\theta\sin(3\theta/2) +$$

$$(\tfrac{1}{2})\sin\theta\cos(3\theta/2)\sin 2\alpha] - 2B_1 r\sin\theta\sin 2\alpha \qquad (20)$$

The two strain-gage readings $(\epsilon_{x'x'})_A$ and $(\epsilon_{x'x'})_B$ and their respective positions r_A, r_B and $\theta_A = \theta_B$ can be used to solve eq (20) for A_0 and B_1. To demonstrate this fact and to show the simplicity of this approach, consider again the aluminum specimen with $\nu = 1/3$ and $\alpha_A = \alpha_B = \theta_A = \theta_B = 60$ deg. With these substitutions, eq (20) reduces to

$$E\epsilon_{x'x'} = (\sqrt{3}/2)A_0 r^{-1/2} - 2B_1 r \qquad (21)$$

Solving eq (21) for A_0 or K_I using data from gages A and B gives

$$K_I = E\sqrt{(8/3)\pi r_A r_B}\left[\frac{(\epsilon_{x'x'})_A r_B - (\epsilon_{x'x'})_B r_A}{r_B^{3/2} - r_A^{3/2}}\right] \qquad (22)$$

If $r_B = qr_A$, then eq (22) becomes

$$K_I = E\sqrt{(8/3)\pi r_A}\,[q(\epsilon_{x'x'})_A - (\epsilon_{x'x'})_B]\left(\frac{\sqrt{q}}{q^{3/2}-1}\right) \qquad (23)$$

The four-parameter solution should be employed with bodies where the crack tip is relatively close to the boundaries and/or loading points, and the strain field requires the fourth term in the series for a more accurate representation.

TABLE 1—ANGLES α AND θ AS A FUNCTION OF POISSON'S RATIO, ν

ν	θ (deg)	α (deg)
0.250	73.74	63.43
0.300	65.16	61.29
0.333	60.00	60.00
0.400	50.76	57.69
0.500	38.97	54.74

Case III: Rectangular Rosette

In some applications in materials testing, the fracture specimen is subjected to a temperature gradient and temperature compensation is important. In these instances it is advisable to employ a stacked rectangular rosette and to place the two gages in adjacent arms of a Wheatstone bridge to achieve temperature compensation. The output from the bridge gives the measurement of $\epsilon_{y'y'} - \epsilon_{x'x'}$ which is shown in eq (13). It is evident from this equation that the coefficient of the B_0 term will vanish if

$$\cos 2\alpha = 0 \quad \text{or} \quad \alpha = \pi/4 \quad (24)$$

With this restriction eq (13) reduces to

$$\mu(\epsilon_{y'y'} - \epsilon_{x'x'}) = -(A_0/2)r^{-1/2}\sin\theta\cos(3\theta/2) + (A_1/2)r^{1/2}\sin\theta\cos(\theta/2) + 2B_1 r \sin\theta \quad (25)$$

Examination of eq (25) shows that θ cannot be selected to eliminate the coefficients of A_1 or B_1 without eliminating the coefficient of A_0. This fact indicates that a single two-element rectangular rosette can only be employed in a two-parameter solution for K_I.

To obtain the two-parameter solution, let $\theta = \pi/2$ and let $A_1 = B_1 = 0$. Equation (25) then gives

$$K_I = \frac{2E}{1+\nu}(\epsilon_{y'y'} - \epsilon_{x'x'})\sqrt{\pi r} \quad (26)$$

It is possible to obtain a three-parameter solution for K_I if the value of A_1/A_0 is known or can be estimated. Again, for $\theta = \frac{\pi}{2}$ with $\alpha = \frac{\pi}{4}$ and $B_1 = 0$, eq (25) becomes

$$\mu(\epsilon_{y'y'} - \epsilon_{x'x'}) = (\sqrt{2}/4)A_0 r^{-1/2}[1 + (A_1/A_0)r] \quad (27)$$

which can be rewritten as

$$K_I = K_{Iapp}/[1 + (A_1/A_0)r] \quad (28)$$

where K_{Iapp} is obtained from eq (26).

An estimate of (A_1/A_0) can be obtained if two or more rosettes are positioned along the line $\theta = \pi/2$. For small values of r, eq (28) can be approximated by

$$K_I \approx K_{Iapp}[1 - (A_1/A_0)r + \ldots] \quad (29)$$

From the measured strain readings a plot of K_{Iapp} versus position is constructed and the limiting value of the slope taken, as illustrated in Fig. 4. From the plotted data, K_I = intercept, and $(A_1/A_0)K_I \approx$ limiting slope.

Strain-Gradient Effects

The error due to strain-gradient effects can be shown by considering a single-element strain-gage positioned in region IIa with $\alpha = \theta = 60$ deg. The gage senses the strain $\epsilon_{x'x'}$ given in eq (18). Its signal represents the average strain over its length which is given by

$$\epsilon_{x'x'}\Big|_{ave} = \frac{k_i}{r_0 - r_i}\int_{r_i}^{r_0} r^{-1/2} dr = 2k_i/(r_0^{1/2} + r_i^{1/2}) \quad (30)$$

where $k_i = K_I/E\sqrt{(8/3)\pi}$ and r_0 and r_i are positions of the active gage element as shown in Fig. 5. The gage output

$$\epsilon_{x'x'}\Big|_{ave}$$

corresponds to the true strain $\epsilon_{x'x'}$ at a specific point r_t along the gage length. It is evident from eqs (18) and (30) that

$$r_t = (r_0^{1/2} + r_i^{1/2})^2/4 \quad (31)$$

The position of the geometric center r_c of the gage is

$$r_c = (r_0 + r_i)/2 \quad (32)$$

Defining Δr as the distance between the geometric center of the gage, r_c, and the true strain point, r_t, gives

$$\Delta r = r_c - r_t = (r_0 - 2r_0^{1/2}r_i^{1/2} + r_i)/4 \quad (33)$$

Noting that the gage length L is

$$L = r_0 - r_i \quad (34)$$

and combining eqs (32), (33) and (34), it is evident that

$$(\Delta r/r_c) = [1 - \sqrt{1 - (L/2r_c)^2}]/2 \quad (35)$$

where $r_c > (L/2)$ to avoid placing the gage over the crack tip. The results from eq (35) which show $(\Delta r/r_c)$ as a

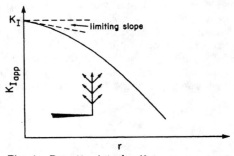

Fig. 4—Rosette data for K_{Iapp} as a function of position r showing the limiting slope, and intercept K_I

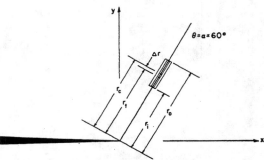

Fig. 5—Definition of radii associated with gage placement near a crack tip

function of (r_c/L) are presented in Fig. 6. The strain-gradient effect is a maximum ($\Delta r/r_c = 0.5$) when the gage is placed as close as possible to the crack tip with $r_c = -L/2$. This placement should be avoided in any event since all or part of the gage is in region I where the plane stress analysis presented here is not valid. For gages placed in region IIa, r_c/L will probably exceed two and the effect of the strain gradient is much smaller. If a correction is required, r_t is determined from eq (31). This value is used in eq (19) to determine K_I. In many applications a correction may not be required. For example consider a gage with $L = 0.030$ in. (0.76 mm) positioned at $r_c = 5L$, and note from Fig. 6 that $\Delta r/r_c = 0.0025$. The correction $\Delta r = 0.000375$ in. (0.0095 mm) is much less than the accuracy which can be achieved in measuring r_c.

Experimental Verification

A compact-tension specimen with $W = 12$ in. (305 mm) was fabricated from a 0.250-in. (6.35-mm) thick plate of aluminum 6061T6 with a machined crack of length $a/W = 0.5$ to verify the theory. Three single-element strain gages with an active grid 0.030×0.030 in. (0.76 × 0.76 mm) were positioned along the line $\alpha = \theta = 60$ deg at $r = 0.192, 0.483$ and 0.783 in. (4.88, 12.27 and 19.89 mm). Three two-element rectangular rosettes were positioned along the line $\theta = 90$ deg, all with $\alpha = 45$ deg and $r = 0.233, 0.500$ and 0.767 in. (5.92, 12.70, 19.48 mm). The specimen was loaded in 200-lb (890-N) increments. The strains were measured at each load increment. The strains $\epsilon_{x'x'}$ and $\epsilon_{y'y'} - \epsilon_{x'x'}$ for the single element and rosettes are given in Table 2 for a load of 2000 lb (8900 N).

The strains were then substituted into eq (19) or eq (28) and the experimental values of K_I determined. For the rosette calculations a value of $(A_1/A_0) = -\frac{1}{3}$ was used.[8] These experimental values were compared to a theoretical value of K_I which was computed from the ASTM formula.[10] Comparisons of the experimental and theoretical results shown in Table 2 indicate that the values of K_I determined with single-element strain gages were consistently less than the theoretical value with differences ranging from 4.7 percent to 12.0 percent. It should also be noted that the largest difference occurred with data from gages close to the tip of the machined crack.

There are two reasons which lead to this difference. Firstly, the crack geometry (its width and sharpness) deviated from the theoretical model and these differences affected the strain field with the largest differences occurring at gage positions close to the crack tip. Secondly, the very large strains predicted by fracture-mechanics theory as r approaches zero cannot be achieved; the material very near the crack tip yields and the stress is redistributed to maintain equilibrium.

To account for the effects of the redistribution of the stress on the strain near the crack tip, Irwin's method of shifting the elastic field by the plastic-zone radius, r_y, to account for the finite stress at the crack tip was used. The r_y correction has two effects on the calculation of K_I from strain-gage data. First, the angle, θ, no longer has the required value to satisfy eq (17). Second, the radius changes. Both of these effects are illustrated in Fig. 7. From the figure it can be shown that

$$r' = r\left[1 - 2\left(\frac{r_y}{r}\right)\cos\theta + \left(\frac{r_y}{r}\right)^2\right]^{1/2} \quad (36a)$$

TABLE 2—STRAIN-GAGE POSITIONS, STRAIN MEASUREMENT, K_I RESULTS AND THE DIFFERENCE BETWEEN THEORETICAL AND EXPERIMENTAL RESULTS FOR 2000-lb (8900-N) LOAD

Gage No.	r in. (mm)	$\epsilon_{x'x'}$ (× 10^{-6})	K_I ksi$\sqrt{\text{in.}}$ (MPa$\sqrt{m}$)	Difference* percent
1	0.192 (4.88)	1533	19.8 (21.8)	−12.0
2	0.483 (12.27)	1038	21.3 (23.4)	−4.7
3	0.783 (19.89)	813	21.2 (23.3)	−5.0
Rosette No.		$\epsilon_{y'y'} - \epsilon_{x'x'}$ (× 10^{-6})	K_I ksi$\sqrt{\text{in.}}$ (MPa$\sqrt{m}$)	
4	0.233 (5.92)	1555	22.1 (24.3)	−0.9
5	0.500 (12.70)	1015	23.4 (25.7)	4.9
6	0.767 (19.48)	727	23.2 (25.5)	4.0

*$K_I = 22.3$ ksi$\sqrt{\text{in.}}$ (24.5 MPa$\sqrt{m}$) ASTM formula

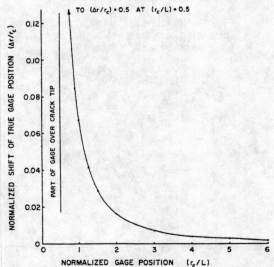

Fig. 6—Normalized shift of true gage position $\Delta r/r_c$ as a function of normalized position r_c/L

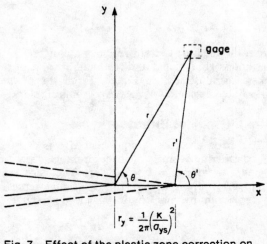

Fig. 7—Effect of the plastic-zone correction on the coordinates of a single-element strain gage

and

$$\theta' = \tan^{-1}\left[\frac{\sin\theta}{\left(\cos\theta - \frac{r_y}{r}\right)}\right] \quad (36b)$$

For single-element gages initially satisfying eqs (16) and (17), eq (12) can be expressed in the form

$$2\mu\epsilon_{x'x'} \approx \frac{A_0}{(r)^{1/2}} f(\theta) \quad (37)$$

where $f(\theta)$ is obtained from the leading term in eq (12). When the plastic-zone corrected coordinates r', θ' are used, eq (12) becomes

$$2\mu\epsilon_{x'x'} \simeq \frac{A_0^c}{(r')^{1/2}} f(\theta') \quad (38)$$

where A_0^c is the plastic-zone corrected estimate of A_0. Equating eqs (37) and (38) yields

$$A_0^c = A_0 \left(\frac{r'}{r}\right)^{1/2} \frac{f(\theta)}{f(\theta')} \quad (39)$$

The combined effects of these errors can, to the first order, be expressed as a function of r_y/r as shown in Fig. 8 for various values of Poisson's ratio. There is an additional correction due to the A_1 term in eq (12); however, for single-element gages approximately satisfying eq (17), the contribution due to this term is less than one percent even for the closest gage. The experimental results for single-element gages, corrected for finite-stress effects, are given in Table 3.

For the case of rosette gages the correction for finite stresses at the crack tip cannot be expressed solely in terms of the ratio r_y/r. Since there is no angle for which the A_1 contribution to the K measurement can be neglected, the influence of this term must be included in the finite stress corrections. An analysis of eq (13) reveals that the correction factor is highly sensitive to estimates of A_1 and r_y/r. As a result, the correction of rosette results is not practical.

Summary

A general method for determining K_I with common commercially available strain gages has been developed in terms of the generalized Westergaard stress functions. A four-parameter solution for K_I was derived from these stress functions to give an experimental approach for measuring K_I with a small number of strain gages.

The area adjacent to the crack tip was divided into three regions. Region I very near the crack tip is invalid because of three-dimensional effects. Region III far from the crack tip is invalid because the truncated series solution does not adequately describe the strain field. Region II located between regions I and III is a valid area where the truncated-series solution represents the strain field to a specified accuracy. The size and shape of region II is presented for the compact-tension geometry. Region II was subdivided into regions IIa and IIb. Region IIb was discarded because the strains in this area are too low for accurate measurement.

Three specific cases are considered in reducing the four-parameter theory to a simple and practical approach. It is shown that a single gage element with proper placement and orientation can be used to provide measurement of K_I while accounting for the effect of the first two non-singular terms. Two in-line single-element gages provide the data necessary to measure K_I with a four-parameter theory. The use of rosettes for the measurement of K_I is inherently less efficient since more gages are required; however, temperature compensation can be achieved on fracture specimens where thermal gradients are required on the specimen. The use of a single rosette gives the

TABLE 3—SINGLE-GAGE RESULTS CORRECTED FOR FINITE CRACK-TIP STRESSES

Gage No.	K_I-Uncorrected ksi$\sqrt{\text{in.}}$ (MPa$\sqrt{\text{m}}$)	K_I-Corrected ksi$\sqrt{\text{in.}}$ (MPa$\sqrt{\text{m}}$)	Error* percent
1	19.8 (21.7)	23.8 (26.1)	6.5
2	21.3 (23.4)	22.7 (24.9)	1.8
3	21.2 (23.3)	22.0 (24.2)	−1.5

*Based on a theoretical value of $K_I = 22.3$ ksi$\sqrt{\text{in.}}$ (24.5 MPa$\sqrt{\text{m}}$), compared to K_I corrected, as determined from strain-gage measurements

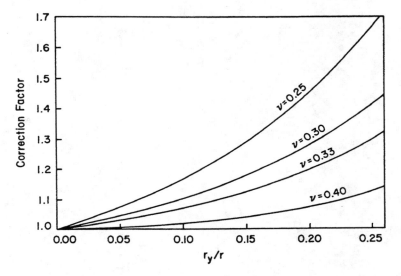

Fig. 8—Finite crack-tip stress correction for single-element gages as a function of r_y/r for various values of Poisson's ratio

data necessary for a two-parameter determination of K_I. The use of two or more rosettes permits a three-parameter determination. A graphical technique is introduced for the rosette approach which permits the experimentalist to determine the need for a higher-order theory. This technique is also applicable for single-element gages if they are placed along a radial line with θ constant.

The effect of strain gradients is considered. A method is shown to locate the true position of the gage, eliminating error due to the strain gradient in the radial direction. It is also shown that the difference between the geometric center of the gage and the true strain position becomes small as r_c/L increases.

Finally, experiments conducted to verify the theory show that K_I determined with single-element strain gages consistently underestimates the theoretical value, with the gages located close to the crack tip exhibiting the largest difference (12 percent). It is believed that these differences are due mainly to the crack-tip geometry of the model and localized yielding at the crack tip. A correction method has been developed to adjust the results for the effects of finite stresses at the crack tip. The results for K_I after correction show differences ranging from -1.5 to 6.5 percent between the theoretical and experimental values. Similarly, the results obtained with rosette gages show acceptable agreement with the theory, i.e., within five percent.

Acknowledgments

The authors are grateful to Mr. Ravi Chona for the computer run necessary to generate Fig. 3 and to Mr. John R. Berger for his assistance in conducting the experiments. Research funding for this project was provided by Martin Marietta Energy Systems (ORNL), Dr. C.E. Pugh, program manager and the National Science Foundation under grant No. MSM-85-13037, Dr. S. Jahanmir, program director.

References

1. Irwin, G.R., "Analysis of Stresses and Strains Near the End of a Crack Traversing a Plate," J. Appl. Mech., 24 (3), (1957).
2. Experimental Techniques in Fracture Mechanics, ed. A.S. Kobayashi, SEM Monograph, Iowa State University Press (1973).
3. Mannog, P., "Schattenoptische Messung der Spezifischen Bruchenergie wahrend des Bruchvorgangs bei Plexiglas," Proc. Int. Conf. on the Physics of Non-Crystalline Solids, Delft, The Netherlands, 481-490 (1964).
4. Theocaris, P.S., "Local Yielding Around A Crack Tip in Plexiglas," J. Appl. Mech., 37, 409-415 (1970).
5. Barker, D.B., Sanford, R.J. and Chona, R., "Determining K and Related Stress-Field Parameters from Displacement Fields," EXPERIMENTAL MECHANICS, 25 (4), 399-406 (1985).
6. Sanford, R.J., "A Critical Re-examination of the Westergaard Method for Solving Opening-Mode Crack Problems," Mech. Res. Comm., 6 (5), (1979).
7. Westergaard, H.M., "Bearing Pressure and Cracks," J. Appl. Mech., 6 (1939).
8. Chona, R., Irwin, G.R. and Sanford, R.J., "Influence of Specimen Size and Shape on the Singularity-Dominated Zone," Fracture Mechanics: Fourteenth Symposium - Volume 1: Theory and Analysis, ASTM STP 791, eds. J.C. Lewis and G. Sines, Amer. Soc. Test. and Mat., I-3-I-23 (1983).
9. Rosakis, A.J. and Ravi Chandra, K., "On Crack Tip Stress States and Experimental Evaluation of Three-Dimensional Effects," Cal. Inst. of Tech. Rep., FM-84-2 (March 1984).
10. ASTM-Standard E399-83, "Standard Test Methods for Plane-Strain Fracture Toughness of Metallic Materials," 1983 Annual Book of ASTM Standards, 03.01, ASTM (1983).
11. Irwin, G.R., "Plastic Zone Near a Crack and Fracture Toughness," Proc. 7th Sagamore Conf., IV-63 (1960).

Author Index

Barsoum, Roshdy S., *361*
Bowie, O.L., *61, 73, 78, 86, 136*
Bradley, W.B., *489*
Brown, Jr., William F., *231*
Cartwright, D.J., *638*
Chan, S.K., *316*
Chell, G.G., *98, 100*
Clark, A.V., *630*
Creager, Matthew, *92*
Crosley, P.B., *590*
Cruse, Thomas A., *405, 423*
Dally, James W., *524, 645*
Dixon, J.R., *497*
Dudderar, T.D., *604*
Dunegan, H.L., *611*
Emery, A.F., *202, 208*
Erdogan, F., *114*
Feddersen, C.E., *159*
Fourney, M.E., *609*
Freese, C.E., *86, 136*
Gorman, H.J., *604*
Grandt, Jr., A.F., *173*
Gross, Bernard, *231, 242, 256, 269*
Harris, D.O., *185, 611*
Henshell, R.D., *374*
Irwin, G.R., *121, 485*
Isida, M., *157*
Johnson, R.L., *502*
Kassir, M.K., *191*
Kobayashi, A.S., *202, 208, 215, 489*
Leven, M.M., *502*
Manogg, Peter, *535*
Marloff, R.H., *502*
McGowan, J.J., *513*
McGregor, J., *497*
McMeeking, R.M., *387*

Mignogna, R.B., *630*
Mostovoy, S., *590*
Mowbray, D.F., *333*
Neal, D.M., *78, 86*
Newman, Jr., J.C., *285, 438, 451*
Paris, Paul C., *3, 92, 114, 121, 387*
Post, D., *461*
Raju, I.S., *438, 451*
Ratcliffe, G.A., *638*
Ravi-Chandar, K., *562*
Rice, J.R., *151*
Ringler, T.N., *502*
Ripling, E.J., *590*
Roberts, Jr., Ernest, *269*
Rooke, D.P., *162, 168*
Rosakis, Ares J., *545, 562*
Sanford, Robert J., *524, 630, 645*
Schroedl, M.A., *513*
Shah, R.C., *215*
Shaw, K.G., *374*
Sih, George C., *3, 114, 121, 151, 191*
Smith, C.W., *513*
Smith, F.W., *202, 208*
Sommer, E., *576*
Srawley, John E., *231, 242, 256, 269, 280*
Strannigan, J.S., *497*
Tada, Hiroshi, *57, 387*
Tatro, C.A., *611*
Tracey, Dennis M., *337, 348*
Tuba, I.S., *316*
Tweed, J., *162, 168*
VanBuren, W., *423*
Walsh, P.F., *351*
Wells, A.A., *461*
Williams, M.L., *145*
Wilson, W.K., *282, 316*

Subject Index

acoustic
    birefringence, *630*
    emission, *611*
alternating method, *208, 215*
anisotropic, *3, 121, 630*

bars, *185*
bi-axial, *3, 61, 173, 285*
bielastic constant, *151*
boundary integral equation, *405, 423*
branching, *489, 545*
brittle fracture, *61, 461, 489, 576*

caustics, *535, 545, 562*
cleavage, *86*
clip gage, *269*
collocation
    boundary, *3, 78, 86, 136, 231, 242, 256, 269, 280, 282, 285, 316, 497*
    local, *524, 630*
compliance, *98, 231, 242, 256, 269, 316, 333, 387, 590, 604, 638*
conformal mapping, *3, 61, 73, 78, 86, 114, 121, 136*
corrosion, *92*
crack
    arc, *3, 114*
    blunt, *92, 489, 502, 513*
    circular, *3, 202, 208, 348, 405, 438, 451*
    edge, *3, 73, 86, 98, 100, 168, 231, 242, 256, 269, 316, 337, 361, 374, 387, 423, 461, 46, 502, 535, 576*
    elliptical, *3, 191, 215, 438, 451*
    interface, *3, 145, 151*
    internal, *3, 57, 78, 86, 98, 100, 114, 121, 136, 151, 202, 215, 285, 316, 387, 502, 513*
    periodic, *3, 57, 73, 151, 387, 576*
    point loads, *3, 114, 121, 151, 576*
    pressurized, *78, 162, 173, 215, 285*
    profile, *3, 98, 100, 173, 202, 208, 269, 387, 405, 451, 576, 590*
    surface, *3, 208, 348, 405, 438, 451*
    three-dimensional, *3, 191, 202, 208, 361, 405, 423, 438, 451, 502, 513, 562*
    velocity, *461, 489, 535, 545*
crack arrest, *92, 489*
crack opening displacement (COD), *269, 405, 576, 590, 611*

disk, *78, 162, 168, 316, 502*
dissimilar media, *3, 86, 145, 151*
dynamic fracture, *3, 461, 485, 489, 535, 545*

element
    isoparametric, *337, 348, 361, 374, 438*
    quadrilateral, *333, 337, 361, 374*
    singularity, *337, 348, 351, 361, 374, 438*
    three-dimensional, *348, 361, 438*
    transition, *351*
    triangular, *316*
estimating, *3*

failure criteria
    Irwin-Orowan, *61, 114, 121, 151, 191*
    strain energy, *3*
    stress intensity factor, *3, 185, 231, 242, 256, 316*
fatigue, *173, 451*
finite element method, *98, 100, 173, 316, 333, 337, 348, 351, 361, 374, 438, 451*
Fredholm equation, *162, 168*
free edge correction factor, *3, 185, 215, 451*

glass, *576*

Green's function, *3, 57, 100, 121, 151, 387, 423*

holes
    circular, *61, 535*
    cracked, *3, 61, 173, 285, 604, 638*
    elliptical, *285, 513*
    fastener, *173*
holography, *604, 609*

interferometry, *461, 485, 576, 590*

J-integral, *316*

Laurent series, *61, 78, 157*
loading
    arbitrary, *191, 202, 524*
    bending, *3, 100, 114, 185, 202, 208, 242, 256, 269, 280, 351, 374, 438, 451, 497*
    crack line, *173, 576*
    gradient, *98, 100*
    point, *57, 114, 151, 280, 285, 316, 423, 638*
    uniaxial, *3, 61, 73, 173, 185, 215, 231, 242, 269, 285, 337, 348, 351, 374, 438, 451, 497, 535, 604, 611*

mesh, *316, 333, 337, 348, 351, 361, 374, 423*
metals, *185, 611, 630*
mode
    anti-plane shear, *3, 92, 121, 185, 191, 387*
    forward shear, *3, 92, 121, 185, 191, 374, 387*
    mixed, *3, 114, 185, 348, 351, 374, 524, 545, 638*
    opening, *3, 92, 121, 185, 208, 374*

near-field equations, *3, 92, 114, 121, 145, 151, 185, 191, 423, 485, 513, 524, 535, 604*
Neuber, *3, 185, 497, 502*
nodal force method, *438*
notches, *3, 185, 497, 502, 513*

orthotropic, *3, 121, 136, 630*

partitioning, *86*
photoelasticity, *461, 485, 489, 497, 502, 513, 524, 630*
plane strain constraint, *202, 423*
plastic zone, *3, 92, 185, 231, 562, 609, 611, 645*
plastics, *3, 451, 461, 489, 497, 513, 524, 535, 545, 562, 604*
plates, *3, 114, 374*
Poisson's ratio, *3, 191, 215, 423, 645*

potential energy, *61, 78*

reciprocal theorem, *387, 405*

shells, *3, 114*
span, *256, 269*
spark gap camera, *461, 489, 535*
specimen
    center cracked tension, CC(T), *100, 136, 157, 159, 405, 513*
    compact tension, C(T), *3, 280, 282, 361, 423, 562, 630*
    double cantilever, DB(T), *590*
    double edge notch, DE(T), *524*
    round bar, R-Bar(T), *3, 185, 337*
    single edge bend, SE(B), *242, 256, 269, 280, 282, 351*
    single edge notch, SE(T), *100, 185, 231, 242, 256, 269, 316, 333, 351, 361, 374, 405, 497, 611*
    tapered double cantilever, DB(T), *269, 590*
strain energy, *3, 98, 162, 168, 374*
strain energy release rate G, *3, 98, 121, 191, 231, 333, 387, 485, 590, 604, 638*
strain gauge, *645*
stress
    concentration factor, *185, 497*
    couple, *3*
    distribution, *461, 497, 502, 535*
    nonsingular, *361, 485, 489, 513, 524, 590, 645*
    residual, *173, 630*
stress function
    complex variables, *3, 61, 73, 78, 86, 114, 121, 151, 157, 285*
    real, *3, 145, 191, 202, 231, 242, 256, 282*
    Westergaard, *3, 57, 121, 387*
superposition, *173, 191, 202, 208, 242, 497*

temperature, *3, 208, 361*
test methods, *185, 231, 611, 630*
thickness, *215, 231, 423, 438, 451, 562*
torsion, *3, 185*

ultrasonics, *611*

viscoelastic, *3*

wedge, *3, 285, 502*
weight function, *100, 173, 387*